Y0-CMH-750

DISCARDED

NON-CIRCULATING MATERIAL

NON-CIRCULATING MATERIAL

AIChE Symposium Series No. 316
Volume 93, 1997

CHEMICAL PROCESS CONTROL-V

Assessment and New Directions for Research

Proceedings of the Fifth International Conference on
Chemical Process Control
Tahoe City, California, January 7-12, 1996

Editors

Jeffrey C. Kantor
University of Notre Dame

Carlos E. Garcia
Shell Chemical Company

Brice Carnahan
University of Michigan

CACHE
American Institute of Chemical Engineers

1997

© 1997
American Institute of Chemical Engineers (AIChE)
and
Computer Aids for Chemical Engineering Education (CACHE)

Neither AIChE nor CACHE shall be responsible for statements or opinions advanced in their papers or printed in their publications.

Library of Congress Cataloging-in-Publication Data

International Conference on Chemical Process Control (5th : 1996 :
    Tahoe City, Calif.)
        Chemical Process Control - V : assessment and new directions for
    research : proceedings of the Fifth International Conference on
    Chemical Process Control, Tahoe City, California, January 7-12, 1996
    / editors, Jeffrey C. Kantor, Carlos E. Garcia, Brice Carnahan.
            p. cm. -- (AIChE symposium series : no. 316)
        Includes bibliographical references and index.
        ISBN 0-8169-0741-2
        1. Chemical process control -- Congresses.   I. Kantor, Jeffrey, C.
    II. Garcia, Carlos, E.   III. Carnahan, Brice.   IV. Title.   V. Title:
    Chemical process control - 5.   VI. Series.
        TP155.75.I55   1996   97-25433
        660' .2815--DC21   CIP

All rights reserved whether the whole or part of the material is concerned, specifically those of translation, reprinting, re-use of illustrations, broadcasting, electronic networks, reproduction by photocopying machine or similar means, and storage of data in banks.
  Authorization to photocopy items for internal use, or the internal or personal use of specific clients, is granted by AIChE for libraries and other users registered with the Copyright Clearance Center Inc., provided that $3.50 per copy is paid directly to CCC, 222 Rosewood Dr., Danvers, MA 01923. This consent does not extend to copying for general distribution, for advertising, or promotional purposes, for inclusion in a publication, or for resale.
  Articles published before 1978 are subject to the same copyright conditions and the fee is $3.50 for each article. AIChE Symposium Series fee code: 0065-8812/1997.

PREFACE

CPC-V

Chemical process control is an area of strong academic and industrial interest. When the first CPC conference was held twenty years ago, process control was a field with an uncertain future and research agenda. That meeting was instrumental in revitalizing academic interest in process control. Since that time, fueled by developments in information systems technology, advanced process control has witnessed unprecedented growth in the breadth and scope of industrial applications, and in underlying theoretical principles. Areas of particular development include predictive control technologies, robust, nonlinear, and adaptive control, identification, and information systems. New companies and a new generation of academic researchers have entered the field in those two decades. The proceedings of CPC-II, CPC-III, and CPC-IV chronicle those developments.

As a consequence of growth and change in the discipline of process control, it seemed timely to meet in the context of a CPC Conference to assess the current status of the field and to discuss new areas for research and development. At least some areas of recent research activity appear to be reaching maturity. Also, the environment for conducting industrial research and development has changed significantly during the recent era of industrial reorganizations. Yet the incentives and opportunities for doing advanced control seem strong.

Chemical Process Control V — the fifth in a series of international conferences held every five years, commencing in 1976 — brought together engineers and scientists from universities, the processing industries, and government laboratories to assess and critique the current and future directions of research in chemical process control. The goals of CPC-V were to:

- promote a vital, interactive discussion among a diverse international group of experts regarding the state of the technology for chemical process control,
- assess current research,
- identify research opportunities for academics, government, and industry,
- promote productive research collaborations.

Conference participants had ample opportunity to interact with peers and experts in their fields. Participants left the conference with a better understanding of current directions that will aid them in formulating their personal research agendas.

Technical Program Committee

Co-Chairmen

Jeffrey Kantor	University of Notre Dame
Carlos Garcia	Shell Chemical Company

Members

Yaman Arkun	Georgia Institute of Technology
James Downs	Tennessee Eastman Company
Francis Doyle	Purdue University
Thomas Edgar	University of Texas
Manfred Morari	ETH Zentrum
John Perkins	Imperial College
James Rawlings	University of Wisconsin
Sigurd Skogestad	University of Trondheim

Sponsors

CACHE Inc.
CAST Division, AIChE

Industrial Contributors

Shell Chemical Company
E.I. duPont de Nemours and Company

Acknowledgments

We wish to acknowledge the assistance of the members of the Technical Program Committee and of the session chairs, whose contributions were instrumental in planning and organizing this conference, and also for the financial assistance of our corporate contributors.

We are grateful to the CACHE Corporation for facilitating the conference arrangements, in particular, to the CACHE Executive Officer, David Himmelblau, and the CACHE office staff in Austin. Special thanks are also due

Robin Craven, who handled on-site registration and took care of many of the details of conference management.

Finally, we acknowledge the outstanding contributions of Prof. Brice Carnahan of the University of Michigan, who assumed full responsibility for setting publication standards and formats, collecting and editing conference papers, assembling (with the able assistance of Matthew Smart, an undergraduate Engineering student at the University of Michigan) the all-electronic version of these proceedings, and supervising the production and initial distribution of this volume.

The CPC-V Co-Chairmen

Jeffery C. Kantor
Carlos E. Garcia

TABLE OF CONTENTS

Industry Assessment — I

Control Technology Challenges for the Future 1
    Brian L. Ramaker, Henry K. Lau and Evelio Hernandez

Adaptive and Nonlinear Control — Fact or Fantasy?

Certainty Equivalence Adaptive Control: What's New in the Gap 9
    B. Erik Ydstie

Nonlinear Process Control — Which Way to the Promised Land? 24
    Frank Allgower and Francis J. Doyle III

Industrial Applications of Nonlinear Control 46
    Babatunde A. Ogunnaike and Raymond A. Wright

Session Summary 60
    Yaman Arkun

Industry Assessment — II

Computer-Aided Process Engineering in the Snack Food Industry 61
    Michael Nikolaou

Batch Food Processing: The Proof is in the Eating 71
    Sandro Macchietto

New Directions for Academic Research

Controller Performance Monitoring and Diagnosis:
Experiences and Challenges 83
    Derrick J. Kozub

Process Analytical Chemical Engineering 97
    Bruce R. Kowalski

Analysis and Control of Combined Discrete/Continuous Systems:
Progress and Challenges in the Chemical Process Industries 102
    Paul I. Barton and Taeshin Park

Session Summary 115
    Francis J. Doyle III

Industry Assessment — III

A Review of Modeling and Control in the Pulp and Paper Industries.............. 117
    Ferhan Kayihan

Case Studies in Equipment Modeling and Control in the
Microelectronics Industry... 133
    Stephanie W. Butler and Thomas F. Edgar

Session Summary 145
    Dale E. Seborg

Impact of Computer Science

Applying New Optimization Algorithms to Model Predictive Control 147
    Stephen J. Wright

Real-time Operations Optimization of Continuous Processes 156
    Thomas E. Marlin and Andrew N. Hrymak

Discrete Events and Hybrid Systems in Process Control 165
    Sebastian Engell, Stefan Kowalewski and Bruce H. Krogh

Session Summary 177
    Lorenz T. Biegler

Past and Future Directions of Process Control Research

CPC I - CPC V: 20 Years of Process Control Research 179
    Manfred Morari

The Future of Process Control — A UK Perspective 192
    Roger Benson and John Perkins

Session Summary 199
    John Perkins

Predictive Control

Recent Advances in Model Predictive Control and Other Related Areas 201
    Jay H. Lee and Brian Cooley

Nonlinear Model Predictive Control: An Assessment 217
    David Q. Mayne

An Overview of Industrial Model Predictive Control Technology 232
    S. Joe Qin and Thomas A. Badgwell

Session Summary 257
    James B. Rawlings

Contributed Papers

Variational Systems: A Basis for New Production-Support Programming 259
    Bruce C. Moore

The Game of Fate — A Modeling Paradigm for Manufacturing Operations 263
    Robert E. Young

Geometric Interpretation of SVD, Rank, Mean Centering and Scaling
in Applying Multivariate Statistical Analysis Methods 267
    Tsewei Wang, Srinivas Vedula and Atul Khettry

Multi-way Analysis in Process Monitoring and Modeling 271
    Barry M. Wise and Neal B. Gallagher

Identification of Paper Machine Full Profile Disturbance Modes
Using Adaptive Principal Component Analysis 275
    Apostolos Rigopoulos, Yaman Arkun, Ferhan Kayihan
    and Eric Hanczyc

Experimental Evaluation of Neural Nonlinear Modeling Techniques 280
    John D. Bomberger, Dale E. Seborg, Gordon Lightbody
    and George W. Irwin

Identification and Control of Industrial-scale Processes via
Neural Networks 284
    Masoud Nikravesh, Anthony R. Kovscek, Tad W. Patzek
    and Masoud Soroush

Design of a Composition Estimator for Inferential Control of High-Purity
Distillation Columns 288
    Joonho Shin, Sunwon Park and Moonyong Lee

On-line Inference System of Tube-wall Temperature for an Industrial
Olefin Pyrolysis Plant 292
    Manabu Kano, Toshihiro Shiren, Iori Hashimoto,
    Masahiro Ohshima, Umetaro Okamo and Shyoji Aoki

Software Sensors and Adaptive Control of Biochemical Processes 297
  Denis Dochain

Control of Nonlinear Distributed Parameter Systems:
Recent Results and Future Research Directions 302
  Panagiotis D. Christofides and Prodromos Daoutidis

Operating Regime-based Controller Strategy 307
  Karlene A. Kosanovich, James G. Charboneau and Michael J. Piovoso

Application of Multi-Unit Control Analysis and Design
to a Reactor/Separation Process 311
  Yudi Samyudia, Peter L. Lee, Ian T. Cameron and Michael Green

Operability of Batch Reactors: Temperature Profile Feasibility 315
  B. Wayne Bequette

Procedural Control of Chemical Processes 319
  A. Sanchez, G. E. Rotstein, N. Alsop, and S. Macchietto

Dynamic Operations in the Planning and Scheduling of
Multi-product Batch Plants 323
  Tarun Bhatia and Lorenz T. Biegler

The Current Status of Sheet and Film Process Control 327
  Richard D. Braatz

On Infeasibilities in Model Predictive Control 331
  P.O.M. Scokaert and James B. Rawlings

Linear Model Predictive Control of Input-Output Linearized
Processes with Constraints 335
  Michael J. Kurtz and Michael A. Henson

Stability of NN-based MPC in the Presence of Unbounded
Model Uncertainty 339
  Alexandros Koulouris and George Stephanopoulos

Mixed Objective Optimization for Robust Predictive Controller Synthesis 343
  Kostas Hrissagis and Oscar D. Crisalle

Mp Tuning of Multivariable Uncertain Processes 347
  Karel Stryczek and Coleman B. Brosilow

Session Summary 352
  Richard D. Braatz

Author Index 353

Subject Index 355

CONTROL TECHNOLOGY CHALLENGES FOR THE FUTURE

Brian L. Ramaker, Henry K. Lau and Evelio Hernandez
Shell Oil Company
Houston, TX 77251-1380

Abstract

If we examine the state of the art in control technology for the last 25 years, we become aware of how much control practices have changed during this time. There have been many remarkable changes beginning with the instruments and equipment that make up the control room, to control implementation and process algorithms, the support tools for control applications, through the skill level of available manpower. Furthermore, indications are that the future rate of change will be even more accelerated. This paper presents a vision of the changes for the next 25 years in the state of the art control room.
We will present a wide variety of topics essential to control development. First, the motivators for change: improved operations, profit, and conservation of the environment will continue to drive improved control. In addition, the workforce is changing, with different skills and different aspirations providing new activities that control will undertake.

This paper provides a vision of the control room of the future, and the activities typically performed by a control engineer. This vision is accompanied with the technical elements that we see at work and the technical challenges that will be encountered. How will we use fundamental modeling in control? How can we enhance the combination of information technologies and classical control technologies? How will we convert the enormous amount of available data into useable information? How can we create control systems that are easier to maintain long term? How do we make the best use of the global computer infrastructure and the communication advances? The technical challenges are presented in a way general enough to maintain a broad view of the problem but detailed enough to provoke specific technical questions that could provide valuable ideas for future work.

Keywords

Control future vision, Control business drivers, Control research challenges, Information research challenges, Process control education, Industry acedemia partnering.

Introduction

Historical Perspective On On-Line Information and Control Systems

In the late 1950s and early 1960s the whole field of on-line process computer systems was just starting to develop, but the visionaries of the time were promoting the potential for advanced control, on-line optimization and information systems to run process units. This was at a time when some process computer equipment did not have any random access memory or compilers. In fact, in many cases, coding was done directly in machine language. Yet these visionaries foretold an implementation close to what we have today. It has taken 35 years to develop the equipment and technology to provide wide spread implementation of their ideas. Our challenge is to be as equally visionary and forecast what will be needed for the next 25 years.

The World in Which We Live

Today's world is changing rapidly as companies are adjusting to global economic competition. Customers continue to demand products with consistent low variability at ever lower prices. Environmental concerns are consuming a large proportion of available capital as efforts are made to eliminate or control all waste streams. Sustainability of the industry into the future is beginning to get attention; natural resources can not be consumed without some thought of the effect on the future. In addition, customers are demanding a wider variety of products and a quicker response to their demands. These factors and the decline in industrial funding for basic research are forcing companies to examine manpower levels and methods to improve efficiency, and to develop strategies to produce tomorrow's technologies at a lower cost.

Business Drivers for Information and Control Systems

Economic competition is forcing every company to understand how all the various segments of their business are contributing to the bottom line. This also applies to the business of developing, installing and maintaining advance information and control systems. If a competitive return on investment can not be demonstrated, then it is likely that the activity will cease to exist. How can process control and information systems contribute to the bottom line? Economic driving forces that exist today and unlikely to change are:

1. Use existing process equipment fully.
2. Deliver the same product consistently.
3. Minimize product variability.
4. Meet safety or regulatory requirements.
5. Increase the operator's span of control.
6. Reduce the cost of implementing and supporting control and information systems.
7. Improve the operating range and reliability of control and information systems.

Control and advisory systems are installed to make money and maintain the plant within safety, regulatory and operating limits. This is achieved by running plants which are sold out closer to the operating limits producing more product or by running not sold out plants closer to the operating limits to produce a given amount of product at the minimum cost. Delivering the same product with reduced variability all the time can also be achieved with these systems and contributes to customer satisfaction which may help win a contract but usually does not permit increased prices. Increasing the operator's span of control may or may not be cost effective depending on how freed-up time is utilized. Reducing the implementation time and improving the performance of the advisory and control systems are other ways to improve profitability.

Technology Requirements

Companies today are faced with the challenge of determining the most cost-effective and profitable technologies for their businesses. Some companies may decide to be technological leaders, but because no company can be a leader in all fields a few fields must be chosen in which to excel. To become a technology leader in a given area a company must have a research activity and a conviction that spending money on research and applying results provides a competitive advantage. Some companies decide that they are technology followers. These companies fund no research and count totally on outside sources to provide technology. Although the approaches are different, each can be profitable.

The cost associated with owning and supporting technologies in a given specialty area is directly related to the number of technologies involved. The larger the number of technologies owned or supported, the higher the costs. For example, consider control algorithms. Using ten different algorithms with individually narrow applicability will cost more to support than using five algorithms with broader applicability that cover the same problem space. Even more costly is using ten technologies with widely overlapping applicability so there is confusion about their use. Introducing a new control algorithm to a company the size of Shell can cost millions of dollars. This means that adding a new technology must offer sufficient benefits to offset the cost. Consequently only a few non overlapping technologies are preferred.

Vision

Future Vision for the Chemical Industry

The future chemical plant will be a flexible manufacturing facility able to produce a wide variety of products just in time and 100 percent of the material produced will be on specification. The chemical production lines will be modified so that no waste or by-product streams are produced. The growing concern over industry sustainability is moving products toward a totally recyclable state in which molecules are disassembled to provide feed stocks or renewable feed stock sources using natural processes. Energy requirements of the processes will be reduced and no longer depend on oil based fuels. Staff at manufacturing facilities will be dramatically reduced but is highly trained. Information and control systems have relieved the board operator of routine tasks.

Vision of Advanced Information and Control Circa: 2020

Rapid development and production of new products is essential to capturing fast changing markets. In order to improve and increase the speed of designing new processes or products, manufacturing sites will be composed of extremely flexible, reusable and interconnectable process and control equipment capable of operating in drastically different regimes. Process design and control system design will occur in a unified step. The control system will use the same reliable dynamic models used for the design and optimization of the process. By including dynamic control considerations during the process design stage and taking advantage of models available at this stage, not only will the development time be reduced, but also the implementation cost of control reduced and the performance will be increased substantially.

Control systems will be economically based, the objective of the controller is to maximize the profitability of the plant while maintaining the highest respect for safety and environmental constraints. Prices and marginal prices of raw materials, utilities, products, and product quality will be maintained on-line and updated continuously to reflect current market conditions. Control systems will use the economic information to dynamically optimize plant-wide regions of operation. On-line control functions will not only maximize the economic benefits of a process by adjusting the existing process equipment, but also by continuously analyzing the process and suggesting changes to the units.

Maintenance of control systems will require minimal human intervention; control systems will be reliable and run during the entire range of operation of continuous or batch plants, from startup to shutdown including major upsets. Dynamic models that accurately represent the entire range of operation of the plant and are continuously validated on-line and updated when necessary will be an integral part of plant operations. Control systems will continuously inspect and validate the integrity of components, from instruments to computers. If incipient failure is detected, the control system turns automatically to backup equipment if available and notifies operation's personnel for assistance with maintenance.

The low maintenance requirements of control systems as well as advances in communications will have made it possible and economic to locate the control room of all units of a manufacturing complex in one safe central location. The control room will use all of the operators' senses to communicate information. Operators will be able to hear and see every part of the complex, including the insides of operating equipment. Operators will communicate via voice with the control room's operations advisor, the decision making interface between the operator and the plant's equipment. The operations advisor will be responsible for many routine tasks assigned to operators today. This will be accomplished by deploying a fleet of robots that can take samples, change equipment as necessary, perform routine maintenance to the equipment, etc. But, routine tasks are only a minor portion of the operation advisor's responsibilities. The advisor will constantly monitor the process and alert the operator when current conditions could lead to a safety or environmental problem or poor plant performance, and provides guidelines and recommends changes to the operator to avoid such situations. The advisor will have access to numerous sources of information: current plant data, historical event data, current models of the systems, and the company's experts on the various processes. The operations advisor will use all of these data to construct a coherent explanation and recommendation. The operator will evaluate the recommendation and order its implementation if acceptable. Otherwise, the operator will probe the advisor on the data and assumptions used to construct the recommendation, may request additional information, and consults the company's experts if necessary to construct an action plan that will be implemented by the operations advisor. The interaction between the operators and the advisor for the particular situation will then be kept in the advisor's database for future reference.

Comparing this future vision with that proposed in the Japanese, European, and American papers presented at the 1991 CPC (Yamamoto and Hashimoto, 1991; Schuler, Allgower and Gilles, 1991; Doss and Downs, 1991) indicates there are few differences. There seems to be not only considerable consistency among groups about how technology will evolve, but also, there has been a consistency over time about this vision. This may indicate that we are all facing the same driving forces and see similar responses to these forces. A consistent industrial future vision should greatly assist efforts to establish research directions for the next five years.

Infrastructure Needed to Support Vision

The future vision is highly dependent on a hardware and software infrastructure that will support the technologies. Computers will need to expand their capability at the present rate into the future in order to not be a bottleneck. Process data must be in digital format and available where needed in a timely fashion. Although today the number of process measurements on new plants is declining because of cost, the future vision requires an increase in measurements using technologies similar to fieldbus to reduce the cost of measurements. Networks must be able to handle the large data loads associated with a totally integrated corporate wide system. Personnel will have a virtual office that provides the same capabilities on the network whether they are at home, on the road, or in an office. Database technologies need to accommodate large volumes of real time, batch and lot, and laboratory data efficiently. Open systems provide the methodology for data to be entered into the system once and to reside in one place only. Interfaces for operators and engineers need to evolve in order to display even more information in an easily understood format, perhaps to the point of using wall size video screens.

Our challenge is the legacy systems of today and how to make the transition to an open completely integrated system. Today, every application has its own database and data structures unique to the application. In most cases these applications are performing well and making money for their users, but the cost of supporting them may be high and opportunities for improved efficiencies may still exist. It has been reported that design engineers working with standalone computer programs spend 50 to 80 percent of their time moving and organizing data between programs (Book et al. 1994). Choices have to be made concerning the large capital cost for infrastructure to enable money making applications versus the hidden fixed costs of supporting and using the existing systems.

Another infrastructure issue is computer network security. As the process control networks are tied into company wide management information systems, security can become a major issue. It is desirable and advantageous to have the data and information from the control systems available in a consistent, easily accessible way to the whole corporation. However it is unacceptable to take the risk of outside sources obtaining access to the systems that control process units. This problem, although it seems to be uncomplicated and easily to solve, is proving to be quite complex.

Challenges in Process Control

To meet the Vision 2020 described in this paper, many technological advances must occur. In this section some of these advances are presented as challenges to the researchers.

Challenge 1: Provide technologies that help operators make better decisions in the control room.

In the control room operators are confronted with many decisions on a daily, hourly, even a moment to moment basis. These decisions may concern scheduling work, maintaining operations, troubleshooting problems, or improving operations. Examples of these decisions are:

- Should this compressor be worked on or should that exchanger leak be fixed first?
- Should the temperature in this column be raised to improve product recovery? Can the downstream unit where the increased product is sent handle the added load?
- The temperature in the reactor is rising. What should we do?
- A unit that provides feed to us is upset. How does this affect us?
- The condition is somewhat abnormal. Should we continue to the next operating phase?

To implement the operations advisor discussed in the 2020 Vision requires technological advances in computing, information, and advisory systems. Specifically, an advisor to help operations make these decisions and run the units closer to their operating objectives must be capable of integrating different forms of information; for example data from the plant, process history, expert statements, statistical analysis, modeling results, and alarm patterns. From this information the advisor forms a coherent picture of where the process is operating and of any potential problem.

Using this picture of the state of the process, the advisor formulates an action plan to meet the quantitative and qualitative operating objectives (profitability, safety, and environmental) of the unit. This action plan includes any corrective actions required to fix or to further understand any potential problem. Displaying the coherent picture of the unit to operations and including the action plan facilitates understanding of the current state of the unit and acceptance of the action plan. Training of operating and technical staff with automated, self-paced, adjustable lessons is part of the advisor's functions.

Innovative technologies in advisory systems affecting plant profitability will be developed. The success of advisory system and robotics leads to remotely operated and maintained plants.

The control community has been working for some time on pieces that will become part of the overall solution. Considerable amounts of work has been devoted to problems such as data monitoring and interpretation (e.g. Cheung and Stephanopoulos, 1990a,b; Bakshi and Stephanopoulos, 1994a,b; Davis et.al., 1995; Nomikos and MacGregor, 1994; Kresta, et. al., 1991; Machresor, et. al., 1994), and diagnosis (e.g. Calandranis et.al., 1990; Leonard and Kramer, 1992) which will surely be pivotal in the solution. The technology involved in these pieces are fairly mature. What we still lack are easy to use and maintain tools for the integration and application of these technologies. These tools would bring together the different sources of information and provide a unifying answer in a cost effective fashion. We are very encouraged by the efforts of the Abnormal Situation Management Consortium administered by Honeywell towards providing such integrated tools (Cochran and Rowan, 1995).

Challenge 2: Develop technologies to implement and maintain a flexible control system

The vision outlined earlier calls for flexible manufacturing systems. A plant will use its operating hardware and software to produce a wide slate of products. This in turn means that the equipment will be used in different physical configurations and over a wider range of conditions. All of this translates to a control system which is required to be highly configurable, and adaptable to the different operating objectives of the different operating modes.

The flexible control system required is divided into two layers, stabilizing control and objectives control. Stabilizing control is the control system that improves the system's ability to reject disturbances and recover from upsets. This control layer runs at a rapid frequency and is typically composed of flow controllers, pressure controllers, and temperature controllers.

The design of flexible control structures begins during the process design phase. Processes are designed to maximize their controllability (here controllability is defined as the "best" dynamic response to setpoint changes and disturbances). Much work has been devoted to this area and significant advances have been accomplished considering the vagueness of the problem. A good review of the work in this area can be found in Morari, 1992 and the references therein. It is envisioned that this work will result in a set of tools and algorithms (as proposed in Seider et al., 1990) which will allow the engineer to arrive at designs with the desired tradeoff between controllability and economics.

The stabilizing layer of the flexible control system has typically posed less problems than the objectives control layer. Drastic changes are not anticipated in the way control is done in the stabilizing. Other than more penetration of self tuning algorithms, there is little else foreseen for this layer. Hardware will change considerably, we will see more control being done in the field and less on Distributed Control Systems (DCS's). With changes in hardware, new information will become available as smart control elements such as valves and sensors continue to be developed and commissioned. Sticking valves and faulty thermocouples will be self diagnosed and the information sent to the operations advisor of the vision.

The objectives control layer completes the flexible control system structure. This layer runs at a slower frequency and pushes the unit to achieve its operating objectives. The new controller at the objective control layer requires no tuning. The speed at which the solution of the algorithm is calculated and the robustness of the algorithm must be such that operations can rely on the objectives control system to run the unit.

This layer takes the place of today's "advanced control layer." The advanced control layer has traditionally been designed by either formulating some optimization problem (Linear Quadratic control: Kwakernaak and Sivan, 1972; Model Predictive Control: Cutler and Ramaker, 1979; H-inf Control: Postlethwaite and Skogestad, 1993), matching some selected response to a pre-defined input (as in input-output linearization: Daoutidis and Kravaris, 1992), or obeying a set of heuristics developed by experts (as done with fuzzy logic: Sugeno, 1985; or expert system control: Tzouanas, 1988). We feel that none of these design approaches have been able to truly achieve their intentions: to obtain the best performance out of the operating equipment.

One fundamental reason for this deficiency is our human incapability of articulating what is the expected performance of the unit. Neither quantitative nor qualitative performance specifications are sufficient to truly define what is desired. For this reason, the algorithm used for the objective layer needs to be able to support both kinds of information. The works of Cagan et. al. (1995) as well as Realff and Kevrikidis (1995) in incorporating logic and other qualitative information along with numerical methods, optimization and quantitative arguments seem to be a foundation that can be used for building the new algorithm. We encourage more work be undertaken in this area.

Challenge 3: Developing a dynamic model as a source of process knowledge

Our vision for control relies heavily on reliable dynamic models of the process. In fact, we see rigorous dynamic models as being a key component of the work tools of many diverse areas of chemical manufacturing: from process and control design, to safety studies, profitability analysis, etc. The increased importance of this model as well as the diversity of uses will bring about numerous technical challenges to the area of dynamic simulation.

A better model development environment is needed to build, modify and maintain dynamic models. Models intended for wide use will be complex but need to be easily built and easily maintained. The level of accuracy required will be different for various uses as well as the speed of solution requirements. It is likely that different interfaces to the same model will be necessary to match up with the intended use. A truly successful modeling environment will allow the user to make the changes to a model in only one place.

Improvements in the basic understanding of chemical and physical phenomena is an important step toward better models that can shadow the processes accurately enough for their applications (e.g. better physical property and reaction kinetics). Improvement in speed of several orders of magnitude must be achieved before a large scale plant dynamic simulation can be implemented on-line. Robust solution techniques are needed to increase the convergence properties of the models and an algorithm is needed to update the models based on process measurements in order to shadow the plant accurately. There have been encouraging reports in the literature of generic tools that have been used to shadow the plant in a dynamic sense.(Riley et. al. 1995) Further investigation of these reports and development of such tools are warranted.

The paradigm of understanding the basic chemical and physical principles that govern the process is relevant and is the philosophical foundation to this approach.

Challenge 4: Providing a relevant undergraduate process control education

The education challenge is to provide students with tools so they can develop in their mind a model of how a process should behave, both in the steady state sense and in the dynamic sense. A solid understanding of the basics of chemical engineering supplemented by process models simulation and experiments will form a foundation that the student can build on when entering industry. An undergraduate in a chemical engineering curriculum process control should be taught using concepts that fit with the rest of the chemical engineering education. For example, a chemical engineer's undergraduate curriculum emphasizes time domain ideas: flowrates, residence times, rate constants, etc. This is contrasted with electrical engineering where a frequency response of a circuit displayed on an oscilloscope is part of their "bread and butter." Thus it would make sense that the electrical engineer be taught control concepts in the frequency domain, while chemical engineers be taught the same concepts in the time domain. This doesn't mean that the Laplace transform cannot be used as a tool to solve differential equations in the undergraduate course. Neither does it mean that frequency domain analysis and design are not useful in chemical engineering. It only means that we feel that frequency domain analysis and design should be taught at a graduate level, maintaining the undergraduate curriculum as closely tied as possible to the time domain.

Industry needs trained personnel to address the future challenges in process control. Undergraduate education should provide bachelor degree engineers with fundamental concepts of process dynamics and its interaction with process variability (Down, Doss,1991). The concept of dynamics and controlling a process should be integrated into the appropriate chemical engineering courses, process design, unit operations, and transportation processes. Process understanding is still the most important ingredient in successfully controlling plants. New personnel need to be better educated and trained to support these systems when they enter the work force.

Challenge 5: Advances in supporting technologies

Advances in other technologies are needed to develop the vision described in this paper and to support the technologies discussed above. Because statistics provide important techniques for estimating and verifying process models, this technology should be considered a fundamental technology to be developed. Statistics provide basic information upon which the other technologies are built, and supply tools that collapse quantities of data to a few dominant factors that represent process behavior. Another statistical contribution is reconciling large amounts of process data and detecting gross errors in the data.

Our emphasis on rigorous nonlinear models of processes places a considerable burden on numerical algorithms. These models will have in the order of ten to one hundred thousand equations. Algorithms and approaches that are used today for data reconciliation, state estimation, parameter optimization, etc. will have to be rethought. What is robust and efficient for hundreds of equations is not necessarily the right approach for one hundred thousand equations. Our vision also has emphasized on the need of combining quantitative and qualitative information. Numerical algorithms will have to be developed to satisfy this vision. We are encouraged by the earlier referenced activity undertaken recently in this area.

Another technology that will have many advances is computing technology, both hardware and software. Software application development environments make applications easy to develop, modify and maintain. Links between different applications and platforms are seamless and connectivity to support communications between faster computers and people is readily available and inexpensive. Software is now capable of searching large databases efficiently and quickly. Knowledge and expert systems support the organization and presentation of information.

How Do We Accomplish These Goals?

A new technology today often takes ten years to progress from initial development to widespread use within a company. Distributed control systems, and model based control are good examples of this time lag. We need to examine what can be done to start shortening delivery time. One factor to consider is legacy systems. Each system is unique and does not support easy access and exchange of information. Open system architecture could make a significant improvement in delivery time, but before starting the transition it is important to ensure that changes are cost-effective.

Companies today are in the midst of a transformation to compete successfully in a global economy. One of the byproducts of this activity is a redirection of research funds into regulation mandated fields and fields that can yield fairly short-term payouts. What is being reduced is the funding of longer term basic research; the cornerstone for future technologies. Companies must consider alternative ways to continue this work or our future technological advancement will be compromised. More extensive partnering between companies, universities, national laboratories and government can provide a significant portion of this basic pre competitive research. A good example of this is Honeywell's consortium on abnormal situation management. This CPC conference was set up to address research needs in the control systems area between industry and academia and we need to take advantage of this opportunity to define industry needs and the cooperation possible to meet those needs.

This is an exciting time and a challenging time to work in the process control area. Advances in computer hardware and software technologies, communication technologies, electronic technologies, etc. can support control tools with capabilities which were only dreamt about a decade earlier. Many more control tools which have not yet been dreamt of will surface as technologies progress.. The span of process control as a discipline is expanding to include all activities inside the control center. Many difficult problems remain to be defined and to be addressed in this environment. The challenge is for industry and academia under tight budget constraint to find ways to more rapidly develop technologies that lead toward the future vision.

Acknowledgments

The authors would like to thank Richard Balhoff and Lena Patsidou of Shell Oil Company for many stimulating discussions.

References

Bakshi, B.R. and G. Stephanopoulos (1994). Representation of Process Trends — Part III. Scale Extraction of Trends from Process Data. *Comp Chem. Eng.* **18**, 267-302.

Bakshi, B.R. and G. Stephanopoulos (1994). Representation of Process Trends — Part IV. Induction of Real Time Patterns from Operating Data for Diagnosis and Supervisory Control. *Comp Chem. Eng.* **18**, 303-332.

Book, N., O. Sitton, R. Motard, M. Blaha, B. Mala-Goldstein, J. Hendrick and J. Fielding (1994). The Road to a Common Byte. *Chem. Eng.*, Sept.

Cagan, J., I. Grossmann and J. Hooker (1995). Combining Artificial Intelligence and Optimization in Engineering Design: A Brief Survey. Presented at Intelligent Systems in Process Engineering Conference, Snowmass, CO.

Calandranis, J., G. Stephanopoulos, and S. Nunokawa (1990). DiAD-Kit/Boiler: On-line Performance Monitoring and Diagnosis," *Chem. Eng. Prog.*, pp. 60-68.

Cheung, J.T. and G. Stephanopoulos (1990). Representation of Process Trends Part I. A Formal Representation Framework. *Comp Chem. Eng.* **14**, 495-510.

Cheung, J.T. and G. Stephanopoulos (1990). Representation of Process Trends Part II. The Problem of Scale and Qualitative Scaling. *Comp Chem. Eng.* **14**, 511-539.

Cochran, E. and D. Rowan (1995). [Human] Supervisory Control and Decision Support: State-of-the-Art. Presented at Intelligent Systems in Process Engineering Conference, Snowmass, CO.

Cott, B. (1995). Introduction to the Process Identifiacation Workshop at the 1992 Canadian Chemical Engineering Conference. *J. Proc. Contr.*, **5**, 67-69.

Daoutidis, P. and C. Kravaris (1992). Dynamic Output Feedback Control of Minimum Phase Nonlinear Processes. *Chem. Eng. Sci.* **47**, 837-848.

Davis, J.F., B.R. Bakshi, K.A. Kosanovich and M.J. Piovoso (1995). Process Monitoring, Data Analysis and Data Interpretation. Presented at Intelligent Systems in Process Engineering Conference, Snowmass, CO.

Downs, J.J. and J.E. Doss (1991). Present Status and Future Needs- A View from North American Industry. In *Proceedings of the Chemical Process Control IV Conference*, Y. Arkun and W.H. Ray Editors, South Padre Island, TX.

Downs, J.J. and E.F. Vogel (1993). A Plant-Wide Industrial Process Control Problem. *Comp. Chem. Eng.*, **17**, 245-255.

Kresta, J.V., J. F. MacGregor, and T. E. Marlin (1991). Multivariable Statistical Monitoring of Process Operating Performance. *Can. J. of Chemistry*, **69**, February.

Kwakernaak, H. and R. Sivan (1972). *Linear Optimal Control Systems*. New York: Wiley-Interscience.

Leonard, J.A and M.A. Kramer (1992). A Decomposition Approach to Solving Large Scale Fault Diagnosis Problems with modular Neural Networks. Presented ant the IFAC Conference On-line Fault Detection and Supervision in the Chemical Process Industry, Newark DE.

MacGregor, J. F., C. Jaeckle, C. Kiparissides, and M. Koutaudi (1996). Process Monitoring and Diagnosis by Multiblock PLS Methods. *AIChE Journal*, **40**(5), May.

Morari, M. (1992). Effect of Design on the Controllability of Chemical Plants. Plenary Address, IFAC Workshop on Interactions Between Process Design and Process Control, London.

Nomikos, P. and J. F. MacGregor (1994). Monitoring Batch Processes Using Multiway Component Analysis. *AIChE Journal*, **40**(8), August.

Postlethwaite, I. and S. Skogestad. Robust Multivariable Control Using H-infinity Methods: Analysis, Design and Industrial Applications.

Prett, D.M. and M. Morari (1987). *The Shell Process Control Workshop*, Butterworth, Stoneham, MA, pp. 355-360.

Realff, M.J. and I. Kevrekidis (1995). Towards Integrated Frameworks of Reasoning and Computation in Chemical Engineering. Presented at Intelligent Systems in Process Engineering Conference, Snowmass, CO.

Riley, G., L. Gdula, T.A. Clinkscales, A. Gokhale and G.W. Hodges (1995). Dynamic Model Supports Control, Engineering and Training for Crude Unit Operations. In *Proceedings of the NPRA Annual Meeting*, Nashville, Tennessee.

Rowan, D.A. (1988). AI Enhances On-Line Fault Diagnosis. *InTech*, May, pp 52-55.

Schuler, H., F. Allgower and E.D. Gilles (1991). Chemical Process Control: Present Status and Future Needs- The View from European Industry. In *Proceedings of the Chemical Process Control IV Conference*, Y. Arkun and W.H. Ray Editors, South Padre Island, TX.

Seider, W.D., D.D. Brengel, A.M. Provost and S. Widagdo (1990). Nonlinear Analysis in Process Design. Why Overdesign to Avoid Complex Nonlinearities. *Ind. Eng. Chem. Res.*, **29**, 805-818

Sugeno, M (Ed) (1985). *Industrial Applications of Fuzzy Control*, North Holland, Amsterdam.

Tzouanas, V.K., C. Georgiakis W.L. Luyben, and L.H. Ungar (1988). Expert Multivariable Control. *Comp. Chem. Eng.*, **12**, 1065-1078.

Yamamoto, S. and I. Hashimoto (1991). Present Status and Future Needs: The View from Japanese Industry. In *Proceedings of the Chemical Process Control IV Conference*, Y. Arkun and W.H. Ray Editors, South Padre Island, TX.

CERTAINTY EQUIVALENCE ADAPTIVE CONTROL: WHAT'S NEW IN THE GAP

B. Erik Ydstie
Department of Chemical Engineering
Carnegie Mellon University
Pittsburgh, PA 15213

Abstract

The paper gives a tutorial overview of the adaptive robust control theory, including a detailed discussion of admissibility and parameter drift. The focus is design of adaptive, discrete time controllers for chemical processes leading up to the Laguerre modeling framework. The Laguerre family has an output error identification structure and is stabilizable. The parameter drift and admissibility problems are therefore easy to deal with in a theoretical and practical manner. The main message of the paper is that certainty equivalence offers significant scope for adaptive control of open loop stable systems with guarantees for stability and robustness with respect to unmodelled dynamics and bounded disturbances. The results can be extended to unstable systems using more complex algorithm design.

Keywords

Adaptive control, Nonlinear system, Process control, Stability, Robustness.

Introduction

In the *certainty equivalence* approach we fit a linear model to plant data and calculate controls as if the estimated model gave an exact representation of the system dynamics. The method is similar to optimal control using a Kalman filter for state estimation. However, the separation principle does not apply – direct implementation of certainty equivalence does not give optimality or even bounded input bounded output stability in general. Research has therefore focused on developing modifications to improve stability properties by solving the so-called admissibility and drift problems that may arise in the direct implementation of certainty equivalence. Significant progress has been made in this direction and solutions to a broad range of adaptive control problems have been obtained. During the past two decades algorithms have been developed achieving stability, robustness with respect to unmodelled terms and good disturbance rejection properties. The theory for bounded input bounded output stability and robustness of adaptive linear control has been completed, some transient performance and optimality issues have been addressed and the number of industrial applications keeps growing. The aim of the paper is to give a tutorial overview over these developments. We focus exclusively on the linear, certainty equivalence based methods and single input single output, discrete time control and estimation. All of results we review have counterparts in continuous time and generalizations have been proposed for multivariable systems.

Excellent reviews, papers and books describe theoretical and practical aspects of certainty equivalence adaptive linear control. An exhaustive review is therefore not warranted here. The historical background of adaptive, robust control and reviews of the local and global stability analysis can be found in Ortega's paper. Ioannou and Sun give an in depth treatment of the robustness analysis of adaptive control. Åström and Wittenmark describe connections between adaptive and robust control and review adaptive and self-tuning algorithms, including the industrially important relay tuning method. A forthcoming volume by Praly and Ydstie focuses on global stability, averaging and performance. Solo describes progress in stochastic adaptive control. Finally, advances in nonlinear adaptive control are reported in Kokotovic (1992) and Åström, et al. (1995).

Recently there has been interest in slow adaptation and iterative design. The averaging analysis shows that this approach gives local stability and robustness provided that a sufficiently reach excitation signal is added to the input or the reference (Anderson, et al., 1986; Sastry and Bodson, 1989; Praly and Ydstie, 1995; Solo and Kong, 1995). Slow adaptation and iterative design has been applied to adaptive LQ control by Bitmead, Gevers and Wertz. Kosut developed the robust calibration methods and recent trends are reviewed by Gevers. These methods are rooted in the certainty equivalence framework. The stability and performance properties can therefore be analyzed using methods developed for adaptive systems provided that the admissibility problem has been addressed.

Åstrøm and Wittenmark provide a recent review of the industrial status of adaptive process control applications. This latter paper adds to the already extensive and important reviews by Åstrøm, Seborg, Edgar and Shah, and Dumont.

The paper is organized as follows. In the first section we review the certainty equivalence paradigm for adaptive linear control. We then give design guidelines to achieve robust stability. In section four we describe the parameter drift issue using tools from nonlinear analysis. In section five we review stability results and in section six we find the opportunity to propose an algorithm for robust adaptive process control. We use this example to show how practical algorithm design and stability analysis go hand in hand.

A Paradigm for Adaptive Control: Certainty Equivalence

"A·dapt, to change so as to be or make suitable for new needs, different conditions, etc."
– Longman dictionary of contemporary English

Certainty equivalence adaptive linear control follows the format

1. Choose initial conditions
2. Estimate a linear model of the process using plant data.
3. Design a linear controller which stabilizes the estimated model.
4. Implement control action (and excitation in some approaches)
5. Return to step 2.

The algorithm model described above is an adaptive controller since step 5 implies retuning to suit new needs. Various permutations of the scheme can be implemented. For example, in the iterative approach we carry out step 3 infrequently since the controller is fixed during the identification phase. Steps 2 and 3 are skipped when the estimator is turned off. The adaptive controller reduces to linear time-invariant control when step 3 is skipped. Adaptive control includes self tuning as a special case. A self tuning controller tunes to one operating point and the decision to retune is made manually.

The objective of our paper is to describe how *adaptive robust control design* can be used to close the gap between theory and application. One of the first discussions of adaptive robust control was given by Goodwin, Hill and Palanaswami. In this paper the adaptive robust control problem was defined to be the combination of a "robust" parameter estimator in step 2 and a "robust" control law in step 3. Robust parameter estimation and robust control design may demand significant computational resources. The adaptive controllers discussed during the last decade therefore appear to be considerably more complex to analyze than their counterparts being discussed at CPC-III in 1986. It turns out that the fundamental issues remain unchanged and that all algorithms falling into the basic certainty equivalence scheme can be analyzed for bounded input bounded output stability and robustness using simple tools. Our paper therefore focuses on the shared and general properties of adaptive certainty equivalence control. Thereby we avoid having to review and explain a plethora algorithm developments and analysis techniques.

In fact, there are only two key issues that must be addressed in certainty equivalence control to achieve bounded input bounded output stability and robustness with respect to disturbances and unmodelled dynamics. The first issue is that the estimated model obtained in step 2 should be well behaved in the sense that the control design applied in step 3 stabilizes the estimated model. This imposes restrictions on the estimated model and controller structures. These should be matched so that it is possible to design a stabilizing controller. For example, consider an adaptive LQ controller. The control equation can be solved if the estimated model is stabilizable. With pole-assignment the model must be controllable to solve the Bezout equation. The problem is referred to as the *admissibility problem*. One early solution to a limited class of admissibility problems was provided by the Nussbaum gain (Nussbaum, 1983). A number of solutions to the admissibility problem have been developed since then. Many of them are complex to implement and some require significant computational resources since it is non-trivial to ensure stabilizability of a general, linear system (Staus, et al., 1995). The simplest and most practical approach for a large number of chemical engineering problems is to use an FIR or Laguerre model which is stable and therefore robustly stabilizable provided that the unmodelled dynamics are stable. An alternative and general approach has been provided by Morse et al.

The second problem is more subtle. Unstructured uncertainty and unmodelled disturbances cannot be compensated for by parameter tuning. Continuous parameter tuning can therefore give *parameter drift* as the estimation algorithm tries to converge the tunable parameters to infeasible or unstable solutions. A number of effective methods have been developed to solve this problem as well. The deadzone approach solves the drift problem by stopping the identification algorithm when the signals are not sufficiently excited to guarantee model improvement. Improvement is defined as a decrease in a Lyapunov like function which measures the size of the parameter error (Middleton, et al., 1988). The method is implemented such that the model changes only when the prediction error is larger than the combined effect of disturbances and unmodelled terms. The drift can be avoided by adding excitation to the setpoint or the manipulated variable to ensure that the signal to noise ratio is sufficiently large in the frequency range of interest. Averaging analysis then shows that the parameter estimates stabilize locally and the parameter drift problem is addressed (Anderson, et al., 1986; Sastry, et al., 1989). Excitation with fixed and finite energy leads to global stabilization and robustness of adaptive control (Radenkovic and Ydstie, 1995). A third method uses parameter projection to ensure that the parameter estimates do not go out of bound. This method was proposed by Egardt and it has been shown that this method also gives bounded input bounded output stability (Ydstie, 1992).

Research into robustness of adaptive linear control systems during the last 15 years focused almost exclusively on developing solutions to the two problems outlined above. A large and diverse collection of methods have been developed. The solutions obtained depend on choice of identification and control algorithms, model parameterization and to some extent the method of analysis.

In order to illustrate more precisely, consider a linear, single input single output system with output $y(t)$, disturbance $v(t)$ and manipulated input $u(t)$ related to each so that

$$y(t) = P(q^{-1})u(t) + G(q^{-1})v(t), \quad t = 1, 2, \ldots \quad (1)$$

Here t is time, normalized so that the sampling time is unity, $P(q^{-1})$ and $G(q^{-1})$ are transfer functions and q^{-1} is the backshift operator defined so that $q^{-1}u(t) = u(t-1)$. The objective is to use the manipulated variable to control the system to a setpoint $y^*(t)$.

In the first step of the certainty equivalence method we fit a nominal model $\{\hat{P}_\theta(q^{-1}), \hat{G}_\theta(q^{-1})\}$, which depends on a tunable parameter θ, to the process data. Without loss of generality we choose the nominal model and noise transfer functions so that

$$\hat{P}_\theta(q^{-1}) = \frac{B_\theta(q^{-1})}{A_\theta(q^{-1})}, \quad \hat{G}_\theta(q^{-1}) = \frac{D_\theta(q^{-1})}{C_\theta(q^{-1})} \quad (2)$$

where

$$\begin{aligned} A_\theta(q^{-1}) &= 1 + a_1 q^{-1} + a_2 q^{-2} + \ldots + a_{n_A} q^{-n_A} \\ B_\theta(q^{-1}) &= b_1 q^{-1} + b_2 q^{-2} + \ldots + b_{n_B} q^{-n_B} \\ C_\theta(q^{-1}) &= 1 + c_1 q^{-1} + \ldots + c_{n_C} q^{-n_C} \\ D_\theta(q^{-1}) &= d_1 q^{-1} + d_2 q^{-2} + \ldots + d_{n_D} q^{-n_D} \end{aligned}$$

The transfer function models in equation (2) are rational functions depending on the parameter set $\theta = \{a_i, b_i, c_i, d_i\}$. The model, often referred to as a Box-Jenkins model, can therefore be tuned by changing the parameters until a good fit has been found.

In most adaptive process applications the noise model is fixed using the internal model principle for rejection of deterministic disturbances (Middleton, et al., 1988). Typically, we set

$$\hat{G}_\theta(q^{-1}) = \frac{1}{1 - q^{-1}}$$

to achieve integral action. The issue of adaptive noise models has been discussed extensively in the literature on stochastic adaptive control and will not be treated here. This literature has had limited impact on adaptive process control up to this point in time. With a fixed noise model we identify one possible parameter vector used for control design as

$$\theta = (a_1, \ldots, a_{n_A}, b_1, \ldots, b_{n_B})$$

One could think of other parameterizations – a Laguerre model is described at the end of the paper. We now define the model error

$$\begin{aligned} e(t) &= A_\theta(q^{-1})y(t) - B_\theta(q^{-1})u(t) \quad (3) \\ &= \phi(t-1)'\theta \end{aligned}$$

where

$$\phi(t-1)' = (y(t-1), \ldots, y(t-n_A), u(t-1), \ldots, u(t-n_B))$$

is called the regression vector. The model is matched to the data by regressing the parameter vector θ against the process data so that the *filtered model error* is small in some average sense. If the l_2 performance is of interest we achieve the objective by minimizing

$$\begin{aligned} J_\theta &= \sum_{i=t-N}^{t} a(i)(F_\theta(q^{-1})e(i))^2 + \quad (4) \\ &\quad (\theta - \theta(t-N))^T P(t-N)^{-1}(\theta - \theta(t-N)) \\ &\text{Subject to: } \theta \in \Theta \\ &\quad P(t-N) \geq \epsilon I > 0 \end{aligned}$$

Θ is the constraint set for the parameter estimates. $F_\theta(q^{-1})$ is a filter chosen to suppress high frequency noise and shape the identification criterion. Typically $F = G_\theta^{-1} F_0$ where F_0 filters out the high frequency noise and the inverse of the noise model has the effect of filtering out periodic signals and bias. There has recently been interest in adaptive noise models. I is the identity matrix and ϵ is a positive number so that the matrix $P(t-N)$ is positive definite. Choosing $P(t-N)$ small slows down the rate of change the parameter estimates by trading off the emphasis placed on the new measurements relative to old estimates. $P(t-N)$ can for example be calculated using a Kalman filter with a forgetting factor. $0 \leq a(t) \leq 1$ is the discount factor (Zhou and Cluett, 1996). Setting $a(t) = 0$ turns the estimator off. Finally, the estimation scheme defined by minimizing J in equation (4) has a recursive solution as described in Goodwin and Sin (1984). However, stability theory applies equally well to batch estimation. There has recently been interest in using different objective functions for estimation. One possibility is to use an l_∞ criterion. The admissibility and drift issues still need to be addressed in order to achieve bounded input bounded output stability. The recursive least squares methods have dominated due to their computational simplicity.

In the second step we use the estimated model to design a feedback controller which satisfies the performance and robustness specifications. Towards this end we define the linear controller

$$R_\theta(q^{-1})u(t) = T_\theta(q^{-1})y(t)^* - S_\theta(q^{-1})y(t) \quad (5)$$

where $\{R_\theta(q^{-1}), S_\theta(q^{-1}), T_\theta(q^{-1})\}$ are polynomials in q^{-1} so that

$$\begin{aligned}
R_\theta(q^{-1}) &= 1 + r_1 q^{-1} + r_2 q^{-2} + \ldots + r_{n_R} q^{-n_R} \\
S_\theta(q^{-1}) &= s_1 q^{-1} + s_2 q^{-2} + \ldots + s_{n_S} q^{-n_S} \\
T_\theta(q^{-1}) &= t_0 + t_1 q^{-1} + \ldots + t_{n_T} q^{-n_T}
\end{aligned}$$

The subscript indicates that the controller depends on the estimated parameter through the application of a control design method like LQ or H$_\infty$.

The closed loop dynamics of the estimated model and the controller can be written by combining equations (3) and (5)

$$\begin{bmatrix} y(t) - y(t)^* \\ u(t) \end{bmatrix} = \frac{1}{(R_\theta A_\theta + B_\theta S_\theta)_t} \begin{bmatrix} B_\theta T_\theta - R_\theta A_\theta - B_\theta S_\theta, & R_\theta \\ A_\theta T_\theta, & -S_\theta \end{bmatrix}_t \begin{bmatrix} y(t)^* \\ e(t) \end{bmatrix} \quad (6)$$

This expression defines the so called "nominal feedback system" (Gevers, 1993). The model error and the setpoint act as an inputs and the tracking error and the manipulated variable are the outputs. In the adaptive case this system is time-varying since the parameter estimates vary as indicated by the subscript t.

When the parameters are fixed we say that the system is *frozen*. The frozen system is bounded input bounded output stable since step 3 of the algorithm gives a characteristic polynomial

$$P(q^{-1}) = R_\theta(q^{-1})A_\theta(q^{-1}) + B_\theta(q^{-1})S_\theta(q^{-1}) \quad (7)$$

with roots inside the unit circle and no roots on the unit circle corresponding to a Jordan block with size larger than one.

The main challenge is to design algorithms and analysis techniques that can be used to show that the model error remains bounded and small and can handle the time-varying (nonlinear) nature of the adaptive system. In order to address these issues we need to introduce the concept of robustness to handle the terms that cannot be modelled by parameter tuning alone. The breakthrough results published around 1980 relied on exact matching, implying that there exists a parameter vector θ so that the model error is equal to zero. One notable exception was the result of Egardt which applies to systems with bounded disturbances. Praly extended this result to unmodelled dynamics using a normalized estimator. In 1989 it was shown that the normalization was not required to achieve robustness with respect to unmodelled terms (Ydstie, 1989).

Imagine that there exists a parameter vector θ^* which is the best possible in terms of meeting the performance and robustness requirements. We call the model corresponding to this parameter for the *centered model*. By comparing equations (1) and (2) we define the additive, *unstructured mismatch* so that

$$\Delta(q^{-1}) = A_{\theta^*}(q^{-1})(P(q^{-1}) - P_{\theta^*}(q^{-1})) \quad (8)$$

It is called the unstructured mismatch since it is the mismatch which remains when the tunable parameter is optimized and set equal to θ^*. The mismatch is not a function of θ. However, it is a function of the excitation since the centered parameter is defined relative to the external input sequences $\{y^*(t), v(t)\}$.

The model error corresponding to the centered model is obtained by combining equations (1), (3) and (8)

$$\gamma(t) = \Delta(q^{-1})u(t) + A_{\theta^*}(q^{-1})G(q^{-1})v(t) \quad (9)$$

where the factor A_{θ^*} on the right hand side in equations (8) and (9) is a consequences of the equation error formulation in equation (3). In the output error formulation this factor does not show up.

The model error, defined in equation (3), can be written so that

$$e(t) = \phi(t-1)'(\theta^* - \theta) + \gamma(t) \quad (10)$$

where $\theta^* - \theta$ is the parameter error. In the robust control literature the parameter error is referred to as being structured. It follows that the frozen system is stable if

- the nominal closed loop is stable, and
- the unmodelled dynamics are small in the frequency range were the controller gain is large. I.e.

$$\Delta(q^{-1})V_\theta(q^{-1})$$

satisfies the Nyquist stability criterion with a margin large enough to take into account the structured error due to parameter error. The transfer function

$$V_\theta(q^{-1}) = \frac{-S_\theta(q^{-1})}{R_\theta(q^{-1})A_\theta(q^{-1}) + B_\theta(q^{-1})S_\theta(q^{-1})}$$

can be viewed as the controller response and it is typically designed to roll off at high frequencies. This allows the mismatch $\Delta(q^{-1})$ to be larger at higher frequencies.

If these conditions are satisfied there exist parameters θ_i in an open neighborhood of the fixed parameters θ_i^* so that the closed loop is stable. The main advantage of adaptive control is that the control algorithm is less sensitive to both structured and unstructured parameter error than an algorithm using fixed parameters when θ^* is not known accurately. The parameter θ in equation (10) is tuned so that the magnitude of the error $e(t)$ is minimized by the parameter identifier. Adaptive control therefore has the potential of finding a parameter θ which gives comparable stability and robustness properties to that which would be achieved using the unknown, centered parameter θ^*.

In the adaptive case the parameters vary and the stability must be analyzed in the context of what is known about stability of time-varying linear or more generally, nonlinear systems. The analysis, which is briefly reviewed in Section 5 of the paper shows that the adaptive system is stable if the unmodelled dynamics are not too large in the frequency range where the control gains are large and the centered parameter does not vary too fast in an average sense. We get the following performance bounds for the adaptive system (Kelly and Ydstie, 1995).

$$\|y - y^*\|_2 \leq k_1 + k_2\|y^*\|_2 + k_3\|v\|_2$$
$$\|y - y^*\|_\infty \leq k_4 + k_5\|y^*\|_\infty + k_6\|v\|_\infty$$

where the k_i's are constants that depend on system parameters and the adaptive design. The l_2- and l_∞- norms of a signal $x(t)$ are defined so that

$$\|x\|_2 = \sqrt{\sum_{1}^{\infty} x(i)^2}, \qquad \|x\|_\infty = \max_{0 < i \leq \infty} |x(i)|$$

Certainty equivalence adaptive schemes therefore give finite gain stability, robustness with respect to small unmodelled dynamics and continuity in the sense that the magnitude of the tracking error tends to zero as the magnitude of the external perturbations tends to zero. Furthermore, the l_2 and l_∞ results can be used to address performance related issues. The results have been applied to one step ahead control and it has been shown that the theory corresponds well with simulation in predicting worst case bounds for the transient. The steady state (l_2) performance of the adaptive system is no more than twice as large as that obtained using the centered parameter (Ydstie, 1992). In the stochastic case it has been shown that the minimum variance controller using least squares parameter estimation converges to the optimal mean square performance (Guo and Chen, 1991). These results do not say anything about convergence of the parameter estimates.

Low level excitation gives stability of the parameter estimator. The excitation must be supplied externally since there is no guarantee, under general conditions, that the adaptive controller generates excitation by itself once the controller has been tuned to give stable performance. The exciation must be "sufficiently strong" to overcome the effects of disturbances and unmodelled dynamics and it must be chosen with the appropriate frequency content so that the unmodelled dynamics are not excited in order to give stable parameter estimation. External excitation does not influence the boundedness properties of the adaptive algorithm and excitation can therefore be used to estimate parameters to a given precison. The stability results cover intermittent adaptation as long as the update algorithm remains active in the sense that the parameter identifier is activated from time to time (Ydstie and Wahlberg, 1994).

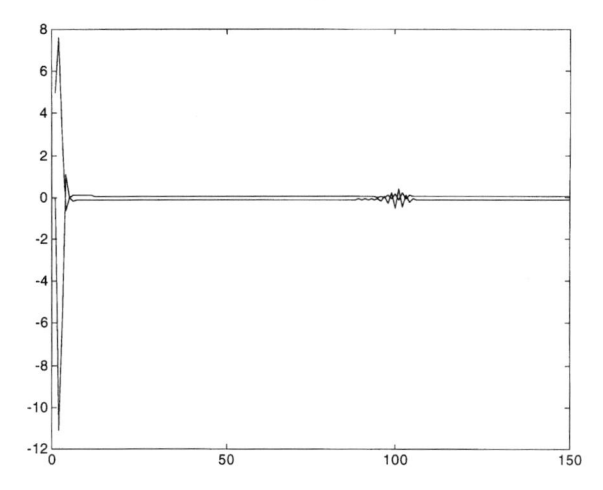

Figure 1. Continual adaptation may lead to small amplitude bursting. The process output (y) initially increases and settles for awhile. Then it starts to oscillate. The process input (u) initially decreases and displays similar behavior during the burst.

Example: Consider adaptive control of the following, second order system

$$y(t) = \frac{q^{-1} + 0.3q^{-2}}{1 - 1.5q^{-1} - 0.5q^{-2}} u(t) + 0.1$$

The system is unstable and stably invertible with poles at $[1.78, -0.28]$ and zeros at $[0, -0.3]$. There is a bias equal to 0.1. We will use an adaptive, certainty equivalence controller based on a first order model

$$y(t) = \frac{\hat{b} q^{-1}}{1 - \hat{a} q^{-1}} u(t)$$

The estimation algorithm tunes the parameters $\hat{a}$ and $\hat{b}$ and uses the estimates for control design. Stable invertibility implies that we can use a one step ahead predictive controller

$$u(t) = \frac{y^*(t+1) - \hat{a}(t)y(t)}{\hat{b}(t)}$$

This controller chooses the control input so that the predicted output is equal to the setpoint. The control law is well defined if $\hat{b} \neq 0$. This is the essence of the admissibility problem. In the Self Tuning Regulator the admissibility problem was solved by assuming that a "workable" lower bound for $|\hat{b}|$ and the sign of $\hat{b}$ was known so that the estimate could be "projected away from zero" to avoid small divisors (Åstrøm and Wittenmark, 1973). The tracking error then satisfies

$$y(t) - y(t)^* = e(t)$$

Thus, we get small tracking error if the model error is small. We will use the least squares algorithm with $N = 1$,

$$P(t) = \begin{pmatrix} 1 & 0 \\ 0 & 1 \end{pmatrix}, \qquad a(t) = \begin{cases} 0 \text{ if } |e(t)| \leq \Delta \\ 1 \text{ otherwise} \end{cases}$$

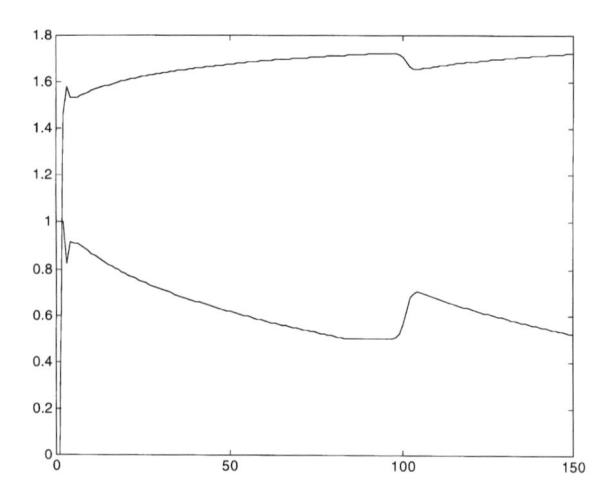

Figure 2. The parameter estimates drift. The estimate of the parameter $\hat{a}$ is shown on top and the parameter $\hat{b}$ on the bottom. Both parameters are constrained. The parameter $\hat{b}$ reaches the constraint around $t = 80$.

and
$$y^* = 0, \quad a_{min} = -1, \quad a_{max} = 2,$$
$$b_{min} = 0.5, \quad b_{max} = 2$$

The analysis developed in Radenkovic and Ydstie (1995) applies and we get bounded input output stability provided that the following sufficient condition is satisfied

$$\left\| \frac{\Delta(z)A_{\theta^*}(z)}{B_{\theta^*}(z) - \Delta(z)(A_{\theta^*}(z) - 1)} \right\|_\infty < 1$$

where $\|\cdot\|$ denotes the H_∞ norm of a bounded operator. This condition is satisfied.

The simulation result for the first 150 time steps is shown in Fig. 1 with no deadzone ($\Delta = 0$). What we notice here is an initial tuning phase and an instability around $t = 100$. At first the output grows to a maximum value, then it decreases and settles to a value close to the setpoint. During the initial transient there is considerable excitation and the parameters quickly converge to nearly stationary values (Fig. 2). There is a small offset due the persistent disturbance. The parameters then drift since the estimator cannot find parameters that exactly match the process data. Around $t = 100$ there is a small burst due to temporary, closed loop instability. The parameters quickly retune. The bursting will persist at regular intervals. However, the magnitude will remain small according to the theory. The offset can be avoided by including integral action in the control law and a deadzone will arrest the drift as shown in Fig. 3. In this case we use $\Delta = 0.2$. Larger deadzone will lead to larger steady state error unless the integrator is included. The bursting will reappear if we set $\Delta < 0.1$. □

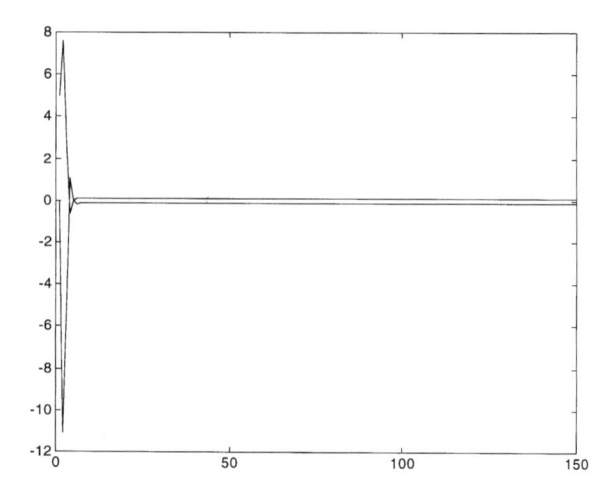

Figure 3. The parameter drift and bursting is arrested by the deadzone.

Algorithm Design Using Certainty Equivalence

"Not much has happened in the area of adaptive control since our review paper was published in 1986"
– Tom Edgar, The CAST dinner lecture, Miami Beach, 1996.

The certainty equivalence approach offers fertile ground for algorithm development. With N linear controllers and M identification algorithms we design $N \times M$ certainty equivalence adaptive controllers. At CPC III, Dumont gave an overview of adaptive techniques and industrial applications. Seborg, Edgar and Shah published an extensive review of the status of adaptive process control around the same time. About 82 experimental applications of adaptive control were listed in the latter paper; 31 were described as "full scale industrial applications". A broad range of control and estimation theory was also reviewed as indicated in Table 1. Indeed, few new adaptive techniques have been invented since then and many of these techniques had in fact been proposed already around 1960 (Åström and Wittenmark, 1995).

A large number of application studies have been published in addition to the ones reviewed about 10 years ago. Fig. 4 reports the number of papers published in Automatica listing adaptive control as a key word since 1985. A listing of recent industrial process applications is given in Table 2. None of the algorithms satisfy the criteria for being bounded input bounded output stable. (The admissibility problem has been ignored in all cases, indicating that the issue may not be of practical concern. The author does not share this view at all). The parameter drift issue is addressed by excitation and/or deadzones in some applications and ignored in other applications. (The author believes that the drift problem has to be addressed in a practical adaptive control application). There are few published applications of adaptive control to cehmical processes in the US. The techniques that are applied do not vary substantially in spirit or implementation from one application to another. Finally, most applications

Table 1. Partial list of techniques surveyed by Seborg, Edgar and Shah (1986). Descriptions and key references are given in this paper.

Estimation algorithms	Control
least squares	deadbeat
maximum likelihood	minimum variance
extended least squares	d-step ahead
approximate maximum likelihood	multivariable linear quadratic
forgetting factor, fixed	predictive
forgetting factor variable	(EHAC, EPSAC MUSMAR, STC)
covariance reset	pole-assignment
pattern recognition	ÅW, WP, Dahlin, VE
square root estimation	stable adaptive
projection	
parameter filter	
instrumental variable	
pseudo linear regression	

Table 2. Representative industrial trial applications of adaptive process control.

Application	Method	Authors
Chip refiner	multiple model, dual adaptive	Allison et al. (1994)
Hot-dip galvanizing	iterative LQ	Partanen and Bitmead (1995)
Glass furnace	predictive control	Zhang et al. (1994)
Bio-reactor	adaptive nonlinear	Pomerleau and Viel (1992)
Alumina calciner	pole assignment	Mills et al. (1991)
Bleach plant	predictive with Laguerre	Zervos and Dumont (1988)
Annealing	predictive and pole assignment	N. Yoshitani (1993)
Supercritical Boiler	predictive	Fujiwara and Miyakawa (1990)

concern processes that are fairly difficult to model using process physics.

The most important development since CPC III is the emergence of a theory for robustness and performance of adaptive systems. Below we outline a design approach for adaptive control systems which takes advantage of these developments. Application results in an in adaptive design with bounded input bounded output stable response, robustness with respect to bounded disturbances and unmodelled dynamics.

- *Model structure:* The most important issue in adaptive control is the choice of model structure. One example is the equation error structure given by equation (3). This choice is common but not usually the best. The admissibility problem is difficult to solve for general autoregressive models and the equation error formulation gives biased estimates. Significant improvements in the performance can be achieved by using *apriori information* to design good noise models to ensure that error minimization leads to unbiased parameter estimation. The following model structures and noise models have been used in conjunction with adaptive control:

  - Auto regressive models
  - State space and instrumental variable representations
  - Finite Impulse Response and Laguerre expansions
  - Internal models for rejection of deterministic disturbances

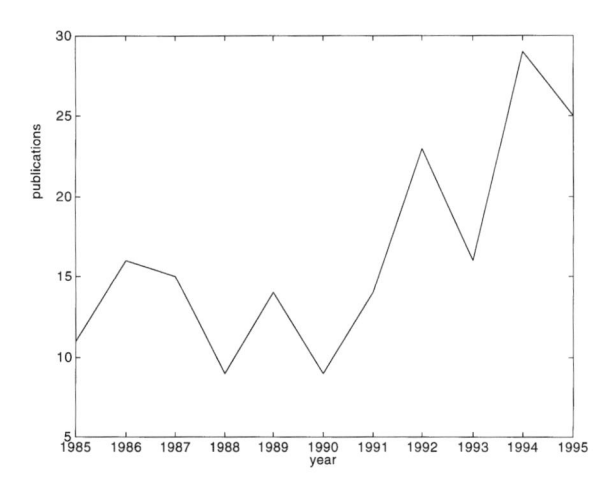

Figure 4. The number of papers published in Automatica, lisiting adaptive control as a keyword, has been on the increase during the last ten years.

– Stochastic noise models

The output error and Box-Jenkins model stuctures are promising candidates for adaptive control. The main difficulty is that the objective function is nonlinear, leading to a significant increase in the computational complexity. These model structures have therefore not played significant roles. The advances made in algorithm design and computer speed make the computational issue less important than it was a decade ago. The output error structure can be maintained without sacrificing linearity and convexity by using the FIR (Finite Impulse Response)/Laguerre modeling framework. Adaptive controllers using these models in conjunction with a controller capable of stabilizing all stable plants do not have singularities and the admissibility problem is a non-issue.

- *Parameter tuning:* Many effective tuning algorithms have been developed using modified versions of least squares. The following modifications have been considered to improve robustness

    – Adaptive deadzone to turn estimator on or off
    – Signal dependent normalization for robustness
    – Square root factorization for numerical robustness
    – Adaptive noise filtering
    – External excitation
    – Forgetting factors and covariance reset to retain adaptivity
    – Parameter projection or leakage (σ-modification)

The only requirement associated with the estimation algorithm are that N is finite and the matrix $P(t - n)$ in the criterion function (4) is positive definite so that

$$\epsilon I \leq P(t - N) \quad \text{for all } t$$

where $\epsilon > 0$ is a positive number and I is the identity matrix. Periodic resetting, condition number monitoring and numerically stable updates for recursive algorithms are effective ways to achieve such properties. $P(t)$ does not need to be bounded from above. However, large $P(t)$ gives poor smoothing and large parameter variations and poor steady state performance. The deadzone provides a way to improve the steady state performance during quiescent periods. The deadzone can be implemented in a number of different ways and it can be adapted to the data as discussed in Ortega, et al. (1992). The only important consideration is that the deadzone should not be too large. It is popular, to leave the estimator running and adding external excitation to the setpoint or the controlled input. Excitation leads to poorer tracking. However, the bounded input bounded output stability properties are not affected and the designer therefore has significant degrees of freedom in choosing the excitation signal. A discussion on excitation is given in (Sastry and Bodson, 1989).

Transient performance analysis has shown that it is very effective to use large $P(t)$ during a transient to achieve rapid parameter estimation. Furthermore, instabilities generate excitation such that poorly estimated parameters are estimated. In this way adaptive control algorithms achieve good transient performance using variable forgetting factors and covariance resetting.

- *Control design:* Any linear controller can be written on the form given by equation (5). A large number of different control design techniques have been suggested for certainty equivalence adaptive control including

    – Model Predictive Control (MPC)
    – Linear quadratic control
    – Model reference control
    – Minimum Variance and Stochastic Control
    – Pole assignment
    – H_∞ robust control
    – Internal Model Control (IMC)

The choice of control law, although a subject of significant and at times heated discussion (Bitmead, et al., 1990), is not as important as it may appear at first sight. The only issues that matter are that the estimated model is stabilized and that the controller does not generate inputs that excite the unmodelled dynamics. These issues are quite well understood in the context of linear control theory and we will not elaborate here.

We now give a more general description of the admissibility problem. Suppose that A_θ and B_θ have a common factor C_θ so that $A_\theta = C_\theta A'_\theta$ and $B_\theta = C_\theta B'_\theta$. We can then write the characteristic polynomial (7) so that

$$C_\theta(R_\theta A'_\theta + B'_\theta S_\theta) = P_\theta$$

The common factor must show up in the denominator of the closed loop. An unstable common factor therefore implies closed loop instability.

A number of different techniques have been developed to deal with the admissibility problem. Some attack the estimator and some the control design. Methods that have been proposed include

– doing nothing and hoping for the best
– stably invertible plant via slow sampling
– hysteresis switching
– modification to estimated parameters
– non-convex optimization
– signal dependent excitation
– parameterization (FIR/Laguerre)

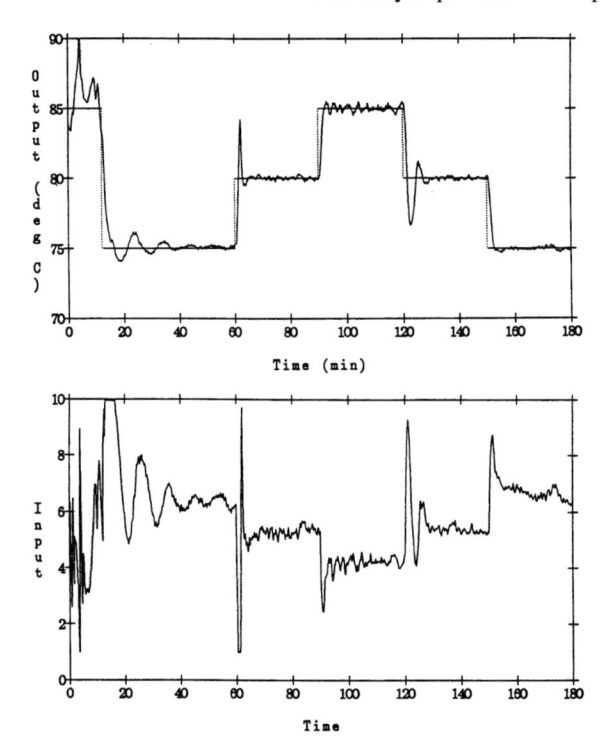

Figure 5. At time $t = 60$ there is a near pole zero cancellation leading to large gains and a spike in the control input. The control input (reflux rate) is shown at the bottom and the controlled output (temperature) is shown at the top. The controller is an LQ controller.

- parallel estimation
- projection into a convex region
- supervisory control

The first solution is by far the most common and is justified on the basis that it is unlikely that the estimated model (3) has unstable poles and zeros that cancel exactly and even less likely that a near common factor will persist – it leads to large control gains and excitation so that the parameters retune to values that hopefully do not give pole zero cancellation. Nevertheless, the performance can be very bad during the transient since the near cancellations gives large control inputs and spikes as shown on Fig. 5.

Many chemical engineering systems are open loop stable, overdamped and have significant delay. Such systems can be modeled with an FIR or Laguerre expansion which does not support unstable pole-zero cancellations. The Laguerre expansion method is therefore recommended for adaptive control of a chemical systems. More recently a method based on non-convex optimization has been developed to handle unstable and oscillatory systems (Staus, et al., 1995).

The adaptive system with least squares estimation behaves almost like a linear system during steady operation. It appears reasonable therefore, to apply the formula for "bias" distribution and other tools from linear identification and control to investigate trade-off issues as discussed in Van den Hof and Schrama (1995). The formula for the bias is derived from the identification cost criterion for the infinite horizon cost and in one form it can be written (Bitmead, et al., 1990; Partanen and Bitmead, 1995).

$$V_N(\theta) = \int_{-\pi}^{\pi} [|P(e^{j\omega}) - \hat{P}_\theta(e^{j\omega})|^2 \Phi_r(\omega) + |1 + \\ + \hat{P}_\theta(e^{j\omega})C_\theta(e^{j\omega})|^2 \Phi_v(\omega)] \qquad (11) \\ \times \frac{|F_\theta(e^{j\omega})|^2}{|1 + P(e^{j\omega})C_\theta(e^{j\omega})|^2|\hat{G}_\theta(e^{j\omega})|^2} d\omega$$

where $C_\theta = -S_\theta/R_\theta$ is the controller transfer function and Φ_r and Φ_v are the reference and disturbance spectra.

Expression (11) highlights trade-offs between, filtering, identification, control and external excitation. We get a good match between the plant and the model when the dither signal is chosen properly and the bandwidth of the controller, noise model and filter respectively are all chosen to suppress the influence of the external disturbances. Unfortunately, the amount of apriori knowledge needed to use expression (11) effectively is quite extensive.

Local Performance of Adaptive Control Systems

"It probably doesn't matter so much what you do as long as you do something"
– Bo Egardt, IFAC World Congress, Budapest, 1984.

Nonlinear analysis, simulation and experiments shows that continual adaptation can give small amplitude chaos even when the admissibility problem is solved and parameter projection is applied to bound the parameter estimates. The amplitude of the tracking error can be shown to be of the order of the external noise and these phenomena do therefore not influence the average performance of the adaptive system in a significant manner. The complexity of these phenomena defy complete analysis in all but the simplest examples.

Example 1: This example is described in Golden and Ydstie (1992). Consider adaptive control of the following system

$$y(t) = bu(t-1) + v$$

v is a constant perturbation and b is the unknown parameter. The recursive estimation algorithm gives

$$\theta(t) = \theta(t-1) + \frac{pu(t-1)e(t)}{r + pu(t-1)^2}$$

where p is the adaptive gain and $e(t)$ is the prediction error. The one-step-ahead predictive controller is defined so that

$$u(t) = \frac{y^*}{\theta(t)} \qquad \theta(t) \neq 0$$

The fixed point $y = y^*$ is globally, uniformly stable if and only if

$$0 < \left(1 - \frac{v}{y^*}\right) p < 1 + \sqrt{1 + 2K^2/y^{*2}} \qquad (12)$$

This double inequality shows important trade-offs between estimation and control and gives hints about possible modifications.

1. We can tune the rate of adaptation p to achieve stability. In particular, from inequality (12) it is clear that we get uniform global stability if

$$p \approx \frac{y^*}{y^* - v}$$

   We need to know the disturbance v to implement this method.

2. We can satisfy inequality (12) by changing the setpoint. This is referred to as excitation. We can for example choose y^* large enough so that $|v/y^*| < 1$. We then get stability provided $p \leq 1$. We require knowledge of v to achieve uniform convergence and we suffer an offset by adding excitation.

Let us see now see what happens when inequality (12) is violated.

1. The lower bound in inequality (12) is violated if $v > y^*$. We then get $\lim \theta(t) = \pm\infty$ where the sign depends on the initial condition. Parameter projection, excitation and/or a deadzone triggered by small errors stops this drift.

2. We observe period doubling bifurcations and chaos when the upper bound is violated. This type of behavior can be avoided by using a smaller adaptive gain. The recursive least squares method is effective since the gain rapidly tends to zero.

Nonlinear analysis shows that the parameter estimate behaves as a random variable when the estimator gain is large. Division with zero in the control equation is therefore a rare event. This explains why the admissibility problem does not occur frequently in a practical application of adaptive control.

Example 2: This example is developed in Espana and Praly (1993) and Golden and Ydstie (1994). We consider adaptive control of the system

$$y(t) + ay(t-1) = u(t) + v$$

The parameter a is estimated and v is a constant perturbation. A local analysis shows that the condition for stability of the fixed point $y = y^*$ is

$$\left|\frac{v}{y^*}\right| < 1$$

A violation leads to a bifurcation and local instability via one of three different routes. Two of these can be classified as high gain instabilities.

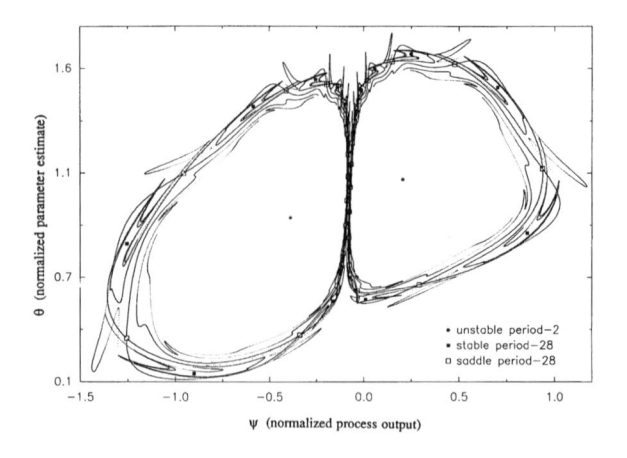

Figure 6. The stable and unstable manifolds associated with a period 28 point describe the burst phenomenon. This figure was prepared by Dr M. P. Golden for her Ph.D thesis at University of Massachusetts. She used software developed at Princeton University in the research group of Prof. I. G. Kevrikidis.

1. The high gain instabilities lead to bifurcation and low amplitude chaos. The nature of this instability is very similar to the one described above and it can be avoided by using small adaptive gain. The least squares method achieves this by reducing the adaptive gain as more information is added.

2. Poor noise to signal ratio gives a global, supercritical Hopf bifurcation and the apperance of invariant circles, frequency-locking and intertwined quasi-periodic (unlocked) and periodic (locked) states. An understanding of this process can be obtained from a picture showing the intersecting stable and unstable manifolds of two neighboring saddles as shown in Fig. 6. A simulation shows the chaotic drift and bursting as shown in Figures 1 and 2 and has been well documented in the adaptive control literature. This problem can be avoided by using a deadzone or low level excitation.

Praly analyzed the effect of modifications, including the deadzone, excitation and disturbance rejection using an internal model. These methods cause some performance loss. At the IFAC World Congress in Sidney a challenge problem was posed for adaptive control. Results were reported in *Automatica* Vol. 30, No.4, 1994. The results are remarkable – very different approaches gave very similar results.

Facts Concerning Stability and Performance

In this section we review very briefly the basis for theoretical analysis of a very general class of adaptive linear controllers. The approaches that have been developed apply to all certainty equivalence schemes and are the result of two decades of research by a number of research groups throughout the globe. Only the highlights will be given here using the notation and style adopted in this paper. First we rewrite the centered model, the estimated model and the control equations so that

$$\begin{bmatrix} A_{\theta^*}(q^{-1}) & -B_{\theta^*}(q^{-1}) \\ A_{\theta}(q^{-1}) & -B_{\theta}(q^{-1}) \\ S_{\theta}(q^{-1}) & R_{\theta}(q^{-1}) \end{bmatrix}_t \begin{bmatrix} y(t) \\ u(t) \end{bmatrix} = \qquad (13)$$

$$\begin{bmatrix} \gamma(t) \\ e(t) \\ T_{\theta}(q^{-1})y^*(t) \end{bmatrix}_t$$

From this expression we derive a (non-minimal) state representation. In the simplest case we use the state vector

$$x(k) = (y_f(t), y_f(t-1), ..., y_f(t-n_y), u_f(t), ...,$$

$$u_f(t-n_u), y_f^*(t), y_f^*(t-1), ..., y_f^*(t-n_{y^*}))$$

where n_y, n_u, n_{y^*} are the appropriate orders and subscript f implies the removal of bias terms by application of an internal model so that the steady state for constant perturbations is given by $x(t) = 0$. We get

$$\begin{aligned} x(t+1) &= F_{\theta}(t)x(t) + G_{\theta}(t)y^*(t) + \eta(t) \\ \begin{pmatrix} y(t) \\ u(t) \end{pmatrix} &= H_{\theta}(t)x(t) \qquad (14) \\ \theta(t) &= \theta(t-1) + K(x(t-s), \theta(t-1), t) \end{aligned}$$

The last equation represents the parameter estimation algorithm. The terms are defined as follows

- x is the state of the nominal closed loop system. This state includes the state of the controller, filters, variables needed for the tuner and the internal model for the disturbances.

- $F_{\theta}(t)$ is a time-dependent matrix describing the closed loop dynamics of the nominal system.

- $G_{\theta}y^*$ is a vector describing the influence of the setpoint on the closed loop.

- $\eta(t)$ is a vector derived from the model error and the so-called swapping terms that result from the fact that time-varying polynomials do not commute.

- H_{θ} describes how the input u and the output y is connected with the state.

- $K(x(t-s), \theta(t-1), t), t \geq s \geq 0$ is called the gain scheduler, tuner or identifier.

Stability and robustness analysis of adaptive systems is greatly facilitated by defining a weighting sequence so that

$$s(t)^2 = \tau^2 s(t-1)^2 + u(t-1)^2 + y(t)^2, \qquad s(0) = 0$$

This sequence provides a bound for all system signals and showing boundedness of $s(t)$ is therefore equivalent to showing boundedness of the adaptive system. The sequence $\{s(t)\}$ is called a *normalization sequence*. An algorithm which uses such a sequence explicitly is called a normalized algorithm. The results in Ioannou and Sun (1996) mostly rely on normalization. The algorithm in Middleton, et al. (1988) uses a slightly different normalization to compute a deadzone. In Ydstie (1992) it was shown that it was not necessary to use normalization to achieve robust stability of adaptive control systems.

We will now state the sufficient conditions for stability.

1. *Nominal stability:* $F_{\theta}(t)$ is (Lipschitz) continuous in θ and has all eigenvalues strictly inside the unit circle. $|F_{\theta}(t)|, |G_{\theta}|$ and $|H_{\theta}|$ are bounded.

2. *Gain schedule:* There exists a constant c_1 and a sequence $0 \leq L(t) \leq L_0$ so that

$$|\eta(t)| \leq c_1 \sum_{i=t-d}^{t} s(t)$$

$$\left(L(t) - L(t-1) + |W(t)| + \frac{\omega(t)}{n(t)} \right)$$

where $n(t)$ is a factor used in the estimation algorithm. In a normalized algorithm we have $n(t) = s(t)$, $\omega(t)$ is related to the bounded disturbance $v(t)$, possibly via a filter and $W(t)$ is related to the rate of change of the parameters describing the centered model.

3. *Smallness of unmodelled dynamics, slow variations in $\theta(t)^*$ and bounded noise:* There exists numbers β_1, β_2 and γ with γ sufficiently small so that for all $N > 0$

$$\frac{1}{N} \sum_{i=t-l}^{t} |W(i)| + \frac{\omega(i)}{n(i)} \leq \frac{\beta_1}{N} +$$

$$\gamma^2 (1 + \sup_{t-N \leq i \leq t} \left(\frac{s(i)^2}{n(i)} + \frac{\beta_2}{n(i)} \right)$$

The Lipschitz property can be relaxed provided that discontinuities do not occur too frequently. There are therefore many ways to implement adaptive controllers. Starting with these conditions we derive stability results for certainty equivalence adaptive linear control. In general what is shown is that there exists constants $\delta, k_i, i = 1, ..., 3$ so that

$$\|x(t)\| \leq \max\{k_1 e^{-\delta t}\|x(0)\|, k_2|v|_{\infty}, k_3|y^*|_{\infty}\}$$

The following insights are drawn.

1. The magnitude of the state vector x depends on the initial conditions in an exponentially decaying manner.

2. The asymptotic performance (bound on the magnitude of the state vector after the transient has died out) is continuous in external disturbances. I.e small changes in disturbance, small model mismatch and small setpoint changes give small excursions.

3. Slowly time-varying and large infrequent jumps in the parameters can be tolerated.

4. Norm bounded (additive and multiplicative) unstructured unmodelled dynamics are tolerated provided these are small in the frequency range where the control gains are large.

5. The magnitude of the constants δ and k_i depend on how the search domain is compactified.

Persistent excitation gives the following property

$$\limsup_{t \to \infty} \|\theta(t) - \theta^*\| \leq d_o$$

where θ^* is the optimal parameter, d_0 is a constant which depends on the power and frequency content of excitation and almost all parameters of the identification and control algorithms as well as the system dynamics (Radenkovic and Ydstie, 1995). Note that θ^* is a function of the excitation power spectrum.

Averaging and other tools from nonlinear analysis are significant because they give precise information about the stability/instability of integral limit sets describing the steady state behavior of the adaptive control system. This information is critical since equation (11) only applies if the steady state is stable. A detailed exposition of these ideas can be found in Anderson, et al. (1986), examples of application are discussed in Espana and Praly (1993) and Bitmead, et al. (1990) and a brief overview is provided in Praly and Ydstie (1995).

In order to present the main idea we will use the representation given by equations (15) and assume that that there are no unmodelled dynamics. This is not a significant limitation since unmodelled dynamics can be viewed as bounded perturbations when the adaptive system is stable. The main idea consists of re-scaling the system by introducing new variables

$$z = \frac{x}{\sqrt{\epsilon}}, \qquad w = \frac{\omega}{\sqrt{\epsilon}}$$

and rewriting (15) so that

$$\begin{aligned} z(t+1) &= A(\theta)z(t) + B(\theta)w(t) \\ \theta(t+1) &= \theta(t) + \epsilon K(z(t-s), \theta(t-1), w(t)) \end{aligned}$$

This system is now a two-time scale system where the parameter update law is viewed as a slow system and the equation for the evolution of z is the fast system. The major conclusion one can draw from the averaging analysis is that there exists an integral manifold which is (locally) attractive if and only if *the persistency of excitation and positive real conditions* are satisfied.

The conditions for bounded input bounded output stability are considerably weaker. These do not require persistent exciation. The bounded input bounded output stability results rely on a rather interesting feature of the certainty equivalence control approach: Excitation is generated when the algorithm gives closed loop unstable control. The parameters therefore retune to optimal values if the instability persists. The certainty equivalence algorithm is therfore *self stabilizing* for this type of instability. However, infinite parameter drift may occur along a manifold which is unobservable from the model error. This drift is associated with small prediction errors and can be stopped by a deadzone and/or low level excitation. In fact, simply constraining the parameter estimates will achieve the desired bounded input bounded ouput stability property.

Adaptive Process Control: A Robust Approach

"Sell, your ideas – they are totally acceptable."
– Chinese cookie # 487.

In this section we develop an algorithm for adaptive process control which gives bounded input bounded output stability, robustness with respect to unmodelled dynamics. We base the approach on the Laguerre modeling concept developed by Wiener and Lee and introduced to adaptive control by Zervos and Dumont.

1. *Parameterization and filtering:* We expand the transfer function $G(q^{-1})$, which is assumed to be stable, in terms of the Laguerre functions

$$L_k(z, a) = \frac{\sqrt{(1-a^2)}}{z-a} \left[\frac{1-az}{z-a}\right]^{k-1}$$

Here $|a| < 1$ is a parameter which is chosen *apriori*. We augment the Laguerre model with FIR terms as described in Finn, et al. (1995) to better handle long timedelays. The key observation is that the base functions

$$\left\{z^{-k}\right\}_{k=1}^{m}, \quad \left\{L_k(z,a)z^{-m}\right\}_{k=1}^{\infty}$$

in discrete time equivalence with the classical continuous Laguerre functions, form an orthonormal and complete set of base functions in $\mathcal{H}_2$. These functions can therefore be used to model any open loop stable system provided that enough terms are included. One will find that the Laguerre approximation converges much faster than the FIR expansion for stable, overdamped systems if the Laguerre pole is chosen consistent with the dominant time constant of the process. The Laguerre based adaptive controller can therfore be designed with few estimated parameters.

We use the moving average noise model and write the FIR-Laguerre model on state space form as described in Finn, et al. (1995). This gives the incremental regression model

$$y(t) = y(t-1) + \phi(t-1)'\theta^* + \gamma(t) \qquad (15)$$

where $\phi(t-1)$ is the state of the Laguerre model. θ^* is the set of parameters to be estimated and γ is the centered model error.

2. *Parameter identification:* Equation (15) defines the model error

$$e(t) = y(t) - y(t-1) - \phi(t)'\theta(t-1)$$

where $\theta(t)$ is the parameter estimate. The least squares objective function is convex and the recursive implementation of least squares reads

$$\theta \mapsto \theta + \frac{P\phi(t-1)F(q^{-1})e(t)}{\lambda r + \phi(t-1)'P\phi(t-1)}$$

$$P \mapsto \frac{1}{\lambda}\left(P - \frac{P\phi(t-1)\phi(t-1)'P}{\lambda r + \phi(t-1)'P\phi(t-1)}\right) + Q$$

The algorithm requires initial conditions for P and θ. λ, r and Q are tuning parameters.

- The tuning parameter Q should be chosen so that

$$0 < \epsilon I \leq P(t-1), \quad \text{for all } t.$$

  The lower bound is needed to ensure adaptivity. Algorithms based on the concept of a variable forgetting factor (λ), covariance resetting (Q) and UDU' factorizations have been shown to work well.

- A fixed deadzone should be implemented. It suffices to turn the estimator off if $|e(t)| \leq \Delta$ where $\Delta > 0$ is an estimate of the noise level. The estimation algorithm then stops when the prediction error is small and the parameters will not drift. Algorithms based on fault detection can be applied as long as the deadzone does not become too large (Hagglund, 1983; Ortega and Garcia, 1992).

- We impose upper and lower bounds on the estimated parameters so that

$$\underline{\theta}_i \leq \theta_i(t) \leq \bar{\theta}_i$$

  where $\underline{\theta}_i$ and $\bar{\theta}_i$ are lower and upper bounds for θ_i^* respectively.

- Low level excitation can be added without destroying the stability properties of the algorithm. The addition of excitation guarantees that the parameters are estimated.

3. *Controller design:* A large class of controllers for model (3) can derived from the following quadratic objective

$$\min_{u_f(t+i)} \sum_{i=0}^{T} q_i(\hat{y}_t(t+1+i) - y^*(t+1+i))^2 + r_i u_f(t+i)^2 \quad (16)$$

The functions $u_f(t+i) = \hat{G}_\theta^{-1}(q^{-1})u(t)$ are the decision variables and the functions $\hat{y}_t$ are predictions based on information at time t. The inclusion of the noise model in the control objective gives rejection of deterministic disturbances. $0 \leq \{T, q_i, r_i\} \leq \infty$ are scalar tuning parameters. A good candidate is the infinite horizon LQ controller obtained by fixing $q = 1, r > 0$ and $T = \infty$. This controller has a linear feedback structure and stabilizes all stabilizable plants. The admissibility problem is not an issue since the Laguerre model is open loop stable and stabilizable for all θ.

We now have an algorithm which is guaranteed to give bounded response for all initial conditions, exponential decay of initial transient, continuity with respect to external perturbations and robustness with respect to unmodelled dynamics. The tradeoff between performance and robustness is given by r. Adaptive feed-forward from measured disturbances can be applied to improve performance without sacrificing stability.

A typical experimental result using the adaptive controller is shown in Fig. 7. The objective is to control temperature on the second plate of a methanol-water distillation column. The experimental set-up is described by Ydstie, Kemna and Liu. The sampling time is 1 minute, there are 20 parameters in the FIR model and an LQ finite horizon controller is used for control. The parameter N_0 in the variable forgetting factor algorithm is set equal to 100. The experiment was carried out by Dr. S. A. Chesna and is described in his PhD work at the University of Massachusetts. The main conclusion we can draw from this and a large number of other experiments is that adaptive certainty equivalence control can work very well indeed when the admissibility and drift issues have been taken care of. In fact, the industrial results in Zervos and Dumont (1988), using a Laguerre formulation of a similar approach, shows that it easily outperforms a well tuned PID controller for an industrial process.

Conclusion

The session chair framed the question "Nonlinear control, fact or fiction?" In this paper we have reviewed very briefly advances that have been made in closing the gap between theory and application in the area of adaptive linear control and we have been able to find a surprisingly precise answer to this question. It is now possible to design a singularity free, globally stable and implementable algorithm for single input single output adaptive control of process systems with slowly varying dynamics and stable overdamped response. The algorithm uses an FIR or a Laguerre type model with parameter projection. The estimated model is always stabilizable so one can use LQ or predictive control based on the estimated model without encountering any discontinuity. The estimation algorithm is based on least squares with a deadzone, covariance resettting and a variable forgetting factor. It is not difficult to check that this particular algorithm possess global stability properties. The performance bounds (maximum transient and steady state error) scale linearly with the magnitude of external perturbations. These are made small by using internal models, filters and all avail-

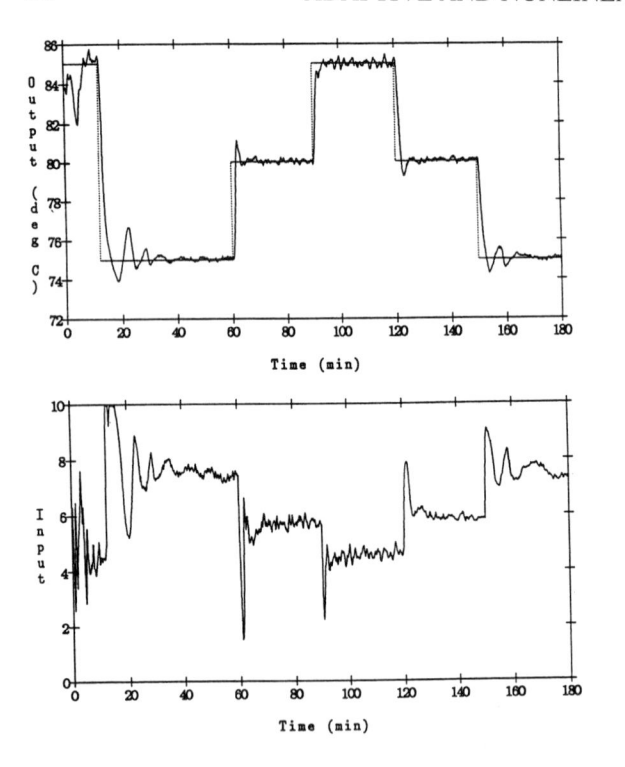

Figure 7. Performance of the adaptive LQ adaptive controller using an FIR model. The admissbility problem does not arise and spikes in the control inputs are avoided.

able apriori information to design estimation and control algorithms using linear theory. Hard bounds for the parameter estimation error can be obtained by using excitation and convergence (local) by application of the deadzone.

The "gap" between theory and application, quite apparent at CPC III and only somewhat narrowed at CPC IV by the use of the GPC, has been closed to a remarkable extent. This may bode well for adaptive control applications and we predict a bright future for this technology. High speed computation will allow the application of considerable more complex algorithms than the ones discussed in this paper. Further research will result in the development of stronger results for multivariable and constraint systems than are currently available. Such results will provide a broader scope for application of adaptive control in the chemical processing industries.

Acknowledgment

Research was supported by the National Science Foundation NSF-CTS 9316572.

References

Allison, B.J., J.E. Ciarniello, P.J.C. Tessier and G.A. Dumont (1995). Adaptive control of chip refiner motor load. *Automatica*, **31**, 1169-1184.

Anderson B.D.O., R.R. Bitmead, C.R. Johnson, P.V. Kokotovic, R.L. Kosut, I.M.Y. Mareels, L. Praly, B.D. Riedle (1986). *Stability of adaptive systems: Passivity and averaging analysis*. MIT Press.

Arkun, Y. and W. H. Ray (1991). *Chemical Process Control IV*, AIChE Publication No. 67, New York.

Åström K.J., B. Wittenmark (1973). On self tuning regulators. *Automatica*, **9**, 185-199.

Åström, K.J. (1983). Theory and application of adaptive control – A survey. *Automatica*, **19**(5), 471-486.

Åström K.J., Wittenmark B. (1995). *Adaptive control*. Addison-Wesley, Reading MA. 2nd edition.

Åström K.J., Wittenmark B. (1996). A survey of adaptive control applications. *Proc. IFAC World Congress*, San Francisco.

Åström, K.J. (1996). Adaptive control around 1960. *IEEE Control Systems*, **16**(3), 44-49.

Åstrøm, K.J., G.C. Goodwin and P.R. Kumar (1995). *Adaptive control, filtering and signal processing*, Springer Verlag, N.Y.

Bitmead R.R., M. Gevers and V. Wertz (1990). *Adaptive optimal control; The thinking man's GPC*. Prentice-Hall, Englewood-Cliffs, N.J.

Chen, D. (1995). Adaptive Control of Hot-dip Galvanizing. *Automatica*, **31**, 715-733.

Guo, L. and H.F. Chen (1991). The Åstrøm Wittenmark self tuning regulator revisited and ELS based adaptive trackers. *IEEE Trans. Automatic Control*, T-AC-36, 802-812.

Chesna, S.A. (1988). *Adaptive predictive control*, Ph.D. Thesis, Univerisy of Massachusetts at Amherst.

Dumont, G. (1986). On the use of adaptive control in the process industries. *Chemical Process Control III*, Morari, M. and T.J. McAvoy (Eds.), AIChE Press, New York. 467-500.

Espana, M. D. and L. Praly (1993). On global dynamics of adaptive systems: A study of an elementary example. *SIAM J. Cont. & Optimization*, **31**(5), 1143-1166.

Egardt, B. (1979). *Stability of Adaptive Controllers*, Lecture Notes in Control and Information Sciences, **20**, Berlin, F. R. G., Springer Verlag.

Finn, C.C. B. Wahlberg and B.E. Ydstie (1995). Quadratic Dynamic Matrix Control using FIR and Markov-Laguerre Models. *AIChE J.* **39**(11), 1810-1826.

Fujiwara, T and H. Miyakawa (1990). Application of predictive adaptive control system for steam temperature control in boiler plants. *Proc. IEEE-CDC*, **29**, Honolulu, Hawaii, 2181-2182.

Hagglund, T. (1983). *New estimation techniques for adaptive control*, Ph.D. thesis, Lund Institute of Technology.

Gevers, M. (1993). Towards joint design of identification and control in *Essays on Control: Perspectives in the theory and its applications*, H.L. Trentelman and J.C. Willems (Eds.), Birkhauser, Boston.

Golden, M. P. and Ydstie, B. E. (1992). Ergodicity and Small Amplitude Chaos in Adaptive Control. *Automatica*, **28**(1), 11-25.

Goodwin G.C., and Sin K.S. (1984). *Adaptive filtering, prediction and control*. Prentice-Hall. Englewood Cliffs, N.J.

Goodwin, G.C., D. Hill and Palanaswami (1985). Towards an Adaptive Robust Controller. *Proceedings of IFAC Identification and System Parameter Estimation*, York, UK., 997-1002.

Golden, M.P. and B. E. Ydstie (1994). Drift instabilities and chaos in forecasting and adaptive decision theory. *PHYSICA D*, **72**, 309-323.

Ioannou, P. and J. Sun (1996). *Robust adaptive control*, PTR Prentice Hall, NJ.

Kelly, J. H. and B. E. Ydstie (1995). Design guidelines for adaptive control with applications to systems with structural flexibility. *Adaptive signal Processing and Control*, P. R. Kumar, G. C. Goodwin and K. J. Åstrøm, (Eds.), Springer Verlag, NY.

Kokotovic, P. (1992). *Foundations of adaptive control*, Grainger lecture at Univ of Illinois, Published by Springer Verlag, NY.

Kosut, R. (1986). Adaptive calibration: An approach to uncertainty modeling and on line robust control design. *Proc. 25th IEEE CDC*, Athens, Greece, Dec.

Middleton R., G.C. Goodwin, D. Hill, D.Q. Mayne (1988). *Design issues in adaptive control*. IEEE Transactions on Automatic Control, **33**(1).

Morse, A.S. D.Q. Mayne and G.C. Goodwin (1992). Applications of hysteresis switching in parameter adaptive control. *IEEE Transactions on Automatic Control*, **37**(9), 343-1354.

Morari, M. and T.J. McAvoy (1986). Chemical Process Control III, AIChE Press, New York.

Mills, P.M., P.L. Lee and P. McIntosh (1991). A practical study of adaptive control of an alumina calciner. *Automatica*, **27**, 441-449.

Nussbaum, R.D. (1983). Some remarks on a conjecture in parameter adaptive control. *System & Control Letters*, **3**, 243-246.

Ortega, R., G. Escobar and F. Garcia (1992). To tune or not to tune?: A monitoring procedure to decide. *Automatica*, **28**(1), 179-184.

Ortega, R. and Y. Tang (1988). Robustness of adaptive controllers - A survey. *Automatica*, **25**, 651-677.

Partanen, A.G. and R.R. Bitmead (1995). The application of iterative identification and controller design to a sugar can crushing mill. *Automatica*, **31**, 1547-1564.

Pomerleau, Y. and G. Viel (1992). Industrial application of adaptive nonlinear control for baker's yeast production. *Proc. IFAC Symposium on Modelling and Control of Biotechnological Processes*, **29**, Keystones, CO, 315-318.

Praly L. (1984). Robust model reference controller - Part I. *IEEE Conference on decision and control*, 1984.

Praly L. (1988). Oscillatory behavior and fixes in adaptive linear control : A worked example. *Proceed. of the 1988 IFAC Workshop on Robust Adaptive Control*, Newcastle, Australia.

Praly L. (1990). Topological orbital equivalence with asymptotic phase for a two time scales discrete time system. *Mathematics of Control, Signals, and Systems*, **3**, 225-253.

Praly, L. and B. E. Ydstie (1995). *Adaptive linear control*, Unpublished research monograph.

Radenkovic, S. R. and B.E. Ydstie (1995). Using persistent exitation with fixed energy to stabilize adaptive controllers and obtain hard bounds for the parameter estimation error. *SIAM J. Cont. and Optimimization*, **33**, 1224 -1246.

Sastry, S. and M. Bodson (1989). *Adaptive control: Stability, Convergence and Robusteness*, Prentice-Hall, Englewood Cliffs, NJ.

Seborg, D.E., T.F. Edgar and S.L. Shah (1986). Adaptive control strategies for process control: A survey. *AIChE J*, **32**, 881-913.

Van den Hof, P.M.J. and R.J.P. Schrama (1995). Identification and control -closed loop issues. *Automatica*, **31**(12), 1725 -1750.

Solo, V. and X. Kong (1995). *Adaptive signal processing algorithms: Stability and Performance*, Prentice-Hall Information and System Sciences Series, Englewood Cliffs, NJ, 1995.

Staus, G.H., L. Z. Biegler and B.E. Ydstie (1995). Adaptive control via non-convex optimization. Nonconvex optimization and its applications. Kluwer Academic Publishers, the Netherlands.

Ydstie, B. E. (1989). Stability of Discrete Model Reference Adaptive Control - Revisited. *Systems and Control Letters*, **13**, 429-438.

Ydstie, B. E. (1992). Transient Performance and Robustness of the Direct Adaptive Control. *IEEE Transactions on Auto. Control*, **37**, 1091- 1105.

Ydstie, B.E., A Kemna and L.K. Liu (1988). Multivariable Extended-Horizon Adaptive Control. *Comp. Chem. Eng.*, **12**(7), 733 -743.

Ydstie, B. E. and B. Wahlberg (1994). Iterative Refinement of Model Predictive Control. *Proc. ADCHEM'94*, Kyoto, Japan.

N. Yoshitani (1993). Modelling and parameter estimation for strip temperature control in continuous annealing process. *IEEE-IECON 93*, 469-473.

Zervos, C.C. and G. Dumont (1992). Deterministic adaptive control based on Laguerre series expansions. *Int. J. Control*, **48**, 2333-2359.

Zhang, H., C.Z. Han, B.W. Wan and R. Shih (1994). Identification and control of a large kinsescope glass furnace. *Automatica*, **30**, 887-892.

Zhou, Quan-Gen and W.R. Cluett (1996). Recursive identification of time-varying systems via incremental estimation. *Automatica*, **32**(10), 1427-1432.

NONLINEAR PROCESS CONTROL – WHICH WAY TO THE PROMISED LAND?

Frank Allgöwer *
Institut für Systemdynamik und Regelungstechnik
Universität Stuttgart
70550 Stuttgart, Germany

Francis J. Doyle III
School of Chemical Engineering
Purdue University
West Lafayette, IN 47907-1283

Abstract

Nonlinear process control systems analysis and synthesis techniques are reviewed with an emphasis on recent results in each area. These include analysis tools for stability, robustness, controllability, observability, and nonlinearity measures. New synthesis results in both the differential algebraic and nonlinear H_∞ frameworks are discussed. Finally, some promising directions for future research pursuits are identified.

Keywords

Nonlinear control, Process control, Nonlinear H_∞, Nonlinearity measure, Differential geometry, Differential algebra, Nonlinear analysis.

Introduction

The interest in nonlinear process control synthesis has remained strong amongst both the academic research community and the industrial practitioners. Many new results have appeared in this area since the last meeting on Chemical Process Control CPC IV, including nonlinear H_∞ and algebraic approaches.

In this paper we limit our discussion on differential geometric control which was thoroughly reviewed at the last CPC meeting (Kravaris and Arkun, 1991), in addition we place less emphasis on nonlinear model predictive control which is reviewed at the present CPC meeting (Mayne, 1996). We also adopt a slightly different viewpoint by emphasizing *analysis tools*, so much so that we begin this review paper with this important area. We attempt to show that such tools are critical in any synthesis development: if one cannot analyze a system property, it will be very difficult to design a control to achieve this property.

Following this section, we review direct synthesis methods, including new results in differential algebraic the-

---

*Currently with: Advanced Process Modeling & Control Group, E.I. DuPont de Nemours & Co., Wilmington, DE 19880

ory, and also cover the emerging field of nonlinear H_∞ control design.

It is important to note that we do not attempt a comprehensive review of nonlinear analysis and synthesis techniques that are relevant to process systems. Instead, we highlight some of the new approaches, the advantages and disadvantages of which we feel to be rather typical for the present state of the art and which we feel bear promise to become successful techniques.

Nomenclature

In the multiple-input multiple-output, time invariant case, the nonlinear state-space model can be written as follows:

$$\begin{aligned} \dot{x} &= f(x) + g(x)u \\ y &= h(x) \end{aligned} \quad (1)$$

where $x \in \mathcal{R}^n$ are state variables, $u \in \mathcal{R}^p$ is the manipulated input, and $y \in \mathcal{R}^q$ is the controlled output. For simplicity, the nonlinear functions f, g, and h are assumed to be sufficiently smooth. The vast majority of chemical processes can be represented in this control-affine state space form due to, among other reasons, the prevalence of flow manipulation which gives rise to the structure above.

A final notational detail is the use of $r = [r_1, \ldots, r_i, \ldots, r_q]^T$ to represent the vector relative degree: the minimum number of differentiations of each process output y_i required to produce an explicit dependence on the process input.

Analysis of Nonlinear Control Systems

Open-loop Nonlinearity Measures

Although most processes are of nonlinear nature, experience clearly shows that not all nonlinear systems require nonlinear analysis and nonlinear synthesis methods. In many cases analysis and controller design based on the Jacobian linearization is sufficient. Nonlinearity measures are a means to systematically determine the severeness of nonlinearity and thus indicate when nonlinear analysis and nonlinear synthesis methods will be needed.

Significant progress has been made in the past couple of years with respect to the quantification of the nonlinearity of I/O-systems. There are in principal two approaches to nonlinearity measures. The first approach, that has its origin in the nonlinear system identification literature (see e.g. (Haber and Unbehauen, 1990)), is based on a quantification of the violation of properties that are known to hold for linear systems. An example for such a property is based on the coherence function

$$\gamma_{u,y}^2(j\omega) := \frac{|S_{uy}(j\omega)|}{|S_{yy}(j\omega)S_{uu}(j\omega)|} \quad (2)$$

where S_{xz} denotes the spectral density (Fourier transformed auto and cross correlation function) (Bendat, 1990). The coherence, $\gamma_{u,y}^2(j\omega)$, is bounded by zero and one, and equal to unity at frequencies where the relationship between $u(t)$ and $y(t)$ is linear. Coherence suppression may actually come from a number of sources: noise in the input and output measurements, unmeasured disturbances, nonlinear relationships between $u(t)$ and $y(t)$ and, if the coherence is estimated and not derived, errors from the coherence estimation process. For linear systems it can be shown that the coherence function is always equal to one. See (Schwarz and Sabotke, 1983; Haber, 1985; Allgöwer and Gilles, 1992; Pearson, 1994b) for a discussion of this approach to nonlinearity measures. Heuristic nonlinearity measures of this kind very often lead to intuitive and meaningful results. There are however cases where a more rigorous definition of the size of nonlinearity is advantageous.

A second approach to quantifying nonlinearity in the I/O-behavior of a dynamical system is derived in an abstract operator setting (Allgöwer, 1995c). General nonlinear multivariable I/O-systems with input u and output y are considered. The I/O-behavior is described by the nonlinear operator N that maps input signals u into output signals y:

$$y = N[u]. \quad (3)$$

We consider multivariable causal operators N, that we require to be BIBO-stable (Vidyasagar, 1993). This setup describes a very general class of systems that comprise finite dimensional systems in continuous and discrete time as well as infinite dimensional systems. Without loss of generality, we assume

$$\lim_{t \to \infty} y(t) = 0 \quad \text{for} \quad u(t) \equiv 0 \; \forall t. \quad (4)$$

Definition 1 *The nonlinearity measure $\phi_N^{\mathcal{U}}$ of a BIBO-stable, causal I/O-system N for inputs signals $u \in \mathcal{U}$ is defined by the nonnegative number*

$$\phi_N^{\mathcal{U}} = \inf_{G \in \mathcal{G}} \sup_{u \in \mathcal{U}} \frac{\|G[u] - N[u]\|_{p_y}}{\|N[u]\|_{p_y}}, \quad (5)$$

with G being a linear I/O-operator, $\|\cdot\|_{p_y}$ being a norm in the output space, and $\mathcal{U}$ being the set of inputs of interest.

The "size" of the nonlinearity of an I/O-system is thus defined as the normalized largest difference between the output of nonlinear system N and a linear system G measured by the norm $\|\cdot\|_{p_y}$. The difference is taken with respect to the worst-case input in $\mathcal{U}$ and with respect to the linear system G^* that approximates N best. The set $\mathcal{U}$ contains the subset of inputs that we consider relevant in the closed-loop.

It is easy to see that the nonlinearity measure $\phi_N^{\mathcal{U}}$ has the value zero for N being a linear operator and from $\phi_N^{\mathcal{U}} = 0$ we can conclude the existence of a linear operator G^* such that

$$\|G^*[u] - N[u]\|_{p_y} = 0 \quad \forall u \in \mathcal{U}. \quad (6)$$

The behavior of nonlinear system N for the input signals considered is thus the same as that of the linear system G^* and therefore N behaves linear in a certain sense. It must be stressed that the value of the nonlinearity measure depends of course on the choice of input signals $\mathcal{U}$ considered.

This is of course not the only sensible definition for a nonlinearity measure. For example a normalization of (5) by $\|u\|$, where the nonlinearity measure has the character of a nonlinear gain, may be useful in some cases (Desoer and Wang, 1981; Sourlas and Manousiouthakis, 1992).

In general the computation of nonlinearity measure $\phi_N^{\mathcal{U}}$ requires the solution of a nonlinear infinite dimensional minimax problem of high complexity. Only for few special cases it can be hoped that an analytical solution of this problem will be feasible. It is however possible to compute $\phi_N^{\mathcal{U}}$ using appropriate approximation schemes. The interested reader is referred to (Allgöwer, 1995b) for an approach based on convex optimization, or to (Allgöwer, 1995c) for the computation of a lower and an upper bound.

A third alternative for the definition and computation of nonlinearity measures is based on a suitably defined operator norm that quantifies the *average nonlinearity* of I/O-systems (Nikolaou, 1994b; Nikolaou, 1994a) and has proven to be useful in many cases.

Another contribution to the area of nonlinearity characterization has been advanced by Guay et al. (1995). In this work they recognize two versions of nonlinearity: open-loop and control law nonlinearity. Control law nonlinearity may be approximately measured by examining the first and second derivative information of the input-output map, although it is recommended to analyze the process inverse map.

If a nonlinearity measure indicates that a system is linear, then linear analysis and synthesis methods will probably suffice. (Exceptions include the case where a nonlinear performance objective is sought.) If considerable nonlinearity is detected then certainly nonlinear analysis tools will be needed in order to determine open- or closed-loop properties. However, a strong nonlinearity does not necessarily imply the need for nonlinear controller design techniques as a linear controller might also be able to achieve the desired closed-loop properties. We will discuss this aspect in more detail in the next section.

Nonlinearity measures have been successfully applied to analyze chemical reactors (Nikolaou, 1994b; Stack and Doyle III, 1995), distillation columns (Allgöwer, 1995a) and to compare different operating points (Allgöwer, 1995c). In the latter reference, various areas of application of nonlinearity measures are described.

Nonlinear Closed-loop Performance

In the context of nonlinearity measures it is of course also of interest to capture the influence that nonlinearities have with respect to achieving certain performance goals in the closed-loop. Sometimes this is referred to as *control-relevant nonlinearity*. For instance, one issue that has received considerable attention in the recent literature is the determination of the appropriate application of nonlinear versus linear control for a particular problem.

From the perspective of stabilization, the relevant question is: does there exist a nonlinear (or linear) controller which guarantees a particular type of stability (*e.g.*, input-output)? A nice review of work that addresses this question is provided by (Poolla, Shamma and Wise, 1990), in which a more difficult problem is also addressed: *robust stabilization*. In an effort to determine whether time-varying nonlinear control is vastly superior, or only marginally superior, to linear time-invariant control, they suggest the *Plant Uncertainty Principle*. A very simplified summary of their findings is that parametric uncertainty favors the application of nonlinear control, while dynamic uncertainty favors the application of linear control.

The next level of analysis would address performance of a controlled system with a typical question being: does a particular plant benefit significantly from the application of nonlinear (versus linear) control to achieve a specific performance objective? For instance, it is well known that the optimal controller to a quadratic objective function for a linear known plant without constraints and subject to additive Gaussian white noise is a linear controller (Kwakernaak and Sivan, 1972). However, if any of these conditions are not satisfied (*e.g.*, nonlinear plant, nonquadratic objective function, or saturation constraints), experience shows that in many cases a nonlinear controller will yield considerable improvement over a linear controller (Bernstein, 1993).

The issues of stabilization and closed-loop performance are clearly tied to the intrinsic characteristics of the particular process in question. Indeed, as demonstrated by Doyle et al. (1993), Morari and Zafiriou (1989), Doyle III et al. (1991), and Shinskey (1988), there exist classes of nonlinear plants that are optimally and robustly controlled by a linear controller.

Clearly, a simple analysis of the open-loop system does not include all the information needed to determine the control relevant nonlinearity of a system. In particular, *it does not include the effects of a performance objective or the costs of control action*.

One preliminary result in this area is given in (Stack and Doyle III, 1995) where an optimal control based method is used to develop a relevant operator for control-relevant nonlinearity assessment. The operator development draws from Lagrangian optimization (for a review with process applications, see the text (Ray, 1990)), and is defined to be the control operator which maximizes the objective $\mathbf{I}$ in (7) subject to the nonlinear model (1) with initial conditions $x(0) = x_0$:

$$\mathbf{I}[\mathbf{u}(t)] = \frac{1}{2} \int_0^{t_f} (\mathbf{x}^2 + \alpha \mathbf{u}^2) dt \qquad (7)$$

In the case without input constraints, this objective is maximized only if the partial derivative of the Hamiltonian with respect to u is equal to zero for unconstrained portions of the path, where the Hamiltonian $\mathbf{H}$ is defined by (8) and is maximized along the constrained portions of the path.

$$\mathbf{H} = \mathbf{F}(\mathbf{x}, \mathbf{u}) + \boldsymbol{\lambda}^T \mathbf{f}(\mathbf{x}, \mathbf{u}) \qquad (8)$$

The time-dependent Lagrange or adjoint multipliers $\boldsymbol{\lambda}$ are defined by $\frac{\partial \boldsymbol{\lambda}^T}{\partial t} = -\frac{\partial \mathbf{H}}{\partial \mathbf{x}}$ and $\lambda_i(t_f) = \frac{\partial G}{\partial x_i}$ for state variables unspecified at $t = t_f$.

From the above discussion, the objective is minimized by $\mathbf{u}$ defined by:

$$\dot{\boldsymbol{\lambda}}^T = -\frac{\partial \mathbf{H}}{\partial \mathbf{x}} = -\mathbf{x} - \boldsymbol{\lambda}^T \left(\frac{\partial \mathbf{f}}{\partial \mathbf{x}} + \frac{\partial \mathbf{g}}{\partial \mathbf{x}} \mathbf{u} \right) \qquad (9)$$

$$\frac{\partial \mathbf{H}}{\partial \mathbf{u}} = \alpha \mathbf{u} + \boldsymbol{\lambda}^T \mathbf{g}(\mathbf{x}) = 0 \qquad (10)$$

$$\lambda_i(t_f) = 0 \qquad (11)$$

Solving equation (10) for u and differentiating yields an explicit form for u (in the scalar case):

$$\dot{u} = \frac{g(x)}{\alpha} x + u \left[\frac{\partial g}{\partial x} \frac{f(x)}{g(x)} - \frac{\partial f}{\partial x} \right] \qquad (12)$$

$$u(t_f) = 0 \qquad (13)$$

Equations (9) and (10) comprise the *"Optimal Control Structure"* (Stack and Doyle III, 1995). This derivation makes no limiting assumptions about the control structure (e.g., MPC, IMC, PID, etc.) that might bias an analysis of the controllability of the original system.

The Optimal Control Structure (OCS) contains the dynamic relationship between the system output and input for the optimal control policy. Therefore, an analysis of its nonlinearity by any open-loop method will give an indication of the control-relevant nonlinearity of the original system. Note that this measure incorporates several key features of the control problem: (i) the objective function (including a weight on u); (ii) region of operation (tied, in a complex way, to the initial conditions in x, and to the ultimate choice of nonlinearity measure); and (iii) the process open-loop model. Finally, note that although the preceding discussion employed the standard quadratic penalty function, the analysis holds for more general nonlinear objective functions which may also include end-point constraints:

$$\mathbf{I}[\mathbf{u}(t)] = \mathbf{G}(\mathbf{x}(t_f)) + \int_0^{t_f} \mathbf{F}(\mathbf{x}, \mathbf{u}) dt \quad (14)$$

Example 1 *The following example is a simplified model of a nonlinear reaction in a batch reactor. The functional form of the temperature dependence of the reaction rate is general: thus arbitrarily nonlinear (open-loop) problems are considered. However, as we will show, a careful analysis of this problem suggests that this system can be optimally controlled by a linear controller.*

The specific example is taken from (Morari and Zafiriou, 1989):

$$C_p \frac{dT}{dt} = (-\Delta H_r) r(T) - UA(T - T_c) \quad (15)$$

The nonlinear reaction term r can be linearized as:

$$r(T) = r^0 + k_T (T - T_0) \quad (16)$$

where $k_T(T) = \partial r / \partial T$, which varies nonlinearly with temperature. Note that this is a local linearization, not a global approximation. This results in the following transfer function between T_c and T:

$$T(s) = \frac{UA/C}{s + a} T_c(s) \quad (17)$$

where

$$a = \frac{UA + \Delta H_r k_T(T)}{C_p} \quad (18)$$

We now consider two approaches to analyzing the control-relevant nonlinearity in this system:

- *The first approach is from (Morari and Zafiriou, 1989) and draws from linear robust control theory. The pole uncertainty in equation (18) arises from the term $k_T(T) = \partial r / \partial T$, which varies nonlinearly with temperature. Letting the possible values of a be represented by a norm-bounded perturbation, Δ_E, weighted by w_E gives*

$$a = \tilde{a}(1 + r_a \Delta_E), \quad |\Delta_E \leq 1|, \quad \Delta_E \text{ real} \quad (19)$$

where $\tilde{a}$ is the nominal pole location and r_a is the relative uncertainty of a. This uncertainty may be expressed as an inverse multiplicative perturbation $(I + w_E \Delta_E)^{-1}$, so that the family of uncertain plants is given by:

$$\frac{1}{s+a} = \frac{1}{s+\tilde{a}} \cdot \frac{1}{1 + w_E \Delta_E} \quad (20)$$

$$\quad (21)$$

where

$$w_E(s) = \frac{r_a}{1 + s/\tilde{a}} \quad (22)$$

This inverse multiplicative uncertainty structure gives rise to a weighted robustness problem which involves the sensitivity function for both stability and performance. As opposed to the typical uncertainty problem (which involves a weighted complementary sensitivity function for robust stability), this uncertain system yields arbitrary robust performance from high gain linear control. It is important to point out that such an analysis ignores process noise as well as high frequency unmodeled dynamics.

- *A second approach to analysis for this problem employs the previously described optimal control structure. In effect, the optimal controller's functional form is constructed, and an open-loop nonlinearity measure can be applied to that operator (although for the purposes of this example, we will only derive the operator). For the simple reactor introduced above, the optimal controller which minimizes a quadratic objective function (including a penalty of the control input with weight α) is given by:*

$$\dot{\lambda} = -x - \lambda \left(\frac{-\Delta H_r k_0 E}{C_p R x^2} e^{\frac{-E}{Rx}} - \frac{UA}{C_p} \right) \quad (23)$$

$$u = \frac{-\lambda UA}{C_p \alpha} \quad (24)$$

where x is the reactor temperature, and u is the cooling jacket temperature.

This controller can be rearranged as follows:

$$\dot{u} = \frac{UA}{C_p \alpha} x + u \left[\frac{-\Delta H_r k_0 E}{C_p R x^2} e^{\frac{-E}{Rx}} - \frac{UA}{C_p} \right] \quad (25)$$

where it is clear that for sufficiently small α (penalty on input), linear compensation will be optimal.

Hence, both approaches confirm that the apparent strong *nonlinearity that is represented by the reaction rate in the open-loop plant presents no hurdle for optimal linear control design provided there is no penalty on the input action.* ◁

Controllability and Observability

Controllability and observability are two important properties that need to be investigated prior to any controller design. Consider again a nonlinear system in input-affine form (1).

Definition 2 *The nonlinear system in (1) is called controllable if for any two points x_1 and x_2 there exists an admissible input trajectory u that drives system (1) from state x_1 to state x_2 in finite time.*

For linear systems (A, B) controllability can be examined using the well known *Kalman rank condition for controllability*

$$\text{rank}\left\{\left[B, AB, A^2B, \cdots, A^{n-1}B\right]\right\} = n, \quad (26)$$

where n is the dimension of the state space. In order to discuss the nonlinear case let us first consider the special class of nonlinear systems without drift term $f(x)$:

$$\dot{x} = \sum_{i=1}^{m} g_i(x) u_i. \quad (27)$$

A system of the form (27) is controllable if the state dependent matrix

$$P(x) = [g_1, \ldots, g_m, [g_i, g_j], [g_i, [g_j, g_k]], \ldots], \quad (28)$$

that consists of all possible Lie brackets of the input vector fields $g_i(x)$, has rank n for all x (Nijmeijer and van der Schaft, 1990), where the Lie bracket is defined as $[g_i, g_j]_{(x)} = \frac{\partial g_j}{\partial x} g_i(x) - \frac{\partial g_i}{\partial x} g_j(x)$:

$$\text{rank}\{P(x)\} = n \quad \forall x. \quad (29)$$

This condition has a nice system theoretic interpretation: It is immediately clear from (31) that all states that lie in the direction of a linear combination of the input vector fields $g_i(x), i = 1, \ldots, m$, can be reached from point x. By an appropriate switching between the different inputs u_i the system can however also be driven in additional directions that are characterized by the higher-order Lie brackets. We show this exemplary for a driftless system with two inputs ($m = 2$). Consider four succeeding time intervals of equal length Δt. During the first interval $(0, \Delta t]$ constant input values $u_1 = 1, u_2 = 0$ are applied. During the second interval $(\Delta t, 2\Delta t]$ $u_1 = 0, u_2 = 1$ is applied, then $u_1 = -1, u_2 = 0$ and finally $u_1 = 0, u_2 = -1$. Computing the solution at time instances $i\Delta t$ as a Taylor series up to order two leads to the following approximation for the state vector x at these time instances:

$$x(\Delta t) = x(0) + \Delta t g_1(x(0)) + \frac{1}{2}\Delta t^2 \left.\frac{\partial g_1}{\partial x} g_2\right|_{x(0)} + \mathcal{O}^{[2+]} \quad (30)$$

$$\vdots$$

$$x(4\Delta t) = x(0) + \Delta t^2 \underbrace{\left[\frac{\partial g_2}{\partial x} g_1 - \frac{\partial g_1}{\partial x} g_2\right]}_{[g_1, g_2]_{x(0)}}\Bigg|_{x(0)}$$
$$+ \mathcal{O}^{[2+]}, \quad (31)$$

where $\mathcal{O}^{[2+]}$ indicates terms of order strictly greater than two. As can be seen from (31) the Lie bracket $[g_1, g_2]_{x(0)}$ describes a control direction into which system (27) can be driven in addition to control directions g_i. It can be easily shown that by choosing more elaborate switching cycles the system can be also steered into the direction of higher-order brackets of g_1 and g_2. This explains condition (33).

Unfortunately for general input affine systems (1) *with drift term f* no generally applicable conditions for controllability like (29) can be given even though there are several approaches to the problem known at present (e.g. (Krener, 1985; Sussmann and Jurdjevic, 1972; Hermann and Krener, 1977)). There is one specific result that is a direct extension of the Kalman rank condition for controllability to the nonlinear case, which is briefly summarized below.

Consider the state dependent matrix

$$P^*(x) = \left[g(x), \text{ad}_f\, g(x), \ldots, \text{ad}_f^{n-1}\, g(x)\right] \quad (32)$$

where $\text{ad}_f^i\, g(x) = \left[f(x), \text{ad}_f^{i-1}\, g(x)\right]$ is the standard notation for higher-order Lie-brackets. If

$$\text{rank}\{P^*(x)\} = n \quad \forall x \quad (33)$$

then system (1) is said to satisfy the *accessibility rank condition*. As $\text{ad}_{Ax}^j\, b_i = A^k b_i$, the accessibility rank condition reduces to the Kalman rank condition for linear systems $(A, [b_1 \ldots b_m])$ and thus implies controllability. For nonlinear systems this is unfortunately not true. Rank$\{P^*(x)\} = n$ only implies *weak controllability* (Vidyasagar, 1986) in the nonlinear case that does not guarantee all states to be reachable (see e.g. (Nijmeijer and van der Schaft, 1990) for a simple example).

The accessibility rank condition is however still an important necessary condition, that needs to be satisfied in order for several nonlinear controller synthesis methods (including feedback linearization) to be applicable.

For the study of observability properties we consider again the nonlinear system in equation (1). The output trajectory depends on the initial condition $x(0)$ as well as on the input trajectory u. We make this explicit in the notation by adding a double subscript to the output y: $y_{(x(0), u)}$.

Definition 3 *System (1) is called observable if for every admissible input function u the equivalence of the output trajectories $y_{(x_1, u)}$ and $y_{(x_2, u)}$ implies equivalence of the initial conditions x_1 and x_2.*

If $y_{(x_1,u)} \equiv y_{(x_2,u)}$ implies $x_1 = x_2$ for one specific input trajectory u, this will also hold for all input trajectories, if the underlying system is linear. This means that for linear systems the input dependency of the system at hand does not have to be taken into account when studying its observability properties (compare the Kalman rank condition (37) which does not depend on the input vector b). For nonlinear systems this is no longer true. For simplicity of exposition we will restrict ourself to *autonomous* single-output systems

$$\begin{aligned} \dot{x} &= f(x) \\ y &= h(x) \end{aligned} \quad (34)$$

in the following, even though the theory is fully developed for the multivariable nonautonomous case (see e.g. (Isidori, 1995; Nijmeijer and van der Schaft, 1990) for an overview).

Consider the following relation between the state x and the output y, and its derivatives, which is usually called the *observability map*:

$$\begin{bmatrix} y \\ \dot{y} \\ \vdots \\ y^{(k-1)} \end{bmatrix} = q(x) = \begin{bmatrix} h(x) \\ L_f h(x) \\ \vdots \\ L_f^{(k-1)} h(x) \end{bmatrix} \quad k \leq n \, . \quad (35)$$

If it is possible to compute the state x from equation (35), then the state x can be reconstructed from information contained in the output y and system (34) is said to be *globally observable*. There is however no simple way to test whether nonlinear equation (35) can be solved for x globally, and in only few cases the explicit inversion of $q(x)$ will be possible. It is however easy to determine the *local* invertibility of (35) at a point x by looking at the rank of the partial derivative of q. If

$$\text{rank}\{Q(x)\} = n \quad \forall x \quad \text{with} \quad Q(x) = \frac{\partial q(x)}{\partial x} \quad (36)$$

holds, then system (34) is called *locally (or weakly) observable* (Nijmeijer and van der Schaft, 1990; Vidyasagar, 1986). Condition (36) is known as the *observability rank condition*, that exactly reduces to the Kalman rank condition for observability

$$\text{rank} \left\{ \begin{bmatrix} c^T \\ c^T A \\ \vdots \\ c^T A^{n-1} \end{bmatrix} \right\} = n \, , \quad (37)$$

when applied to linear systems (A, b, c^T). In general local observability does however not imply observability. In other words, although a system may satisfy the observability rank condition for all x, it may not be possible to reconstruct the state from the information contained in the output. An in depth treatment of nonlinear observability can be found in (Birk, 1992).

Summarizing, the structure of the nonlinear controllability and observability problem is well studied at present and there is also a clear understanding of the "obstacles" that make the nonlinear case more difficult. However, although researchers have studied this problem for over 20 years, there are still no simple tests available that allow the practicing engineer to determine controllability or observability for a given nonlinear control system. The main reason for our inability to solve these fundamental problems is most probably the large class of nonlinear systems (arbitrary control affine smooth systems) that we are considering. For more restricted classes of nonlinear systems, it may be possible to derive results. We tried to make this explicit by showing a controllability criterion for the limited class of driftless nonlinear systems that are rather common in mechanical engineering, but unfortunately not in chemical engineering. For a process example, consider the work of Bastine and Levine (1993) in which they characterize the state accessibility rank for a generic class of chemical reaction systems. The class they consider has the structure

$$\dot{x} = Cr(x) - Dx + Bu \quad (38)$$

where r is the vector of reaction rate kinetic laws. They successfully develop a theory for the direct computation of the (strong) accessibility rank of systems of this class.

Stability

The notion of stability is of central importance for the analysis and synthesis of control systems. There is, however, no single definition of stability and particularly in the nonlinear case it is important to clearly state what type of stability is meant in each context. Space doesn't permit us to go into the details here, but we comment on a few of the approaches. The most common notion of stability goes back to the work of Ljapunov (Ljapunov, 1907) and is concerned with the problem of analyzing the stability of state space representations of dynamic systems. The second common approach, that has attracted a lot of attention especially in the nonlinear context, is the notion of input-output stability that goes back to the work of Sandberg and Zames (Sandberg, 1964; Zames, 1966a; Zames, 1966b). The interested reader is referred to an in depth treatment of these notions in the following texts (Hahn, 1967; Willems, 1970; Atherton, 1981; Vidyasagar, 1993). Assessment of either type of stability is a routine task for linear systems. For nonlinear systems this is no longer the case. By Ljapunov's first method (Hahn, 1967) it is usually easy to determine the *local* stability characteristics of a given system near an equilibrium by looking at the stability of the Jacobian linearization around this equilibrium. This holds however only if the Jacobian linearization has no poles on the imaginary axis. For those critical cases Center Manifold Theory (Carr, 1981) can help to reduce the complexity of the problem of determining stability. In the linear case stability results are always of a global nature: If a linear system is locally stable it is also globally stable. For nonlinear systems this is, of course, not true. In general it is in fact a very difficult task to determine the region of attraction of a given stable equilibrium

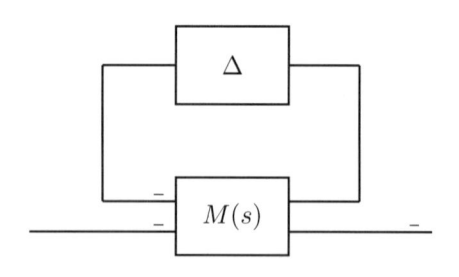

Figure 1. Nonlinear closed-loop system.

point in the nonlinear case. This is, as of today, still one of the most challenging and important research questions in the area of nonlinear control. Input-output approaches, like the Popov- or Circle Criterion (Vidyasagar, 1993) are more useful in this respect (if the assumptions for their application are fulfilled) compared to Ljapunov type approaches.

With these few remarks we emphasize that in a nonlinear context, even the problem of defining and proving stability is a nontrivial one and as yet still widely open. It is important to retain this perspective when considering the notions of performance and robustness in the following sections.

Robustness

Tools for robust stability analysis of nonlinear systems are still in the early stages of research development. Some of the results of linear theory can be carried over to the nonlinear context. This is for example true for the notion of unstructured model uncertainty as it is common to use for example in the linear H_∞- or μ-approaches to control (Doyle, Francis and Tannenbaum, 1992). In this section we review this approach, in an effort to point out the computational and theoretical difficulties to this problem. In a later section, we will discuss the extension of this particular type of uncertainty description to the nonlinear case and will demonstrate potential problems of this approach in the nonlinear context with a simple example.

A generic representation for a nonlinear closed-loop system is given in Fig. 1. Here, the Δ and M blocks are general, possibly nonlinear and/or time-varying operators. Later in the discussion we will assign specific roles to Δ and M. The stability of this feedback system can be analyzed using the small-gain theorem (Zames, 1966a). Before stating the small gain result, we need to introduce the notion of an operator gain:

Definition 4 *The incremental gain of an operator M, denoted by $g(M)$, is:*

$$g(M) = \sup \frac{\|M(x_1) - M(x_2)\|}{\|x_1 - x_2\|} \quad (39)$$

where the supremum is taken over all x_i in the domain of M, such that $M(x_i)$ is in the range of M, and $x_1 \neq x_2$.

This incremental gain can be defined for various L_p norms (see for example, (Vidyasagar, 1993)). Having defined the differential gain of an operator, we can state the main result in (Zames, 1966a):

Theorem 1 *If $g(M) \cdot g(\Delta) < 1$, then the closed-loop is internally input-output stable.*

Clearly the burden in this analysis is the calculation of the nonlinear operator norm. This remains an area of active research (see, e.g., (Choi, Nikolaou and Manousiouthakis, 1994)), and trends in computing algorithms may lead to more tractable solutions in the next few years.

The so-called M-Δ diagram in Fig. 1 also represents a general framework for analysis of *linear* uncertain systems. In that context, M represent the nominal process operator (possibly closed-loop), and Δ represents the uncertainty (structured or unstructured) associated with that nominal system. It is straightforward to show (Morari and Zafiriou, 1989) that problems in which the uncertainty is of an additive, multiplicative, or inverse multiplicative nature can be cast in the structure in Fig. 1. In (Packard, 1987), (Doyle III, Packard and Morari, 1989), there were extensions presented to include classes of nonlinear systems. This allows, for example, the application of the structured singular value (SSV) to nonlinear problems in which the nonlinearity can be represented by a linear fractional transformation (LFT). Furthermore, one can express particular families of uncertain nonlinear systems as LFTs, leading to the possibility of robust analysis of nonlinear systems using the SSV.

It was shown in (Packard, 1987) that for norm bounded nonlinear operators, constant D-scalings could be used in the structured singular value analysis (Doyle, 1982) to give a conservative small gain condition for robust stability; an application to a CSTR was also presented. Furthermore, it is possible to calculate a Ljapunov function from the D-scalings, thus linking the input-output stability results to the Ljapunov stability. One final technical note, the preceding discussion concerns the stability of the nonlinear closed-loop, however the *performance* of the system can also be addressed in this context using a "virtual" uncertainty block to represent performance requirements between input-output pairings.

A class of nonlinear systems is now described which fits into the proposed M-Δ structure framework. A conic sector is defined as:

$$\text{Cone}(C, R) \equiv \{(u, y) : \|y - Cu\| \leq \|Ru\|\} \quad (40)$$

where (u, y) is the input/output pair for the operator. A nonlinear operator enveloped tightly by a conic sector is most accurately approximated linearly by the cone center C. In general, the cone center will not coincide with the plant described by the Jacobian of the nonlinear model evaluated at the operating point. Note that because we have replaced a potentially highly nonlinear function by two linear time-invariant operators, this simplification is likely to be conservative. The $\mathrm{Cone}(C, R)$ describes many input-output pairs, some of which may yield poorer performance than the original operator. There is a direct correspondence between a nonlinear cone-bounded operator and a time-varying gain. From the conic sector definition, the plant can be interpreted as being equal to the nominal value (C) which is perturbed by a time varying gain of magnitude R. R and C can be absorbed into the system to arrive at the general uncertainty structure in Fig. 1, where Δ is a time-varying gain of norm one. Consider the nonlinear saturation nonlinearity:

$$\mathrm{sat}(u) = \begin{cases} u^{max} & \text{if } u > u^{max} \\ u & \text{if } u^{min} \leq u \leq u^{max} \\ u^{min} & \text{if } u < u^{min} \end{cases}$$

One can represent this nonlinearity as the following conic sector operator ($C = 0.5$, $R = 0.5$):

$$\mathrm{sat}(u) = 0.5u + 0.5\delta(u)u \quad (41)$$

where

$$\delta(u) = 2\frac{\mathrm{sat}(u)}{u} - 1 \quad (42)$$

which is easily shown to be a nonlinear operator with gain equal to one ($-1 \leq \delta(u) \leq 1$).

An advantage of representing nonlinear operators by structured decomposition operators (as in Fig. 1) is the fact that operator inverses are easily calculated from such an additive construction. Consider, for example the additive uncertainty formulation below, where Δ_{nl} represents a specific structured nonlinear operator; P is a linear plant approximation (not necessarily the Jacobian), and the combination $P + \Delta_{nl}$ represents the true or uncertain plant. The inverse

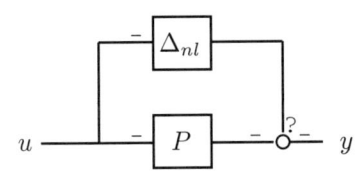

Figure 2. Hierarchical nonlinear model.

of this operator can be constructed analytically from the inverse of the linear sub-operator through the arrangement depicted in Fig. 3. Such an approach to operator inversion has

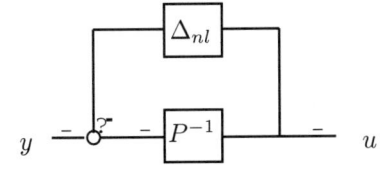

Figure 3. Inverse of hierarchical nonlinear model.

been pointed out by several researchers (Economou, Morari and Palsson, 1986; Hernández, 1992). It is possible to extend this idea one step further by considering a generalized inverse of the linear operator. This arises, for example, in a weighted least-squares solution to the unconstrained MPC problem. It is still possible to represent analytically the generalized nonlinear inverse through a simple block-diagram construction as shown in Fig. 4. A rigorous derivation of such a construction is given in (Doyle III, 1995). The utility

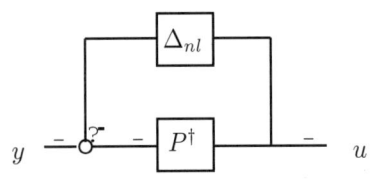

Figure 4. Generalized inverse of hierarchical nonlinear model.

of such an approach goes beyond a realization of a nonlinear operator inverse; the real strength of such a hierarchical operator representation is the ability to apply robustness analysis tools to nonlinear closed-loop operators.

A potentially less conservative approach to the nonlinear closed-loop system analysis questions (robust stability and robust performance) is addressed using a different norm formulation for the closed-loop signals. Using the same diagram (Fig. 1), one can utilize l^{∞} norm metrics, which result in the induced-l^{∞} norm or $\mathcal{A}$-norm ($\equiv l^1$-norm). Khammash and Pearson (1991) give necessary and sufficient conditions for robust stability and robust performance which carry over to systems with nonlinear and/or time-varying uncertainty, Δ. It should be noted that although the conservatism is reduced in the formulation of the stability result (cf. the small-gain theorem), there may be considerable conservatism in the conic sector uncertainty formulation for the nonlinear operator. In particular, several classes of practical nonlinearities are known to give extremely poor characterization as conic sectors (e.g., input saturation nonlinearity). An attractive feature of the result are the closed-form expressions available for the $\mathcal{A}$-norm for three or fewer structured uncertainty elements. For example, the necessary and sufficient conditions for two structured nonlinear Δ blocks are:

$$\|P_{22}\|_{\mathcal{A}} < 1$$

$$\|P_{11}\|_{\mathcal{A}} + \frac{\|P_{12}\|\|P_{21}\|}{1-\|P_{22}\|} < 1$$

Hence the robust analysis computation is straightforward in this case, once a nonlinear uncertainty description has been derived.

Synthesis Approaches

Model-Based Direct Synthesis

In this section, we give a brief review of the various "direct synthesis" nonlinear control design methodologies which

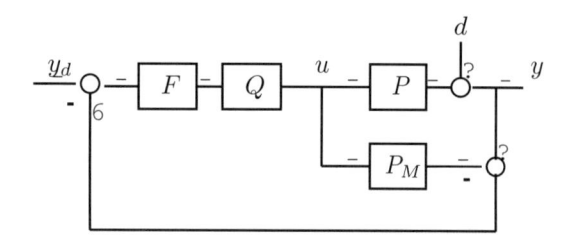

Figure 5. IMC control structure.

make explicit use of a nonlinear process model. The somewhat arbitrary use of the words "direct synthesis" will connote control design in which a prescribed closed-loop behavior is sought (for example, linear relationship between the setpoint, y_d and the output, y). These ideas are most clearly motivated in the context of IMC design, with the corresponding architecture depicted in Fig. 5. Here, P represents the nonlinear plant, P_M is the process model, Q is the IMC controller, and F is the filter which is required for causality and imparts robustness properties (Morari and Zafiriou, 1989). We consider in this section the case where $P = P_M$; robustness is postponed for later discussion. If one takes as a suitable objective, the design of a controller QF that yields a linear response from y_d to y (which is denoted, in the linear case, the complementary sensitivity function, $\eta(s)$), then the controller QF represents the product of $\eta(s)$ and the inverse of the nonlinear plant model. Hence, a key component of any such approach is the derivation (implicit or explicit) of the nonlinear plant operator inverse. Limitations, such as right-half plane zeros, restrict the choices of $\eta(s)$.

One of the earliest approaches to nonlinear direct synthesis was the paper by Economou et al. (1986) in which the nonlinear inverse is solved by numerical methods. A Newton method is recommended, which yields the following discrete numerical implementation for inverse computation:

$$u^{n+1} = u^n - \frac{y - h(x)}{-\frac{\partial h}{\partial x}\Gamma}$$

$$\dot{\Gamma} = \frac{\partial f}{\partial x}\Gamma + g(x)$$

Another approach to the problem (which was reviewed thoroughly at CPC IV (Lee, 1991) and, hence, will not be detailed here) is the generic model control (GMC) or reference system synthesis approach (Bartusiak, Georgakis and Reilly, 1989), in which an implicit computation is set up for the control law (assuming a relative degree 1 plant):

$$\frac{\partial h}{\partial x}(f(x) + g(x)u) = k_1(r - y) + k_2 \int_0^t (r - y)d\tau$$

Here, the parameters k_1 and k_2 determine the shape of the complementary sensitivity function, $\eta(s)$.

In the general case (relative degree not equal to 1), more powerful machinery from differential algebra may be employed to implicitly solve for the control inverse using an expression of the form (Fliess, 1990):

$$\mathcal{C}(\mathbf{q}, u, \dot{u}, \ddot{u}, \ldots, u^{(\nu)}) = -\sum_{i=1}^{n} \lambda_{i-1} q_i + v$$

where $\mathcal{C}$ represents the input-output nonlinear plant operator, and the λ_i represent the coefficients of the pole polynomial for $\eta(s)$.

Formal machinery has been available for the analytical construction of state-space nonlinear model inverses for some time (Hirschorn, 1979), with the possible critique being the computational burden associated with the derivation of such structures for high order dynamic systems. In this case, a minimal order right inverse is realized by the following system:

$$\dot{\zeta}_1 = q_1\left(y, \dot{y}, \ldots, \frac{d^{r-1}y}{dt^{r-1}}, \zeta_1, \ldots, \zeta_{n-r}\right)$$

$$\vdots$$

$$\dot{\zeta}_{n-r} = q_{n-r}\left(y, \dot{y}, \ldots, \frac{d^{r-1}y}{dt^{r-1}}, \zeta_1, \ldots, \zeta_{n-r}\right)$$

$$u = \frac{\frac{d^r y}{dt^r} - \alpha\left(y, \dot{y}, \ldots, \frac{d^{r-1}y}{dt^{r-1}}, \zeta_1, \ldots, \zeta_{n-r}\right)}{\beta\left(y, \dot{y}, \ldots, \frac{d^{r-1}y}{dt^{r-1}}, \zeta_1, \ldots, \zeta_{n-r}\right)}$$

This basic structure has driven the development of many of the differential geometric approaches (Kravaris and Kantor, 1990; Isidori, 1995). Differential geometry was an excellent tool for posing and solving this problem. There are, however, many drawbacks, see for example (Allgöwer and Gilles, 1995). For the last several years, many researchers have worked on extensions to the theory to overcome the drawbacks, however, these were not developed in the framework of geometric control.

Recent extensions of the differential geometric approaches for process applications include: (i) the design of anti-windup controllers for multivariable process systems subject to input saturation constraints (Calvet and Arkun, 1988; Balchen and Sandrib, 1994; Kendi and Doyle III, 1995); (ii) observer design (van Dootingh, Viel, Rakotopara, Gauthier and Hobbes, 1992); (iii) incorporation of adaptation schemes (Dochain, 1994); (iv) extension to two-time scale systems (Christofides and Daoutidis, 1994); (v) input-output linearization of nonminimum phase systems (Doyle III, Allgöwer and Morari, 1996; Allgöwer, Doyle III and Morari, 1996); (vi) input-output linearization by output feedback (Soroush, 1995; Groebel, Allgöwer, Storz and Gilles, 1994); and (vii) unmeasured disturbance estimation using Kalman filtering (Kurtz and Henson, 1995).

Several industrial applications of these synthesis methods as well as academic pilot plant studies have been reported in the literature. These include an extrusion process at Dow (Wassick and Camp, 1988) in which the dominant nonlinearity was the pumping curve for the extruder. A nonlinear IMC algorithm was also reported in an application in the hydrometallurgical processing industry for the

Bayer process which involved the control of digested slurry (Colombe, Dablainville and Vacarisas, 1987). Several experimental studies at Texas Tech involving a plasma reactor pressure control problem (Subawalla and Rhinehart, 1994) and a heat exchanger problem (Paruchuri and Rhinehart, 1994) indicated that nonlinear model-based control outperformed scheduled linear IMC with feedforward compensation for setpoint tracking, but both controllers fared equally well for measured disturbance rejection. Lévine and Rouchon (1989) demonstrated the application of geometric control to an industrial distillation column using an aggregated, low-order column model. Lee (1993) presents a number of impressive industrial applications of GMC including distillation column control, particulate dryer control, furnace control, and chemical reactor control. Finally, Groebel et al. (1994; 1995) show that input-output linearity can be achieved robustly for a distillation column on a pilot plant scale.

Differential Algebraic Approach

The differential algebraic approach to nonlinear control design offers a unified treatment for linear and nonlinear systems, lumped and distributed parameter systems (Fliess and Glad, 1993). Indeed many of the "open" problems in nonlinear control theory have been solved using this approach: decoupling, invertibility, model-matching, and realization. The resulting canonical forms suggest dynamic feedback for input-output linearization in which the transformations may depend on inputs and their derivatives (Fliess, 1990; Sira-Ramirez and Lischinsky-Arenasr, 1991; Sira-Ramirez, Lischinsky-Arenas and Llanes-Santiago, 1993). As the name implies, this methodology makes use of *algebraic* relationships between system variable derivatives, as opposed to systems of differential equations as in the geometric setting.

We briefly summarize the canonical form results from (Fliess, 1990). Assume that we start with a general single-input nonlinear state-space model (possibly non-control-affine):

$$\dot{x} = f(x, u)$$

The key idea from differential algebra is that of the differential primitive element. Simply stated, we seek an element, z, for our original nth order state model such that the system dynamics can be represented as n algebraic equations of $z, \dot{z}, \ddot{z}, \ldots, z^{(n)}$ as well as $u, \dot{u}, \ddot{u}, \ldots, u^{(\nu)}$. (The dependence of the transformations on u and its derivatives distinguishes this approach from the traditional differential geometric approaches). If this is satisfied, we can define $z \equiv q_1$, and derive the global generalized controller canonical form for the original system:

$$\begin{aligned}
\dot{q}_1 &= q_2 \\
\vdots &= \vdots \\
\dot{q}_{n-1} &= q_n \\
0 &= \bar{\mathcal{C}}(\dot{q}_n, \mathbf{q}, u, \dot{u}, \ddot{u}, \ldots, u^{(\nu)})
\end{aligned} \quad (43)$$

Finally, if we start with a scalar system output $y = h(x)$, and y qualifies as a differential primitive element, *i.e.*, (Sira-Ramirez et al., 1993):

$$\left\{ \text{rank} \frac{(y, \dot{y}, \ddot{y}, \ldots, y^{(n-1)})}{\partial x} \right\} = \left\{ \text{rank} \frac{(y, \dot{y}, \ddot{y}, \ldots, y^{(n)})}{\partial x} \right\}$$

then we can derive the global generalized canonical form:

$$\begin{aligned}
\dot{q}_1 &= q_2 \\
\vdots &= \vdots \\
\dot{q}_{n-1} &= q_n \\
0 &= \hat{\mathcal{C}}(\dot{q}_n, \mathbf{q}, u, \dot{u}, \ddot{u}, \ldots, u^{(\nu)}) \\
y &= q_1
\end{aligned} \quad (44)$$

Locally, we can solve $\hat{\mathcal{C}}$ for $\dot{q}_n$ to yield the local generalized observability canonical form:

$$\begin{aligned}
\dot{q}_1 &= q_2 \\
\vdots &= \vdots \\
\dot{q}_{n-1} &= q_n \\
\dot{q}_n &= \mathcal{C}(\mathbf{q}, u, \dot{u}, \ddot{u}, \ldots, u^{(\nu)}) \\
y &= q_1
\end{aligned} \quad (45)$$

Given this final structural form (which is related to the external differential form in (Nijmeijer and van der Schaft, 1990)), one can easily construct a dynamic feedback linearizing controller:

$$\mathcal{C}(\mathbf{q}, u, \dot{u}, \ddot{u}, \ldots, u^{(\nu)}) = -\sum_{i=1}^{n} \lambda_{i-1} q_i + v \quad (46)$$

where the λ_i are chosen for pole placement. In the case where the original problem has an affine dependence on u in the state equations, then one can solve for $u^{(\nu)}$ from $\mathcal{C}$, *i.e.*:

$$u^{(\nu)} = \mathcal{C}_\nu(v, y, \dot{y}, \ldots, y^{(n-1)}, u, \dot{u}, \ldots, u^{(\nu-1)}) \quad (47)$$

from which it is easy to realize a state-space representation for the controller in equation (46) for implementation purposes:

$$\begin{aligned}
\dot{\xi}_1 &= \xi_2 \quad (48) \\
\vdots &= \vdots \\
\dot{\xi}_\nu &= \xi_{\nu+1} \\
\dot{\xi}_{\nu+1} &= \mathcal{C}_\nu(v, y, \dot{y}, \ldots, y^{(n-1)}, u, \dot{u}, \ldots, u^{(\nu-1)}) \\
u &= \xi_1
\end{aligned} \quad (49)$$

A process example which includes a realization procedure for such a control architecture is given in (Doyle III and Hobgood, 1995). The limitations to such an input-output linearizing design are nonminimum phasedness, and impasse points which yield singular control laws. The latter points correspond to places where $\frac{\partial \mathcal{C}}{\partial u^{(\nu)}}$ is singular. The nonminimum phase property can be checked from the unforced zero dynamics, which are conveniently given by the following expression:

$$\mathcal{C}(0, u, \dot{u}, \ddot{u}, \ldots, u^{(\nu)}) = 0 \qquad (50)$$

In addition, the structure of $\mathcal{C}$ may suggest terms to be included in a semi-empirical model for nonlinear input-output identification (as will be shown in the following example).

Example 2 *Consider the following CSTR model (from (Uppal, Ray and Poore, 1974)):*

$$\begin{aligned}
\dot{x}_1 &= -x_1 + \mathcal{D}a(1 - x_1)e^{\frac{x_2}{1+x_2/\gamma}} \\
\dot{x}_2 &= -x_2 + B\mathcal{D}a(1 - x_1)e^{\frac{x_2}{1+x_2/\gamma}} + \beta(u - x_2) \\
y &= x_2
\end{aligned} \qquad (51)$$

To obtain the Fliess canonical form, one must differentiate the output two times:

$$\begin{aligned}
\dot{y} &= -x_2 + B\mathcal{D}a(1 - x_1)e^{\frac{x_2}{1+x_2/\gamma}} + \beta(u - x_2) \\
\ddot{y} &= -\dot{x}_2 - B\mathcal{D}a\dot{x}_1 e^{\frac{x_2}{1+x_2/\gamma}} + \\
&\quad B\mathcal{D}a(1 - x_1)e^{\frac{x_2}{1+x_2/\gamma}} \frac{1}{(1+x/\gamma)^2} + \beta(\dot{u} - \dot{x}_2)
\end{aligned}$$

Using the inverse transformation:

$$\begin{aligned}
x_1 &= 1 - \frac{\dot{y} - \beta(u - y) + y}{B\mathcal{D}a e^{\frac{y}{1+y/\gamma}}} \\
x_2 &= y
\end{aligned}$$

one can obtain the generalized observability canonical form for the CSTR problem:

$$\begin{aligned}
\ddot{y} &= (-2 - \beta)\dot{y} + (-\beta - 1)y + \beta\dot{u} + \beta u \\
&\quad + \frac{1}{(1+y/\gamma)^2}(\dot{y} - \beta(u-y) + y) + \\
&\quad \mathcal{D}a e^{\frac{y}{1+y/\gamma}}(B - \dot{y} + \beta(u-y) + y)
\end{aligned} \qquad (52)$$

(Note that for complex, higher order state dynamics, the analytical derivation of the transformation from the inputs, outputs, and their derivatives to the states may require symbolic manipulation software, or may not be globally solvable.)

This resultant structure is useful for several reasons: (i) the expression has an affine dependence on both u and $\dot{u}$, indicating that a dynamic feedback linearizing controller is easily constructed; and (ii) the structural form of the y nonlinearities suggests terms that might be included in a nonlinear input/output semi-empirical model structure for identification (e.g., terms such as $\frac{1}{(1+y/\gamma)^2}$ which are not immediately evident from the state-space fundamental model.) ◁

A recent research development in the differential algebraic literature is the concept of a differentially flat output for a controlled dynamical system (Fliess, Lévine, Martin and Rouchon, 1995). Although the flat output is a "virtual" process variable, it lends insight into several structural features of the original problem. Here "flat" refers to the "flat coordinates" which result from application of the Frobenius Theorem in the differential geometric approaches (Isidori, 1995). Such an output possesses the following properties:

- The components of the flat output can be expressed as a function of the original process states, **x**, the process inputs, **u**, and a finite number of derivatives of the inputs.

- The elements of the flat output vector are differentially independent.

- Each of the inputs and process states (as well as their derivatives and functions thereof) can be expressed as a function of the flat outputs and its derivatives.

The key connection between flat outputs and nonlinear control design is that systems which possess a flat output are guaranteed to yield a dynamic feedback linearizing control synthesis, as well as a quasi-static feedback linearization (Delaleau and Fliess, 1992). In fact the conditions are necessary and sufficient in both cases. Consequently, the flat output is quite useful for constructing tracking controllers for a complex process system. These principles have been applied to the analysis and tracking controller design for a simple chemical reactor in (Rothfuss, Rudolph and Zeitz, 1995), and recent results in (Fliess, Lévine, Martin, Ollivier and Rouchon, 1995) show that certain classes of singularities may be avoided by suitable flat-output-based controller design.

The following example illustrates the notion of flat outputs.

Example 3 *Consider the unstable biological reactor described in (Hoo and Kantor, 1986):*

$$\begin{aligned}
\dot{x}_1 &= \mu_1(S_f - \frac{x_1}{k_1} - \frac{x_2}{k_2})x_1 - u_1 x_1 & (53) \\
\dot{x}_2 &= \mu_2(S_f - \frac{x_1}{k_1} - \frac{x_2}{k_2}, x_3)x_2 - u_1 x_2 & (54) \\
\dot{x}_3 &= -p x_1 x_3 - u_1 x_3 + u_2 & (55)
\end{aligned}$$

where μ_1 and μ_2 are specified growth functions in the original reference, for the present context they are left as general functions with the proviso that μ_2 is invertible with respect to x_3. The following is proposed as a flat output vector for the process:

$$\begin{aligned}
y_1 &= x_1 & (56) \\
y_2 &= \ln(\frac{x_1}{x_2}) & (57)
\end{aligned}$$

Note that this choice of flat output has physical significance, as y_2 represents the relative extent of the inhibitor-resistant cells to the inhibitor-sensitive cells.

To verify this choice of a flat output, we need to show that each state and each input can be represented as a function of **y** *and its derivatives. (The other two conditions are* *obviously satisfied). First, note that:*

$$x_1 = y_1 \tag{58}$$
$$x_2 = y_1 e^{-y_2} \tag{59}$$

One can solve for u_1 from equation (53):

$$u_1 = \frac{\dot{y}_1 - \mu_1(S_f - \frac{y_1}{k_1} - \frac{y_1 e^{-y_2}}{k_2})y_1}{y_1} \tag{60}$$

Next, using equation (60), one can substitute for u_1 and solve equation (54) for x_3 (details avoided for simplicity of presentation):

$$x_3 = f_1(y_1, y_2, \dot{y}_1, \dot{y}_2) \tag{61}$$

Finally, one can solve equation (55) for u_2 to yield:

$$u_2 = f_2(y_1, y_2, \dot{y}_1, \dot{y}_2, \ddot{y}_1, \ddot{y}_2) \tag{62}$$

Hence, we see that the flat outputs have physical significance, are functions of the physical states, and one can solve for the process inputs and states as functions of the flat output and its derivatives. While this might be considered a trivial example at first glance (the outputs are equal to the appropriate distinguished outputs *from static feedback linearization*), it is noteworthy that the present results are independent of the specific structure of the growth functions μ_1 and μ_2. Hence, arbitrarily complex functional forms may be used, and the present results do not change. This generalizing property has been shown for several other chemical engineering examples as well (Rouchon, 1994; Rudolph, 1995; Rothfuss et al., 1995). ◁

Nonlinear H_∞ Control

For several years linear H_∞ control theory has experienced remarkable popularity in engineering applications. The main reasons for this are the possibility to include robustness considerations explicitly in the design and the fact that meaningful physical performance objectives can be expressed as H_∞ design specifications. Linear H_∞ control proved as a very successful controller design method and there are many reports of successful applications ((Skogestad, Morari and Doyle, 1988; Allgöwer and Raisch, 1992; Geddes and Postlethwaite, 1994), to mention a few). Most practical control problems are however of nonlinear nature. Therefore the obvious advantage of *nonlinear H_∞* theory over linear H_∞ theory is that system nonlinearities can be taken into account and thus stability and performance can be achieved over a larger operating region.

Any H_∞ controller design for a practical control problem consists of three major design steps: (i) formulation of the physical control objectives as H_∞ design specification, (ii) transformation of this H_∞ problem into the so-called standard form (see Fig. 6), and (iii) solution of the resulting H_∞ standard problem. For many interesting *linear* control problems there exists reasonable knowledge and rich experience to facilitate the first two steps. In the third step a

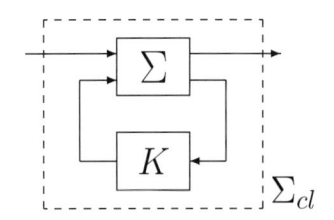

Figure 6. H_∞ standard problem.

controller K is then sought such that (a) the closed-loop is asymptotically stable and (b) the H_∞ norm of the transfer matrix between external inputs d and external outputs e is smaller than some prescribed number γ. The computation of such a controller involves the solution of two uncoupled Riccati equations in the linear case, for which reliable numerical tools exist.

Even though linear H_∞ control has its clear deficiencies (see for example (Morari and Zafiriou, 1989) for a discussion), it can be said that it is a mature controller design technique that allows the solution of many practically meaningful control problems in a transparent and numerically sound way.

During the last few years a theory for solving the nonlinear equivalent of the standard problem from linear H_∞ control has been developed (e.g. (Ball, Helton and Walker, 1993; Isidori and Astolfi, 1992; van der Schaft, 1992)). For a nonlinear system Σ:

$$\Sigma : \begin{array}{rcl} \dot{x} & = & f(x) + g_1(x)d + g_2(x)u \\ e & = & h_1(x) + k_{11}(x)d + k_{12}(x)u, \end{array} \tag{63}$$

where x represents the state, u the control input, d the external input and e represents the external output (see Fig. 6), a state feedback controller K:

$$K : \quad u = \gamma(x) \tag{64}$$

is sought such that (a) the closed-loop is asymptotically stable and (b) the L_2-gain from external input d to external output e of the closed-loop Σ_{cl} is smaller than some prescribed number γ:

$$\int_0^T \|e\|^2 dt \leq \gamma^2 \cdot \int_0^T \|d\|^2 dt, \quad \forall d(\cdot), T. \tag{65}$$

By an abuse of notation we will succinctly express the L_2-gain condition (65) as

$$\|\Sigma_{cl}\|_\infty \leq \gamma. \tag{66}$$

An interpretation of equation (65) is that the "gain" between the energy of the L_2-input-signal $d(\cdot)$ and the energy of the L_2-output-signal $e(\cdot)$ is smaller or equal to γ for all possible input-signals $d(\cdot)$. Thus, this holds for the worst-case input $d^*(\cdot)$. In the linear setup the smallest γ for which eq.(66) holds is the H_∞ norm of the linear transfer matrix Σ_{cl} and thus the nonlinear H_∞ standard problem is the direct extension of the linear problem to the nonlinear case.

Under the so called *standard assumptions*

$$\begin{aligned} \boldsymbol{k}_{11}(\boldsymbol{x}) &= 0 \\ \boldsymbol{h}_1^T(\boldsymbol{x})\boldsymbol{k}_{12}(\boldsymbol{x}) &= 0 \\ \boldsymbol{k}_{12}^T\boldsymbol{k}_{12}(\boldsymbol{x}) &= I \end{aligned} \quad (67)$$

the optimal state-feedback law that achieves closed-loop stability and L_2-gain less than γ is then given by (van der Schaft, 1993):

$$\boldsymbol{u}(\boldsymbol{x}) = -\frac{1}{2}\boldsymbol{g}_2^T(\boldsymbol{x})V_x^T(\boldsymbol{x}) \quad (68)$$

where V_x is a solution of the Hamilton-Jacobi inequality

$$V_x \boldsymbol{f}(\boldsymbol{x}) + \boldsymbol{h}_1^T(\boldsymbol{x})\boldsymbol{h}_1(\boldsymbol{x}) + \quad (69)$$
$$V_x \left(\frac{1}{4\gamma^2} \boldsymbol{g}_1(\boldsymbol{x})\boldsymbol{g}_1^T(\boldsymbol{x}) - \frac{1}{4}\boldsymbol{g}_2(\boldsymbol{x})\boldsymbol{g}_2^T(\boldsymbol{x}) \right) V_x^T \leq 0 \ .$$

Even though the solution is nicely characterized by (68) and (69) the actual computation of the controller involves the solution of the nonlinear partial differential equation/inequality (69). A global analytical solution is of course not feasible except for simple problems. However a numerical solution is sometimes possible. One noteworthy approach, which is mostly used in connection with nonlinear H_∞ control, is Lukes' method (Lukes, 1969). This method is based on a series expansion of the problem and leads to a local approximate solution. For problems with less than ten states, an approximate solution with terms up to about tenth order can be calculated using this method if the linearized problem is not degenerate. The obvious disadvantage is the exclusively local character. Promising new approaches are given in (James and Yuliar, 1995b), (Lu and Doyle, 1995) and (Flockerzi and Knobloch, 1994), and (Rehm and Allgöwer, 1996).

One of the big advantages of linear H_∞ control is that the output feedback case can be nicely treated within this framework. Unfortunately in nonlinear H_∞ control the output feedback case is much harder. Despite considerable progress during the last years (Isidori, 1994; Krener, 1994; van der Schaft, 1993; James and Yuliar, 1995a; Ball et al., 1993) the output feedback case is still not fully understood.

In addition to the computational difficulties and open theoretical problems with respect to output feedback the stringent assumptions (67) on the generalized plant (63) may also lead to serious problems in the nonlinear case. It turns out that the so-called 'standard' assumptions (67) are rarely satisfied in practical applications. In the linear setup, assumptions (67) can always be satisfied, for example, by various 'loop-shifting' transformations (Safonov, Limebeer and Chiang, 1989). For example, assumption $\boldsymbol{k}_{22} = 0$ can be assured simply by applying the change of variables

$$\boldsymbol{y}_{new} = \boldsymbol{y} - \boldsymbol{k}_{22}\boldsymbol{u} \quad (70)$$

to the original plant. In the nonlinear setup there is yet no simple equivalent to the linear 'loop-shifting' transformations. Some assumptions (e.g. $\boldsymbol{k}_{12}^T\boldsymbol{k}_{12} = I$) can be satisfied by a simple input transformation. Also $\boldsymbol{k}_{11} \neq 0$ constitutes no severe problem as a general solution can still be obtained. However, the case $\boldsymbol{k}_{22} \neq 0$ is as yet unsolved. The trivial change of variables (70) used in the linear case, fails because $\boldsymbol{k}_{22}$ is state-dependent in the nonlinear case.

Although the standard problem can be formulated and, with some reservations, solved in the nonlinear case, this does not imply that nonlinear H_∞ control incorporates both the practicality of the linear theory plus the possibility to explicitly consider (and possibly make use of) nonlinearities. The biggest problem lies in the formulation of meaningful and broadly applicable H_∞ design specifications in the nonlinear case. We will motivate this difficulty with an example of the practicality of robustness criteria involving unstructured uncertainty in the linear versus nonlinear case. The possibility of expressing conditions for robust stability to unstructured uncertainty as a bound on the H_∞ norm of certain transfer matrices is the main reason for choosing the H_∞ framework for linear control problems.

Unstructured uncertainty can be assumed to "lie" at different location in the loop (e.g. at the plant input or at the plant output). A key restriction is that unstructured uncertainty appears at only one "location" in the loop. Typical H_∞ robustness specifications are thus directed for example towards multiplicative output uncertainties or towards additive uncertainties, but never explicitly to both (see for example Fig. 7). Necessary and sufficient robust stability condi-

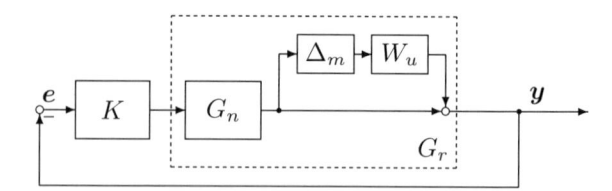

Figure 7. Closed-loop with multiplicative output uncertainty. G_r represents the real plant, G_n stands for the nominal plant, Δ_m stands for the uncertainty, W_u for the uncertainty weight and K represents the controller.

tions with respect to unstructured uncertainties can be formulated and expressed as bounds on the H_∞ norm of certain transfer matrices (see for example (Morari and Zafiriou, 1989)). Thus the assessment of robust stability depends both on the size of the respective uncertainty (usually described as a bound on its maximal singular value) *and* on the assumed location of the uncertainty (type of uncertainty).

In practical applications it is usually not known which "type" (multiplicative output uncertainty, additive uncertainty, etc.) of unstructured uncertainty is present in the system. In fact the "true" uncertainty will almost certainly be of a different (more structured) class. For linear systems, it is possible to express stability with respect to one type of uncertainty with a robustness condition on another type. This is potentially very conservative, indicating that, for example, robustness with respect to a large uncertainty at the system input is needed in order to achieve robustness with respect to a small uncertainty at the system output. The larger the condition number of the nominal system G_n, the more conservative the robust stability condition will be in general. For practical applications this is usually still a reasonable approach: One can employ engineering judgement to determine the principle source of uncertainty. If robust stability is assured for the assumed uncertainties, the closed-loop also exhibits at least some robustness with respect to other possible locations of uncertainty.

Based on the small gain theorem similar robustness conditions can also be proved for the nonlinear case. Nonlinear unstructured uncertainties can be defined in an analogous way as in the linear case where the nominal system G_n and the uncertainty weight W_u are nonlinear I/O-operators and the unstructured uncertainty Δ_m is an arbitrary *nonlinear* I/O-system with (nonlinear) L_2-gain smaller than some bound. A sufficient, but *not* necessary condition for robust stability can be given. We summarize this with the following result for nonlinear multiplicative output uncertainty (compare with Fig. 7):

Theorem 2 *Under certain technical assumptions the nonlinear closed-loop system as in Fig. 7 with nonlinear I/O-operators G_n, K, W_u and Δ_m, is asymptotically stable for all stable nonlinear perturbations Δ_m satisfying the L_2-gain condition $\|\Delta_m\|_\infty < 1$ if the nominal closed-loop is asymptotically stable and*

$$\|W_u \cdot G_n K(I + G_n K)^{-1}\|_\infty \leq 1 \ . \quad (71)$$

The same type of robustness theorem can also be proven for nonlinear additive uncertainties, nonlinear multiplicative input uncertainties, etc. As with linear control problems, the mathematical uncertainty rarely matches the exact type described in practical applications. This is particularly true in the nonlinear case where there are many more possibilities of uncertainty locations. But for nonlinear systems different locations of the unstructured uncertainty cannot be "shifted" to other locations in general, meaning that a robustness guarantee for one particular uncertainty description does not imply *any* robustness with respect to other uncertainty types.

Example 4 *We consider the simple nonlinear nominal system*

$$\begin{aligned} y &= \sin x \\ \dot{x} &= -x + u \end{aligned} \quad (72)$$

and its linearization around the steady state $x_s = 0$

$$\begin{aligned} y &= x \\ \dot{x} &= -x + u \ . \end{aligned} \quad (73)$$

Both the linear as well as the nonlinear system have an L_2-gain of one. If we assume a multiplicative uncertainty with $W_u = 1$ and $\|\Delta_m\|_\infty < 1$ for linear system (73), then it is easy to show that robustness with respect to this multiplicative uncertainty also guarantees robustness with respect to additive uncertainties satisfying $\|\Delta_a\|_\infty < 1$. For the nonlinear system robustness with respect to multiplicative output uncertainty $\|\Delta_m\|_\infty < 1$ does not give rise to any robustness to additive uncertainties in this example. This can be seen immediately by the following argument: The output of nominal system (72) is always constrained to lie between minus one and plus one. A multiplicative output uncertainty can never lead to a change of sign between the real and nominal output y. If we consider additive uncertainties with $\|\Delta_a\|_\infty < \epsilon$, then for any $\epsilon > 0$ there exists always an input function $u(\cdot)$ large enough and an uncertainty Δ_a^* satisfying $\|\Delta_a^*\|_\infty < \epsilon$ such that the sign of the nominal and real output differ. Therefore this system cannot tolerate any additive uncertainty. ◁

This example demonstrates that a robustness guarantee with respect to one unstructured uncertainty does not imply robustness with respect to another uncertainty class in the nonlinear case. Once more, results from linear theory can be nicely extended to the nonlinear case, but they unfortunately loose their merits and general applicability. Moreover, condition (71) by itself can be arbitrary conservative as it is only a sufficient condition in the nonlinear case, whereas it is also necessary in the linear case.

A certain level of robustness can, however, always be expected when applying nonlinear H_∞ theory. Any linear or nonlinear H_∞ state-feedback law guarantees an "infinite gain margin and 50% gain reduction tolerance" to static input uncertainties (Glad, 1987). Furthermore if feedback (68) renders the nominal closed-loop exponentially stable, then solution V of eq. (69) is a Ljapunov-function for the closed-loop and thus guarantees stability also for the closed-loop with $f = f_n + \Delta f$ and $g = g_n + \Delta g$ etc. where Δg, Δf have to satisfy a Lipschitz condition.

Similar remarks with respect to restricted applicability also hold for the definition of H_∞ performance objectives in the nonlinear setting (Allgöwer, Rehm and Gilles, 1994).

Nonlinear H_∞ control is one of the areas in nonlinear control that received considerable attention during the last five years. This attention was however focused mainly on the development of the theory but not on applications. Only a very few nonlinear H_∞ applications involving real or at least realistic plants are known (Feng and Postlethwaite, 1993; Astolfi and Lanari, 1994; Kang, 1995). The authors are only aware of one single process control application (Rehm and Allgöwer, 1996). We tried to explain the reasons for this with the discussion above. Three areas can be defined in which there are still major deficiencies as compared to the popular linear case. The first area concerns important open problems in the theoretical development. From an application point of view, open problems in connection with the output-feedback case and the restrictive standard assumptions are of principle interest. The second area is related to the computational solution of the standard problem. At the moment few practical solutions are known. Because of the rapid development of computing power, there is however the hope that a numerical (on-line) solution of the Hamilton-Jacobi equations is feasible. The third problem area, that will be the most crucial one, concerns the formulation of meaningful nonlinear H_∞ problems. For linear systems there is a wealth of practical problems that can be formulated in the H_∞ framework. For nonlinear systems this is no longer the case. We tried to point this out through the preceding example and the ensuing discussion regarding robustness to unstructured uncertainty. It is clear that for nonlinear systems a generally valid design procedure cannot be expected. However for specific practical problems nonlinear H_∞ theory can and will become a useful tool for achieving robustness and desired performance.

Which Way to the Promised Land?

With the discussion of some selected control system analysis and synthesis methods we wanted to convey the state of the art in nonlinear process control. The coverage is by far neither complete nor exhaustive, and our selection is certainly not meant to imply that the methods and techniques not mentioned are inferior to the ones discussed. For example, we did not discuss nonlinear model predictive control or adaptive control even though we think that both approaches have clear and distinct advantages over other control schemes — as they also have disadvantages. These two approaches are covered in detail and are critically discussed in the papers by Mayne (1996) and Ydstie (1996) in these proceedings. In addition, we will not elaborate on the important issue of benchmark studies for nonlinear control systems; though it is important to develop both: (i) simple, yet dynamically rich theoretical benchmarks (e.g., based on the van de Vusse kinetic scheme (Van der Vusse, 1964)), and (ii) practical benchmarks for evaluating the applicability of various nonlinear methodologies (e.g., the Eastman problem (Downs and Vogel, 1993)).

The deficiencies of the methods we discussed here are rather typical and representative for present nonlinear control theory. It is taken as accepted that there is no single nonlinear control paradigm at present that proves to be rich enough to inspire and allow the development of a practically satisfying nonlinear control theory. Rather, there is a selection of more or less capable analysis and synthesis approaches, each having its own strengths and weaknesses and each being suitable for some (perhaps limited) class of problems but not for others. Nevertheless our understanding of nonlinear control has improved considerably during the last decade. In particular, the geometric framework has led to an understanding of many of the peculiarities of nonlinear control and has helped to develop insight into the difficulties, complications and also into the differences to the linear case. We are also able to (partly) classify nonlinear control problems, for example with respect to the type of feedback required to achieve certain closed-loop properties or with respect to the difficulty one will encounter when designing controllers. In fact, many fundamental tools have been developed that will certainly facilitate the development of new analysis and synthesis methods in the future.

None of the present paradigms are completely mature, so it will be impossible to answer in a specific way the question posed by the title of this section: *Which way to the promised land?* Instead, we point out some of the important research challenges which might serve as milestones to the development of more capable methods which lead, ultimately, in the "right direction".

Synthesis and Analysis

Any practitioner can tell that the analysis step prior to the actual controller design is very often the most crucial one, that saves, when done properly, time and effort and leads in most cases to the decision on which controller synthesis approach to take. Control system analysis is, however, a field that appears to be neglected by the nonlinear process control community when compared to the many efforts in the area of controller synthesis. There are many significant unresolved problems, including stability, controllability and observability, and robustness and performance analysis. But analysis is not only important because of its direct practical importance but also because it is often the first step in the development of synthesis approaches: If you cannot analyze a property, you will in many cases not be able to synthesize a controller that achieves it. Therefore, we believe that it is most advantageous to put more effort into the (sometimes even simpler) field of nonlinear control system analysis. As outlined earlier, there are preliminary analysis results available that can serve as a starting point. Extensions of the structured singular value theory has provided some initial pointers, albeit conservative, for research directions in determining both nominal stability (of plant operators and their inverse), and robustness properties. Of particular importance is the analysis of achievable performance in the nonlinear case. This will help in developing insight into when the application of nonlinear control is worthwhile when compared to linear control. The quantification of the size of nonlinearity of open-loop plants and of controller design problems is a first step in that respect (see also the paragraph on performance below).

On-line vs. Off-line Controller Computation

Most linear controllers can be synthesized using concepts from linear algebra and the necessary numerical computations can be performed off-line in the vast majority of cases. Nonlinear analysis and synthesis methods usually draw from different mathematical concepts than linear algebra. The current understanding and the state of the art is that the solution of nonlinear problems requires symbolical computations (for example to compute the various forms of Lie derivatives). Symbolical computations are performed using computer algebra packages like MATHEMATICA, MACSYMA or MAPLE. Any user of such packages knows their inherent limitations from painful experience: Only rather small problems with few states can be solved (if there is no special problem structure that can be exploited). Even if one extrapolates the current status of computing power into the future with faster CPUs, more memory, and improved software tools, it is clear that this alone will not solve the problem. We believe that there is (unfortunately!) no way around massive numerical computations that most certainly will have to be performed *on-line*. Nonlinear MPC, and to a certain extent nonlinear H_∞ control, are steps in that direction that have already proven to be very successful. We believe that if we throw out the natural bias toward off-line controller computation, we might be able to open up a whole new world of possibilities. We want to stress that we certainly do *not* want to advocate numerical computation over analytical thinking. As much as can be done off-line should be done off-line. But if the goal is worthwhile we should not dismiss ideas because they seem to require computationally daunting calculations

A useful current approach of sufficient generality, that avoids on-line computations, is a series approximation based approach. While these approaches are inherently suboptimal, they do allow (via a series expansion) a decomposition of difficult nonlinear problem into several (provably) easier problems. The weakness of such an approach is that the resultant solution tends to be optimal in a local region, or semi-global region at best (provided that the linear problem is nondegenerate). However, the strengths are the computational ease of solution as well as the intuitive extension of the linear methodologies. Representative examples of this approach include: (i) Volterra series approach for IMC and MPC (Doyle III, 1995); (ii) Taylor series approach for feedback linearization (Krener, Karahan, Hubbard and Frezza, 1987; Champetier, Reboulet and Mouyon, 1984; Doyle III and Hobgood, 1995; Allgöwer et al., 1996); and (iii) series solution to nonlinear H_∞ synthesis (Isidori and Kang, 1995).

Nonlinear Uncertainty Descriptions

There is no doubt that a "good" controller design method has to be able to incorporate robustness specifications that guarantee proper control action despite uncertainties in the underlying model description. When one assumes a linear nominal model and also that the real plant is linear, there are many approaches to analyzing robust stability (given a controller) and also to synthesizing a controller that achieves the required robustness of stability (Kharitonov, 1978; Doyle and Stein, 1981; Chapellat, Dahleh and Bhattacharyya, 1990). In the linear case there are even methods that allow analysis and design of controllers that achieve *robust performance* (Doyle, 1982). We have pointed out some promising approaches to developing nonlinear robustness theories earlier in the paper. However, these results have yet to prove their utility. The main reason for this is not necessarily the lack of proper mathematical tools for deriving such theorems but rather a lack of appropriate *uncertainty descriptions*. Unstructured uncertainty, that has proven very successful in the linear case, is by far not as useful in the nonlinear case. Also its "structured" extension (Doyle, 1982) will probably not be appropriate. As we required a frequency domain representation in the linear case to develop practical uncertainty descriptions and robustness theorems, we will likely have to go to alternative system representations in the nonlinear case as well. The behavioral approach of Willems (Willems, 1995) or the differential algebraic approach of Fliess (Fliess and Glad, 1993) might lead in the right direction.

Another obvious uncertainty description that we did not mention so far, but that makes a great deal of sense in the nonlinear case, is parametric uncertainty. When considering parametric uncertainties *adaptive techniques* might turn out to be much more valuable in the nonlinear case.

Output Feedback

In chemical process control – more so than in most other control areas – the complete state cannot be measured in general and therefore state feedback approaches are not directly suited for practical applications. However most current nonlinear controller design methods assume that the whole state is measurable. Our feeling is that five years ago the difficulty and the implications of this drawback of most methods were not completely recognized or have been partly underestimated. One of the achievements since then is the current lively discussion of the output feedback case.

As is well known there are two approaches to output feedback. The first class of controllers explicitly consist of a static or dynamic state feedback plus an observer or filter that reconstructs the states. The underlying idea is the so called *certainty equivalence principle*. The control performance depends of course heavily on the observer used and, due to a lack of a nonlinear *separation principle*, stability of the closed-loop has to be proven in each case even though each component by itself might be stable/stabilizing. There has been significant progress in the area of nonlinear observer design during the last years (Walcott, Corless and Zak, 1987; Gauthier and Kupka, 1994; Misawa and Hedrick, 1989; Birk, 1992; Soroush, 1995; Daoutidis and Kravaris, 1994) but there is still no completely satisfying solution. This is not surprising as the nonlinear observer design problem is as difficult, or more difficult, than the nonlinear stabilization problem.

The second approach is the direct output feedback synthesis were the actual states are not explicitly reconstructed. This direct approach will probably prove much more valuable than the indirect approach because stability and performance of the closed-loop will be assessed directly. Except for the nonlinear H_∞ output feedback case – which is also still largely unresolved – there are no major approaches in this direction at the moment.

Without achieving significant progress in the output feedback case, a successful practical application of nonlinear control theory for industrial control problems will not be possible.

Performance Limitations and Performance Objectives

The design of nonlinear controllers will most certainly always be rather involved and will most certainly always require significant on-line and off-line computations and significant time on the side of the designing engineer. In order to justify this effort, the achievements over robust linear control or simple PI-control must be worthwhile. The quantification of achievable performance and the (quantitative) assessment of performance limitations imposed by certain open-loop plant properties should, therefore, be seen as key research areas. Preliminary results can for example be found in (Shamma, 1991; Russo and Bequette, 1992; Sourlas, Choi and Manousiouthakis, 1994).

Another question that falls under this category is the definition of *meaningful* performance objectives. In the linear case the use of quadratic performance costs (e.g. H_2) makes sense because this allows one to retain the linear structure of the problem. In the nonlinear case, it makes sense to exploit the generality afforded in the nonlinear setting to employ objective functions that accurately reflect complex quantities such as profitability or environmental consciousness. These quantities are not easily expressible in terms of simple linearly weighted norms of systems variables.

The notion of performance is strongly coupled to process nonlinearity. In particular, the issues raised earlier concerning control-relevant nonlinearity deserve further research attention. At the present time, there are only preliminary research results in this area, certainly compared to the level of maturity of other aspects of process characterization and their corresponding tools (e.g., multivariable interaction–RGA).

Classes of Nonlinear Systems

The difference in dynamic behavior that nonlinear systems can display is striking. It would therefore be wildly optimistic to hope to find a theory that is equally well applicable to *any* nonlinear system. Even the class of control affine smooth systems (eq. 1), that is the starting-point for most current theories, is much too broad to allow the solution of challenging practically relevant problems. We should, therefore, actively search for specific, but broad enough classes which satisfy two principle criteria: (i) exhibit an accurate match to physical behavior; and (ii) facilitate both control system synthesis and closed-loop analysis. Three examples, that have already been studied in detail, are Hammerstein and Wiener model descriptions and Volterra series models (Eskinat, Johnson and Luyben, 1991; Ouarti and Edgar, 1993; Zhu and Seborg, 1994; Doyle III, 1995; Fruzzetti, Palazoglu and McDonald, 1995). System classes can be defined with respect to mathematical properties/structures (like for example Hammerstein models, etc.) or with respect to specific physical processes (like for example distillation, crystallization, etc.). Particularly from an application point of view, a stronger process orientation will be most beneficial and can lead to an improved appreciation of nonlinear process control. The results by Bastin and Lévine (Bastin and Lévine, 1993) mentioned earlier are exemplary in that respect.

If we are able to define mathematical classes of systems/problems that we can solve successfully, we have to be able to find system representations of physical processes that fit into these classes. We therefore have to find means of representing given physical systems as members of certain mathematical classes of systems. There are two ways of approaching this problem: Nonlinear system identification and nonlinear model reduction, that both have a rather broader scope.

Nonlinear System Identification

In model-based process control applications (e.g., MPC, IMC, IOL, etc.), one must balance model complexity and the associated computational burden against model accuracy. Invariably, detailed fundamental models consist of numerous coupled differential algebraic equations, which pose too great a burden for control synthesis using the methodologies described in this paper.

One class of models which achieves a reasonable compromise on accuracy versus complexity are nonlinear empirical models (for a recent survey, see (Pearson, 1994a)). Many possibilities exist, including polynomial ARMAX models, Volterra series models, and neural networks.

These classes of empirical nonlinear models have been utilized in process model-based control applications (e.g., NARMAX – (Hernández and Arkun, 1993), Volterra series – (Doyle III, 1995), polynomial models – (Ogunnaike, Pearson, Samardzija and Bomberger, 1994), neural networks – (Eaton, Rawlings and Ungar, 1994)), however, there are still many open research issues. To name a few:

- What nonlinear basis functions (i.e., model structure) are most appropriate for particular process applications?

- What nonlinear model structures give rise to computationally tractable control analysis *and* synthesis problems?

- What are the most efficient algorithms for initial parameter identification and subsequent on-line updating?

While recent work (Rhodes and Morari, 1995) indicates that the first issue may admit a formal mathematical solution, the practicality of such an approach has yet to be demonstrated for a complex problem.

Nonlinear Model Reduction

The empirical identification approach to nonlinear model reduction has some history in theoretical development, however, a principle shortcoming of this approach is the non-physical nature of the resulting variables. There is a need for *fundamental* approaches to process model development. For a motivating example of this principle, consider the work done by Skogestad and Morari (1988) on linear reduced models for distillation columns. Physical principles drove the development of the two-time constant models, which were readily implemented in model-based control schemes such as IMC.

A similar set of tools (possibly application dependent) is required for nonlinear process models. It will be extremely useful from a control synthesis perspective to have a well defined methodology for fundamental model reduction which yields physically intuitive, and computationally tractable models.

A representative example of this approach for a specific class of nonlinear systems is the traveling wave phenomena for certain distributed parameter systems (for a review, see (Marquardt, 1990; Hwang, 1995)). This approach reduces the complex fundamental models which describe packed bed reactors and distillation columns to one on the order of two to four nonlinear differential equations which describe the underlying transport phenomena leading to wave propagation.

Open research issues in this area include:

- What mechanisms exist for reducing fundamental models of complex lumped parameter systems (e.g., multiple reactions in a CSTR)?

- How does one extend the ideas for wave propagation to more complex distributed parameter systems involving multiple components and concurrent reaction/separation?

In addition, there is also the possibility of formal mathematical reduction of nonlinear models using balanced realizations (Scherpen, 1994). Although the physical meaning of states is lost, the input-output behavior is preserved in a meaningful way.

Multiple-model and Scheduling

Another research area receiving considerable attention in the literature is the multiple-model or scheduled approach to control synthesis. Strictly speaking, such an approach might be deemed nonlinear only because of the fact that the controller cannot be represented by any single linear controller. Early work in this area included the so-called *extended* approaches to model predictive control which relied on updated Jacobian models (García, 1984). The theoretical framework for the scheduling approaches involving a parametrized linear model (*e.g.*, along the equilibrium manifold) have been developed by Rugh (1991), Packard (1994), and Apkarian and Gahinet (1995). Another theoretical approach to the multiple-model description involves a Bayesian formulation in which probabilities are used to weight individual models (Banerjee, Arkun, Ogunnaike and Pearson, 1994). In the process control community, these approaches have employed a variety of modeling and synthesis approaches including fuzzy logic-based scheduling (Ling and Edgar, 1992), H_∞ design (Banerjee et al., 1994), adaptive control (Schott and Bequette, 1995), and dynamic scheduling (Lawrence and Rugh, 1995), (Kwatra and Doyle III, 1995). Experimental studies in the process area include pH reactor control using static nonlinear gain scheduling (Klatt and Engell, 1995).

Conclusions

With this paper we tried to give a brief assessment of the state of the art of nonlinear process control and tried to point out some promising and broad research directions. Despite many promising results, and also quite a number of practical applications that demonstrate the advantages of nonlinear control, this field is still largely unexplored and wide open. Only few practically meaningful problems can be posed and solved for systems that typically arise in chemical process control. However we want this to be understood in an optimistic rather than a pessimistic way: Nonlinear control is a vital area, offering many possibilities and advantages over linear control. Also many of the necessary fundamentals have been established, and it can be expected that there will be significant progress in the near future. We do not, however, see one direct answer to the question concerning the route to the "promised land" that we posed in the title. Nonlinear control is a difficult and fairly young field. There will certainly be not *one* way but there will be *many* ways.

We would like to make one more remark as with respect to the judgment of progress in this field. There should be two goals that we are heading for: One is the long-term development of the basic theory and the other is the "short-term" development of practically useful theory for *today's* industrial problems. As with any challenging and difficult field we should not merely judge new developments with respect to the second goal (namely their immediate practical relevance). Theoretical results that do not appear to have direct practical impact might impart important insight into a problem and might serve in the long run as an important intermediate step necessary for the development of truly practically meaningful and useful methods.

Acknowledgments

It is important to acknowledge a number of very thorough reviews in both nonlinear systems theory and nonlinear process control methodologies. In the general area of nonlinear systems theory, Vidyasagar (1986) has a thorough treatment of results up through 1986, and more recently Corless summarized results in the areas of analysis and control of nonlinear uncertain systems (Corless, 1993).

In the process control area, Kantor and Kravaris have an excellent review and tutorial on nonlinear differential geometric approaches (Kravaris and Kantor, 1990), McLellan et al. completed a review on error trajectory based approaches (1990), Kravaris and Arkun surveyed the geometric nonlinear control field at CPC IV (1991), and Bequette has a very nice overview of nonlinear control with an emphasis on process applications (1991).

A number of excellent textbooks are also available for

the reader who seeks a thorough tutorial on particular aspects of nonlinear control theory. From a pure systems perspective, there are the texts by Vidyasagar (1993), Isidori (1995), Nijmeijer and van der Schaft (1990), Slotine and Li (1991) and Marino and Tomei (1995). An upcoming monograph addresses aspects of nonlinear process control with applications (Henson and Seborg, in press, 1995).

One of the authors (FJD) gratefully acknowledges funding from the National Science Foundation in the form of an NYI Award (CTS 9257059). The other author (FA) would like to thank Dr. W.D. Smith Jr. of the DuPont Company for providing an excellent research environment for the one year stay at DuPont in which course this paper was written

References

Allgöwer, F. (1995a). Analysis and controller synthesis for nonlinear processes using nonlinearity measures, *AIChE Annual Meeting*.

Allgöwer, F. (1995b). Definition and computation of a nonlinearity measure, *3rd IFAC Nonlinear Control Systems Design Symposium*, Lake Tahoe, CA, 279–284.

Allgöwer, F. (1995c). Nichtlinearitätsmaße — Ein Werkzeug zur Analyse und Synthese nichtlinearer Regelkreise, *in* S. Engell (ed.), *Entwurf Nichtlinearer Regelungen*, Oldenbourg Verlag, München, 309–331.

Allgöwer, F., Doyle III, F. and Morari, M. (1996). Approximate input-output linearization of nonminimum phase nonlinear systems. In preparation.

Allgöwer, F. and Gilles, E. (1992). Approximate input/output-linearization of nonlinear systems, AIChE Annual Meeting, Miami, FL, **126f**.

Allgöwer, F. and Gilles, E. (1995). Einführung in die exakte und näherungsweise Linearisierung nichtlinearer Systeme, *in* S. Engell (ed.), *Entwurf Nichtlinearer Regelungen*, Oldenbourg Verlag, München, 23–52.

Allgöwer, F. and Raisch, J. (1992). Multivariable controller design for an industrial distillation column, *in* N. Nichols and D. Owens (eds), *The Mathematics of Control Theory*, Clarendon Press, Oxford, 381–406.

Allgöwer, F., Rehm, A. and Gilles, E. (1994). An engineering perspective on nonlinear H_∞ control, *Proc. 33rd IEEE Conf. Decision Contr.*, Orlando, FL, 2537–2542.

Apkarian, P. and Gahinet, P. (1995). Convex characterization of gain-scheduled H_∞ controllers, *IEEE Trans. Autom. Control*, **40**, 853–864.

Astolfi, A. and Lanari, L. (1994). Disturbance attenuation and setpoint regulation of rigid robots via H_∞ control, *Proc. 33rd IEEE Conf. Decision Contr.*, Orlando, FL, 2578–1583.

Atherton, D. (1981). *Stability of nonlinear systems*, John Wiley & Sons.

Balchen, J. and Sandrib, B. (1994). Input saturation in nonlinear multivariable processes resolved by nonlinear decoupling, *Proc. IEEE Conf. on Control Applications*, 39–44.

Ball, J., Helton, J. and Walker, M. (1993). H_∞ control for nonlinear systems with output feedback, *IEEE Trans. Automat. Contr.*, **AC-38**(4), 546–559.

Banerjee, A., Arkun, Y., Ogunnaike, B. and Pearson, R. (1994). Robust nonlinear control by scheduling multiple model based controllers, *AICHE Annual Meeting*, **230a**.

Bartusiak, R., Georgakis, C. and Reilly, M. (1989). Nonlinear feedforward/feedback control structures designed by reference system synthesis, *Chem. Eng. Sci.*, **44**, 1837–1851.

Bastin, G. and Lévine, J. (1993). On state accessibility in reaction systems, *IEEE Trans. Autom. Control*, **38**, 733–742.

Bendat, J. S. (1990). *Nonlinear System Analysis and Identification from Random Data*, John Wiley & Sons, New York, NY.

Bequette, B. (1991). Nonlinear control of chemical processes: A review, *Ind. Eng. Chem. Res.*, **30**, 1391–1413.

Bernstein, D. (1993). Nonquadratic cost and nonlinear feedback control, *Int. J. Rob. Nonl. Control*, **3**, 211–229.

Birk, J. (1992). *Rechnergestützte Analyse und Lösung nichtlinearer Beobachtungsaufgaben*, VDI Verlag, Düsseldorf.

Calvet, J. P. and Arkun, Y. (1988). Feedforward and feedback linearization of nonlinear systems and its implementation using IMC, *Ind. Eng. Chem. Res.*, **27**(10), 1822–1831.

Carr, J. (1981). *Applications of center manifold theory*, Springer-Verlag, Berlin.

Champetier, C., Reboulet, C. and Mouyon, P. (1984). A new approach to linearize nonlinear systems: The pseudolinearization. comparison with classical methods, *Proceedings of the IFAC 9th Triennial Congress*, 393–397.

Chapellat, H., Dahleh, M. and Bhattacharyya, S. (1990). Robust stability under structured and unstructured perturbations, *IEEE Trans. Automat. Contr.*, **AC-35**(10), 1100–1107.

Choi, J., Nikolaou, M. and Manousiouthakis, V. (1994). Nonlinear system gains and incremental gains over ball. IEEE Trans. Auto. Control, submitted.

Christofides, P. and Daoutidis, P. (1994). Feedback synthesis for two-time-scale nonlinear systems with uncertainty, *Proc. IEEE Conf. Dec. Control*, 805–807.

Colombe, P., Dablainville, R. and Vacarisas, J. (1987). Simulation and Development of Internal Model Control Applications in the Bayer Process, *Proceedings of Light Metal Session at AIME Annual Meeting*, 27–32.

Corless, M. (1993). Control of uncertain nonlinear systems, *Trans. ASME*, **115**, 362–372.

Daoutidis, P. and Kravaris, C. (1994). Dynamic output feedback control of minimum-phase multivariable nonlinear processes, *Chem. Eng. Sci.*, **49**, 433–447.

Delaleau, E. and Fliess, M. (1992). Algorithme de structure, filtrations et découplage, *C.R. Acad. Sci. Paris*, **315**, 101–106.

Desoer, C. and Wang, Y.-T. (1981). Foundations of feedback theory for nonlinear dynamical systems, *IEEE Trans. Circ. Syst.*, **CAS-27**(2), 104–123.

Dochain, D. (1994). Design of adaptive linearizing controllers for non-isothermal reactors, *Int. J. of Control*, **59**, 689–710.

Downs, J. J. and Vogel, E. F. (1993). A plant-wide industrial process control problem, *Computers Chem. Engng.*, **17**(3), 245–255.

Doyle III, F., Allgöwer, F. and Morari, M. (1996). A Normal Form Approach to Approximate Input-Output Linearization for Maximum Phase Nonlinear SISO Systems, *IEEE Trans. Autom. Control*, **41**, 305–309.

Doyle III, F. and Hobgood, J. (1995). Input-output linearization using approximate process models, *J. Proc. Cont.*, **5**, 263–275.

Doyle III, F. J. (1995). An Anti-windup Scheme for Input-Output Linearization, *submitted to Automatica*.

Doyle III, F. J., Morari, M. and Doyle, J. C. (1991). Some practical considerations in the selection of "linearizing" control over linear control, *AIChE Annual Meeting*, **149d**.

Doyle III, F., Packard, A. and Morari, M. (1989). Robust controller design for a nonlinear CSTR, *Chem. Eng. Sci.*, **44**, 1929–1947.

Doyle, J. C. (1982). Analysis of feedback systems with structured uncertainties, *IEEE Proc.*, D, **6**, 242–250.

Doyle, J., Francis, B. and Tannenbaum, A. (1992). *Feedback control theory*, Macmillan Publishing Comp., New York, NY.

Doyle, J., Georgiou, T. and Smith, M. (1993). The parallel projection operators of a nonlinear feedback system, *Syst. Control Lett.*, **20**, 79–85.

Doyle, J. and Stein, G. (1981). Multivariable feedback design: Concepts for a classical/modern synthesis, *IEEE Trans. Automat. Contr.*, **AC-26**(1), 4–16.

Eaton, J., Rawlings, J. and Ungar, L. (1994). Stability of neural net based model predictive control, *Proceedings of the American Control Conference*, 2481–2485.

Economou, C., Morari, M. and Palsson, O. (1986). IMC. 5. Extension to nonlinear systems, *Ind. Eng. Chem. Process Des. Dev.*, **25**, 403–411.

Eskinat, E., Johnson, S. H. and Luyben, W. L. (1991). Use of Hammerstein models in identification of nonlinear systems, *AIChE J.*, **37**(2), 255–268.

Feng, W. and Postlethwaite, I. (1993). Robust nonlinear H_∞ control adaptive control of robot manipulator motion, *Proc. 12th IFAC World Congress*, Sydney, Australia, 31–34.

Fliess, M. (1990). Generalized controller canonical forms for linear and nonlinear dynamics, *IEEE Trans. Autom. Control*, 994–1001.

Fliess, M. and Glad, S. (1993). An algebraic approach to linear and nonlinear control, *Essays on Control: Perspectives in the Theory and its Applications*, Birkhäuser, 223–267.

Fliess, M., Lévine, J., Martin, P., Ollivier, F. and Rouchon, P. (1995). Flatness and dynamic feedback linearizability: Two approaches, *Proceedings of European Control Conference*, 649–654.

Fliess, M., Lévine, J., Martin, P. and Rouchon, P. (1995). Flatness and defect of nonlinear systems: introductory theory and examples, *Int. J. of Control*, **61**, 1327–1361.

Flockerzi, D. and Knobloch, H. (1994). Integral manifolds in H_∞-theory, *Proc. 33rd IEEE Conf. Decision Contr.*, Orlando, FL, 2559–2564.

Fruzzetti, K., Palazoglu, A. and McDonald, K. (1995). Nonlinear Model Predictive Control Using Hammerstein Models, *J. Proc. Cont.*. Submitted.

García, C. E. (1984). Quadratic dynamic matrix control of nonlinear processes. An application to a batch reaction process. *AIChE Annual Meeting*, San Francisco, CA.

Gauthier, J. and Kupka, I. (1994). Observability and observers for nonlinear systems, *SIAM J. Contr. Optim.*, **32**(4), 975–994.

Geddes, E. and Postlethwaite, I. (1994). Multivariable control of a high performance tandem cold rolling mill, *Proc. of Int. Conf. on CONTROL '94*, 389–397.

Glad, S. (1987). Robustness of nonlinear state feedback - a survey, *Automatica*, **23**, 425–435.

Groebel, M., Allgöwer, F., Storz, M. and Gilles, E. (1994). Nonlinear control of a high-purity distillation column, *AIChE Annual Meeting*, **126c**.

Groebel, M., Allgöwer, F., Storz, M. and Gilles, E. (1995). Asymptotic exact I/O Linearization of an industrial high-purity distillation column, *Proc. Amer. Control Conf.*, 2648–2652.

Guay, M., McLellan, P. and Bacon, D. (1995). Measurement of nonlinearity in chemcial process control systems: The steady state map, *Can. J. Chem. Eng.*, **73**, 868–882.

Haber, R. (1985). Nonlinearity tests for dynamic processes, *Proc. 7th IFAC/IFIP Symp. on Identification and System Parameter Estimation*, 409–413.

Haber, R. and Unbehauen, H. (1990). Structure identification of nonlinear dynamic systems - A survey on input/output approaches, *Automatica*, **26**(4), 651–677.

Hahn, W. (1967). *Stability of motion*, Springer-Verlag, Berlin.

Henson, M. and Seborg, D. (eds) (in press, 1995). *Nonlinear Process Control*, Prentice Hall, Englewood Cliffs, NJ.

Hermann, R. and Krener, A. (1977). Nonlinear controllability and observability, *IEEE Trans. Automat. Contr.*, **22**, 728–740.

Hernández, E. (1992). *Control of Nonlinear Systems Using Input-Output Information*, PhD thesis, Georgia Institute of Technology, Atlanta, GA. Chemical Engineering.

Hernández, E. and Arkun, Y. (1993). Control of nonlinear systems using polynomial ARMA models, *AIChE J.*, **39**(3), 446–460.

Hirschorn, R. M. (1979). Invertibility of multivariable nonlinear control systems, *IEEE Trans. Autom. Control*, **AC-24**(6), 855–865.

Hoo, K. and Kantor, J. (1986). Linear feedback equivalence and control of an unstable biological reactor, *Chem. Eng. Commun.*, **46**, 385–399.

Hwang, Y.-L. (1995). On the nonlinear wave theory for dynamics of binary distillation columns, *AIChE J.*, **41**(1), 190–194.

Isidori, A. (1994). A necessary condition for nonlinear H_∞ control via measurement feedback, *Syst. Contr. Lett.*, **23**, 169–177.

Isidori, A. (1995). *Nonlinear Control Systems*, third edn, Springer-Verlag, New York, NY.

Isidori, A. and Astolfi, A. (1992). Disturbance attenuation and H_∞ control via measurement feedback in a nonlinear system, *IEEE Trans. Autom. Control*, **37**, 1283–1293.

Isidori, A. and Kang, W. (1995). H_∞ control via measurement feedback for general nonlinear systems, *IEEE Trans. Autom. Control*, **40**, 466–472.

Isidori, A. and Tarn, T. (1995). Robust regulation for nonlinear systems with gain-bounded uncertainties, *IEEE Trans. Autom. Control*, **40**, 1744–1754.

James, M. and Yuliar, S. (1995a). A nonlinear partially observed differential game with a finite-dimensional information state, *Syst. Contr. Lett.*, **26**(2), 137–145.

James, M. and Yuliar, S. (1995b). Numerical approximation of the H_∞ norm for nonlinear systems, *Automatica*, **31**(8), 1075–1086.

Kang, W. (1995). Nonlinear H_∞ control and its application to rigid spacecraft, *IEEE Trans. Automat. Contr.*, **40**(7), 1281–1285.

Kendi, T. and Doyle III, F. (1995). An Anti-Windup Scheme for Input-Output Linearization, *Proc. of European Control Conference*, 2653–2658.

Khammash, M. and Pearson, J. B. (1991). Robust synthesis for discrete-time systems with structured uncertainty, *Proceedings of the American Control Conference*. preprint.

Kharitonov, V. (1978). Asymptotic stability of an equilibrium position of a family of linear differential equations, *Differential'nye Uraveniya*, **14**(11), 2086–2088.

Klatt, K. and Engell, S. (1995). Gain scheduling trajectory control of neutralization processes, *Proceedings of Eurpoean Control Conference*, 2665–2670.

Kravaris, C. and Arkun, Y. (1991). Geometric nonlinear control - an overview, *Proceedings of the 4th International Conference of Chemical Process Control*, 477–516.

Kravaris, C. and Kantor, J. C. (1990). Geometric methods for nonlinear process control. 1. Background 2. Controller synthesis, *Ind. Eng. Chem. Res.*, **29**, 2295–2323.

Krener, A. (1985). (ad_f g), (ad_f,g) and locally (ad_f,g) invariant controllability distributions, *SIAM J. Contr. Optim.*, **23**, 523–549.

Krener, A. (1994). Necessary and sufficient conditions for nonlinear worstcase H_∞ control and estimation, *Journal of Mathematical Systems, Estimation, and Control*, **4**(4), 485–488.

Krener, A. J., Karahan, S., Hubbard, M. and Frezza, R. (1987). Higher order linear approximations to nonlinear control systems, *Proceedings of the 27th Conference on Decision and Control*, 519–523.

Kurtz, M. and Henson, M. (1995). Disturbance estimation for input-output linearizing controllers, *AIChE Annual Meeting*, **180i**.

Kwakernaak, H. and Sivan, R. (1972). *Linear Optimal Control Systems*, Wiley, New York.

Kwatra, H. S. and Doyle III, F. J. (1995). Dynamic gain scheduled control, *AIChE Annual Meeting*, **179h**.

Lawrence, D. and Rugh, W. (1995). Gain scheduling dynamic linear controllers for a nonlinear plant, *Automatica*, **31**, 381–190.

Lee, J. H. and Ricker, N. L. (1993). Extended Kalman filter based nonlinear model predictive control, *Proceedings of the American Control Conference*, 1895–1899.

Lee, P. (1991). Direct use of nonlinear models for process control, *Proceedings of the 4th International Conference of Chemical Process Control*, 517–542.

Lévine, J. and Rouchon, P. (1989). Quality control of binary distillation columns based on nonlinear aggregated models, *Automatica*, **27**(3).

Ling, C. and Edgar, T. (1992). A new fuzzy gain scheduling algorithm for process control, *Proceedings of American Control Conference*, 2284–2290.

Ljapunov, A. (1907). Problème générale de la stabilité de mouvement, *Ann. Fac. Sci. Toulouse 9*, 203–474. Engl. translation in: Int. J. Contr. 55 (1992) pp.531-773.

Lu, W.-M. and Doyle, J. (1995). H_∞ control of nonlinear systems: a convex characterization, *IEEE Trans. Automat. Contr.*, **40**(9), 1668–1675.

Lukes, D. (1969). Optimal regulation of nonlinear dynamical systems, *SIAM J. Contr.*, **7**(1), 75–100.

Marino, R. and Tomei, P. (1995). Nonlinear control design.

Marquardt, W. (1990). Traveling waves in chemical processes, *Int. Chem. Eng.*, **30**(4), 585–606.

Mayne, D. (1996). Nonlinear model predictive control: An assessment, *Proceedings of the 5th International Conference of Chemical Process Control*.

McLellan, P., Harris, T. and Bacon, D. (1990). Error trajectory descriptions of nonlinear control designs, *Chem. Eng. Sci.*, **45**, 30.

Misawa, E. and Hedrick, J. (1989). Nonlinear observers – a state-of-the-art survey, *ASME J. Dyn. Syst. Meas. Cntrl.*, **111**, 344–352.

Morari, M. and Zafiriou, E. (1989). *Robust Process Control*, Prentice-Hall, Englewood Cliffs, NJ.

Nijmeijer, H. and van der Schaft, A. (1990). *Nonlinear Dynamical Control Systems*, Springer-Verlag, New York, NY.

Nikolaou, M. (1994a). The 2-norm of nonlinear processes: Application to modeling and control problems, *Proc. PSE'94, Kyongju, Korea*, 971–976.

Nikolaou, M. (1994b). How nonlinear is a "nonlinear" system? Old and new results under a unifying theory. preprint.

Ogunnaike, B. A., Pearson, R. K., Samardzija, N. and Bomberger, J. D. (1994). Low order emprical modeling for nonlinear systems, *ADCHEM'94*, 41–46.

Ouarti, H. and Edgar, T. (1993). The use of approximate models and exact linearization for control of nonlinear processes, *Proc. Amer. Control Conf.*, 2268–2273.

Packard, A. (1987). *What's New with μ: Structured Uncertainty in Multivariable Control*, PhD thesis, University of California, Berkeley, CA. Engineering.

Packard, A. (1994). Gain scheduling via linear fractional transformations, *Syst. Control Lett.*, **22**, 79–92.

Paruchuri, V. and Rhinehart, R. (1994). Experimental demonstration of nonlinear model-based-control of a heat exchanger, *Proceedings of American Control Conference*, 3533–3537.

Pearson, R. (1994a). Nonlinear input/output modeling, *ADCHEM'94*, 41–47.

Pearson, R. K. (1994b). Nonlinear input/output modeling, *ADCHEM'94*, IFAC, 1–15.

Poolla, K., Shamma, J. and Wise, K. (1990). Linear and nonlinear controllers for robust stabilization problems: A survey, *Proceedings of IFAC World Congress*, 176–183.

Ray, W. (1990). *Advanced Process Control*, McGraw-Hill, New York.

Rehm, A. and Allgöwer, F. (1996). Nonlinear H_∞ control of a high purity distillation column. Accepted for Control'96, Exeter, U.K.

Rhodes, C. and Morari, M. (1995). Determining the model order of nonlinear input/output systems directly from data, *Proceedings of the American Control Conference*, 2190–2194.

Rothfuss, R., Rudolph, J. and Zeitz, M. (1995). Flatness based control of a chemical reactor, *Proceedings of European Control Conference*, 637–642.

Rouchon, P. (1994). Necessary condition and genericity of dynamic feedback linearization, *J. Math. Sys. Est. Control*, **4**, 257–260.

Rudolph, J. (1995). A canonical form under quasi-static feedback, *Systems and Networks: Mathematical Theory and Applications*, 445–448.

Rugh, W. (1991). Analytical framework for gain scheduling, *IEEE Control Syst. Magazine*, **11**, 79–84.

Russo, L. and Bequette, B. (1992). CSTR performance limitations due to cooling jacket dynamics, *in* J. Balchen (ed.), *Selected Papers from the 3rd IFAC Symposium DYCORD+ '92, College Park, MD*, Pergamon Press, 149–154.

Safonov, M., Limebeer, D. and Chiang, R. (1989). Simplifying the H_∞ theory via loop-shifting, matrix-pencil and descriptor concepts, *Int. J. Contr.*, **50**(6), 2467–2488.

Sandberg, I. (1964). On the L_2-boundedness of solutions of nonlinear functional equations, *Bell Sys. Tech. J.*, **43**, 1581–1599.

Scherpen, J. (1994). *Balancing for nonlinear systems*, PhD thesis, University of Twente. Chemical Engineering.

Schott, K. and Bequette, B. (1995). Control of chemical reactors using multiple-model adaptive control (mmac), *Proceedings of DYCORD'95*, 345–350.

Schwarz, J. and Sabotke, J. (1983). Die Kohärenzfunktion als Beurteilungskriterium der Linearität von Übertragungswegen, *messen + prüfen /automatik*, 193–198.

Shamma, J. (1991). Performance limitations in sensitivity reduction for nonlinear plants, *Syst. Contr. Lett.*, **17**(1), 43–47.

Shinskey, F. (1988). *Process-Control Systems: Application, Design and Adjustment*, McGraw-Hill, Inc., New York.

Sira-Ramirez, H., Lischinsky-Arenas, P. and Llanes-Santiago, O. (1993). Dynamic compensator design in nonlinear aerospace systems, *IEEE Trans. Aero. Elec. Sys.*, **29**, 364–378.

Sira-Ramirez, H. and Lischinsky-Arenasr, P. (1991). Differential algebraic approach in non-linear dynamical compensator design for d.c.-to-d.c. power converters, *Int. J. of Control*, **54**, 111–133.

Skogestad, S., Morari, M. and Doyle, J. C. (1988). Robust Control Of Ill-Conditioned Plants: High-Purity Distillation, *IEEE Trans. Autom. Control*, **33**(12), 1092–1105.

Slotine, J. J. E. and Li, W. (1991). *Applied Nonlinear Control*, Prentice-Hall, Englewood Cliffs, NJ.

Soroush, M. (1995). Nonlinear output feedback control of a polymerization CSTR, *Proceedings of the American Control Conference*, 2672–2676.

Sourlas, D., Choi, J. and Manousiouthakis, V. (1994). Best achievable control system performance: The saturation paradox, *Proc. 33rd IEEE Conf. Decision Contr.*, 3816–3818.

Sourlas, D. and Manousiouthakis, V. (1992). Development of linear models for nonlinear plants, AIChE Annual Meeting, Miami, FL, **125c**.

Stack, A. and Doyle III, F. (1995). The optimal control structure approach to measuring control-relevant nonlinearity, *Computers Chem. Engng.*. submitted.

Subawalla, H. and Rhinehart, R. (1994). Experimental comparison of model-based and conventional pressure control for a plasma reactor, *Proceedings of the American Control Conference*, 3122–3126.

Sussmann, H. and Jurdjevic, V. (1972). Controllability of nonlinear systems, *J. Diff. Eqns.*, **12**, 95–116.

Uppal, A., Ray, W. H. and Poore, A. B. (1974). On the dynamic behavior of continuous stirred tanks, *Chem. Eng. Sci.*, **29**, 967–985.

van der Schaft, A. (1992). Complements to nonlinear H_∞ control by state feedback, *IMA Journal of Mathematical Control & Information*, **9**, 245–254.

van der Schaft, A. (1993). Nonlinear state space H_∞ control theory, *in* H. Trentelman and J. Willems (eds), *Essays on Control: Perspectives in the Theory and its Applications*, Birkhäuser, 153–190.

Van der Vusse, J. G. (1964). Plug-flow type reactor versus tank reactor, *Chem. Eng. Sci.*, **19**, 994–997.

van Dootingh, M., Viel, F., Rakotopara, D., Gauthier, J. and Hobbes, P. (1992). Nonlinear deterministic observer for state estimation: Application to a continuous free radical polymerization reactor, *Computers Chem. Engng.*, **16**, 777–791.

Vidyasagar, M. (1986). New directions of research in nonlinear system theory, *Proc. IEEE*, **74**, 1060–1091.

Vidyasagar, M. (1993). *Nonlinear Systems Analysis*, second edn, Prentice Hall, Englewood Cliffs, NJ.

Walcott, B., Corless, M. and Zak, S. (1987). *Int. J. Contr.*, **45**, 2109–2132.

Wassick, J. and Camp, D. (1988). Internal model control of an industrial extruder, *Proceedings of American Control Conference*, 2347–2352.

Willems, J. (1970). *Stability theory of dynamical systems*, Nelson, London.

Willems, J. (1995). The behavioral approach to systems and control, *Journal of the Society of Instrument and Control Engineers*, **34**(8), 603–612.

Ydstie, E. (1996). Certainty equivalence adaptive control: Paradigms, puzzles, and switching, *Proceedings of the 5th International Conference of Chemical Process Control.*

Zames, G. (1966a). On the input-output stability of time-varying nonlinear feedback systems - Part I: Conditions derived using concepts of loop gain, conicity, and positivity, *IEEE Trans. Automat. Contr.*, **AC-11**(2), 228–238.

Zames, G. (1966b). On the input-output stability of time-varying nonlinear feedback systems - Part II: Conditions involving circles in the frequency plane and sector nonlinearities, *IEEE Trans. Automat. Contr.*, **AC-11**(3), 465–476.

Zhu, X. and Seborg, D. E. (1994). Nonlinear model predictive control based on Hammerstein models, *in* E. S. Yoon (ed.), *Proceedings of the 5th International Symposium on Process Systems Engineering*, Vol. II, Korean Institute of Chemical Engineers, 995–1000.

INDUSTRIAL APPLICATIONS OF
NONLINEAR CONTROL

Babatunde A. Ogunnaike
E.I. du Pont de Nemours and Company
Experimental Station
Wilmington, Delware 19880-0101

Raymond A. Wright
Dow Chemical Company
1400 Building
Midland, MI 48667

Abstract

As a result of increased customer demand for consistent attainment of high product quality, coupled with increasingly stringent safety and environmental regulations, and intensified global competition, the current drive in the chemical and allied industries has been towards more efficient utilization of exisiting assets (especially capacity and energy) rather than new capital expenditure. The result is that a growing number of industrial processes must now operate under conditions that emphasize their inherent nonlinearities. Nonlinear control is thus becoming more important in industrial practice. This paper assesses the current status of nonlinear control applications in the chemical industry; discusses some of the most pertinent issues of, and barriers to, practical implementation; and presents *two* complementary industrial applications of nonlinear control to illustrate the main points.

Keywords

Industrial process control, Nonlinear control, Model predictive control, Neural networks, pH control.

Introduction

It is well known that virtually all processes of practical importance exhibit some degree of nonlinear behavior. Nevertheless, the vast majority of well-established controller design techniques are for *linear* systems. Such techniques typically work well *in practice* for processes that exhibit only mildly nonlinear dynamic behavior. More recently, increasingly stringent requirements on product quality and energy utilization, as well as on safety and environmental responsibility, demand that a growing number of industrial processes operate in such a manner as to emphasize their inherent nonlinearity even more. There is therefore increased industrial and academic interest in the development *and implementation* of controllers that will be effective when process nonlinearities cannot be ignored without serious consequences.

The growing interest of the process control community in nonlinear control is reflected in several reviews of currently available nonlinear controller design techniques (see for example Kravaris and Kantor, 1990; Kravaris and Arkun, 1991; Bequette, 1991; and Rawlings, Meadows, and Muske, 1994). To be sure, many significant theoretical and practical issues remain unresolved; nevertheless, the impact of the available theory on industrial practice is becoming more noticeable. First, observe that it has become standard industrial practice to use certain simple nonlinear elements to improve performance in some control loops — for example, square root correction in flow control (see Shinskey, 1979). But beyond such simple applications, there is a growing number of more complex nonlinear control applications that have appeared in the open literature — for example, Wassick and Camp, (1988): model-based control of an industrial extruder; Labossiere and Lee, (1991): generic model control of an industrial blast furnace stove; Levine and Rouchon (1991): geometric nonlinear model-based control of a binary distillation column; Dore, *et al*. (1994): geometric nonlinear control of an industrial CO_2 adsorption/desorption pilot plant process; Singstad *et al*., (1992): multivariable nonlinear decoupling control of industrial low density polyethylene autoclave reactors; Wright *et al.*, (1992), and Wright and Kravaris (1995): nonlinear control of industrial pH processes; Temeng *et al.* (1995): nonlinear model predictive control of an industrial packed-bed reactor; Berkowitz and Gamez (1995): nonlinear model predictive control for economic optimization and control of gas processing plants. (See also Newell and Lee, 1989.) However, while the number of industrial applications of nonlinear control is growing, a careful consideration of the current opportunities *vis à vis* the currently available theory indicates that such applications are, in fact, not as widespread as they could be. This paper has a twofold overall objective:

1. to discuss some of the main issues involved in implementing nonlinear control in industry: assessing the current status (the problems and challenges) and identifying the means by which the impact of nonlinear control on industrial practice can be improved;

2. to use *two* complementary industrial case studies (a) to demonstrate the potential impact of nonlinear control, appropriately applied; and (b) to illustrate the main issues involved in successful industrial implementations of nonlinear control.

The paper is organized as follows: a contemporary perspective of the role of nonlinear control in industrial practice will be presented in the next section. This will be followed by a discussion of two process applications. The first involves nonlinear model predictive control of an industrial "Spent Acid Recovery Converter," while the second involves nonlinear model-based, adaptive control of an industrial pH process. The example applications complement each other in several ways: one is based on an empirical model obtained from input/output data correlation, the other is based on a "gray-box" model (derived by augmenting a first-principles model with empirical correlations); one involves a MIMO process, the other, a SISO process; one employs a predictive technique, the other, model-based transformations; one employs on-line optimization with a fixed model structure, the other, on-line adaptation with a fixed controller structure. The discussion of the applications will be followed by a summary and some conclusions.

Applying Nonlinear Control to Industrial Processes

A significant proportion of the demands placed on the typical industrial production facility translates into one, or more, of the following:

1. the need to increase capacity (to meet overall market demands); or,

2. the need to improve product quality (to meet individual customer demands); or,

3. the need to reduce environmental emmisions (to meet safety and environmental regulatory demands).

Traditionally, it has been customary to adopt the "capital expenditure" approach in solving these problems: for example, building new production facilities to handle the "capacity problem;" adding blending facilities to handle the "quality problem;" and redesigning and retrofitting processing units to handle the "environmental problem." More recently, however, increasing global competition has dictated the current trend towards finding alternative solutions requiring little or no capital expenditure. This almost invariably implies seeking effective control solutions first, wherever possible. But when most processes are operated under the conditions dictated by these stringent market, customer and environmental demands, the tendency is for the inherent process nonlinearities to become more pronounced — making it more difficult to obtain acceptable solutions from traditional linear controller design techniques. The prevailing global economic conditions thus continue to create opportunities for the application of nonlinear control techniques.

Given the current potential for nonlinear control to contribute significantly to industrial productivity, we now consider some of the issues that must be addressed for such potential to be realized fully. In this regard, it is convenient to consider these issues in two categories: (i) *Quantitative needs assessment* (*When* does one actually need to use nonlinear control?) (ii) *Appropriate Technology and its Implementation* (Even when the need is clear, *What* strategy should be used, and *How* should they be used?)

Quantitative Needs Assessment

It is widely accepted that only about 10–20% of industrial control problems require the application of so-called "advanced control;" it is also accepted that processes in which such problems are encountered account for close to 80% of the revenue. Of the industrial control problems in need of advanced control applications, there is now an increasing realization that a certain proportion cannot be solved effectively by linear techniques, which constitute the bulk of the most widely applied of these advanced techniques. However, the application of nonlinear techniques requires incrementally greater investments in implementation effort and costs, and such costs must therefore be economically justifiable. Thus, being able to answer the following questions as objectively as possible will increase the impact of nonlinear control in industrial practice:

1. **Identifying Opportunities:** For which problem is the application of nonlinear control critical to the achievement of the desired operational objectives? (and which of the available tools is most appropriate for the specific application?)

2. **Economic Viability:** How does the cost of implementation compare to the potential benefits to be derived from the application?

For many of the documented applications of nonlinear control, these questions were relatively straightforward to answer. When the process nonlinearity is obvious, and severe enough (as with the applications soon to be discussed) the need for nonlinear control is usually clear. By the same token, if a critical process is virtually inoperable with linear controllers, it will be relatively straightforward to quantify the benefit of nonlinear control. For other processes, however, identifying the need for nonlinear control can often be quite difficult. In general, the industry has traditionally had more practitioners skilled at implementing the "capital expenditure" solutions. By contrast, recognizing opportunities for the application of nonlinear control requires a specialized background typically possessed by a comparatively small group of industrial practitioners.

The vast, virtually untapped — and currently difficult to quantify — potential for nonlinear control lies with the class of problems for which linear control methods are applicable, but for which nonlinear methods will result in significant process performance improvements. In this regard, observe that theoretical tools for quantifying the degree of process interaction (and process conditioning) have been useful in assessing the applicability of multivariable control and have thereby promoted industrial application. Similar tools for measuring the degree of process nonlinearity could conceivably play a commensurate role in promoting the industrial application of nonlinear control methods.

To undertake any industrial application, it is often necessary to demonstrate beforehand that the potential benefits outweigh the implementation costs; the same is true for nonlinear control. Many of the costs associated with nonlinear control implementation will be discussed in the next section. We simply note here the obvious fact that reducing the costs of application will make nonlinear control more economically viable for a significantly larger number of industrial processes and thereby promote increased industrial application.

Technological and Implementation Issues

Even in the cases where the need is obvious, and the potential benefit is known to be substantial, the following are some factors that currently constitute barriers to the widespread use of nonlinear control:

1. **Control Technology:** The typical analytical tools required for rigorous nonlinear systems analysis and controller design still remain largely inaccessible to all but a few researchers concerned with such problems. Nonlinear control techniques thus tend to be more complicated than linear ones, and are more likely to involve *unique* development and implementation, directly raising costs.

2. **Model Development:** Virtually all high performance controllers are model-based; and nonlinear controllers in general require nonlinear process models. Developing adequate nonlinear models for control can be time-consuming and costly.

3. **Implementation and Support:** Most nonlinear controller design techniques give rise to complex controllers that can require unique, specialized software and hardware resources for real-time implementation, and in turn, require extensive training and support for long term effectiveness.

These issues arise primarily because of the intrinsic characteristics of nonlinear systems. First, because *nonlinearity* is an intrinsically more complex phenomenon to analyze than *linearity*, nonlinear systems require a different set of analysis and design tools that are more involved and are currently not as well developed and widely accessible as their linear counterparts.

Second, because of all the nice properties enjoyed by linear systems (additivity, homogeneity, superposition, etc) linear model development is relatively straightforward, in concept, if sometimes tedious in practice. The literature on linear model identification from empirical plant data in particular, is vast, and essentially complete; and industrial practice of linear empirical modeling is reasonably well-developed. When the desired process model is to be nonlinear, however, many additional issues immediately arise by virtue of this departure from linearity — the most important of which has to do with what modeling approach to adopt: the theoretical (or first-principles) approach, the empirical approach, or the "hybrid" approach?

The first-principles approach is often not employed because it requires a significant amount of process knowledge which may not always be available; and when such knowledge is available, the resulting model may simply be too difficult to be useful for controller design purposes. The empirical approach has the advantage of depending strictly on data, but it requires an à-priori choice of model structure (itself a very difficult task); in addition it requires a very careful design of the input sequence to be used for the identification. Also, the fact that empirical model parameters are usually not related to fundamental process properties has contributed to increased cost of long-term support in certain applications. An increasingly promising approach is the so-called "gray-box" or hybrid approach in which basic first-principles information is augmented with empirical data, thereby taking advantage of the benefits of each approach. This is illustrated by the pH control case study in which dynamic modeling was accomplished by using fundamental chemical engineering knowledge to obtain a model structure containing only a small number of parameters that can be estimated from process data. For some additional hybrid modeling applications, see, for example, Psychogios and Ungar, 1992; Tulleken, 1993; Lindskog and Ljung, 1994; Ogunnaike, 1995.

Finally, by definition, and intrinsically, nonlinear systems tend to defy classification: they are *all* characterized by the property they lack — linearity. Each nonlinear control application thus tends to be unique and specialized, making it difficult to employ any generalized approach, or tools or implementation platforms.

Taken together, all the foregoing factors argue strongly for the research that will lead to the development of commercial nonlinear control packages in the same spirit as those available for (linear) Model Predictive Control (MPC). Observe that even though (i) linear MPC analysis and design techniques, obviously less complicated than nonlinear techniques, are still complicated enough compared to classical methods; and (ii) linear model development for MPC applications is still not a trivial task; commercial packages such as DMC and IDCOM (see Ogunnaike and Ray, 1994, Chapter 27 for a summary of other commercial MPC packages) have made the implementation of this technology much more widely accessible than would otherwise be possible.

Despite the obvious difficulties regarding "standardization" of model forms and design techniques, Continental Controls, Inc. has commercialized one nonlinear control package — MVC — with the claim that it could potentially do for nonlinear control what IDCOM and DMC did for the linear Model Predictive Control. One of the reported applications of this technology may be found in Berkowitz and Gamez (1995).

In addition to initial implementation costs, it is also important to consider the training and long term support costs associated with nonlinear control applications. At the plant level, the engineers need to understand the controller and what it is intended to achieve, or the performance of the control scheme will quickly degenerate. A significant effort is required to provide the basic understanding of any advanced control technique industrially applied, and the starting point is generally from an undergraduate control course. Communication and education targeted at the gap between an undergraduate level of understanding and the theoretical level of the open literature can significantly reduce the effort in training when advanced control methods are industrially applied. Additionally, industrial processes generally do not usually remain static. The drive to operate more efficiently will lead to changes, in aspects other than the control scheme, that will have a secondary effect on controller performance. The performance of methods that require a significant effort to update model or controller parameters usually tend to degrade more with time and have a higher support cost associated with them. Nonlinear techniques resulting in a small number of parameters that can be intuitively understood can generally reduce long term support costs.

In the next two sections, we will discuss the development and on-line performance of two different nonlinear control systems for two industrial processes, to illustrate how the problems noted above — control technology, modeling, and control system implementation — were addressed in each specific case.

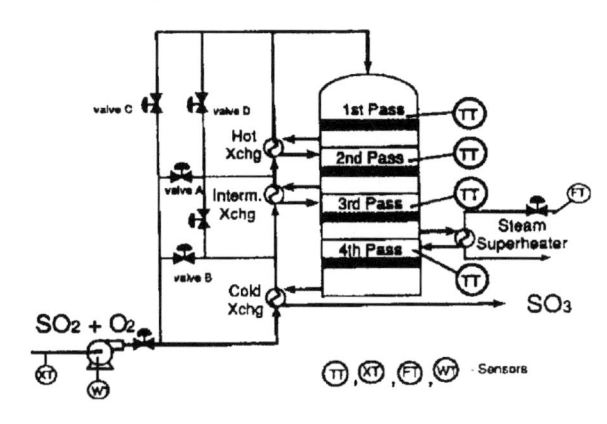

Figure 1. The spent acid recovery converter.

Model Predictive Control of a Spent Acid Recovery Converter

The Process

The process in question is the "Spent Acid Recovery" converter shown schematically in Fig 1. It consists of a series arrangement of four Vanadium Pentoxide fixed-bed reactors used to convert a cold feed of sulfur dioxide, (SO_2), Oxygen, (O_2), and some inerts, into SO_3. Because the reaction is highly exothermic, interstage cooling is provided primarily via heat exchange with the incoming cold feed, except between stages 3 and 4 where cooling is achieved via heat transfer to steam in a superheated steam generator.

Process Operation Objectives:

Safe, reliable and economic process operation requires close regulation of the inlet temperatures of the first, second, and third stages. In general, there is an "optimum" inlet temperature for each stage (or pass) that will give rise to optimum conversion. These desired target values are determined by "gas strength" (SO_2 concentration), production rate, and the conversion achieved in the preceding passes. In addition, these temperatures must not fall below 410° (otherwise the reaction will be quenched) or rise above 600° (otherwise the catalyst active life will be shortened considerably).

Frequent fluctuations in feed conditions — the blower speed, gas strength (SO_2 concentration), and O_2 concentration — constitute the main obstacles to smooth process operation. Primarily to minimize yield losses, and to comply with strict environmental regulations on SO_2 emmissions, these persistent disturbances must be rejected effectively and quickly. Ineffective process control has been responsible for low conversions; and low conversions result in *both* high SO_2 emmission rates and high yield losses.

The indicated network of pipings, baffles and valves A, B, and C provide the means for controlling the inlet temperatures through by-pass feeding of cold reactants. (For reasons that will soon become clear, only the valve openings (or "valve loadings") for valves A, B, and C are available for manipulation; the valve loading of valve D is not.) For example, observe that increasing by-pass flow through valve

C will *reduce* the first pass inlet temperature.

The dynamic characteristics induced by the network of valves can be quite complex. First, observe that the valves merely *redistribute* the feed, sending a portion directly as cold feed, and the rest through the various heat exchangers. A change in a single valve loading therefore affects not just the feed flow rate through that valve; it affects the flow rate through *all* the other valves. These manipulated variables are therefore *not* entirely independent. Observe therefore that only 3 of the 4 valves can be manipulated independently. Next, consider — for the purpose of illustration — the effect of an increase in the valve C loading. The initial direct response will be a *decrease* in the first pass inlet temperature (as a result of increased cold feed by-pass to this stage); but because the increased by-pass through valve C causes a concurrent decrease in the amount of cold feed distributed to the interstage heat exhangers, this action also results in an *increase* in the second and third pass inlet temperatures. This otherwise "normal" process interaction is then complicated by *secondary* effects resulting from the fact that a reduction in the first pass inlet temperature ultimately causes a reduction in the exit temperature, which in turn causes a reduction in the inlet and outlet temperatures in the succeeding stages. The reduced temperature in all the stages then produces a *tertiary* effect in which the amount of the first stage feed preheating provided by the three interstage heat-exchangers is reduced, further reducing the first pass inlet temperature. This now starts another round of inlet temperature reductions with the potential for open-loop instability induced by the progressive cooling, and the possibility of quenching the reaction outright. Finally, as a result of the nonlinearity induced by the chemical reaction kinetics and the heat exchanger characteristics, a "mirror image" decrease in the valve C loading will not give rise to a precise, "mirror image" reverse net effect in inlet temperatures. To keep the process away from potentially unstable operating regimes, a lower constraint of 30% is imposed on the valve loadings; the upper constraint of 100% is physical.

The overall process objective may therefore be stated as follows:

> *In the face of persistent process disturbances, control the inlet temperature for each of the first three passes to their respective prespecified desired target values; maintaining them between the operating constraints of 410°, and 600° at all times; with the loadings for Valves A, B, and C constrained to lie between 30% and 100%.*

A Control Perspective of the Process:
The process variables may be categorized as follows:

- **Output (controlled) variables**:
    1. First pass inlet temperature;
    2. Second pass inlet temperature;
    3. Third pass inlet temperature.

- **Input (manipulated) variables**:
    1. Valve A loading;
    2. Valve B loading;
    3. Valve C loading.

- **Disturbance variables**:
    1. SO_2 concentration;
    2. O_2 concentration;
    3. Blower speed;
    4. Valve D loading.

As summarized above, the main control problems are caused by the persistent disturbances; the strong interactions among the process variables; the constraints on both the input and output variables; and the process nonlinearities due to the reaction kinetics, heat transfer characteristics, and the flow distribution network.

The specific objective of the application is to develop an effective control system for this process; but the broader objective in this section is to use this specific application to illustrate various aspects of how nonlinear control can be applied on an industrial process.

Overall Control Strategy

The multivariable nature of the process, along with the process operating constraints, make this an ideal candidate for model predictive control (MPC); however, the severity of the process nonlinearities argue strongly for the application of nonlinear MPC instead of the more popular standard, linear version. The most important implications of this decision are as follows: Technologically, this boils down — in principle — to obtaining a reasonable, nonlinear process model, and a reliable nonlinear optimization routine for performing the optimization that lies at the heart of MPC. In practice, however, unlike with linear MPC, very few theoretical results are available to guide the critical design choices such as the prediction horizon, the control move horizon, and the various weights in the objective function. The nonlinear optimization will thus have to be carried out with extra care. Also, unlike with linear MPC, no standard commercial packages were available at the time of this application (1991/92).

At the heart of the nonlinear model predictive control technique is the nonlinear process model; and based on the following three main points, the decision was made to obtain this model via input/output data correlation.

1. Not enough is known about certain critical details of the process to generate a first-principles model having sufficient integrity.

2. Even if the required fundamental process knowledge were available, the resulting first-principles model will be far too complicated for on-line optimization-based control. Observe that, *at the very least*, such a model will consist of a combination of individual models for each subprocess making up the overall process: a gas

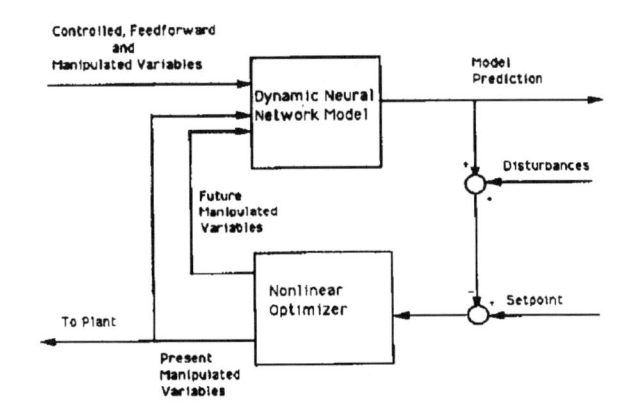

Figure 2. The control strategy.

distribution network model; a heat transfer model for the four heat exchangers; and a kinetic model for the four fixed-bed catalytic reactors. Each contributing model could conceivably consist of a system of several, coupled nonlinear partial differential equations; and the overall combination will clearly be far too complex for controller design.

3. From a process control perspective, the process is a 3 × 3 process with 4 disturbances; this process dimensionality is actually not so high as to render empirical modeling prohibitively time-consuming.

The issue of model structure selection in empirical nonlinear modeling is not trivial, and many factors influence each individual choice (see for example Pearson, 1994; Pearson and Ogunnaike, 1996). For this particular application, a recurrent neural network representation was chosen because of the flexibility of the neural network paradigm in general for representing arbitrary nonlinear input/output maps; the recurrent structure (as opposed to the standard feedforward structure) was chosen in particular for improved long range prediction (cf. Temeng et al. 1995), a critical requirement for model predictive control.

The overall control strategy is therefore to represent the process dynamics with a recurrent neural network, and to use this in a model predictive control framework in conjunction with a nonlinear optimizer. This control structure is shown in Fig 2.

Process Model Development

A systematic procedure for nonlinear empirical model development involves the following steps (cf. Pearson and Ogunnaike, 1996):

1. Model structure selection;

2. Model identification; *(Input sequence design; data collection and preconditioning; model parameter estimation)*;

3. Model validation.

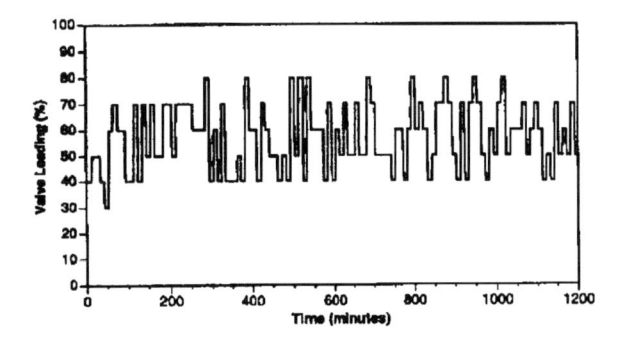

Figure 3. Identification input sequence for Valve A loading.

In this specific application, the selected model structure — a recurrent neural network — and the reasons for the choice have been presented. The next step — actual identification of the neural network model for the spent acid recovery converter — involves making decisions about the input sequences to be used for the model identification, implementing these input changes, collecting the sets of process response data, and analysing the collected input/output data sets.

The theoretical issues concerning input sequence design for nonlinear model identification remain largely unresolved (cf. for example Pearson and Ogunnaike, 1996); much of what is done in practice is influenced mostly by sensible, but vague heuristics. For example, it is generally recommended that the *magnitude* of the inputs must be such that the desired region of operation is "adequately covered;" and that the "frequency content" must be such that those aspects of the process that must be captured in the model are "adequately excited." Such heuristics and available theoretical results immediately rule out the typical inputs employed in industrial practice for linear model identification, i.e. single steps, single pulses, and the PRBS; but there is as yet no comprehensive theory regarding "optimum" input sequences for *general* nonlinear model identification.

In this specific case, therefore, the decision was to employ 6-level, pseudo random sequences (as opposed to the *binary*, i.e. 2-level, sequences employed for linear systems) that span the "normal" input range. From process operation data, and process knowledge, this "normal" range was determined to be 30% – 80% valve loadings. Because the "dominant time constant" for the process is known to be approximately 40 mins, the duration of each "step change" in the sequence was fixed at 5 mins, at the end of which the valve loading was switched to a different randomly drawn level. The total duration for each input sequence was at least 12 hours.

Fig 3 shows the valve A loading input sequence and Fig 4 shows the observed responses in the first, second, and third pass inlet tempratures, respectively. Similar responses were obtained from similar input changes in valves B and C.

Each process data set acquired during the plant tests was partitioned into two: one part for model development (the "training set"), the other for model validation (the "val-

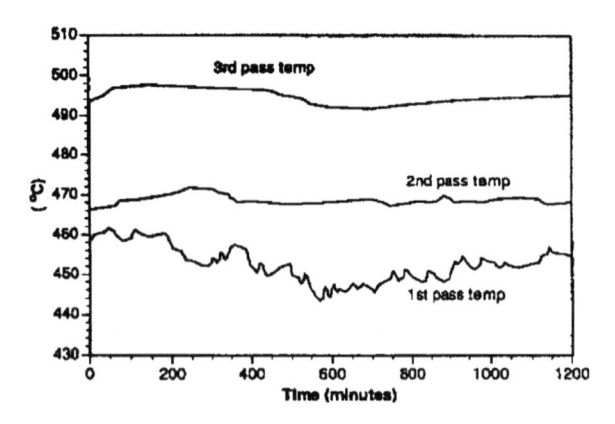

Figure 4. Temperature responses to changes in Valve A loading.

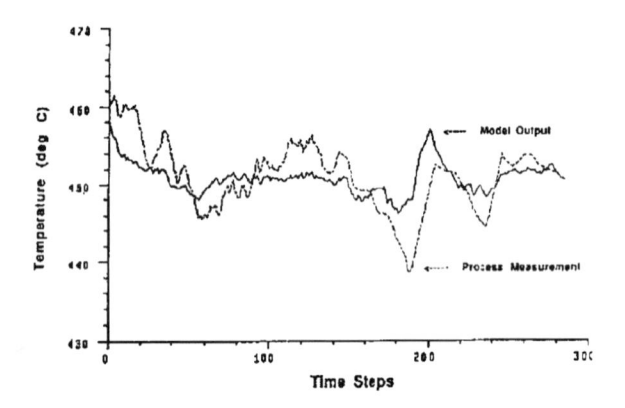

Figure 5. First pass inlet temperature prediction and validation data.

idation set"). The backpropagation-through-time algorithm was used to obtain the seven-input, three-output recurrent NN model from the plant data in the "training set." The final NN model architecture consisted of three layers and four nodes in the hidden layer, with unit time-delayed output feedback connections to the input layer. For additional details about the model development, see Temeng et al., 1995. The performance of the resulting model is illustrated in Fig 5 where the long range, pure prediction of the first pass inlet temperature is compared with corresponding validation data. Comparable performance was observed from the other parts of the model.

Control System Design and Implementation

Conceptually, the nonlinear model predictive controller was implemented as shown in Fig 2: the NN model provided the long-range prediction, and "ADS," a public domain nonlinear optimization routine (obtained from the Naval Postgraduate School in Monterey, California) was used to determine optimal control action sequences. The model prediction, and control sequence horizon lengths were chosen to be 20 and 5 respectively; with $\Delta t = 10$ minutes. Additional details about the optimization routine are available in Temeng et al., 1995.

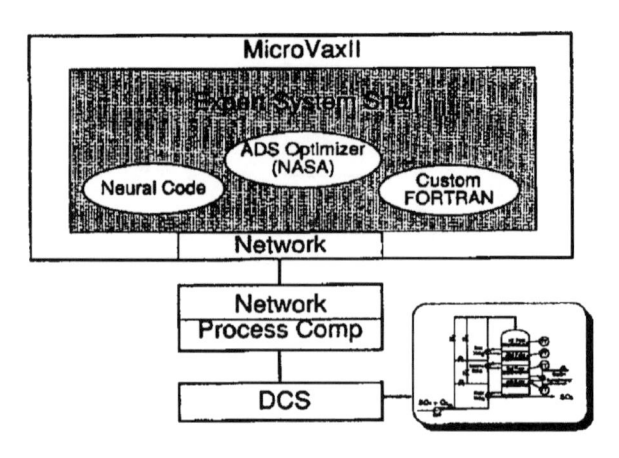

Figure 6. Control system implementation architecture.

The actual implementation of this nonlinear MPC scheme requires a few additional hardware and software considerations. Process operation data were collected and archived by a PDP 11/85 host computer interfaced to a dedicated DCS (Distributed Control System) through vendor-supplied software running on a MicroVAX system. The NN process model and the optimizer were deployed within an in-house expert system shell on the same MicroVAX computer. Apart from providing a convenient environment for integrating all the FORTRAN routines used to execute the modeling and the optimization functions of the nonlinear MPC scheme, the expert system also performed two additional relatively simple, but critical, tasks: (i) it determined when it was time to execute the controller; (ii) it checked the availability and validity of process data, and the "reasonableness" of the computed control action.

At each control cycle, the desired setpoints computed for the valve loadings were sent from the expert system (in the microVAX) to the host computer; this was then communicated to the DCS, from where it was implemented on the actual process. The implementation hardware/software architecture is shown in Fig 6.

Control System Performance

Figures 7, 8, and 9 are representative of the actual closed loop performance of the control system. Fig 7 shows the process output variables over a 24 hour period during which the process was subject to the disturbances indicated in Fig 8. Between $t = 500$ and $t = 900$, the SO_2 concentration dropped by more than 15 % — by process operation standards, a significant feed disturbance; the indicated change in the blower speed (related to the process throughput) is also significant. In responding to these disturbances, the control scheme successfully maintained the inlet temperatures close to their respective desired setpoints, as shown in Fig 7, by implementing the control action sequences shown in Fig 9.

Compared to standard process operation prior to the implementation of this controller (not shown) the controller performed remarkably well. Observe that the 30% – 100% constraint range was enforced for each of the valves during

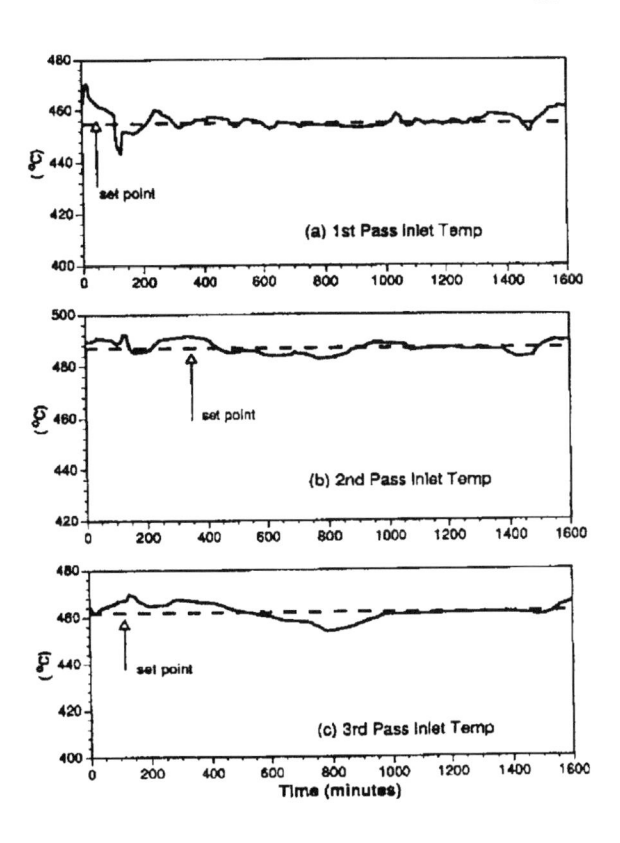

Figure 7. Closed-loop temperature responses.

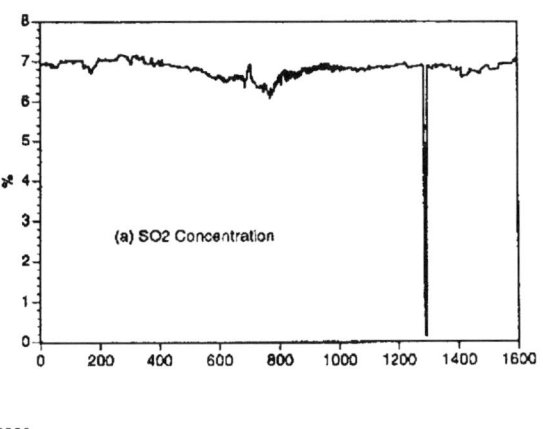

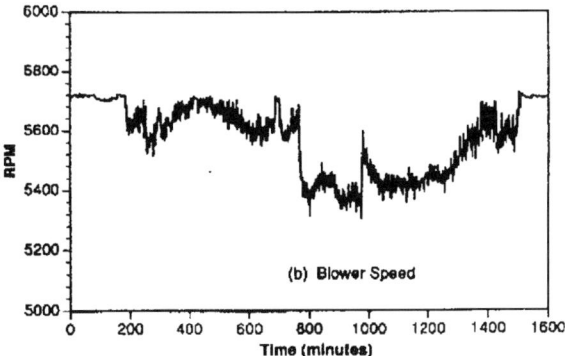

Figure 8. Process disturbances during closed-loop operation.

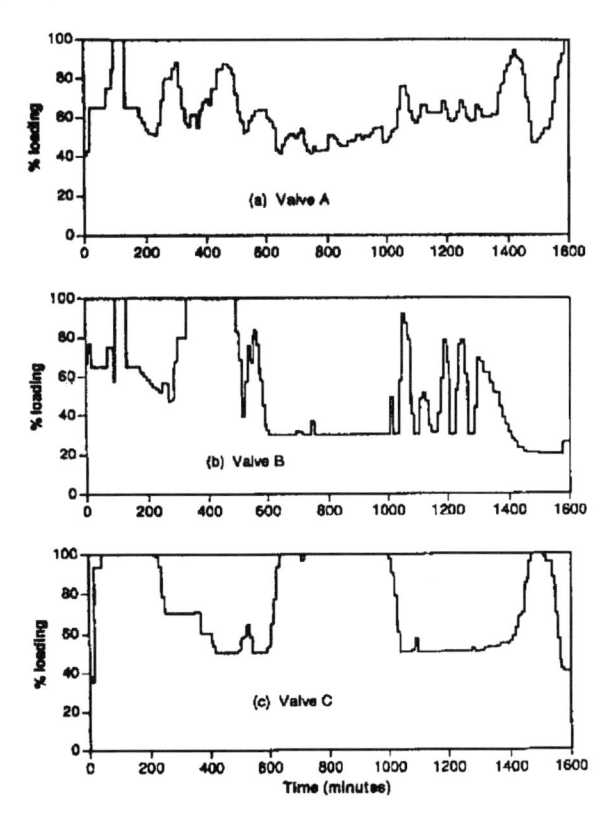

Figure 9. Implemented control actions.

the entire period. The SO$_2$ concentration "spike" that occured at $t = 1300$ was due to the daily scheduled analyzer calibration; observe however that such a clearly anomalous measurement did not affect the controller performance. This illustrates the effectiveness of the expert system in checking and validating process measurements before they are used in computing corrective control action. For additional details on the performance of the controller and a comparison to conventional control approaches, see Temeng *et al*. 1995.

On-line Identification and Model-Based Control of an Industrial pH Process

The control of pH processes is one classic example frequently used in the literature to demonstrate nonlinear control algorithms. In this section, a nonlinear controller developed and implemented for a particularly difficult industrial pH process will be presented. The emphasis will be from the process and plant point of view in relation to the issues mentioned previously in this paper. A theoretical development of the algorithm employed will not be presented here; a detailed development can be found in Wright and Kravaris (1995).

Identifying an Opportunity

A plant was having difficulty maintaining process equipment handling a particular process stream. Upstream of this equipment is a tank where lime is added based on a mea-

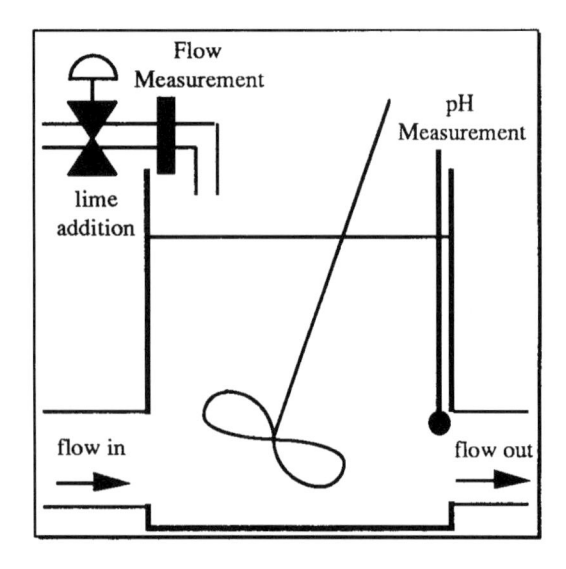

Figure 10. Overview of industrial pH process.

sured pH value. Prolonged excursions at low pH values were felt to be a major contributing factor to the maintenance problems. Additionally, it was determined that significant deviations at either low or high pH values were affecting the operation of processes even further downstream. It was quickly recognized that a control problem with a single control loop was a major part of the plant problems encountered.

The process consists of a tank with an acidic stream of unknown composition flowing into it. The process stream is usually acidic with large changes in pH and base demand observed. The location of the process and the upstream processes makes installation of upstream pH probes impractical. A lime slurry is added at the tank in order to control the effluent pH. The flow rate of the lime slurry and the pH at the exit point of the tank are measured. The process stream flow rate and tank volume are not directly measured, but can be reliably estimated from other process measurements. The lime slurry and the process stream are vigorously mixed together by an agitator in the tank. The effectiveness of the mixing in the tank can be evaluated by pH measurements further downstream. An overview of this industrial pH process is shown in Figure 10.

An example of the types of changes in pH of the process stream is given in Figure 11. During the time shown in this figure, the lime addition was turned off. The change in pH is, therefore, totally attributable to the changes in the upstream process. The spikes in pH from a low pH value to near neutral frequently occur around changes from one operating region to another.

The time-varying nature of the base demand on the system is further illustrated in Figure 12. Samples taken from the process stream at five different times, during different upstream conditions, were taken and titrated in the lab. The results illustrate changes in both concentration and buffering of the process stream. The two curves on the left of the figure show very little base demand and have a very high

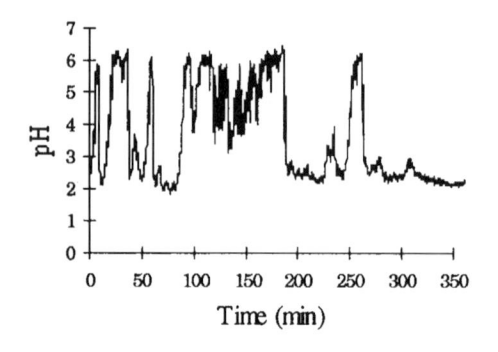

Figure 11. Example of pH changes from upstream processes.

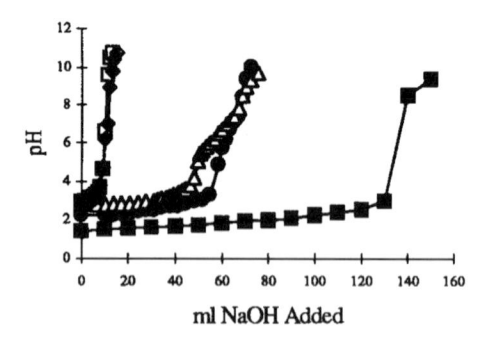

Figure 12. Five lab titration curves of the process stream.

slope at pH values near 7. The third and fourth curves reflect about the same overall base demand, but have a significantly flatter slope, especially the third curve, through the neutral region. The final curve shows a very large base demand with little change in pH as base is added at low pH values. To solve the plant problems, large pH deviations, especially in the acidic region, in the presence of this variability, must be avoided. Clearly, an opportunity existed to improve performance of the lime addition controller.

Economic Viability

The maintenance costs associated with the poor performance of this control loop, were significant. It was also clear that this problem could not be significantly improved with the existing control structure. For this reason, it was rather easy to determine the need for applying a more sophisticated control structure and the benefits associated with solving the plant's problem.

The nonlinear nature of the process suggested pursuing nonlinear methods to develop a controller. The time-varying nature of the process stream suggested a parameterization of the nonlinearity and on-line identification of those parameters will be required. The costs associated with this approach will be given in more detail below. In general, it was desired to provide a solution which would run on the plant's existing computer. Costs associated with this application would include the initial development and implementation, but the long term support cost would be reasonably low. It was therefore decided that if initial development and implementation costs were reasonable, the benefits would outweigh the costs and the application of nonlinear control was economically justifiable.

Model Development

The general approach used to develop a nonlinear model for this process was developed in Wright and Kravaris (1993). This method is a gray-box approach that results in a small number of parameters, each of which has a physical significance acssociated with it. The following discussion will illustrate this approach.

A "white" or completely first-principles model with detailed chemical information would involve balances for the analytical concentrations of all chemical species involved (McAvoy *et al.*, 1972, or Gustafsson and Waller, 1983). It has been shown in the literature (Wright and Kravaris, 1991) that the input/output behavior of such a model, when the assumption that the concentration of the titrating stream is much larger than that of the process stream ($u \ll F$), is made, collapses to:

$$V\frac{dX}{dt} = u - FX \quad (1)$$

$$T(pH) = X \quad (2)$$

where: F is the flow rate of the process stream
V is the volume of the mixing tank
u is the flow rate of the lime stream
X is the model state
$T(pH)$ is the inverse titration curve of the process stream

The dynamics given in (1) depends primarily on quantities for which either measurements exist or can easily be inferred. The process nonlinearity is completely within the inverse titration curve, $T(pH)$, which depends on the analytical concentrations and dissociation constants of all chemical species present. Furthermore, $T(pH)$ can be represented as a ratio of two expressions: the numerator depending on the analytical concentrations and dissociation constants of the process stream, and the denominator depending on those of the titrating stream.

If detailed chemical information for the process stream is unavailable, one could still use the same representation, (1-2), but with empirically fitting the pH-dependent expression for the process stream:

$$T(pH) = \frac{\Phi(pH;\theta)}{a_L(pH)\alpha_L} \quad (3)$$

where:

$\Phi(pH;\theta)$ depends on the process stream analytical concentrations and dissociation constants

θ are parameters to be fitted empirically

α_L is the analytical concentration of lime in the titrating stream

$a_L(pH)$ is a function of the dissociation constants

for lime, and is given by

$$a_L(pH) = \frac{2[H^+]^2 + \frac{K_w}{K_{b2}}[H^+]}{[H^+]^2 + \frac{K_w}{K_{b2}}[H^+] + \frac{K_w^2}{K_{b1}K_{b2}}} \quad (4)$$

K_w is the dissociation constant for water
K_{b1}, K_{b2} are the dissociation constants for lime

For this particular industrial process, the parameterization of the nonlinearity was chosen as:

$$\Phi(pH;\theta) = 10^{-pH} + 10^{pH-14} + s + c\log_{10}\left[\frac{1 + 10^{(pH-m_1)}}{1 + 10^{(pH-m_2)}}\right] \quad (5)$$

This parameterization corresponds to a pulse function between pK values m_1 and m_2 (for details, see Wright and Kravaris, 1993). An impulse function at a particular pK value corresponds to the same function obtained from the chemical equilibria for a species with one weak dissociation. The pulse representation allows the inverse titration curve to be affected over a wider pH range with a smaller number of parameters that would be possible using impulse functions. For the process at hand, a pulse from 3.5 to 5.5 was chosen. This choice was made partially due to the titration curves shown in Figure 12 and partially to potential buffering in the acidic region which could lead to low pH excursions for prolonged periods.

The s parameter in (5) represents a shift in the entire curve. This can be physically interpreted as representing strong acid or strong base concentrations in the process stream. With the choice of m_1 and m_2, the last term in (5) will have a nonlinear impact on $T(pH)$ only in the acidic region. The parameter, c, roughly corresponds to the analytical concentration of a weak acid. Changing the relative contributions of s and c corresponds to changing the buffering of the system.

In the present application, the additional complication of the time-varying nature of the nonlinearity means that a static fit of s and c would not be particularly useful. It is therefore necessary to identify these parameters on-line. This will force the empirical fit of the inverse titration curve as the process evolves with time. Integrating (1) on-line, X can be estimated and then (see Wright and Kravaris, 1993, for details):

$$\Phi(pH;\theta) = Xa_L(pH\alpha_L) \quad (6)$$

is linear in the unknown parameters and therefore in standard form for off-the-shelf identification algorithms. For this industrial application, the identification method developed in Papadoulis, *et al.* (1989) was used to obtain on-line estimates of s and c.

The nonlinear model developed for this process represents the application of gray-box modeling principles. Simple and directly applicable dynamics and a macroscopic representation of the process nonlinearity in terms of the inverse titration curve of the process stream results from first

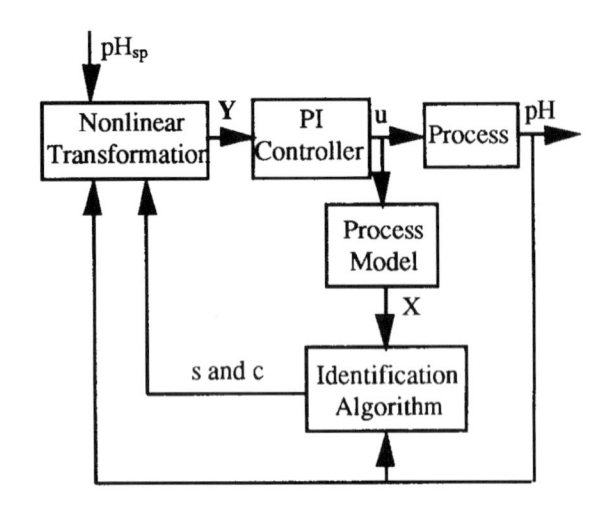

Figure 13. Complete control structure.

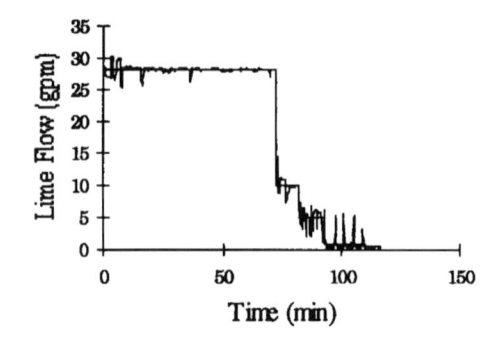

Figure 14. Tracking of lime flow rate to set point and corresponding valve position.

principles. Where practical, i.e. for lime, chemical equilibria information is directly included. The nonlinearity is represented using a small number of parameters that can be fitted empirically. Due to the time-varying nature of this process, it is necessary to fit the parameters on-line using an identification algorithm. This model was quickly developed and ready for implementation. The model development cost for this application was, therefore, resonably small.

Nonlinear Control Technology

The control technology used here is a direct application of theory available in the literature (Wright and Kravaris, 1991,1993), uniquely adapted to the particular industrial process being controlled.

The difference between the inverse titration curve evaluated at the measured pH and evaluated at a set point pH value, pH_{sp}, is a nonlinear transformation on the process output. The quantity:

$$Y = T(pH_{sp}) - T(pH) \qquad (7)$$

is a dynamically equivalent output (Wright and Kravaris, 1993b) to pH, since controlling pH to pH_{sp} is equivalent to controlling Y to 0, as long as there is a one-to-one relation between pH and $T(pH)$. Reformulating the control problem in terms of this output results in a linear problem for which a linear PI controller in terms of the dynamically equivalent output:

$$u = K_c \left[Y + \frac{1}{\tau_I} \int_0^t Y\, dt \right] \qquad (8)$$

where K_c and τ_I are the controller parameters, can be used. The complete nonlinear control structure is shown in Figure 13.

Implementation and Support

The controller was implemented directly on the control computer for this process with calculations being performed once every second. The largest cost associated with the implementation was the time to develop the initial code installed in the plant. The set point pH value for this process is 4.5. The two parameters, s and c, are adapted on-line using a forgetting factor parameter of 0.2 and a minimum forgetting factor of 0.1. A slave PI controller is used to control the measured lime flowrate to output of the master pH controller by adjusting the valve position on the control valve. All results that follow are actual plant data obtained at a sampling frequency of ten seconds.

To fully commission the controller in the plant, normal and abnormal operating conditions must be evaluated with the new control structure in place. This begins with checking the performance of the slave controller. The tracking of the lime flow rate to its set point value during step tests is shown in Figure 14. The response of the slave is reasonable, but the figure illustrates other points that will make pH control of this process difficult. At lower flow rates solids in the slurry can partially plug the valve making control to set point difficult. Even at higher flow rates, significant deviations occur from set point. Given the time constant of the pH process, these deviations will clearly be visible in the measured pH. In addition, the dynamics of this loop will force a relatively slow tuning of the master pH controller.

Downstream pH probes are used to check the validity of the pH measurement in the vessel. Since there are no further additions between the probes and no additional mixing, it is expected that the downstream probe will give a value differing only by a time shift. This is usually observed in practice unless a probe fails or becomes coated. During start up of the algorithm, the probes did become coated giving a response shown in Figure 15. Note the sluggish response of the probe in the vessel compared to the downstream probe. The periods in Figure 15 with relatively flat pH values near 7 correspond to no lime addition, as can be seen in Figure 16. The difference at these times is the calibration error between the probes.

The level in the vessel is not directly controlled. If the volume is outside of certain bounds, the agitator must be shut off. During these times, the controller must cope with a much poorer measurement until the level returns to the normal operating range. Figure 17 shows the pH response

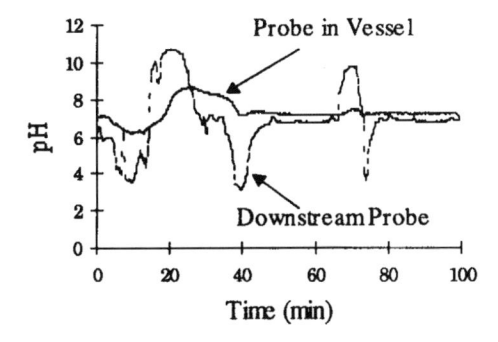

Figure 15. pH response with coated probe.

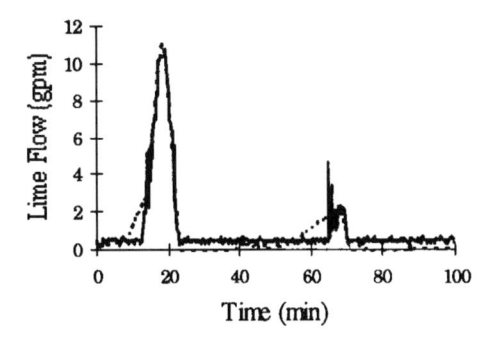

Figure 16. Lime flow, flow set point, and valve position corresponding to Figure 15.

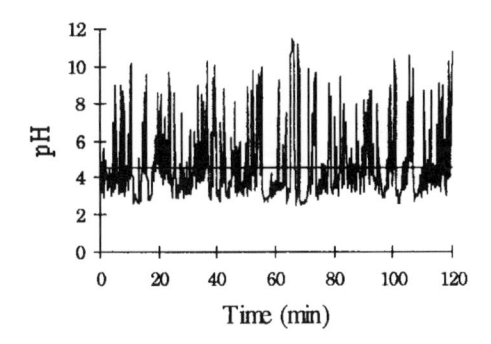

Figure 17. pH response with no agitation in vessel.

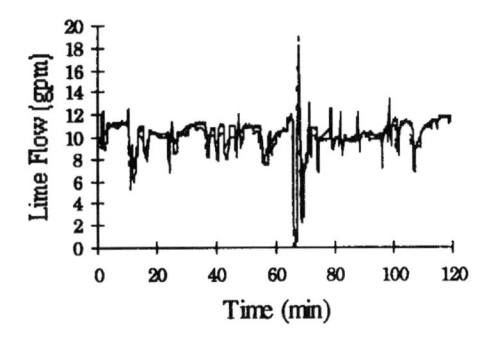

Figure 18. Lime flow, flow set point, and valve position corresponding to Figure 17.

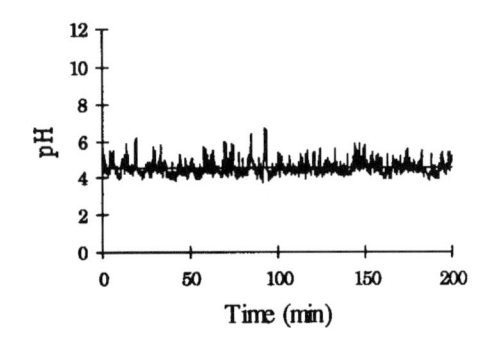

Figure 19. pH response with no process upsets.

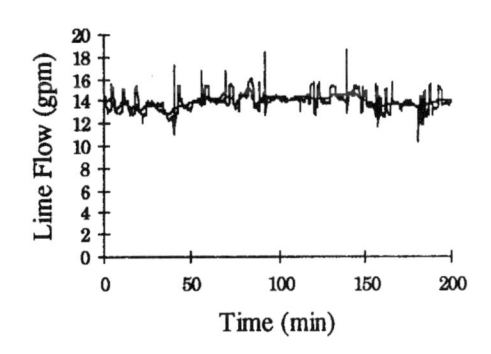

Figure 20. Lime flow, flow set point, and valve position corresponding to Figure 19.

without agitation. While the pH control is not good, the lime flow rate set point, as shown in Figure 18, did not reflect the large changes in pH. This is considered a reasonable response to loss of agitation The downstream pH is higher than would normally be expected due to the poor mixing in the vessel.

When the problems mentioned previously have been resolved and the upstream processes are in stable operation, the resulting pH control can be quite good, as shown in Figure 19. During this time period the lime flow rate did not vary much, as is shown in Figure 20.

The response of the system to a change in acid demand is shown in Figures 21 and 22. In the early parts of these plots, the lime flow rate is relatively stable, even though the pH is slightly high. These pH values are on the horizontal part of the $T(pH)$ curve and, consequently, do not generate large changes in flow rate. When a disturbance enters the process, the flow rate is doubled even though the deviation in terms pH is not that large. Notice the decrease in flow rate when the pH reaches 10.

The application of nonlinear control to this process has proven beneficial to the plant. The result is that harmful pH deviations for prolonged periods are avoided and the overall level of control of this process is significantly improved. The long term support costs of this installation have been minimal, while the savings in maintenance and improved process operation are significant.

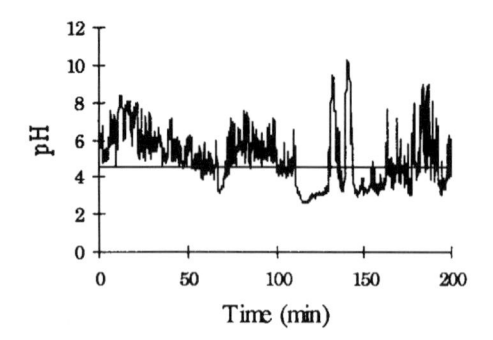

Figure 21. pH response to change in inlet acid.

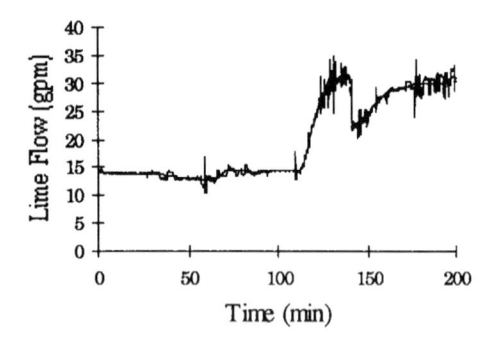

Figure 22. Lime flow, set point, and valve position in response to change in inlet acid.

Conclusions

We have presented here one perspective of the "many-sided" issues involved in the industrial application of nonlinear control, using the Spent Acid Recovery process and the pH process as illustrative case studies of the successful design and implementation of two such industrial nonlinear control systems.

Clearly, nonlinear control is becoming ever more relevant to industrial practice; the key issue now is essentially one of how best to identify and capture the stake presented by the ever-increasing demands on process operation. In this regard, by making the inevitable comparison with (linear) model predictive control and what has been primarily responsible for the significant impact it has had on industrial practice to date, it is not difficult to arrive at the following conclusion: the commercialization of nonlinear control packages similar in spirit to those available for linear MPC will significantly increase the impact of nonlinear control on industrial practice. There are several obstacles to the widespread development — and application — of such packages; and some of the most important have been noted. Nevertheless, that one such package is in fact already available is an encouraging sign that the potential exists for a significant increase in the application of nonlinear control techniques on many more actual industrial cases.

References

Bequette, B. W. (1991). Nonlinear Control of chemical processes: A review. *Ind. Eng. Chem. Res.*, **30**, 1391-1413.

Berkowitz, P. N. and Gamez, J.P. (1995). Economic On-line Optimization for Liquids Extraction and Treating in Gas Processing Plants. Presented at the Gas Processors Association 74th Annual Convention, San Antonio.

Dore, S. D., Perkins, J. D., and Kershenbaum, L.S. (1994). Application of Geometric Nonlinear Control in the Process Industries – A Case study. Proceedings of ADCHEM '94, Kyoto, Japan.

Gustafsson, T. K., and Waller, K. V. (1983). Dynamic Modeling and Reaction Invariant Control of pH. *Chemical Engineering Science*, **38**(3), 389.

Kravaris, C. and J. C. Kantor (1990). Geometric methods for nonlinear process control. 1. Background. *Ind. Eng. Chem. Res.*, **29**, 2295-2310; 2. Controller Synthesis. *Ind. Eng. Chem. Res.*, **29**, 2310-2323.

Kravaris, C. and Arkun, Y. (1991). Geometric Nonlinear Control – An Overview. *Chemical Process Control – CPCIV*, CACHE, 477.

Labossiere, G.A. and Lee P. L. (1991). Model-based Control of a Blast Furnace Stove Rig. *J. Proc. Cont.*, **1**(4), 217.

Lee, P. L. and Sullivan, G. R. (1988). Generic Model Control (GMC). *Comp. and Chem. Eng.* **12**(6), 573.

Levine, J., and Rouchon, P. (1991). Quality Control of Binary Distillation Columns via Nonlinear Aggregated Models. *Automatica*, **27**(3), 463.

Lindskog, P. and Ljung, L. (1994). Tools for semi-physical Modelling. *Preprints IFAC Symposium on Sysytems Identification*, **3**, 237.

McAvoy, T. J., Hsu, E., and Lowenthal, S. (1972). Dynamics of pH in a Controlled Stirred Tank Reactor. *Ind. Eng. Chem. Process Des. Dev*, **11**(1), 68.

Newell, R. B. and Lee, P.L. (1989). *Applied Process Control: A Case Study*, Prentice-Hall, Englewood Cliffs, NJ.

Ogunnaike, B.A. and Ray, W.H. (1994). *Process Dynamics, Modeling, and Control*, Oxford University Press, N.Y.

Ogunnaike, B.A. (1995). Application of hybrid modeling in control system analysis and design for an Industrial Low-Boiler column. *Proceedings European Control Conference*, Rome, 2239.

Papadoulis, A. V., Tsiligiannis, C. A., and Svoronos, S. A. (1987). A Cautious Self-Tuning Controller for Chemical Processes. *AIChE J.*, **33**(3), 401.

Pearson, R.K. and Ogunnaike, B.A. (1996). Nonlinear Process Identification. Chapter 4 in *Nonlinear Process Control*, Ed. Henson, M.A. and Seborg, D.E. Prentice-Hall, Englewood Cliffs, N.J.

Pearson, R.K. (1995). Nonlinear Input/Output Modelling. *J. Proc. Cont.* **5**(4), 197.

Psychogios, D. C. and Ungar, L. H. (1992). A hybrid neural network – first principles approach to process modeling. *A.I.Ch.E.J.*, **38**, 1499.

Rawlings, J. B., E. S. Meadows, and K. R. Muske (1994). Nonlinear Model Predictive Control: A Tutorial and Survey. Proceedings of ADCHEM'94, Kyoto, Japan.

Shinskey, F. G. (1979). *Process Control Systems*, 2nd Ed., McGraw-Hill, N.Y.

Singstad, P., K. Nordus, K. Strand, M. Lien, L. B. Lyngmo, and O. Moen (1992). Multivariable Nonlinear Control of Industrial LDPE Autoclave Reactors. *Proc. ACC*, Chicago, 615.

Temeng, K.O., Schnelle, P.D., and McAvoy, T.J. (1995). Model Predictive Control of an Industrial Packed Bed Reactor using Neural Networks. *J. Proc. Cont.* **5**(1), 19.

Tulleken, H.J.A.F. (1993). Grey-box modeling and identification using physical knowledge and Bayesian techniques. *Automatica*, **29**, 285.

Wassick, J. M. and Camp, D. T. (1988). Internal Model Control of an Industrial Extruder. *Proceedings ACC*, Atlanta, 2347.

Wright, R. A., and Kravaris, C. (1991). Nonlinear Control of pH Processes Using the Strong Acid Equivalent. *Ind. Eng. Chem. Res.*, 30, 1561.

Wright, R. A., Kravaris, C., Camp, D. T. and Wassick, J. M. (1992). Control of an Industrial pH process using the Strong Acid Equivalent. *Proceedings ACC*, Chicago, 620.

Wright, R. A., and Kravaris, C. (1993). On-line Identification and Nonlinear Control of pH Processes. Proc. Amer. Contr. Conf., San Francisco, 1167.

Wright, R. A., and Kravaris, C. (1993b). Dynamically Equivalent Outputs and Their Use in Nonlinear Controller Synthesis. *Chem. Eng. Sci.*, **48**, 3207.

Wright, R. A., and Kravaris, C. (1995). On-line Identification and Nonlinear Control of an Industrial pH Process. *Proceedings ACC*, Seattle, 2657.

SESSION SUMMARY: ADAPTIVE AND NONLINEAR CONTROL — FACT OR FANTASY?

Yaman Arkun
School of Chemical Engineering
Georgia Institute of Technology
Atlanta, GA 30332-0100

Summary

Chemical processes are nonlinear and their physical parameters often vary with time. Therefore nonlinear and adaptive control hold much promise for advanced process control, yet the promise has largely not been realized. There has been significant academic research in these two areas with limited but notable industrial applications. The purpose of this session was to present an industrial assessment, review the current status in research and point out the outstanding theoretical and practical issues.

The first paper by two industrial participants motivates the need for nonlinear control by noting that in recent years increasingly stringent control objectives have been imposed on chemical plants, resulting in more pronounced nonlinear process dynamics and difficult control problems. The authors describe a pH and a spent acid recovery system to elucidate the problems associated with the design and implementation of nonlinear controllers. In particular, they discuss the need assessment (when is nonlinear control required?), economic viability (cost vs. benefit analysis) and technological issues such as the nonlinear model development and control approaches used. They conclude by stating that the challenge lies in the identification of the appropriate processes that call for nonlinear techniques and commercialization of nonlinear control packages that can have widespread industrial impact.

The second paper reviews the state of the art of nonlinear analysis and controller synthesis methods and points out the research issues involved. They make certain statements that may help to identify the route to the "promised land," some of which are:

- We need to pay significantly more attention to nonlinear systems analysis in the process control community.

- We need to be able to identify limiting plant properties and quantify the achievable performance for nonlinear systems.

- We need to quantify benefits from nonlinear control vs. robust linear and vs. PI control.

- It is wildly optimistic to assume that there is a "silver bullet" for all nonlinear system classes.

- We will most likely have to go to alternative system representations in the nonlinear case.

- We believe that there is (unfortunately) no way around massive numerical on-line computations.

The last paper discusses the certainty equivalence approach to adaptive linear control. In doing so, new theoretical results are given to analyze the stability and transient properties of adaptive control. Next, two "puzzles" concerning the certainty equivalence approach are discussed: when to stop estimation? and how to handle the controllability problem associated with pole zero cancellation in the estimated models? Finally, switching between different models and controllers is presented as a new paradigm for adaptive control.

COMPUTER-AIDED PROCESS ENGINEERING IN THE SNACK FOOD INDUSTRY

Michael Nikolaou
Chemical Engineering Department
Texas A&M University
College Station, TX 77843-3122

Abstract

Food science and engineering traditionally have been the basis for the development of the process technology related to the manufacture of snack foods. While these disciplines remain important, new challenges for the snack food industry have created an incentive to explore the potential of recent advances in computer-aided process engineering (CAPE). Such challenges include consistent quality, productivity, safety, and environmental and consumer friendliness. In this paper, we briefly describe basic snack food production processes, and compare them to processes with which most chemical engineers are more familiar. A number of snack food process engineering tasks for which computers and chemical engineering principles can provide powerful aids are outlined, and the fundamental and practical problems associated with these tasks are identified. Examples of our experience with theoretical and applied developments for specific snack food processes are provided.

Keywords

Computer aided process engineering, Snack food.

Introduction

Snack foods figure eminently in America's dietary and business profiles. In 1993, Americans consumed 3.48 billion *lb* of salty snacks — potato chips, corn and tortilla chips, extruded and fabricated snacks, and multigrain chips (Snack Food, June 1994, p. 42) — an amount corresponding to 13.5 *lb* per capita. The 1993 ex-factory snack food sales were $42 billion (Snack Foods, 1994), the retail value of which was close to $60 billion. While the US snack food market is dominated by giant corporations, many small companies are flourishing by concentrating on specialty snacks.

New developments are taking place, given the benefits of new knowledge in nutrition science and new process technologies (Shukla, 1994; Tettweiler, 1991). Snack foods and processes are being thoroughly revamped or redesigned, either under regulatory mandate or under the food industry's self-imposed guidelines for healthful public nutrition. This is a formidable activity, necessitating the use of talent from several disciplines, including chemistry, biology, food science, agricultural and food engineering, and chemical and process engineering. Computer-aided process engineering (CAPE), in particular, is an important tool that can be used effectively in the development of flexible, safe, cost-effective, high quality snack food production processes. CAPE can contribute solutions to a broad range of snack food process technologies. For salty snacks such technologies include ultrafiltration and reverse osmosis for concentration of fresh fruits and vegetables; extrusion for blending, mixing, cooking, reacting, in-process flavor production, forming, and texturizing; and microwave processing for drying, dehydration, proofing, and puffing (Shukla, 1994).

This paper is an attempt to give an overview of the snack food industry from a CAPE perspective. The author's experience with the snack food industry during the last four years has been both fruitful[1] and enjoyable. Hoping to broaden the discussion between the CAPE and snack food communities, the author discusses open

---

[1]Snackful.

problems and possible future directions throughout the text.

Snack Food Products

There is a multitude of products that the snack food industry produces. The list is growing steadily, as competition forces companies to introduce snacks with refined features, such as new raw material basis, improved texture, shape, color, flavor, and nutritional content. Basic product categories are potato chips (http: //www. fritolay. com/ chips. html); meat-based snacks; popcorn-based snacks; puffed snacks; corn/tortilla chips and simulated potato chips; nut-based snacks.

Snack Food Processes and Equipment

There is a relatively small number of processes that appear most frequently in snack food production plants. A brief description of the most important processes follows. The discussion emphasizes elements that are important from a CAPE viewpoint.

Extrusion

Food extrusion simultaneously achieves three functions: (a) mixing of raw materials — e.g. various ground grains and water, (b) shearing/kneading, and (c) heating and/or cooking, using the heat released by friction inside the extruder (Harper, 1981; Frame, 1984). For some extrusion processes, the last function may be followed by puffing, i.e. abrupt expansion of high-temperature plasticized and gelatinized cornmeal from a pressurized chamber into the atmosphere[2] (Matz, 1975). The control of extrusion cooking processes is a formidable challenge, because of variability of raw materials, process modeling difficulties, frequent wear of equipment, and various external and internal disturbances. A case study is discussed in a subsequent section of this paper.

Frying

This is a process in which a snack is cooked by floating or being immersed in hot oil that provides heat at high transfer rates. Continuous fryers are used for large scale operations (~5,000 *lb/hr* throughput), while batch (kettle-style) fryers are experiencing a comeback in small scale enterprises (<200 *lb/hr*). The frying fat or oil becomes a significant component of the end product, varying from as little as 10% by weight in breaded fish sticks to 40% in potato chips (Matz, 1975). Responding to consumer preferences, the snack food industry is now emphasizing good-tasting low-fat snacks. The design and operation of manufacture processes for such snacks poses a plethora of exciting CAPE problems.

Baking

In baking processes snacks are cooked by heat transferred through the air by convection, conduction, or radiation. The effectiveness of each of these mechanisms varies with oven and product design, and is generally lower than that of heat transfer based on liquids such as oil. Baking does not contribute any added fat to the finished snack. This may result in more healthful products, a fact that explains the recent surge in popularity of baked snacks.

Drying

Drying is required for several snacks to develop the right crispness. Puffed snacks, for example, must be dried after extrusion in order to obtain moisture content sufficiently low (to ~4%) to insure good texture and storage stability. Because a puffed snack with moisture level above or below normal feels stale or spongy, there is a need for accurate control of the drying process..

Packaging

Challenges include accurate control of the snack weight contained in each package, and safe extension of shelf storage life.

Miscellaneous

(a) Nut Processing equipment: Sorters, blanchers, roasters, coolers. (b) Oil, Powder, and Granule Applicators: Oil and cheese sprayers, powder dispensers, electrostatic salters, coating tumblers. (c) Transfer and Storage Equipment. (d) Measuring and Weighing Equipment.

Comparison with Other Industries

Raw materials Variability

Because most raw materials for the snack food industry are natural products, they can exhibit substantial variability. This may be due to the variety of a particular crop, location of cultivation, weather conditions during crop growing and harvesting, and storage conditions. The chemistry of natural products can be particularly complex. Similar concerns exist for other process industries, such as the petroleum, metal, and pulp and paper industries, that rely on the primary production sector for raw materials (e.g., oil, metal ores, and timber).

Unit Operations

While the majority of the traditional chemical processes deal mostly with fluids, snack food processes deal mostly with solids or highly viscous fluids. Heat transfer is common in snack food production. Mechanical or mass

---

[2]This phenomenon was discovered more or less by accident (Matz, 1975).

transfer based separations are not uncommon. Cleaning, peeling, slicing, washing, crushing, and seasoning are common solids handling unit operations.

Snack Food Challenges and Relevant CAPE Tools

Product and Process Development

Computer modeling of new snack products and processes can substantially speed up their development. Examples are (a) the design of healthful snacks; (b) the development of flexible processes that can reliably operate on a multitude of raw materials and are capable of producing a variety of products; (c) the design of environmentally benign snack food processes. Examples of environmental issues are (a) the treatment of wastewater containing considerable amounts of starch and peel fragments resulting from potato peeling, slicing, and washing operations; (b) the containment of oil or particulate solid containing fumes resulting from frying. In addition to design tasks, process models can be used for personnel training, and process control.

Scheduling

Unlike oil and petrochemical products, snacks can be stored for a relatively limited time only. In addition, demand may vary unexpectedly and quite rapidly. This creates the need for just-in-time production methods. Given that production equipment is limited, the selection of what snack to produce, at what quantity, and in which production line, become particularly important scheduling issues.

Process Control

Variables to control include temperature, pressure, moisture, oil content, color, shape, and texture. Image-based control is not uncommon. It should be stressed that, unlike oil and petrochemicals, snack food blending is hardly ever possible, a fact that renders it imperative for the product to be within specifications during as large a percentage of the production time as possible. Controlled variables may be difficult or impossible to measure on-line, thus creating a need to infer values from secondary measurements, and, ultimately, a need for new sensors. Quality control laboratory results, produced at ordinary but infrequent intervals, need to be integrated with on-line process control methods. During frying, for example, one needs to adjust the amount of fresh oil exchanged with used oil, to counter the effects of free fatty acids (released from the oil triglycerides and imparting a foamy or even soapy feel to the oil) and rancid-tasting compounds produced from oil oxidation (Robards et al., 1988). Control systems that compensate for equipment wear and can indicate when equipment maintenance is necessary need to be perfected. In extrusion cooking, for example, there is a need to know when screw wear can no longer be compensated for by feedback control and equipment maintenance is needed.

An adverse result of effective control systems is that they may offer few opportunities for novice operators to understand the dynamics of a process in depth. This situation might jeopardize the smooth operation of a process when circumstances are encountered that cannot be handled autonomously by the process control system, thus requiring human intervention (Brooks, 1995).

Statistical Methods

Statistics can be used to extract workable knowledge from an abundance of process data available from real time measurements or in databases. One such case, as previously mentioned, is integration of statistical quality control results from the quality control laboratory with on-line process control. Statistics are also used extensively in translating subjective sensory judgments about the quality of a snack to numerical values of appropriate variables (Vickers, 1987; Defreitas and Molins, 1988; Prinyawiwatkul, 1993).

Software Engineering and Artificial Intelligence

An idea is to use as many forms of information available in the plant as possible, such as past operator experience, heuristics, complex food chemistry, and qualitative objectives and constraints. Computer system integration, from the factory floor to the corporate headquarters, is a task that is showing signs of becoming feasible.

Case Studies

Extrusion Process

Twin screw extrusion has found widespread application in extrusion cooking, because of several advantages, most important of which is the possibility of good process control (Straka, 1985). However, varied usage of the term *control* in the past seems to have contributed to misunderstandings, and to have stifled progress in research on extrusion cooking control, both in industry and academia. What most studies refer to as extrusion *control*, means, in reality, *design* of an extrusion process with the ability to operate at certain steady state operating conditions (Matz, 1975, p. 135; Falcone and Phillips, 1988). For example, numerous studies have concentrated on controlling pressure inside an extruder. While an extruder must be able to reach high enough pressures, regulation of pressure does not guarantee product quality, because the latter is affected by numerous other factors, not all of which can be taken into account. Direct feedback (or feedback-feedforward) control based on direct measurement of variables associated with product quality clearly provides better opportunities for regulatory control.

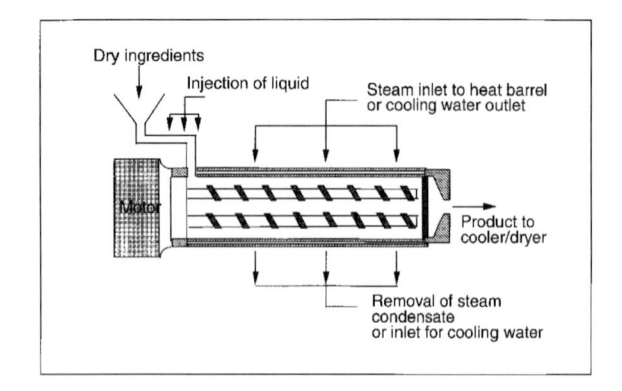

Figure 1a. Twin-screw extruder.

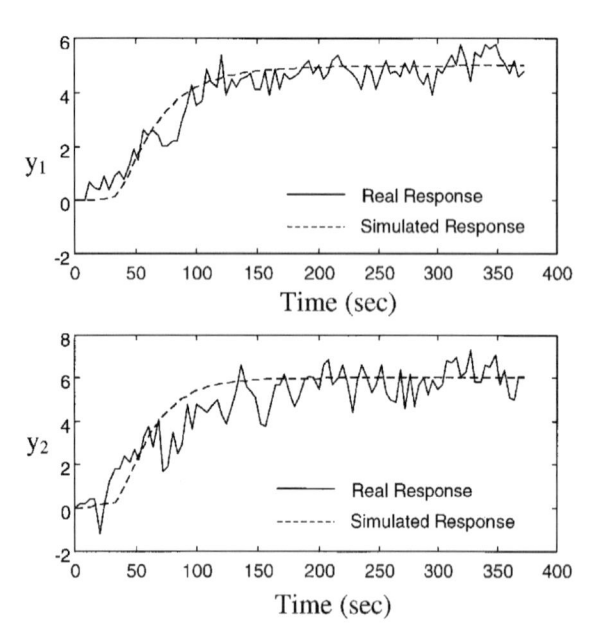

Figure 1b. Response of product quality variables to setpoint changes, for a closed loop with a 3x2 controller.

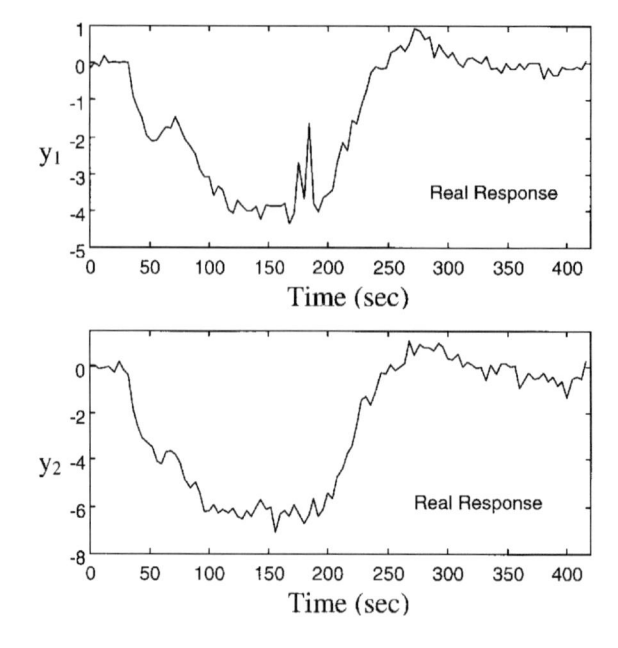

Figure 1c. Response of product quality variables to change in the raw material, for a closed loop with a 3x2 controller.

The measurement of variables associated with product quality raises two fundamental issues:

1. What variables can be used to infer product quality?
2. How can these variables be measured on-line?

The first question can be addressed using statistical approaches, that relate human judgment of the product quality to number-valued variables. The second question is related to the development of sensors and analytical methods for product evaluation. Experience has shown that common sense does not always provide reliable solutions to the above problems (Vickers, 1987).

In our experience we dealt with an extrusion cooking system for puffed snacks. We first collaborated with our industrial sponsor to address the two questions raised above. Subsequently, we identified the following fundamental questions to address:

- How can one configure a control system?
- How much accuracy is needed in a process model for control purposes? When do I stop identification experiments?
- Are constraints important for the controller, thus requiring an on-line optimization based control system?
- Are nonlinear modeling and control required?
- How can a model be refined under closed-loop control, with minimum penalty on product quality?

Addressing the above questions requires good understanding of both the underlying physical process and the current state of the art in control theory. It is evident, however, that the above questions are, by no means, confined to snack food process control, but emerge naturally in a wide range of process control problems.

For the particular problem, we decided to work in a model predictive control (MPC) framework. Excellent reviews on MPC and related issues are provided by Lee (1995) and Mayne (1995) elsewhere in these proceedings. Therefore, we do not attempt to present the state of the art or review related work by other investigators in this particular area. Instead, we summarize below our contributions to research in this field, and explain the significance and usability of our results.

- *Robust stability of constrained MPC* — On the basis of the seminal work of Rawlings and Muske (1993), theoretical results that can be used for the design of constrained MPC systems with robust stability and performance were obtained in Genceli and Nikolaou (1993) for MPC with l_1 objective, and Vuthandam et al. (1995) for MPC with quadratic objective. These results were extended to nonlinear MPC systems with second-order Volterra models (Genceli and Nikolaou, 1995a) and to non-square MPC systems (Sarimveis et al., 1996). A comparison between simulation and experiment for a puffed snack extrusion cooking system is shown in Fig. 1.

- *Performance of constrained MPC* — A common theme of the above results is that they create a connection among (a) a process model and its accuracy, expressed as upper and lower bounds in the model's parameters; (b) the optimization and control horizon lengths; (c) the minimum values of the input move suppression coefficients in the on-line objective function. For overdamped or "slightly" underdamped systems, the smallest values of the input move suppression coefficients allow the most aggressive control action, thereby producing the tightest control. This realization has the following implications:

    ◊ Control system configurations that result in the smallest values of the input move suppression coefficients, for the same level of modeling uncertainty, are preferable.
    ◊ Identification experiments should produce modeling accuracy (e.g., bounds for model parameters) such that the resulting closed-loop performance bound can be satisfactory.
    ◊ To decide whether to use a linear or nonlinear model in constrained MPC, one can compare the closed-loop performance bounds corresponding to the different uncertainty levels of the two classes of models.

- *Simultaneous MPC and Identification (MPCI)* — A first attempt for simultaneous constrained MPC and process identification is presented in Genceli and Nikolaou (1995b) and Shouche et al. (1995). The idea is to conduct the on-line optimization over a set of process inputs that satisfy a persistent excitation condition, in addition to all standard MPC constraints. The resulting on-line optimization problem is non-convex. An approach to its solution, based on the solution of a series of semi-definite programming problems is shown in Genceli and Nikolaou (1995b). The importance of MPCI is that it performs process identification under closed-loop MPC, without resorting to external excitation of the closed loop. Therefore, the perturbation of the process output is kept at a minimum level, while, at the same time, process identification remains feasible.

- *Multiscale MPC* — This is a recent effort to unify phenomena, control objectives, and constraints that occur at different time scales. The goal is to handle situations where frequency- or time-domain analyses alone cannot capture the essential features of signals that have time-varying frequency spectra. A first attempt is described in Feng (1996), where a wavelet-domain framework is employed for external disturbance prediction and subsequent incorporation in the objective function of an MPC system.

Potato Chip Frying Process

Frying is one step of an integrated potato chip manufacture line that includes a sequence of operations: cleaning, peeling and spotting, slicing, washing, frying, seasoning, and packaging. The exposure of a potato chip to hot oil during frying has several effects (Mottur, 1989):

- Starch, the predominant food component of potato (up to 99.5% of dry weight), is gelatinized, thus rendering the potato more easily digestible.
- The water content of the slice is reduced from about 80% to about 1-2%, resulting in a desirable crisp texture and stability from microbiological spoilage.
- Cooking oil is absorbed into the chip, enhancing flavor and texture.
- The level and variety of flavor compounds are greatly increased over those present in the raw potato.

The design and operation of fryers poses a variety of CAPE problems such as the following:

- How can the process be optimized through better design and/or operation (e.g., for minimum oil uptake by the potato chip)?
- At what level of detail should a first-principles process model be developed for design and optimization purposes?
- At what level of detail should a process model be developed for control purposes?

- How accurate does a process model have to be?
- How can on-line and quality control laboratory measurements, obtained at different time scales, be used effectively in process control?
- What control strategy provides adequate control, without excessive design and maintenance requirements?

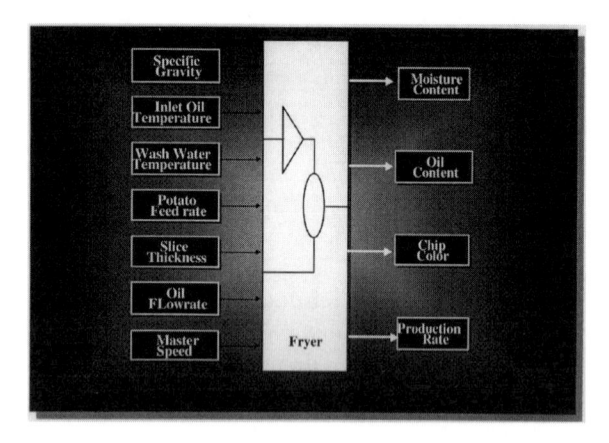

Figure 2a. Overview of the fryer model.

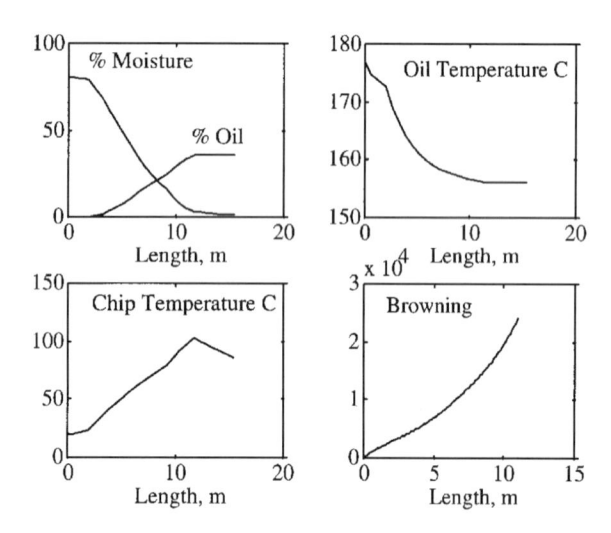

Figure 2b. Steady state profiles along the fryer predicted by the model.

Tight process control of snack chip frying offers significant economic incentives. It is estimated that reduction of potato chip losses by 1% of throughput would reduce production costs by over $3 million per year for the single largest US snack producer alone (Brooks, 1995). There are interesting process control problems associated with potato chip frying, as outlined below:

- *Moisture control* — It is important to end the frying process so that final moisture will be in the critical 1-2% range. A 1.4 mm thick slice at 180 C will reach this range after about 2 minutes of fry time. If moisture is above 2%, crispness will suffer. If it is below 1% it will result in excessive oiliness, dark color, and scorched flavor (Mottur, 1989).
- *Oil content control* — Typical finished potato chips contain 30-40% oil. Since potato tissue contains only 0.1% fat, nearly all of the oil in the finished chip is due to absorption from the frying medium. If the oil content rises above 40% the chip becomes unpleasantly greasy with an oil-soaked appearance. For chips designated as low-fat, maintaining oil content below the level advertised on the product's package poses a constraint that must be satisfied throughout production.
- *Color control* — There are two components in the color of the finished chip: (a) background color, and (b) dark spots. Background color, mostly due to starch caramelization, is associated with the length of frying period, oil temperature, and slice thickness (Talburt and Smith, 1975). Background color is relatively straightforward to control at a golden yellow level, using feedforward/feedback control. Dark spots, a feature undesirable to consumers, are due to the Maillard reaction of reducing sugars (mostly sucrose, fructose, and glucose (Yada et al., 1985)) with aminoacids, that occurs during frying. While the extent of the Maillard reaction can be partially controlled, it is mostly the sugar content of the potato that determines whether dark spots will appear or not. Sugar concentration may increase during potato storage at low temperature (2-4 C) due to slowing of the breakdown of sugars to carbon dioxide and water (Snackfood Association, 1987; Blankson et al., 1988). Sugar content can be lowered back to an acceptable level (<0.2-0.4%) by reconditioning (storage at >13 C for several weeks prior to use). It is clear that knowledge of the sugar concentration before frying can be valuable for use in feedforward-feedback control schemes.

The above process control tasks are challenging because of factors such as the following:

- large dead-time (a few minutes, required for passage of chips through the fryer);
- time-varying dynamic behavior (e.g., due to oil degradation, or variation of raw materials due to different sources and/or storage

conditions (Evensen et al., 1988; Louwes and Neele, 1987; Sieczka and Maatta, 1986));
- nonlinearity (e.g., due to chemical reaction kinetics);
- strong coupling of variables (e.g. moisture and oil content of the cooked chip);
- constraints on process variables (e.g., specifications for "low-fat" product designation).

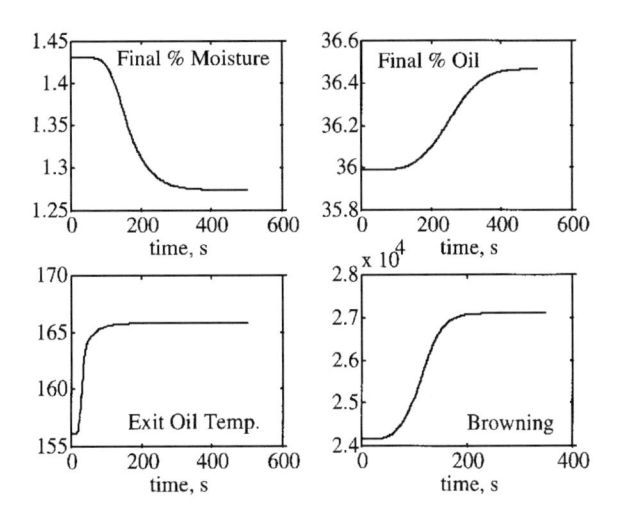

Figure 2c. Fryer dynamics predicted by the model.

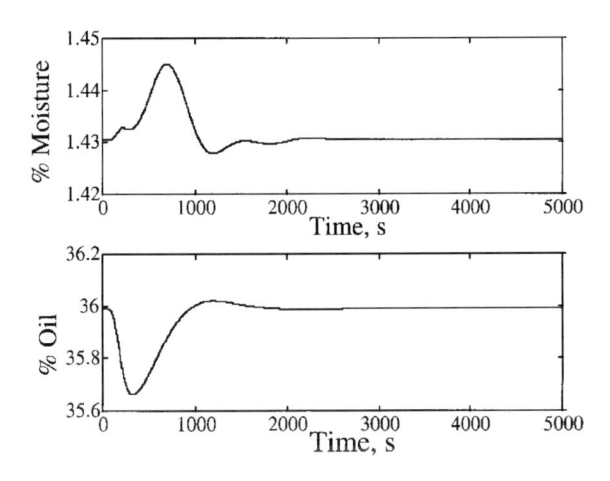

Figure 2d. Simulation of closed-loop response to step change in the thickness of potato slices.

In addressing the above issues, we started with the development of a first-principles model for a continuous fryer (Feng et al., 1995). The objective was to develop a model usable in the design of control systems for several variations of continuous fryers, without requiring excessive experimentation for each particular fryer. The challenge for such a model is to decide what physical and chemical phenomena to incorporate, and at what level of detail. There have been several studies on individual phenomena that occur during frying, such as

- chip texture formation (Pravisani and Calvelo, 1986) ;
- chip color formation (Buera et al., 1987; Coffin et al., 1987; Dahlenburg, 1982; Habib and Brown, 1957; Lee et al., 1984; Leszjowiat et al., 1990; Lyman and Mackay, 1961; Marquez and Anon, 1986; Picha, 1986; Pravisani et al., 1986; Roe and Faulks, 1991; Roe et al., 1990; Talburt and Smith, 1975);
- oil penetration into the chip (Farkas, 1994; Keller et al., 1986; Lamberg et al., 1990);
- oil uptake by the chip (Gamble et al., 1993; Miller et al., 1975; Nonaka et al., 1977; Pinthus et al., 1993; Sayer et al., 1975; Shouche and Feng, 1995; Talburt and Smith, 1975);
- heat and mass transfer in the chip (Keller et al., 1986; Mittelman et al., 1984; Rice and Gamble, 1989.

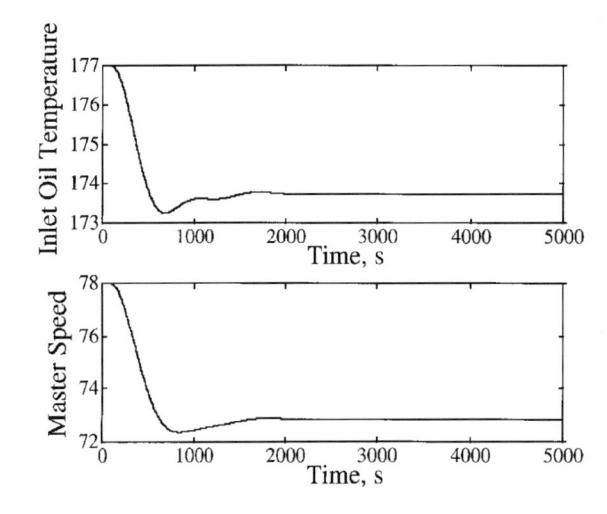

Figure 2e. Simulation of response of manipulated inputs to step change in the thickness of potato slices.

A list of models was constructed and a heuristic sensitivity analysis was conducted, to determine the effects of various model parameters on the prediction of controlled variables. A model was selected that showed very good agreement between predicted and measured values for controlled variables. The computer implementation of the model was done in a Matlab-Simulink environment. Simulations produced by the model are shown in Fig. 2. Testing of the model in an MPC system is underway.

Conclusions

There is hardly an economic commodity more important for our lives than food. Snack foods, in particular, are a substantial part of Americans' diet. Salty snack consumption could triple in a decade, as eating on the run and grazing are becoming more common (Shukla, 1994). This makes the improvement of snack foods and processes a high-priority economic opportunity and social responsibility. The US snack food industry is hard at work addressing the marketing and technological challenges associated with the production of better snacks. This is a multifaceted endeavor, requiring talent and tools from several disciplines. The academic CAPE community, strongly represented in chemical engineering, can be a substantial contributor to that effort. Cross fertilization of ideas common in the chemical and food industries can offer to each industry a fresh view of related challenges and possible solutions. Collaboration between academia and the snack food industry can identify exciting research problems to work on. This can steer research to fruitful directions for the production of better snacks.

Acknowledgment

This paper is based on work partially supported by NSF, Shell Development, Frito-Lay, and the Texas Advanced Technology Program. The author gratefully acknowledges that support.

References

Blankson, J. E., Coffin R. H., Yada R., Lougheed E. C. (1988). The Effects of Moderate Carbon-Dioxide Levels in Storage upon Potato-Chip Color, *Canadian J. of Plant Science*, **68**(2), 567-568.

Brooks, A. (1995). Personal Communication with the Author.

Buera, M.P., J. Chirife, S.L. Resnik and R.D. Lozano, (1987). *J. Food Sci.*, **52**(4), 1059.

Coffin, R.H., R.Y. Yada, K. L. Parkin, B. Grodzinski and D.W. Stanley, (1987). *J. Food Sci.*, **52**(3), 639.

Dahlenburg, A.P., (1982). *Food Tech. in Australia*, **34**(11), 544.

Defreitas, Z., Molins R. A. (1988). Development of Meat Snack DIPS — Chemical, Physical, Microbiological and Sensory Characteristics, *Journal of Food Science*, **53**(6), 645-1649.

Evensen, K. B., Russo J. M., Braun H. (1988). Predicting Potato-Chip Quality and Yield, *HortScience*, **23**(3), 728-728.

Falcone, R. G., Phillips R. D. (1988). Effects of Feed Composition, Feed Moisture, and Barrel Temperature on the Physical and Rheological Properties of Snack-Like Products Prepared from Cowpea and Sorghum Flours by Extrusion, *Journal of Food Science*, **53**(5), 1464-1469.

Shouche, M. S., H. Genceli, and M. Nikolaou (1995). Simultaneous Model Predictive Control and Identification of DARMA Processes, submitted to *Automatica*.

Farkas, B. E., (1994). *Ph.D. Dissertation*, UC, Davis.

Feng, W Y., M. Shouche, and M. Nikolaou (1995). Modeling and Predictive Control of a Continuous Fryer, *paper 173j, AIChE Annual Meeting*, Miami Beach, FL.

Frame, N. D. (1994). *The Technology of extrusion cooking*, 1st ed., Blackie Academic & Professional.

Gamble M.H, P. Rice, and J.D. Selman, (1987). *Int. J. of Food Sci. and Tech.*, **22**, 233-241.

Genceli, H., and M. Nikolaou (1993). Robust Stability Analysis of Constrained l1-Norm Model Predictive Control, *AIChE J.*, **39**(12), 1954-1965.

Genceli, H., and M. Nikolaou (1995a). Design of Constrained Model-Predictive Controllers with Volterra Series, *AIChE J.*, **41**(9), 2098-2107.

Genceli, H., and M. Nikolaou (1995b). A New Approach to Simultaneous Model Predictive Control and Identification, *AIChE J.*, **42**(10), 2857-2869.

Habib, A. T, and Brown, H. D, (1957). *Food Technol.*, **11**(85).

Harper, J. M. (1981). *Extrusion of foods*, CRC Press, Boca Raton, Fla.

Keller C. F., Escher., and J. A. Solms, (1986). *Lebensmittel-Wissenschaft & Technologie*, **19**(4), 346-348.

Feng, W. (1996). *Application of Time-Frequency Techniques to Process Control and Identification*, Ph. D. Thesis.

Lamberg, I., B. Hallstrom, and H. Olsson, (1990). *Lebensmittel-Wissenschaft & Technologie*, **23**(4), 245-251.

Lee, C.M., B. Sherr and Y.N. Koh, (1984). *J. Agric. Food Chem.*, **32**, 379-382.

Lee. J. H. (1995). Recent Advances in Model Predictive Control and Other Related Fields, *CPC V Proceedings*.

Leszjowiat, M.J., V. Barichello, R.Y. Yada, R.H. Coffin, E.C. Lougheed, and D.W. Stanley, (1990). *J. Food Sci.*, **55**(1).

Louwes, K. M., Neele A. E. F. (1987). Selection for Chip Quality and Specific-Gravity of Potato Clones — Possibilities for Early Generation Selection, *Potato Research*, **30**(2), 241-251.

Lyman, S. and A. Mackay, (1961). *American Potato J*, 38:51.

Marquez, G and M. C. Anon, (1986). *J. Food. Sci.*, **51**(1), 157.

Matz, S. A. (1975). *Snack food technology*, 1rd ed., Van Nostrand Reinhold, New York; 3rd ed. 1993.

Mayne, D. (1995). Optimization in Model Based Control, *CPC V Proceedings*.

Miller, R.A., J.D. Harrington and G.D. Kuhn, (1975). *Am. Potato J.*, **52**, 379.

Mittelman, N., S. Mizrahi, and Z. Berk, (1984). *Engineering and Food*, Ch. 12, B. M. McKenna (Ed.), Elsevier 109-116.

Mottur, G. P. (1989). A Scientific Look At Potato-Chips — The Original Savory Snack, *Cereal Foods World*, **34**(8), 620-626.

Nonaka, M., R. N. Sayre., and M. L. Weaver, (1977). *Am. Potato J.*, **54**, 151-159.

Orr, P.H., Janardan K. G. (1990). A Procedure to Correlate Color Measuring Systems Using Potato-Chip Samples, *American Potato Journal*, **67**(9), 647-654

Picha, D. H. (1986). Influence of Storage Duration and Temperature on Sweet-Potato Sugar Content and Chip Color, *Journal of Food Science*, **51**(1), 239-240.

Picha, D.H., (1986). *J. Food Sci.*, **51**(1), 239.

Pinthus, E. J., P. Weinberg, and I. S. Saguy, (1993). *J. Food. Sci.*, **58**(1), 204.

Pravisani, C. I., and A. Calvelo, (1986). *J. Food Science*, **51**(3), 614.

Prinyawiwatkul, W., Beuchat L. R., Resurreccion A. V. A. (1993): Optimization of Sensory Qualities of an Extruded Snack Based on Cornstarch and Peanut Flour, *Food Science and Technology-Lebensmittel-Wissenschaft & Technologie*, **26**(5), 393-399.

Rawlings, J. B., and K. R. Muske (1993). The Stability of Constrained Receding Horizon Control, *IEEE Trans. AC*, **38**(10), 1512.

Rice, P., M. H. Gamble, (1989). *Int. J. of Food Sci. and Tech.*, **24**, 183-187.

Robards, K., Kerr A. F., Patsalides E. (1988). Rancidity and Its Measurement in Edible Oils and Snack Foods — A Review, *Analyst*, **113**(2), 213-224.

Roe, M.A., and R.M. Faulks, (1991). *J. Food Sci.*, **56**(6), 1711.

Roe, M.A., R.M. Faulks and J.L. Belson, (1990). *J. Sci. Food Agric.*, **52**, 207-214.

Sarimveis, H., H. Genceli, and M. Nikolaou, (1996). Design of Robust Non-Square Constrained Model Predictive Control, *AIChE J.*, **42**(9), 2582-2593.

Sayre, R. N., M. Nonaka, M. L. Weaver, (1975). *Amer. Potato J.*, **52**, 73-82.

Shukla, T. P. (1994). Future Snacks and Snack Food-Technology, *Cereal Foods World*, **39**(9), 704-705.

Sieczka, J. B., Maatta C. (1986). The Effects of Handling on Chip Color and Sugar Content of Potato-Tubers, *American Potato Journal*, **63**(7), 363-372.

Snackfood Association, (1987). *Fifty Years: A Foundation for the Future*, Alexandria, VA.

Sowokinos, J. R., Knoper J. A., Orr P. H., Varns J. L. (1987). Influence of Potato Storage and Handling Stress on Sugars, Chip Quality and Integrity of the Starch (Amyloplast) Membrane, *American Potato Journal*, **64**(5), 213-226

Straka, R. (1985). Twin-Screw and Single-Screw Extruders for the Cereal and Snack Industry, *Cereal Foods World*, **30**(5), 329-332.

Strock, H., C. O. Ball, S. S. Chang, and E. F. Stier, (1966). *Food Technology*, **20**(4), 193-196.

Talburt, W. F., and O. Smith (1975). *Potato Processing*, 3rd Ed., AVI Publishing, Westport, Connecticut.

Tettweiler, P. (1991). Snack Foods Worldwide, *Food Technology*, **45**(2), 58-62.

Vickers, Z. M. (1987). Sensory, Acoustical, and Force-Deformation Measurements of Potato-Chip Crispness, *Journal of Food Science*, **52**(1), 138-140.

Vuthandam, P., H. Genceli, and M. Nikolaou (1995). Performance Bounds for Robust Dynamic Matrix Control with End-Condition, *AIChE J.*, **41**(9), 2083-2097.

Warner, K., Orr P., Parrott L., Glynn M. (1994). Effects of Frying Oil Composition on Potato-Chip Stability, *Jounal of the Americal Oil Chemists Society*, **71**(10), 1117-1121.

Yada, R. Y., Stanley D. W., Fitts M., Coffin R. H., Leszlowiat M. J. (1985). Effect of Sucrose, Glucose and Fructose Content of Ontario Grown Potato-Tubers on Chip.

BATCH FOOD PROCESSING: THE PROOF IS IN THE EATING

Sandro Macchietto
Centre for Process Systems Engineering
Imperial College
London SW7 2BY, UK

Abstract

Some of the characteristics of manufacturing in the food industry are reviewed and the major business drivers are highlighted. Based on this analysis, the major control needs are identified. It is argued that the largest opportunities lie in a broad view of control, to include plantwide control of operations for maximum resource utilisation, operational flexibility and responsiveness to market and product innovation. The second part of the paper presents the activities of a consortium project aimed at the development of an on-line scheduling system suitable for large scale, multipurpose food manufacturing, to meet some of the above requirements. The project goals, its organisation, technology choices and system design utilised are briefly described. Results of two demanding industrial applications from the Dairy and Brewery industries are presented and discussed.

Keywords

Food manufacturing control, Business drivers, Manufacturing execution system, On-line scheduling, Dairy production, Beer production.

Introduction

By any account, food and drink are BIG. In the UK, for example, they are the largest of the manufacturing sectors, accounting for 2.9% of Gross Domestic Product (GDP) in 1992 (14.1% of total manufacturing output) and 2% of employment. To put things in perspective, this is a bigger contribution to GDP than those of the chemicals, motor vehicles, defence and aerospace, or materials industries. The above figures exclude food and drink wholesaling and retailing, which represent an additional 2.5% of UK GDP and 4.7% of employment. Food and drink are among the UK's largest exports, at £8.2bn in 1993 (Technology Foresight Report, 1995a, 1995b).

By business sector, the activities involved are production and/or processing of meat, starch and other miscellaneous foods, dairy, bread, biscuit and flour confectionery, brewing and malting, ice cream, cocoa and chocolate confectionery, animal feeding stuffs, soft drinks, spirit distilling, ready meals, oils and fats, fruit and vegetables, sugars, fish, grain milling, and wine and cider.

The industry is characterised by a small number of very large players, with global, integrated product development, manufacturing and distribution networks and strong product brands. In Europe, for example, there are but three food giants, Nestle, Unilever and Danone. Nestle employs 200,000 people, has 490 factories in 60 countries, and sells 15,000 products in nearly every country of the globe. Unilever has approximately 600 companies (not all in food), operates in just as many countries and generated £1.2 billion operating profit from a £15 billion turnover in foods in 1994. Of these, food sales in emerging markets accounted for £4.6, twice the figure for 1992, and about a quarter of the total. At a slightly lower level, a company like Heinz had sales of $8 billions in 1995 (Company Reports).

There are also very much larger numbers of intermediate and small manufacturers with more localised production and distribution capabilities. Approximately 90% of UK food and drink manufacturers employ less than 100 people, although the other 10% of large companies control about 90% of the market value (Technology Foresight Report, 1995b).

In the western world at least, retailing of food and beverage products is dominated by large companies with multiple outlets and integrated operations. For example, in the UK 5 big retailers share approximately 80% of the market. They have very large buying muscles and command much attention from their manufacturing suppliers. The retailing end of the business is characterised by high volume, low inventory, short lead times, wide variety of products and strong competition on price. The consumer preference for freshness is leading to fewer additives being used and shorter shelf life of many products. Sunday trading has been recently introduced in the UK and is making inroads in the rest of Europe. Deregulation within the European Community and due to GATT is leading to substantially increased competition. Retailers use sophisticated techniques for point-of-sale tracking and inventory update, while Electronic Data Interchange (EDI) is increasingly being used to link retailers, distributors and manufacturers, offering substantial opportunities for optimising the whole supply chain. Weekly exchange of forecasted orders through EDI between retailers and their suppliers is becoming more commonplace (Booty, 1993), but is still not adequate to deal, for example, with perishable products which are produced to order on a daily basis, with direct delivery to the stores rather than warehouses.

Product Innovation

The rate of new product introduction in the food industry is very high. In the USA, over 12,000 new products are introduced every year (Food Technology, December 1994). The average cost of a product launch is about $20 million. Most products show some short term success but five out of six then disappear (typically within six months). Less than 1% of products ever achieve more than $15 million annual sale (Food Technology, March 1995).

Product life is also becoming shorter. Bruin (1992) reports that the half life of products, approximately ten years in 1970 and reduced to five by 1990, is expected to decrease further to 2-3 years by the end of the century. Consequently, "the high bonus of being first with a product innovation of substance is getting increasingly difficult to achieve. Speed up of the product/process development cycle is therefore of paramount importance".

It should be observed that product innovation in the food industry often means innovation in marketing, packaging, and basic food science leading to new products (e.g. replacement of fats by other functional ingredients, new flavours, textures), not just process innovation.

Food Manufacturing

The manufacturing end of food has become much more difficult recently. As a simple example, consider the manufacturing of cheddar cheese, a typical dairy industry product. Making cheese is pretty straightforward, however the raw milk price has recently been deregulated in the UK. A medium size cheddar cheese manufacturer calculates that the farmers' profit amounts to 12p (pence), the retailer about 10p and the cheese producer about 3p per liter of milk (Financial Times, Sept. 1, 1995, p.15). Smaller operators are being squeezed hard, and as a consequence, much rationalisation is happening in the industry. Economies of scale from high volumes however are no longer sufficient to keep costs down, since the emphasis is on much more responsive production. Synergies arise from combining manufacturing with related products, such as yoghurt and liquid milk, thus sharing raw material processing, intermediates, plant services and distribution. Some of these products (e.g., yoghurt) are typically made in multipurpose plants with several end products made daily, and are characterised by high rate of product innovation (160 new yoghurts were introduced in the UK in 1994). The resulting large, integrated dairies produce in some cases up to 2-300 products. Thus pretty quickly making even a boring, commodity product like cheddar cheese becomes a rather more lively and demanding activity.

The above observations are well supported by a systematic study by the Process Industries Manufacturing Advantages (PRIMA) Project carried out under the auspices of the European Union (summarised in Technology Foresight Report, 1995b, Appendix E). A survey of European manufacturers was carried out to establish, amongst other things, current and future business and technology drivers. The wide spectrum of foods and drink companies amongst the respondents identified as top factors for competitive advantage the rapid development of new products/processes, the rapid response to change, flexible production and rapid customer response.

Manufacturing Processes

Focusing in more detail on the processes involved, food processes have been classified into four broad categories (Bruin, 1992):

i. Separation processes (e.g. diffusional extraction, concentration of juices, mechanical separation of flower).
ii. Assembly or texturising processes (e.g. emulsion processes for margarine and ice cream, foaming of whipped cream, extrusion processes, dough making).
iii. (Bio)conversion processes (e.g. sugar fermentation to alcohols, roasting).
iv. Preservation processes (e.g. retorting to eliminate microbial, enzymatic or chemical spoilage)

This author claims that all food processes can be described by something in the order of 120 "unit operations" (as opposed to approximately 30 used in chemical processes). Indeed some operations, such as pasteurisation, ohmic or microwave heating, are hardly known outside the food industry. However it may be argued that, apart from some minor details, chemical and food industries are not all that different, since both are concerned with the transformation of raw materials into a set of products, utilising chemical, physical and biological transformations. Some of these may be slightly unfamiliar to a traditional chemical engineer, but they can all be described fundamentally in terms of mass, energy and momentum balances, and all analysis and design skills from a chemical engineering background can be easily ported to food applications.

With regards to the mode of operation, Sawyer (1993) quotes a 1987 study by the Automation Research Corporation indicating that 65% of the food industry is batch. Although it is not mentioned whether this is by product value, no. of processes, etc., the figure is higher than for chemicals and other process industries (except pharmaceuticals) and indicates a pervasive presence of batch operations.

Control of Food Processes

There are only a few publications specifically dedicated to food engineering and control (e.g. Food Engineering in a Computer Climate, IChemE Symp. Series No. 126, Trans. IChemE, Part C, Bimbenet et al, 1994) but they report a number of interesting modelling and control applications, typically centered on individual unit operations and ranging from classical extrusion cooking control, to control of spray dryers, fermenters, etc. Control methods span from linear and nonlinear model based control to neural nets. A review by Bimbenet and Trystram (1992) identifies four broad objectives of process control in the food industry, shared with other sectors:

- to increase equipment productivity and yields from raw materials.
- to increase human productivity through automation.
- to control product quality.
- to increase operations flexibility.

It also identifies two food industry specific control aspects, arising from: i) the need to meet hygienic requirements (this requires that equipment be periodically shut down for cleaning and sterilization, hence turning even otherwise continuous processes into batch) and ii) the need to account for product perishability, and hence limits on residence and wait times, which dictates segregation into batches and makes production scheduling much more difficult. I would add another one, related to the need for accurately tracking and recording individual sources and operation history of materials that end up into the final products, and not just in a statistical average sense, but for individual packaged items (increasingly this is becoming a legal requirement). Again, this sometimes results in an otherwise continuous process, such as coffee roasting, being artificially split into "batches" (Gallagher, 1995).

Some problems in the control of food operations are posed by the lack of adequate sensors, a limited understanding of the often very complex phenomena involved and the difficulty in predicting physical and other properties, as required for closed loop control.

A second obstacle is posed by the intrinsic complexity of the materials and transformations involved in food processes, and the difficulty in describing them in terms of fundamental models.

Nonetheless, in my personal view many of these are not insurmountable, and proper application of the standard tools of the trade offer vast opportunities both for the control community and the food industry.

In my experience, a further difficulty arises from the cultural background rooted in the industry, which is dominated by biological and food science aspects at the research end, and by the sparse presence and general lack of weight of engineers at the operations end. Manufacturers often rely on equipment suppliers, contractors and systems integrators for design and control expertise.

Type of Control Problems

Given all the above, it is clear that the food manufacturing focus is expected to be for the foreseeable future on safe, efficient and responsive operation of integrated plants producing a variety of products. The pressure on margins requires minimum wastage and high resource utilisation. This demands maximum flexibility from manufacturing operations, hence the ability to handle short product runs, variable batch sizes, frequent, high speed changeovers, robust control of individual batch operations and the ability to quickly introduce new product formulations.

In this environment, control problems which need to be addressed include the design, validation automated implementation and commissioning of control for sequential (discrete and hybrid) systems, the robust optimisation and tracking of batch control profiles in individual unit operations, techniques for monitoring state and performance of individual batch unit operations and overall plant, supervisory batch management, with very much tighter integration of operations planning, scheduling and control into closed loop control of overall manufacturing operations.

Of course, it is imperative that each and any of the activities above should be driven by suitably generic models, and supported by easy to use tools, so that developments can be ported between applications, and investment in new technologies can be discounted over many projects.

The need for tighter integration between manufacturing planning and execution functions is increasingly being recognised (the current banner being Manufacturing Execution Systems — MES). An important point to note is that such integration is not just an exercise in data communication, but that there are serious, challenging and largely unresolved control problems. For example, it is not clear whether there is an optimal degree of decentralisation of control functions (raising the important issues of central vs. distributed planning/scheduling/control), or how autonomous these decentralised "agents" should be.

Current developments on smart control devices suggest that a great deal of self monitoring, retuning, etc. can be distributed much more than at present, but what about the ability of an equipment module, a plant section, or an entire plant, to monitor and (re)organise itself autonomously? Such issues are very interesting both practically and intellectually, and there are hardly any theories around enabling us to deal sensibly with the above questions.

Some work is under way, however, and one such project is described next.

On-line Scheduling Project

This section describes a collaborative project carried out over the last three years, aimed at developing and demonstrating in practice the application of on-line scheduling techniques in the food industry, so as to address some of the above problems. The goal is to provide a tactical capability for production planners, plant supervisors and process operators to accurately schedule the production of flexible batch plants, to implement such schedules in conjunction with existing control systems, and to dynamically respond to process variations or demand changes as they arise.

Project Partners

The project partners are APV Regional Engineering, Northern Foods, Scottish Courage, Safeway Stores and Imperial College. It is perhaps indicative of the rate of change in the industry that three out of five of the partners had a name change since the start of the project due to major restructuring, merger and takeover.

APV, the overall manager of the project, is a leading UK engineering and equipment manufacturing company with worldwide presence in the food industry. It builds turnkey plants for dry and liquid foods, and has substantial in house equipment build capability (e.g. heat exchangers, UHT systems, distillation columns). It was the winner of the 1995 McRoberts award, the UK top prize for innovation (piping on the line our own Speedup entry) for its novel ohmic heating process for aseptic processing of ready meals, dessert products, etc. without the need for additives, freezing or chilling. Plants built by APV produce 80% of the UK milk, 60% of the world butter and 40% of the world ice cream.

APV has about 400 specialist engineers worldwide in automation, supplying control from process units to complete integrated factories. It provides about 1000 of its own ACCOS control systems per annum, about 20% of which were last year integrated in a network. It also operates as a systems integrator for third party control systems.

Northern Foods is a large food manufacturer, with a strong presence in dairy and fruit juices. In particular, the partner in the project is their Eden Vale dairy subsidiary, which operates a number of large, integrated dairy plants. Eden Vale provides to the project one of these plants as a test bed for full scale on-line scheduling demonstration.

A second food manufacturer in the project is Scottish Courage (formerly Scottish and Newcastle), the largest UK brewer. It supplies approximately 30% of British beer production from a number of large plants, in a great variety of products. Scottish Courage brings to the project one of their plants for a modelling and off-line scheduling exercise.

Safeway Stores is one of the big 5 UK food retailers. It has a large number of outlets and a sophisticated distribution system. A 16% increase in the number of their 5000 own-brand products posted in a single year (Argyll annual report, 1993) gives an indication of the rate of change they experience. Safeway's participation to the project stems from their interest in developing closer links with their suppliers (of which Northern Food and Scottish Courage are two) so as to better integrate the planning and scheduling of their respective operations along the supply chain. They already operate sophisticated logistics systems connecting manufacturers, regional distribution centers and stores.

Imperial College is a research and technology supplier to the project, bringing to it its experience and algorithms on batch production modelling, scheduling and control. We also provide a fully automated batch pilot plant for testing and demonstration.

The project is funded mainly by the industrial partners, with a substantial contribution by DTI/MAFF under its LINK Programme for Food Processing Sciences.

The major components of the project include:

- a detailed analysis of the industrial requirements for on-line scheduling, with input from all project partners.
- the further development of data models and scheduling algorithms to meet those requirements.
- the development of a detailed specification of system architecture and design, databases, interfaces, etc.
- software implementation and testing.
- installation and demonstration of the on-line system on the College's pilot plant.
- demonstration of the off-line system on a large beer plant.
- installation and demonstration of the on-line system on a busy dairy plant.
- analysis of management changes required for best use of the technology and of benefits.

The project started in April 1993 and we are presently several months away from its end date. Of the above list, the top 3 items have been completed, the next three are essentially completed, and the on line demonstration on the dairy plant has been initiated. On the last item, a number of aspects are already apparent but a proper assessment is planned nearer the end of the project.

It is worth noting that both manufacturing companies in the consortium have a long experience with production scheduling software. Scottish Courage utilise in particular ProScheduler, a DEC product (Benwell, 1994), while Eden Vale have used for many years Numetrix's Schedulex.

Production Models and Scheduling Technology

The basic approach utilised in the project builds on earlier research work described in Cott and Macchietto (1989). It postulates a three level control hierarchy, consisting of: i) a batch control layer (i.e. one or more standard Distributed or Programmable Logic Controllers) for execution of discrete and continuous control phases on the plant; ii) a supervisory batch control and management layer, responsible for initiating control activities by the lower layer controller according to a current schedule, monitoring their execution in relation to the desired schedule, and updating the schedule to take into account any disturbances; iii) a production planning/optimisation layer, where some of the decision variables relating to the planned production are taken and which communicates with other business systems.

The work in this project specifically addresses the provision of a robust system for the intermediate, supervisory batch control layer. It relies on a rather generic set of modelling primitives. These can be utilised to define a production model and problem for specific applications, in a similar way that a flowsheet program is utilised to define a simulation case. A set of algorithms is also supplied which can take such model as an input and produce an operation schedule.

The modelling primitives essentially include the definition of materials, plant topology networks, product recipes, control resources available at the lower level and procedures required to make the desired materials in the available plant utilising the available control resources. Additional definitions permit describing the initial state of plant/processes, future availability of all resources and a desired production plan. From these an operations schedule is calculated.

The set of material states involved is defined, with generic material classes and specific materials within a class. Materials may have attributes such as an identifier (e.g. on a supplier/customer database), stability (with a time limit), compatibility with other materials, etc. Equipment units represent actual equipment. Units of the same functional type are grouped into equivalent equipment resources. A plant topology is defined as networks of equipment (units or resources) and connections, each with suitable attributes. Equipment resources may be defined as able to carry out process or storage activities, with attributes (batch exclusive, phase exclusive, non-exclusive) defining whether a piece of equipment can be shared or not by different operations (interlocking features). Equipment may also have associated states, which can be used to define/check feasible sequences of operations and compatibility with materials. Common resources, such as operators and utilities may be defined. These can be simultaneously shared between processing tasks. Equipment and Common resources have attributes such as capacity.

Recipes define in an equipment independent way the generic materials, proportions and any resources required for producing a reference quantity of product. The interlock mode for each step may also be indicated. Materials and resource requirements may be parameterized and made batch size dependent, according to a simple linear relation. Similarly, the duration of a step may be specified simply as a constant time or based on constant rate (nominal estimates). Recipes consists in a network of steps, indicating their precedence structure, with additional attributes defining, for example, the wait interval allowed (if any) between steps.

The control resources available to implement procedural control at the lower control level are explicitly defined in terms of "phases". Generic "resource phases" are used to declare how the plant resources can be used to carry out the product recipes using the available controls. For actual execution, such resource phases are specifically instantiated by associating them to specific control "unit phases" running onto specific equipment. Again, full use is made of parameterizations to define which lower level control sequence must be executed, which parameters are to be exchanged for specifying and monitoring the sequence execution and the mapping to corresponding control items in the target control system. This results in generality and flexibility in the definition of the interfaces to lower level control systems.

A desired production is specified in terms of an ordered sequence of procedures (each producing a batch of material), with associated equipment assignment and batch size. The problem solved by the scheduling algorithms is to find the start and end times of all activities required to produce a feasible schedule. This is clearly a subset of a more general schedule optimisation problem, since in principle there are large numbers of alternative assignments, batch orders, and sizes to be explored. Even the limited "feasibility" problem, however, is still rather difficult, in particular in view of the very large number of constraints to be taken into account. The advantage is that solution times are small even for very large problems, which allows the on-line use of the algorithms.

The particular choice outline above is clearly a compromise, but also a deliberate design choice. Optimality is left for the upper decision level, while detailed operational feasibility is aimed at here. Earlier experience indicated that while the number of potential decisions left for the upper level is theoretically very large, in practice it is often reduced, for example, by the common practices of running full batches whenever possible (this is usually also optimal), and of favouring retaining a current sequence order unless there are major disruptions. Similarly, our experience indicates that it is often rather difficult for our industrial friends to formulate hard optimisation objective functions reflecting soft and sometimes shifting "preferences" (for example, in the value to assign to specific customers' satisfaction). On the other hand, particularly in a dynamic operations environment, it rather helps to be able to assure detailed operational feasibility (an imperative nobody argues with) while a few of the more interesting choices are explored interactively and in useful time.

The operation scheduling decisions left for the upper decision level can in fact be taken algorithmically, using mathematical programming or other types of algorithms, if available. Indeed some of our current research addresses this very point.

Finally, the modelling constructs described above broadly conform in substance and where applicable to the standards proposed for batch control by the Instrument Society of America (ISA-SP88).

System Design and Software Aspects

A schematic outline of the SUPERBATCH II system design is given in Fig. 1. It consists of two parts, an off-line scheduler and an on-line version of the same (the Monitor). The current state of plant and production can be used to initialise the off-line scheduler, in conjunction with the plant and production models described above. The off-line scheduler can be used to just import from the Monitor and view the current schedule, to produce one or more altogether new schedules or just variations from that initial schedule (perhaps by including/eliminating batches, defining new equipment, utilities and services, their availabilities, new recipes, etc.). Several users can do this simultaneously and independently. When a satisfactory scenario is produced, a new production schedule can be posted (subject to authorisation) to the Monitor. The latter processes any new requests submitted to it, makes extensive checks for compatibility with the current state of the production, and if this is successful, adopts the new schedule. The Monitor also interacts with one or more control systems, and sends instructions and parameters to them to initiate the required control phases. These are actually executed by the controllers themselves. The Monitor checks the progress of the schedule being executed by scanning status and other defined parameters associated with the control phases. The current schedule is then updated, enforcing all constraints defined in its production models. The Monitor operates non-stop in closed loop at an update frequency of once every minute. The Monitor and the off-line scheduler are fully driven by the (shared) production models and utilise the same scheduling algorithms. The system can be used in off-line scheduling mode alone, if so desired or prior to connection with a control system.

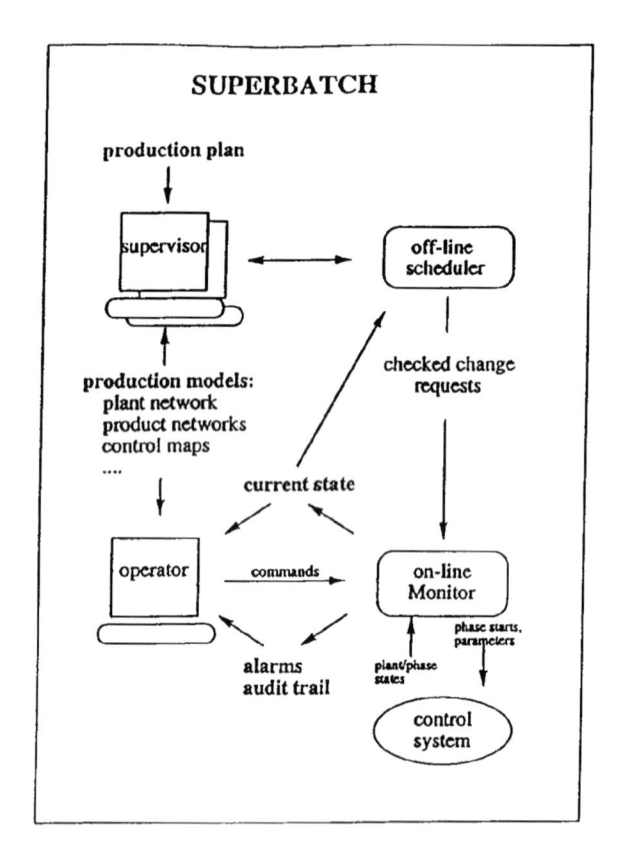

Figure 1. Schematic structure of SUPERBATCH software.

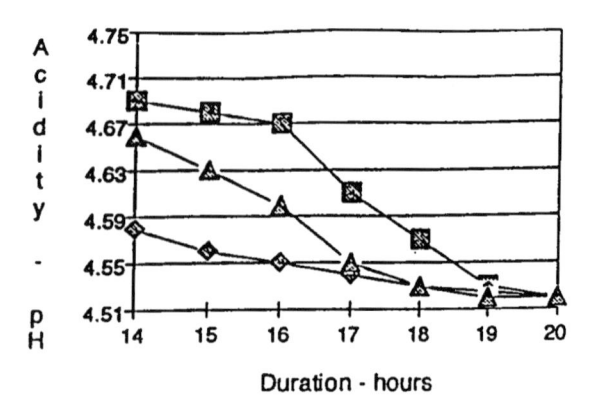

Figure 2. Various incubation times for samples of yoghurt base which matures at pH = 4.55.

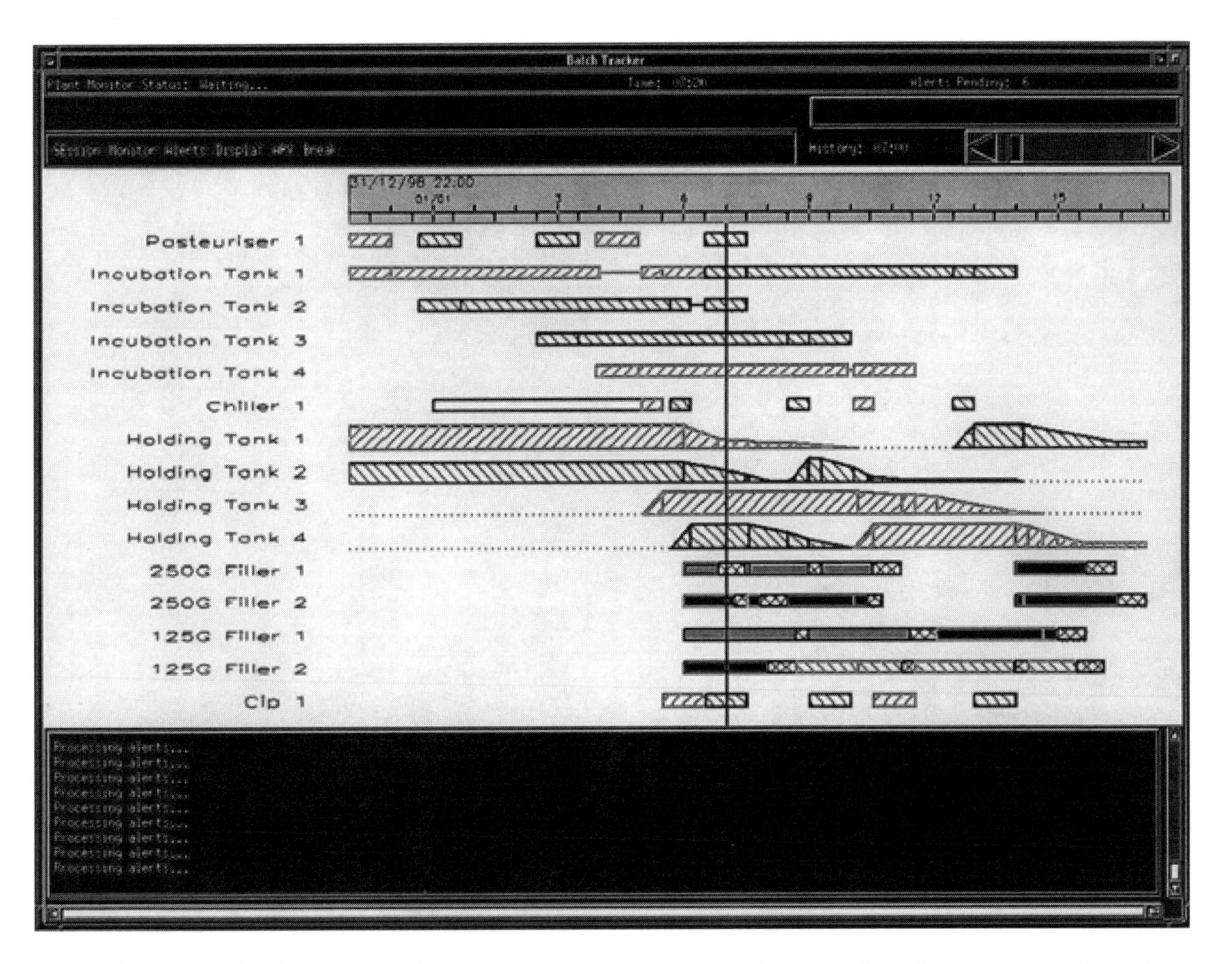

Figure 3. a) Graphical definition of plant equipment and manufacturing procedures for a sample yoghurt plant. b) Schedule for yoghurt plant production, with actual event history before 7:00 am and predicted operation schedule thereafter.

Although the nominal operations models are indeed simple, their parameters may be updated on-line at every update interval. For example, the termination condition for a yoghurt incubation step is that its acidity has reached a certain value (pH = 4.55 for the case in Fig. 2) and a nominal average time may be initially assigned to this step. During the fermentation samples are taken (even manually) and the pH is measured. Based on the lab measures a new time estimate for completing the step can be calculated (even from simple historical charts such as Fig. 2). This can be picked up by the Monitor at its next pass (i.e. within a minute of becoming available) and used immediately to update the entire future schedule). Of course this estimation can also be done algorithmically, especially in the presence of an on-line pH meter. And of course, a proper controller of the fermentation could also be developed to eliminate such variability in the first place!

The supervisory control software communicates with control systems (and other applications) in a client-server mode utilising a specially developed piece of software, called Control Link. Each application is interfaced to Control Link through a well defined protocol. This way

SUPERBATCH can also be utilised to run dynamic simulations of a plant running elsewhere in a network, or carry out sophisticated numerical computations (e.g. to improve on-line parameter estimates as for the yoghurt example above, for more demanding optimizations, for model based control), while sharing data amongst clients. Some details and examples are given by Liu and Macchietto (1995).

The basic SUPERBATCH models and software engine have been incorporated by APV in a proprietary application, called BATCH MANAGER, which features a number of extensions. These include highly interactive graphical user interfaces for problem definition, schedule viewing, powerful model and version management, etc. (some of the features and a small example are described in Jakeman, 1995). They also include facilities for accepting orders, breaking them down into individual batches, and posting them to the scheduler while tracking their origin and association to customers, etc. A Gantt chart display is utilised to describe a schedule (Fig. 3, from Jakeman, 1995). The on-line version has a moving vertical bar separating historical information (on the left of the current time) from the future schedule (on the right). Each item on the chart is an object and its characteristic attributes (e.g. type of control phase, material being processed, start times, transfer rate, etc.) may be easily viewed. The plant component attributes on the equipment network diagram may be similarly interrogated, while other facilities permit tracing material movements, the relation between materials and recipes, etc.

Behind all this, all plant, recipe, materials, production network and other data are stored in an SQL database, which permits the easy configuration of reports.

All SUPERBATCH software is implemented in C and runs on UNIX workstations (SUN and IBM RS6000), supporting multiuser access over a network, in a client-server mode. The Batch Manager user interfaces can also be run on networked PCs.

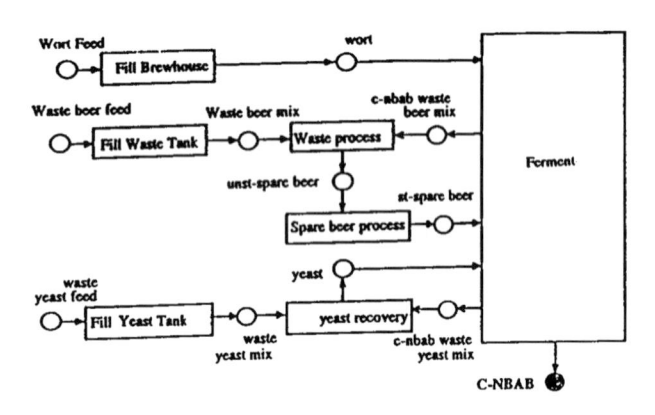

Figure 4. Schematic representation of process for making beer product C-NBAB.

Applications

Earlier off-line versions of SUPERBATCH and Batch Manager have been used at APV for routine engineering work for over four years. Much of this use has been for developing multipurpose food plant designs, while taking into account detailed operations schedules. Over 100 large scale projects have been completed to date, including both grassroot designs and retrofit projects.

The extended versions of the software produced in this project are at present under consolidation and testing by the project partners.

A multipurpose batch pilot plant facility (Macchietto, 1992) has been built and utilised to demonstrate the on-line version of the software operating in closed loop. This application is discussed in some detail in Liu and Macchietto (1995). Much development work has been done utilising a detailed dynamic model of the plant, implemented in the gPROMS simulator (Pantelides, 1996) and interfaced to the control system itself.

The two industrial applications developed as part of this project are rather challenging and are therefore discussed in some detail here.

Industrial Case 1 — Beer Manufacturing

Beer is made in a number of main processing stages, including wort preparation, fermentation, maturation and packaging, of which fermentation is the most critical. Wort prepared in a brewhouse (lasting a few hours total) is blended with yeast and fed to a large number of very large fermenters, where the yeast causes the alcoholic fermentation (lasting several days). The main product is then removed from the fermenters, centrifuged, and sent to maturation vessels, prior to packaging in bottles or cask. Depending on the type of fermenter used and type of beer produced, a mixture of beer and yeast (waste beer) may also be produced, from which the beer is recovered. Upon pasteurisation in a sterilizer, this "spare beer" is available for blending in various percentages with some of the main products. Some of the yeast, which grows during fermentation, is recovered and reused for subsequent batches. There are therefore two main recycle loops (spare beer and yeast), which are very important for a smooth operation. The process up to the maturation stage is shown schematically in Fig. 4.

Different yeasts are utilised to produce different beers, the quality of which is also determined by the fermentation and cooling profiles in the fermenters, according to the product recipe. There is a large amount of variability here, and several corrective actions can be taken to ensure strict constraints are met on flavour, colour, and alcohol content. These range from small changes in the cooling profiles, to rather larger changes involving significant extra processing time, to redirecting the beer to a different product in more serious cases, likely requiring re-blending with other beers. Different yeasts must be carefully segregated to avoid cross contamination, and so must spare beers for different products. Over 25 different types of beer are produced in this plant.

In the plant studied here, the raw material supply does not present a problem, and sufficient capacity is also available for the maturation stage, so these two steps are not a bottleneck and it was sufficient to model them in a simplified way. However, due to the very large number of operational constraints posed by the recipes, plant size and structure, storage and recycles of different materials, quality requirements, and frequency of variations arising, scheduling and managing the fermentation operations is extremely complex. The critical fermentation, spare beer and yeast recovery portions of the process were therefore modelled in detail. The total number of number of major equipment considered is over 100. The overall plant diagram is confidential, but a small portion of it is shown in Fig. 5. The equipment is located in several rooms, is connected by a network of shared pipework, and shares a number of services, including staff and CIP. These constraints were also modelled in detail.

The plant is well instrumented (accurate record keeping being mandated by Tax and Excise regulations) and already fully computer controlled. However all major operations at the supervisory control level are currently initiated and monitored by expert brewers.

Equipment, recipe and production models were developed in conjunction with company personnel, including planners, master brewer and operations personnel and checked in site visits. Actual orders were also collected for a typical period of operation, and a complete production schedule developed using Batch Manager. A sample schedule is shown in Fig. 6 (kept deliberately small for confidentiality reasons). The good accuracy of the model (for nominal conditions and allocation choices as in the actual operation) was confirmed through discussions with company personnel.

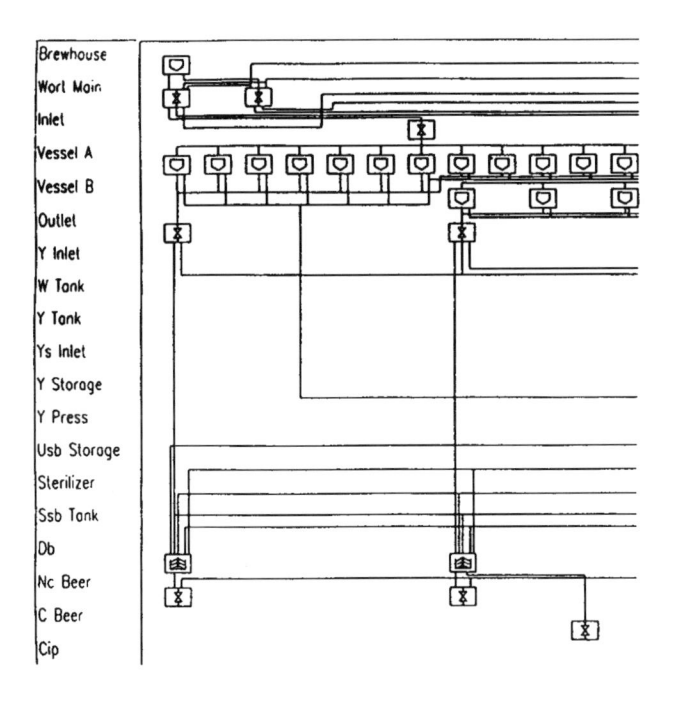

Figure 5. Partial flowsheet of beer plant.

The validated model is now being used to highlight bottlenecks, to explore the effects on production of common operational problems and to develop and assess alternative operation choices at the supervisory level (e.g. with respect to yeast recovery and recycle, fermentation problems, etc.). Finally, some structural changes are also being examined, to assess the viability and economic return of capital investment decisions. For example, the model has shown that the same production as in the reference case can be scheduled in approximately 9 days instead of 11 if the piping network for use of available Cleaning In Place stations is modified only slightly.

The study has shown that the Batch Manager model can handle well production scheduling problems of large scale and complexity. The study was carried out over three months, mainly by a post-doc with experience in modelling but no previous experience in Batch Manager or beer (other than as an end user). Most of the effort was spent on data collection and initial model building and validation.

Industrial Case 2 – Dairy Manufacturing

The plant selected for the on-line demonstration is a fromage frais plant. Such exotic continental dairy products have been introduced in the UK only fairly recently and have rapidly become popular. The fromage frais plant is one of four in an integrated dairy facility processing about 1 million liters per day. An upstream milk plant produces skim milk, which feeds a large cottage cheese plant (covering a rather large fraction of the UK market), the fromage frais plant and a cream plant. This arrangement permits balancing wild seasonal swings in individual product demands while retaining an overall constant milk intake.

The fromage frais process is straightforward in its broad terms. Proteins are first concentrated in the skim milk feed by ultrafiltration to give a "sweet retentate," followed by incubation lasting several hours in incubation vats and by other heat treatment and separation steps to make a base. Late product differentiation is achieved by blending the base with cream and fruit in various qualities and proportions. Finally the fromage frais is filled in various size pots and tray combinations and kept in cold storage until despatch. Approximately 100 distinct packaged products are produced (the manufacturer's plus retailers' brands), including 25 different flavours of strawberries! Each brand and retailer jealously guards its own recipe and there are strong pressures for even further differentiation.

The specific process used is novel and was pioneered on this plant, so details of the equipment and production cannot be given for confidentiality reasons. Nonetheless, a few general aspects which can be mentioned show that this is a very interesting scheduling application.

Particular difficulties arise from the natural variability in the properties of the milk (both in type and protein content), which makes incubation performance rather variable. The filling machinery is mechanically complex and subject to breakdown and stoppages. Holding tanks are used flexibly to store different intermediate materials at different stages of processing, with some recycles. As with beer, the connectivity of the plant imposes significant constraints, as do the cleaning requirements. Finally, the intermediate base is highly perishable and can be stored for no more than a very few hours. On the other hand, the lead times for supplying the fruits are up to three weeks and shipments do sometimes get delayed. Most interestingly, the overall processing times and commercial pressures are such that production must be committed, based on best forecasted estimates, well before orders arrive. The production schedule in progress is then adjusted on the fly when actual orders are confirmed. This makes scheduling of the plant very dynamic and challenging.

The plant already had a full set of control systems, but in preparation for the on-line trials, a full check of instrumentation and measurements was made. New instrumentation was installed to provide additional information on the state of plant and processes (e.g. levels of tanks, number of product trays filled) and to permit a much more accurate control of critical ingredient flows.

A scheduling model of the plant and its operation analogous to that described for the beer application was developed. This has started to be used in off-line mode at the plant to aid re-scheduling decisions at the daily operations review meetings, in particular to schedule the fillers operation.

Current development work includes testing for on-line operation in closed loop but in "simulation" mode. The current on-line scheduler has been interfaced to an exact copy of the actual full plant control software running on a duplicate control system at APV. Interactions between the on-line scheduler and the distributed control system are thus as in the final application, however signals from the controller are sent to a screen rather than to the actual control devices. From the screen the system response (e.g. completion of a control step, abnormal conditions, etc.) can be entered manually. This permits a thorough step-by-step testing and familiarization with concepts and software by the plant planners and operators in a realistic but safe environment. This activity is ongoing at present.

A parallel step is the development of a model predicting the duration of incubations from laboratory analysis data of samples taken manually approximately every hour during this process phase. The improved phase duration estimates will be used for the early update of schedules, as described for the yoghurt incubation example in a previous section.

Full commissioning of the on-line scheduler on the actual plant is planned for the first quarter 96 and all results so far are very encouraging. In the meanwhile, use of the Schedulex software for this plant has been discontinued by the company.

Conclusions

The food and drink industry is larger than other sectors where control ideas are typically developed and exercised. There are many opportunities for control applications at all levels, from the control of individual unit operations, to the control of entire supply chains.

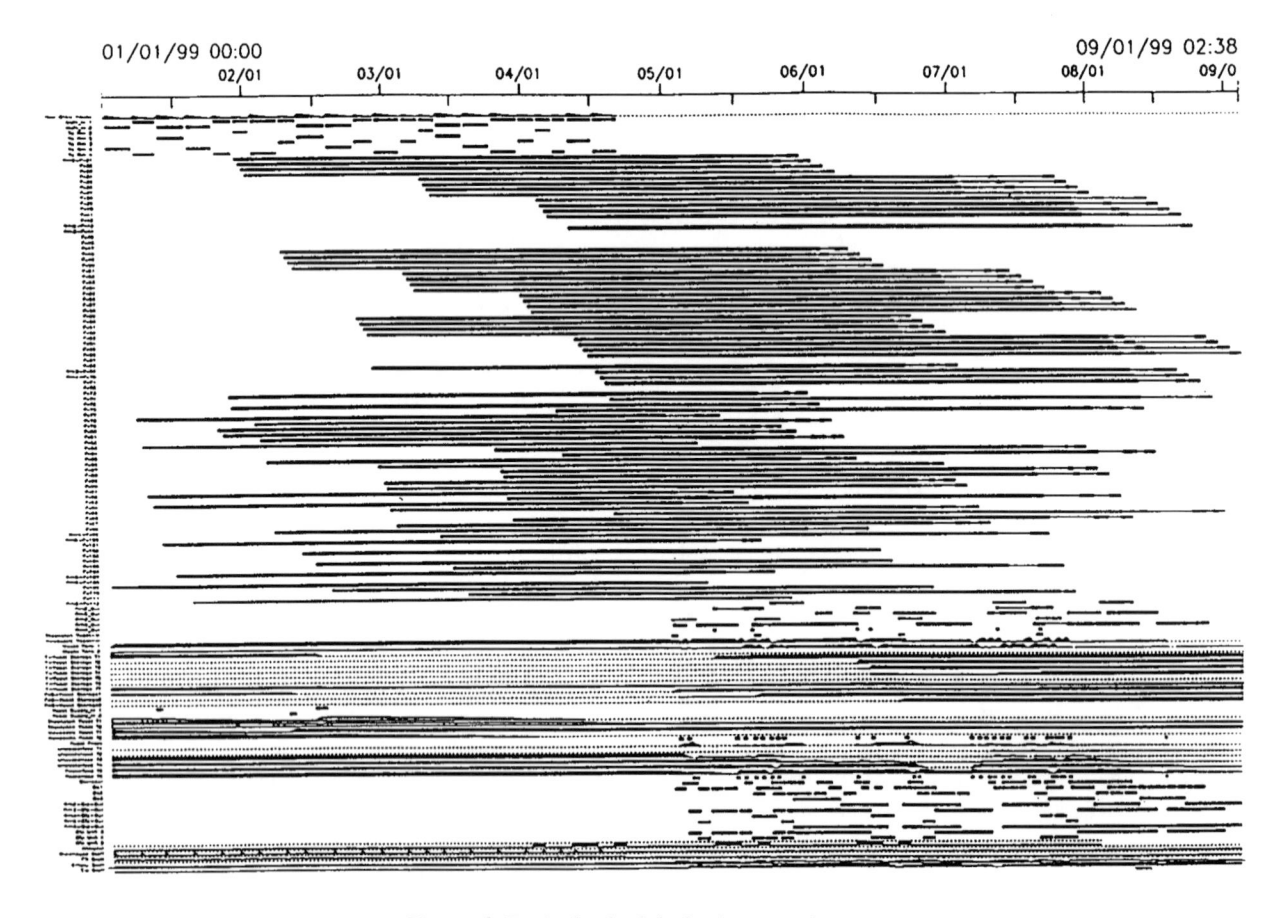

Figure 6. Typical schedule for beer production.

Some particular problems at the unit operation level are posed by the lack of adequate sensors, the limited understanding of the often very complex phenomena involved and the difficulty in predicting physical and other properties, as required for closed loop control. Some further problems arise from the cultural background rooted in the industry, which is dominated by biological and food science aspects at the research end, and by the sparse presence and lack of weight of engineers at the operations end.

There are very strong business drivers demanding to view the control of food manufacturing facilities in a broad rather than narrow sense, as plantwide, integrated systems. Specific requirement are for manufacturing facilities to respond to rapid product and market innovation, to produce flexibly a variety of products, while making efficient use of material and equipment resources.

Plantwide food production control requires the normalization and automation of supervisory batch control functions so far performed manually, and the development of what are now being called Manufacturing Execution Systems, to support effective production planning, on-line scheduling and control.

A project was presented aimed at developing such a system suitable for large scale multipurpose batch plants. The consortium-based project is providing a demonstration of the concepts and software tools on two diverse and challenging industrial applications, which were described in some detail.

Although the study is not yet completed, results so far show that models, algorithms, system designs, and software implementations are applicable to industrial production facilities of large size and complexity, and that their use leads to significant benefits.

Acknowledgments

The work described was supported, financially and otherwise by EPSRC, DTI/MAFF (LINK Programme for Food Processing Sciences) and the industrial partners of the On-line Scheduling project: APV, Scottish Courage, Northern Foods and Safeway. I am particularly indebted to Bill Kirkland and Chris Jakeman, APV for many useful discussions and challenges over the years, and for kindly passing on many of the insights on the food industry contained in this paper. The SUPERBATCH and brewing modelling work were very professionally carried out more recently by Terrence Crombie, Edwin Hutton and Zhenhai Liu, with previous contributions from many others. Their contribution is gratefully acknowledged.

References

Benwell, N. (1994). Organising a Brewery. *DEC User*, January, 37-38.

J.J. Bimbenet, E. Dumoulin and G. Trystram (Eds) (1994). Automatic Control of Food and Biochemical Processes, Developments in Food Science, **36**, Elsevier.

Bimbenet, J.J. and G. Trystram (1992). Process Control in the Food Industry. *Trans. IChemE*, **70**, Part C, 115-125.

Booty, F. (1993). Missing in Action. *Computer Weekly*, September, 32-33.

Bruin, S. (1992). Integrated Process Design: Issues and Opportunities. *Trans. IChemE*, **70**, Part C, 126-130.

Cott, B. J. and S. Macchietto (1989). An Integrated Computer Aided System for the Operation of Batch Plants. *Comput. and Chem. Eng.*, **13**(11/12), 1263-1272.

Gallagher, J. (1995). Nestle, Personal Communication.

Jakeman, C. (1995). On-line Scheduling in the Food Industry. Proceedings, 7th IFAC/IFORS/IMACS Symposium on Large Scale Systems: Theory and Applications, London, July 11-13.

Liu, Z. H. and S Macchietto (1993). Cleaning in place policies for a food processing batch pilot plant. *Food and Bioproducts Processing Journal*, September issue, 194-196.

Liu, Z.H. and S. Macchietto (1994). Model based control of a multi-purpose batch reactor with a variable control structure. In J.J. Bimbenet, E. Dumoulin and G. Trystram (Eds), *Automatic Control of Food and Biochemical Processes*, Elsevier, 337-344.

Liu, Z. H. and S. Macchietto (1995). An advanced operation support system for batch plants. Workshop on Analysis and Design of Event-Driven Operations in Process Systems, London, April 10-11.

Macchietto, S. (1992). Automation research on a food processing pilot plant. In Food Engineering in a Computer Climate, *IChemE Symp. Series No. 126*, Hemisphere Publishing Co., 179-189.

Pantelides, C. C. (1996). gPROMS — An Advanced tool for process modelling, simulation and optimisation. Proc. Chemputers 96 Conf., Frankfurt, McGraw Hill.

Sawyer, P. (1993). Computer Controlled Batch Processing. IChemE, Rugby, UK.

Technology Foresight Report (1995a). Manufacturing, Production and Business Processes. HMSO Publications Centre, London, UK.

Technology Foresight Report (1995b). Food and Drink. HMSO Publications Centre, London, UK.

CONTROLLER PERFORMANCE MONITORING AND DIAGNOSIS: EXPERIENCES AND CHALLENGES

Derrick J. Kozub
Shell Development Company
Houston, TX 77251-1380

Abstract

The goal of achieving reliable and profitable automated control applications in the chemical process industries requires controller performance monitoring and diagnosis technology. Research in process control has mainly focused on design techniques while giving little attention to troubleshooting and maintenance of installed systems. Due to the many assumptions and compromises made during design, and the many changes that occur in our plants, controllers seldom work as designed. Performance monitoring and diagnosis technology is a way of guiding engineers in real time as to the need for corrections and the type of corrections needed. In this paper, the industrial goals and requirements for technology in this area are defined. A review of some of the significant research in this area is presented, along with a list of suggestions for future work. Examples with industrial data are also presented to illustrate our experience using existing theory from past research, and the technical challenges that still remain. The paper demonstrates that there exist tremendous research opportunities related to controller performance monitoring and diagnosis, and that advancements could have a major impact in the highly automated chemical process industries.

Keywords

Controller performance monitoring and diagnosis, Time series analysis, Stochastic control.

Introduction

The goal at Shell is to arrive at 100% reliable process operation achieving maximum profitability with minimal cost. The application of several key technologies have been identified in order to meet this goal. Included in these technologies is Automated Process Control (APC).

A plethora of technologies related to the design of automatic feedback control algorithms are available. Traditional approaches require the use of identified transfer function models relating manipulated inputs to the controlled outputs, and models for the disturbance characteristics (either identified, specified, or implicitly assumed). Based on this modeling information, a controller design strategy is chosen, and tuning parameters are selected which meet some closed-loop performance criteria for a set of simulation case studies. The controlled output may yield on-line performance characteristics that are significantly different from what is ultimately desired. Reasons for this discrepancy include:

- errors in the input/output transfer function models.
- errors in the assumed magnitude and dynamic trend characteristics associated with the loop disturbances.
- process nonlinearities and changing operating conditions.
- poor choice of tuning parameters relative to the process performance objectives.

Process control engineers have the responsibility of ensuring that control applications are performing according to specifications. It is often difficult, if not impossible, to effectively monitor the performance and diagnose problems associated with these applications from raw data trends. The data trends are often characterized by complicated response patterns that can result from the presence of persistent and multiple sources of measurable and unmeasurable disturbances, process noise, time variant response phenomena, and nonlinearities. To further complicate the matter, control engineers are often responsible for hundreds of applications, making the analysis of raw data virtually unmanageable. As a consequence of this, poorly performing applications can often go unnoticed for a lengthy period of time before some corrective action is taken.

As the discussion above indicates, controller performance monitoring and diagnosis plays a crucial role in the success, utilization, and maintenance of control strategies, and is one of the most frequently carried out activities by plant location engineers. It is an activity which influences implementation costs the most. Given the importance of the task, it is surprising that very little attention has been given to this area by the research community with respect to developing tools and methodologies that are effective in both meeting the goals in this area, and dealing with the complicated characteristics of industrial response data.

The goal of this paper is to provide a vision and requirements for automated controller performance monitoring and diagnosis technology in the Chemical Process Industries (CPI). A review and demonstration of significant published results with industrial data will be presented. Both the value and limitations of these approaches are discussed along with areas for further development. The paper concludes by indicating that there exist tremendous research opportunities and challenges in the area of controller performance monitoring and diagnosis in the CPI.

Performance Monitoring and Diagnosis: Goals and Requirements

The goal is to develop automated, on-line technology that provides the necessary information for determining if performance targets are being met by the controlled process variables. The technology should be easy to use and interpret. Example operating targets might include:

- ± limits of error from set point deviation.
- ± limits on manipulated input moves.
- standard deviation (or variance) specifications
- closeness of the output error standard deviation(or variance) to the best that can be achieved through feedback compensation given the existing disturbance and process characteristics. This is essential for differentiating between controller and process problems.
- effectiveness in constraint enforcement.

- for the case of surge volume control, effectiveness of the controller in making use of the available surge volume capacity.

It is also important to determine if the dynamic response characteristics associated with the controlled output are satisfactory. Useful information would include:

- speed of response of the controlled outputs returning to their set points.
- time response characteristics of the output error and input manipulations.
- extent of underdamped, oscillatory, and overshoot behavior.
- extent of dynamic interactions for the case of multivariable and/or multi-loop control configurations

The monitoring technology should also serve as a useful aid for the prioritization of control problems, for assessing the effect of controller design or tuning modifications, and predicting the economic incentive for controller applications.

The goals for diagnosis are to analyze process data, and assimilate this information along with process knowledge and models (if available) to arrive at root cause problems, and recommendations for corrective actions. The technology should provide information on the relative contributions and dynamic response trends associated with both set point changes (if present) and measurable disturbances. It would be very useful for the technology to isolate problems associated with the performance of the control algorithms from those associated with the process. Example root cause problems associated with control would include:

- poor tuning.
- poor / no feedforward compensation.
- input saturation & constraint problems.
- manipulated input and disturbance transfer function error in the models used to design the controller
- poor choice for the sampling interval.
- poor selection of input/output pairings

Example root cause problems that lead to poor output response characteristics but are not the fault of the control algorithm may include:

- abnormal characteristics associated with the disturbance upsets such as a severe oscillatory trend.
- faulty sensors and/or actuators.

For the monitoring and diagnosis technology to be useful, it must, in general, be robust and applicable to the characteristics that can be exhibited by industrial data. Some of the characteristics normally encountered in practice include:

- data that is difficult to analyze by visual inspection due to complicated response trend characteristics.
- the presence of significant process noise (not necessarily white nor Gaussian).
- the existence of both stochastic and event driven, deterministic disturbances, some, or all of which may not be measurable.
- time variant response characteristics resulting from changing process conditions.
- multivariate systems with the possibility for significant colinearity

The technology should make use of both routine operating data and any data generated from planned testing. Indication of poor performance should first come from routine operating data. As much diagnosis as possible should first be carried with the routine data. If a root cause(s) cannot be determined from the routine data then the action of introducing intrusive plant testing may be justified. Plant testing can be carried out by the introduction of random dithering at set points, manipulated inputs, and disturbances (if possible). It is important to note that one should not rely solely on planned test data, especially generated from set point changes, to evaluate controller performance. The ultimate performance of a controller depends on the dynamic characteristics associated with the dominating loop disturbances, which could deviate substantially from deterministic upsets introduced from a planned test.

The long term goal of meeting the objectives stated above will likely require the integration and enhancement of technologies such as robust control design, stochastic control, Statistical Process Control (SPC), time series analysis, dynamic process modeling, and artificial intelligence technologies. In this paper, we will focus mainly on SISO control for the sake of simplicity in introducing the basic problem, and because most of our applications experience so far has been related to technology developed for SISO control. It is important for the reader to bear in mind that meeting all of the objectives for the more general multivariate, and multi-loop problem is the ultimate aim, and where many research challenges lie ahead.

A Review of Controller Performance Monitoring and Diagnosis Technology

In this work we consider the case where the process is described by a linear discrete time transfer function model

$$y_t = P(z^{-1})u_t + \sum_i D_i(z^{-1})d_{i,t} + v_t \tag{1}$$

where, at sampling interval t, y_t is the output, u_t is the manipulated input, $d_{i,t}$ is the i'th measured disturbance, and v_t represents the net additive effect of noise and unmeasured disturbances at the output. $P(z^{-1})$ and $D_i(z^{-1})$ are rational discrete time transfer functions corresponding to the manipulated input and measurable disturbance i respectively, and z^{-1} is the backwards shift operator. The manipulated input is determined by the feedback/feedforward controller

$$u_t = C(z^{-1})e_t + \sum_i C_{f,i}(z^{-1})d_{i,t} \tag{2}$$

$C(z^{-1})$ and $C_{f,i}(z^{-1})$ are the feedback and feedforward controller transfer functions respectively, and e_t is the output error from set point (s_t), given by

$$e_t = s_t - y_t \tag{3}$$

From these equations, the closed loop output error response can be shown to be

$$e_t = \frac{s_t - \sum_i (D_i(z^{-1}) + P(z^{-1})C_{f,i}(z^{-1}))d_{i,t} - v_t}{1 + P(z^{-1})C(z^{-1})} \tag{4}$$

The net dynamic response trend for e_t (4) can be represented by an Autoregressive Moving Average (ARMA) or high order Moving Average (MA) time series model (Box and Jenkins, 1976) of the forms

$$e_t = \frac{\theta(z^{-1})}{\phi(z^{-1})}a_t = (1 + \psi_1 z^{-1} + \psi_2 z^{-2} + \psi_3 z^{-3} + ...)a_t \tag{5}$$

The identification of the dynamic response characteristics associated with (4) and (5) is one of the main goals for monitoring and diagnosing variables where feedback/feedforward control is applied.

So far a minimal, yet significant, research effort has been carried out in developing statistical, or SPC technology, for monitoring process variables where automated process control is applied (DeVries and Wu, 1978; Harris 1989; Desborough and Harris, 1992; Stanfelj et al., 1993; Desborough and Harris; 1993; Harris et al. 1995; Tyler and Morari, 1995; Rhinehart, 1995). In the literature cited, time series analysis and/or stochastic control technologies were used to arrive at SPC tools to monitor performance. The next sections will provide a review and discussion on the value of some of this work. Feedback on where technology advancements are needed will also be provided.

Performance Analysis using the Auto-correlation Function

Astrom (1970), Harris (1989), and Stanfelj et al. (1993) have reported the use of the estimated auto-correlation function (Box and Jenkins, 1976) for the characterization of the dynamic response characteristics associated with the output error from set point trend, and for the evaluation of controller performance. The motivation for the use of the auto-correlation function is that it can be easily estimated from plant response data, and that the dynamic response characteristics for data trends, such as output error from set point, can be inferred without having to resort to the more complicated tasks associated with the identification and interpretation of time series models.

The use of the auto-correlation function by the researchers stated above focused mainly on testing if the conditions for minimum variance feedback control (Box and Jenkins, 1976; Astrom, 1970) were being satisfied, or the nearness of controller performance to this condition. The output error from set point (e_t) response under the conditions of minimum variance feedback control is described by the moving average time series model of order f (Harris, 1989)

$$e_t = \left(1 + \psi_1 z^{-1} + \psi_2 z^{-2} + \ldots + \psi_f z^{-f}\right) a_t \qquad (6)$$

where f is the sampling intervals of delay between the manipulated input and output (obtained from the model identified for $P(z^{-1})$). The moving average coefficients ψ_i are determined solely by the open loop dynamic characteristics of the net effect of disturbances and noise at the output, and the set point change trend (Harris, 1989). The condition for minimum variance feedback control can be detected when $r_e(k)$, the auto-correlation coefficient for the sampled output error (e_t) at sampling lag k, is statistically equal to zero at lags k>f, where f is equal to whole discrete sampling intervals of delay. Hence, auto-correlation analysis on e_t is a very straightforward test to check for the condition of minimum variance feedback control. If this condition is satisfied, no modification to the feedback controller can be made which would yield a lower output error variance with respect to rejecting the net effect of the disturbance and set point change trends in the data. If the variance must be reduced in order to meet the process target, it must be accomplished by reducing the source of disturbance variability and/or through the introduction of feedforward compensation if the disturbance delays are equal or larger than the manipulated input delay.

In many practical circumstances, depending on the process and the disturbance characteristics, the rating of output error trend characteristics relative to the condition of minimum variance is neither practical, nor achievable. In such cases, a different auto-correlation pattern, $\rho_e(k)$, can be used as an alternative to the one represented by the condition of minimum variance for judging the effectiveness of the control strategy. For example, suppose one would consider a first order exponential output error decay trend given by

$$e_t = \frac{1}{1 - \lambda z^{-1}} a_t \qquad (7)$$

to be acceptable, with λ given by

$$\lambda = \exp\left(-\frac{T}{\tau}\right) \qquad (8)$$

T is equal to the sampling frequency, and τ is equal to the first order response time constant. The time series model (7) can be shown (Box and Jenkins, 1976) to have an auto-correlation pattern given by

$$\rho_e(k) = \lambda^k \qquad (9)$$

$\rho_e(k)$ can be compared to $r_e(k)$ to determine if the trend for e_t is acceptable. This procedure can be extended to establish auto-correlation function targets and acceptable bounds for any closed-loop output error response transfer function specification. This approach can be used as an SPC test for performance monitoring in practice.

Based on our experiences using auto-correlation analysis in practice, we've found that the tool can simplify the analysis of complicated output error response trends. Underdamped behavior, overshoot, and speed of response are just some of the characteristics that are easily inferred from the auto-correlation patterns. The ability of this tool to detect closeness to the condition of minimum variance feedback has also been valuable. We've often run into cases where excessive variability from operating targets were wrongly attributed to a poor controller design when in fact the controller was yielding performance close to minimum variance feedback. Hence, the auto-correlation function has been found to be helpful for separating problems associated with a poor controller design from bad process operating conditions that cannot be corrected by better feedback control.

We've also had some negative experiences using auto-correlation analysis to monitor output error trends, especially when the performance specifications were not close to minimum variance. In general, the output error auto-correlation pattern is significantly different from the output error time response trend, particularly at early lags. Unfortunately the two trends have often been mistaken to be equal by engineers in practice. Another problem is that we've run into cases where significantly different time response characteristics yielded auto-correlation functions which look very similar. Hence, we've had some experiences where an incorrect conclusion was drawn by engineers relying solely on auto-correlation patterns. Examples will be shown at the end of the paper to illustrate these points.

In practice we have found the auto-correlation function to be more useful than just characterizing the response characteristics associated with e_t. Application of the auto-correlation function to input move response data can be very helpful in determining if the manipulated input dynamic response is well tempered, or unfavorable, as for example the cases of excessive cycling or ringing. With respect to diagnosis of controller problems, the auto-correlation function has been valuable when applied to measured disturbance input trends. For example, auto-correlation analysis can be used to diagnose unfavorable dynamic response characteristics associated with measured disturbances (e.g. severe periodicity). This type of problem would normally be corrected at the source and not through controller modification. Moreover, auto-correlation analysis is very useful for testing if the dynamic trend characteristics associated with measured disturbances are suitable for effective feedforward compensation. We've had many experiences where the introduction of feedforward compensation was harmful because of the incorrect application of the classical deterministic random step assumption for the measured disturbance trend.

Spectral analysis approaches have also been proposed as alternatives to auto-correlation analysis (Desborough and Lane, 1992; DeVries and Wu 1978; Kendra and Cinar, 1996). Spectral methods essentially provide the same information as the time domain auto-correlation analysis (Box and Jenkins, 1976) but with the dynamic response characteristics expressed in the frequency domain. Our experiences using spectral approaches have been less successful than the time domain characterizations since the latter is far more easily interpreted by practicing process control engineers.

Time Series Modeling of the Output Error Trend

The fitting of an ARMA time series model of the form (5) directly to measured closed loop output error response data can also be carried out to identify the dynamic response characteristics associated with e_t. The pulse response of the fitted time series model provides an estimate of the average time response of the output error from set point. From the pulse response, information related to dynamic response characteristics, such as settling time, amount of overshoot, amount of cycling, and closeness to the condition of minimum variance control, can be easily extracted. The e_t pulse response from the estimated model can be compared to the pulse response for a specified desired response trend for e_t, such as (7), in order determine if the output error response characteristics are acceptable.

Relative to auto-correlation analysis, we've found the time series modeling approach more informative and easier to interpret since explicit time response information is provided. The drawback with the model based approach is that expertise is required to select a suitable ARMA model structure, and estimate parameters. This task must be automated for this approach to be useful in industrial practice.

Single Number Controller Performance Statistics

ISE and output error variance have been commonly used statistics for monitoring the effectiveness of a control strategy. Such statistics are, without doubt, important with respect to rating the overall process performance. However, from the perspective of controller performance monitoring such statistics can be misleading for two reasons. First, unacceptable output error variance from set point can be a result of a large magnitude associated with the disturbance upsets. Secondly, with respect to the quality of feedback control, a variance or ISE statistic is essentially meaningless unless it can be compared to the best that can be potentially achieved through modification of the existing feedback / feedforward strategy.

In light of the above, several researches (DeVries and Wu, 1987; Desborough and Harris, 1992; Stanfelj et al., 1993; Kozub and Garcia, 1993) have proposed alternative statistics which basically amount to the normalization of the closed-loop output error variance. This is accomplished by defining some ratio of the output error variance with the potentially best achievable variance for the measured conditions under minimum variance control. In arriving at such a monitoring statistic, we shall adopt in this paper the convention of DeVries and Wu (1978), and will name it the Closed-Loop Potential (CLP) Factor defined as

$$CLP = \frac{\sigma_{mv}^2}{\sigma_e^2} \qquad (10)$$

where σ_{mv}^2 is the variance that can be potentially achieved by the application of minimum variance feedback control with respect to the disturbance characteristics in the measured data, and σ_e^2 is the existing output error closed-loop variance. CLP can take on a number from 0 to 1. When CLP is equal to 1 the best achievable feedback control is being applied relative to both the process and disturbance characteristics. When CLP is equal to 0, no effective feedback control is being applied to drive the output error to zero. Typically, CLP will take on some value between 0 to 1, and its nearness to zero, along with the value for σ_e^2, provides an indication for the incentive to improve the feedback controller design. If CLP is close to 1 and σ_e^2 is unacceptably large, the statistic indicates that improvements need to be made by either the reduction of the source of variability, and/or through the introduction of feedforward compensation.

The estimation of CLP requires an estimate for σ_{mv}^2. The procedure for estimating σ_{mv}^2 given the input/output sampling intervals of deadtime (f) is provided by Harris (1989), and is equal to

$$\sigma_{mv}^2 = (1 + \psi_1^2 + \psi_2^2 + \ldots + \psi_f^2)\sigma_a^2 \qquad (11)$$

where the ψ_i's and variance of a_t (σ_a^2) are obtained by fitting a time series model of the form (5) to e_t response data.

As was the case with auto-correlation analysis, the rating of controller performance against the condition of minimum variance feedback may not be practical in many circumstances. In such cases a minimum bound for acceptable performance can be specified for CLP and compared against the estimated one as a straightforward SPC type test. For example suppose that an acceptable closed-loop response bound was given by (7). It can be shown that (Box and Jenkins, 1976)

$$\sigma_e^2 = \frac{1}{1-\lambda^2} \sigma_a^2 \qquad (12)$$

and that σ_{mv}^2 is equal to σ_a^2 with f=0. Hence, a bound limit for CLP in this example would be given by

$$CLP = 1 - \lambda^2 \qquad (13)$$

If the estimated CLP falls below this bound the performance of the controller would be considered poor. The approach taken in this example could be easily generalized to any type of response specification made for e_t.

So far we've found CLP to be a useful first pass indicator that a problem may exist with a controlled output. However, some cases were observed where the information provided by CLP was misleading. This is expected given that two very different output error time response patterns, such as a cycling and an overdamped error response decay, can yield similar variances, and likewise, CLP's. Hence, one should not solely rely on CLP to monitor controlled variables. Given that the calculation of CLP depends on the specification of the input/output deadtime (f), it is important to have a good estimate for f to prevent the result from being misleading.

Harris (1995) has indicated that the class of statistics similar to CLP can be viewed from the general perspective as the ratio of the variance of the e_t k step ahead forecast error to the variance of e_t. The generalized version of CLP, which we shall refer to as CLP_k, would then be given by

$$CLP_k = \frac{(1+\psi_1^2+\psi_2^2+\ldots+\psi_k^2)\sigma_a^2}{\sigma_e^2} \qquad (14)$$

One useful specification for k would be the desired settling sampling intervals for the output error response. If the settling time specification is satisfied, the estimated value for CLP_k would be very close to one. We have found such a statistic to be useful when we are not interested in being close to minimum variance, and all that matters to us is achieving some settling time specification. However, we have seen cases where reliance on CLP_k alone to monitor performance is misleading as a result of bad initial transients associated with the output error response trend.

Another time series based statistic for monitoring the performance of controlled outputs has been proposed by Rhinehart (1995). The statistic, referred to as the watchdog for controller performance, is defined to be the ratio of twice the variance of e_t to the variance of the differenced e_t trend. If the output error from set point response trend is given by $e_t=a_t$ (white noise), the watchdog will take on a value of one. If the output error trend becomes more highly positively correlated the watchdog will be significantly larger than one, and will indicate a poor output error response trend, according to Rhinehart (1995). The assumptions taken to derive the watchdog statistic are not sound, in general, from the perspective of fundamental stochastic theory. The condition $e_t=a_t$ can only be met at minimum variance control for deadtime free processes (refer to equation 6). Although it is trivial to extend the watchdog concept to set targets for specified e_t response targets, such as (6) or (7) for example, our experience has led us to conclude that it provides little useful information about e_t relative to the variance ratio tests discussed above.

Tyler and Morari (1995) have proposed several pass/fail likelihood ratio tests for determining if the output error response characteristics are acceptable based on specified dynamic response performance bounds. The approaches were demonstrated to be useful for detecting poor e_t response trend characteristics using simulated data, and industrial data provided by Shell (Kozub and Garcia, 1993). One big drawback with their monitoring approach is that it is both conceptually and computationally very sophisticated to apply relative to other methods described in this paper which could have been employed to arrive at similar conclusions. In their work, a settling time specification was illustrated for constructing the likelihood ratio tests. As stated before, reliance on settling time alone can be misleading for error responses with bad initial transients.

The use of a single number statistic, such as CLP for example, can hardly be as effective as time series model based impulse response estimation or auto-correlation function analysis for controller performance monitoring given that these approaches offer far more detail concerning the e_t trend characteristics. However, the latter approaches typically require a time commitment for analysis, and if a control engineer is responsible for several hundred loops, the time commitment relying solely on these tools may become too burdensome unless the necessary research can be developed for automating the analyses. Based on our experience, we have found monitoring a group of simple single number statistics useful to a least provide a first pass indication that a problem exists. If a problem is detected, further analysis should be carried with tools providing more detailed dynamic response information.

Cross-correlation Analysis

Stanfelj et al. (1993) proposed the use of the estimated cross-correlation function (Box and Jenkins, 1976) for testing the dynamic significance of measured disturbances and set point changes with respect to the closed-loop output error response. Cross-correlation analysis serves as a useful pass / fail test to determine if measurable loop disturbances and set point changes have a significant association with e_t. Both the magnitude and extent of statistically significant cross-correlation coefficients provide some indication of the relative importance of different measured disturbances with respect to their relationship with e_t. The procedure was also advocated for testing the effectiveness of feedforward compensation.

Based on our experiences, we have found the cross correlation function to be of limited value. In most of our work we often would like to know not only if a loop disturbance has a significant relationship with e_t, but also the dynamic response characteristics associated with the measured disturbance trend and e_t. This is the case when we are willing to allow the disturbance to have some transient effect on e_t provided that the dynamic response characteristics meet some performance specification. The estimated cross-correlation function, in general, exhibits very complicated patterns which, even for some of our more technically skilled people, are very difficult to interpret when statistically significant coefficients are estimated.

In order to facilitate cross-correlation analysis, prewhitening of the disturbance/set point trends has been attempted (Kozub and Garcia; 1993). This should have the effect of making the cross-correlation coefficients at sampling lag k proportional to the corresponding impulse response weights of a moving average transfer function model describing the net average dynamic response of a causal measured disturbance or set point trend on e_t. (Box and Jenkins, 1976). This approach has led to a marginal improvement in our applications. The problem may be due to inadequate prewhitening of the disturbance trends in our industrial data, perhaps because a fixed linear, time invariant time series model is applied for prewhitening.

Stanfelj et al. (1993) have also advocated the use of cross-correlation analysis for testing if model mismatch exists in the models used for both the feedback and feedforward controller designs. In this case, cross-correlation is carried out between the model prediction error and either the set point or the measured feedforward disturbance trends. Based on our experience, we have also found that cross-correlation for model mismatch to be of limited use in practice. One limitation that we often face is the lack of significant random set point changes given that most of our controllers are regulatory. This often necessitates the introduction of intrusive random set point changes in order to satisfy the conditions for the test. Once sufficient set point excitation is present, the test provides some indication that model mismatch is present. However, the analysis falls short of providing any useful quantitative information on the nature of the model mismatch, such as gain error or deadtime, and if the extent of modeling error is sufficient enough to prevent meeting specifications for the control system. The latter issue is extremely important because in practice, low order transfer function model approximations of the true process often yield acceptable performance despite the presence of modeling error. Given these limitations, we have found that closed-loop cross-correlation and transfer function model identification approaches, such as advocated by Box and MacGregor (1974), are more useful since more informative, structural information concerning model mismatch can be inferred through these procedures. Furthermore, the information gained by these procedures allows one to directly make model adjustments to the controller design.

Model-based Approaches and ANOVA

Performance assessment and diagnosis with respect to the effect of set point changes (s_t) and measurable disturbances ($d_{i,t}$) on the output error from set point trend can be carried out by the identification of input/output transfer functions for these individual contributions using response data. The identification would yield a model of the form

$$e_t = S(z^{-1})s_t + \sum_i R_i(z^{-1})d_{i,t} + n_t \qquad (15)$$

where $S(z^{-1})$ and $R_i(z^{-1})$ represent rational, discrete time transfer functions, and n_i represents the effect of unmeasured disturbances and noise. Standard empirical input/output identification procedures (e.g. Box and Jenkins, 1976; Ljung, 1987) can be applied to arrive at these models provided that the loop inputs aren't perfectly correlated. These models can be used to predict e_t responses for any s_t and $d_{i,t}$ input trends such as a step change. If a time series model is fit to the disturbance or set point trend, the net average trend effect of these loop upsets on e_t can be predicted by generating s_t and $d_{i,t}$ input trends from the impulse response of their fitted time series models. Hence, the response characteristics associated with any loop upset can be predicted with these models and used for controller dynamic response validation, diagnosis, and the prediction for incentive of feedforward control.

Desborough and Harris (1993) have shown that the identified models can be useful for carrying out an analysis of variance (ANOVA) of e_t with respect to the disturbances. Their results are based on the assumption of causal, independent loop disturbance trends. Their ANOVA procedure can yield valuable information on the relative contributions of the different measured disturbances to the variance of e_t, and the contribution due to unmeasured effects. Their approach also provides information on variance effects resulting from the feedforward design in order to assess its performance.

More general nonlinear model approaches, such as Neural Networks or nonlinear ARMAX model structures could also be applied as an alternative to (15). The benefits of using these more advanced model forms in practical problems in an issue that needs to be researched.

Our success in applying model identification and ANOVA in practice has been somewhat limited. Although the model identification approach offers a wealth of information, it demands expertise in model identification and estimation, which is usually lacking in the field. Hence, for the approach to find widespread application in the field, automation of the model fitting procedure is essential. Another significant problem that we've frequently encountered is the presence of significant associations between process disturbance and set point trends. This condition can invalidate the ANOVA results, and yield biased, non-causal models. If both of these obstacles can be overcome by further research, the model based/ANOVA procedure will be very beneficial for carrying out performance monitoring and diagnosis in industry.

General Concluding Comments on Time Series Methods

The time series based theories discussed above have so far been useful in some of our work related to controller performance monitoring and diagnosis. Our experiences have indicated that no single statistic, or statistical analysis, by itself is sufficient for effective performance monitoring and diagnosis. The best results are often obtained by the collective application of several methods. With respect to our ultimate goals stated earlier, there still remain significant problems and enhancements that need to be addressed by further research. Some issues which we believe to be important are discussed below.

The success of these methods requires that the analyses be applied to what can be considered representative data from the perspective of controller analysis. The data set(s) should contain information on important disturbance upsets, and not be corrupted by abnormal process operating conditions, such as temporary equipment shutdown for maintenance for example. Furthermore, the data set(s) should be neither too short nor too long. With too short data sets the variability of statistical estimates will be too high for them to be of any use. Too long data sets can also lead to misleading results when many different response characteristics are juxtaposed into one long time series. In fact, we've run into cases where an analysis carried out on a very lengthy data set suggested high performance, when in fact an analysis at a more local time scale looking at smaller subsets of data, revealed consistently very poor performance, but with significantly different response characteristics. We believe that some rigorous research needs to be carried for selecting data window sizes which take into account both statistical distributional properties as well as the historical properties observed in the data.

The theory discussed so far pertains only to SISO control problems. Although SISO controllers are most frequently applied, a significant number of cost incentive MIMO applications exist. Hence, research is needed to generalize the concepts to these problems, and arrive at practical solutions. So far, we are aware of only one significant paper that has been recently published by Harris et al. (1995) which extends the auto-correlation and variance ratio concepts to the unconstrained MIMO control problem. Also, theory needs to be developed to analyze multi-loop systems where adverse interactions can take place. We've seen cases, such as processes where recycles exist, where one malfunctioning loop can lead to the detection of bad performance in other loops as result of interactions.

It would be useful to have single number type statistics, such as CLP, for both detecting and determining the extent of underdamped, or cyclical, response trend characteristics. Without examining either the estimated auto-correlation function or the times series model response prediction, the problem of output error oscillation can go undetected.

With some important control applications, constraint enforcement prevents set points from being reached in a steady state sense. It would be useful to extend the previously shown theory to carry out dynamic performance assessment under conditions where nonzero offset from steady state needs to be enforced.

With respect to the measured disturbances $d_{i,t}$ and s_t, the assumption has been made that they are causal and independent. As stated earlier, we've encountered many data sets where strong colinearity was observed in the measured disturbance trends. Cases have also been observed, particularly with plants that have recycles, where feedback effects introduce correlation with respect to the relationship between the output and the disturbances. The theory needs to be extended to address this type of data characteristics, or at least provide the necessary alarms for preventing the arrival at erroneous conclusions.

All of the statistical analyses discussed so far require interpretation. This task can be difficult with respect to relating the statistical information provided to the user to both the controller and process economic objectives. Some results in simplifying this task would be significant.

We've often observed a substantial degree of time variant response characteristics in our data. Changes in the dominant disturbance characteristics and the process operating conditions have been seen to be the cause of this, with the former often being the predominant factor. The time series based approaches make the implicit assumption of time invariant response characteristics, which clearly violates the data characteristics described above. Research results in the area of modeling time variant systems could yield some potential improvements.

The time series based approaches assume linear response characteristics. At this point it is unclear whether nonlinearities are significant in most of our practical applications. Nevertheless, tests for significance of nonlinear effects, and extension of the theory to handle such scenarios may have potential value.

Industrial Examples

In this section we shall illustrate the application and value of some of the statistical time series / modeling approaches using Shell industrial response data. Both examples will make use of routine operating data.

Example 1

The first example concerns the regulation of a variable associated with one of Shell's distillation columns. There is no manipulated input deadtime. The control engineer has specified an output error standard deviation limit of 1, and would like an error response similar to (7) with a time constant of 100 minutes and settling time of about 300 minutes. Fig. 1 shows one week of response data sampled at a rate of one minute. The measured output and set point are indicated by (-) and (—) respectively.

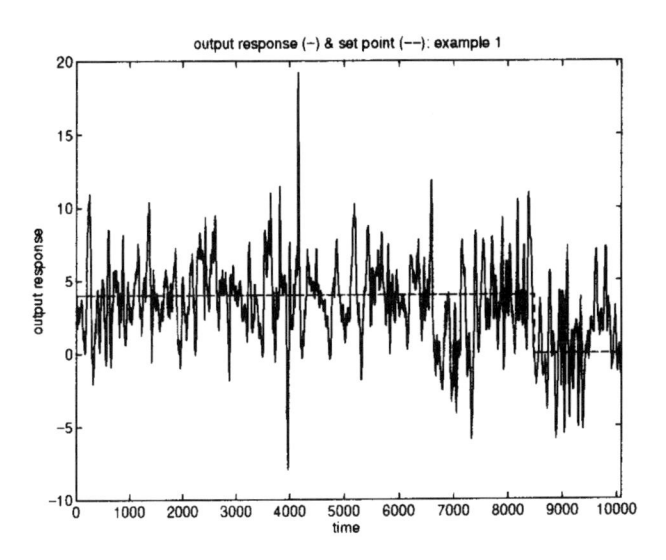

Figure 1. Example 1 output (-) and set point (—) trends.

Fig. 2 shows three daily average sliding window statistics. Fig. 2(a) shows the estimated error standard deviation (STD) (o), the specified limit for the error STD (-), and the estimated lowest STD possible by minimum variance control (+) as determined by (5) and (6). This plot indicates that the standard deviation limit is violated on the average for each day. It also indicates that the specified limit for the standard deviation is theoretically achievable on all days by feedback control since the estimated minimum variance (STD) always lies below the target.

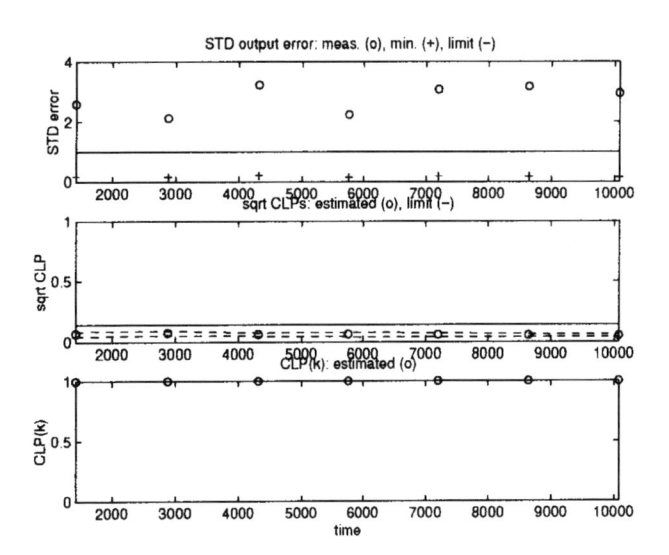

Figure 2. Daily average sliding window statistics.
a) output error STD (o); STD Limit (-); MVC STD estimate (+).
b) estimated square root CLP (o); specified square root CLP target (-).
c) estimated CLP_k (o) with k=300.

Fig. 2(b) shows the estimated square root CLPs (o) along with its desired limit (-) based on (13). 95% confidence intervals for the estimated CLPs are indicated by (—). Since all the estimated CLPs lie below the limit, the data indicates that the output response trend with respect to variance rejection is poorer than desired relative to reference model (7). The average of the weekly square root CLPs is 0.061 which is less than half the desired calculated value of 0.14. Using the average estimate of 0.166 for the lowest possible error STD from the data in Fig. 2(a) (+) along with the specified square root CLP target of 0.14, equation (10) predicts that the desired error STD of 1 would have been closely met (about 1.19) if model (7) were valid. This result suggests that the output error time response characteristics might be significantly different from (7).

Fig. 2(c) shows the CLP_k daily average time response trend with k set to 300. The specification of k=300 was based on the settling time specification of 300 minutes. All the daily average CLP_k are equal to 1, indicating that the settling time specification is being met. Hence, the CLP and CLP_k results taken together suggests that some poor initial transient might be associated with the error response for this loop.

Fig. 3(a) shows the estimated auto-correlation function (vertical bars) together with its 95% confidence interval (- -) and desired pattern (solid line) based on (7) for the entire week of data. The auto-correlation pattern indicates that the loop is not yielding minimum variance control since there are significant lags beyond k=f=0. The auto-correlation plot suggests that the output error trend settles in about 60 sampling periods, which is significantly faster than specified. The auto-correlation results lead one to conclude that the output error response trend characteristics are better than desired.

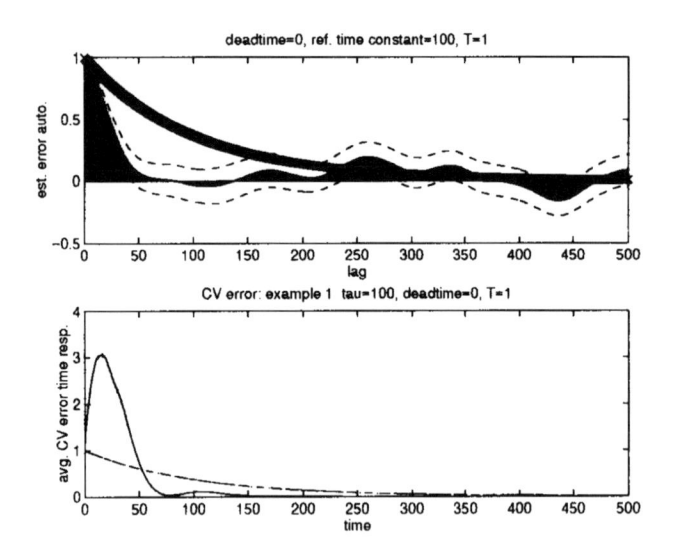

Figure 3. Average dynamic response characteristics for one week of data.
a) Auto-correlation function.
b) Fitted time series model impulse response.

Fig. 3(b) shows the average estimated output error trend (-) as determined from the impulse response of a model of the form (5) fitted to data. The impulse trend corresponding to (7) is indicated by (—). The average estimated output error response indicates a large initial excursion from set point that recovers quickly in about 60 minutes relative to reference trend (7). The large initial excursion is consistent with both CLP and CLP_k results, and is the source of the higher than desired output error variance. Note that the time series modeling result is not inconsistent with the estimated auto-correlation function data since it can be shown that the fitted model does yield a theoretical auto-correlation function which is similar to the one observed in 3(a).

The following conclusions may be drawn from this example:

- The engineer should attempt to modify the controller design to eliminate the high initial excursion in the error response.
- If a controller design improvement is not possible, either the performance specifications for the loop should be relaxed to permit the initial transient, or the disturbance source contributing to this characteristic should be eliminated.
- One should make use of several statistical analyses when carrying out controller performance monitoring and diagnosis in order to arrive at correct conclusions. In this case, the information provided by the auto-correlation function was somewhat misleading by itself.

Example 2

The second example also concerns the regulation of a variable associated with one of Shell's distillation columns. The manipulated input deadtime is 3 minutes and the sampling rate is 1 minute. The control engineer would like an output error standard deviation limit of 1, and would like an error response close to first order plus delay with a time constant of about 3 minutes and a settling time of about 12 minutes. Fig. 4 shows one week of response data. The measured output and set point are indicated by (-) and (—) respectively. There is evidence of a time variant output trend with sudden bursts of cycling.

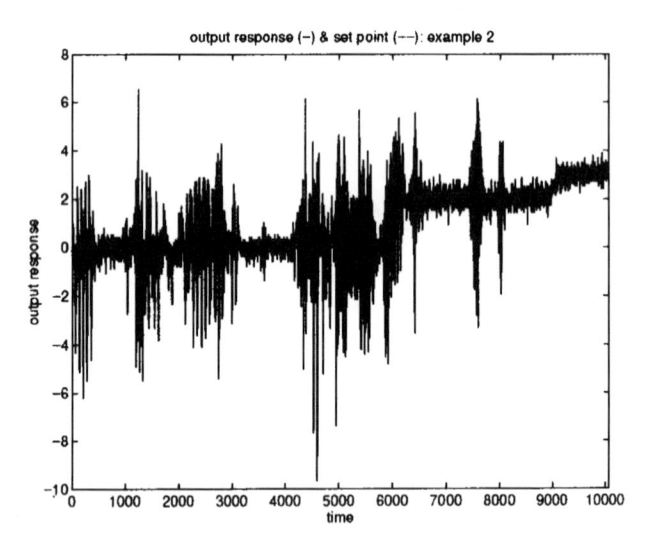

Figure 4. Example 2 output (-) and set point (—) trends.

Fig. 5 shows the output error STD, square root CLP, and CLP_k (k=12) estimates based on a sliding window of 360 minutes. Fig. 5(a) shows that the estimated output error STD (o) is frequently larger than desired (-). There are also periods where the desired STD error specification (-) is lower than the estimated minimum variance error STD (+), making this specification impossible to achieve by feedback control alone. In Figures 5(b) and 5(c), both the estimated CLP and CLP_k (o) frequently fall significantly below their limits (-). This indicates a poor output error response trend which may potentially be improved by better feedback control for the measured conditions. All three statistics exhibit trends which vary considerably, suggesting time variant response characteristics.

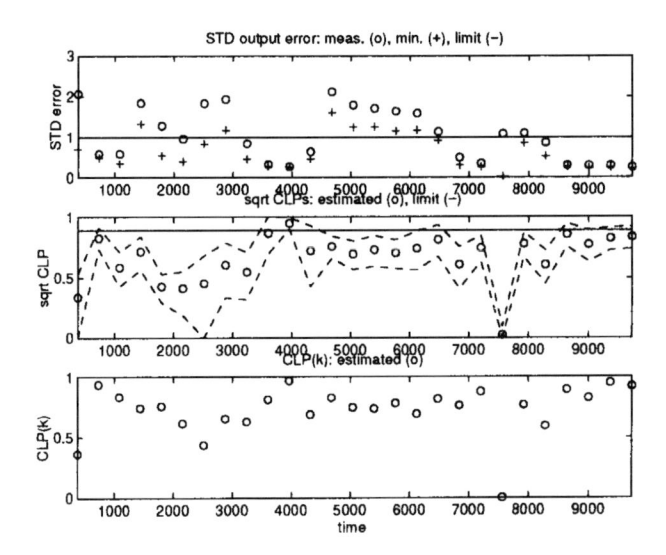

Figure 5. Daily average sliding window statistics.
a) output error STD (o); STD Limit (-); MVC STD estimate (+).
b) estimated square root CLP (o); specified square root CLP target (-).
c) estimated CLPk (o) with k=12.

Fig. 6 shows the estimated average output error time response obtained from a time series model fitted to the full week of output error data. Relative to the desired response (-) the estimated average output trend oscillates, and is slow to settle.

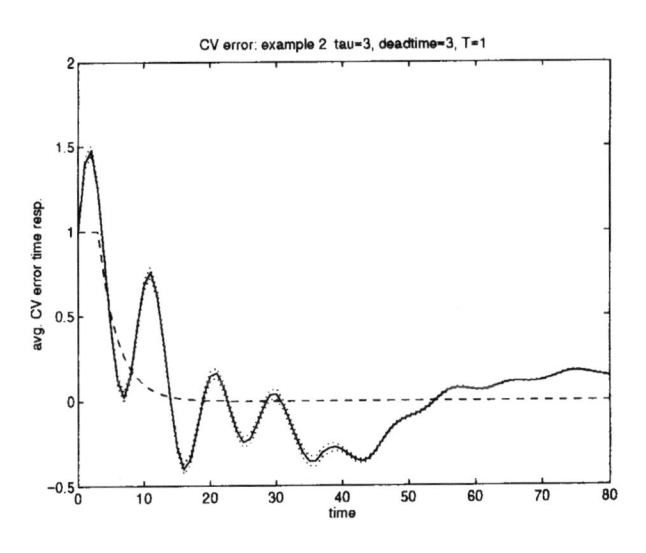

Figure 6. Time series model estimated average error response. Estimated (-); Desired (—).

The information gathered so far indicates that the output error response trend is poor but may not be a result of a bad feedback control alone. In order to gain a clearer picture of the situation, measured loop disturbance trends are examined. Two measured disturbance trends are shown in Fig. (7). Feedforward compensation was not applied. Time variant, erratic behavior can be observed in these plots, particularly with disturbance 2 (Fig. 7(b)). There is evidence of some correlation between the disturbance trends.

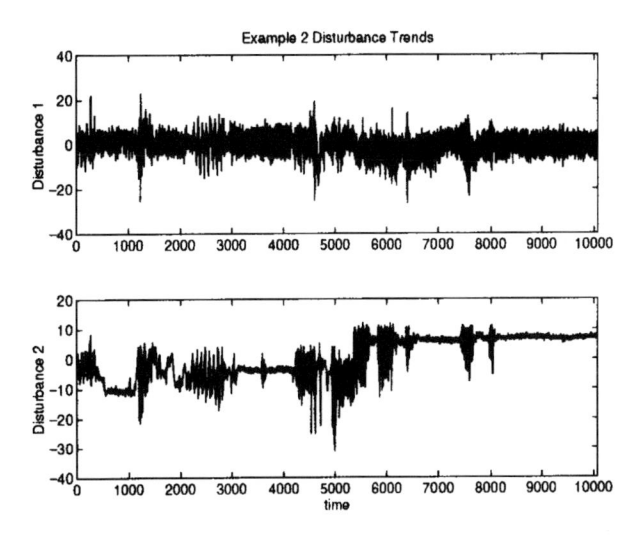

Figure 7. Measured disturbance response trends.
a) Disturbance 1.
b) Disturbance 2.

Auto-correlation function plots of the differenced disturbance trends are shown in Fig. 8. The auto-correlation function for disturbance 2 shows a significant high frequency cycling pattern which may be the cause of the poor output error response.

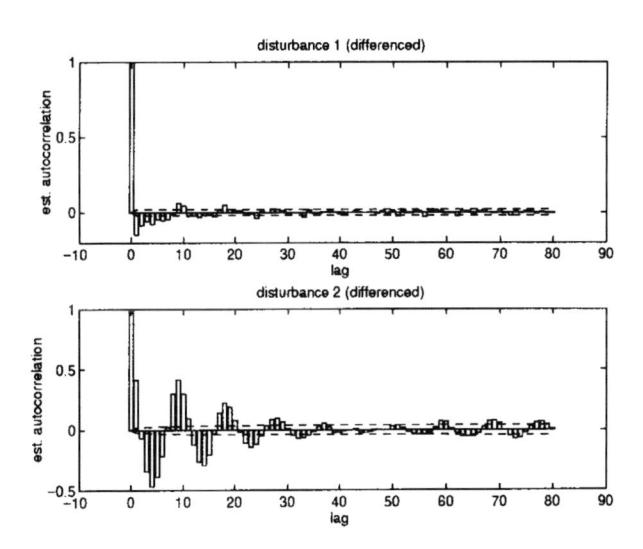

Figure 8. Estimated disturbance auto-correlation functions.
a) differenced disturbance 1.
b) differenced disturbance 2.

Fig. 9 shows the estimated cross-correlation functions between the output error and the prewhitened disturbance trends. Both disturbances are shown to be statistically significant. Both disturbances appear to have some cyclical associations with the output error. Fig. 9(b) shows that disturbance 2 is more strongly associated with the output error given that it has large cross-correlation coefficients.

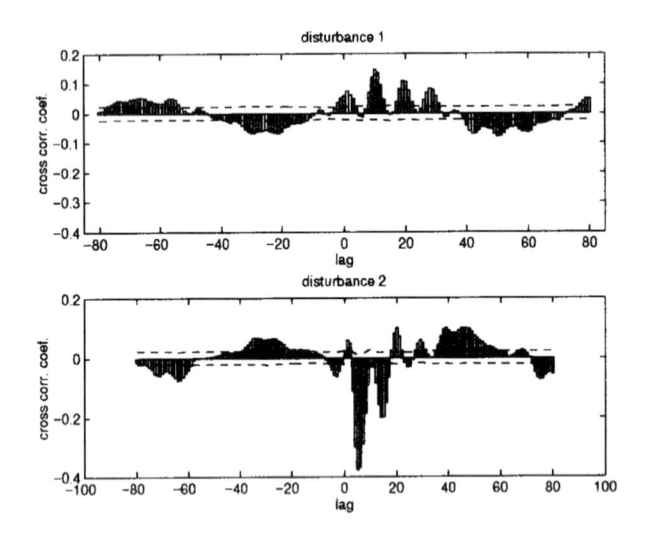

Figure 9. Estimated output error and disturbance cross-correlation functions.
a) Prewhitened disturbance 1.
b) Prewhitened disturbance 2

A dynamic model of the form (15) was fitted between the output error trend and the two disturbance trends. Fig. 10(a) shows the measured output error trend. Fig. 10(b) shows the model predicted contribution of the first disturbance to the output error trend. The plot shows that the contribution of disturbance 1 is small. The model predicts that this disturbance contributes to about 3.5% of the output error variance. Fig. 10(c) shows the model predicted contribution of the second disturbance. The plot shows that the contribution of disturbance 2 is large. The model predicts that this disturbance contributes to about 74% of the output error variance. It is important to note that the variance contributions are rough estimates in this case study given that there is some correlation between the disturbance trends.

Figs. 11(a) and 11(b) show the model predicted unit step response and net average response respectively corresponding to disturbance 2. The net response was obtained by fitting a time series model to the second disturbance trend and then generating an average disturbance 2 input from the time series model impulse response. The results in Fig. 11 show the net response to be dynamically far worst than the predicted unit step. This suggests that the time response trend characteristics associated with the second disturbance input is the cause of the poor output error response trend. This is confirmed in Fig. 12 where an analysis of the model residuals is shown. The autocorrelation function of the residuals, once the modeled contributions of both disturbances are eliminated, indicates that a good output error response would be obtained.

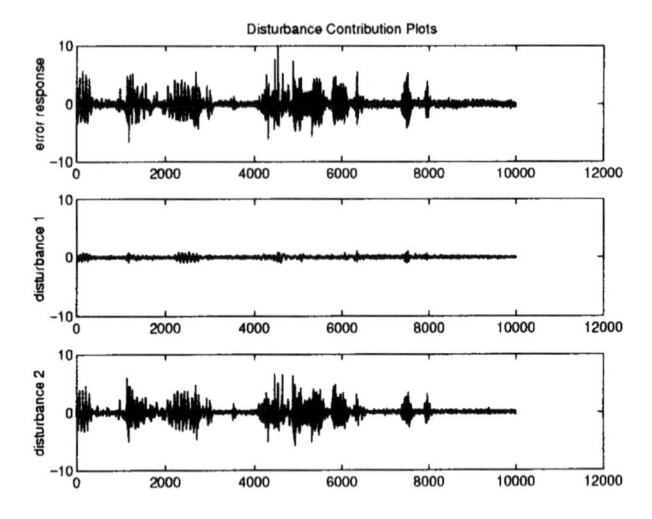

Figure 10. Model predicted output error trend contributions.
a) Output error trend.
b) Disturbance 1 contribution.
c) Disturbance 2 contribution.

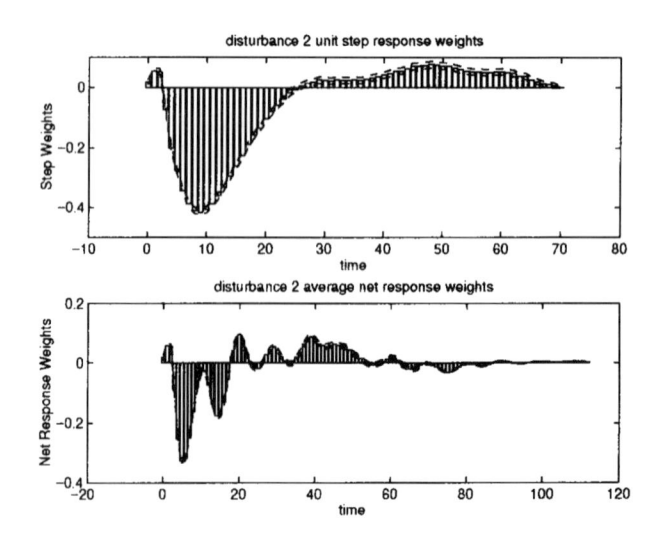

Figure 11. Model predicted disturbance 2 response.
a) Step.
b) Net average response.

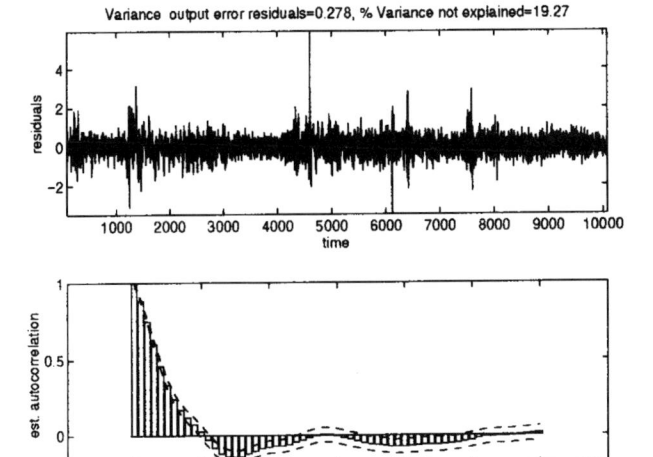

Figure 12. Residual analysis of the output error model fit.
a) Model residuals.
b) Estimated auto-correlation function of residuals.

From the results of these analyses, it can be concluded that the output error trend characteristics are poor as a result of the time trend associated with disturbance input 2. Improvements should be focused at eliminating adverse time response trend characteristics associated with this disturbance. The introduction of feedforward control for disturbance 2 would not correct the problem because of both the time variant and cycling trend characteristics associated with this disturbance. Although much of the conclusions in this example could have been arrived at by visual data inspection, the example is nevertheless valuable for a straightforward application of the theory discussed in the paper which will be useful for automating the tasks of monitoring and diagnosis.

Artificial Intelligence Technologies

As was shown in the previous section, the tasks of data analysis and statistical interpretation can require both a high degree of expertise and time commitment. The current economic climate has led to the necessity of reducing both the manpower and expertise available at plants for carrying out such tasks. Hence, a high degree of automation will almost be essential if any of the technologies developed are to have a major impact in the field. We believe that Artificial Intelligence (AI) technologies can play an important role in meeting this goal. The goal for the AI technology would be to carry out the necessary interpretation of the results, and present the conclusions in straightforward manner for a nonexpert user in the field.

AI technologies could play a very important role in the task of diagnosis. Statistical information, such as correlation analysis and regression model fitting, can provide useful information on associations between different variables, but not necessarily any direct causal information that can help isolate a root cause problem. Our experience has shown that effective diagnosis requires the assimilation of all process measurement information, statistical information, process knowledge from experience, and detailed process modeling information (if available) in order to arrive at root causes for problems. Furthermore, cases often arise when some event might have a simultaneous adverse effect on many process variables and controllers, making the analysis virtually unmanageable without more sophisticated technologies to interpret the information.

In conclusion, the challenge of automating controller performance monitoring and diagnosis is an important new frontier for the AI research community. Positive results in this area have the potential for making a tremendous impact in the highly automated chemical process industries.

Conclusions

The task of controller performance monitoring and diagnosis plays an important role in arriving at control systems which are both reliable and profitable. It is one of the most frequently carried out activities by plant location engineers, and the one which influences cost the most. Given the importance of this task, it surprising that very little research has been devoted to addressing the industrial challenges in this area.

In order to stimulate activity and provide direction to the research community, a problem definition along with a list of goals and requirements were provided. A review of past research related to the subject was presented together with feedback based on the author's industrial experience using the various technologies. Industrial examples were demonstrated to illustrate the value of existing technologies.

In conclusion, tremendous research opportunities exist in the area of controller performance monitoring and diagnosis. Advancements in this area can have a major impact in the highly automated chemical process industries.

References

Astrom, K.J. (1970). Introduction to stochastic control theory. Academic Press.

Box, G.E.P. and G.M. Jenkins (1976). Time series analysis forecasting and control. Holden-Day.

Box, G.E.P. and J.F. MacGregor (1974). The analysis of closed-loop dynamic — Stochastic systems. *Technometrics*, **16**(3).

DeVries, W.R. and S.M. Wu (1978). Evaluation of process control effectiveness and diagnosis of variation in paper basis weight via multivariate time-series analysis. *IEEE Transactions On Automatic Control*, **AC-23**(4).

Desborough, L. and T.J. Harris (1992). Performance assessment measures for univariate feedback control. *The Canadian Journal Of Chemical Engineering*, **70**.

Desborough, L. and T.J. Harris (1993). Performance assessment measures for univariate feedforward/feedback control. *The Canadian Journal Of Chemical Engineering*, **71**.

Harris, T.J. (1989). Assessment of control loop performance. *The Canadian Journal Of Chemical Engineering*, **67**.

Harris, T.J. Boudreau, F., and J. F. MacGregor (1995). Performance assessment of multivariate feedback controllers. *Submitted to Automatica*.

Harris, T.J. (1995). Personal communication.

Kendra, S.J. and A. Cinar (1996). Controller performance assessment by frequency domain techniques. Submitted to *Journal Of Process Control*.

Kozub, D.J. and C. E Garcia (1993). Monitoring and diagnosis of automated controllers in the chemical process industries. AIChE Meeting, St. Louis.

Ljung, L. (1987). System identification. Theory for the user. Prentice-Hall.

Rhinehart, R. R. (1995). A watchdog for controller performance monitoring. *ACC proceedings*.

Stanfelj, N., Marlin, T.E. and J.F. MacGregor (1993). Monitoring and diagnosing process control performance: The single-loop case. *Ind. Eng. Chem. Res.*, **32**.

Tyler, M. L. and M. Morari (1995). Performance monitoring of control systems using likelihood methods. *Submitted to Automatica*.

PROCESS ANALYTICAL CHEMICAL ENGINEERING

Bruce R. Kowalski
Center for Process Analytical Chemistry
University of Washington
Seattle, WA 98195

Abstract

A paradigm shift is currently underway in the field of Analytical Chemistry that opens new opportunities for both academic and industrial Chemical Engineering. This paper describes the changes taking place and invites chemical engineers to join with analytical chemists to modify the academic curricula in both fields and form research and application partnerships in the interdisciplinary areas of Process Analytical Chemistry *and* Process Analytical Chemical Engineering.

Keywords

Process analytical chemistry, Process analyzer.

Introduction

Analytical Chemistry is the science that provides quantitative chemical information on chemical systems. It has developed as a laboratory science whereby processes and systems are physically sampled and samples are brought to the analytical laboratory for analysis. The emphasis is on analysis of the samples and not the process from which they originated. For the chemical engineer this means that real-time chemical information is scarce. The result is that control engineers rely primarily on data from on-line gauges measuring temperature, pressure and flow to monitor and control processes. As processes become more complicated and plant engineers are pressured to lower production costs in all ways (energy, raw materials, effluents, etc.) it is clear to some that more advanced control is needed. Either more powerful computer hardware and software is needed and/or control systems just need better input.

In response to this need a few analytical chemists began to rethink the science of analytical chemistry at its foundation recognizing that the focus of analysis should be the entire process and not just the samples. At first, laboratory instruments were "interfaced" to chemical processes leading to both positive and negative results. As the new field of Process Analytical Chemistry (Callis, et al., 1987) emerged in the early 1980's new research lead to in-line and non-invasive chemical sensors (e.g. arrays of coated quartz crystal microbalances) and analyzers based on fiber optic coupled near infrared, mid infrared, Raman and other spectroscopy, that could operate in real-time and thus shorten the delay in providing monitoring and control systems accurate concentrations of some or all process components. As process analytical chemists brought new in-line analyzers to plant engineers and supervisors several problems surfaced. As might be expected, these new analyzers were more complex and less rugged than the traditional gauges familiar to instrument engineers. Sample ports clogged, windows became coated and calibration was problematic. In short, the new analyzers could not survive the mandatory "direct hit from a fork-lift truck." These problems are in the technical domain and can and are being solved. However, there is a much more serious problem that needs to be addressed. Simply stated analytical chemists and chemical engineers are not trained to work together. The jargon is different and their interests are not in common. Most chemical engineers learn that chemical analysis is performed in a laboratory. Even those that do take a course in analytical chemistry are introduced to pipettes, burettes and instruments far too delicate to be of use outside the laboratory. Most analytical chemists receive no engineering training of any kind. Now, as any engineer knows when two systems, each designed without the other in mind, are brought together there is usually a problem at the interface. The problem in this case is in the educational domain and can be solved by a cooperation between educators in the two fields.

An effort to solve this problem has already begun at a local level at the National Science Foundation's Center for Process Analytical Chemistry (CPAC) where a graduate course (Chemistry 525: Process Analytical Chemistry) has been taught for eight years. In the course analytical chemists and a few brave engineering students learn about modern chemical sensors and analyzers as well as the basics of control engineering and then work in teams to apply what they learn to solve actual problems given to them by guest lecturers from industry. These case studies (Figure 1) are crucial to their education and the solutions they devise are often of real interest to our industrial visitors. Now, following the introduction of a new textbook on the subject (McLennan and Kowalski, 1994) plans are underway to expand the course material and create a new undergraduate level course offered jointly by the Departments of Chemical Engineering and Chemistry. The course (Figure 2) that will be partially sponsored by CPAC will involve other NSF Centers as well as a number of process analyzer manufacturers. It will be offered to universities and corporations via satellite.

In chemical process control, multivariate control algorithms require a mathematical model be available that, hopefully, describes the process being controlled. Process models can be classified as theoretical or empirical depending on the type of data used to develop the model and the interpretability of the estimated model parameters. Theory-based models attempt to describe the process based on the laws of thermodynamics, kinetics, mass flow, etc. Empirical or "data-based" models use generalized basis sets (e.g. singular vectors in principal components analysis or sigmond functions in neural networks) fit to historical process data (temperature, pressure and flow-rate, predominantly) and, while empirical models may possess an accurate predictive capability, they can only be trusted to interpolate. Also, examination of estimated model parameters from empirical models to gain process understanding is problematic.

Due to the recent development of powerful process analyzers, a new opportunity has been opened that, with a strong collaborative R & D effort, promises to add a new paradigm to chemical process modeling, control and optimization. These new analyzers are capable of monitoring the concentrations of process components in real-time and at multiple locations along a process stream. Given the wealth of chemical information that can be fused with physical data it should be possible to capture the essential dynamics of today's complex chemical processes in a process model. Using systems engineering jargon, it is becoming possible to "fully" identify a process (system). From a mathematical perspective, this means that the process vector space stands a better chance to be spanned by the process measurement vector space and the model be "determined" or, better yet, "overdetermined." Herein lies the challenge for the future that this paper addresses. Research and education that aims to develop in-line, real-time chemical analyzers and integrate them into process design, modeling and control must be conducted as a partnership between process analytical chemists and chemical engineers.

CPAC has had analytical chemists working with chemical engineers on process monitoring and control studies since it began operation in 1984. Perhaps the earliest study employed an array of quartz crystal microbalance sensors to monitor solvent evaporation in a pharmaceutical process (Cary and Kowalski, 1988). The next study employed flow injection analysis (FIA) to monitor the concentrations of metal ions in order to control electrolytic plating bath chemistry (Whitman, et al., 1988). Next, the first application of multivariate statistical process control (MSPC) was undertaken to monitor a glass melter process during vitrification (Wise, et al., 1988). More recently, process control has been used to control fermentation processes using several novel process analyzers (Cavinato, et al., 1990). Currently work is being completed that uses a noninvasive near infrared spectrometer to control the particle size distribution during the emulsion polymerization of styrene. These and other studies have resulted in the training of control engineers familiar with modern process analyzers as well as analytical chemists familiar with control theory and practice. These students are formally educated by taking Chemistry 525 and then employ and extend what they learn in interdisciplinary research projects leading to their advanced degrees.

Two additional studies conducted by CPAC investigators deserve special attention as they show what can be done using modern process analyzers that significantly advance Systems Identification. These studies were conducted by the Laboratory of Chemometrics which uses multivariate statistics and mathematics to develop novel instrument calibration, standardization and error detection and correction algorithms. Chemometrics (Brown, et al., 1994) is the natural bridge between getting the best data available from process analyzers and then using the data for process monitoring, modeling, control and optimization.

The first study was conducted in collaboration with a process analytical chemist at du Pont who used an on-line Fourier Transform Infrared Spectrometer (FTIR) to monitor a manufacturing process (Tauler, et al., 1993). At one point in the process the reaction was quenched and no change in the readings from the FTIR instrument were expected. Mysteriously, in a region of the spectrum where no signal was expected a number of peaks appeared and changed their intensities over the time the reaction was believed to be completely stopped. Data was sent to CPAC where multivariate curve resolution and evolving factor analysis were used to determine that up to three unexpected (and unwanted) reaction intermediates were produced and changed their concentrations over time. Analysis of historical data from eight separate runs showed that the three intermediates were not always present together depending on the process conditions used for the particular run. Since the pure spectra (which lead to the identity of the components) as well as their relative concentration versus time profiles were estimated, a prediction algorithm was developed that could be used in future runs to monitor the formation and time behavior of the unwanted process states and thereby eliminate them if desired.

The important point of this application is that a process cannot be controlled if all states are not identified. While this work was done using data from a manufacturing process, the real benefit would arise if the study was conducted at the pilot plant level so that production was not burdened by an incomplete knowledge of all possible process states, thereby forcing control algorithms to deal with process upsets in an underdetermined condition.

In another study, researchers monitored a complex polymerization process via passive mid-IR emission spectrometry (Pell, et al., 1991). The experiment was simple: Mix the process components and deposit the mixture (a urethane paint product) on a surface; acquire several infrared spectra at a suitable time interval (every two minutes) until the process was complete (two hours). From the 60 spectra acquired, a process model was constructed and used to determine the following:

1. The process consisted of three physical states; reactants, intermediate and final product.
2. The three resolved spectra allowed the concentration versus time plots to be made which then yielded the complete kinetics of the reaction; initial liquid, semi-crosslinked gel state and multiple cross-linked (cured) polymer product.

It was later determined that all states were not identified. Since the initial reactants were initially of equal concentration and reacted stoichiometrically a mathematical ambiguity existed baring observation of the individual reactants. Further experiments using unequal proportions and different temperatures allowed more states to be identified which then lead to better process understanding.

During meetings with industrial engineers the author has learned that some engineers believe that all processes can be controlled using only temperature, pressure and flow data as inputs to process models. At the same time the author has seen many manufacturing processes shut down due to unexpected upsets that lead to production delays and high control costs (the cost associated with reprocessing, destroying or selling off-spec product for a lower price). While process control algorithms can be somewhat robust to weak or inaccurate input data, there is little hope for optimal control without complete system identification. With a strong collaboration between a new breed of process analytical chemists and process analytical chemical engineers, powerful analyzers will be made commercially available (and hopefully inexpensive as well) capable of providing all reactant, intermediate, product and impurity concentrations in real-time.

While many research challenges lie ahead, the most pressing problem is the interdisciplinary cultural problem mentioned earlier that must be solved by changes in education. Just as engineers have enjoyed the result of computation moving from centralized to distributed they will also enjoy and hopefully participate in the movement of analytical chemistry out of the centralized laboratory and distributed to where quantitative chemical information is needed most; the process.

Analytical chemists, particularly in industry, are beginning to seize the opportunity to move out of the laboratory and closer to manufacturing and other processes (i.e. environmental and medical). This is evidenced by the recent additional of a review on Process Analytical Chemistry (Blaser, 1995) in the Application Review issue of the most prestigious journal in the field, Analytical Chemistry. It is interesting to note that the review was authored by our industrial colleagues and not academicians. As the review shows, more and more university scientists are engaging in research on in-line non-invasive analyzers and with the availability of a book on Process Analytical Chemistry (McLennan and Kowalski, 1994), it is hoped that a greater number of universities will offer a course on the subject in the future.

Chemical Engineers, particularly those at universities, can begin to understand how this paradigm shift provides new opportunities in chemical engineering research and education. At the fundamental level the belief held by some that most chemical processes are well understood should be reexamined in light of the real-time chemical analyzers that are becoming available. Also, questions such as what chemical information is needed, what the optimal sampling rates should be and even, where should analyzers be placed in a given process are largely unanswered. Some progress is being made as, for example, C. Moore at the University of Tennessee (Moore, 1987) showed the optimal number and placement of in-line analyzers for a distillation tower. This is a good example of how engineering skills are needed in chemical process analysis and how a collaboration between engineers and process analytical chemists is the key to a successful application.

Chemical engineering departments currently add physical chemists to their faculty so it is conceivable that process analytical chemists could be added as well provided that research in the field could be recognized as an outgrowth of process monitoring and control activities currently in place. These new analyzers need to be developed as integral parts of chemical processes so it seems natural that this activity become a subdiscipline in chemical engineering.

Regardless of whether or not chemical engineering as a field decides that process analytical chemical engineering should become part of the field the educational problem addressed in this paper is still an issue. Through my work at CPAC I see little doubt that research and development of chemical process analyzers is definitely a high priority need in the chemical industry. The Council on Chemical Research, the Chemical Manufacturing Association, and other organizations recently identified it as one of five priority needs in their Vision 20/20, Road Map for the Chemical Industry. Therefore, at the very least, chemical engineering students will require some formal training in their development and application. The new course being developed at the University of Washington is a start. It is now up to chemical engineers to examine the opportunity and decide whether or not to participate in the changes that will take place in the future as chemical analysis moves from centralized to distributed.

Acknowledgments

The author wishes to thank Ray Chrisman of Dow, Mel Koch, formerly with Dow and now with CPAC, and the investigators and industrial sponsors of CPAC for stimulating discussions leading to this paper. Professor Larry Ricker's help with Chemistry 525 as well as this paper is also very much appreciated.

Examples of Case Studies

Chemistry 525
University of Washington

On-line measurement of Fe(II) and Fe(III) and control of the Suferox Process. Shell Development Company.

Continuous emission monitor system for CO, NO_x, and SO_2 in the effluent of a hazardous waste incinerator. The Dow Chemical Company.

Measurement and control of acid addition in an alkylation reactor. Amoco.

Multipoint monitoring of multiple analytes in boilers for the Navy nuclear fleet. Savannah River Laboratory.

Monitoring and control of butadiene polymerization (% monomer, % solid, catalyst concentration and polymer properties). Goodyear.

Continuous analysis of H_2O and CO_2 in an amine stream. Amoco.

Determination of "gum" content in butane/butene stream. Amoco.

On-line analysis of multiple layer thickness' and active ingredient in medical patches during production. 3M.

Tentative Outline of New Course on Process Analytical Chemical Engineering

University of Washington

I. **Process Analysis in Perspective**
   - Introduction and History
   - Terminology
   - Safety, Maintenance and Support
   - Literature Sources

II. **Sampling Systems**

III. **Physical Property Analyzers**

IV. **Process Chromatography**
   - GC
   - LC
   - SCF
   - Ion Chromatography

V. **Flow Injection Analysis**

VI. **Process Spectroscopy**
   - IR, Near-IR and Raman
   - Mass Spectrometry
   - Other Topics

VII. **Electroanalytical Techniques**

VIII. **Environmental Monitoring**
   - Pollution Control
   - Pollution Prevention

IX. **Non-Invasive Methods**

X. **Chemical Sensors and New Analyzer Systems**

XI. **Monitoring, Modeling and Control**
   - SPC and MSPC
   - Multivariate Calibration and Instrument Standardization
   - Process Modeling
   - Analyzer Selection and Location
   - Loop Configurations
   - Optimization
   - Control (PID and model-based)

References

Blaser, W.W. (1995). Application Review; Process Analytical Chemistry. *Analytical Chemistry*, **67**, 47R.

Brown, S.D., T.B. Blank, S.T. Sum and L.G. Weyer (1994). Fundamental Review; Chemometrics. *Analytical Chemistry*, **66**, 315R.

Callis, J.B., D.L. Illman and B.R. Kowalski (1987). Process Analytical Chemistry. *Analytical Chemistry*, **59**, 624A.

Cary, W.P. and B.R. Kowalski (1988). Monitoring a Dryer Operation using an Array of Piezoelectric Crystals. *Analytical Chemistry*, **60**, 541.

Cavinato, A.G., D.M. Mayes, Z. Ge and J.B. Callis (1990). A Noninvasive Method for Monitoring Ethanol in Fermentation Processes Using Fiber Optic Near-Infrared Spectroscopy. *Analytical Chemistry* **62**, 1977–1982.

McLennan, F. and B.R. Kowalski (1994). *Process Analytical Chemistry*. Blackie Academic and Professional, London.

Moore, C. (1987). Determining Analyzer Location and Type for Distillation Column Control. Manufacturing and Control Engineering Center, University of Tennessee, Knoxville, TN.

Pell, R.J., J.B. Callis and B.R. Kowalski (1991). Noninvasive Polymer Reaction Monitoring by Infrared Emission Spectroscopy with Multivariate Statistical Modeling. *Applied Spectroscopy*, **45**, 808.

Tauler, R., S. Fleming and B.R. Kowalski (1993). Multivariate Curve Resolution Applied to Spectral Data from Multiple Runs of an Industrial Process. *Analytical Chemistry*, **65**, 2040.

Whitman, D.A., G.D. Christian and J. Ruzicka (1988). Spectrophotometric Determination of Nickel(I), Iron(II), Boric Acid and Chloride in Plating Baths by Flow Injection Analysis. *The Analyst*, **113**, 1821–1826.

Wise, B.M., D.J. Veltkamp, B. Davis, N.L. Ricker and Kowalski, B.R. (1988). Principal Components Analysis for monitoring the West Valley Liquid Fed Ceramic Melter. Waste Management 88 Proceedings, Tucson, AZ.

Wise, B.M., N.L. Ricker, D.J. Veltkamp and B.R. Kowalski (1990). A Theoretical Base for the Use of Principal Component Models for Monitoring Multivariate Processes. *Process Control & Quality* **1**(1), 41–51.

ANALYSIS AND CONTROL OF COMBINED DISCRETE/CONTINUOUS SYSTEMS: PROGRESS AND CHALLENGES IN THE CHEMICAL PROCESSING INDUSTRIES

Paul I. Barton and Taeshin Park
Department of Chemical Engineering
Massachusetts Institute of Technology
Cambridge, MA 02139

Abstract

System behavior during many transients of interest in the chemical processing industries exhibits significant discrete aspects in addition to the more familiar continuous behavior. A typical chemical plant can be viewed as a combined system of interacting continuous-state and discrete-state subsystems. Continuous behavior arises from phenomena such as mass, energy and momentum conservation, and discrete behavior arises due to control actions from intelligent control systems, sequence control systems, automatic protective devices, etc., and due to physical phenomena such as phase changes.

Analysis and control of combined discrete/continuous (or hybrid) systems has recently become an active area for research in several communities. Typically, continuous-state behavior is modeled in terms of differential equations, whilst models of discrete-state behavior rely on the notion of transition systems such as finite state machines. Although theoretical and analytical tools have been developed for each subsystem in isolation, the current challenge lies in similar developments for hybrid systems in which both aspects of behavior are important and must be integrated. Further, the issues of control and/or design, which appear to be closely related, are even more difficult.

The current state of hybrid systems analysis and control is reviewed, focusing on potential applications and challenges in the chemical processing industries. Current progress and problems are commented on.

Keywords

Combined discrete/continuous systems, Hybrid systems, Modeling and simulation, Control, Verification, Phase equilibrium.

Introduction

There are many examples in the chemical processing industries of transients in which discrete phenomena superimposed on the more familiar continuous system dynamics make a significant contribution to the overall system response. In these situations, the behavior of a typical chemical plant is probably more correctly viewed as that arising from a series of coupled and interacting discrete and continuous subsystems. In particular, the strong coupling of these two facets of process behavior in many problems of practical interest demands the development of analytical technologies that can address discrete and continuous behavior appropriately and simultaneously. This paper is devoted to the progress and challenges in the development of such technologies from a chemical processing industry perspective.

Early work and reviews in this area focused primarily on the issues of modeling and simulation (Fahrland, 1970; Ören, 1977), and the term combined discrete/continuous system was adopted to avoid confusion with hybrid digital/analogue approaches to computation. In recent years, the study of such systems has become popular again, and now encompasses the issues of modeling and simulation, analysis, control, verification, and even design. In this latter work, the term hybrid system has become the vogue. In the interests of brevity we shall adopt the term hybrid system for the remainder of this paper.

The study of hybrid systems in the chemical processing industries is motivated by the fundamentally hybrid nature of process transients. While continuous behavior arises in a familiar manner from phenomena such as mass, energy, and momentum conservation, it is useful to classify discrete behavior into two broad categories (Barton and Pantelides, 1994): physico-chemical or autonomous discontinuities, and discrete controls and/or disturbances.

Physico-chemical discontinuities arise as an integral part of the physical behavior of most systems, especially if system behavior is studied over a large enough region of state space. Numerous examples include phase changes, flow reversals, shocks and transitions (e.g., laminar to turbulent to choked), discontinuities in equipment geometry, internal features of vessels (e.g., weirs), etc.. In all of these cases, discrete phenomena occur purely as a consequence of the system moving through state space.

On the other hand, discrete controls and/or disturbances are most frequently encountered in the study of process operations. Examples include digital regulatory control, the use of hybrid controllers to regulate highly nonlinear continuous dynamics (Friedrich et al., 1995), process upsets and the action of passive protective devices (Park and Barton, 1994), and planned operational changes such as start-up, shut-down and feedstock changeovers (Sédès, 1995). Further, there are whole classes of processes such as batch, semi-continuous and periodic processes that rely on discrete control actions to implement normal operation, hence even the design of such systems can be viewed as a hybrid problem (Allgor et al., 1996).

From the above, we can see that dynamic modeling and simulation in the chemical process industries is most conveniently viewed as a hybrid problem (Barton and Pantelides, 1994), and hybrid process simulators such as gPROMS (Barton, 1992) and ABACUSS[1] are beginning to be used in a widespread manner. Hence, the current challenge lies in extending this hybrid world view to other problems where a simultaneous treatment of discrete and continuous phenomena can yield significant benefits. Although there are few tangible examples at present, we can imagine several potential areas of application:

- hybrid regulatory control, which can be subdivided into the control of systems exhibiting autonomous discontinuities, and the use of hybrid control laws for highly nonlinear continuous systems.
- qualitative analysis of systems exhibiting autonomous discontinuities in some form of analogy to the study of nonlinear dynamics.
- design, optimization and verification of sequential controls (e.g., operating procedure synthesis, recipe design in batch processes).
- design and verification of logic based (implicit) control systems that embed multiple functionalities, for example the permissive and shut-down functionalities of a safety interlock system.

While this list of potential applications is impressive, the underlying theory supporting the study of hybrid systems is currently very limited, there is no consensus on how to model these systems, and very few applications have been demonstrated. In light of the immature status of this field of study, this paper will attempt to classify phenomena inherent in the class of hybrid systems of primary interest to chemical engineers, generalize certain notions concerning these systems, review the relevant results currently reported in the literature, and study in detail simple multi-phase systems, which represent a class of hybrid systems of particular interest to chemical engineers.

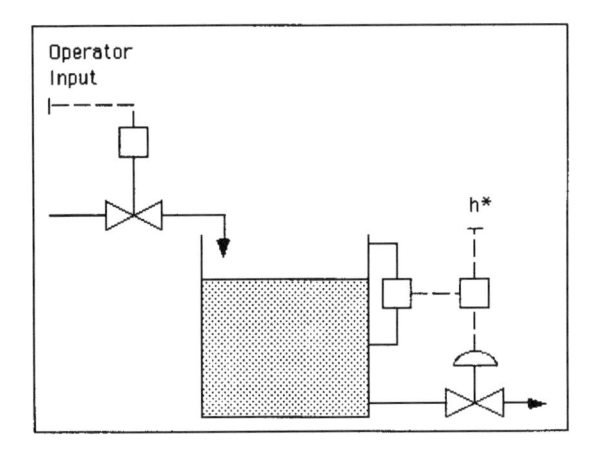

Figure 1. Buffer tank and related instrumentation.

Hybrid Systems

Fig. 1 shows a simple buffer tank and its associated instrumentation which will used to illustrate some of the properties of hybrid systems discussed in this section. Since it is difficult to discuss an example without reference to a model, a straightforward model of this system would include three ODEs, representing the mass balance on the tank and the stem position dynamics for the two valves, coupled with four algebraic equations, one relating the level in the tank to the pressure at the tank outlet, two relating the pressure drop, stem position and flow through

---

[1] ABACUSS (Advanced Batch and Continuous Unsteady-State Simulator) Process Modeling Software, a derivative work of gPROMS Software, Copyright 1992 by the Imperial College of Science, Technology and Medicine.

each valve, and finally the proportional control law. The inputs to this system will then be the upstream, downstream and atmospheric pressures, the level set point, and the on/off operator input to the inlet valve.

Definitions

Hybrid systems inherently combine coupled discrete and continuous sub-systems interacting with each other. In our example, the operator input to the on/off valve might be one of the discrete sub-systems, and the overall DAE model near to the nominal steady-state will be one of the continuous sub-systems. Depending upon a particular application, various classes of hybrid systems can arise with respect to the configuration and types of interaction between continuous and discrete sub-systems. However, at the lowest level hybrid systems are considered as interacting networks of finite automata (Grossman, et al., 1993; Nerode and Kohn, 1994).

For the purpose of the paper, we employ the notion of hybrid systems focused on transitions between continuous sub-systems. Fig. 2 depicts this notion of hybrid systems, where each node represents a continuous sub-system and each arc represents a transition between continuous sub-systems. Transitions can be *autonomous* or *controlled* by discrete sub-systems. It is usually assumed that these transitions are instantaneous, and the instant at which these transitions occur is called the *transition time*, which is not in general known *a priori* and is defined implicitly by *transition conditions*.

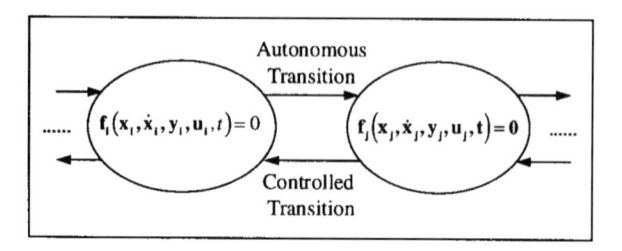

Figure 2. Autonomous/Controlled transitions in hybrid systems.

With this notion of hybrid systems, they experience a repeated sequence of two steps: a *continuous state transformation* and a *discontinuous state transformation* where a state is a point on the trajectory determined by the current continuous sub-system. During a continuous state transformation, the current continuous sub-system is evolving over time continuously according to the set of governing equations. At transition times, a discontinuous state transformation occurs autonomously or by control actions exerted by discrete sub-systems. Discontinuous state transformations can be simple parameter or control settings, or changes in the functional form of the governing equations, including dimensional change of the state space. Discontinuous state transformations result in a new current continuous sub-system, and continuous state transformation is resumed. In summary, a discontinuous state transformation changes the *mode* of a hybrid system (active continuous sub-system at a given time).

To illustrate these definitions, consider the response of the buffer tank (initially at steady-state) to a major upset that leads to a dramatic and prolonged increase in the inlet flow. This can be modeled by, for example, a step change in the upstream pressure. A continuous state transformation will immediately follow, and time will advance as the level rises and the controller attempts to compensate by increasing the signal to the control valve, until the valve stem becomes fully open. This marks a transition time. Note that this transition time is defined implicitly by a condition that the variable representing the stem position reaches an upper bound, and the actual time can only be determined by playing out a particular scenario. This triggers an autonomous discontinuous state transformation: the current continuous-subsystem is switched to a mode in which the differential equation coupling the stem position to the control signal is replaced by an algebraic equation setting the stem position to the fully open value. Following this transition, the system reverts to a continuous state transformation in this new mode (new functional form for the governing equations). It should be noted that there is an asymmetry in this transition: while the *stem position* reaching a threshold marks a transition to modes not containing the valve dynamics, it is the *control signal* (which is not bounded) dropping back below this same threshold that marks a return to modes containing the valve dynamics. Hence, the pending transition conditions can be a function of the current mode.

This new continuous state transformation begins, the level continues to rise as time advances, until eventually the level in the tank reaches the top of the vessel. This transition time is also implicitly defined by the level in the vessel crossing a threshold, and leads to another autonomous discontinuous state transformation to a mode in which the mass balance has a term for material overflowing the vessel in addition to those for the inlet and outlet streams, and the outlet valve is still shut. Continuous state transformation begins again until the operator notices liquid on the plant floor, at which point he pushes a button to close the on/off valve. This can be considered a controlled discrete transformation: the operator is a discrete controller that monitors the continuous state and implements discrete control actions such as pushing a button. Alternatively, a passive protective system may have detected an overflow and implemented the same action. This controlled transformation in the discrete signal to the inlet valve immediately triggers an autonomous discontinuous state transformation to a mode in which the inlet valve dynamics are now present. During the continuous state transformation that follows, the variable representing the inlet valve stem position will decrease until it closes (another discontinuous transformation), and the system will then experience a final continuous state transformation until the controller closes the outlet valve.

It should also be noted that this *sequence* of continuous and discontinuous state transformations experienced by a hybrid system is a function of the specific scenario. In general, there may be a very large number of possible sequences a system may experience. For example, in response to a larger upstream disturbance, the tank may overflow before the control valve saturates, leading to another distinct possible sequence.

There are three main issues in the study of hybrid systems: continuous systems, discrete systems, and an interface between them. While the continuous behavior is effectively modeled in terms of differential equations, the modeling of the discrete behavior depends upon the application domain and there is currently no universal framework; current practice relies on the notion of transition systems (e.g., finite automata, Petri-nets, Grafcet, synchronous modeling language, etc.). In this paper, the modeling of continuous behavior or discrete behavior will not be treated in detail because we are mainly interested in *hybrid phenomena*, which are not observed in purely continuous systems nor purely discrete systems. However, it should be noted that the degree of model complexity for each sub-system and the nature of the interface between them depend on both the particular application and the characteristics of the problem (e.g., modeling and simulation, control, verification, etc.) because there is a direct trade-off between model complexity and the strength of the results that can be derived.

Discontinuous state transformations or mode changes are hybrid phenomena that can occur in the class of hybrid systems we have defined. The hybrid system changes its mode based on the partition of the state space (not necessarily disjoint — see the valve saturating above), which is determined by the inherent dynamics of the embedded continuous systems (for autonomous transitions) and/or by the interface between the continuous sub-systems and input symbols of discrete sub-systems (for controlled transitions). In the example above, there are two input symbols for the operator: liquid is spilling on the floor, or not. In any case, we know that *events* occur, which is the satisfaction of the corresponding transition conditions at the boundaries of the current partition. For controlled transitions, output symbols of the discrete sub-systems are associated with modes of the hybrid system. For example, the on/off output from the operator categorizes the system into two distinct sets of modes.

There can be various classes of mode changes depending upon the nature of the embedded continuous system. The next section discusses classes of hybrid phenomena for ODE-embedded hybrid systems and DAE-embedded hybrid systems.

Hybrid Phenomena in ODE-embedded Hybrid Systems

The continuous dynamics of this hybrid system are governed by a system of ordinary differential equations of the form:

$$\dot{\mathbf{x}}(t) = \mathbf{g}(\mathbf{x}(t), \mathbf{u}(t), t) \qquad (1)$$

where $\mathbf{x}, \dot{\mathbf{x}} \in \mathbf{R}^n$, $\mathbf{u} \in \mathbf{R}^l$, and $\mathbf{g}: \mathbf{R}^n \times \mathbf{R}^l \times \mathbf{R} \mapsto \mathbf{R}^n$. The unknowns $\mathbf{x}$ are differential variables, $\mathbf{u}$ are the known system inputs, and t is the independent variable time. Hybrid phenomena due to discontinuous state transformation can be partitioned into disjoint two classes: *jumps* and *switching*, both of which can be autonomous or controlled by discrete sub-systems (Branicky et al., 1994). Jumps refer to the case where the differential variables change discontinuously in an autonomous manner or in response to an external control, and switching refers to the case where the functions $\mathbf{g}$ and/or $\mathbf{u}$ change discontinuously in an autonomous manner or in response to an external control. Examples are given by Branicky et al. (1994).

Note that the number of independent initial conditions is always n, and in the absence of autonomous or external actions leading to discontinuities in $\mathbf{x}$, these n initial conditions can be used to preserve continuity of the n differential variables because the differential variables represent conserved quantities or are directly related to them. Hence, jumps can only be caused by autonomous or control actions directly causing discontinuities in $\mathbf{x}$, and $\mathbf{x}$ is continuous at switching transitions even if there are discontinuities in $\mathbf{u}(t)$ or $\dot{\mathbf{x}}(t)$. In other words, $\dot{\mathbf{x}}(t)$ can experience discontinuities even though $\mathbf{x}(t)$ is continuous. For the same reason, it is always possible to impose n jumps in $\mathbf{x}$ at a transition.

Hybrid Phenomena in DAE-embedded Hybrid Systems

The continuous dynamics of this hybrid system are governed by a system of differential-algebraic equations of the general form:

$$\mathbf{f}(\mathbf{x}(t), \dot{\mathbf{x}}(t), \mathbf{y}(t), \mathbf{u}(t), t) = 0 \qquad (2)$$

where $\mathbf{y} \in \mathbf{R}^m$ and $\mathbf{f}: \mathbf{R}^n \times \mathbf{R}^n \times \mathbf{R}^m \times \mathbf{R}^l \times \mathbf{R} \mapsto \mathbf{R}^{n+m}$. The unknowns $\mathbf{y}$ are referred to as algebraic variables. Similar to the ODE-embedded hybrid systems, there are two classes of hybrid phenomena: *jumps* and *switching*. Jumps refer to the case where the differential variables change discontinuously in an autonomous manner or in response to an external control, and switching refers to all other autonomous or controlled discontinuities.

Unlike ODE-embedded hybrid systems, other discontinuous changes (e.g., discontinuities in $\mathbf{u}(t)$) can also cause jumps, and the number of jumps that can be imposed independently at a transition is not necessarily n. This complication is caused by the fact that a system of

DAEs requires *consistent* initial conditions in order to start continuous state transformations (Pantelides, 1988). Consistent initial conditions are determined by the original system (Eqn. (2)), first and high order differentials with respect to time, and a number of additional independent specifications on the initial condition that can be expressed in general (Barton and Pantelides, 1994) by:

$$\mathbf{c}(\mathbf{x}(t^*), \dot{\mathbf{x}}(t^*), \mathbf{y}(t^*), \mathbf{u}(t^*), t^*) = 0 \quad (3)$$

where t^* is a transition time. The number of these additional specifications (or degrees of freedom γ for consistent initialization) is not necessarily n and can be smaller than n. Depending upon the value of γ and the classification of DAEs according to the *differential index*, care should be taken to analyze hybrid phenomena properly. We present definitions of differential index and degrees of freedom for consistent initialization before considering each case separately.

> *Definition 1* : The *(differential) index* ν of a system of DAEs (Eqn. (2)) is the minimum number of times that all or part of it must be differentiated with respect to time in order to define $\dot{\mathbf{x}}$ and $\dot{\mathbf{y}}$ uniquely as functions of $\mathbf{x}, \mathbf{y}$, and $\mathbf{u}$ (and its time derivatives), and t (Brenan et al., 1989).

> *Definition 2* : The number of degrees of freedom γ for consistent initialization is the number of additional independent specifications required to determine a set of consistent initial values. Note that $\gamma \leq n$ and that equality holds if and only if

$$\text{rank}\left[\mathbf{f_x f_y}\right] = n + m \quad (4)$$

Case 1 : $\text{rank}\left[\mathbf{f_x f_y}\right] = n + m$

This equality holds for all index 0 systems (ODEs) and most index 1 systems. In this case, n degrees of freedom can be used to preserve continuity of all n differential variables in the absence of autonomous or external control actions that may deplete some degrees of freedom. If this is the case, jumps can be introduced only by discontinuous changes in $\mathbf{x}$. For example, consider the case in Eqn. (5) with $\nu = 0$ and $\gamma = 2$:

$$\begin{aligned} \dot{x}_1 &= x_0 - x_1 \\ \dot{x}_2 &= x_1 - x_2 \\ x_0 &= u_0(t) \end{aligned} \quad (5)$$

Any discontinuous change in $u_0(t)$ cannot cause jumps in x_1 or x_2 because two degrees of freedom can always be used to preserve continuity of x_1 and x_2. Note that Eqn. (5) is a simple model for a cascade of two buffer tanks with the inlet flow to the first tank specified. This case is discussed by Barton (1992), and for the more restricted class of linear-implicit DAE-embedded systems subject to discontinuities in the controls, this is case 1 of Majer et al. (1995).

Case 2 : $\text{rank}\left[\mathbf{f_x f_y}\right] < n + m$

This includes some index 1 systems and all index 2 or greater systems. In this case, γ is strictly less than n because all n differential variables are not independent and there are some hidden relationships between them. Therefore, continuity of all n differential variables cannot be guaranteed. Furthermore, the selection of γ differential variables to be continuous among n differential variables may not be unique, and a discontinuous change in $\mathbf{u}(t)$ may cause jumps depending upon the particular problem. However, this problem can be resolved uniquely in certain special cases. Below, we discuss existing results and introduce some new cases that can be resolved uniquely.

Special Index 1 Problems

This case corresponds to index 1 problems whose Jacobian with respect to the highest order derivatives is singular. For example, consider the DAEs in Eqn. (6) with $\nu = 1$ and $\gamma = 1$:

$$\begin{aligned} \dot{x}_1 + 2\dot{x}_2 &= 1 \\ x_1 + x_2 &= u(t) \end{aligned} \quad (6)$$

The system of equations constraining the initial values of $\mathbf{x}$ and $\dot{\mathbf{x}}$ are:

$$\begin{aligned} \dot{x}_1 + 2\dot{x}_2 &= 1 \\ x_1 + x_2 &= u(t) \\ \dot{x}_1 + \dot{x}_2 &= u'(t) \end{aligned} \quad (7)$$

Since $\gamma = 1$, the choice of one differential variable to be continuous is not unique, and only one out of $\{x_1, x_2\}$ can be made continuous in the presence of a step change in $u(t)$. Clearly either choice will lead to a different initial state, and this would appear to introduce an ambiguity.

Brüll and Pallaske (1991) study index 1 linear-implicit DAEs and introduce the concept of a genuine initial value corresponding to a jump in the controls. This can be used to define jumps uniquely for all such systems with $n + m \leq 2$ (e.g., Eqn. (6)). However, they note that for $n + m > 2$ there will exist systems for which a genuine initial value following a jump in the controls does not exist. Majer et al. (1995) also study index 1 linear-implicit DAE embedded systems subject to discontinuities in the controls from a slightly different perspective, and identify several different cases, some of which can be resolved without ambiguity.

The case of changes in the general functional form of Eqn. 2, for example as a result of an autonomous switching, has not been investigated.

Problem with No Degrees of Freedom, $\gamma = 0$

There are no degrees of freedom available for consistent initialization and the new state of the system at the transition is uniquely determined. In this case, an arbitrary introduction of jumps or switching is impossible. However, a discontinuous change in $\mathbf{u}(t)$ may cause jumps. For example, consider the DAEs in Eqn. (8) with $v = 3$ and $\gamma = 0$:

$$\begin{aligned} \dot{x}_1 &= x_0 - x_1 \\ \dot{x}_2 &= x_1 - x_2 \\ x_2 &= u_2(t) \end{aligned} \quad (8)$$

which is equivalent to specifying the outlet flow from the second tank. The system of equations constraining the initial values of $\mathbf{x}$ and $\dot{\mathbf{x}}$ are:

$$\begin{aligned} \dot{x}_1 &= x_0 - x_1 \\ \dot{x}_2 &= x_1 - x_2 \\ x_2 &= u_2(t) \\ \dot{x}_2 &= u_2'(t) \\ \ddot{x}_2 &= u_2''(t) \\ \ddot{x}_2 &= \dot{x}_1 - \dot{x}_2 \end{aligned} \quad (9)$$

There are no degrees of freedom in Eqn. (9) and the initial state of the system is uniquely determined by $u_2(t)$. A step change in $u_2(t)$ causes jumps in x_1 and x_2.

Unique Equivalent Index 1 Problem

A family of index 1 systems of DAEs that preserve the invariants of the original DAEs can be obtained for any system that does not satisfy Eqn. (4) by index reduction techniques (e.g., index reduction using dummy derivatives (Mattsson and Söderlind, 1993)). If this family of equivalent index 1 systems, which satisfies Eqn. (4) by construction, contains only one member, we will have the same situation as in Case 1. For example, consider the DAEs in Eqn. (10) with $v = 2$ and $\gamma = 1$:

$$\begin{aligned} \dot{x}_1 &= x_0 - x_1 \\ \dot{x}_2 &= x_1 - x_2 \\ x_1 &= u_1(t) \end{aligned} \quad (10)$$

which is equivalent to specifying the outlet flow from the first tank. Applying index reduction method using dummy derivatives (Mattsson and Söderlind, 1993), we derive the unique equivalent index 1 system of DAEs:

$$\begin{aligned} x_1' &= x_0 - x_1 \\ \dot{x}_2 &= x_1 - x_2 \\ x_1 &= u_1(t) \\ x_1' &= u_1'(t) \end{aligned} \quad (11)$$

where x_1' is a dummy algebraic derivative. Clearly, in this case, continuity of x_2 is implied without ambiguity.

Autonomous Index Transition

Consider a vessel which is partially full of an incompressible[2] liquid initially, whose dynamics are described by the system of DAEs in Eqn. (12):

$$\begin{aligned} \dot{N}_1 &= -x_1 F_{out} \\ \dot{N}_2 &= -x_2 F_{out} \\ \dot{H} &= \dot{Q}(t) - F_{out} C_p (T - T_{ref}) \\ x_1 &= N_1 / (N_1 + N_2) \\ x_2 &= N_2 / (N_1 + N_2) \\ H &= C_p (N_1 + N_2)(T - T_{ref}) \\ hA &= (N_1 + N_2)/\rho(T) \\ &\text{if } h \le h_{max} \text{ then } F_{out} = 0 \\ &\text{else } h = h_{max} \end{aligned} \quad (12)$$

The index of the system increases from one to two when the liquid has expanded sufficiently for h to become equal to h_{max}, since the previously independent molar component holdups, N_1 and N_2, become constrained by the fixed volume of the vessel and the incompressibility of the liquid. Applying the index reduction algorithm (Mattsson and Söderlind, 1993) to the index 2 problem, the family of equivalent index 1 systems in this case is composed of two members: one with N_1 selected as a dummy algebraic variable, and one with N_2 selected. In either case, continuity of the algebraic molar holdup is redundant rather than inconsistent with the initial condition defined by assuming continuity of the differential molar holdup. Clearly, this redundancy eliminates any ambiguity in the new initial state. Further work is required to identify these classes of DAEs and switchings for which this redundancy property holds, and hence simulation and analysis can proceed in an unambiguous fashion.

---

[2] Here we use the term incompressible to refer to a fluid with temperature and concentration dependent density, but negligible pressure dependence.

Current Status of Hybrid Systems Study

Modeling and Simulation

Despite the expected benefits of modeling and simulation of hybrid systems, this technology is not yet widely used in the chemical processing industries because most existing tools primarily focus on modeling continuous behavior alone. For a modeling tool for hybrid systems to be useful, it should have the capability to model a variety of discrete behavior (e.g., logic based control system, sequence control, etc.) with a well defined interface between continuous and discrete sub-systems, and its solution algorithms should locate events accurately and handle hybrid phenomena properly.

The current trend is to pose the hybrid system simulation problem as a sequence of initial value problems, described by DAEs, interspersed by autonomous or controlled state transitions (Cellier et al., 1993; Barton and Pantelides, 1994). Most existing tools rely on simple *if-then-else* constructs for the modeling of discrete behavior, although Barton and Pantelides (1994) also provide systematic modeling of sequences of control actions. This is implemented in the gPROMS and ABACUSS systems.

Further work is required to develop a general-purpose modeling language for hybrid systems that can express a variety of discrete sub-systems such as logic-based controllers, sequence controllers, etc.

Regarding solution algorithms, guaranteed and efficient location of events defined by implicit conditions is now available in ABACUSS (Park and Barton, 1996). In addition, the ABACUSS system is capable of automated consistent initialization and numerical integration of a broad class of nonlinear differential-algebraic equations of arbitrary index (Feehery and Barton, 1996a). Hybrid simulations with fluctuating index may also be solved, provided that for high index systems any jumps are properly defined because there are either no degrees of freedom for consistent initialization, or there is a unique equivalent index 1 problem (see above). Further work is required to handle jumps properly in a broader class of high index systems.

Literature examples of hybrid system modeling and simulation applied to the chemical processing industries include batch process simulation (see Barton (1994) for a review), simulation of process start-up and shut-down (Barton, 1992; Sédès, 1995), simulation of complex pressure swing adsorption cycles (Barton, 1992), and simulation of phase changes (Barton, 1992; Gopal and Biegler, 1994).

Analysis and Control

Even though theory for the analysis and control of purely continuous or purely discrete systems exists, similar techniques for hybrid systems do not exist at present primarily due to the difficulty of extending the usual definitions of stability, controllability, etc. to such systems. The current approach is to attempt to extend the theory for dynamic systems described by state-space models. All the hybrid models developed in the literature for analysis and control are ODE-embedded, and most of them are closely related to each other. They are slightly different with respect to the classes of hybrid phenomena supported, the way to partition state space, etc.. These hybrid system models may be categorized into two classes depending upon the existence of *explicit* discrete sub-systems. No complete review of all the models is attempted here. Instead, we will present a hybrid system model from each class. See Grossmann et al. (1993) and Branicky et al. (1994) for others.

Back et al. (1993) developed an autonomous hybrid system model without explicit discrete sub-systems. The dynamics of this model is described by $\dot{\mathbf{x}}(t) = \mathbf{g}(\mathbf{x}(t), \mathbf{p}(t))$ where $\mathbf{p}(t) \in \mathbf{R}^k$ is a vector of parameters. The problem domain V in the state space is partitioned into the form, $V = \bigcup_{q \in I} V_q$ where I is a finite index set and V_q is an open, connected subset of $\mathbf{R}^n$. Each V_q is associated with one particular system of ODEs, and has an open set U_q such that $\overline{U}_q \subset V_q$ and piecewise smooth ∂U_q where $\overline{U}_q$ is the closure of U_q and ∂U_q is the boundary of U_q. The boundary ∂U_q is defined by a set of boundary functions such that $\mathbf{x} \in U_q$ if and only if $h_{q,i}(\mathbf{x}) - C_{q,i} > 0 \; \forall i$. The transition function $T_q : S_q \rightarrow V \times I$ is defined on transition boundary S_q where $S_q \subseteq \partial U_q$. Autonomous jumps and autonomous switching are allowed through transition functions which are evaluated when a transition boundary is hit. The model of Tavernini (1987), the so called *differential automata*, is very similar to this model, but does not support autonomous jumps. Tavernini (1987) proves the uniqueness of the solution of the initial value problem with a set of assumptions on ODEs, boundary functions, and discrete transitions, and analyzes the effect of small perturbations in the initial condition upon the sequence of discontinuous state transformation. This class of models would seem an appropriate basis to develop qualitative tools for understanding hybrid dynamics. For example, in general simple multi-phase systems cannot be modeled by a single system of equations throughout large enough domains of state space. Modeling of simple multi-phase systems is therefore one application of Tavernini-type hybrid models that support autonomous switching between a set of continuous subsystems. This will be discussed in detail below.

While the models of Back et al. (1993) and Tavernini (1987) are developed for simulation, Nerode and Kohn (1993) and Antsaklis et al. (1993) developed hybrid systems models for controlling continuous sub-systems with high-level discrete sub-systems. The model of Antsaklis et al. (1993) consists of three basic parts: the plant, the controller, and the interface. The plant (continuous sub-systems) is modeled as a time-invariant state-space form:

$$\dot{\mathbf{x}}(t) = \mathbf{g}(\mathbf{x}(t), \mathbf{u}(t))$$
$$\mathbf{z}(t) = \mathbf{h}(\mathbf{x}(t)) \quad (13)$$

where $\mathbf{z} \in \mathbf{R}^p$. The controller (discrete sub-systems) is modeled as a deterministic finite automaton, which can be specified by a quintuple, (Q, I, O, δ, η), consisting of the set of states, the set of plant input symbols, the set of controller output symbols, transition functions, and output functions, respectively. The action of controller is described by

$$q(t) = \delta(q(t^-), i(t))$$
$$o(t) = \eta(q(t)) \quad (14)$$

where $q(t) \in Q$, $i(t) \in I$, and $o(t) \in O$. The plant and controller communicate through an interface consisting of two memoryless maps, DA and AD. The first map $DA: O \rightarrow \mathbf{R}^l$ converts a control symbol to a constant plant input, $\mathbf{u}(t) = DA(o(t))$. The second map $AD: \mathbf{R}^n \rightarrow I$ maps the state space of the plant to the set of plant input symbols, $i(t) = AD(\mathbf{x}(t))$. The map AD is based upon a partition of the state space, where each region of the partition is associated with one plant input symbol. The model of Nerode and Kohn (1993) is similar in its concept. Both models support only controlled switching. One application of this type of models is the control of a batch crystal growth process (Friedrich et al., 1995). In order to control the crystal structure, it is necessary to move the concentration of a key component in the solution along a prescribed trajectory during crystal growth. Due to the highly nonlinear nature of the process dynamics, moving the process to another operating point, or along a trajectory, requires switching between different control schemes. This is because a conventional control scheme does not work well throughout state space, even though it can provide satisfactory performance at a single operating point.

For both types of hybrid system models, a *correct* partition of the state space is essential for the performance of the models. For example, autonomous switchings, which are determined by the partition, usually represent discontinuous phenomena inherent in the physical system being modeled (e.g., phase changes). Therefore, a correct partition is a prerequisite for correct modeling of autonomous switching. However, correct partitioning is a very difficult task because it requires identification of all possible behaviors of the continuous sub-systems in the state space in advance. For example, see the section on multi-phase systems. Hybrid systems with controlled transitions also require a proper partition of the state space

by the controller (discrete sub-systems) to be able to achieve control specifications (e.g., stability, controllability, etc.). In this case, a partition of the state space represents the symbolic partitioning of the continuous sub-systems that the controller recognizes and utilizes to decide upon the appropriate control action. This partition effectively determines the interface between the state space of the continuous sub-systems and input symbols of the discrete sub-systems, or simply an *AD* map. Lemmon and Antsaklis (1994) define transition stability (*T*-stability) in terms of whether or not a specified transition in the continuous sub-systems can be controlled in a stable manner, and propose an algorithm to determine the partition that ensures *T*-stability when the continuous dynamics is described by $\dot{\mathbf{x}}(t) = \mathbf{g}_0(\mathbf{x}(t)) + \sum_{i=1}^{l} u_i(t) g_i(t)$ where $g_i(t)$ is a family of Lipschitz continuous functions and $u_i(t)$ is linear. Most hybrid system models in the literature employ a set of boundary functions in the form of logical propositions to represent a partition of the state space. Further work is required to investigate alternative way to represent the partition, especially those that characterize the discontinuites resulting from inherent physical phenomena.

A systematic and general way to design hybrid control systems satisfying given control objectives does not exist. Bencze and Franklin (1994) propose a design method based on control engineering practice, which is to design a set of controllers for each operating mode in state space and then to combine them designing a meta-controller (logic to switch between controllers) and interface. Lemmon and Antsaklis (1994) approach the design of hybrid control systems from the viewpoint of discrete event systems (DES). They obtain a DES plant model corresponding to continuous sub-systems by adding an interface through a partition of state space as described above, then synthesize controllers for the DES plant model using a DES controller design method (Antsaklis et al., 1993).

Hybrid System Verification

For hybrid systems implementing sequence and/or logic control, correctness must be formally verified. Currently, only a very limited class of hybrid systems (with respect to model complexity) can be verified in a formal manner. A typical sub-problem in the verification problem is *reachability analysis* — given a final discrete state, determine whether or not the hybrid system has a sequence of computation terminating at that final state from a given initial state. This reachability problem is undecidable already for very restricted classes of hybrid systems, e.g., constant slope hybrid systems, in which the right hand side of all differential equations is an integer constant (Alur et al., 1993; Kestne et al., 1993). Phase transition systems (Manna et al., 1993) and extended state transition graphs

(Nicollin et al., 1993) provide a similar framework for the verification of a limited class of hybrid systems using a temporal logic.

Since the formal verification of general hybrid systems is a formidable task, Lygeros et al. (1994) advocates the use of simulation to validate the performance of the hybrid system, even though it cannot replace formal proof techniques, because no formal analytical tools are currently available.

A major application of hybrid system verification in the chemical processing industries is verification of discrete control systems embedded in continuous systems. For example, consider the sequence control system embedded in the solid transport system of an aluminum smelting process (Probst and Powers, 1994). The sequence controller regulates the movement of crushed solids from storage tanks to a distribution bucket. In principle, a hybrid model should be developed and verified to confirm overall system functionality against specifications. However, the current state of hybrid system verification cannot address this problem. Instead, Probst and Powers (1994) abstracted the entire system as a set of discrete subsystems and applied symbolic model checking to verify the system.

Simple Multi-Phase Systems at Equilibrium

A broad class of hybrid dynamic systems of interest to chemical engineers is a well mixed region of known volume or pressure containing a multi-component mixture of fluids distributed between one or more thermodynamic phases (a *simple* multi-phase system (Modell and Reid, 1983)). Typically such a system will be one part of a larger system under study, such as a tray in a distillation tower. Certain transients of such a system will exhibit autonomous switching in the form of changes to the number and/or nature of the phases present (e.g., vapor, aqueous liquid, organic liquid, crystal) at well defined points in time (*phase transitions*). For the purposes of this discussion, we will restrict ourselves to so-called "equilibrium" models in which it is assumed that phase equilibrium is reached instantaneously in comparison to the other system time constants. We will also assume that the models are derived from balances on a control volume that encompasses the entire simple multi-phase system, which will typically lead to index-1 DAEs (Ponton and Gawthrop, 1991). This section will discuss the challenges in dynamic modeling and analysis of simple multi-phase systems.

As a simple example, consider the autonomous model for isobaric simple distillation of a mixture exhibiting at most two liquid phases proposed by Pham and Doherty (1990a) as shown in Fig. 3:

$$\frac{dz_i}{d\xi} = z_i - y_i \quad \forall i = 1 \ldots NC - 1 \quad (15)$$

$$\sum_{i=1}^{NC} z_i = 1 \quad (16)$$

$$Py_i = P_i^{SAT}(T)\gamma_i^I(T,\mathbf{x}^I)x_i^I \quad \forall i = 1 \ldots NC \quad (17)$$

$$\gamma_i^I(T,\mathbf{x}^I)x_i^I = \gamma_i^{II}(T,\mathbf{x}^{II})x_i^{II} \quad \forall i = 1 \ldots NC \quad (18)$$

$$\sum_{i=1}^{NC} y_i = 1 \quad (19)$$

$$\left(x_{NC}^I - x_{NC}^{II}\right)z_i = \left(z_{NC} - x_{NC}^{II}\right)x_i^I + \left(x_{NC}^I - z_{NC}\right)x_i^{II} \quad (20)$$

$$\forall i = 1 \ldots NC - 1$$

$$\sum_{i=1}^{NC} x_i^I = 1 \quad (21)$$

where $\mathbf{z}$ are the overall mole fractions for the liquid residue, $\mathbf{x}^I$ and $\mathbf{x}^{II}$ are mole fractions in liquid phases *I* and *II* respectively, and $\mathbf{y}$ are the mole fractions in the vapor phase. Any initial mixture lying within the projection of the heterogeneous liquid boiling surface onto the composition simplex will result in a residue composed of two liquid phases (Fig. 4). As such an orbit is traced forward in time it will eventually intersect the boundary of the heterogeneous liquid boiling surface, at which point the second liquid phase will disappear and the system will exhibit simple vapor-liquid equilibrium behavior for the remainder of the transient.

Eqn. (15)-(21) can be viewed as a member of the class of differential-algebraic equations for which, given values for the differential variables $z_i(\xi) \ \forall i = 1 \ldots NC - 1$, the algebraic Eqn. (16)-(21) can be solved to yield values for all the algebraic variables. The first problem arises from the fact that equality of temperature, pressure and chemical potential between all phases (Eqn. (17)-(18)) is only a necessary condition for phase equilibrium, and that in general multiple solutions satisfying this condition exist in and around multi-phase regions (Baker et al., 1982). For example, an examination of Eqn. (15)-(21) indicates that given the residue mole fractions $z_i \ \forall i = 1 \ldots NC - 1$, the trivial solution $\mathbf{x}^I = \mathbf{x}^{II}$ will satisfy these equations[3] throughout the entire state space, and within the heterogeneous boiling region there must exist at least one other solution corresponding to an unequal distribution of the components between the two liquid phases. Hence, we are faced with the question: which of the solutions to the necessary conditions corresponds to the state exhibited by the physical system under these conditions? A popular approach is to select that solution to the necessary conditions corresponding to a global extremum of the thermodynamic function appropriate to the specified thermodynamic restraints on the simple multi-phase

---

[3] It should also be noted that the Jacobian of Eqn. (16)-(21) will always be singular at the trivial solution, creating further complications for some numerical methods.

system. For example, for a system restrained to a given temperature and pressure, the solution corresponding to the global minimum of Gibbs free energy of the overall system is thermodynamically stable (Baker et al., 1982). Although this approach will be adopted for the remainder of this discussion, it should be noted that there is no guarantee that a physical system will actually reach the stable equilibrium state in the time horizon of interest, instead a metastable equilibrium state may be observed in certain regions of state space. This property of multiple solutions poses problems for consistent initialization, numerical solution, and analysis of these hybrid dynamic systems.

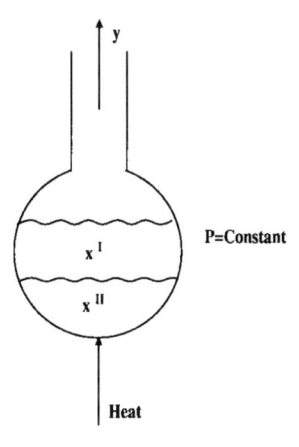

Figure 3. Heterogeneous simple distillation still.

The consistent initialization problem can be stated as given the DAE model of the simple multi-phase system and the appropriate number of additional equations (Eqn. (3)) to fully define the initial condition, find that equilibrium state corresponding to a global extremum of the appropriate thermodynamic function. If the additional equations explicitly give the initial mole numbers, temperature and pressure of the system, then Gibbs free energy should be minimized, and a recent review of approaches to this specific problem is given by McDonald and Floudas (1994). These authors also describe an approach that will guarantee location of the global minimum for an ideal vapor phase and a selection of liquid phase activity coefficient models (McDonald and Floudas, 1994; McDonald and Floudas, 1995). However, much work remains to solve general consistent initialization problems in a satisfactory manner. For example, consistent initialization of the model in Eqn. (15)-(21) occurs under restraints of given pressure and vapor fraction (i.e., the solution must lie on the bubble point surface) rather than temperature. Pham and Doherty (1990b) describe an algorithm that applies for mixtures limited by an upper critical solution temperature, but there is no discussion of how the solution found relates to a global extremum of the appropriate thermodynamic function. There are at least three significant unresolved research issues here. First,

given an arbitrary specification of the initial condition for a simple multi-phase system, how can the appropriate thermodynamic function be found, preferably in an automated manner? For example, in a dynamic system (where both temperature and pressure are rarely specified) what is the appropriate thermodynamic function for the very common specification of steady-state as the initial condition? Second, given the appropriate criterion, efficient and correct algorithmic approaches are required to find the global extremum for arbitrary chemical potential model. One approach might be an embedded global optimization in which the inner problem minimizes Gibbs free energy at a given mole number, temperature and pressure, and the outer problem adjusts the mole number, temperature and pressure to match the specified initial conditions (Gibbs free energy will be minimized at the temperature and pressure implied by the initial conditions). However, this approach is likely to be computationally prohibitive. Finally, consistent initialization of a typical process model will require the simultaneous initialization of many coupled simple multi-phase systems. Efficient approaches to this problem that properly take into account the coupling are required.

Given a consistent initial condition, it is then necessary to advance the system in the time domain. Here the appropriate thermodynamic function is clearer, at least for an explicit integration scheme. Two typical systems are one at a known volume (for example, a flash vessel) or a known pressure (for example, fluid constrained by a piston). In the constant volume case, function evaluations are performed at known mole numbers, volume and system energy, implying maximization of system entropy as the appropriate criterion. Further thought is required in the case of an implicit integration scheme. Moreover, it is possible for the restraints on a system to change at events during a transient. For example, an action that locks the piston above would correspond to a transition from a constant pressure to a constant volume system.

One approach to numerical integration (see, for example, Pham and Doherty (1990a)) is to eliminate the algebraic Eqn. (16)-(21) implicitly, and solve the ODE system:

$$\frac{dz_i}{d\xi} = z_i - y_i(\mathbf{z}, T(\mathbf{z})) \quad \forall i = 1...NC-1 \qquad (22)$$

This has the advantage in this case that $\mathbf{z}(\xi)$ are C^1 functions throughout state space, hence eliminating the discrete aspect of the problem. However, this type of elimination is not in general possible, and even if it is possible, leads to prohibitively expensive function evaluations in the numerical integration (each is, potentially, a global optimization), and in general does not guarantee elimination of discontinuities in time derivatives or algebraic variables. Another approach is to solve the original DAE system, and view the algebraic Eqn. (16)-

(21) as an algebraic system parameterized by $z_i(\xi)$ $\forall i = 1...NC-1$. For an initial condition within the heterogeneous boiling region, this system will exhibit multiple solutions and the consistent initialization should be designed to lock onto the appropriate branch in the parameter space. As the parameters $z_i(\xi)$ $\forall i = 1...NC-1$ vary along a trajectory this branch can probably be traced reliably by standard numerical integrators, until the boundary of the heterogeneous boiling region is encountered, which will correspond to a bifurcation point in parameter space. At this boundary, the branch being traced (vapor-liquid-liquid) will disappear, leaving only the trivial solution (vapor-liquid), and many variables will not be C^1. Furthermore, we know from the physical nature of this system that the two branches intersect at this point in parameter space, hence the need to switch to the vapor-liquid branch is clear. Further work is required to establish the theoretical basis for the use of bifurcation points in this fashion, and to develop a general, correct and efficient algorithm that could be used in a process simulator. Moreover, although simple multi-phase systems provide a good physical example, a preliminary investigation of the current literature has only found one brief discussion of DAEs exhibiting bifurcations in the algebraic equations. (Hairer and Wanner, 1991). Clearly research is required to elucidate the theoretical properties of such systems.

The other major complication in modeling of simple multi-phase systems is that in general one single system of equations cannot be used to describe the system behavior throughout the subset of state space of interest, and therefore we must accommodate autonomous switching. In fact, the model in Eqn. (15)-(21) is a very special case and only applies throughout the composition simplex as a result of both restraining the system to the bubble point surface and the existence of the trivial solution to Eqn. (18). For example, consider an isobaric model of a closed system exhibiting the same phase phenomena, with temperature as a time varying input. If the initial state is a vapor phase with a composition lying in the heterogeneous boiling region, and temperature is dropped monotonically throughout the simulation, the system will experience transitions from vapor phase only to vapor and one liquid phase to vapor and two liquid phases to finally just two liquid phases. The conditions (17) and (18) can be enforced in the vapor-liquid-liquid regime, but Eqn. (18) must be dropped in the vapor phase regime, and Eqn. (3) in the liquid-liquid regime. Further, in a particular regime certain variables may be undefined (e.g., vapor phase mole fractions in the liquid-liquid regime) and must be either dropped from the model or assigned dummy values.

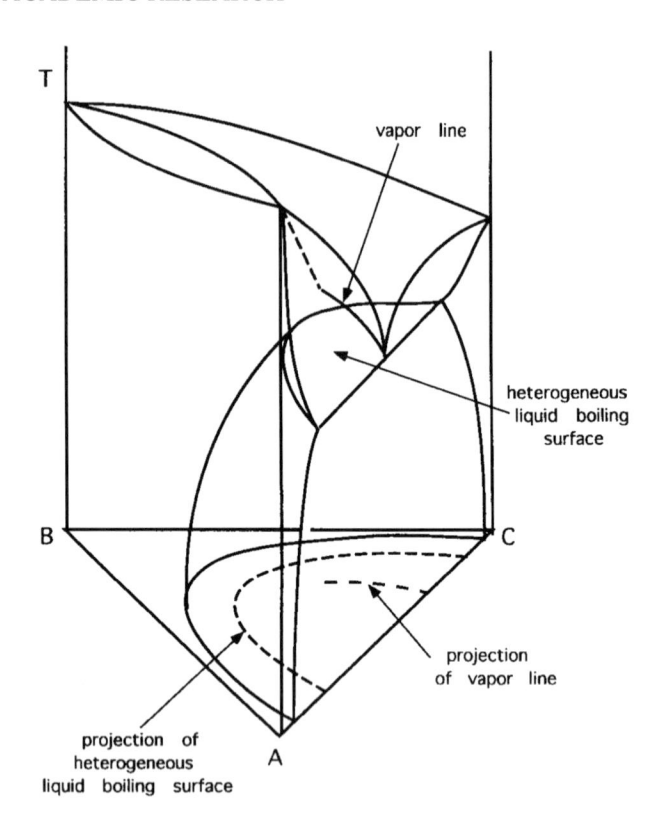

Figure 4. VLL equilibrium diagram of three component mixture.

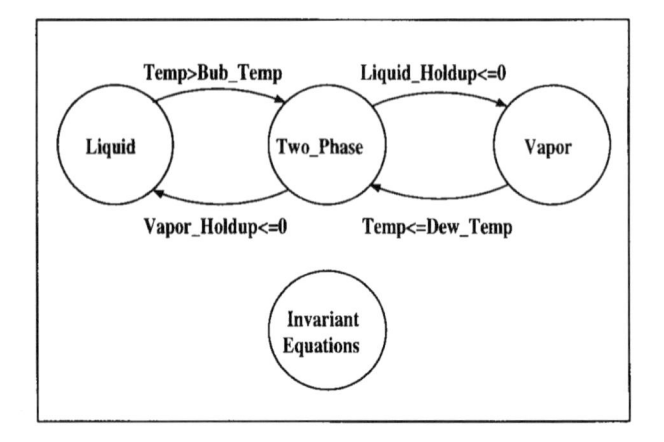

Figure 5. Finite automaton model for flash vessel.

One approach to autonomous switching in simple multi-phase systems is to employ a general finite automaton formalism (Barton and Pantelides, 1994) as now widely used in practice by both the ABACUSS and gPROMS (Barton, 1992) process simulators. For example, Barton (1992) models a flash vessel (only one liquid phase possible) with the finite automaton shown in Fig. 5. In general, this approach requires some form of a priori analysis of state space in order to identify the possible phase regimes (i.e., define partition of state space) and the logical propositions that define the boundaries between these partitions. For such an analysis to be appropriate, it must guarantee that the automaton always selects the stable phase regime at any point in state space. Further work is required to determine if, in general, such an a priori analysis is possible. From the point of view of analysis and control of simple multi-phase systems, this approach is promising because it is most compatible with the existing results discussed in previous section.

A recent publication (Gopal and Biegler, 1994) has sketched in a preliminary manner an alternative based on the introduction of slack variables in the equations defining equality of chemical potentials coupled with a set of inequalities that augment the dynamic system and implicitly define phase transitions through the slack variable values. Although elegant, this approach only solves the necessary conditions for phase equilibrium, and in its present form relies on the costly solution of an iterated linear program at each corrector iteration in a numerical integration. Feehery and Barton (1996) describe a potentially much more efficient approach to the simulation of dynamic systems coupled with inequality constraints.

Conclusions

Many applications in the chemical processing industries can be posed as a hybrid system problem because most transient behavior in chemical plants exhibits both continuous and discrete behavior. The notion of hybrid systems focused on transitions between continuous sub-systems is clarified. Based on this notion of hybrid systems, hybrid phenomena in DAE-embedded hybrid systems are classified and examined according to the class of embedded DAEs. In particular, it has been shown that analysis of hybrid phenomena in this case is closely related to consistent initialization problems and the differential index of the DAEs.

The current status of hybrid systems study is surveyed. The complexity of the hybrid system model (the complexity of each constituent sub-system and the degree of interaction) depends upon the particular domain of application since there is a direct trade-off between model complexity and/or generality and the ability to derive a specific result. For instance, models for simulation should include all the relevant details of continuous and discrete behavior, while models for formal verification purposes should be simple to yield strong results.

A general purpose modeling language for hybrid systems that can express various discrete behaviors is necessary to solve a variety of simulation problems along with the development of a solution algorithm to resolve the non-unique choice of differential variables to be continuous. It has been shown that simple multi-phase systems form a complex class of hybrid dynamic systems of particular interest to chemical engineers, and that they exhibit mathematical properties not addressed by the current theory. Further, there is no approach available at present that can capture all the issues associated with modeling of simple multi-phase systems in an appropriate manner.

From the point of view of analysis, control and design of hybrid control systems, a systematic method to express a proper partition of the state space for general cases, which is essential to analyze hybrid systems, is not available. The design of hybrid control systems satisfying given constraints is a formidable unresolved task along with the formal verification of hybrid systems except for limited cases.

Acknowledgments

This work was supported by the National Science Foundation under grant No. CTS-9321863 and by the Exxon Education Foundation.

References

Allgor, R.J., M.D. Berrera, P.I. Barton and L.B. Evans (1996). Optimal batch process development, *Computers chem. Engng.*, **20**, 885-896.

Alur, R., C. Courcoubetis, TA Henzinger, and P. Ho (1993). Hybrid automata: An algorithmic approach to the specification and verification of hybrid systems, In R.L. Grossman, A. Nerode, A.P. Ravn, and H. Rischel (Ed.). *Hybrid Systems*, Lecture Notes in Computer Science, Vol.736, Springer-Verlag, New York.

Antsaklis, P.J., J.A. Stiver, and M. Lemmon (1993). Hybrid system modeling and autonomous control systems, In R.L. Grossman, A. Nerode, A.P. Ravn, and H. Rischel (Ed.). *Hybrid Systems*, Lecture Notes in Computer Science, Vol. 736, Springer-Verlag, New York.

Back, A., J. Guckenheimer, and M. Myers (1993). A dynamical simulation facility for hybrid systems, In R.L. Grossman, A. Nerode, A.P. Ravn, and H. Rischel (Ed.). *Hybrid Systems*, Lecture Notes in Computer Science, Vol.736, Springer-Verlag, New York.

Baker, L.E., A.C. Pierce, and K.D. Luks (1982). Gibbs energy analysis of phase equilibria, *Soc. Petrol. Engrs. J.*, **22**, 731-742.

Barton, P.I. (1992). *The Modeling and Simulation of Combined Discrete/Continuous Processes*, Ph.D. thesis, University of London.

Barton, P.I. (1994). Batch process simulation — why and how, *Proc. of Aspen World*, Boston, Massachusetts.

Barton, P.I. and C.C. Pantelides (1994). Modeling of combined discrete/continuous processes, *AIChE Journal*, **40**, 966-979.

Bencze, W.J. and G.F. Franklin (1994). A separation principle for hybrid control system design, *Proc. of IEEE/IFAC Joint Symposium on Computer-Aided Control System Design*, Tucson, Arizona, 327-332.

Branicky, M., V.S. Borkar, and S.K. Mitter (1994). A unified framework for hybrid control: background, model, and theory, *Proc. of the 33rd IEEE Conference on Decision and Control*, Lake Buena Vista, Florida.

Brenan, K.E., S.L. Campbell, and L.R. Petzold (1989). *Numerical Solution of Initial-Value Problems in Differential-Algebraic Equations*, North-Holland.

Brüll, L., and U. Pallaske (1991). On consistent initialization of differential-algebraic equations with discontinuities, *Proc. 4th European Conference Mathematics in Industry*, 213-217, Kluwer Academic Publishers.

Cellier, F.E., H. Elmqvist, M. Otter, and J.H. Taylor (1993). Guidelines for modeling and simulation of hybrid systems, *Proc. of IFAC 12th Triennial World Congress*, Sydney, Australia, 1219-1225.

Fahrland, D.A. (1970). Combined discrete event continuous systems simulation, *Simulation*, **14**, 61-72.

Feehery, W.F., J.R. Banga, and P.I. Barton (1995). A novel approach to dynamic optimization of ODE and DAE systems as high-index problems, AIChE Annual Meeting, Miami.

Feehery, W.F. and P.I. Barton (1996). Dynamic simulation and optimization with inequality path constraints, *European Symposium on Computer Aided Chemical Engineering 6*, Rhodes, Greece.

Friedrich, M. and R. Perne (1995). Design and control of batch reactors — an industrial viewpoint, *Computers chem. Engng.*, **19**, Suppl., S357-S368.

Gopal, V. and L.T. Biegler (1994). A novel approach to nonsmooth dynamic process simulation, AIChE Annual Meeting, San Francisco.

Grossman, R.L., A. Nerode, A.P. Ravn, and H. Rischel (1993). *Hybrid Systems*, Lecture Notes in Computer Science, Vol.736, Springer-Verlag, New York.

Hairer, E. and G. Wanner (1991). *Solving Ordinary Differential Equations II Stiff and Differential-Algebraic Problems*, Springer-Verlag.

Kestne, Y., A. Pnueli, J. Sifakis, and S. Yovine (1993). Integration Graphs: A class of decidable hybrid systems, In R.L. Grossman, A. Nerode, A.P. Ravn, and H. Rischel (Ed.). *Hybrid Systems*, Lecture Notes in Computer Science, Vol.736, Springer-Verlag, New York.

Lemmon, M. and P. Antsaklis (1994). A computationally efficient framework for hybrid controller design, *Proc. of IEEE/IFAC Joint Symposium on Computer-Aided Control System Design*, 327-332, Tucson, Arizona.

Lygeros, J., D. Godbole, and S. Sastry (1994). Simulation as a tool for hybrid system design, *Proc. of IEEE Conference on AI, Simulation and Planning in High-Autonomy Systems*, Gainsville, Florida.

Manna, Z. and A. Pnueli (1993). Verifying hybrid systems, In R.L. Grossman, A. Nerode, A.P. Ravn, and H. Rischel (Ed.). *Hybrid Systems*, Lecture Notes in Computer Science, Vol.736, Springer-Verlag, New York.

Mattsson, S.E. and G. Söderlind (1993). Index reduction in differential-algebraic equations using dummy derivatives, *SIAM J. Sci. Stat. Comput.*, **14**, 677-692.

McDonald, C.M. and C.A. Floudas (1994). Global optimization for the phase equilibrium problem using the NRTL equations, *IChemE Symposium Series*, **133**, 273.

McDonald, C.M. and C.A. Floudas (1994). Global optimization and analysis for the gibbs free energy function using the UNIFAC, Wilson, and ASOG equations, *Ind. Eng. Chem. Res.*, **34**, 1694.

Modell, M. and R.C. Reid (1983). *Thermodynamics and Its Applications*, 2nd ed., Prentice-Hall.

Nerode, A. and W. Kohn (1993). Models for hybrid systems: automata, topologies, controllability, observability, In R.L. Grossman, A. Nerode, A.P. Ravn, and H. Rischel (Ed.). *Hybrid Systems*, Lecture Notes in Computer Science, Vol.736, Springer-Verlag, New York.

Nerode, A. and W. Kohn (1994). Hybrid systems as a substrate for DSSA, *Proc. of IEEE/IFAC Joint Symposium on Computer-Aided Control System Design*, Tucson, Arizona, 359-361.

Nicollin, X., A. Olivero, J. Sifakis, and S. Yovine (1993). An approach to the description and analysis of hybrid systems, In R.L. Grossman, A. Nerode, A.P. Ravn, and H. Rischel (Ed.). *Hybrid Systems*, Lecture Notes in Computer Science, Vol.736, Springer-Verlag, New York.

Ören, T.I. (1977). Software for simulation of combined continuous and discrete systems: a state-of-the-art review, *Simulation*, **12**, 33-45.

Pantelides, C.C. (1988). The consistent initialization of differential-algebraic systems, *SIAM J. Sci. Stat. Comput.*, **9**, 213-231.

Park, T. and P.I. Barton (1994). Towards dynamic simulation of a process and its automatic protective system, *International Symposium and Workshop on Safe Chemical Process Automation*, AIChE/CCPS, New York.

Park, T. and P.I. Barton (1996). State event location in differential-algebraic models, *ACM Trans. Modeling and Computer Simulation*, **6**, 2.

Pham, H.N. and M.F. Doherty (1990). Design and synthesis of heterogeneous azeotropic distillations-I. heterogeneous phase diagrams, *Chem. Eng. Sci.*, **45**, 1823-1836.

Pham, H.N. and M.F. Doherty (1990). Design and synthesis of heterogeneous azeotropic distillations-II. residue curve maps, *Chem. Eng. Sci.*, **45**, 1837-1843.

Ponton, J.W. and P.J. Gawthrop (1991). Systematic construction of dynamic models for phase equilibrium, *Computers chem. Engng.*, **15**, 803-808.

Probst, S.T. and G.J. Powers (1994). Automatic verification of control logic in the presence of process faults, *AIChE Annual Meeting*, San Fransisco, USA.

Sédès, D. (1995). Modelling, simulation and process safety analysis. A case study: the formaldehyde process, Internal Report, MIT.

Tavernini, L. (1987). Differential automata and their discrete simulators, *Nonlinear Analysis, Theory, Methods & Applications*, **11**, 6, 665-683.

SESSION SUMMARY:
NEW DIRECTIONS FOR ACADEMIC RESEARCH

Francis J. Doyle III
School of Chemical Engineering
Purdue University
West Lafayette, IN 47907-1283

Summary

This session, New Directions for Academic Research, addressed several key topics of interest to academics in the process control research community. The authors reflect a varied perspective: an industrialist, a chemist, and an academic chemical engineer. So too, are their topics varied, including controller performance monitoring, chemical process analyzers, and hybrid process systems.

The first speaker, Derrick Kozub from the Shell Development Company, addressed the challenges in process monitoring. The abnormal behaviors which require monitoring include modeling errors in model-based controllers, unmeasured disturbances, nonlinearity across the operating region, and poor selection of controller tuning parameters. Kozub made a persuasive argument in favor of automated approaches to determine if performance targets are being met (steady-state and dynamic), as well as automated diagnosis of root causes when these conditions are not met. In his paper he reviews a number of research methods for controller performance monitoring and diagnosis, including correlation methods, time series methods, and model-based methods.

The second speaker, Bruce Kowalski from the University of Washington, gave a talk which was entitled Process Analytical Chemical Engineering. His thesis was that chemists and chemical engineers must work together to integrate the process control, process instrumentation, and analytical chemistry functions for industrial processes. Just as the design and control interface benefits from collaborative endeavors, Kowalski proposes that engineers and chemists must each be educated using a common nomenclature and tool base. Representative of the trend that must emerge is the current emphasis on novel approaches to non-invasive measurement of key process characteristics. Kowalski pointed out that measurement needs top the lists of most industrial vision plans, and argued that process analytical chemistry should be incorporated into chemical engineering as a sub-discipline which represents a logical extension of process monitoring and control areas.

The final speaker, Paul Barton of MIT, addressed the emerging research area of hybrid systems — combinations of discrete and continuous subsystems. While most control engineers are familiar with the continuous domain and an artificially induced discrete domain from sampling, Barton pointed out several examples of actual discrete events which occur in chemical processes. These include human events such as operating decisions, and natural events such as phase changes. While the early work in this area dates back more than 30 years, the last five years have witnessed an increased research interest in this area, probably coupled to the emerging power of computational algorithms for mixed-integer programming problems. The current efforts, as described by Barton, lie primarily in the modeling and simulation environment with tools such as gPROMS and ABACUSS. One can easily imagine the impact of such tools on complex control problems such as optimal operation of batch recipes, as well as combined control and fault detection algorithms.

A REVIEW OF MODELING AND CONTROL IN THE PULP AND PAPER INDUSTRIES

Ferhan Kayihan[1]
Weyerhaeuser Company
Tacoma, WA 98477

Abstract

Pulp and paper industry converts a unique natural resource into everyday consumer products. The conversion processes are generally continuous and exhibit complex dynamic behavior. Fundamental dynamic models are not available for some processes. Environmental requirements, product quality targets and competitive pressures have increased the industry's desire for better process management practices including advanced process control implementations. A review of the last fifteen years of modeling and process control literature reveals that many of the industry specific problems have been worked on and solutions to some of the complex problems have emerged. Continuous digesters and paper machines represent a significant fraction of a mill's capital investment and have naturally attracted their share of attention so far. In light of their pivotal roles in intermediate and final product qualities, and due to process complexities and limited measurement capabilities, both of these processes need continuous improvements in modeling and control. The pulp and paper industry is striving to take full advantage of the technical advancements to stay competitive.

Keywords

Pulp and paper control, Dynamic models, Chips, Digesters, Kappa number, Paper machines, Profile control, CD control, Headbox, Refiners, Lime kilns.

Introduction

Pulp and paper industry faces three challenges derived from the distinct nature of the industry: One, it is the most capital intensive commodity product industry. Two, the main raw material is a natural product which retains its non-uniform characteristics and behavior through the manufacturing processes all the way to the final consumer products. Three, like other manufacturing industries, it has to comply with increasingly stringent environmental requirements and be responsive to customers' demands for using environmentally friendly processes.

The first challenge makes it more important than ever before for rigorous engineering approaches to maximize the return from capital investments. The second underlines an eye-opening reality which the industry has clearly recognized during the past two decades. It is no longer acceptable to claim that since we are dealing with a naturally variable material, pulp or paper-making are unique art forms reserved to be exercised only by the experts. Science and engineering fundamentals and disciplines are being implemented more and more to provide the missing links in our understanding of major manufacturing steps. The third challenge accelerates the industry's push to improve its accountability for overall environmental impact and specific product targets that need to be maintained for both short and long term periods. An expected by-product of this effort is the development and implementation of comprehensive production tracking and process management systems.

---

[1] present address: *IETek*®, 5533 Beverly Ave NE, Tacoma WA 98422, e-mail: *fkayihan_ietek@msn.com*

The technical common denominator for all three challenges is the improved understanding of process behavior through mathematical modeling. The industry has seen, and continues to see, modeling activities covering a wide range from empirical (statistical, time-series, neural networks, etc.) to first principles based fundamental developments. Major processes like digesters, recovery boilers, pulp and paper machines, evaporators and lime kilns attract the most attention because of their critical importance and high capital costs.

Process control follows a parallel route to modeling in the degree of its recognized importance and emphasis during the last two decades. As in other industries, high level of customer awareness and requirements of product quality are clearly identifying the targets and the role of reliable process control systems. Pulp and paper processes are typically very similar to systems which chemical engineers have been accustomed. Physical separations, reactions and combinations of the two in mixed or plug flow transport conditions with intermediate storage units cover the general nature of processes in a pulp or paper mill. Therefore, appropriate and successful control approaches are also similar in principle to what have already been established through applications in petroleum and chemical industries. However, there is much more to be done.

Some of the unique process modeling and process control challenges of the pulp and paper industry arise from the stochastic nature of the raw material, highly interactive multivariate process behavior, long time delays, incomplete measurements and grade change transients.

In what follows, an overview of an integrated pulp and paper mill operations will be presented. Then, modeling and control literature for the last fifteen years will be reviewed with respect to major process areas.

Process Background

Fig. 1 shows a simple schematics of the Kraft process for pulp and paper manufacturing. Only a brief overview will be presented here as the comprehensive details are available in recent publications by Grace et al. (1989), Smook (1990, 1992), Mims et al. (1993), Kline (1991), Lavigne (1993), Hagenmyer et al. (1992), and Thorp and Kocurek (1991).

On dry weight basis, wood is about 70% cellulosic material, 25% lignin and 5% extractives. Lignin serves as the binding glue for the cellulosic matrix. The main objective of a pulp mill is to remove lignin and recover as much of the cellulosic fibers as possible. Usually part of the fibers are lost in the process and the total yield is about 50% of the original dry weight. A typical mill capacity is 1000 dry tons/day of fiber output.

Trees are cut, debarked and chipped in preparation for the conversion processes. Chips are stored in outdoor piles with enough holdup to supply at least several weeks of production capacity. Chipping exposes the porous structure of wood for efficient mass transfer and reactions in the digester. Wood has a relatively structured but a very intricate porous matrix. In fresh chips about 60% of the total volume is void space which is mostly filled with liquid water.

Kraft process relies on alkaline cooking with an aqueous solution of NaOH and Na_2S called white liquor. Chips are screened to separate extreme sizes and steamed to drive off air trapped in pores before being introduced to the digester along with white liquor. Although there are both batch and continuous digesters in the industry, recently the latter is more commonly encountered. The digester is the main reaction unit for lignin removal. It operates under pressurized conditions reaching to about 800 kPa to prevent liquor from evaporating at operating temperatures which may reach to 170 °C. Delignification reactions occur inside the chip porous matrix and are naturally affected by the diffusion rates of chemicals through the entrapped liquor in pores.

Main lignin removing reactions take place in the first co-current section of the digester called the cook zone. Both chips and white liquor flow downwards in the column where each may initially occupy equal volume. As reactions proceed and chips soften, they pack more efficiently and occupy more of the reactor volume. Chip compaction is an important concern in maintaining constant operating conditions. After the cook zone the spent white liquor, now called weak black liquor, is extracted while the chips continue through the digester this time against the upflow of first mildly concentrated liquor and then dilution water. During the counter-current stages additional lignin is removed at slow rates and the reaction liquor, containing dissolved solids, is washed off the pulp. Total chip residence time in the digester may vary from two to ten hours depending on reaction conditions and production rates. The extent of reactions is measured by Kappa number which reflects the mass fraction of lignin still present in the pulp.

From this point on pulp is processed to wash out the remaining chemicals, to separate incompletely reacted chips, to remove more lignin and to bleach as necessary in order to achieve the fiber qualities required for specific product targets. The separated spent liquor goes through a recovery process with recycling of essential chemicals. During the chemical recovery the combustion of lignin and other dissolved organics from wood provides a substantial energy source to the mill.

The recovery operations start with the thickening of black liquor through a series of multi-effect evaporators to about 65% solids. Black liquor contains dissolved organics originating from wood and inorganic coming from the reaction chemicals. Strong black liquor is sprayed into the recovery boiler where the organic content is burned to generate steam and the inorganic portion is recovered as molten salt or smelt. The molten chemicals are dissolved to generate the green liquor containing Na_2CO_3 and Na_2S. In a series of stirred causticizing reactors Na_2CO_3 reacts with CaO to produce NaOH needed for the white liquor. The causticizing product $CaCO_3$ is converted back to CaO in a gas fired rotary lime kiln.

Combustion energy in the recovery boiler is converted to superheated steam which is then used to run the turbogenerators to generate electricity and process steam. Additional sources of mill energy are bark, off-size chips from screening and the partially converted chips from the digester. These are burned separately in a power boiler which also generates superheated steam.

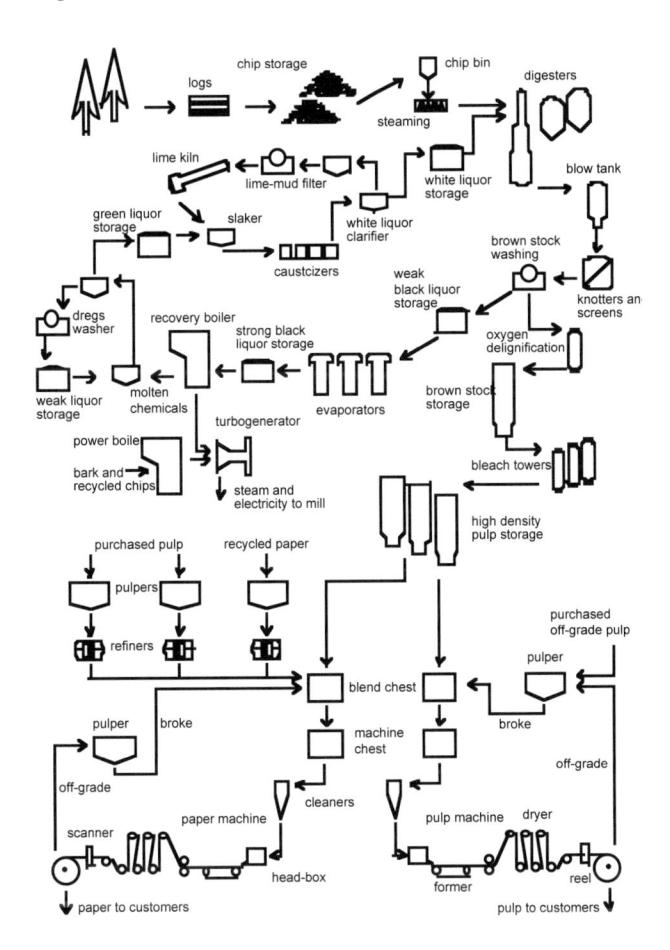

Figure 1. Simplified flow diagram of a pulp and paper mill.

Product from the digester is called the brown stock. Its pressure coming out of the digester is reduced by releasing it into the blow tank which also disperses the fibers. After the separation of partially unreacted chips (or knots), pulp goes through a washing stage to remove all of the reaction chemicals. At this point there is usually another delignification process in a vertical reactor with counter-current O_2 to remove more lignin with minimum yield loss. For final products such as office paper and personal care items bleaching is a necessary step to achieve the required brightness. Other products like corrugated boards and newsprint do not require full bleaching.

Pulp available for final product is stored in high-density chests which may each hold up to eight hours of stock at about 10% fiber mass fraction or consistency. Stock preparation starting from the high density chest is a sequence of dilution and cleaning operations aimed at providing a uniform mixture of fiber and water to the headbox of the pulp machine. Blend chest mixes all fiber sources together and reduces the consistency to about 4%. Auxiliary sources of fiber are the recycled off-grade product from the operations and purchased pulp. These additional sources of clean fiber are mixed with water in an agitated chest called pulper or re-pulper and fed to the blend chest. Machine chest dilutes the fiber level to 3.5% and dampens consistency fluctuations. After the machine chest the fiber slurry is diluted again and cleaned before going into the headbox. Headbox consistency is about 0.75%. Immediately after the head-box the fiber slurry looses most of its water reaching to about 10% solids on the wire where a well defined mat is formed. Further water removal takes place through pressing and long steam heated drying stages. Product coming out of the dryer is in the form of a sheet and it has only about 6% moisture. A typical production rate for a pulp machine is 3m/s. Final quality is monitored by a traversing scanner which measures basis weight and moisture. In traditional units, basis weight is the equivalent weight of approximately 1300 ft^2 of product in pounds. The preferred international unit is grammage (gm/m^2). From a pulp mill the final product is shipped to the customer in special size rolls or in bales.

On the paper production side the stock preparation operations are essentially the same. Differences come from the additional fiber sources as shown in Fig. 1 and from the chemical additives introduced after the blend chest. Stock or furnish which is carefully prepared and supplied to the paper machine is a prescribed combination of different fiber contents to give the desired product properties. Different paper grades are obtained by changing furnish composition and additives. Paper machines operate at speeds of up to 20 m/s.

A particular feature of the mill setup as depicted in Fig. 1 is the liberal use of surge or storage capacity. These are necessary to isolate various parts of the mill from the effects of short term stoppages and to manage production schedules and grade changes.

Modeling and Control

System Simulation

Traditional steady-state process simulation packages for the pulp and paper industry are recently reviewed by Syberg and Wilde (1992). The major simulators are: GEMS, developed at the University of Idaho; MAPPS, developed at the Institute of Paper Science and Technology in Atlanta; and MASSBAL, developed by SACDA in Ontario. These are sequential modular simulators which are customized for the specific needs of the industry. There are also proprietary software developments like PROSPR of Weyerhaeuser and successful use of other established softwares like ASPEN (Erickson et al., 1994).

System-wide dynamic simulation is a growing area of activity especially in the last ten years. Shewchuk (1991) describes SACDA's dynamic simulator designed for engineering analysis, control system design and operator training. Simons-Eastern Consulting in Atlanta has developed an object oriented package for detailed design analysis, distributed control system (DCS) development and testing, advanced control design, and operator training (Meincer and Andersen, 1991). Haynes, Scheldorf and Edwards (1988, 1990) describe the real-time dynamic version of GEMS or PCGEMS as it is used in an industrial case for operator training in conjunction with a DCS. Clarke and Bialkowski (1986) have developed an extended dynamic simulator as a teaching tool to train mill engineers for process analysis and controller design.

General State of Process Control

An acknowledged driving force for better process control is the combination of the need for overall performance improvement for profitability and the increasing pressure to adopt international quality standards (Ingman, 1992a, 1992b). There is a strong incentive for improvement through process control. At the same time, there is a significant need to increase mill staff competency to implement and maintain process control systems (Bialkowski and Elliot, 1995; Bialkowski, 1991, 1992; Bialkowski and Weldon, 1994; Dumont, 1986, 1988; Elliot, 1992; Leffler, 1993; Walsh, Bialkowski and Kooi, 1987). Recently, a few control textbooks were written specifically for the practitioners in the industry (Ostroot, 1993; Rao, Xia and Ying, 1994; Sell, 1995).

Advanced control methods are being continuously tested against the performance of PI controllers and simple dead-time compensation methods. Adaptive control, IMC, MPC, DMC, Generalized Predictive Control, neural networks and expert systems have all been tried for their suitability in time-varying, nonlinear and multivariable applications (Dumont, 1992; Haggman and Bialkowski, 1994; Gough and Kay, 1993a; Gough, Kay and Seebach, 1994; Perrier, Gendron and Legault, 1992; Rao and Ying, 1990; Rao, Ying and Corbin, 1991; Xia et al., 1995; Zhou et al., 1991). Data reconciliation (Isaksson and Kaul, 1992) and loop performance analysis (Jofriet et al., 1995; Eriksson and Isaksson, 1994; Perrier and Roche, 1992) are also recognized as integral parts of modern control systems. Looking to the future Fadum (1987), Kaunonen and Paunonen (1993), and Mangin (1992) expect information systems to become the central part of the next generation advanced control systems in pulp and paper mills.

Digesters

The first effort in quantifying digester operations is due to Vroom (1957) who expressed the extent of delignification reaction through the "H-factor" which combines time and temperature effects of reactions through an assumed first order mechanism. Later on, Hatton (1973) correlated the results of numerous laboratory batch cooking experiments and proposed a universal empirical relationship for Kappa number and yield as functions of effective alkali charge and H-factor. Hatton (1984) also correlated residual alkali and H-factor for Kappa number prediction.

Fundamental approaches to digester modeling starts with the efforts Smith and Williams (1974) with further improvements and validation by Christensen, Albright and Williams (1982), and Butler and Williams (1988). Operations of continuous digesters were described through a combination of transport and kinetic relationships between the three phases of the system: wood chips, entrapped liquor and free liquor. Different flow regimes of the digester where chips and free liquor phases are first co-current and then counter-current are appropriately incorporated into the model. Gustafson et al. (1983, 1984) and Agarwal, Gustafson and Arasakesari (1994) developed another fundamental digester model with particular attention to the diffusion effects of liquor in chips. Model predictions show the sensitivity of Kappa number to the size distribution of chips and indicate how process variability could be affected by the raw material variations alone.

In the models, the kinetic relationships that are used for assumed delignification reactions are at best good estimations. Various approximations are available for selection. Model builders rely on laboratory and field data for their choices and parameter adjustments. Efforts continue to simplify and to improve the accuracy of reaction kinetics (Datta et al., 1994). Another aspect of modeling which requires more attention is the prediction of chip compaction. Harkonen (1987) worked on the problem and developed a model describing the expected behavior of the two-phase flow system. Present models either use an empirical compaction relationship or predict it based on Harkonen's work.

Recent work by Funkquist (1995), Michelsen (1995), Wisnewski (1995) and Kayihan et al. (1996) are all targeted towards the development of reliable continuous digester models for dynamic analysis, model reduction and robust controller design. These models also incorporate the recent digester design improvements present in all of the new installations.

Continuous digesters are essentially large vertical reactors in which both chips and liquor form a moving packed column. Operating conditions must be kept as uniform as possible to guarantee uniform packing properties and continuous chip movement. A major concern is to keep chip levels constant in the digester and, if present, in the impregnation vessel. Level control algorithms require special attention due to the potential dynamic variations in packing and to the time delays in level change after flow adjustments. Belanger et al. (1986), Petrus (1990), and Allison, Dumont and Novak (1991) have developed and implemented a self-tuning approach and an adaptive scheme based on Generalized Predictive Control.

Kappa number control is a relatively difficult problem for digesters. Raw material variability, long residence times between inputs and output, and infrequent manual sampling for Kappa number measurement create a serious challenge in designing effective control systems. The classical approach to Kappa number control is to manually adjust the reaction temperature to compensate for deviations from the target. Adjustments take two to ten hours to affect the results and often interact with other corrections which were previously made and forgotten. Overcompensation and cycling as a result of control action is very common. Therefore, the typical tendency is to be very conservative in taking any corrective action.

Armstrong (1991) developed and implemented an SPC approach to prevent both over- and under-control. Wells and Hassler (1990), Obradovic et al. (1993) and Rudd (1994) worked with neural networks to augment feedback control decisions. All of these approaches worked better than unaided decision making. Dayal et al. (1994) developed and compared both neural networks and partial least squares methods to build empirical Kappa number models using historical data from an industrial digester. Both methods worked equally well in correlating most of the process variables.

On-line Kappa number measurement has been a possible technology for the last ten years and implementations into feedback control systems show distinct advantages. Horner and Rasmusson (1990) described some commercial applications of the STFI Opti-Kappa sensor. DeWitt, Wang and Ulinder (1991) used the sensor to control an industrial oxygen delignification reactor. Agneus and Damlin (1992) describe the operating principles of a similar optical sensor for on-line lignin detection.

There are also indirect approaches to monitoring the degree of delignification during digester operations. However, these are limited to laboratory batch operations and are not yet available for commercial continuous digesters. Alen et al. (1991) developed a method to monitor hydroxy carboxylic acids as reaction by-products and showed that, in addition to estimating the progress of reactions, the method also differentiates between softwood and hardwood species. Paulonis and Krishnagopalan (1988, 1990) and Liaw and Krishnagopalan (1992) developed an on-line liquor analysis method for alkali and sulfide concentrations to predict Kappa number from material balance calculations. Vanchinathan and Krishnagopalan (1993) extended the method to include the measurement of dissolved total solids and lignin in liquor.

Other advanced control applications use a simplified reference model for Kappa number predictions and update model parameters on-line with each measurement or as needed. Beller et al. (1988) developed a simple two dimensional model using input temperature and effective alkali strength to predict Kappa number and residual alkali. Christensen et al. (1990) and Michaelsen et al. (1994) used a simple mechanistic model compensated by an optimal state estimator as the basis for a model predictive control algorithm. Venkateswarlu and Gangiah (1992) compared a nonlinear observer to the extended Kalman filter in estimating the predictions of a model based on laboratory batch experiments and found the latter to be more effective. Lee and Datta (1994) used the extended Kalman filter with on-line liquor measurements to control the cooking time of laboratory batch digesters with a finite-horizon model predictive method.

The opportunities of Kappa number control improvement for a continuous digester are far from being exhausted (Lundqvist, 1990). On-line chip property measurements (Pettersson, Olsson and Lundqvist, 1988; Pettersson, 1989), spatial and temporal tracking of liquor properties in the digester and more reliable lignin analysis (Agneus and Damlin, 1990; Alen et al., 1990; Roman and Jeffers, 1992; and Jeffers and Roman, 1994) will open new horizons. Meanwhile, we have to be more creative with the available capabilities in present installations.

Washing and Bleaching

Dissolved solids and remaining cooking chemicals are removed in washers and pulp brightness is obtained in multistage bleach towers. Both operations strive for property improvements at uniform levels without excessive use of chemicals, loss of fiber or fiber strength.

Han (1989) has developed a model for brown stock washing, thereby predicting mat formation, washing and water removal behavior and the sensitivity of washing efficiency to holding tank level and vacuum. Harris and MacGregor (1987) and Turner, Allison, and Oei (1992) developed and implemented multivariable level in cascaded washer filtrate tanks. Bender, Richard and Dorn (1988), Wigsten (1988) and Noel et al. (1993) have implemented optimized complete washer control system based on dynamic model predictions and operating curves relating dissolved solid to washing cost. Patrick (1991) has successfully used the neural networks approach to supplement existing mill controllers to maintain uniform dissolved solids in washed pulp.

Germgard, Teder and Tormund (1985) and Barrette and Perrier (1995) have modeled various bleaching stages to develop strategies for control and process optimization. Mill data was used for modeling where sequential stirred tanks approximated the plug flow behavior. A cascade control strategy with time-delay compensation was found to be appropriate. Adaptive-predictive PID (Dumont, Martin-Sanchez and Zervos, 1989) and Laguerre based adaptive-predictive (Zervos, Dumont Pageau, 1989; Dumont, Zervos and Pageau, 1990) approaches were also used for bleach plant applications and shown to be robust with respect to time delay variations. Gough and Kay (1993b) used the universal adaptive controller for brightness control in a mill with clear improvements over the prior performance of PID with Smith predictor. Tham and Jones (1985) worked with multivariable self-tuning controllers and Rankin and Bialkowski (1984) used nonlinear compensation for process interactions to improve bleach plant performances. Xia, Rao, Farzadeh et al. (1993) combined neural networks with intelligent monitoring and fault detection to propose a control approach for a peroxide bleaching stage.

Evaporators, Recovery Boiler and Causticizing

Concentration of black liquor, recovery of chemicals and generation of steam are the basic functions of the recovery cycle. Steady-state and dynamic behavior of evaporators and causticizers are relatively well established. Lime kiln modeling is progressing for process analysis and controller design. However, the development of mechanistic recovery boiler models, especially simple ones to use with controllers is a continuing challenge.

Niemi (1989) developed an evaporator model from radioactive tracer tests and implemented an optimal feedforward algorithm to a two-stage evaporator plant. Through a series of efforts Ricker and coworkers (Cheng and Ricker, 1985; Erickson, Sim and Ricker, 1985; Ricker, Subrahmanian and Sim, 1989) and Cheng (1989) have developed dynamic models from mill data and designed LQ, IMC and MPC approaches including constraints on both manipulated and output variables.

Johnson (1988), Adams (1994) and Grace (1995) provide a clear perspective of the importance and physical principles of recovery boilers, and the challenges of predicting their behavior.

Harrison and Ariessohn (1985) developed an optical smelt-bed imaging system to provide early information to operators about operational changes. Hosti, Sutinen and Leiviska (1994) combined a knowledge based system with a burner image analyzer to evaluate performance and to advise operators for appropriate actions. Ozaki et al. (1994) proposed a qualitative modeling approach with neural networks and fuzzy logic to control recovery boiler operations. Kemna et al. (1992) used the canonical variate analysis technique to develop a model from historical mill data. Fundamental modeling approaches for recovery boilers are provided by Salcudean, Nowak and Abdullah (1992), Wessel, Parker and Akan-Etuk (1993) and Yang, Horton and Adams (1994).

Sethuraman, Krishnagopalan and Krishnagopalan (1995) developed a kinetic model for causticizing without diffusion limitations using a laboratory system. Wand, Englezos and Tessier (1994) reported a dynamic simulation system which describes the effects of causticizer residence times and lime specific surface on reaction efficiency. Lime kiln modeling both for steady-state (Mumford and Edwards, 1986; Gorog and Adams, 1987; Correia et al. 1990) and for dynamic behavior (Smith and Edwards, 1991) have provided a reasonable starting point for process analysis and control. However, mill applications have so far tended to be more pragmatic. Youngblood, Curtis and Vachtsevanos (1990) established a rule-based system and used fuzzy logic to demonstrate good control performance through simulations. Charos, Arkun and Taylor (1991) successfully implemented an MPC algorithm to an industrial lime kiln using step response data for modeling. Ribeiro, Dourado and Costa (1993) proposed a multivariable adaptive neural networks approach and showed applicability through simulations. Osmond, Tessier and Savoie (1994) have obtained a transfer function model from an industrial kiln and designed decoupled PI controllers and feedforward compensation to predict improved performance through simulations.

Refiners

Refiners are used to separate the fibers of steamed or chemically treated chips when lignin does not need to be removed for certain applications. Refiners are rotating disks where rotation speed, energy consumption, pressure drop across the plates, and output fiber properties are important aspects of the process. For fine paper applications, refiners are also used to create uniform fibers after the repulping of recycled fiber sources.

For a chip refiner, Di Ruscio and Balchen (1992) developed a dynamic model for fiber size distribution resulting from refining and showed that unknown model parameters can be recursively estimated. Karlstrom and Koebe (1993) modeled two serially linked refiners with a linear parametric approach to get two multi-input single-output dynamic systems. Miles and Kay (1993) worked on a constitutive model to predict the degree of refining from refiner geometry and operating conditions. Strand (1993) used a statistical approach to relate fiber properties after the refiners to both operating conditions and raw material changes. Qian and Tessier (1993) developed a model to relate sheet properties to refining operating conditions. Subsequently, Tessier and Qian (1994) modeled the dynamic behavior of the system from the chip silos to the secondary refiners in order to study the impacts of process disturbances and operating conditions on performance. Qian, Tessier and Dumont (1994) have developed a fuzzy logic model for the performance optimization of wood chip refiners.

Dahlqvist, Munster and Hill (1993) examined options of controlling either the refiner gap or the hydraulic pressure for a twin-refiner setup and concluded that pressure is a better control objective for pulp quality. Partanen and Kovio (1984) directly measured pulp freeness on-line and developed a non-linear PI algorithm to control a thermomechanical plant. Toivonen and Tamminen (1990) applied a minimax robust LQ design for freeness control. They identified multiple process models at various operating conditions and designed a controller that is optimized for the complete range.

Dumont and Astrom (1988) paid special attention to the possibility of sudden changes in process gain between motor load and plate gap due to fiber pad collapse. They developed various control approaches including an adaptive method with fault detection. Fu and Dumont (1988) used a Laguerre filter based dynamic model with an adaptive approach to control the motor load of a chip refiner. The method was found to have fast convergence in estimation and accurate prediction. Allison et al. (1994) used a linear dynamic system approach with fast time-varying parameters to detect pulp pad collapse behavior and to adaptively control refiner load.

Kooi and Khorasani (1992) developed static and dynamic neural networks based controllers to replace self tuning regulators for chip refiners. They found the dynamic neural networks to be more effective. Di Ruscio et al. (1994) used a nonlinear model based strategy to estimate freeness between two refiners to control overall performance. Gresham (1994) combined a rule based system with PI controllers to successfully reduce load and freeness variations in a mill refiner.

To control the performance of high-consistency refiners, specifically to reduce both short and long term freeness variations, Blanchard and Fontebasso (1993) used an on-line freeness analyzer in a cascade mode to adjust specific energy set-point. Freeness is a measure of the ability of fibers to release water from a fiber water mixture, especially on the forming wire.

Stock Preparation and Wet-end

Stock or furnish preparation is a sequence of steps for the careful dilution, mixing, final cleaning and chemical modification of the fiber. In principle, process control issues in this area are straightforward. In practice however, they require a lot of attention, mainly due to process and loop interactions, sensor problems, changes in operating conditions (Dumdie, 1988) and imperfect mixing in tanks. Stock preparation area plays an important role in grade changes. Design improvements for better process management is of continued interest (Goldenberg, 1994). From this point within the process to the end of sheet forming, total material balance control is crucial as the final product is highly sensitive to consistency and flow variations in fiber delivery.

Wet-end of the machine refers to the management of fiber and water between the machine chest and the end of the former (wire) where fiber mat is formed. Significant amounts of water are added and then removed from the fiber mixture in the wet-end. Most of this water, called white-water, is recycled through various holdup systems and can accumulate fiber fines and chemical additives. Behavior of fiber mat on the wire like retention, formation, and drainage are directly affected by white-water characteristics (Guest, 1990). Residence time of fibers in the wet-end is on the order of ten seconds while white-water average residence time is about an hour.

White-water systems have been modeled for dynamic behavior and process sensitivity analysis (Kaunonen et al., 1985; Virtanen, 1988; and Bussiere, Roche and Paris, 1992). Wet-end chemistry, consistency and retention control (Edwards, 1987; Farley, 1988; Kaunonen et al., 1991; and Rantala, Tarhonen and Koivo, 1994a, 1994b) were shown to be necessary for white-water management.

The headbox receives a very dilute mixture of fiber and water and uniformly distributes it through a slice opening onto the moving wire for formation. Liquid level in the headbox and air pressure above it determines the flowrate and velocity of the jet through the slice opening. Constant jet behavior requires tight control of liquid level and air pressure. These are interacting loops for a closed headbox.

Ibrahim (1985, 1987) developed a model for the headbox appropriate for design and optimization studies. Bialkowski (1991) implemented a decoupled control design for a multi-ply headbox system. Piirto and Koivo (1992) used a self-tuning SISO controller for retention with self-tuning MIMO for headbox control. Rao and coworkers developed fault tolerant (Xia et al., 1992), mill DCS bump test and decoupling (Xia, Rao, Shen et al., 1993), decoupling through quadratic performance index (Xia, Rao, Sun et al., 1993), state estimation through adaptive fading Kalman filter (Xia, Rao, Ying et al., 1994), and bilinear decoupling (Ying et al., 1994) approaches to control headbox performance. Sbarbaro and Jones (1994) used a non-linear model with MPC to show promise through simulations.

Knowledge based systems are also used for performance evaluation. Lindholm et al. (1989), Pokela et al. (1994) and Ritala (1993) described the development and the benefits of the WEDGE system which uses available data around the wet-end to help diagnose the sources of process variability. Estevez-Reyes and Dumont (1995) presented a mathematical model to be used in the diagnosis of sudden faults in pressure screens.

Further developments and reliable on-line applications of low consistency range optical sensors (Jack, Bentley and Barron, 1990) and wire drainage rate sensors (Woodard, 1991) are likely fill a serious gap in process measurement capability.

Paper Machines

The product of the paper machine is a continuous sheet for which all of the properties like basis weight, moisture, caliper, color, smoothness and strength are expected to be on target and uniform everywhere. Basis weight and moisture are the primary properties to regulate. Traditionally sheet characteristics are described in terms of temporal (machine direction, MD) and spatial (cross direction, CD) components. Flow of fiber into the headbox, adjusted by the basis-weight valve, is used to control MD basis weight and overall steam pressure in the drying section is used to control m.d. moisture. Headbox slice lip and individual steam hood adjustments are then used for CD basis weight and moisture control respectively. Final sheet properties are measured by a scanning sensor which traverses the sheet width in about 30 seconds. Headbox to scanner residence time may be as short as about 20 seconds for high speed newsprint machines or as long as ten minutes for low speed pulp machines.

Comprehensive first principles modeling of paper machines is not straightforward. This is mainly due to the complex hydrodynamics of the fiber-water mixture during formation (Westermeyer, 1987; and Clos, Edwards and Gunawaran, 1994). Approximate dynamic models for MD variations and static CD gain models are usually obtained through experiments and are used for controller design (Chen, 1995a). Slice adjustments affect neighboring positions and essentially give an interaction matrix for expected CD behavior. This simple approach ignores the simultaneous nature of temporal and spatial properties (Holik, 1986; Kastanakis and Lizr, 1991). Scanner signals are separated into m.d. and c.d. components for controller action. Differences in controllers performances arise as much from the identification of MD/CD variations as from the controller design (Nuyan, 1986; Heaven et al., 1993, 1994).

Bond and Dumont (1988) used the Dahlin algorithm with gain scheduling and recursive model identification for a color control scheme. Vincent et al. (1994) developed an adaptive approach using both linear and non-linear predictions to control color during grade transitions. Chen (1995b) applied color control successfully to more than 50 commercial machines.

Hagberg and Isaksson (1994) and Isaksson, Jonsson and Hagberg (1994) provided a dynamic paper machine model as a benchmark to test different control approaches to regulate basis weight and filler retention. They compared the results provided by various groups (Bozin and Austin 1994; Chow, Kuznetsov and Clarke 1994; Fu and Dumont, 1994; Makkonen et al., 1994; and Piipponen and Ritala, 1994) which used one or more of the techniques among the Smith predictor, IMC, DMC, and GPC.

Sikora et al. (1984) designed and implemented a self-tuning moisture control strategy. Brown and Millard (1993) developed an MD horizon multivariable control (HMPC) approach for mill implementation and found it to be more effective than Smith predictors. Rao and coworkers proposed basis weight and moisture control schemes using bilinear suboptimal control (Ying, Rao and Sun, 1991), MIMO adaptive (Xia, Rao, Shen and Zhu 1993, 1994), and model algorithmic control based on an impulse response model (Xia, Rao and Qian, 1993) approaches.

Bergh and MacGregor (1987) used a linear-quadratic-gaussian controller for both MD and CD variations. Balakrishnan and McFarlin (1985) showed that the elastic nature of the slice lip is an important part of profile control and incorporated a slice lip model into their controller design. Sutcliffe, Swanson and Balakrishnan (1993) proposed an improved profile controller using a multiple frequency actuator design. Graeser (1995) considered the interactions between disturbances, control action coupling and slice lip bending to precalculate control actions. Matsuda (1990) used the interaction matrix as a weighing function to control thickness profile.

Dumont and coworkers (Natarajan, Dumont and Davies, 1989; Dumont et al., 1991; Dumont et al., 1993; Wang, Dumont and Davies, 1993a, 1993b, 1993c) developed a control method combining Kalman filtering approach for m.d. with least squares estimation for c.d. properties. Kristinsson and Dumont (1993) used discrete Gram polynomials to approximate the constrained shaping of the slice lip under physical limitations which conditioned the interaction matrix for more robust control behavior. Chen and Wilhelm (1986), Chen (1988, 1992), and Chen and Adler (1990) developed commercial controllers using dual Kalman filtering approach to estimate both temporal and spatial variations including the use of quadratic penalty functions to handle constraints on the control action. Morari and coworkers (Laughlin and Morari, 1989; Braatz and Morari, 1991; Braatz et al., 1992a,b; Laughlin, Morari and Braatz, 1993; and Tyler and Morari, 1995) developed an MPC approach to control spatial variations using Kalman filtering with the lifting method. Campbell and Rawlings (1994) and Rawlings and Chien (1996) formulated a time-varying Kalman filter to estimate full sheet properties and designed an MPC approach that can handle hard constraints. Rigopoulos et al. (1996a,b and 1997) used a particular principal component analysis technique, Karhunen-Loève expansion, for spatial disturbance modeling and showed that an effective MPC formulation can be implemented in a state-space format. Braatz and Ogunnaike (1995) prepared a comprehensive review of the recent control efforts in this area.

Development of model identification and controller design for film forming processes while simultaneously considering both dimensions has also attracted considerable attention. Rogers and Owens (1994) reviewed the recent developments in what is called the two-dimensional approach. Efforts of Wellstead, Duncan and coworkers are significant in this area (Kjaer, Waller and Wellstead, 1994; Wellstead and Heath, 1994; Kjaer, Heath and Wellstead, 1995; and Duncan and Bryant, 1994, 1995).

A recent design alternative offered in headboxes is going to ease some of the cross directional interactions encountered in the traditional slice design. The new design offers consistency profiling by mixing controlled dilution water at closely spaced slice actuator locations (Vyse, King and Heaven, 1995). Other recent advancements which are likely to have a significant impact on profile control are the instantaneous non-scanning nuclear weight (Francis, Stenbak and Kleinsmith, 1988) and optical weight and moisture measurement technologies, and the medium-frequency (0.5 sec) MD weight measurement capability (Wallace 1990, 1992).

Conclusions

The pulp and paper industry uses a variety of complex conversion and separation processes and demonstrates a clear need for the effective implementation of modeling, identification and control technologies. The most challenging processes include continuous digesters and paper machines. For both of these processes, focused research is necessary in fundamental modeling, model reduction, adaptive model and disturbance identification, state estimation, and robust model based controller design.

Industry and academia must collaborate to test the new control methodologies under realistic conditions and simplify them enough to be accepted and to be confidently incorporated into the existing systems by the practitioners. The pulp and paper industry is striving to take full advantage of the new developments in the process systems technologies in order to stay competitive.

References

Adams, T. N. (1994). Recovery boiler physical principles. In *1994 Kraft Recovery Short Course*. TAPPI Press, Atlanta, GA, USA. 239-244.

Agarwal, N., R. Gustafson, and S. Arasakesari (1994). Modeling the effect of chip size in kraft pulping. *Pap. Puu*, **76**(6/7), 410-416.

Agneus, L. and S.-A. Damlin (1990). On-line kappa-number analysis. In *Pulp Technology Energy: New Available Techniques and Current Trends*. SPCI, Stockholm. 234-241.

Agneus, L. and S.-A. Damlin (1992). Optimal process control with on-line kappa number analysis. *Pulp Pap. Can.*, **93**(2), 20-24.

Alen, R., P. Hentunen, E. Sjoestroem and L. Sundstrom (1991). New approach for process control of kraft pulping. *J. Pulp Pap. Sci.*, **17**(1), J6-J9.

Alen, R., A. Makela, E. Sjostrom, T. Hartus, L. Paavilainen and O. Sundstrom (1990). Analysis of lignin degradation products as a new tool for monitoring the kraft pulping process. In *Pulping Conference*. TAPPI Press, Atlanta. 55-60.

Allison, B. J. et al. (1994). Dual adaptive control of chip refiner motor load: Industrial results. In *Control Systems 94*. Swedish Association of Pulp and Paper Engineers, Stockholm, Sweden. 289-297.

Allison, B. J., G. A. Dumont and L. H. Novak (1991). Multi-input adaptive-predictive control of Kamyr digester chip level. *Can. J. Chem. Eng.*, **69**(1), 111-119.

Armstrong, M. K. (1991). Statistical process control of digester permanganate and kappa numbers: closing the loop with SPC. *Tappi J.*, **74**(6), 244-248.

Balakrishnan, R. and D. McFarlin (1985). Application of a model of the slice lip as an elastic beam or weight profile control. In *Process Control Symposium*. TAPPI Press, Atlanta. 105-109.

Belanger, P. R. et al. (1986). Self-tuning control of chip level in a Kamyr digester. *AIChE Journal*, **32**(1), 65-74.

Barrette, M. and M. Perrier (1995). Modeling and control of a hypochlorite bleaching stage for dissolving pulp. In *Process Control, Electrical and Information Conference*. TAPPI Press, Atlanta. 11-22.

Bergh, L. G. and J. F. MacGregor (1987). Spatial control of sheet and film forming processes. *Can. J. Chem. Eng.*, **65**(1), 148-155.

Beller, J., R Kammerer, A. Kaya and M. A. Keyes (1988). Kappa number estimation, control and management in the pulping process. In *American Control Conference*. American Automatic Control Council, Green Valley, AZ. 1951-1958.

Bender, G., L. Richard, and W. Dorn (1988). Advanced control for brown stock washers. In *Engineering Conference*. TAPPI Press, Atlanta. 247-251.

Bialkowski, W. L. (1991). Multi-ply headboxes have special control requirements. *Am. Papermaker*, **54**(11), 41-42.

Bialkowski, B. and R. Elliot (1995). Competency in process control — industry guidelines. In *Annual Meeting — Technical Section, Canadian Pulp and Paper Association, Preprints B*, 81st. CPPA, Montreal. B175-B182.

Bialkowski, W. L. (1991). Recent experiences with advanced control aimed at optimizing product uniformity. In *Process Control Conference*. TAPPI Press, Atlanta. 127-136.

Bialkowski, W. L. (1992). Process variability, control, standards and competitive position. In *ISA 92 Canada General Program*. ISA, Research Triangle Park, NC. 395-417.

Bialkowski, W. L. and A. D. Weldon (1994). Digital future of process control; possibilities, limitations, and ramifications. *Tappi J.*, **77**(10), 69-75.

Blanchard, P. and J. Fontebasso (1993). Mill implementation of a freeness-specific energy control strategy for high-consistency refiners. In *International Mechanical Pulping Conference: Poster Presentations*, Session: Process Measurement and Control. EUCEPA, . 215-244.

Bond, T. and G. Dumont (1988). On adaptive control of color for paper machines. In *American Control Conference*. American Automatic Control Council, Green Valley, AZ. 1965-1970.

Bozin, A. S. and P. C. Austin (1994). Dynamic matrix control of a paper machine benchmark problem. In *Control Systems 94*. Swedish Association of Pulp and Paper Engineers, Stockholm, Sweden. 242-248.

Braatz, R. D. and M. Morari (1991). mu-sensitivities as an aid for robust identification. In *American Control Conference*. American Automatic Control Council, Green Valley, AZ. 231-236.

Braatz, R. D. and B. A. Ogunnaike (1995). Control of sheet and film processes — a critical review. Submitted to *Automatica*.

Braatz, R. D., M. L. Tyler, M. Morari, F. R. Pranckh and L. Sartor (1992a). Identification and cross-directional control of coating processes: Theory and experiments. In *American Control Conference*, **2**. American Automatic Control Council, Evanston. 1556-1560.

Braatz, R. D., M. L. Tyler, M. Morari, F. R. Pranckh and L. Sartor (1992b). Identification and cross-directional control of coating processes. *AIChE Journal*, **38**(9), 1329-1339.

Brown, T. and D. Millard (1993). New model-based machine-direction control algorithm. In *Process Control Conference*. TAPPI Press, Atlanta. 15-20.

Bussiere, S., A. A. Roche, and J. Paris (1992). Analysis and control of white water network perturbations in an integrated newsprint mill. *Pulp Pap. Can.*, **93**(4), 41-44.

Butler, A. C. and T. J. Williams (1988). *A Description and User's Guide for the Purdue Kamyr Digester Model*. Technical report 152. Purdue University, West Lafayette, Indiana.

Campbell, J. C. and J. B. Rawlings (1994). Estimation and control of film forming processes. Presented at the SIAM Symposium on *Control Problems in Industry*. San Diego, CA. June 22-23.

Charos, G. N., Y. Arkun, and R. A. Taylor (1991). Model predictive control of an industrial lime kiln. *Tappi J.*, **74**(2), 203-211.

Chen, S. C. (1988). Kalman filtering applied to sheet measurement. In *American Control Conference*. American Automatic Control Council, Green Valley, AZ. 643-647.

Chen, S. C. (1992). Full-width sheet property estimation from scanning measurements. In *Control Systems 92*. CPPA, Montreal. 123-130.

Chen, S. C. (1995a). Modelling of paper machines for control: theory and practice. *Pulp Pap. Can.*, **96**(1), 44-48.

Chen, S. C. (1995b). Color control system for paper machines. In *Process Control, Electrical and Information Conference*. TAPPI Press, Atlanta. 3-10.

Chen, S.-C. and L. Adler (1990). Cross machine profile control for heavy weight paper. In *Control Maintenance Environment: New Available Techniques and Current Trends*. SPCI, Stockholm. 77-94.

Chen, S. C. and R. G. Wilhelm Jr. (1986). Optimal control of cross-machine direction web profile with constraints on the control effort. In *American Control Conference*. IEEE, New York. 1409-1415.

Cheng, Chun-M. (1989). Linear quadratic-model algorithm control method with manipulated variable and output variable set points and its applications. *Industrial & Engineering Chemistry Research*, **28**(2), 187-192.

Cheng, Chun-M. and N. L. Ricker (1985). Optimal Controller for Pulp Mill Applications. In *American Institute of Chemical Engineers, National Meeting*. AIChE, New York. .

Chow, C. M., A. Kuznetsov, and D. Clarke (1994). Application of generalised predictive control to the benchmark paper machine. In *Control Systems 94*. Swedish Association of Pulp and Paper Engineers, Stockholm. 249-255.

Christensen, T., L. F. Albright and T. J. Williams. (1982). *A Mathematical Model of the Kraft Pulping Process*. Technical report 129. Purdue University, West Lafayette, Indiana

Christensen, T., M. P. Ivan, R. Michaelsen, G. Lundman and U. Zetterland (1990). Advanced model applications to digester control at SCA/Nordliner, Munksund, Sweden. In *Pulp Technology Energy: New Available Techniques and Current Trends*. SPCI, Stockholm. 265-274.

Clarke, F. R., W. L. Bialkowski (1986). Dynamic process control simulation package. *Pulp & Paper Canada*, **87**(5), 90-95.

Clos, R. J., L. L. Edwards, and I. Gunawaran (1994). A limiting-consistency model for pulp dewatering and wet pressing. *Tappi J.*, **77**(6), 179-187.

Correia, A. D., R. Correia, J. L. Amaral and M. Amaral (1990). Simulation and performance analysis of a pulp mill rotary kiln. In Z. Jacyno (Ed.), *ISMM*

International Conference: MICRO 90. Acta Press, Anaheim. 177-180.
Dahlqvist, G., H. Munster, and J. Hill (1993). Advances in fundamental refiner control. In *International Mechanical Pulping Conference: Poster Presentations*, Session: Process Measurement and Control. EUCEPA, 208-214.
Datta, A. K., J. H. Lee, S. Vanchinathan and G. A. Krishnagopalan (1994). Model-based monitoring and control of batch pulp digester. In *American Control Conference*. American Automatic Control Council, Green Valley, AZ. 500-504.
Dayal, B. S., J. F. MacGregor, P. A. Taylor, R. Kildaw and S. Marcikic (1994). Application of feedforward neural networks and partial least squares regression for modelling Kappa number in a continuous Kamyr digester. *Pulp Pap. Can.*, **95**(1), 26-32.
DeWitt, S. A., P. H. Wang, and J. D. Ulinder (1991). Oxygen-stage advanced control using on-line kappa sensors. In *Pulping Conference*, Book 1. TAPPI Press, Atlanta. 325-336.
Di Ruscio, D. and J. G. Balchen (1992). State space model for the TMP process. In *Control Systems 92*. CPPA, Montreal. 107-114.
Di Ruscio, D. et al. (1994). Experience with a nonlinear model based control strategy applied to a two stage tmp plant. In *Control Systems 94*. Swedish Association of Pulp and Paper Engineers, Stockholm, Sweden. 282-288.
Duncan, S. R. and G. F. Bryant (1994). Design of dynamics for cross-directional controllers in papermaking. In *International Conference on CONTROL '94*. IEE, Stevenage, Engl. 618-623.
Dumdie, D. P. (1988). Systems approach to consistency control and dry stock blend. *Tappi J.*, **71**(7), 135-139.
Dumont, G. A. (1986). Application of advanced control methods in the pulp and paper industry — survey. *Automatica*, **22**(2), 143-153.
Dumont, G. A. (1988). Challenges and opportunities in pulp and paper process control. In *American Control Conference*. American Automatic Control Council, Green Valley, AZ. 1959-1964.
Dumont, G. A. (1992). High technology for process control. In *Annual Meeting — Technical Section, Canadian Pulp and Paper Association, Preprints A*, 78th. CPPA, Montreal. p. A90.
Dumont, G. A. and K. J. Astrom (1988). Wood chip refiner control. *IEEE Cont. Syst. Mag.*, **8**(2), 38-43.
Dumont, G. A., M. S. Davies, K. Natarajan, C. Lindeborg, F. Ordubadi, Y. Fu, K. Kristinsson and I. Jonsson.(1991). An improved algorithm for estimating paper machine moisture profiles using scanned data. In *IEEE Conference on Decision and Control*, 30th. IEEE, Piscataway, NJ. 1857-1862.
Dumont, G. A.,I. M. Jonsson, M. S. Davies, F. Ordubadi, Y. Fu, K. Natarajan, C. Lindeborg and E. M. Heaven (1993). Estimation of moisture variations on paper machines. *IEEE Trans. Cont. Syst. Tech.*, **1**(2), 101-113.
Dumont, G. A., J. M. Martin-Sanchez, and C. C. Zervos (1989). Comparison of an auto-tuned PID regulator and an adaptive predictive control system on an industrial bleach plant. *Automatica*, **25**(1), 33-40.
Dumont, G. A., C. C. Zervos, and G. L. Pageau (1990). Laguerre-based adaptive control of pH in an industrial bleach plant extraction stage. *Automatica*, **26**(4), 781-787.
Duncan, S. R. and G. F. Bryant (1995). The spatial bandwidth of cross-directional control systems for web processes. Submitted to *Automatica*.
Edwards, K. R. (1987). Wet-end chemical control reduces recycled coated stock problems. *Pulp Pap.*, **61**(2), 52-54.
Elliot, R. (1992). Process control optimization skills: need for competency standards. In *Annual Meeting - Technical Section, Canadian Pulp and Paper Association, Preprints B*, 78th. CPPA, Montreal. B273-B279.
Erickson, V. B., T. Sim, and N. L. Ricker (1985). Internal model control of two simulated multieffect evaporators. In *American Institute of Chemical Engineers, National Meeting*. AIChE, New York. .
Eriksson, P. G. and A. J. Isaksson (1994). Some aspects of control loop performance monitoring. In *IEEE Conference on Control Applications*, v 2. IEEE, Piscataway, NJ. 1029-1034.
Eriksson, G., U. Gren, C. Hanson, Y. Liu and H. Theliander (1994). Some comments on the simulation of a Kraft pulp process. In *Institution of Chemical Engineers Symposium Series*, 133: *European Symposium on Computer Aided Process Engineering*. Rugby, Engl.: Inst of Chemical Engineers. Dublin, Ireland. 191-198.
Estevez-Reyes, L. W. and G. A. Dumont (1995). Dynamic model of pressure screens for fault detection. In *Process Control, Electrical and Information Conference*. TAPPI Press, Atlanta. 47-52.
Fadum, O. (1987). Process information and control systems: a technology overview. *Tappi J.*, **70**(3), 62-66.
Farley, C. E. (1988). Practical application of charge analysis and control. In *Papermakers Conference*. TAPPI Press, Atlanta. 387-388.
Francis, K., M. Stenbak, and C. Kleinsmith (1988). Stationary bone-dry weight sensor. In *Engineering Conference*. TAPPI Press, Atlanta. 561-567.
Fu, Y. and G. A. Dumont (1993). Chip refiner motor load adaptive control using a nonlinear Laguerre model. In *IEEE International Conference on Control and Applications*, v 1. IEEE, Piscataway, NJ. 371-376.
Fu, Y. and G. A. Dumont (1994). A generalised predictive control design for the benchmark. In *Control Systems 94*. Swedish Association of Pulp and Paper Engineers, Stockholm, Sweden. 256-261.
Funkquist, J. (1995). *Modeling and Identification of a Distributed Parameter Process: The Continuous Digester*. Dr. of Tech. Thesis. Royal Institute of Technology, Stockholm.
Graeser, A. (1995). Cross profile control - predetermination of the control results regarding slice lip bending. *Pulp Pap. Can.*, **96**(1), 35-39.
Grace, T. M. (1995). Challenges in modeling recovery boilers. *PIMA Mag.*, **77**(2), 50-51.

Germgard, U., A. Teder, and D. Tormund (1985). Mathematical models for simulation and control of bleaching stages. In *Pulping Conference*, Book 3. TAPPI Press, Atlanta. 529-536

Goldenberg, P. H. (1994). Stock prep's future: Evolution toward improved quality and operating efficiency. In *Papermakers Conference*. TAPPI Press, Atlanta. 213-217.

Gorog, J. P. and T. N. Adams (1987). Design and performance of rotary lime kilns in the pulp and paper industry: part 1 - a predictive model for a rotary lime reburning kiln. In *Kraft Recovery Operations Seminar*. TAPPI Press, Atlanta. 41-47.

Gough, B. and J. Kay (1993a). Minimum-effort adaptive control of effluent treatment, kilns, and pulp brightness. In *Process Control Conference*. TAPPI Press, Atlanta. 207-218.

Gough, W. and J. Kay (1993b). Process automation: adaptive control of pulp brightness. *Can. Mill Prod. News*, **4**(1), 13-14.

Gough, B., J. Kay, and G. Seebach (1994). Adaptive control applications in pulp and paper. *Pulp Pap. Can.*, **95**(6), 51-53.

Grace, T. M., B. Leopold, E. W. Malcolm, and M. J. Kocurek, eds. (1989). *Pulp and Paper Manufacture, Volume 5: Alkaline Pulping*. 3rd ed. Joint Textbook of the Paper Industry. Tappi Press, Atlanta, GA. and CPPA, Montreal, Que.

Gresham, G. (1994). Rule-based PI control of a TMP refiner's motor load. *Tappi J.*, **77**(11), 195-199.

Guest, D. A. (1990). Wet-end chemistry control. *Paper*, **213**(10), 23-25.

Gustafson, R. R., C. A. Sleicher, W. T. McKean and N. L. Ricker (1983). Theoretical Model of the Kraft Pulping Process. *Ind. Eng. Chem. Process Des. Dev.* **22**(1), 87-96.

Gustafson, R. R., C. A. Sleicher, W. T. McKean and N. L. Ricker (1984). Applications of a theoretical kraft pulping model. In *Applications of Chemical Engieering in the Forest Products Industry*. AIChE, New York. 56-65.

Hagberg, M. and A. J. Isaksson (1994). Benchmarking for paper machine md-control part 1: the organisation and the simulation model. In *Control Systems 94*. Swedish Association of Pulp and Paper Engineers, Stockholm, Sweden. 207-213.

Hagenmyer, R. W., D. W. Manson and M. J. Kocurek, eds. (1992). *Pulp and Paper Manufacture, Volume 6: Stock Preparation*. 3rd ed. Joint Textbook of the Paper Industry. Tappi Press, Atlanta, GA. and CPPA, Montreal, Quebec.

Haggman, B. C. and W. I. Bialkowski (1994). Performance of common feedback regulators for first-order and deadtime dynamics. *Pulp Pap. Can.*, **95**(4), 46-51.

Han, Y. S. (1989). *Modeling and Simulation of Wood Pulp Washers* [PhD Thesis]. University of Idaho.

Harrison, R. E. and P. C. Ariessohn (1985). A smelt-bed imaging system. *Tappi J.*, **68**(8), 62-66.

Harkonen, E. J. (1987). Mathematical model for two-phase flow in a continuous digester. *Tappi J.*, **70**(12), 122-126.

Harris, T. J. and J. F. MacGregor (1987). Design of Multivariable Linear-Quadratic Controllers Using Transfer Functions. *AIChE Journal*. **33**(9), 1481-1495.

Hatton, J. V. (1973). Development of Yield Prediction Equations in Kraft Pulping. *Tappi*. **56**(7), 97-100.

Hatton, J. V. (1984). Significance of residual effective alkali/H-factor relationships in kraft pulping process control. In *Pulping Conference*, Book 3. TAPPI Press, Atlanta. 469-478.

Haynes, J., J. Scheldorf and L. Edwards (1988). Modular simulator interfaces to distributed control for training. *Pulp & Paper*. **62**(9), 192-195.

Haynes, J., J. Scheldorf and L. Edwards (1990). Toward Better and More Economical Process Trainers. *Pulp & Paper Canada*. **91**(5), 45-48.

Heaven, E. M., T. M. Kean, I. M. Jonsson, M. A. Manness, K. M. Vu and R. N. Vyse (1993). Applications of system identification to paper machine model development and controller design. In *IEEE International Conference on Control and Applications*, v 1. IEEE, Piscataway, NJ. 227-232.

Heaven, E. M., I. M. Jonsson, T. M. Kean, M. A. Manness and R. N. Vyse (1994). Recent advances in cross machine profile control. *IEEE Cont. Syst. Mag.*, **14**(5), 35-46.

Holik, H. (1986). Improving paper quality by optimum cd profile control. In *Control Systems '86: Symposium on Control Systems in the Pulp and Paper Industry*. Swedish Forest Products Lab, Stockholm. 87-94.

Horner, M. G. and L. Rasmusson (1990). The use of on-line kappa measurement in Kamyr digester control. In *Pulp Technology Energy: New Available Techniques and Current Trends*. SPCI, Stockholm. 242-252.

Hosti, T., R. Sutinen, and K. Leiviska (1994). Burning expert system for recovery boilers. In *Control Systems 94*. Swedish Association of Pulp and Paper Engineers, Stockholm, Sweden. 313-317.

Ibrahim, A. A. (1985). Computer analysis of headbox/slice flow dynamics. *PIMA Mag.*, **67**(2), 32-33,36-37.

Ibrahim, A. A. (1987). How headbox design and operating criteria affect machine performance and sheet structure. *Pulp Pap. Can.*, **88**(12), 33-34,36-37.

Ingman, L. C. (1992a). Managing total quality. In *Tappi 92*. TAPPI Press, Atlanta, GA, USA. 235-236.

Ingman, L. C. (1992b). Process control and quality control: integral parts of ISO 9000. *In Process & Prod. Quality Conf. (Appleton, WI) Proc.* TAPPI Press, Atlanta, GA, USA. 43-48.

Isaksson, A. J., L. E. Jonsson, and M. Hagberg (1994). Benchmarking for paper machine md-control part 2: simulation results. In *Control Systems 94*. Swedish Association of Pulp and Paper Engineers, Stockholm. 262-272.

Isaksson, A. J. and V. Kaul (1992). Survey of missing data identification techniques. In *Control Systems 92*. CPPA, Montreal. 95-99.

Jack, J. S., R. G. Bentley, and R. L. Barron (1990). Optical pulp consistency sensors. *Pulp Pap. Can.*, **91**(2), 59-62,64.

Jeffers, L. A. and G. W. Roman (1994). Continuous on-line measurement of lignin concentration in wood pulp.

In *Industrial Energy Technology Conference*. Department of Energy, Washington. P6.

Jofriet, P. J. C. T. Seppala, D. M. Harvey, B. W. Surgenor and T. J. Harris (1995). Experience with implementation of an expert system for control loop performance analysis. In *Advances in Instrumentation and Control*, pt 1, Vol. 50. ISA, Research Triangle Park, NC. 21-30.

Johnson, R. K. (1988). Design consideration for advanced control of the modern recovery boiler. In *Engineering Conference*. TAPPI Press, Atlanta. 57-66.

Karlstrom, A. and M. Koebe (1993). Modeling of wood-chip refining process. *Nord. Pulp Pap. Res. J.*, 8(4), 384-388,404.

Kastanakis, G. and A. Lizr (1991). Interaction between the md and cd control processes in papermaking and plastics machines, optimal tuning criteria. *Tappi J.*, 74(2), 77-83.

Kaunonen, A., H. Kovio, P. Virtanen and P. Sumanen (1985). Modeling and simulation of paper machine's wet-end process. In B. Wahlstrom and K. Leiviska (Eds.), *IMACS World Congress on Systems Simulation and Scientific Computation*, v 3. North-Holland Publishing, Amsterdam, New York. 229-231.

Kaunonen, A, K. Lehmikangas, J. Nokelainen and P. Tikkanen (1991). Practical experiences of how to control wet-end operations using continuous retention measurement. In *Papermakers Conference*. TAPPI Press, Atlanta. 39-45.

Kaunonen, A. and H. Paunonen (1993). Trends in pulp and paper process control. *Pap. Puu*, 75(8), 550-552,554,557-558.

Kayihan, F., M. S. Gelormino, E.M. Hanczyc, F.J. Doyle III and Y. Arkun (1996). A Kamyr Continuous Digester Model for Identification and Controller Design. In *13th IFAC World Congress*, June 30-July 4, San Francisco, CA.

Kemna, A. H., D. A. Mellichamp, D. E. Seborg and N. Sweerus (1992). Dynamic models of recovery boilers from input-output data. In *International Chemical Recovery Conference*. TAPPI Press, Atlanta. 305-315.

Kjaer, A. P., W. P. Heath, and P. E. Wellstead (1995). Identification of cross-directional behaviour in web production: Techniques and experience. *Control Engineering Practice*, 3(1), 21-29.

Kjaer, A., M. Waller, and P. Wellstead (1994). Headbox modelling for cross-direction basis weight control. *ISA Transactions*, 33(3), 245-254.

Kline, J. E. (1991). *Paper and Paperboard, Manufacturing and Converting Fundamentals*. 2nd ed. Miller Freeman Publications, San Francisco, CA.

Kooi, S. B. L. and K. Khorasani (1992). Control of wood-chip refiner using neural networks. *Tappi J.*, 75(6), 156-162.

Kristinsson, K. and G. A. Dumont (1993). Paper machine cross directional basis weight control using Gram polynomials. In *IEEE International Conference on Control and Applications*, v 1. IEEE, Piscataway, NJ. 235-240.

Laughlin, D. L. and M. Morari (1989). Robust performance of cross-directional basis-weight control in paper manufacturing. In *American Control Conference*, v 3. IEEE, Piscataway, NJ. 2122-2127.

Laughlin, D. L., M. Morari, and R. D. Braatz (1993). Robust performance of cross-directional basis-weight control in paper machines. *Automatica*, 29(6), 1395-1410.

Lavigne, J. R. (1993). *Pulp and Paper Dictionary*. Miller Freeman Publications, San Francisco, CA.

Lee, J. H. and A. K. Datta (1994). Nonlinear inferential control of pulp digesters. *AIChE Journal*, 40(1), 50-64.

Leffler, N. (1993). Process control today and tomorrow. In *Pulp and Paper 2000*, vol. 2. EUCEPA, . 275-280.

Liaw, S.-J. and G. A. Krishnagopalan (1992). On-line measurement of sulfide and alkali concentrations during kraft pulping. *Tappi J.*, 75(9), 219-224.

Lindholm, P, H. Paulapuro, I. Penttinen and R. Ritala (1989). WEDGE — intelligent tool kit speeds up paper-machine diagnostics. *Pap. Puu*, 71(5), 463-466.

Lundqvist, S.-O. (1990). Recent Developments in Continuous Digester Control. In *Pulp Technology Energy: New Available Techniques and Current Trends*. SPCI, Stockholm. 222-233.

Makkonen, A., R. Rantanen, A. Kaukovirta, H. Koivisto, J. Lielehto, T. Jussila, H. N. Koivo and T. Nuhtelin (1994). The paper machine benchmark for control systems 94. In *Control Systems 94*. Swedish Association of Pulp and Paper Engineers, Stockholm. 219-225.

Mangin, H. V. (1992). Control systems in the year 2010. In *Engineering Conference*. TAPPI Press, Atlanta. 37-38.

Matsuda, M. (1990). A new decoupling matrix method for paper basis weight profile control. In *IEEE Conference on Decision and Control*, 29th. IEEE, Piscataway, NJ. 1592-1594.

Meincer, W. P. and J. P. Andersen (1991). Real time dynamic process simulation — a spectrum of choices. In *Engineering Conference*, (Nashville, TN; 1991 Sep 30 - Oct 3). Tappi Press, Atlanta. 421-434.

Michaelsen, R., T. Christensen, G. G. Lunde, G. Lundman and K. Johansson (1994). Model predictive control of a continuous Kamyr digester at SCA-Nordliner, Munksund, Sweden. *Pulp Pap. Can*, 95(12), 146-149.

Michelsen, F. A. (1995). *A Dynamic Mechanistic Model and Model-Based Analysis of a Continuous Kamyr Digester*. Dr. Ing. Thesis. The Norwegian Institute of Technology, Trondheim.

Miles, K. B. and W. D. May (1993). Predicting the performance of a chip refiner. a constitutive approach. *J. Pulp Pap. Sci.*, 19(6), 268-274.

Mims, A., M. J. Kocurek, J. A. Pyatte, and E. E. Wright, eds. (1993). *Kraft Pulping: a Compilation of Notes*. Tappi Press, Atlanta, GA.

Mumford, W. and L. Edwards (1986). Lime-kiln simulation — predicted versus observed mill performance. In *American Institute of Chemical Engineers, National Meeting*, Paper no. 81d. AIChE, New York.

Natarajan, K., G. A. Dumont, and M. S. Davies (1989). Algorithm for estimating cross and machine direction moisture profiles for paper machines. In *Identification and System Parameter Estimation*, v 2. Pergamon Press, Oxford. 1091-1096.

Niemi, A. J. (1989). Optimal control of evaporator and washer plants. *Isotopenpraxis*, **25**(4), 171-175.

Noel, A. et al. (1993). Advanced brown stock washer control: successful industrial implementation at James Maclaren. In *IEEE International Conference on Control and Applications*, v 1. IEEE, Piscataway, NJ. 351-357.

Nuyan, S. (1986). Requirements for the cross-machine direction basis weight control of the paper web. In *American Control Conference*. IEEE, New York. 1416-1422.

Obradovic, D., G. Deco, H. Furumoto and C. Fricke (1993). Neural networks for industrial process control: applications in pulp production. In *International Conference on Neural Networks and their Industrial and Cognitive Applications*. EC2, Nanterre, France. 25-32.

Osmond, D. R., P. J. C. Tessier, and M. Savoie (1994). Control of an industrial lime kiln operating close to maximum capacity. *Tappi J.*, **77**(2), 187-194.

Ostroot, G. F. (1993). *Consistency Control Book*. TAPPI Press, Atlanta.

Ozaki, N., S. Yamazu, K. Tateoka, W. Shinohara and S. Hayashi (1994). Recovery boiler intelligent control. In *Control Systems 94*. Swedish Association of Pulp and Paper Engineers, Stockholm, Sweden. 307-312.

Partanen, K. and H. Koivo (1984). On-line freeness control at a thermomechanical pulping plant. In *Advances in Instrumentation*, pt 2, Vol. 39. ISA, Research Triangle Park, NC. 877-884.

Patrick, K. L. (1991). Neural networks keeps BSW filtrate solids at maximum, uniform levels. *Pulp Pap.*, **65**(3), 55-58.

Paulonis, M. A. and A. Krishnagopalan (1988). Application of cooking liquor analysis to on-line estimation of pulp yield and Kappa number in a kraft batch digester. In *Pulping Conference*, Book 3. TAPPI Press, Atlanta. 555-560.

Paulonis, M. A. and A. Krishnagopalan (1990). Kraft batch-digester control based on pulping-liquor analysis. In *AIChE Forest Products Symposium 1989 and 1990*. TAPPI Press, Atlanta. 49-60.

Perrier, M., S. Gendron, and N. Legault (1992). On the choice of algorithms for time-varying processes. In *Annual Meeting - Technical Section, Canadian Pulp and Paper Association, Preprints B*, 78th. CPPA, Montreal. B249-B254.

Perrier, M. and A. A. Roche (1992). Towards mill-wide evaluation of control loop performance. In *Control Systems 92*. CPPA, Montreal. 205-209.

Petrus, C. J. (1990). On the application of generalized predictive control to Kamyr digester chip level. In *Pacific Paper Expo Technical Conference, Program 2: Process Control*. Maclean Hunter, Vancouver. 2/2/0-2/2/15.

Pettersson, A. (1989). Size and thickness distribution of wood chips - now measured on-line. In *Pulping Conference*. TAPPI Press, Atlanta. 471-475.

Pettersson, A., L. Olsson, and S.-O. Lundqvist (1988). On-line measurement of wood chip size. *Tappi J.*, **71**(7), 78-81.

Piipponen, J. and R. Ritala (1994). Basis weight and filler control with static decoupling and smith predictor. In *Control Systems 94*. Swedish Association of Pulp and Paper Engineers, Stockholm, Sweden. 226-229.

Piirto, M. and H. Koivo (1992). Advanced control of a paper machine wet end. In *Advanced Control of Chemical Processes: ADCHEM '91*. Pergamon Press, Tarrytown, NY. 47-52.

Pokela, J., E. Kukkamaki, H. Paulapuro, R. Ritala and M. Vuori (1994). Practical experiences of process analysis on a modern paper machine. In *Control Systems 94*. Swedish Association of Pulp and Paper Engineers, Stockholm, Sweden. 159-165.

Qian, X. and P. J. Tessier (1993). Modeling and simulation of a chip refining process. In *Pacific Paper Expo Technical Conference*. Maclean Hunter, Vancouver. 116-122.

Qian, X., P. J. Tessier, and G. A. Dumont (1994). Fuzzy logic modeling and optimization of a wood chip refiner. *Tappi J.*, **77**(2), 181-186.

Rankin, P. A. and W. L. Bialkowski (1984). Bleach plant computer control: design, implementation, and field experience. Part II: control engineering, performance monitoring, and human resources. *Tappi J.*, **67**(7), 66-70.

Rantala, T., P. Tarhonen, and H. N. Koivo (1994a). Adaptive retention control in a paper machine. *Pulp Pap. Can.*, **95**(8), 20-23.

Rantala, T., P. Tarhonen, and H. N. Koivo (1994b). Control of paper-machine wire retention. *Tappi J.*, **77**(12), 125-132.

Rao, M. and Y. Ying (1990). Intelligent control: a new decade for pulp and paper processes. In *IEEE International Symposium on Intelligent Control 1990*. IEEE, Piscataway, NJ. 1095-1099.

Rao, M., Y. Ying, and J. Corbin (1991). Intelligent engineering approach to pulp and paper process control. In *Annual Meeting - Technical Section, Canadian Pulp and Paper Association, Preprints A*, 77th. CPPA, Montreal. A195-A199.

Rao, M., Q. Xia and Y. Ying (1994). *Modeling and Advanced Control for Process Industries, Applications to Paper Making Processes*. Advances in Industrial Control. Springer-Verlag, London.

Rawlings, J. B. and I. L. Chien (1996). Gage control of film and sheet forming processes. *AIChE Journal*, **42**(3), 753-766.

Ribeiro, B., A. Dourado, and E. Costa (1993). Lime kiln process identification and control: a neural network approach. In R. F. Albrecht, C. R. Reeves, and N. C. Steele (Eds.), *International Conference on Artificial Neural Nets and Genetic Algorithms*. Springer-Verlag, Berlin. 117-124.

Ricker, N. L., T. Subrahmanian, and T. Sim (1989). Case studies of model-predictive control in pulp and paper production. In T. J. McAvoy, Y. Arkun, and E. Zafiriou (Eds.), *Model Based Process Control: Proceedings of the IFAC Workshop*. Pergamon Press, Oxford. 13-22.

Rigopoulos, A., Y. Arkun and F. Kayihan (1996a). Identification of paper machine full profile disturbance modes using adaptive principal component analysis. CPC-V conference, Jan 7-12, 1996, Tahoe City, CA.

Rigopoulos, A., Y. Arkun and F. Kayihan (1996b). Control relevant disturbance modeling of paper machine full profile properties using adaptive PCA. In *Control Systems '96*, April 30 — May 2, Halifax, Nova Scotia. CPPA 35-39.

Rigopoulos, A., Y. Arkun and F. Kayihan (1997). Identification of full profile disturbance models for sheet forming processes. *AIChE Journal*, in press.

Ritala, R. (1993). New tool for powerful process analysis. *Pap. Puu*, **75**(1), 488-490.

Rogers, E. and D. H. Owens (1994). 2D systems theory and applications-a maturing area. In *International Conference on CONTROL*, vol.1. IEE, London. 63-69.

Roman, G. and L. Jeffers (1992). Development of a prototype lignin concentration sensor. In *DOE/Industry Advanced Sensors Technical Conference*. US Department of Energy, . p.44.

Rudd, J. B. (1994). Prediction and control of pulping processes using neural network models. In *Annual Meeting — Technical Section, Canadian Pulp and Paper Association, Preprints B*, 80th. CPPA, Montreal. B169-B173.

Salcudean, M., P. Nowak, and Z. Abdullah (1992). Mathematical modeling of recovery furnaces. In *International Chemical Recovery Conference*. TAPPI Press, Atlanta. 197-208.

Sbarbaro, D. G. and R. W. Jones (1994). Multivariable nonlinear control of a paper machine headbox. In *IEEE Conference on Control Applications*. IEEE, Piscataway, NJ. 769-774.

Sell, N. J. (1995). *Process Control Fundamentals for the Pulp and Paper Industry*. Tappi Press, Atlanta.

Sethuraman, J., J. Krishnagopalan, and G. Krishnagopalan (1995). Kinetic model for the causticizing reaction. *Tappi J.*, **78**(1), 115-120.

Shewchuk, C. F. (1987). Massbal mk11: New process simulation system. *Pulp & Paper Canada*. **88**(5), 76-82.

Sikora, R. F., W. L. Bialkowski, J. F. MacGregor and P. A. Taylor (1984). Self-tuning strategy for moisture control in papermaking. In *American Control Conference*, v 1. IEEE, Piscataway, NJ. 54-61.

Smith, D. B. and L. L. Edwards (1991). Dynamic mathematical model of a rotary lime kiln. In *Engineering Conference*. TAPPI Press, Atlanta. 447-456.

Smith, C. C. and T. J. Williams (1974). *Mathematical Modeling, Simulation and Control of the Operation of a Kamyr Continuous Digester for the Kraft Process*. Technical report 64. Purdue University, West Lafayette, Indiana

Smook, G. A. (1990). *Handbook of Pulp and Paper Terminology*. Angus Wilde Publications, Vancouver, B.C.

Smook, G. A. (1992). *Handbook for Pulp and Paper Terminologists*. Angus Wilde Publications, Vancouver, B.C.

Strand, B. C. (1993). Analyzing the effects of raw material variation on mechanical pulp properties using integrated factor networks. In *Annual Meeting — Technical Section, Canadian Pulp and Paper Association, Preprints A*, 79th. CPPA, Montreal. A215-A220.

Sutcliffe, L., D. Swanson, and R. Balakrishnan (1993). Optimization of multiple cross-direction actuator control. *Appita J.*, **46**(1), 27-30.Syberg, O. and N. W. Wild (1992). *Introduction to Process Simulation*, 2nd ed. TAPPI Press, Atlanta.

Tessier, P. and X. Qian (1994). Modeling and simulation of a ctmp process. In *Annual Meeting — Technical Section, Canadian Pulp and Paper Association, Preprints A*, 80th. CPPA, Montreal. A143-A149.

Tham, M. T. and R. W. Jones (1985). Advanced control of multivariable chemical processes. In *Advances in Process Control*. Institution of Chemical Engineers, Rugby, Engl. P8.1-P8.19.

Thorp B. A., and M. J. Kocurek, eds. (1991). *Pulp and Paper Manufacture, Volume 7: Paper Machine Operations*. 3rd ed. Joint Textbook of the Paper Industry. Tappi Press, Atlanta, GA. and CPPA, Montreal, Que.

Toivonen, H. T. and J. Tamminen (1990). Minimax robust lq control of a thermomechanical pulping plant. *Automatica*, **26**(2), 347-351.

Turner, P. A., B. J. Allison, and J. K. Oei (1992). Brown stock washer control. Part II. Filtrate tank level control. In *Annual Meeting — Technical Section, Canadian Pulp and Paper Association, Preprints B*, 78th. CPPA, Montreal. B109-B115.

Tyler, M. L. and M. Morari (1995). Estimation of cross directional properties: scanning versus stationary sensors. *AICHE Journal*. **41**, 846-854.

Vanchinathan, S. and G. A. Krishnagopalan (1993). Kraft delignification kinetics based on liquor analysis. In *Pulping Conference*. TAPPI Press, Atlanta. 431-441.

Venkateswarlu, C. and K. Gangiah (1992). Dynamic modeling and optimal state estimation using extended Kalman filter for a kraft pulping digester. *Industrial & Engineering Chemistry Research*, **31**(3), 848-855.

Vincent, J.-P, T. Nguyen-Minh, W. Hagen, B. Lebeau, J. M. Dion and L. Dugard (1994). Advanced control strategies for the dyeing of white and colored papers. *Pulp Pap. Can.*, **95**(9), 34-38.

Virtanen, P. (1988). Computer simulator for paper machine wet-end. In *Process Simulation Symposium*. CPPA, Montreal. 85-89.

Vroom, K. E. (1957). The "H" factor: a means of expressing cooking times and temperatures as a single variable. *Pulp and Paper Magazine of Canada*. Convention Issue.

Vyse, R., J. King and M. Heaven (1995). Cross direction basis weight control on multiply board machines. *Pulp Pap. Can.*, **96**(4), 29-34.

Wallace, B. (1990). Challenge of higher-frequency MD and CD control. *PIMA Mag.*, **72**(2), 56-57.

Wallace, B. W. (1992). Advances in on-line measurement for tighter control of paper quality. *Appita J.*, **45**(2), 74-77.

Walsh, R. A., W. L. Bialkowski, and S. B. L. Kooi (1987). Process control study in the pulp and paper industry. *Pulp Pap. Can.*, **88**(3), 121-127.

Wang, L., P. Englezos, and P. Tessier (1994). Dynamic modeling and simulation of the recausticizing plant in a kraft pulp mill. *Tappi J.*, **77**(12), 95-103.

Wang, X. G., G. A. Dumont, and M. S. Davies (1992). On-line estimation of a dynamic model for basis weight variations in paper machines. In *IEEE Conference on Control Application*, vol.2. IEEE, New York. 1082-1087.

Wang, X. G., G. A. Dumont, and M. S. Davies (1993a). Adaptive basis weight control in paper machines. In *IEEE International Conference on Control and Applications*, v 1. IEEE, Piscataway, NJ. 209-216.

Wang, X. G., G. A. Dumont, and M. S. Davies (1993b). Estimation in paper machine control. *IEEE Cont. Syst. Mag.*, **13**(4), 34-42.

Wang, X. G., G. A. Dumont, and M. S. Davies (1993c). Modelling and identification of basis weight variations in paper machines. *IEEE Trans. Cont. Syst. Tech.*, **1**(4), 230-237.

Wells, C. H. and J. C. Hassler (1990). Deep-knowledge expert system for batch digester control. In *Pulping Conference*, Book 2. TAPPI Press, Atlanta. 581-595

Wellstead, P. E. and W. P. Heath (1994). Two-dimensional control systems: application to the CD and MD control problem. *Pulp Pap. Can.*, **95**(2), 48-51.

Wessel, R. A., K. L. Parker, and A. Akan-Etuk (1993). Three-dimensional flow and combustion model of a kraft recovery furnace. In *Engineering Conference*. TAPPI Press, Atlanta. 651-664.

Westermeyer, W. N. (1987). Modeling flow in the headbox slice and on the fourdrinier wire with regard to cross-machine basis weight profile of the paper web. *Papier*, **41**(11), 591-601.

Wigsten, A. L. (1988). Conductivity-based shower water control in brown stock washing. In *Pulping Conference*, Book 1. TAPPI Press, Atlanta. 1-14.

Wisnewski, P. A. (1995). *Model Reduction, State Estimation and Model Predictive Control for the Cook Zone of a Kamyr Digester*. MS Thesis. Purdue University, Indiana.

Woodard, E. R. (1991). Measurement and control of the fourdrinier drainage process. In *Papermakers Conference*. TAPPI Press, Atlanta. p. 11.

Xia, Q. and M. Rao (1992). Fault-tolerant control of paper machine headboxes. *Journal of Process Control*, **2**(4), 171-178.

Xia, Q., M. Rao, H. Farzadeh, K. Danielson, C. Henriksson and j. Olofsson (1993). Integrated intelligent control system for peroxide bleaching processes. In *IEEE International Conference on Control and Applications*, v 2. IEEE, Piscataway, NJ. 593-598.

Xia, Q., M. Rao, C. Henriksson and H. Farzadeh (1995). Case-based reasoning for intelligent fault diagnosis and decision making in pulp processes. In *Annual Meeting — Technical Section, Canadian Pulp and Paper Association, Preprints B*, 81st. CPPA, Montreal. B195-B198.

Xia, Q., M. Rao, and J. Qian (1993). Model algorithmic control for paper machines. In *IEEE Conference on Control Applications*, v 1. IEEE, Piscataway, NJ. 203-208.

Xia, Q, M. Rao, X Shen and Y. Ying (1993). New technique for decoupling control. *International Journal of Systems Science*, **24**(2), 289-300.

Xia, Q, M. Rao, X. Shen, Y. Ying and J. Zurcher (1993). Systematic modeling and decoupling control of a pressurized headbox. In V. K. Bhargava (Ed.), *Canadian Conference on Electrical and Computer Engineering*, vol. 2. IEEE, New York. 962-965.

Xia, Q, M. Rao, X. Shen and H. Zhu (1993). Adaptive control of basis weight and moisture content for a paperboard machine. *Journal of Process Control*, **3**(4), 203-209.

Xia, Q., M. Rao, X, Shen and H. Zhu (1994). Adaptive control of a paperboard machine. *Pulp Pap. Can.*, **95**(5), 51-55.

Xia, Q, M. Rao, Y. Ying and X. Shen (1994). Adaptive fading kalman filter with an application. *Automatica*, **30**(8), 1333-1338.

Yang, W., R. R. Horton, and T. N. Adams (1994). Effects of boundary geometries on cfd simulations of recovery furnace char beds. *Tappi J.*, **77**(8), 189-199

Ying, Y, M. Rao, X. Shen and Q. Xia (1994). Bilinear decoupling control and its industrial application. *Control - Theory and Advanced Technology*, **10**(1), 97-109.

Ying, Y. Q., M. Rao, and Y. X. Sun (1991). A new design method for bilinear suboptimal systems. In *American Control Conference*. American Automatic Control Council, Green Valley, AZ. 1820-1821.

Youngblood, R., J. W. Curtis, and G. Vachtsevanos (1990). Fuzzy logic control. New approaches and applications. In *Advances in Instrumentation*, pt 3, Vol. 45. ISA, Research Triangle Park, NC. 1129-1136.

Zervos, C. C., G. A. Dumont, and G. L. Pageau (1989). Laguerre-based adaptive control of pH in an industrial bleach plant extraction stage. In *Adaptive Control of Chemical Processes: Selected Papers from the 2nd International IFAC Symposium*. Pergamon Press, Oxford. 135-140.

Zhou, Wei-W., G. Dumont, and B. Allison (1991). Prototype expert system based on Laguerre adaptive control. In *Intelligent Tuning and Adaptive Control*. Pergamon Press, Tarrytown, NY. 319-325.

CASE STUDIES IN EQUIPMENT MODELING AND CONTROL IN THE MICROELECTRONICS INDUSTRY

Stephanie W. Butler
Texas Instruments
Dallas, TX 75265

Thomas F. Edgar
Department of Chemical Engineering
University of Texas
Austin, TX 78712

Abstract

Major advances in modeling and control will be required to meet future technical challenges in microelectronics manufacturing. This paper discusses some of the technological improvements under development and the challenges faced by the industry. Because of the tutorial nature of this paper, a brief overview is given of how an integrated circuit is made. Sources of variation requiring control and diagnosis are discussed, and current control practices in the industry are described. Recent efforts involving joint university-industry research and development are making significant progress in developing standardized equipment designs, models, and control/instrumentation systems in unit operations such as lithography, chemical vapor deposition, rapid thermal processing, and plasma etching. Several industrial case studies are presented showing how advanced process modeling and control have been successfully applied to rapid thermal processing.

Keywords

Microelectronics manufacturing, Rapid thermal processing, Batch control, Equipment models.

Introduction

During the next ten years major improvements in design, modeling and control of process equipment will be required to remain globally competitive in semiconductor manufacturing. Recently advanced process modeling and control has been identified as a key technology to receive increased emphasis in the microelectronics industry. In the Semiconductor Industry Association (SIA) roadmap (1994), the microelectronics industry has forecasted R&D needs beyond the year 2000; in this document they have adopted a broad definition of advanced process control (APC), which is a set of automated methodologies to achieve desired process goals (Butler, 1995a) in operating individual process steps. APC is considered to include four components:

- fault detection
- fault classification
- fault prognosis
- process control

Fault detection includes the function of statistical process control (SPC). APC uses information about the material to be processed, measured data, and the desired results (target values) to select which model and control strategy to employ. APC also determines if the desired results are achievable on the machine and notifies the user of necessary machine maintenance. In its most basic mode, APC monitors the process and determines the necessary manipulated variable action. The controller also monitors the adjustments to ensure they satisfy operating constraints and generates the necessary alarms. Recently, the definition of APC has been expanded to include not just a single machine, but to encompass the entire fab (Butler, 1995b).

At CPC-IV Larrabee (1991) discussed the microelectronics factory of the future. This paper mainly focused on the financial aspects of the semiconductor industry, especially device technology and fab costs, and was based on the 1992 SIA roadmap. The current SIA roadmap (1994) has accelerated a few areas compared to the 1992 version. One key area is wafer size, which will increase to 300mm (from the current 200mm) before the end of the decade, requiring a new generation of equipment. Also needed are flexible factories, which can produce a variety of specialty products, each in small volume and very quickly. A prototype of flexible manufacturing cited by Larrabee was the Microelectronics Manufacturing Science and Technology Program (MMST), at Texas Instruments, sponsored by the Department of Defense and completed in 1993 (Schaper et al., 1994; IEEE, 1994; Solid State Technology, 1994).

MMST was very successful in that it introduced new concepts, including real-time control, to the semiconductor industry. Based upon this program and other process control work in the industry, improvements in the plant control system are expected to beneficially impact throughput, cycle time, yield, maintenance scheduling, flexibility, local and global process understanding, and time to market. The use of APC should also increase the life of the processing equipment in the plant. The 1994 SIA Roadmap underscores the need for productivity improvement in order to maintain the industry's historical 30% per-year per-function reduction in cost. In the past these cost reductions were attainable through yield improvement, but yields are currently very high. Hence future improvements must come in increased capital equipment utilization, which translates to maximizing throughput of product wafers with reduced set-up costs.

Manufacturing of an Integrated Circuit

Integrated circuits are created by going through a sequence of unit processes, also known as a "flow." These unit processes are:

- oxidation
- deposition (dielectrics, silicon, metals)
- doping/diffusion/implant
- lithography (photoresist pattern)
- etching (pattern transfer)
- resist removal (ash, part of etching or lithography)
- clean
- metrology

As many as 300 unit process steps with an overall duration of two months may be required in manufacturing a given product. In contrast to the chemical industry, products cannot be blended to rectify processing errors, and rework is usually not possible.

On each wafer are formed many dies (or bars or chips). The dies are delineated by scribe lines. As the wafer progresses through the flow, it acquires topography (vertical features) on the dies; scribe lines also create topography. The smallest horizontal feature (linewidth) on the wafer is known as the critical dimension. The lowest critical dimension in production currently is 0.35 μm (1/100 the thickness of a human hair).

There are several terms which relate to the number of wafers processed together. A lot is a quantity of wafers (24) processed together through the flow, and the lot is carried in a boat. There is usually an extra slot in the boat for a pilot wafer, which is used for metrology reasons. Most new processes are single wafer, which denotes that wafers are processed in a reactor one at a time, but the wafer is not taken to the next process until the entire lot is finished processing. Cluster tool refers to equipment which has several single wafer processing chambers. The chambers may carry out the same process or different processes; some vendors base their chamber designs on series operation, while others utilize parallel processing schemes. The term batch denotes wafers that are processed more than one at a time. Only furnaces (usually oxidation, and some deposition), clean-up hoods, and some ash equipment treat more than one wafer at a time in a modern line. A batch may be comprised of more than one lot.

In traditional chemical engineering terminology, most processes are semi-batch. For example, in a single wafer processing tool the following steps are carried out:

1. the operator loads the boat of wafers
2. the machine transfers a single wafer into the processing chamber
3. gases flow continuously and reaction occurs
4. the gas flow is stopped and the machine removes the wafer
5. the next wafer is processed.

Thus, control of semiconductor processes can be viewed as a batch reactor control problem. When all wafers are finished processing, the operator takes the boat of wafers to the next machine. All of these steps are carried out in a clean room designed to minimize device damage by particulate matter.

The recipe consists of the regulatory setpoints and parameters for the real-time controllers on the equipment. The equipment controllers are normally not capable of receiving a continuous setpoint trajectory. Only furnaces and rapid thermal processing (RTP) tools are able to ramp up, hold, and ramp-down the temperature or power supply. A recipe can consist of several steps; each step processes a different film and/or uses a different chemistry. The process step may be automatically terminated when the endpoint is reached, indicated by a change in a process variable. The same recipe on two chambers of the same brand may produce different results. This lack of repeatability across chambers is a big problem with cluster tools or when a fab has multiple machines of the same type because it requires that a fab keep track of different recipes for each chamber. In microelectronics manufacturing, APC represents a new paradigm of "control to target" because specifications will consist of desired results (e.g., target of 2000 A) rather than a recipe based on machine settings (e.g., 200 mT), which have been used historically. The controller translates the desired results specification into a machine recipe. Thus the fab supervisory controller must only keep track of the product specifications.

This paper contains a description of how integrated circuits are manufactured along with relevant terminology, followed by a discussion of the principal sources of variation that must be addressed by process control in the unit operations utilized. Three types of variations arise due to physical process changes, operational and control practices and the limitations of metrology currently used. We illustrate some of the key modeling and control challenges by reviewing several cases where advanced process control has been applied to Rapid Thermal Processing (RTP).

Sources of Variation in Manufacturing

The sources of variation during manufacturing can be grouped into three categories:

1. physical process changes
2. operational/control practices
3. metrology limitations.

Current control practice in the industry, while eliminating one source of variation in the product, may induce another. We discuss each category in more detail below.

Physical Process Changes

In other batch processing industries, process dynamics within a single batch or setpoint trajectories may be considered the major control issue. However, except for furnaces and rapid thermal processing, these are not the critical issues in semiconductor manufacturing today. Disturbance rejection, feedforward control, and product target changes (setpoint changes) are the main control issues. Disturbances include equipment aging, machine maintenance, chamber build-up, and unmeasured incoming wafer state changes. Feedforward values which may change are measured incoming wafer state (different products or different process units in the flow), as well as measurable machine states (such as tube age). Product target changes are due to the same machine being used several times during a single flow (for different processes or for different desired wafer state changes) or the same machine being used for several different products. However, dynamic control may require more attention in future processes because quickly igniting and extinguishing plasmas and on/off process changes between steps may cause wafer damage.

One source of variation due to physical phenomena is called loading. Loading denotes the consumption of large amounts of the reactant, especially in a non-uniform manner. Macroloading is non-uniform depletion across the wafer. Sometimes the nonuniformity is due to the non-axisymmetric placement of the pump port. Often the available manipulated variables do not influence uniformity, i.e., there is a controllability problem. Microloading occurs when actual features on the device cause depletion or impact mass/ion transport; features on a wafer are affected differently depending on the feature size/topography. The only solution in this case is to try to find operating points where the process is insensitive to device features/topography or to request that VLSI engineers change the product design. It would be desirable to quantify how specific features influence the processing behavior of different products. In theory feedforward control could be used if the controller has such knowledge. However, this knowledge is hard to gain without first having the actual device to process. In addition, process control has only a gross effect and the wafer can suffer from microloading.

Table 1. Events Affecting Time Scales for Process Variation/Control

- Within the run (real-time)
  - startup and shutdown of the process
  - different materials present (clearing)
  - dynamic target (set point changes)
  - processing times (several seconds to several minutes)
- Wafer to wafer (run to run)
  - Long term machine drift
- Within a lot
  - process warmup (first wafer effect)
- Lot to lot
  - different devices and/or step in flow (setpoint changes)
  - variations around targeted incoming wafer state
  - long term machine drift
  - interactions between different processes run sequentially
- Machine events
  - gas bottle changes, chamber cleans, equipment part replacements (preventive maintenance)

There are several time scales over which variations occur and a control system could operate; see Table 1. Clearing indicates the macro- and micro-scale removal of the film across the wafer, which results in "islands" of material present across the wafer at particular moments in time. Due to nonuniformities and microloading effects, device damage can occur during clearing. As mentioned above, damage to the wafer may also occur during startup and shutdown of the process. Due to the short time scale of the process, startup effects can occupy a large fraction of the process run time, resulting in run-to-run variability. When a chamber has been idle for a long time, a different result may be obtained compared to when there is no idle time. The difference in results is attributed to the "first wafer effect," which may last for more than one wafer (Stefani et al., 1995).

Figure 1 shows the variation of the optical emission signal and the exhaust valve position as a function of time within a single run of a plasma etch process. The optical emission shows the variation due to start-up and the presence of various materials in the stack being etched. The chemistry does not change during the entire process, although two of the materials in the stack are endpointed; analysis of the emission signal determines when to stop the process. The overetch, the amount of time the wafer is etched past clearing, is required due to the topography. It is not possible currently to endpoint the overetch, hence test wafers must be used. Too much overetch will damage the wafer, while too little etch will result in material remaining, which will also cause the device to function incorrectly. Figure 2 shows the typical variation within a lot (wafer-to-wafer) and lot-to-lot. Different amounts of idle time occurred between the lots, as well as the running of other processes.

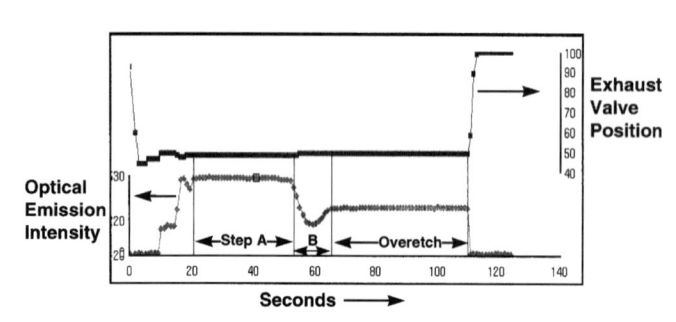

Figure 1. Variations of emission signal (lower trace) and exhaust valve position (upper trace) for a single wafer as a function of time.

There are several issues that arise if two processes are run in the same chamber. Because of operational flexibility and cost considerations, it is desirable to run as many processes as physically possible on the same machine. However, sometimes interactions between processes run sequentially may occur, i.e., the results of process A are dependent on the previous process (A or B) and the sequence of processes run in a given chamber may affect results. This interaction occurs because of the dependence of process variables upon the chamber state, and unfortunately the nature of many processes impacts the condition of the chamber walls. If process variables must be changed to make process A perform to a desired specification (i.e., different controller actions were required during consecutive runs of Process A), one must determine which changes should be made to process B to make its first wafer perform successfully. The changes that were required during process A could have been due to lot material (incoming wafer state) or due to chamber state changes. If they were due to chamber state changes, the amount of compensation for process A must be correlated to the amount required for process B. Because the chamber state is not directly measured, measurements of the process and product (and mathematical models) must be used to infer the chamber state and its effect on various processes/products.

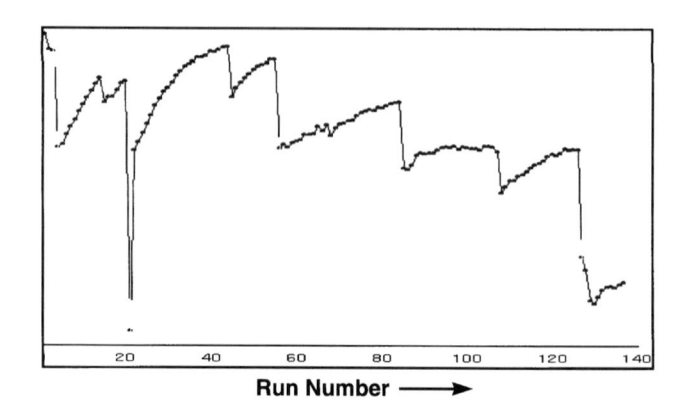

Figure 2. Wafer to wafer variation of an emission signal, averaged over an etching step. The abscissa is the wafer number.

The issue of how control actions taken during process A relate to required control actions for process B are also of extreme importance in operating flexible factories. It is quite likely that some time has passed since a given process was run on the chamber of interest. The only recent experience may be for a different process. Some method for modeling/controlling across several processes and products is necessary to eliminate pilot wafers, which is similar to the small sample statistical process control (SPC) problem. This may entail development of more complex models in order to treat the process variables involved.

Machine and chamber aging are the main reasons for long term drift, although a drift in a machine processing the incoming wafer previous to the current process sometimes is the culprit. Build-up on chamber parts, erosion of chamber materials (anodized layer), wearing of parts used for power delivery, and loss of calibration of sensors used for regulatory controllers (such as the baratron gauge used for pressure measurement) all contribute to aging. The windows used for optical emission spectroscopy also cloud (etch/build-up), which attenuates the signal and creates noise. It is important to know whether a signal change is due to window clouding or chamber aging, because only aging may affect process/product results.

Operational/Control Practices

Current control practices are aimed at removing the sources of variation caused by the physics of the process. The first practice is to decide what time scale is used for control (Table 1). Real-time control (within a run) is mainly focused on the machine, although real-time control of the process pressure is quite common. Conventional process control is also employed in furnaces and rapid thermal processes. However, the current focus of semiconductor manufacturers (rather than equipment vendors) is run-to-run control because it does not require hardware changes or vendor involvement. Wafer to wafer, lot to lot, or queue to queue are all called run to run control (a queue is where several lots are loaded together on a single wafer tool and the same recipe is used for all the lots in the queue). Run to run control allows feedforward and feedback information between consecutive processes and is naturally compatible with ex situ metrology, which has been the normal industrial practice. Figure 3 depicts two controllers and their time scales. The real-time controller is provided by the vendor and is usually a regulatory controller for a machine setting (although it could be a process variable, such as pressure). The run to run controller is usually provided by the user and is a supervisory controller for the setpoints and settings for the regulatory controller (i.e., the machine recipe). The supervisory controller runs at the factory level, as part of the computer integrated manufacturing (CIM) system, or is a cell controller which controls and downloads the recipe electronically to several similar machines. Supervisory controllers provided by outside or third party vendors are becoming more common. Another recent development is the supplying of real-time control software to equipment vendors who traditionally have written their own code.

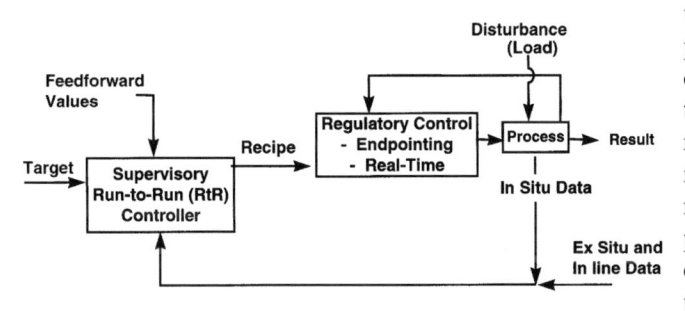

Figure 3. Regulatory (Real-Time) vs. supervisory (Run-to-Run) control.

Most unit processes are multivariable in nature. Unfortunately, the real-time regulatory controllers provided on the equipment are a system of single input-single output (SISO) controllers. Figure 4 shows the interacting SISO controllers on an etch tool that oscillate from startup to shutdown of the process. Usually, the controllers are heavily de-tuned to prevent this interaction. One benefit of the supervisory run to run controllers is that they perform multiple input-multiple output calculations.

Statistical Process Control (SPC) is typically performed on a run to run basis using ex situ metrology. The standard response to a SPC failure is to "fix" the machine. However, many times the machine is not broken, only slightly aged, and so the engineer adjusts the process. If a formal automatic supervisory controller does exist, in most cases it will only adjust the time of the process to control thickness (many processes cannot be endpointed with available on-line measurements). However, more advanced supervisory controllers other than the thickness-time loop are starting to be placed in production. The control law used in the industry is similar to minimum variance control. However, the estimator on the feedback loop usually has SPC as a gating function for feedback. In addition, no filtering of setpoint changes occurs.

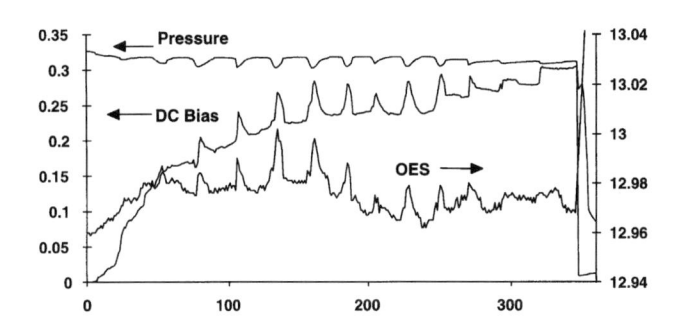

Figure 4. Variations in process variables for interacting SISO controllers in a reactive ion etcher.

Due to the lack of controllers that perform coherently across several processes/products, a qualification (set-up) pilot wafer (with or without topography) or a "lookahead" wafer from the lot is used to check and optimize the process. Because of process changeover and first wafer effects discussed above, pilot or lookahead wafers determine the chamber state and decide the amount of compensation required. Pilot or lookahead wafers are also used after machine maintenance to determine the correct machine recipe, to obtain the desired chamber state for successful processing, and as warm-up or conditioning wafers. It is estimated that 20 to 40% of wafers processed in a fab are pilot wafers. Because productivity improvement is required for microelectronics companies to remain competitive, capital equipment utilization must be maximized by increasing throughput while maintaining existing high yields. The principal way to achieve this is to reduce the number of pilot and lookahead wafers. Besides requiring non-value added time, the use of pilot wafers causes logistical problems because the CIM system has to keep track of the different type of runs (pilot, production, lookahead, warm-up) along with their different machine recipes, data sampling plans, and analysis techniques.

Metrology Limitations

Metrology can be carried out in situ (in the chamber) or ex situ, which can be inside the fab (metrology tool) or outside the fab (possible contamination problem). It may not be compatible with the topography of product wafers and may be destructive. Thus, pilot wafers are frequently used, but non-topographic or minimal topographic pilots give results which may not represent actual product wafers. Another constraint caused by the ex situ metrology is its slowness, because the measurement is slow and the metrology tool may be physically located some distance from the processing tool as well. The slowness of the measurement causes fabs to start a lot before receiving the measurements from the last lot. This measurement delay can obviously cause problems for run to run control.

In situ sensors are very limited, even for measurement of the machine state, such as valve position. There are even fewer process sensors. However, in the past few years, considerable work has been done in this area and new sensors from commercial suppliers are beginning to appear on the market. Some of the more common new sensors are rf sensors and various optical sensors for both the process (optical emission spectroscopy or OES) as well as the wafer (interferometry, ellipsometry, FTIR). Cheap mass spectrometers are also showing promise. Only recently have machine measurements been obtainable from the equipment using the semiconductor equipment communication standard (SECS-II) port. Data for regulatory controller setpoints and process variable values (such as flow rate, pressure, power), data for manipulated variables for the regulatory controllers (throttle valve, capacitors), and some additional process measurements (such as impedance) may be obtained. One result of additional sensors and SECS port data is data overload. Because run to run control and SPC are done at the CIM level or at a cell controller level, a large amount of data must be sorted to obtain process-specific information.

Several companies are attempting to predict wafer results from process and machine data and thereby replace the metrology pilots. Understanding how the short and long term variations impact the models is necessary, especially when considering that variation in normal operating conditions is relatively small. One of the main issues with respect to these models is the ability to update the model when no forcing function is present, i.e., dynamic testing concepts are not accepted in the industry today. Thus, with the slow dynamics and the constraint of running 24 wafers with a single product setpoint (lot size), difficulties can be encountered with model uncertainty when changing product setpoints or to obtain unique classification of a fault. Other issues include how to determine which faults should be compensated by the controller, the amount of data required to uniquely identify a fault, and when a virtual (soft) wafer state sensor is no longer valid. This task is extremely difficult when there is no ex situ data for comparison.

Case Studies on Control of Rapid Thermal Processing

Having discussed the general features of process control in microelectronics processing, we now review its application to a particular unit operation, namely Rapid Thermal Processing (RTP). As mentioned earlier, RTP control systems are the best analog in this industry to conventional process control. A number of equipment/instrumentation designs have been proposed and demonstrated, but this is a rapidly changing technology. Universities, equipment vendors, control software companies, and manufacturing companies are all involved in the evolution of RTP systems. These developments are discussed in detail in this section.

A typical temperature cycle for RTP is shown in Figure 5. This cycle shows a ramp from room temperature to 950 °C in 10 seconds and a steady state portion for 20 seconds at 950 °C. Temperature measurement and control have proven to be major challenges for RTP (Roozeboom, 1992; Roozeboom and Parekh, 1990), and the development of more accurate temperature measurement systems is still a critical research area. So far the optical pyrometer has been the most widely used wafer temperature measurement technique in commercial RTP applications. The physical quantity that is measured by optical pyrometer is spectral radiosity, not temperature. The conversion of spectral radiosity measurements to temperature measurements is not a trivial task. The conversion requires knowledge of the spectral emissivity of the surface (which depends on surface roughness, surface temperature, the types of films on the surface and their thicknesses, and wafer doping characteristics). Because these parameters may not be quantified sufficiently, commercial RTP applications have employed pyrometer calibration techniques using SensArray thermocouple wafers. Recent advances in SensArray's wire bonding technology have improved the accuracy and reproducibility of water temperature measurements in RTP equipment. However, there are two fundamental problems with this approach. First, in a typical semiconductor manufacturing facility, wafer-to-wafer variations are common, as discussed earlier. These variations include spectral emissivity variations, which introduce systematic bias errors in the optical pyrometer temperature measurement. The second problem has to do with the overlap between the narrow wavelength band operating regime of a typical pyrometer and the electromagnetic spectrum of the radiant heat sources. Interference of radiant heat source reflections must be taken into account when using pyrometers for temperature measurement in RTP systems. Closed-loop control of the radiant heat sources also signifies that the interference is not constant during a typical RTP process cycle.

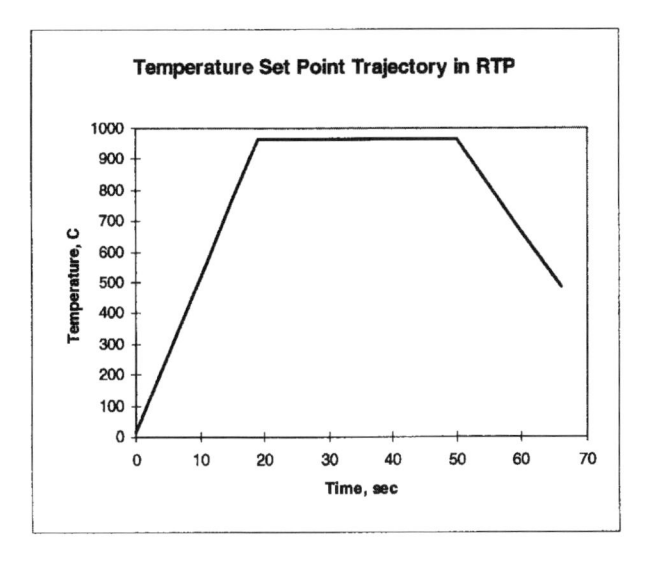

Figure 5. Typical RTP cycle.

Another concern in single wafer RTP systems is the temperature non-uniformity across the wafer surface, especially for larger wafers. This is more of a problem during the transient period, when the temperature on the wafer surface is being ramped from room temperature to the processing temperature, in contrast to the steady state (constant temperature) part of the processing. Different parts of the wafer exchange heat differently with the surroundings, thus creating non-uniform regions on the wafer. These temperature non-uniformities in turn cause slip dislocations and non-uniform reaction on the wafer surface. The resulting defects and non-uniformities then result in poor yields. Temperature non-uniformity problems can be addressed in two ways: (1) design of reactors with inherently better uniformity (2) application of control algorithms to maximize uniformity. For reactors that are already being used commercially, the second method is probably the most practical solution to meet performance standards.

SEMATECH has sponsored research teams at MIT, Sandia, ISI, CVC, and AG Associates to develop detailed three-dimensional time-dependent mathematical models for RTP based on rigorous application of the principles of heat transfer, mass transfer, and fluid flow. Even at this level of detail, it is important to remember that these models do not incorporate all of the features, both macroscopic and microscopic, of the real process. One focus of this research has been to evaluate how design decisions affect controllability of the resulting system (Aral et al., 1994). However, these global models are not useful for real-time control, even given the capabilities of the latest workstations. Real-time in the context of RTP would mean a sampling time in digital control of seconds, with a total run time of minutes. On the other hand, computationally-intensive process models are of utility for off-line computer simulations of the process to investigate process behavior.

Most model-based controllers developed so far use gain scheduling to treat the nonlinearities. The process model is linearized about several operating points, and then linear controllers are designed for different regions of operating temperatures. This approach was utilized in the Applied Materials RTP control system described by Elia (1994) and the Texas Instruments-MMST reactor control system developed by Schaper et al. (1994) at Stanford University. A similar approach was proposed by Chatterjee et al. (1993), where step response curves were fitted to a first order dynamic model for a RAPRO RTP system. They found that the gain and time constant of this reactor varied as much as 50% as the power level for a given lamp bank was increased from 15% to 30% of full scale.

Elia (1994) of Integrated Systems Inc. (ISI) reported the successful commercialization of optimal multivariable control for the Applied Materials RTP chamber for both annealing and oxidation. The AM-RTP has twelve independently actuated groups of lamps and eight temperature measurements (the specific type of sensor was not reported) across the width of the chamber. The number of manipulated variables was reduced to eight so that the number of inputs and outputs were equal. A simple discrete-time model was developed (first order) based on input-output testing using pseudo-random binary sequences (PRBS). Model parameters were fitted depending on the temperature range covered. The specifications on the controller included

a. a uniform temperature profile in the chamber with less than +2.5°C error band during ramps
b. steady state error of less than 1.5°C peak and 0.5°C average
c. less than 5°C overshoot at the end of the ramp with a fast settling time.

A robust controller design using MATRIX-X was employed with some additional logic needed to prevent reset windup and lamp voltage saturation. Integral action was also included to minimize offset. No numerical details were provided in the paper, so it is difficult to compare the characteristics of this control system to any other for RTP. For a 200 mm wafer, Elia reported that six temperatures were successfully controlled from 750°C to 1050°C with a 50°C/sec ramp rate. Slower (25°C/sec) ramp rates were used for 125 mm and 150 mm wafers.

Investigators at Texas Instruments and Stanford University (Schaper et al, 1994b) used a low-order nonlinear model in the design of multivariable feedforward/feedback control strategies for the TI-MMST reactor shown in Figure 6. They employed the model in an open-loop optimal computation based on the desired trajectory. Linear multivariable feedback based on internal model control (IMC) combined with gain-scheduling was used to compensate for modeling errors and disturbances. The experimental work used a four-inch silicon wafer with three radial thermocouple measurements. Model identification experiments were used to obtain parameters of the low-order model. Using gain scheduling, a controlled ramp was achieved from 20°C to 900°C at a rate of 45°C per second with less than 15°C nonuniformity during the ramp and less than 1°C average nonuniformity during the hold at 900°C. The TI MMST design and control system are currently being modified for a five zone reactor by CVC, Inc.

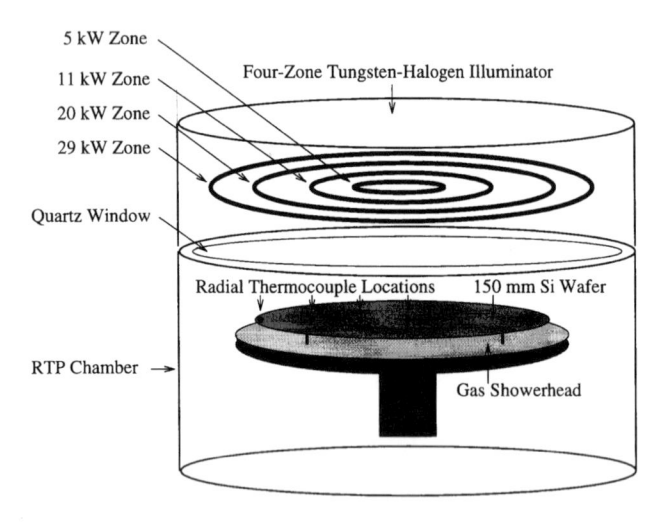

Figure 6. Drawing of the MMST RTP system.

In a cooperative project between Texas Instruments and the University of Texas, Breedijk et al. (1994) reported an enhanced nonlinear model predictive control scheme using successive model linearization and QDMC for the TI-MMST RTP reactor, which gives improved experimental control over linear model-based schemes discussed above. Breedijk et al. obtained a simplified model for the energy equation in the semiconductor wafer and were able to fit this model to different lamp configurations and chamber geometries through parameter estimation of configuration factors. Hence the model structure can probably be generalized to other RTP systems. In developing the multivariable control system for the four-zone reactor (see Figure 6), Breedijk et al. recognized the ill-conditioned nature of the 4 x 4 control system as seen in the condition number of the gain matrix. Instead of controlling the four temperatures directly, they controlled the average and variance of the four temperatures using the QDMC algorithm for model predictive control (4 x 2 transformed system). A plot of gain matrix condition numbers indicated that the transformed system was much better conditioned (by a factor of over 100), and therefore easier to control than the original system, which was verified experimentally using instrumented wafers.

The University of Texas is also working with AG Associates on the development of a model-based control scheme for the HeatPulse system that is widely used in industry for annealing (see Figure 7). The dynamics of the system can be explained using a macroscopic lumped parameter model that is based completely on physical parameters. The model utilizes the input-output relationship between the radiant energy inputs from the lamps and the thermal response of the wafer. Based on this model, nonlinear controllers are being developed that will be used to control the thermal response of the wafer when it is subjected to a typical setpoint trajectory as shown in Figure 5.

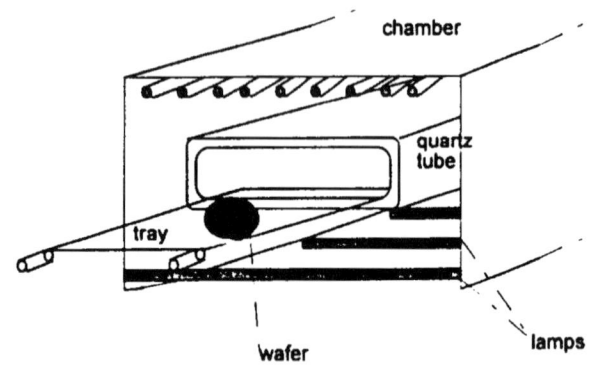

Figure 7. (a) Schematic of the AG Associates; (b) Arrangement of lamps. Heatpule reactor above and below the wafer

The model state variables include the temperatures of the quartz, the lamps, the chamber walls and the wafer in the reactor chamber. The model takes into account both radiation and convection in the chamber. The wafer is assumed to be at uniform temperature and thus the conduction in the wafer is neglected. The model accounts for varying thermal emissivities of the various components in the reactor during the temperature cycle. The short and long wavelength portions of the radiation from the various components in the reactor chamber are treated separately and solved in the energy balance equations.

Experimental results show that the thermal response of the wafer is approximately linear when subjected to small changes in lamp power. The nonlinear model is linearized around different operating points in those regions. The gain and time constants predicted by the model and computed from experimental data are within $\pm 20\%$. The macroscopic model shows good agreement with the linear empirical models determined from experimental data. Linear controllers are being developed based on the above model and results will be published shortly. These linear controllers will be gain-scheduled over the temperature range and will be used to control the wafer temperature when the wafer is subjected to temperature trajectories similar to the one shown in Figure 5.

An interesting alternative to commercial RTP designs has been developed as a prototype test bed at SEMATECH. This design is the basis for new commercial equipment being developed by Matrix Integrated Systems, and its control system is being configured by Matrix and the University of Texas in a joint research project. As shown in Figure 9, the wafer is heated from above by a linear bank of 16-2 kW tungsten-halogen lamps surrounded by a gold plated reflector, and the light passes through a hexagonal kaleidoscope on its way to the wafer. This arrangement provides uniform illumination entering the top of the reaction chamber. Inert gases flow through the quartz bell jar surrounding the wafer under atmospheric pressure. The edge of the reaction chamber is far from the edge of the wafer, hence the wafer edge is cooler than the wafer center when illuminated from above. This effect is compensated by a ring of 36 tungsten-halogen lamps (each 0.5 kW) located concentrically in the lower zone around the edge of the wafer (equally-spaced) and backed by a gold plated reflector. The power output from the lamps is regulated by controlling the voltage across the lamps. The lamp resistance varies with temperature such that each lamp draws an approximately constant current. Thus the voltage control is also a power control. Two Accufiber ripple pyrometers have also been installed and will be used as the primary temperature sensors when test wafers are processed with a typical RTP cycle (Stuber et al., 1995). A parallel development is underway at Stanford University to develop a three pin acoustic thermometer for use in this system.

shown in Figure 8 and a TI-MMST three-zone reactor (TI3) illustrated in Figure 6. For the STB the upper and lower lamp zones are defined as u_1 and u_2 respectively, and the center temperature and edge temperatures are T_1 and T_2. Let $y_1 = T_1$, $y_2 = T_2$ and $y_3 = T_2 - T_1$. The first STB configuration uses (u_1, u_2) and y_1, y_2). The second configuration uses (y_1, y_3). RGA analysis clearly shows that the second configuration will have less interaction when both loops are closed. In addition, singular value analysis was performed and the condition number (CN) tabulated for both cases. For the first configuration, RGA numbers range from 3.5 to 7.0 (depending on the temperature levels) and the CN ranges from 30-60. For the second configuration, RGA numbers range from 0.80 to 0.90 and the CN ranges from 15-25. Thus the second configuration has fewer multivariable interactions and is apparently much better conditioned than other RTP systems currently being studied, as discussed below.

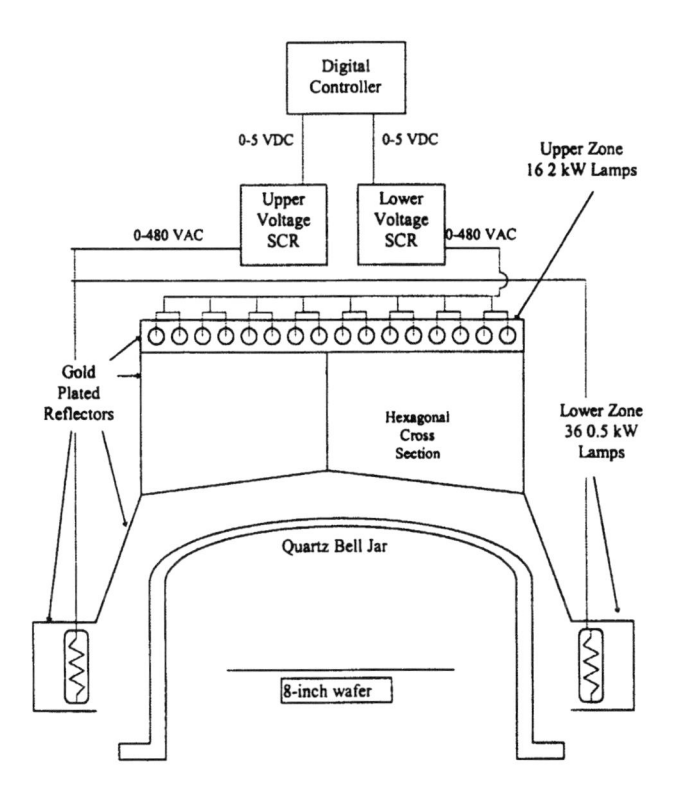

Figure 8. Drawing of the SEMATECH Test Bed (STB).

The SEMATECH test system has some controllability advantages over the TI and AG multi-zone configurations. One can compare the RGA's and the condition numbers for the SEMATECH test bed (STB)

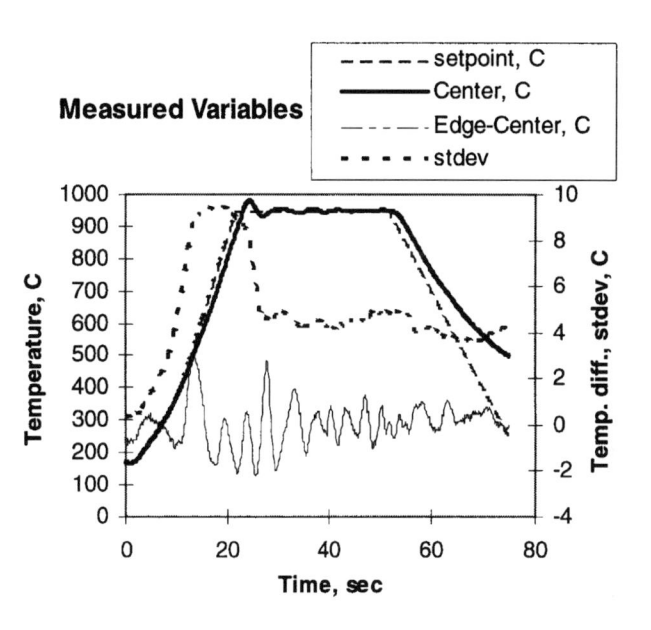

Figure 9. Gain-scheduled PID controllers with standard set point ramp from 400°C to 950°C at 50°C/s.

In contrast, the TI3 system yields an RGA of

$$\Lambda = \begin{bmatrix} 112.27 & -108.76 & -2.51 \\ -205.05 & 211.68 & -5.63 \\ 93.78 & -101.92 & 9.14 \end{bmatrix}$$

and the condition number is 185. Because of the size of the diagonal elements, we expect severe interactions in this RTP design, which was experimentally demonstrated by Schaper et al. (1994a) and Breedijk et al. (1994). In fact, changes in the second decimal point of the gain matrix can cause significant changes in the RGA. In the above expression the inputs are lamp zone voltages (u_1 = inner zone, u_2 = middle, u_3 = edge) and the outputs are temperatures measured by an instrumented wafer (T_1 = center, T_2 = middle, T_3 = edge). The conditioning of this system can be improved by transforming the variables and using $y_1 = T_1$, $y_2 = T_2 - T_1$, $y_3 = T_3 - T_1$. As mentioned earlier, Breedijk et al. (1994) also showed that a significant improvement in conditioning resulted by transforming the control system for the four zone reactor into a 4 x 2 system with y_1 = avg (T_1, T_2, T_3, T_4), y_2 = var (T_1,T_2,T_3,T_4).

Based on the above results one can conclude for design of an RTP system that the lamp zones should be configured with one high power zone which heats the entire wafer and other zones that compensate for non-uniformities. This is the case in the STB which delivers a nearly uniform, high intensity light from its upper zone and an edge-concentrated, low intensity light from the lower zone. Chamber symmetry forces the wafer temperature distribution to lie between the temperatures at the center and the edge. The temperature uniformity across the entire wafer can therefore be controlled with only two zones.

Recent joint work between the University of Texas and SEMATECH has focused on the development of model-based controllers for RTP tools that can be implemented in real time as a "piggy-back" or add-on component to an existing tool; the STB is the system currently being studied (Figure 8). A nonlinear, multiple-input multiple-output (MIMO) model can be used to describe the temperature at multiple points on a wafer in response to inputs from multiple lamps or lamp zones. If used for Model Predictive Control (MPC), the models must be executable in real time. Practically, this means that the data acquisition (temperature measurement), control calculation, and adjustment of the output must occur in about 1/10 of the dominant time constant in the system. If the RTP recipe calls for a 50°C/s ramp, then the control cycle must be completed in 0.10 seconds in order to limit the temperature error during the ramp to 5.0°C. A controller with this cycle time has been implemented using Labview® software running on a Quadra 950 computer which takes data from a 17-point thermocouple wafer. Two analog signals are sent to the SCR's powering the two lamp zones in the STB. The STB processes 200 mm silicon wafers and is physically located at SEMATECH in Austin. Stuber et al. (1995) have presented a transfer function matrix description of the STB system by analytically solving only the radiation heat transfer equation. In the transfer function matrix the gains and time constants are shown to be inversely proportional to the third power of temperature.

The implementation of the temperature cycle is problematic for several reasons. First, optical temperature sensors (e.g. pyrometers) cannot receive enough radiation to make a reliable temperature measurement if the wafer temperature is less than 400°C. This necessitates the use of an open-loop sequence which heats the wafer to at least 400°C. Second, the profile calls for a discontinuous change in the temperature velocity, $\frac{dT}{dt}$, at the transition from ramp to hold. Since the wafer has a positive thermal mass $\rho V C_p$, the thermal momentum $\rho V C_p \frac{dT}{dt}$ cannot be abruptly changed from a positive value to zero by an abrupt change in the heat input. This non-physical temperature trajectory is therefore infeasible. Third, the maximum rate of cooling obtainable by the wafer is dependent on the chamber reflectivity, thermal inertia of the quartz window, and wall temperature. While a high cooling rate is desirable, it is generally not possible to cool the wafer as quickly as it is heated because of the inherent limitations in the system. This asymmetry in process dynamics influences the choice of control strategy.

With these limitations in mind, experiments were performed to demonstrate the processing capabilities of the STB. In each of the three cases shown here, two independent SISO control loops were implemented for the STB where temperatures were sampled and control moves implemented every 0.10 sec. One SISO loop used a standard PID controller to control the difference between the wafer edge and center temperatures by manipulating the lower zone voltage (lower loop). The other loop paired the average temperature and the upper zone voltage. The PID controller parameters (gain, integral time, and derivative time) for both loops were changed during the process depending on the value of the wafer temperature and whether or not the set point was being ramped or held constant. There were six regions between the threshold measurement temperature (400°C) and the annealing temperature (950°C). The controller parameters were set according to ITAE tuning rules (Seborg et al., 1989) based on the model parameters determined at the average temperature of each region.

Several experiments were run to determine the limits of multi-loop control. The results of a closed-loop experiment with a standard 50°C/s ramp from 400°C to 950°C and a 30 sec hold at 950°C are shown in Figure 9. The temperature exceeded the desired processing temperature by 30°C due to reset windup during the ramp and the infeasible set point trajectory. The standard deviation of the eight measured temperatures on the wafer peaked at 9°C during the ramp and leveled out around 4°C during the high temperature hold. Multivariable interactions caused the manipulated variables to oscillate out of phase with one another.

The results of a second closed-loop experiment with a 40°C/s ramp are shown in Figure 10. In this experiment the deviation between the set point and the measured temperature in the upper loop during the ramp was limited. If the set point of the previous sample period was more than a fixed amount away from the measured temperature of the current sample period, the set point was not incremented but was instead held constant. This prevented a large error from accumulating during the ramp sequence and greatly reduced the overshoot at the hold temperature. It also smoothly changes from 40°C/s to zero which prevented the upper zone voltage from having to make a large adjustment as the set point "turned the corner." The peak in the standard deviation in temperature across the wafer was also moved to a lower temperature. This is beneficial in preventing wafer slip because the wafer is stiffer when it is cooler. However, this did increase the processing time by approximately 20%, which would not be acceptable for a commercial process.

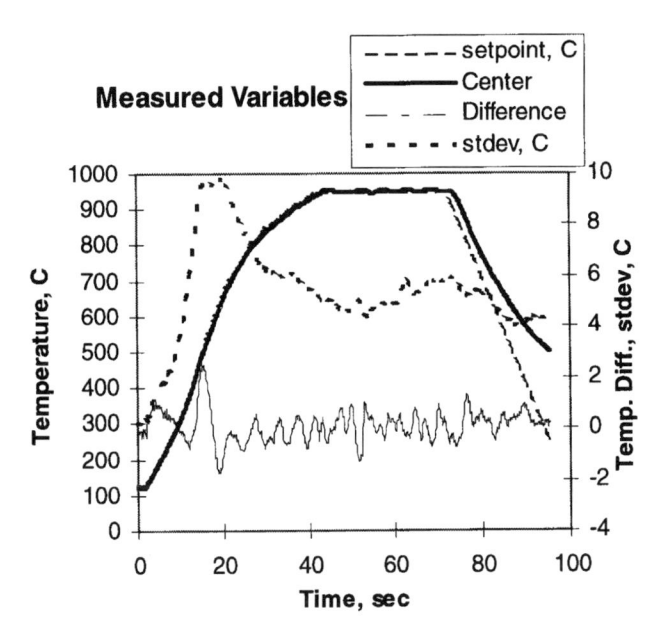

Figure 10. Gain-scheduled PID controllers with 40°C/s ramp, set point lead limited to 10°C.

In order to eliminate the large correction in the upper zone voltage at the peak of the ramp sequence, a second technique anticipated the "corner" in the set point trajectory by inserting an open loop model predictive control sequence into the upper zone lamp power profile. During this sequence, the upper zone power was set to a constant value while the error in the upper loop was set equal to zero. Figure 11 shows how this technique can limit the overshoot to 6°C while increasing the processing time only 1 sec out of 75 and still maintaining a low temperature non-uniformity. The open loop sequence can be updated in a run-to-run fashion where the value of the open loop power is updated after each run. The shape of the curve showing the temperature standard deviation changed because a different thermocouple wafer was used, this one measuring 16 points across the wafer.

Future tests of closed-loop control at SEMATECH will implement multivariable model predictive control in order to determine its benefits over control based on SISO design. Other control issues to be addressed include the treatment of variations in incoming wafers and integration with run-to-run control (and use of pilot-wafers). A related instrumentation issue involves use of temperature measurement for control rather than direct film property measurement such as dopant distribution and whether the backside temperature measurement is sufficient for predicting film properties, due to topographic effects.

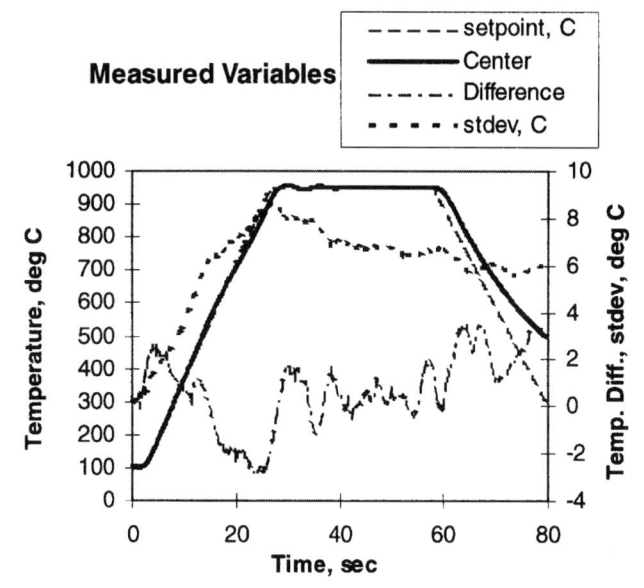

Figure 11. Gain-scheduled PID control, 35°C/s ramp, with open loop set point anticipation at the "corner."

Conclusions

There are many opportunities to apply advanced modeling and control techniques in microelectronics manufacturing. Improvements in supervisory (run-to-run) control are likely to have a major impact, especially in reducing the number of test wafers that must be used. The development of fundamental mathematical models for single wafer reactors has reached a fairly high level of sophistication and should provide a means of evaluating various advanced control techniques for this type of equipment. Mathematical models should also be helpful in analyzing how design parameters affect the quality of control for single wafer reactors, because precise control for larger wafers will be mandatory. However, control strategies (multivariable, model-predictive, and possible adaptive) need to be developed for such reactors, especially in RTP. The use of feedback control for RTP has been hindered by the lack of real-time measurements; additional research on accurate and relatively inexpensive non-invasive measurement techniques should be a high priority in order to implement real-time process control techniques.

Acknowledgments

Support for this work was provided by the National Science Foundation, Texas Instruments, AG Associates, and SEMATECH. John Stuber served as the facilitator for intermittent communications between the co-authors.

References

Aral, G., T. Merchant, J.V. Cole, K. Knutson, and K. Jensen (1994). Concurrent engineering of an RTP reactor: design for control. *Proceedings of RTP'94 Conf.*, 288-296, Monterrey, CA.

Breedijk, T., T.F. Edgar, and I. Trachtenberg (1994). Model-based control of rapid thermal processes. *Proceedings of Amer. Cont. Conf.*, 887- 891

Butler, S.W. (1995a). Process control in semiconductor manufacturing. *J. Vac. Sci. Tech. B.*, **13**(4), 1917-1923.

Butler, S.W. (1995b) Control of plasma processes in semiconductor manufacturing. *Proceedings Volume A of Workshop on Industrial Application of Plasma Chemistry, Low Pressure Non-Equilibrium Plasma Applications*, in conjunction with the 12th International Symposium on Plasma Chemistry, Aug 21-25.

Butler, S.W., J. Stefani, M. Sullivan, S. Maung, G. Barna, and S. Henck (1994). An intelligent model based control system employing in situ ellipsometry. *J. Vac. Sci. Tech*, A., **12**(4), 1984-1991.

Butler, S.W. and J. Stefani (1994). Supervisory run-by-run control of polysilicon gate etch using in situ ellipsometry. *IEEE Trans Semi. Manufact.*, **7**(2), 193- 201.

Chatterjee, S., I. Trachtenberg, and T.F. Edgar (1993). Modeling and control of RTCVD of polysilicon. *Proceedings of RTP'93 Conf.*, Scottsdale, AZ. 386-392.

Edgar, T.F. and T. Breedijk (1994). Overview of process control isues in rapid thermal processing. *Proceedings of RTP'94 Conf.*, Monterrey, CA. 267-277.

Elia, C.F. (1994). RTP multivariable temperature controller development. *Proceedings of Amer. Cont. Conf.*, 907-911

IEEE Trans. Semicond. Manufact. (1994). Issue on MMST Program, **7**(2).

Larrabee, G.L. (1991). The microelectronics factory of the future. Proceedings of. *Chem. Proc. Cont. IV Conf.*, AIChE, New York, 673-684.

Lie, K.H., W. Cole, T.P. Merchant, and K.F. Jensen (1993) Simulation of rapid thermal processing equipment and processes. *Proceedings of RTP'93 Conf.*, 376-385, Scottsdale, AZ.

Miner, G., C. Gronet, B. Peuse, and J. Grilli (1994). Rapid thermal processor with dynamic spatial control and emissivity-independent temperature. *Proceedings of RTP'94 Conf.*, 94-101, Monterrey, CA.

Roozeboom, F. (1992) *Manufacturing Equipment Issues in Rapid Thermal Processing*. Academic Press, New York.

Roozeboom, F. and N. Parekh (1990). Rapid thermal processing system: A review with emphasis on temperature control. *Journal Vac. Sci. Technol. B.*, **8**(6), 1249-1259.

Schaper, C.D., M. Moslehi, K. Saraswat, and T. Kailath (1994a). Control of MMST RTP: Repeatability, uniformity, and integration for flexible manufacturing. *IEEE Trans. Semicond. Manufact.* **7**(2), 202- 219.

Schaper, C.D., M. Moslehi, K. Saraswat, and T. Kailath (1994b). Modeling, identification and control of rapid thermal processing systems. *J. Electrochem. Soc.* **141**(11), 3200-3215.

Seborg, D.E., T.F. Edgar, and D.A. Mellichamp (1989). *Process Dynamics and Control*. Wiley, New York.

Semiconductor Industry Association (SIA) (1994). *The National Technology Roadmap for Semiconductors*, SIA, San Jose, CA.

Solid State Technology, (1994). Issue on MMST Program.

Stefani, J., L. Loewenstein, and M. Sullivan (1995). On-line diagnostic monitoring of photoresist ashing. *IEEE Trans. Semicond. Manufact.*, **8**(1), 2-9.

J.D. Stuber, T.F. Edgar, and T. Breedijk (1995). Model-based control of rapid thermal processes. Proceedings of Electrochemical Society, Reno, NV. Vol. 95-4, 113-147.

SESSION SUMMARY: INDUSTRY ASSESSMENT — III

Dale E. Seborg
Chemical Engineering Department
University of California
Santa Barbara, CA 93106-5080

Summary

The papers in this session survey the status of process modeling and control in two industries, pulp and paper and microelectronics. Although control problems in these industries have historically received little attention at AIChE meetings and control engineering conferences, advanced process modeling and control have become important enabling technologies for both industries in recent years. The two papers introduce control problems and summarize the current state of the art in each of the two industries.

APPLYING NEW OPTIMIZATION ALGORITHMS TO MODEL PREDICTIVE CONTROL

Stephen J. Wright
Mathematics and Computer Science Division
Argonne National Laboratory
Argonne, IL 60439

Abstract

The connections between optimization and control theory have been explored by many researchers, and optimization algorithms have been applied with success to optimal control. The rapid pace of developments in model predictive control has given rise to a host of new problems to which optimization has yet to be applied. Concurrently, developments in optimization, and especially in interior-point methods, have produced a new set of algorithms that may be especially helpful in this context. In this paper, we reexamine the relatively simple problem of control of linear processes subject to quadratic objectives and general linear constraints. We show how new algorithms for quadratic programming can be applied efficiently to this problem. The approach extends to several more general problems in straightforward ways.

Keywords

Optimization, Model predictive control, Interior-point methods.

Introduction

In this paper we apply some recently developed techniques from the optimization literature to a core problem in model predictive control, namely, control of a linear process with quadratic objectives subject to general linear constraints. To describe our algorithms, we use the following formulation:

$$\min_{x_j,u_j} \sum_{j=0}^{N-1} \tfrac{1}{2}(x_j^T Q x_j + u_j^T R u_j) + q^T x_j + r^T u_j$$
$$+ \tfrac{1}{2} x_N^T \tilde{Q} x_N + \tilde{q}^T x_N,$$
$$x_{j+1} = A x_j + B u_j, \quad j = 0, \ldots, N-1, \quad (1)$$
$$x_0 \text{ fixed},$$
$$G u_j + J x_j \leq g, \quad j = 0, \ldots, N-1,$$

where Q and R are positive semidefinite matrices and

$$u_j \in \mathbb{R}^m, \quad x_j \in \mathbb{R}^n, g \in \mathbb{R}^{m_c}.$$

This problem is well known from the optimal control literature, but it has been revived recently in the context of model predictive control (MPC). In MPC applications such as receding horizon control (Rawlings and Muske, 1993) and constrained linear quadratic regulation (Scokaert and Rawlings, 1995), controls are obtained by solving problems like (1) repeatedly. As we describe later, the methods we outline here can be extended easily to more general forms of (1), which may contain outputs $y_j = C x_j$, penalties on control jumps $u_{j+1} - u_j$, and so on.

The two approaches we consider in detail involve the infeasible-interior-point method and the active set method. Both methods are able to exploit the special structure in the problem (1), as they must to obtain a solution in a reasonable amount of time.

The techniques discussed here represent just one of many potential contributions that optimization can make to MPC. Developments in MPC have created a demand for fast, reliable solution of problems in which nonlinearities, noise, and constraints on the states and controls may all be present. Meanwhile, recent algorithmic developments in areas such as interior-point methods and stochastic optimization have produced powerful tools that are yet to be tested on MPC problems. By no means do we expect optimization algorithms to be a panacea for all the problems that arise in MPC. In many cases, more specialized algorithms motivated by the particular control problem at hand will be more appropriate. We do expect, however, that some MPC problems will benefit from the optimization viewpoint and that interactions between optimizers and engineers are the best way to realize these benefits.

In the next section, we sketch the way in which algorithmic research in optimization relates to applications, illustrating the point with a problem from optimal control. Next, we present the interior-point algorithm and show how it can be applied efficiently to the problem (1). We then move to the active set approach and again outline its application to (1).

Applications and Paradigms

The field of optimization was founded as a separate academic discipline during the 1940s. Its emergence was due to a number of factors. On the "demand side," there was a desire to approach the huge logistical problems posed by the wartime economy in a more systematic way, and a realization that the same techniques also could be applied to the logistical problems faced by industry and commerce during peacetime. On the "supply side," Dantzig's development of the simplex method and the appearance of digital computers were two factors that played an important role.

Connections between optimization and other mathematical disciplines, such as the calculus of variations and game theory, were recognized by the earliest researchers in the field. Today, research in optimization continues to give impetus to other areas of mathematics, such as nonsmooth analysis, linear algebra, and combinatorics. It has found applications in operations research, industrial engineering, and economics; in experimental sciences and statistics (the problem of fitting observed data to models is an optimization problem (Seber and Wild, 1989)); and in the physical sciences (for example, meteorological data assimilation (National Research Council, 1992) and superconductor modeling (Garner, Spanbauer, Benedek, Strandburg, Wright and Plassmann, 1992)).

Most researchers in optimization work with a set of standard *paradigms*, each of which is a mathematical formulation that is supposed to represent a large class of applications. Examples include linear programming, convex quadratic programming, unconstrained nonlinear optimization, and nonlinear programming. These paradigms and a few others were proposed in the early days of optimization, and they are still the focus of most of the research effort in the area. Optimization paradigms are an interface between optimization research and optimization applications. They focus the efforts of theoreticians and software developers on well-defined tasks, thereby freeing them from the effort of becoming acquainted with the details of each individual application. Linear programming is possibly the most successful paradigm of all, because a vast range of linear programs can be solved with a single piece of software, with little need for case-by-case interactions between software developers and users. More complex paradigms, such as nonlinear programming, are not so easy to apply. General software for these problems often is unable to take advantage of the special features of each instance, resulting in inefficiency. The algorithms often can be customized, however, to remove these inefficiencies.

Optimal control and model predictive control illustrate the latter point. Many problems in these areas fit into one of the standard optimization paradigms, but it is often unclear how the optimization algorithms can be applied efficiently. In some cases, special-purpose algorithms have been devised (for example, differential dynamic programming (Jacobson and Mayne, 1970)); in other cases, standard optimization algorithms such as Newton's method, the conjugate gradient method, and gradient projection algorithms have been adapted successfully to the optimal control setting (Polak, 1970; Bertsekas, 1982; Dunn and Bertsekas, 1989).

We close this section with the example of the classical discrete-time optimal control problem with Bolza objectives, a problem that arises frequently in the MPC literature (Rawlings, Meadows and Muske, 1994). This problem provides a nice illustration of the potential impact of optimization on control. It shows, too, that naive application of optimization algorithms to control problems can lead to gross inefficiencies, which can be remedied by a little customization and adaptation. The problem is

$$\min_{x_j, u_j} \sum_{j=0}^{N-1} L_j(x_j, u_j) + \hat{L}_N(x_N),$$
$$x_{j+1} = f_j(x_j, u_j), \quad j = 0, \ldots, N-1, \quad x_0 \text{ fixed}, (2)$$

where $x_j \in \mathbb{R}^n$, $u_j \in \mathbb{R}^m$. It can be viewed as an unconstrained optimization problem in which the unknowns are $(u_0, u_1, \ldots, u_{N-1})$; the states $(x_1, x_2, \ldots, x_N)$ can be eliminated through the state equations $x_{j+1} = f_j(x_j, u_j)$. However, it would be highly inefficient to solve (2) with a general implementation of Newton's method for unconstrained optimization. Such a code usually requires the user to evaluate the function, gradient, and Hessian on request at any given set of variable values. The Hessian for (2) with respect to $(u_0, \ldots, u_{N-1})$ is dense, so the code would require $O(N^3 m^3)$ operations simply to compute the Newton step. The paper by (Dunn and Bertsekas, 1989) shows how the same step can be obtained through a specialized calculation that takes advantage of the structure in (2) and requires only $O(N(m^3 + n^3))$ operations. Hence, the Newton algorithm must be tailored to the special form of (2) if we are to have any hope of solving this problem efficiently.

The problem (2) also can be viewed as a nonlinear programming problem in which the variables are $(u_0, \ldots, u_{N-1}, x_1, \ldots, x_N)$ and the state equations are viewed as equality constraints. The special structure becomes transparent in this formulation, since the Jacobian of the constraints and the Hessian of the objective function are both sparse (block-banded) matrices. Therefore, a nonlinear programming code that implements some variant of sequential quadratic programming may perform quite efficiently, provided that it uses exact second derivatives and exploits the sparsity. The disadvantage is that nonlinear programming algorithms tend to have weaker global convergence properties than do unconstrained optimization algorithms. In the special case of f_j linear and L_j convex quadratic, the problem (2) is a convex programming problem, and global convergence is attained easily with either formulation.

When constraints on the controls u_j are added to (2), we still have the choice of eliminating the states x_j or not, though both formulations yield a nonlinear programming problem. (The formulation in which the states are eliminated will have simpler constraints and a more complicated objective function.) When constraints on the states are introduced, however, elimination of the x_j becomes more problematic, and there is little choice but to view the problem as a nonlinear programming problem in which the unknowns are $(u_0, u_1, \ldots, u_{N-1}, x_1, x_2, \ldots, x_N)$. We consider the linear-quadratic form of this problem in the next two sections.

Interior-Point Methods for Linear-Quadratic Problems

In this section, we consider the linear-quadratic problem (1). It has frequently been noted that this problem is, in optimization terms, a convex quadratic program. Two successful methods for addressing this class of problems are the active set method described by (Fletcher, 1987) and the interior-point method described by (Wright, 1996). However, the special structure of (1) means that we must take care in applying either approach to this problem. A naive application of a quadratic programming code based on the active-set method (for instance, QPOPT (Gill, Murray, Saunders and Wright, 1991)) will give poor results, typically requiring $O(N^3(m+n+m_c)^3)$ operations, where m_c is the number of rows in the matrices G and J. In this section, we show how the interior-point algorithm can be applied to (1), while in the next section we examine the adaptations that are needed to make the active set approach more efficient.

We state at the outset that the interior-point approach we describe here can be adapted to various modifications and generalizations of (1) without significant loss of efficiency. For instance,

- the matrices Q, R, A, B, G, and J can vary with j;

- an output vector $y_j = Cx_j$ can be incorporated into the objective and constraints;

- we can incorporate constraints and objective terms that involve states/controls from adjacent stages; for example, a penalty on the control move $(u_{j+1} - u_j)_i$ for some component $i = 1, 2, \ldots, m$.

The last generalization is useful when the problem is obtained as a discretization of the continuous problem, since many discretization schemes for ordinary differential equations and differential algebraic equations lead to relationships between states and controls at a number of adjacent time points.

The rest of this section is organized as follows. We define the *mixed monotone linear complementarity problem* (mLCP), a powerful paradigm that generalizes the optimality conditions for linear and quadratic programs and is a convenient platform for describing interior-point methods. We then outline an infeasible-interior-point algorithm for the mLCP and discuss its properties. Finally, we customize this algorithm to convex quadratic programming and the linear-quadratic problem (1).

Mixed Linear Complementarity and the Infeasible-Interior-Point Framework

The mLCP is defined in terms of a square, positive semidefinite matrix $M \in \mathbb{R}^{n \times n}$ and a vector $q \in \mathbb{R}^n$. The problem is to find vectors z, x, and s such that

$$\begin{bmatrix} M_{11} & M_{12} \\ M_{21} & M_{22} \end{bmatrix} \begin{bmatrix} z \\ x \end{bmatrix} + \begin{bmatrix} q_1 \\ q_2 \end{bmatrix} = \begin{bmatrix} 0 \\ s \end{bmatrix}, \quad (3)$$

$$x \geq 0, \ s \geq 0, \ x^T s = 0. \quad (4)$$

Here, M_{11} and M_{22} are square submatrices of M with dimensions n_1 and n_2, respectively, and the vector q is partitioned accordingly.

The infeasible-interior-point algorithm for (3),(4) starts at point (z^0, x^0, s^0) for which $x^0 > 0$ and $s^0 > 0$ (*interior* to the nonnegative orthant) but possibly *infeasible* with respect to the constraints (3). All iterates (z^k, x^k, s^k) retain the positivity properties $x^k > 0$ and $s^k > 0$, but the infeasibilities and the *complementarity gap* defined by

$$\mu_k = (x^k)^T s^k / n_2 \quad (5)$$

are gradually reduced to zero as $k \to \infty$. Each step of the algorithm is a modified Newton step for the system of nonlinear equations defined by the feasibility conditions (3) and the complementarity conditions $x_i s_i = 0, i = 1, 2, \ldots, n_2$. We can write this system as

$$F(z, x, s) \stackrel{\text{def}}{=} \begin{bmatrix} M_{11}z + M_{12}x + q_1 \\ M_{21}z + M_{22}x - s + q_2 \\ XSe \end{bmatrix} = 0, \quad (6)$$

where we have used the notation

$$X = \text{diag}(x_1, x_2, \ldots, x_{n_2}), \quad S = \text{diag}(s_1, s_2, \ldots, s_{n_2}).$$

The algorithm has the following form:

Algorithm IIP
Given (z^0, x^0, s^0) with $(x^0, s^0) > 0$;
for $k = 0, 1, 2, \ldots$
    for some $\sigma_k \in (0, 1)$, solve

$$\begin{bmatrix} M_{11} & M_{12} & 0 \\ M_{21} & M_{22} & -I \\ 0 & S^k & X^k \end{bmatrix} \begin{bmatrix} \Delta z \\ \Delta x \\ \Delta s \end{bmatrix} = \begin{bmatrix} -r_1^k \\ -r_2^k \\ -X^k S^k e + \sigma_k \mu_k e \end{bmatrix},$$
$$(7)$$

to obtain $(\Delta z^k, \Delta x^k, \Delta s^k)$, where

$$\begin{aligned} r_1^k &= M_{11}z^k + M_{12}x^k + q_1, \\ r_2^k &= M_{21}z^k + M_{22}x^k - s^k + q_2, \\ e &= (1, 1, \ldots, 1)^T. \end{aligned}$$

set

$$\begin{aligned}(z^{k+1}, x^{k+1}, s^{k+1}) & \quad (8) \\ = (z^k, x^k, s^k) + \alpha_k(\Delta z^k, \Delta x^k, \Delta s^k),\end{aligned}$$

for some $\alpha_k \in (0,1]$ that retains
$$(x^{k+1}, s^{k+1}) > 0;$$
end(for).

Note that (7) differs from the pure Newton step for (6) only because of the term $\sigma_k \mu_k e$ on the right-hand side. This term plays a stabilizing role, ensuring that the algorithm converges steadily to a solution of (3),(4) while remaining inside the positive orthant defined by $(x,s) > 0$.

The only two parameters to choose in implementing Algorithm IIP are the scalars σ_k and α_k. The convergence analysis leaves the choice of σ_k relatively unfettered (it is often confined to the range $[\sigma, 0.8]$, where σ is a fixed parameter, typically 10^{-3}), but α_k is required to satisfy the following conditions.

(i) The reduction factor for the infeasibility norms $\|r_1\|$ and $\|r_2\|$ should be smaller than the reduction factor for μ; that is, the ratios $\|r_1^k\|/\mu_k$ and $\|r_2^k\|/\mu_k$ should decrease monotonically with k;

(ii) The pairwise products $x_i s_i$, $i = 1, \ldots, n_2$ should all approach zero at roughly the same rate; that is, the ratios $x_i^k s_i^k / \mu_k$ should remain bounded away from zero for all i and all k. (Note that μ_k represents the average values of the terms $x_i^k s_i^k$.)

(iii) We require sufficient decrease in μ, in the sense that the decrease actually obtained is at least a small fraction of the decrease predicted by the linear model (7). (See (Dennis and Schnabel, 1983) for a discussion of sufficient decrease conditions.)

(iv) The chosen value of α_k should not be too much smaller than the largest possible value for which (i)–(iii) are satisfied.

For details, see (Wright, 1993b; Wright, 1996). When σ_k and α_k satisfy these conditions, global convergence to a solution of (3),(4) is attained whenever such a solution exists. A slightly enhanced version of the algorithm, in which σ_k is allowed to be zero on some of the later iterations, exhibits superlinear convergence under some additional assumptions. The property that excites many of the theoreticians working on interior-point methods— polynomial complexity—is also attained when the starting point (z^0, x^0, s^0) has sufficiently large x^0 and s^0 components, relative to the initial residuals r_1^0 and r_2^0 and the solutions of (3),(4).

In practical implementations of Algorithm IIP, α_k often is chosen via the following simple heuristic. First, we set $\alpha_k^{\max}$ to be the supremum of the following set:
$$\{\alpha \in (0,1] \mid (z^k, x^k, s^k) + \alpha(\Delta z^k, \Delta x^k, \Delta s^k) > 0\}. \quad (9)$$

Then we set
$$\alpha_k = \min(1, 0.995 * \alpha_k^{\max}). \quad (10)$$

That is, we forget about the theoretical conditions (i)–(iv) above and simply choose α_k to step almost all the way to the boundary of the nonnegative orthant. This tension between theory and practice existed for a long time during the development of primal-dual methods. However, recent work has reconciled the differences. There exist relaxed versions of conditions (i)–(iv) that are satisfied by the "practical" choice of α_k from (9),(10). Hence, the parameters σ_k and α_k and be chosen to make Algorithm IIP both practically efficient and theoretically rigorous.

The major operation to be performed at each step of Algorithm IIP is the solution of the linear system (7). The matrix in this system obviously has a lot of structure due to the presence of the zero blocks and the diagonal components I, S^k, and X^k. Additionally, the matrix M is sparse in most cases of practical interest, including our motivating problem (1), so sparse matrix factorizations are called for. In general, these are fairly complex pieces of software, but problems of the form (1) require only banded factorization code, which is comparatively simple.

The first step in solving (7) is to eliminate the Δs component. Since the diagonal elements of X^k are positive, we can rearrange the last block row in (7) to obtain
$$\begin{aligned}\Delta s &= (X^k)^{-1}(-X^k S^k e + \sigma_k \mu_k e - S^k \Delta x^k) \\ &= -s_k + (X^k)^{-1}(\sigma_k \mu_k e - S^k \Delta x^k).\end{aligned}$$

By substituting into the first two rows of (7), we obtain
$$\begin{bmatrix} M_{11} & M_{12} \\ M_{21} & M_{22} + (X^k)^{-1}S^k \end{bmatrix} \begin{bmatrix} \Delta z \\ \Delta x \end{bmatrix} \quad (11)$$
$$= \begin{bmatrix} -r_1^k \\ -r_2^k - s_k + \sigma_k \mu_k (X^k)^{-1} e \end{bmatrix}.$$

In most cases, some of the partitions M_{11}, M_{12}, M_{21}, or M_{22} are zero or diagonal or have some other simple structure, so further reduction of the system (11) is usually possible. This phenomenon happens, for instance, when (3),(4) is derived from a linear or quadratic program, as we show below.

Since the factorization of the coefficient matrix in (7) comprises most of the work at each iteration, we may be led to ask whether it is really necessary to compute a fresh factorization every time. A set of heuristics in which the factorization is essentially re-used on alternate steps was proposed by (Mehrotra, 1992). Mehrotra's algorithm has proved to be successful in practice and is the basis for the vast majority of interior-point codes for linear programming.

Linear and Quadratic Programming as mLCPs

We now show how linear and convex quadratic programming problems can be expressed in the form (3),(4) and solved via Algorithm IIP. Consider first the linear program in standard form:
$$\min_x c^T x \text{ subject to } Ax = b, x \geq 0, \quad (12)$$
where c and x are vectors in R^n, $b \in \mathbb{R}^m$, and $A \in \mathbb{R}^{m \times n}$. The dual of (12) is
$$\max_{\lambda, s} b^T \lambda \text{ subject to } A^T \lambda + s = c, s \geq 0, \quad (13)$$

where $\lambda \in \mathbb{R}^m$ are the dual variables (or, alternatively, the Lagrange multipliers for the constraints $Ax = b$) and $s \in \mathbb{R}^n$ are the dual slacks. The Karush-Kuhn-Tucker (KKT) conditions for (12),(13) are as follows:

$$\begin{aligned} A^T\lambda + s &= c, \\ Ax &= b, \\ x \geq 0, \; s \geq 0, \quad x^T s &= 0. \end{aligned} \quad (14)$$

Because (12) is a convex programming problem, the KKT conditions are both necessary and sufficient. Hence, we can find a primal-dual solution for the linear program by finding a vector (x, λ, s) that satisfies the conditions (14). We can verify that (14) has the form (3),(4) by making the following identifications between these two systems:

$$\begin{bmatrix} M_{11} & M_{12} \\ M_{21} & M_{22} \end{bmatrix} = \begin{bmatrix} 0 & A \\ -A^T & 0 \end{bmatrix}, \begin{bmatrix} q_1 \\ q_2 \end{bmatrix} = \begin{bmatrix} -b \\ c \end{bmatrix},$$

$$z \leftarrow \lambda, \; x \leftarrow x, \; s \leftarrow s.$$

Hence the KKT conditions (14) are an mLCP, and we can obtain solutions to the linear program (12) and its dual (13) simultaneously by applying Algorithm IIP to (14).

Next, we consider the following general convex quadratic program:

$$\min_z \tfrac{1}{2} z^T Q z + c^T z \;\; \text{s.t.} \; Hz = h, \, Gz \leq g, \quad (15)$$

where Q is a symmetric positive semidefinite matrix. The KKT conditions for this system are

$$\begin{aligned} Qz + H^T\zeta + G^T\lambda &= -c, \\ -Hz &= -h, \\ -Gz - t &= -g, \\ t \geq 0, \; \lambda \geq 0, \quad t^T \lambda &= 0. \end{aligned} \quad (16)$$

The following identifications confirm that the system (15) is an mLCP:

$$M_{11} = \begin{bmatrix} Q & H^T \\ -H & 0 \end{bmatrix}, \; M_{12} = \begin{bmatrix} G^T \\ 0 \end{bmatrix},$$

$$M_{21} = \begin{bmatrix} -G & 0 \end{bmatrix}, \; M_{22} = 0,$$

$$q_1 = \begin{bmatrix} c \\ h \end{bmatrix}, \; q_2 = g,$$

$$z \leftarrow \begin{bmatrix} z \\ \zeta \end{bmatrix}, \; x \leftarrow \lambda, \; s \leftarrow t.$$

The reduced form (11) of the linear system to be solved at each iteration of Algorithm IIP is

$$\begin{bmatrix} Q & H^T & G^T \\ -H & 0 & 0 \\ -G & 0 & (\Lambda^k)^{-1}T^k \end{bmatrix} \begin{bmatrix} \Delta z \\ \Delta \zeta \\ \Delta \lambda \end{bmatrix} \quad (17)$$

$$= \begin{bmatrix} -r_c^k \\ -r_h^k \\ -r_g^k - t^k + \sigma_k \mu_k (\lambda^k)^{-1} e \end{bmatrix}.$$

Here, μ_k is defined as $\mu_k = (t^k)^T \lambda^k / m$, where m is the number of inequality constraints in (15). It is customary to multiply the last two block rows in (17) by -1, so that the coefficient matrix is symmetric indefinite. We then obtain

$$\begin{bmatrix} Q & H^T & G^T \\ H & 0 & 0 \\ G & 0 & -(\Lambda^k)^{-1}T^k \end{bmatrix} \begin{bmatrix} \Delta z \\ \Delta \zeta \\ \Delta \lambda \end{bmatrix} \quad (18)$$

$$= \begin{bmatrix} -r_c^k \\ r_h^k \\ r_g^k + t^k - \sigma_k \mu_k (\lambda^k)^{-1} e \end{bmatrix}.$$

Since $(\Lambda^k)^{-1}T^k$ is diagonal with positive diagonal elements, we can eliminate $\Delta\lambda$ from (18) to obtain an even more compact form:

$$\begin{bmatrix} Q + G^T \Lambda^k (T^k)^{-1} G & H^T \\ H & 0 \end{bmatrix} \begin{bmatrix} \Delta z \\ \Delta \zeta \end{bmatrix} \quad (19)$$

$$= \begin{bmatrix} -r_c^k + G^T[\Lambda^k (T^k)^{-1} r_g^k + \lambda^k - \sigma_k \mu_k (T^k)^{-1} e] \\ r_h^k \end{bmatrix}.$$

Factorizations of symmetric indefinite matrices have been studied extensively in the linear algebra literature; see (Golub and Van Loan, 1989) for references to the major algorithms. Standard software is available, at least in the dense case (Anderson, Bai, Bischof, Demmel, Dongarra, Du Croz, Greenbaum, Hammarling, McKenney, Ostrouchov and Sorensen, 1992). Algorithms for the sparse case have also received considerable attention recently (Ashcraft, Grimes and Lewis, 1995) and have been implemented in the context of interior-point methods by (Fourer and Mehrotra, 1993).

In solving the linear systems (18) or (19), we are free to reorder the rows and columns of the coefficient matrix in any way we choose, before or during the factorization. As we see below, the problem (1) benefits dramatically from such a reordering, since the coefficient matrix in this case becomes a narrow-banded matrix, which is easily factored via existing linear algebra software such as LAPACK (Anderson et al., 1992).

"Soft" Constraints and Penalty Terms

The cost of introducing slack variables and dummy variables into the formulation (15) is often surprisingly small when the quadratic program is solved by the technique outlined above, even though the total number of variables in the problem may increase appreciably. The reason is that the new variables can often be substituted out of the linear system (11), as in (18) and (19), so that they may not affect the size of the linear system that we actually solve. This comment is relevant when we are adding norm constraints or "soft" constraints to the problems (15) or (1), as in (Scokaert and Rawlings, 1996). Suppose, for instance, that we wish to include the soft constraint $G_s z \leq g_s$ in (15), and we choose to do this by including a term $\nu \|(G_s z - g_s)_+\|_1$ to the objective function. (The subscript "+" denotes the positive part of a vector, obtained by replacing its negative components by zeros.) To restore the problem to the form of a standard convex quadratic program, we introduce the "dummy" vector v, and

- add the term $\nu e^T v$ to the objective, where $e = (1, 1, \ldots, 1)^T$;
- introduce the additional constraints $v \geq 0, v \geq G_s z - g_s$.

By computing the mLCP form (3),(4) of the expanded problem, applying Algorithm IIP, and reducing its step equations as far as we can via simple substitutions, we find that the linear system is ultimately no larger than (19). The details are messy and we omit them from this discussion.

In the case of soft constraints and penalty terms, the consequence of this observation is that the amount of work per iteration is not really sensitive to whether we choose a 1-norm penalty or an ∞-norm penalty, though the latter option adds fewer dummy variables to the formulation of the problem.

Solving LQR Efficiently via the mLCP Formulation

The linear-quadratic problem (1) is obviously a special case of (15), as we see by making the following identifications between the data:

$$Q \leftarrow \begin{bmatrix} R & & & & & \\ & Q & & & & \\ & & R & & & \\ & & & \ddots & & \\ & & & & Q & \\ & & & & & R \\ & & & & & & \tilde{Q} \end{bmatrix},$$

$$G \leftarrow \begin{bmatrix} G & J & & & \\ & G & J & & \\ & & \ddots & \ddots & \\ & & & G & J \end{bmatrix},$$

$$H \leftarrow \begin{bmatrix} B & -I & & & \\ A & B & -I & & \\ & & \ddots & & \\ & & & A & B & -I \end{bmatrix},$$

$$z \leftarrow \begin{bmatrix} u_0 \\ x_1 \\ u_1 \\ \vdots \\ u_{N-1} \\ x_N \end{bmatrix}, \quad c \leftarrow \begin{bmatrix} r \\ q \\ r \\ \vdots \\ r \\ \tilde{q} \end{bmatrix},$$

$$g \leftarrow \begin{bmatrix} g \\ g \\ \vdots \\ g \end{bmatrix}, \quad h \leftarrow \begin{bmatrix} -Ax_0 \\ 0 \\ \vdots \\ 0 \end{bmatrix}.$$

As suggested earlier, the matrices in this problem are all *block-banded*; their nonzeros are clustered around a line connecting the upper-left and lower-right corners of the matrix. When we formulate the linear system (17), the nonzeros seem to become widely dispersed. However, by interleaving the states x_j, the controls u_j, the adjoints p_j and the Lagrange multipliers λ_j for the constraints $Gu_j + Jx_{j+1} \leq g$, we can return the coefficient matrix in (17) to banded form. We order the unknowns in the system as

$$(\Delta u_0, \Delta \lambda_0, \Delta p_1, \Delta x_1, \Delta u_1, \ldots, \Delta \lambda_{N-1}, \Delta p_N, \Delta x_N)$$

and rearrange the rows and columns of the matrix accordingly to obtain

$$\begin{bmatrix} R & G^T & B^T & & & & & & \\ G & -D_0^k & 0 & & & & & & \\ B & 0 & 0 & -I & & & & & \\ & & -I & Q & 0 & J^T & A^T & & \\ & & & 0 & R & G^T & B^T & & \\ & & & J & G & -D_1^k & 0 & & \\ & & & A & B & 0 & 0 & -I & \\ & & & & & & -I & Q & \ddots \\ & & & & & & & \ddots & \ddots \end{bmatrix}. \quad (20)$$

(We have used D_j^k to denote sections of the diagonal matrix $(\Lambda^k)^{-1}T^k$ from (17).) This system can be reduced further by using the diagonal entries D_j^k to eliminate $\Delta \lambda_j, j = 0, 1, \ldots, N-1$.

The computational savings obtained by recovering bandedness are significant. If we assume the dimensions

$$u_j \in \mathbb{R}^m, \quad x_j \in \mathbb{R}^n, \quad \lambda_j \in \mathbb{R}^{m_c},$$

the banded matrix has total dimension $N(2n+m+m_c)$ and half-bandwidth of $(2n+m+m_c-1)$. Therefore, the time required to factor this matrix is $O(N(2n+m+m_c)^3)$, compared with $O(N^3(2n+m+m_c)^3)$ for a dense matrix of the same size. In the absence of constraints (that is, $m_c = 0$), this cost has exactly the same order as the cost of solving a linear-quadratic version of (2) by using dynamic programming techniques.

Stagewise ordering of the equations and variables in (17) is the key to obtaining a banded structure. As mentioned above, we can maintain the banded structure when outputs $y_k = Cx_k$ and other constraints are introduced into the model, provided that we continue to order the variables by stage.

The techniques outlined above are quite similar to those described in (Wright, 1993a). The difference lies in the use of infeasible-interior-point methods above, in contrast to the techniques of (Wright, 1993a) which combined feasible interior-point methods with an embedding of the original problem into an expanded problem for which a feasible initial point is easy to compute. The new approach is cleaner, more practical, and more efficient, while remaining theoretically rigorous.

Active Set Methods for Linear-Quadratic Problems

The structure of the problem (1) can also be exploited when we use an active set method in place of the interior-point method described above. Again, we find that the linear algebra at each step can be performed in terms of banded matrices rather than general dense matrices. The details are different from (and somewhat more complicated than) the interior-point case. We start by sketching a single iterate of the active set approach. For a more complete description, see (Fletcher, 1987).

Active set methods for the general convex program (15) generate a sequence of feasible iterates. At each iterate, a certain subset of the constraints $Gz \leq g$ are *active* (that is, hold as equalities). On each step, we choose a subset of the active set known as the *working set*. (Typically, the working set either is identical to the active set or else contains just one less constraint.) We then compute a step from the current point z that minimizes the objective function in (15) while maintaining activity of the constraints in the working set, and also ensuring that the original equality constraints $Hz = h$ remain satisfied. If we denote by $\bar{G}$ the subset of rows of G that make up the working set, the step Δz is obtained by solving the following system:

$$\min_{\Delta z} \tfrac{1}{2}(z+\Delta z)^T Q(z+\Delta z) + c^T(z+\Delta z)$$
$$\text{subject to } H\Delta z = 0, \quad \bar{G}\Delta z = 0.$$

Equivalently, we have

$$\min_{\Delta z} \tfrac{1}{2}\Delta z^T Q \Delta z + \tilde{c}^T \Delta z, \quad \text{s.t.} \quad H\Delta z = 0, \quad \bar{G}\Delta z = 0, \tag{21}$$

where $\tilde{c} = c + Qz$. The KKT conditions for (21) are that Δz is a solution of this system if and only if there are vectors $\Delta \zeta$ and $\Delta \bar{\lambda}$ such that $(\Delta z, \Delta \zeta, \Delta \bar{\lambda})$ satisfies the following system:

$$\begin{bmatrix} Q & H^T & \bar{G}^T \\ H & 0 & 0 \\ \bar{G} & 0 & 0 \end{bmatrix} \begin{bmatrix} \Delta z \\ \Delta \zeta \\ \Delta \bar{\lambda} \end{bmatrix} = \begin{bmatrix} -\tilde{c} \\ 0 \\ 0 \end{bmatrix}. \tag{22}$$

We can obtain Δz from this system and then do a line search along this direction, stopping when a new constraint is encountered or when the minimum of the objective function along this direction is reached.

Note the similarity between (22) and (17). The coefficient matrices differ only in that there is no diagonal term in the lower left of (22), and some rows are deleted from G. For the problem (1), the matrices Q, H, and $\bar{G}$ are banded, so a stagewise reordering of the rows and columns in (22) again produces a banded system. The matrix will not be quite as regular as (20), because of the missing columns in $\bar{G}$, but similar savings can be achieved in factoring it.

The banded matrix is best factored by Gaussian elimination with row partial pivoting, as implemented in the LAPACK routine `dgbtrf` and its affiliated solution routine `dgbtrs` (Anderson et al., 1992). To the author's knowledge, there is no software that can exploit the fact that the matrix is symmetric in addition to being banded.

Updating Factorizations

We have noted that the matrix in (22) is banded for the problem (1), so we can use software tailored to such matrices to obtain significant savings over the dense linear algebra that is usually employed in standard quadratic programming software. The story is not quite this simple, however. The systems (22) that we solve at each iteration are closely related to one another, differing only in that a column is added and/or deleted from $\bar{G}$. Therefore, it does not make sense to solve each system "from scratch"; we try instead to modify the matrix factorization that was computed at an earlier iteration to accommodate the minor changes that have occurred since then. In the general case of dense matrices, updating of factorizations has been studied extensively and is implemented in software for many kinds of optimization problems. (Simplex algorithms for linear programming, for instance, depend heavily on efficient updating of the basis matrix at each iteration.) The question we face here is: Can we perform efficient updating of the factorization while still exploiting bandedness? We answer this question in the affirmative by sketching a technique for re-solving (22) after a row has been added to or deleted from $\bar{G}$.

We start with addition of a column to the system (22). Changing the notation for convenience, we denote the original coefficient matrix by M and the new row/column by a. We assume without loss of generality that a is ordered last. Then the updated matrix $\bar{M}$ is

$$\bar{M} = \begin{bmatrix} M & a \\ a^T & 0 \end{bmatrix}.$$

We assume that the LU factorization of the original matrix M is known, so that there is a permutation matrix P, a lower triangular L and upper triangular U such that

$$PM = LU. \tag{23}$$

(If M is banded, then the factors L and U also have nonzeros only in locations near their diagonals.) We can easily modify L and U to accommodate the new row/column by adding an extra row and column to each factor to obtain

$$\begin{bmatrix} P & 0 \\ 0 & 1 \end{bmatrix} \begin{bmatrix} M & a \\ a^T & 0 \end{bmatrix} = \begin{bmatrix} L & \\ a^T U^{-1} & 1 \end{bmatrix} \begin{bmatrix} U & L^{-1}Pa \\ 0 & \alpha \end{bmatrix},$$

where

$$\alpha = -a^T U^{-1} L^{-1} Pa.$$

Hence, the factorization can be updated at the cost of triangular substitution with each of L and U. For (1), the cost of this process is $O(N(m+n+m_c)^2)$, so it is less expensive than refactoring $\bar{M}$ from scratch, unless $(m+n+m_c)$ is very small. Since the new row/column does not participate in the pivoting, the new row of L and column of U are dense in general, so the factors are not as sparse as they would be in a factorization from scratch. Stability issues may arise, since the diagonal element α may be small. A good strategy for dealing with these problems is to compute a fresh factorization whenever α becomes too small by some measure, or when $O(m + n + m_c)$ iterations have passed since the last refactorization.

We turn now to the case in which a row and column are deleted from M. We assume that the factorization (23) is known and that we obtain a new coefficient matrix $\hat{M}$ by deleting row i and column j from the matrix PM. (Note that the indices of the deleted row and column will be identical in the original matrix M by symmetry, but row pivoting may cause them to differ in the permuted matrix PM.) Obviously, we can modify L and U so that their product equals $\hat{M}$ by simply deleting the ith row of L and the jth column of U. Unfortunately, these deletions cause L and U to become nontriangular; the entries $L_{i+1,i+1}, L_{i+2,i+2}, \ldots$ now appear *above* the diagonal of the modified version of L. We can restore triangularity by removing the ith *column* of L as well as the ith row, and similarly removing the jth *row* of U as well as the jth column. The modified matrix $\hat{M}$ can be expressed in terms of these modified L and U factors as follows:

$$\hat{M} = \hat{L}\hat{U} + vw^T, \qquad (24)$$

where $\hat{L}$ is the matrix L with ith row and column removed and $\hat{U}$ is U with the jth row and column removed, and we have

$$v = \begin{bmatrix} 0 \\ \vdots \\ 0 \\ L_{i+1,i} \\ \vdots \\ L_{n,i} \end{bmatrix}, \quad w = \begin{bmatrix} 0 \\ \vdots \\ 0 \\ U_{j,j+1} \\ \vdots \\ U_{j,n} \end{bmatrix}.$$

The expression (24) shows that the product $\hat{L}\hat{U}$ of two triangular matrices differs from $\hat{M}$ by a rank-one matrix, which can be accounted for by using the Sherman-Morrison-Woodbury formula (Golub and Van Loan, 1989, page 51). From this expression, we have

$$\begin{aligned} \hat{M}^{-1} &= (\hat{L}\hat{U} + vw^T)^{-1} \\ &= (\hat{L}\hat{U})^{-1} + \frac{[(\hat{L}\hat{U})^{-1}v][w^T(\hat{L}\hat{U})^{-1}]}{1 + w^T(\hat{L}\hat{U})^{-1}v}. \end{aligned}$$

Hence, the solution $\hat{x}$ of the system

$$\hat{M}\hat{x} = \hat{c}$$

can be written as

$$\begin{aligned} \hat{x} &= \hat{M}^{-1}\hat{c} \\ &= (\hat{L}\hat{U})^{-1}\hat{c} + \frac{(\hat{L}\hat{U})^{-1}v}{1 + w^T(\hat{L}\hat{U})^{-1}v} w^T (\hat{L}\hat{U})^{-1}\hat{c}. \end{aligned}$$

In computing $\hat{x}$ via this formula, the main operations are to perform two pairs of triangular substitutions with $\hat{L}$ and $\hat{U}$: one pair to compute $(\hat{L}\hat{U})^{-1}\hat{c}$ and another to compute $(\hat{L}\hat{U})^{-1}v$.

Hot Starts

In MPC, the problem (1) is not usually encountered in isolation. On the contrary, we usually need to solve a sequence of these problems in which the data A, B, Q, etc, and/or the starting point x_0 vary only slightly from one problem to the next. It is highly desirable that the algorithms should be able to take advantage of this fact. The information could be used, for instance, to choose good starting values for all the variables or to make a good guess of the active constraint set (that is, the matrix $\bar{G}$ in (21)). The process of using this information is called *hot starting*.

Sequences of similar linear-quadratic problems can arise in the control of nonlinear systems. When we apply the sequential quadratic programming algorithm to a constrained version of (2), we obtain a search direction at each iteration by solving a problem like (1). (Actually, we solve a slightly more general problem in which the data A, B, Q, R varies with stage index j and linear terms may appear in the objectives and constraints.) As the iterates converge to a solution of the nonlinear problem, the data matrices become more and more similar from one iteration to the next. A starting guess of $u_j \equiv 0$ and $x_j \equiv 0$ is best, because the subproblem is obtained by approximating the nonlinear problem around the current iterate. We can however make an excellent guess at the active set and the initial working set, particularly on later iterations. Active set methods will typically require just a few steps to identify the correct active set from this good initial guess.

Model predictive control also gives rise to sequence of similar linear-quadatic problems. The usual procedure is to solve a problem like (1) using the current state of the system as the initial value x_0, and then apply the control u_0 until the next time point is reached. The process is then repeated (Scokaert and Rawlings, 1995; Rawlings and Muske, 1993). In the absence of disturbances, the problems (1) are very similar on successive steps, sometimes differing only in the initial value x_0. An excellent starting point can be obtained from the solution at the previous set by setting

$$(x_0, x_1, \ldots, x_N) = (x_0^{\text{new}}, x_2^-, \ldots, x_N^-, x_N^{\text{new}}),$$
$$(u_0, u_1, \ldots, u_{N-1}) = (u_1^-, u_2^-, \ldots, u_{N-2}^-, u_{N-1}^{\text{new}}),$$

where u_j^-, x_j^- are the solution components at the previous step, x_0^{new} is the new initial state, and x_N^{new} and u_{N-1}^{new} are some well-chosen estimates for the final stages. In the case of the active set approach, an excellent starting guess can also be made for the active constraint matrix $\bar{G}$.

In some situations, particularly when disturbances are present, the starting point chosen by the obvious techniques may not be feasible even though it is close to a solution. This represents no problem for the infeasible-interior-point approach, although it is desirable in Algorithm IIP for the initial complementarity to be comparable in size to the initial infeasibilities. The active set method assumes a feasible starting point. One remedy is to use a two-phase approach, in which we solve a "Phase I" problem to find a feasible point, then a "Phase II" problem to find the optimum. A second option is to introduce penalty terms into the objective for the infeasibilities and then obtain a solution in a single phase (provided that a heavy enough penalty is imposed.)

Active set methods typically gain more from hot starting than do interior-point methods, for reasons that are not yet fully understood. On linear programming problems, the best interior-point codes gain about a factor of three in compute time when they are hot started, in comparison with a "cold" (i.e., no prior information) start. The relative savings for simplex/active set methods are significantly higher. It is difficult to predict how much the situation will change when we consider the problem class (1). Numerical testing is the only way to find out.

Acknowledgments

This work was supported by the Mathematical, Information, and Computational Sciences Division subprogram of the Office of Computational and Technology Research, U.S. Department of Energy, under Contract W-31-109-Eng-38.

References

Anderson, E., Bai, Z., Bischof, C., Demmel, J., Dongarra, J., Du Croz, J., Greenbaum, A., Hammarling, S., McKenney, A., Ostrouchov, S. and Sorensen, D. (1992). *LAPACK User's Guide*, SIAM, Philadelphia.

Ashcraft, C., Grmies, R.L. and Lewis, J.G. (1995). Accurate symmetric indefinite linear equation solvers. in preparation.

Bertsekas, D.P. (1992). Projected Newton methods for optimization problems with simple constraints, *SIAM J. on Control and Optimization*, **20**, 221-246.

Dennis, J.E. and Schnabel, R.B. (1983). *Numerical Methods for Unconstrained Optimization*, Prentice-Hall Englewood Cliffs, NJ.

Dunn, J.C. and Bertsekas, D.P. (1989). Efficient dynamic programming implementations of Newton's method for unconstrained optimal control problems, *J. of Optimization Theory and Applications*, **63**, 23-38.

Fletcher, R. (1987). *Practical Methods of Optimization*, 2nd ed., John Wiley and Sons, New York.

Fourer, R. and Mehrotra, S. (1993). Solving symmetric indefinite systems in an interior-point method for linear programming, *Mathematical Programming*, **62**, 15-39.

Garner, J., Spanbauer, M., Benedek, R., Strandburg, K.J., Wright, S.J. and Plassmann, P.E. (1992). Critical fields of josephson-coupled superconducting multilayers, *Physical Review B*, **45**, 7973-7983.

Gill, P.E., Murray, W., Saunders, M.A. and Wright, M.H. (1991). Inertia-controlling methods for general quadratic programming, *SIAM Review*, **33**, 1-36.

Golub, G.H. and Van Loan, C.F. (1989). *Matrix Computations*, 2nd ed., The Johns Hopkins University Press, Baltimore.

Jacobson, D.H. and Mayne, D.Q. (1970). *Differential Dynamic Programming*, American Elsevier, New York.

Kall, P. and Wallace, S.W. (1994). it Stochastic Programming, John Wiley and Sons.

Mehrotra, S. (1992). On the implementation of a primal-dual interior point method, *SIAM J. on Optimization*, **2**, 575-601.

National Research Council (1991). *Four-Dimensional Model Assimilation of Data*, National Academy Press.

Polak, E. (1970). *Computational Methods in Optimization*, Academic Press, New York.

Rawlings, J.B., Meadows, E.S. and Muske, K.R. (1994). Nonlinear model predictive control: A tutorial and survey, *Proc. of ADCHEM*, Tokyo, Japan.

Rawlings, J.B. and Muske, K.R. (1993). The stability of constrained receding horizon control, *IEEE Transactions on Automatic Control*, **38**(10), 1512-1516.

Scokaert, P.O.M and Rawlings, J.B. (1995). Constrained linear quadratic regulation, *Technical Report*, Department of Chemical Engineering, University of Wisconsin-Madison.

Scokaert, P.O.M and Rawlings, J.B. (1996). On infeasibilities in model predictive control, *Chemical Process Control: CPC-V*, CACHE.

Seber, G.A.F. and Wild, C.J. (1989). *Nonlinear Regression*, John Wiley and Sons, New York.

Vandenberghe, L. and Boyd, S. (1994). Semidefinite programming, *Technical report*, Electrical Engineering Department, Stanford University, Stanford, CA 94305. To appear in *SIAM Review*.

Wright, S.J. (1993a). Interior point methods for optimal control of discrete-time systems, *Journal of Optimization Theory and Applications*, **77**, 161-187.

Wright, S.J. (1993b). A path-following interior-point algorithm for linear and quadratic optimization problems, *Preprint MCS-P401-1293*, Mathematics and Computer Science Division, Argonne National Laboratory, Argonne, Ill. To appear in *Annals of Operations Research*.

Wright, S.J. (1996). Primal-Dual Interior-Point Methods. Book in preparation. See URL http://www.mcs.anl.gov/home/wright/ippd.html.

REAL-TIME OPERATIONS OPTIMIZATION OF CONTINUOUS PROCESSES

Thomas E. Marlin and Andrew N. Hrymak
Department of Chemical Engineering
McMaster University
Hamilton, Ontario, Canada L8S 4L7

Abstract

Real-time operations optimization (RTO) is a feedback control system that maximizes a calculated, inferred estimate of the plant profit by adjusting selected optimization variables within specified bounds. An RTO system has three components: 1) model updating for data validation and parameter estimation, 2) optimization of a model using the updated parameters, and 3) results analysis to decide whether the results should be implemented. This paper discusses the current technology applied in these components as well as future technology needs.

Keyword

Economic optimization, Data reconciliation, Flowsheeting, Parameter estimation, Results analysis, Optimizing Control, Real-time optimization, Modeling.

Introduction

Process control, involving both closed loop automatic control and advisory systems, contributes to the safe and profitable production of materials with consistently high product qualities. In many processes considerable opportunity exists for further improvement of plant operations beyond conventional process control methods, and one of the most important opportunities is the increase of profit that is addressed by real-time operations optimization (RTO). This paper has three goals in discussing real-time optimization. First, the great potential for RTO will be discussed and an overall system structure described. Second, some key factors in successful RTO design and implementation will be presented. Finally, some of the salient questions and challenges will be summarized to suggest areas for future technology development.

The RTO Opportunity

Increased plant profit is possible via real-time optimization when the following criteria are satisfied: (1) adjustable optimization variables exist after higher priority safety, product quality and production rate objectives have been achieved, (2) the profit changes significantly as values of the optimization variables are changed, (3) disturbances occur frequently enough for real-time adjustments to be required, and (4) determining the proper values for the optimization variables is too complex to be achieved by selecting from several standard operating procedures. Some small scale optimizations can be addressed using direct search methods (e.g. Box and Chanmugan, 1962; Bamberger and Isermann, 1978). This paper emphasizes the important topic of plant-wide optimization, which involves a large number of variables throughout a slowly responding plant. For this situation, a model-based method is required to approach the best conditions within reasonable times.

Current applications of real-time optimization address complex plants. For example, an ethylene plant involves many parallel chemical reactors, a multiproduct fractionator, five or more distillation towers, several refrigeration systems, and extensive heat integration. A typical plant contains about 50 optimization variables with the majority associated with the reactors, but with several in the separations and refrigeration areas.

Many successful industrial applications of RTO have been reported, most prominently the impressive optimizations implemented at the SUNOCO Sarnia Canada refinery, that was recognized by a 1995 Computerworld Smithsonian Award for innovative information technology in manufacturing. The following examples with economic benefits suggest the wide range of processes on which RTO has been successful.

- Ethylene plant, payback in less than one year (Fatora et al., 1992)
- Ethylbenzene/styrene, 1-2.6 M$/y (Divekar and Lepore, 1991)
- Ethylene oxide, 50+ percent return (Larmon et al., 1977)
- Textiles refrigeration, 30k$/y (May et al., 1979)
- Industrial steam and power, 1.5 M$/y (Foster, 1987)
- Petroleum crude distillation, 1-2 M$/y (Cronkwright, 1994)
- Hydrocracker, 1 M$/y (Pedersen et al., 1995)

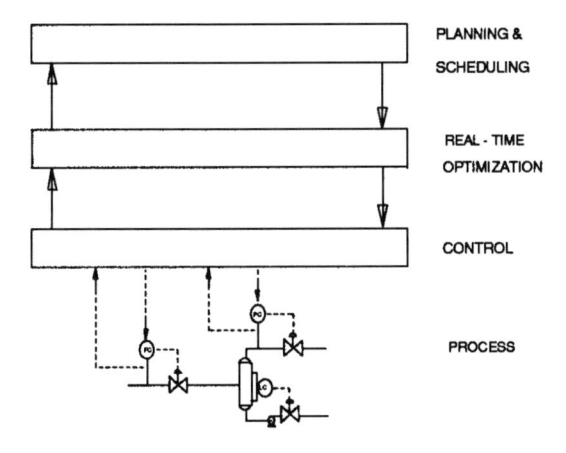

Figure 1. Automation decision hierarchy.

Today's Plant Automation Hierarchy

The typical plant automation system involves a hierarchy of several levels as shown in Fig. 1. This structure, which is essentially a cascade design, is generally appropriate because of the differences in the plant dynamics and disturbance frequencies associated with the decisions at each level. The process control level achieves safety, product quality, and production rate goals, and all closed-loop technology including multivariable control (e.g. QDMC) is implemented at this level. The Real-time Optimization (RTO) level is the first level at which economics is considered explicitly in operations decisions, i.e. in control calculations. This level addresses the short term decisions on a time scale of hours to a few days. The third level addresses longer term issues like material inventories and production rate targets. All levels utilize process measurements as inputs to their feedback loops. Each higher level provides guidance to the subsequent lower level; for example, the RTO outputs could be set points to the process controllers.

The structure in Fig. 1 has advantages and disadvantages. One important advantage is the assurance that the performance of lower levels is not compromised by decisions from upper levels. Thus, the safety and product quality control are not adversely affected by a properly designed RTO level. Another advantage is the distribution of computing intensity that generally matches the time available to make the decisions. Typically, acceptable process control performance can be achieved with algorithms that are fast and converge reliably. On the other hand, model-based RTO often involves large models that may require long computing times, on the order of one hour, and may (infrequently) fail to converge. A disadvantage of the structure in Fig. 1 is the potential conflict among levels leading to difficulty in selecting the proper communication among the levels; this issue is discussed later.

Components in the RTO Loop

This paper deals primarily with the real-time optimization level, which typically contains the elements in Fig. 2. In this section, the major RTO loop components are introduced, beginning with model updating. One key aspect of model updating is the validation of process measures, usually involving steady-state detection and data reconciliation. Since the models used in most RTO applications are based on steady-state assumptions, it is important to ensure that that the plant is near steady state before a data reconciliation or parameter estimation is undertaken. In the simplest case, the mean values of a given set of measurements are the same between two time periods used for averaging. Other indications of steady state include constancy of time series coefficients, small rate of change of a variable over a period, or measurements remaining within prescribed bounds about a mean. The problem is made more difficult by the fact that different sections of a plant may be in different states, e.g. the front section may have come to a steady state, but significant transients are affecting the downstream units. Narashimhan et al. (1987) provide a number of statistical tests which can be used to compare means between periods. A too rigorous test for steady state may mean that there would be few opportunities to do an RTO calculation; thus, the engineer must select reasonable time periods, measurements and decision thresholds

Gross errors include instrument malfunctions, miscalibrations, leaks, poor sampling, or information from units off-line. It is important to identify and remove gross errors before the parameter estimation phase, and Crowe (1994) has reviewed a number of statistical tests that have been developed for gross error detection. A global test for the appearance of gross errors within a data set was developed by Reilly and Carpani (1963). Methods that help locate gross errors are the constraint test of Reilly and Carponi (1963) and the measurement test of Mah and Tamhane (1982). Recall that gross error detection assumes a valid model, so that departure from steady state and model mismatch may appear as gross errors which may not be readily removable by deletion of measurement values

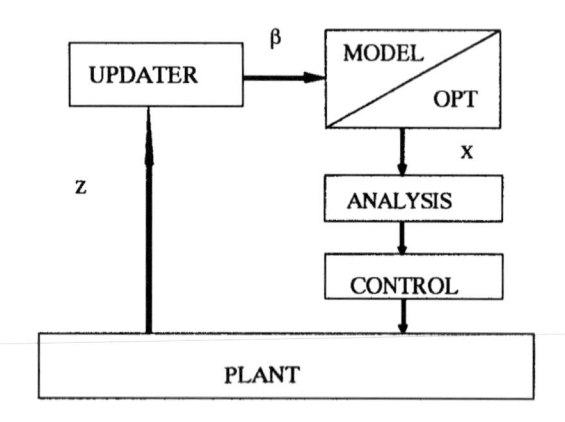

Figure 2. Detailed schematic of RTO loop components.

Data that has satisfied the validation checks, after appropriate filtering, provide feedback for use in the updating of model parameters. Some important characteristics of the estimation problem are 1) the variables appear implicitly in the models, 2) significant errors can appear in all variables, and 3) the equations are generally nonlinear in the parameters. An appropriate method for such an estimation problem minimizes the lack of closure in the algebraic model equations representing energy and material balances (Box, 1970; Sutton and MacGregor, 1977). The problem can be stated as

Problem 1 $$\min_{\beta, a} \varepsilon^T V^{-1} \varepsilon \quad (1)$$

with $f_i(x,\beta,a)=\varepsilon_i$ (with $\varepsilon_i=0$ representing closure), ε a vector containing ε_i, and V the covariance matrix of the residuals which is diagonal when the residuals are independent. Both the updated parameters (β) and adjustments to the measurements (a) are the variables in the updating minimization problem. An alternative formulation for parameter estimation which has the advantage of simultaneous gross error detection is given by Tjoa and Biegler (1991). The resulting β is transmitted for use in the economic optimization.

The economic optimization involves a model of the process, limits on process variables, ranges on feed availabilities and production rates, and economic values. In general, this involves the following problem.

Problem 2 $$\max_{x} P \quad (2)$$
s.t.
$$h(x,\beta) = 0$$
$$g(x,\beta) \geq 0$$
$$x_{min} \leq x \leq x_{max}$$

The calculated profit, P, is maximized by adjusting selected optimization variables (x) within specified limits.

The optimization variables are selected so that they can be transmitted to the process control level; however, the use of large amounts of process data, inevitable model mismatch, and the complex calculations makes the RTO system susceptible to errors. Therefore, a results analysis component evaluates the results and decides whether they should be implemented. The process control system enforces the operating policy supplied by the RTO system until the next RTO execution.

Today's commercial RTO systems have components that perform the functions in Fig. 2. The successes of many RTO systems attest to the insight of practicing engineers, although a few failures have also been reported (e.g., Hanley, 1985). The next section discusses the design and implementation issues for RTO systems, covering fundamental issues in RTO, an overview of some current practices, and proposed improvements.

Design and Implementation Methods for RTO

The design and implementation methods should be reviewed with the three goals for an RTO system in mind. First, the RTO system should predict operating conditions that closely track the true plant optimum. The RTO feedback system controls a calculated inference of profit; therefore, the RTO system will not exactly achieve the true optimum conditions because of inevitable errors in models and measurements. The model is optimized, while the profit in the true plant is increased. Second, the RTO results vary because of measurement noise which is propagated through the updater and optimizer to the operating conditions. This variability should be minimized to reduce both the unnecessary changes to plant operation and the loss of profit due to deviations from good operation. Third, because of the RTO system's complexity, it should provide diagnostic analysis of its performance. This will prevent incorrect actions on bad inputs and speed corrective actions.

Parameter Updating

Model updating provides essential feedback information for RTO, and data validation is required for successful, long term RTO operation. The importance of accurate measurements at proper locations is perhaps not considered sufficiently in designing RTO systems. Recall that data reconciliation provides an important check on measurements; however, it is not possible without some redundant measurements! In addition, RTO may require high accuracy from specific measurements, e.g., when reactor yields are updated using flow rates.

In general, reliable measurements satisfying the validity checks should be used for model updating. However, Krishnan et al (1992) point out that some measurements may not contribute substantially to reducing the offset from the true plant optimum and may greatly increase the variability of the RTO results due to sensor noise. This is an important issue which can be addressed, as proposed by Krishnan, by eliminating some measurements or by properly weighting the noisy measurements with large variances in a combined data reconciliation and parameter update. The updated parameters can be constrained in Problem 1, but if a parameter encounters a constraint, the situation should be evaluated to determine if this outlier is the result of faulty data or unexpected process behavior.

Key design decisions for model updating include measurements used, parameters updated, and the parameter estimation method, and these decisions should be made in a coordinated manner. By far, most of the parameters in the detailed RTO models will not be updated. Some criteria for selecting the parameters for updating are that 1) they are observable , 2) they represent actual changes in the plant, and 3) they contribute to the approach to the plant optimum. Observability was addressed by Stanley and Mah (1977); actual changes are problem specific; and the final issue is referred to as adequacy (Forbes et al, 1994) and discussed next.

The goal of the closed-loop system is to achieve the true plant optimum operating conditions (x_p) in spite of inevitable structural and parameter model errors. For the RTO model-based optimizer to yield the plant optimum, the proper inequality constraints must be active, and the first and second order optimality conditions must be satisfied in the reduced space; in other words, the KKT conditions for optimizing the **model** are valid at the optimum plant conditions. Adequacy requires that this criterion be achieved for some allowable values of the selected adjustable parameters (β), which is stated below.

Problem 3
$$\nabla P_r(x_p, \beta) = 0 \qquad (3)$$
$$x^T \nabla^2 P(x_p, \beta) x < 0$$
$$h(x_p, \beta) = 0$$
$$g_A(x_p, \beta) = 0$$
$$g_I(x_p, \beta) > 0$$

with the subscript r denoting the reduced space, x_p the true plant optimum, and A and I denoting active and inactive constraints, respectively. Finding at least one β that satisfies problem 3 is a necessary, but not sufficient, condition for the RTO system to find the true plant optimum, and this quantitative test can be applied to candidate model structures and updated parameters. Equally important, it provides insight for the engineer; adequacy requires that the updated model parameters affect the gradient and curvature, but not the value, of the profit. This makes sense because the optimizer uses the derivative information in maximizing profit. Thus, the engineer can select candidate parameter sets that affect the key requirements: constraints and profit derivatives.

One important conclusion from the adequacy criterion is the effectiveness of updating model biases, which is a common industrial practice. Since the bias parameters often do not appear in the gradient or Hessian of the profit, updating only bias parameters generally will not provide an adequate system for RTO. This is particularly important conclusion in evaluating the potential for economic optimization within a linear model predictive controller (MPC). Since the standard MPC design provides feedback as bias updating, the sensitivity of profit to values of the optimization variables is not corrected by updating, and the uncorrected model must be quite accurate to provide successful RTO (Forbes and Marlin, 1995).

Model-Based Economic Optimization

In current practice RTO systems involve fundamental models. This practice has the advantages of providing the best accuracy possible, realistic model parameters for updating (e.g., feed impurity), and easy modification as equipment performance changes, while encountering the challenge of optimizing large, non-linear models reliably in reasonable computing times. The models used in current RTO implementations are written as a sparse set of nonlinear equations generally in the form f(x) = 0 (Gallun et al., 1992). Component and overall material balances and energy balances are written for each unit. Thermodynamic and physical property equations, and their derivatives, are supplied either through regressed equation forms imbedded within the equation set or from separate physical property subroutines. Typically, standard models are used for process units like mixers, splitters, heat exchangers, and so forth. The range of operations introduces modeling challenges; for example, flash equations are difficult because the correct phase of the system is not known until the final solution has been found. Changes in the number of phases introduce discontinuities into the solution path, which must be anticipated and dealt with. Distillation columns may be described by short-cut methods such as Fenske-Underwood-Gilliland, but usually the accuracy is poor. It is better, in terms of model accuracy, to imbed the full tray-by-tray material and energy balances within the optimization problem for the theoretical number of trays within the column (Bailey et al, 1993). For the special case of columns where there are many trays with little change in composition between the trays, it is possible to achieve smaller, robust models using a collocation approach (Seferlis and Hrymak, 1994).

A difficult modeling problem occurs with units that are described by ordinary differential equations (ODEs) such as reactors. The approach of integrating the ODE set repeated times to obtain gradients requires substantial computational effort. Another approach is to rewrite the ODEs as nonlinear algebraic equations through the application of finite difference or collocation methods (Renfro et al., 1987; Vasantharajan and Biegler, 1988). The resulting equations are imbedded within the nonlinear constraints of the optimization problem and most importantly are readily analytically differentiable. One of the major challenges is making sure that the reaction kinetics are adequately represented within the model to capture the important effects on the optimization problem without introducing so many reactions as to make the resulting problem parameters unobservable with the available data.

Many nonlinear programming algorithms (Fletcher, 1987) are available for the optimization problems. In practice, the two types of algorithms most commonly used are augmented lagrangian methods (Murtagh and Saunders, 1995) or reduced sequential quadratic programming (RSQP) (e.g. Schmid and Biegler, 1994). Both are infeasible path methods, but MINOS in general allows greater control to maintain feasibility along the solution path which may be important in some problems that are difficult to converge. The RSQP strategy is an eminently suitable method for problems which are well posed, especially in the case where the reduced degrees of freedom for a problem can be identified in advance and where there are relatively few degrees of freedom compared to the number of variables in the problem

The reliable optimization of a high order model requires great care in formulation. Such considerations as variable and equation scaling, analytical first derivatives (where possible), and enforcing step bounding at each iteration are important factors. A good starting point improves convergence, and the parameter estimate solution provides a converged flowsheet with the most up-to-date parameter estimates: this is an excellent starting point for the optimization. In addition, typical practice enforces a maximum change in plant operation per RTO execution which prevents unacceptably large changes from being transmitted to the process control system for implementation. However, this practice limits the optimizer to a region around the current operating conditions. A preferred approach would search the entire allowable region for the operation that maximizes profit. If the new operating conditions involve too large a change from the current state during one RTO cycle, a second optimization problem could be solved to determine the best path toward the optimal operation.

The model represents the response of the process to changes in the optimization variables. Thus, it must include the (steady-state) effects of the process control system and any routine actions by the operators. For example, the response of a distillation tower to a change in the distillate composition set point depends on the control of the bottoms of the tower, even if the bottoms is not adjusted by the optimizer. Thus, the process control system *implemented in the plant* must be modeled. This is straightforward for simple controllers which never experience saturation; the controlled variables are fixed to their set point values. However, the typical situation involves complex control designs for non-square systems with saturation occurring due to operation near constraints. Methods are available for modeling controllers in open-form solution approaches (Kassidas and Marlin, 1993).

Results Analysis

Results analysis involves additional evaluation of the optimization results before the values are transmitted. Some of the checks evaluate whether the plant situation is the same as when the RTO calculations began by determining if 1) the plant operation is not currently upset, i.e., not in a dynamic transient, 2) the optimization variables remain available for manipulation, and 3) bounds of the optimization variables are unchanged If a significant change has occurred, the RTO cycle must be restarted at either the update or optimization phase, as appropriate.

Also, the plant operator may be given an opportunity to accept or reject the entire solution. It is important to recognize that implementing only part of the solution. i.e., implementing some values of x while setting others to independent values, is not appropriate; this could lead to poor economic performance or even infeasible process operation. If the results are not deemed acceptable, additional bounds can be defined and Problem 2 solved again.

Results analysis can evaluate the (estimated) change in profit from the current to the new operation; if the increase is not significant, the change is not implemented in the plant. Also, guidelines for optimum conditions can be compared with the RTO results; if they disagree, an engineer should evaluate the performance of the RTO system (or the guidelines).

A more fundamental approach for results analysis can be based on concepts from statistical process control to determine when the results represent a meaningful change. Even when no significant change occurs in the plant and the data is filtered, the measurements will contain variability due to sensor noise and high frequency, stationary disturbances, and this variability is transmitted by the RTO system to the calculated values of the optimization variables. Thus, the calculated optimization variables contain a common cause variability (Koninckx et al, 1988), which can be characterized using the measurement covariance and the linear sensitivity of the RTO updater and optimizer components (Wolbert et al, 1994). A hypothesis test based on the Hotelling T^2 statistic can be applied to the multivariate deviation between the current plant variables and calculated variables. If the change is within the common cause variability at a specified level of significance, the results would not be output; if the change is significant, the results would be output. An example of this approach applied to the Williams-Otto reaction system is presented in Figures 3a (without test) and 3b (with test) (Miletic and Marlin, 1996) for the two optimization variables. The hypothesis test reduces the variability in the optimization variables, tracks the change in operation, and yields higher average profit.

Real-time diagnosis of the RTO system components is critical because of changing situations encountered by the system. An important example is the behavior of the parameter updater as plant operation and data change. An example is the real-time optimization of the Williams-Otto reaction system with updates of both the frequency factor and activation energy terms for one of the reaction rate constants. When using a single set of steady-state, noisy plant data the updating calculation is very ill-conditioned; the parameter estimates will likely contain substantial error; and the optimization results will likely be far from the true optimum. If the hypothesis test described in the previous paragraph were employed, the common cause variability would be large and the incorrect values likely not implemented; otherwise, the erroneous results would be implemented. A real-time diagnostic check can evaluate the conditioning of the parameter estimation problem using singular value analysis and could define a reduced, well conditioned updating problem (Miletic and Marlin, 1996).

Process Control

The results analysis component transmits the solution in a format the can be implemented by the process control system. Generally, the transmitted information includes the set points of controlled variables and targets for some manipulated variables. Naturally, the values sent to the control system must be measured or inferred by the system, while the optimization model can contain some product quality variables which are neither measured nor inferred. This discussion leads to an important conclusion: a tight integration of the optimization, results analysis, and control are required because the proper bounds must be implemented and values *calculated in the optimization*; the bounds should not be imposed after the optimization.

The process control system implements RTO results, which include product qualities, production rates, and active constraints. Typically, more manipulated variables than controlled variables exist, and some controlled variables can vary within specified limits. Clearly, the RTO results should include not only the optimum operating point, but also the manner for responding to disturbances, i.e., the key constraints to remain active, variables to be maximized or minimized, priority for adjusting manipulated variables, and so forth. Many process control designs under RTO employ multivariable, model-predictive, constraint-handling process controllers, and for these, the RTO system could output the controlled variable set points, manipulated variable targets, and weights for variable deviations from their predicted optimum values (set point or target, as appropriate) at the end of the control horizon. In short, the RTO results should be transmitted as an operating policy, rather than an operating point.

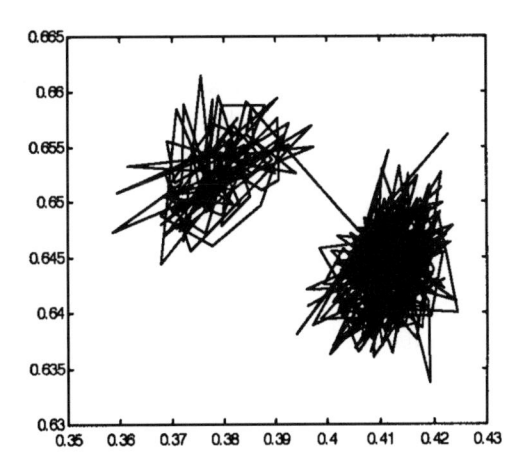

Figure 3a. RTO conditions without hypothesis test.

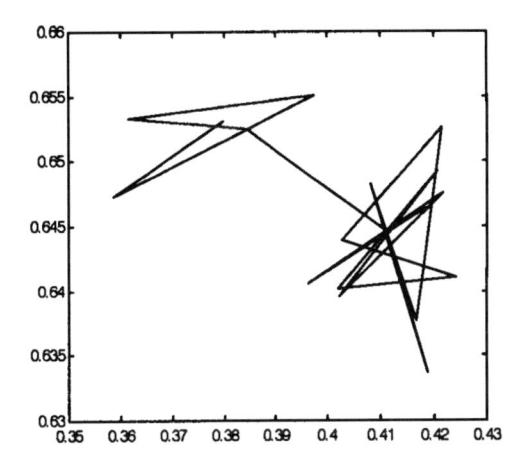

Figure 3b. RTO conditions with hypothesis test.

RTO Productivity: People and Tools

Real-time optimization represents a major step-out in technology for control engineers. The scope of the system becomes the entire plant (if not multiple plants), model fidelity involves broad ranges of operating conditions, and calculations involve thousands of nonlinear equations. The engineer must have an understanding of 1) the entire plant process, product qualities and economics, 2) design-quality process models and their limitations, 3) the fundamentals of statistics and optimization as applied to a closed-loop system, and 4) issues involving the solution of large sets of nonlinear equations. Given the range and complexity of skills required, training is needed for most engineers entering the field.

Building and solving large systems of equations for parameter estimation and optimization requires excellent software tools. Engineers should be able to define plant models as a connection of standard unit models with options for common process configurations. One would expect that the RTO system have graphical interfaces and model management tools comparable to current flowsheeting packages for plant design. Since optimization algorithms require accurate derivatives, the analytical derivatives should be provided automatically, without programming by the engineer. Physical properties should provide smoothed values for the properties and their derivatives. Finally, the package should have the flexibility to perform diagnostics after each RTO component and based on the diagnostic, redefine the problem and restart anywhere in the RTO loop. An open architecture would allow the inclusion of statistical and conditioning analysis and knowledge-based heuristics to be integrated with the equation solvers.

Finally, the plant decision-making organization must respond to the automated tools. The planning and scheduling group should recognize the need for accurate and up-to-date economic values, as well as timely information on inventories and production rates. A consistent set of updated plant models for use in scheduling and process analysis as well as RTO is a major side benefit of optimization projects. In some plants, the RTO system displays results which can be accepted or rejected by the operator, while in many applications the RTO system is entirely closed-loop. In both situations the plant operator role changes to the "supervisor" of the RTO system

Needs for Future RTO Systems

Today's RTO system design and implementation represents the best approach given the current computing resources, mathematical algorithms, and engineering understanding. Some of the key potential improvements in economic optimization are discussed briefly below. All of these issues have been investigated intensively, but further technology is required for routine, real-time economic plant optimization.

Scheduling — The temporal decomposition in Figure 1 is never entirely satisfactory because time scales of disturbances and plant dynamics do not segregate neatly into three categories. In fact, decisions made at the RTO level influence equipment performance and inventory issues which are the major concerns at the scheduling level. Thus, the scheduling (longer range plan) and RTO (current operating decisions) should be made simultaneously or at least be well coordinated. If the hierarchy were retained, it appears as though some time-dependent behavior would be required at the RTO level.

Process Control — To properly establish a control design as the output of the RTO system, the RTO calculations would have to evaluate the economic performance for a range of likely disturbances, such as feed compositions and cooling water temperature. Then, the RTO system would design a control system that would yield the highest profit for a range of these disturbances until the next RTO execution.

Modeling — The required model fidelity for economic optimization remains an open issue. No procedure exists for establishing the model structure a priori, nor do many useful design heuristics exist. Currently, simulation studies of the closed-loop RTO system are required to evaluate alternate models.

Discrete Variables — In many plants economic performance is strongly influenced by discrete variables like feed type and equipment selection. The mathematical optimization involves mixed integer nonlinear programming which might require computing times too lengthy for real-time application. In addition, it is not clear how to automate the results when considering the switching costs and personnel requirements for implementing changes in integer variables.

Uncertainty — All of the information used for economic optimization, economic values, production targets, process models, and so forth, have uncertainty. Therefore, a more proper approach would be to optimize the expected value of the profit over the distributions of all parameters.

Unsteady-state Operation — The currently applied approaches to data reconciliation and parameter estimation require steady-state data, but it would be advantageous to perform these components during normal dynamic plant operation, even if the optimization were restricted to steady-state. Naturally, further optimization improvement is at least theoretically possible by operating the process in unsteady-state.

Model Updating — The use of typical operating data provides a limitation to the accuracy of parameters estimates. Experimental design is a mature technology and could be integrated with sensitivity analysis of the RTO system to improve the estimates of key parameters by selective plant experiments tailored to the current plant scenario.

Results Analysis — There is a large literature on sensitivity analysis; however, many methods require the Hessian which is difficult to evaluate for large-scale industrial problems. Also, while results analysis of the final solution with regards to infinitesimal parameter changes are possible (e.g. Wolbert et al., 1994), approaches are required for the analysis of finite changes from the solution (e.g., Seferlis and Hrymak, 1995). In addition, there has been little work done on developing aids to diagnose causes and remedies for numerical difficulties, such as singular sets of equations, ill-posed problems or degenerate sets of equations. While it is possible to develop such aids for the off-line model-building exercise (Chinneck, 1995), it is still an open question about what diagnosis tools will be needed for on-line closed loop optimization implementations.

The greatest demand from optimization users is a system that can be easily understood: they want "why" and "how accurately known" to be transmitted with the recommendation for operating condition.

RTO Conclusions: A Profitable Endeavor

Real-time economic optimization of plant operations, long a desired goal, has been achieved recently through progress in digital computing, optimization methods, and software support tools. The topic provides opportunities for both the industrial practitioner and academic researcher.

For the practitioner, the major benefit is the opportunity to substantially increase plant profit. The challenge is to design and implement the hierarchical automation system in Fig. 1 with coordination that represents the plant economics. The monitoring and enhancing of the RTO system will become an integral part of the continual improvement program in the plant. This activity provides an increased scope for process control as some traditional topics reach maturity and require less engineering effort.

For the academic, closed-loop economic optimization provides a wealth of research topics that combine fundamental issues with the broad potential application. These topics include data evaluation, parameter estimation, scheduling, optimization, and process control. The key feature is that these technologies are integrated in a closed-loop, feedback system that produce plant operating conditions to be implemented with little or no human intervention.

References

Bailey, J.K., A.N. Hrymak, S.S. Treiber and R.B. Hawkins (1993). Nonlinear Optimization of a Hydrocracker Fractionation Plant, *Comp. Chem. Eng.*, **17**, 123-138.

Bamberger, W. and R. Isermann (1978) Adaptive On-line Steady-State Optimization of Slow Dynamic Processes, *Automatica*, **14**, 223-230.

Box, G. and J. Chanmugan (1962) Adaptive Optimization of Continuous Processes, *IEC Fund.*, **1**(1), 2-16.

Chinneck, J.W. (1995). Analyzing Infeasible Nonlinear Programs, *Computational Optimization and Applications*, **4**, 167-179.

Cronkwright, M. (1994). Experience with Online Optimization, *Proceed. Can. Chem. Confer. Annual Meet.*, Calgary, 477-478.

Crowe, C.M., (1994). Data Reconciliation — Progress and Challenges, *Proceedings of PSE 94*, 111-121, Seoul, Korea.

Divekar, S. and J. Lepore (1991). Operational Improvements in Ethylbenzene/Styrene Plant Due to Advanced Process Control and Online Optimization, EB/SM Technology Conference, October 20-25, 1991.

Fatora, F., G. Gochenour, B. Houk and D. Kelly (1992). Closed-Loop Real-Time Optimization and Control of a World Scale Olefins Plant, AIChE Spring Meet., New Orleans, March 1992.

Fletcher, R., *Practical Methods of Optimization, Second Edition*, John Wiley & Sons, Toronto, 1987.

Forbes, F., T. Marlin and J. MacGregor (1994). Model Selection Criteria for Economics-Based Optimizing Control, *Comp Chem Engng*, **18**, 497-510.

Forbes, F. and T. Marlin (1995). Model Accuracy for Economic Optimizing Controllers: The Bias Update Case, *IEC Fund.*, **33**(8), 1919-1929.

Foster, D. (1987). Economic Performance Optimization of a Combined heat and Power Station, *Proceed. IMechE*, **201**(A3), 201-206.

Gallun, S.E., R.H. Lueke, D.E. Scott and A.M. Morshedi (1992). Use open equations for better models, *Hydrocarbon Processing*, July, 78-90.

Hanley, J. (1985). Experience with Ethylene Plant Optimization, ISA Meet Paper no. 85-1411.

Kassidas, A. and T. Marlin (1993). Predictive Controller Design with Implicit Economic Criteria, Auto. Contr. Conf., San Francisco, June 1993.

Koninckx, J., T. McAvoy and T. Marlin (1988). Online Steady-state Optimization of Continuous Chemical Plants, AIChE Nat. Meet., New Orleans, March 6-10, Paper 27e.

Krishnan, S., G. Barton and J. Perkins (1992). Robust Parameter Estimation in Online Optimization: Part 1, Methodology and Simulated Case Study, *Comp Chem Engng*, **16**, 6, 545-562.

Larmon, F., J. Van Reusel and L. ViVille (1977) On-Line Optimization of an Ethylene Oxide Unit, in Van Nauta Lemke (ed.), *Digital Computer Applications to Process Control*, IFAC and North-Holland Publishing, 59-77.

Mah, R.S.H., and A.C. Tamhane (1982). Detection of Gross Errors in Process Data, *AIChE J.*, **28**, 828.

Martin, G., P. Latour, and L. Richard (1981). Closed-Loop Optimization of Distillation Energy, *CEP*, 33-37.

May, D., Norden, B. Andreasen, C. and Cho, C. (1979). Optimizing Plant Refrigeration Costs, *ISA Transactions*, **18**, 1, 71-79.

Miletic, I. and T. Marlin, (1996) Results Analysis for Real-Time Optimization, Escape-6, Rhodes, May 26-29, 1996.

Murtagh, B.A. and M.A. Saunders (1995). *MINOS 5.3 User's Guide*, Systems Optimization Laboratory, Stanford University

Narasimhan, S., C.S. Kao and R.S.H. Mah, (1987). Detecting Changes of Steady State Using the Mathematical Theory of Evidence, *AIChE J.*, **33**, 1930.

Pedersen, C., D. Mudt, J.K. Bailey and J. Ayala (1995). Closed Loop Real Time Optimization of a Hydrocracker Complex, Nat. Petro. Refiners Compter Conf., Nashville, Paper CC-95-121.

Reilly, P.M. and R.E. Carpani (1963). Application of Statistical Theory of Adjustment to Material Balances, 13th Can. Chem. Eng. Conf., Montreal, Quebec.

Renfro, J., A. Morshedi and O. Asbjorsen (1987). Simultaneous Optimization and Solution of Systems Described by Differential/Alegebraic Equations, *Comp. Chem Engng.*, **11**, 503-517.

Schmid, C. and L.T. Biegler, (1994). Quadratic Programming Methods for Reduced Hessian SQP, *Comp. Chem. Eng.*, **18**, 817-832.

Seferlis, P. and A.N. Hrymak (1994). Optimization of Distillation Units Using Collocation Models, *AIChE J.*, **40**, 813-825.

Seferlis, P. and A.N. Hrymak (1996). Sensitivity analysis for chemical process optimization, *Comp Chem Eng.* **20**, 1177-1200,

Shara, L., A. Chontos and D. Hatch (1990). On-line Optimization at the Painter Complex Gas Processing Plant Improves Profitability, Nat. Petro. Refiners Computer Conf., Seattle, Paper CC-90-121.

Stanley, G. and R. Mah (1977). Estimation of Flows and Temperatures in Process Networks, *AIChE J*, **23**, 642-650.

Sutton, T. and J. MacGregor, J. (1977). The Analysis and Design of Binary Vapour-Liquid Equilibrium Experiments Part I: The Design of Experiments, *Can. J. Chem. Eng.*, **55**, 609-613.

Tjoa, I. and L. Biegler (1991). Simultaneous Strategies for Data Reconciliation and Gross Error Detection of Nonlinear Systems, *Comp. Chem. Engng.*, **15**, 10, 679-690.

Vasantharajan, S. and L. Biegler (1988). Simultaneous Solution of Reactor Models within Flowsheet Optimization, *Chem Engr. Res. Des.*, **66**, 396-407

Wolbert, D., X. Joulia, B. Koehret and L.T. Biegler (1994). Flowsheet Optimization and Optimal Sensitivity Analysis Using Analytical Derivatives, *Comp Chem Eng*, **18**, 1083-1095.

DISCRETE EVENTS AND HYBRID SYSTEMS IN PROCESS CONTROL

Sebastian Engell and Stefan Kowalewski
Process Control Group, Department of Chemical Engineering
University of Dortmund
D-44221 Dortmund, Germany

Bruce H. Krogh
Department of Electrical and Computer Engineering
Carnegie Mellon University
Pittsburgh, PA 15213

Abstract

Process control systems typically perform a large number of logical functions for supervisory and safety control, in addition to closing the continuous-variable control loops which have been the traditional focus of process control theory. The interaction of discrete control actions with the continuous process dynamics, and the discrete phases that can arise in the physical processes themselves, lead to *hybrid dynamical systems*. In this paper we identify the sources of hybrid dynamics in chemical process control systems and present a general framework for modular modeling of interacting continuous and discrete dynamic elements. We then review recent research themes in hybrid systems as they relate to process control, including simulation, extensions of tools for analysis and synthesis of logical (discrete event) systems, and general methods for analysis and optimization.

Keywords

Hybrid systems, Discrete event systems, Modeling, Simulation, Finite state machines, Formal verification, Controller synthesis

Introduction

Traditionally, most of the work in process control theory is concerned with the control of continuous dynamic processes described by ordinary differential equations (ODEs) or differential-algebraic equations (DAEs). However, a large portion of process control software is devoted to processing and generating discrete events as, e.g., alarms and discrete control actions. Even for the continuous control loops, monitoring variable limits, handling exceptions, and switching control regimes — predominantly logical functions — usually require much more complex code than the control algorithm itself.

There are various sources of discrete phenomena in chemical processes:

- discrete changes in the processes (e.g., a vessel becomes empty, a liquid is completely evaporated, a column is flooded)
- instruments which give discrete rather than continuous outputs (especially for levels and flows)
- actuators with discrete settings/positions (e.g., on/off valves, pumps, heatings with constant current, motors without speed control) and the associated binary feedback signals
- continuous measurements trigger thresholds (e.g. alarms) or the plant is classified by discrete states ("normal," "exceptional," "dangerous / shutdown necessary")

- steps in recipes, transitions between batches or cuts in batch distillation
- startup and shutdown of continuous plants, product changeover, etc.

In process control systems, various layers of the automation hierarchy can be distinguished:

- safety related control functions
- basic control functions (regulation of process variables, such as temperatures and pressures; basic sequence control, e.g. to switch on a pump or a vacuum system)
- advanced control functions (multivariable control, startup and shutdown, setpoint optimization, recipe driven production, etc.)
- logistic control functions (e.g., assignment of equipment, sequencing of batches, selection of batch sizes and campaigns).

Classical continuous control occurs only on two of these four layers. But on all levels, there are numerous control tasks where logical (discrete) control functions interact with continuous or piecewise continuous dynamic processes. Such constellations are nowadays classified as *hybrid systems*.

The interaction of switching controls with continuous plants may occasionally lead to very undesirable consequences. An example is an accident which occurred at a chemical plant in Germany some years ago and led to major investigations and public discussions.

The accident occurred at a manually operated plant where two substances had to be put into a reactor while stirring. Under normal conditions this leads to a reaction which starts slowly but is strongly exothermic and is controlled by the cooling system. However, when the second substance was added, the stirrer was not on, the second substance accumulated on top of the first and almost no reaction took place. The operators realized that something was wrong because the temperature did not rise, and discovered the problem with the stirrer. They anticipated that much heat could be created soon, so they started the stirrer to improve the heat transfer to the cooling system. This discrete action led to an immediate strong reaction, and then to a discrete and irreversible event: the distribution of the reactor contents in the (inhabited) neighborhood of the plant.

This accident would not have occurred if a computer-based control system had been used which would have blocked the addition of the second substance when the stirrer did not turn. But it also points to a more general problem: especially in plants where many different products are made, there is a large number of possible situations which may occur due to some malfunction, and the correct reactions in such exceptional situations require relatively deep understanding of the processes (turning the stirrer on in a different situation would have been the right reaction).

In modern plants, the majority of the control tasks are performed by computer programs in a distributed control system (DCS) or programmable logic controller (PLC). Even for safety related functions, redundant computer-based systems are used now in many new plants in place of the traditional hardwired control logic. This allows a more flexible reaction and the implementation of complex emergency shutdown procedures, thus minimizing the potential dangers and the economic losses. Even if the safety of the environment and the humans in the plant or neighboring vicinity is maintained by a separate hardwired safety system, a malfunction of the discrete controls may cause severe economic losses. It is therefore extremely important to guarantee the correct reaction of the control software in all critical situations.

To date, however, this is not possible. The state of the art is that the potentially critical situations in the plant are identified by process experts and the desired function of the code is determined. Sometimes, parts of the discrete control software are tested offline or against simple models to check their behavior. The lack of verification methods also leads to extensive tests on the site and long commissioning periods.

So there exists a strong motivation to apply formal methods to the analysis and, as a long term goal, to the synthesis of process control programs beyond checking the stability and performance of continuous control loops rigorously or by simulations. Whereas this is not possible for realistic problems at the moment, research activities have started in this direction which are very relevant for, and often directly related to, process control. This research can build upon theoretical results and software tools from computer science and discrete dynamical systems theory which were developed in recent years. In this paper, we provide an overview of interesting approaches to deal with the interaction of discrete controls and continuous or piecewise continuous processes. We concentrate on the three lower levels of the automation hierarchy, i.e. we do not discuss the logistic aspects.

Complementary Views of Hybrid Systems

In the scientific discussion, two complementary views of hybrid systems have emerged. In the first view, the interest focuses on the continuous dynamical behavior, whereas in the second the logical aspects are dominant.

From a process engineering point of view, hybrid systems naturally arise as generalizations of continuous dynamical systems. Many processes have structurally different dynamics in different situations (cf. Pantelides, 1995). A simple example is a liquid in a vessel which is heated and then evaporates when the temperature reaches the boiling temperature. The most natural description of such phenomena is by different sets of differential and algebraic equations, and switching conditions which govern the transition from one description to the next.

In simple cases, it is sufficient to describe the boundaries where the dynamics change as functions of the real-valued state variables of the continuous system. But often a switching logic is required which has discrete internal states which store the past of the system. For the evaporator it is appropriate to introduce two discrete states, heating/cooling (hc) and boiling (b) where the condition for the transition from hc to b is that the boiling temperature is reached, whereas the condition for the transition in the other direction is that the amount of heat added becomes negative.

Quantitatively, the most important reason for discrete changes in chemical processing plants is the instrumentation with discrete sensors and actuators. The number of discrete valves, motors with constant speed, etc. usually is much higher than the number of continuous control inputs. The operation of these discrete inputs, manually or automatically, requires a discrete decision which may be a static logical expression of binary variables (from binary sensors or from discretizations of continuous measurements), or depend on the discrete state of the control logic, i.e. past measurements.

Another important source of discrete changes is the sequence control logic. This is obvious in batch processes where recipes are executed step by step, parts of the plant are activated and deactivated, vessels are filled and emptied, batches of material are created, united, separated, undergo transformations and finally leave the system as a number of discrete units of product. The same is true for startup and shutdown of continuous plants. Such processes require adequate models where complex switching sequences can be expressed in a transparent fashion.

The interaction of logical systems with even very simple continuous dynamical systems can lead to surprising behaviors and complex dynamics.

An illustration of this fact is the switched flow system described in (Chase et al., 1993). There, three tanks with continuous constant outlet flows are filled by one constant inlet flow which can be switched instantaneously between the tanks. Surprisingly, for a very simple and intuitively reasonable (discrete) control strategy (always switch the server to the tank which becomes empty), this system will show chaotic behavior although the continuous dynamics are simply integrators. In the reverse situation where the consumer is switched and the tanks have constant inflow, the system becomes periodic after a few switchings.

When hybrid systems are modeled from the point of view of continuous dynamics, a standard description of the continuous dynamics by systems of ODEs or DAEs is used in most cases and the discrete aspects are captured by simple switching terms or small state machines. Such models are used primarily for the simulation and the optimization of plant operations. Simple systems as the switched tank system can be analyzed using concepts and tools from nonlinear dynamics as deterministic chaos and entropy (Chase et al., 1993, Schürmann and Hoffmann, 1995).

If a large number of discrete variables is present, complex logical expressions must be evaluated and the interesting phenomena are determined strongly by the discrete behavior. Typical examples are safety control issues (e.g. for burners) and the operation of batch plants with multiple products and shared resources. In such cases, modeling in terms of continuous variables is no longer adequate. To model the predominantly discrete dynamics, the continuous aspects are simplified by approximating the continuous dynamics either by a purely discrete behavior (so-called *qualitative* dynamics) or by discrete models with timers (so-called *quantitative* dynamics). Such models can be used to specify a desired behavior and, based on this, to verify or to synthesize discrete controllers.

In the logical systems framework, aspects of the behavior of the system which are qualitatively different from those which naturally arise in continuous systems can be treated. The simplest question which can be asked (and answered with modern computer tools even for relatively large problems) is whether a set of forbidden (discrete) states of the system will ever be reached, e.g. whether two combustible products will occupy a transport pipeline at the same time. This analysis can be extended to more complex logical expressions and forbidden sequences of states. In contrast to the usual system-theoretical approach in which one argues about properties of models and mathematical formulae and puts all implementation aspects outside the scope of the discussion, the goal of the verification of logical systems is to provide a formal correctness proof of the actual implementation.

One approach to control logic synthesis using formal methods is sequential refinement. First a basic specification is formulated and a basic algorithm is verified using a simple model. Then additional features of the model and details of the implementation are added and shown to imply the basic specification. This is iterated until the logical control program is sufficiently detailed to be automatically translated into executable code. This approach is successfully used in communication protocol verification (Herrmann and Krumm, 1994).

In logical systems, the state space grows combinatorially with the number of binary or discrete variables considered in the model. How to handle this combinatorial explosion is a major theoretical and practical problem. There are tools available to perform analysis functions for systems with some 10^{20} to 10^{30} states, but clearly an unstructured model of this size is as likely to be faulty as the control software which is to be verified. Adequate modeling paradigms and techniques which support modularity and hierarchical models are therefore indispensable for progress towards the analysis of real systems (Kowalewski et al., 1995).

The Continuous Dynamics Approach

The first step towards an understanding of hybrid systems is adequate modeling and reliable and efficient numerical simulation. An early attempt to provide a formal framework for the modeling and simulation of such phenomena is the differential automaton introduced by Tavernini (1987). The differential automaton is an autonomous dynamical system which lives in a state space $S = \mathbb{R}^n \times Q$ where Q is a finite set of discrete states. The continuous trajectories are governed by ODEs

$$\dot{x} = f(x, m)$$

and on the *switching manifolds* $g_{q,p}(x) = 0$, the discrete variable m switches from q to p. The trajectory of the x-component is continuous in this framework, and the discrete part m is piecewise constant. Under certain conditions on the switching manifolds, the system has a unique solution and infinitely fast switching cannot occur.

This model was generalized in various directions. In (Back et al., 1993), local state spaces and jumps of the state in the case of switching were introduced as well as control inputs. Although one can argue that often discontinuous changes of the states of physical systems on the macroscopic scale are the result of simplifications in modeling, it is often helpful to include such abrupt changes (equivalent to impulsive inputs), e.g. to avoid problems with the numerical treatment of extremely different time-scales (cf. Barton and Pantelides, 1994). The models of Nerode and Kohn (1993) and Antsaklis et al. (1993) include control automata which determine the discrete dynamics and interfaces to the continuous part which are generalizations of DA- and AD-converters. A comprehensive model which includes all those mentioned was proposed by Branicki et al. (1994).

A General Block-oriented Model

In (Krogh, 1993), a modular modeling paradigm for logic controlled continuous systems was introduced. It consists of ODE-subsystems of the Tavernini-type with an additional "event input" which can reinitialize the state vector. The logical part is described by *condition/event systems* (Sreenivas and Krogh, 1991a). A c/e-system can be described by a finite state automaton with two types of input and output signals.

- piecewise constant right continuous "condition signals" which assume values from a finite alphabet
- "event signals" which are almost always zero and assume values from a (different) alphabet.

The use of two types of signals permits a clear and transparent distinction between enabling and enforcing of state transitions. Similar signal types are used in logic programming languages.

A c/e-system has a condition input $c(t)$ and an event input $d(t)$, a condition output $p(t)$ and an event output $q(t)$.

The internal behavior is described by the *state transition function F*:

$$s(t) \in F(s(t^-), c(t), d(t)),$$

where $s(t^-)$ denotes the limit from the left (which corresponds to the "previous" state) satisfying $\forall s \in S, c \in C$: $s \in F(s,c,0)$, (the system can always remain in the same state if no event occurs); and the *condition and event output functions G and H*

$$p(t) = G(s(t), c(t)),$$

$$q(t) = H(x(t), s(t), d(t)).$$

satisfying $\forall s \in S$: $0 = H(s,s,0)$ (event outputs only are generated if the internal state changes or an event is received).

In contrast to other models, this is a modular input-output description in which all types of interactions of logical systems can be expressed conveniently, and all external quantities are defined over time. Hence the coupling of c/e-systems and continuous systems is straightforward.

To obtain a general modular modeling paradigm for hybrid systems, c/e-systems can be coupled with switched DAE-systems which include threshold functions to generate condition and event outputs from continuous variables (see Fig. 1).

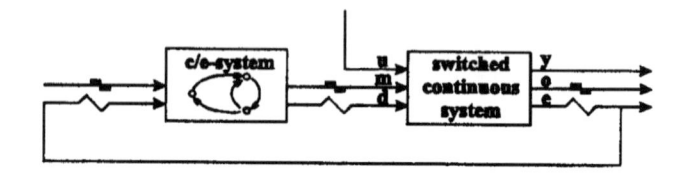

Figure 1. An example of a feedback connection of a c/e-system and a continuous system.

The switched continuous block is a DAE-system with dynamical state vector x, algebraic state vector z, continuous input vector u, continuous output vector y, mode input vector m, quantized output vector o, event input vector d and event output vector e.

The vectors x and z are real-valued, as well as u and y. m and o are condition signals, d and e are event signals. For constant mode input vector m, the system is a conventional DAE-system

$$f(\dot{x}, x, z, u, m) = 0, x(t_0) = x_0, z(t) = z_0$$

with continuous output

$$y = h(x, z, u).$$

A mode change leads to a change of the dynamics (including changes of the dimensions of the variables x and z), but x and z cannot jump due to mode changes, similar to the Tavernini model. Jumps are enforced by the event input d. If d is nonzero, the system is reinitialized

$$(x(t), z(t)) = g(x(t^-), z(t^-), u(t^-), m(t^-), d(t)),$$

where $x(t^-)$ etc. are the limits from the left of $x(t)$, etc.

The condition output is generated by quantization functions:

$$o_i = 1 \text{ if } k_i(x, z, u, m) \geq 0, \text{ else } o_i = 0,$$

and component e_i of the event output e is nonzero iff o_i changes, and e_i indicates the direction of the change.

The switched continuous block may represent the physical part of the (sub)system, i.e. the (sub)process together with the discrete actuators and sensors. It is connected to the environment by physical streams, continuous or discrete (usually binary) signals and messages (events).

In the simplest case, the condition inputs and the condition outputs of the continuous block are directly coupled. If the direction in which the boundary is crossed determines the next mode, an event output must be connected to a c/e-system which generates the mode input vector of the continuous block. Then we have the DAE-equivalent of Tavernini's differential automaton where the dynamics change when a function of the continuous variables reaches a predefined threshold. To introduce jumps, the event output and input can be coupled such that for some or all changes of the quantized variables, a reinitialization which may depend on the internal variables and the external inputs, takes place. Such discontinuous (sub)systems can be coupled via the continuous inputs and outputs, but also the discrete outputs of one subsystem can be connected to the discrete inputs of another one.

With the logical blocks, more complex logical behaviors can also be modeled. Mode changes and state resets then can depend on the inner state of the logical system, and on arbitrary conditions and events in the overall system. The logical systems can be organized in layers where only the lower layer communicates with the continuous system, e.g. to represent automation hierarchies. Discrete dynamical changes can be represented locally by individual logical blocks associated with the continuous subsystems, whereas the global recipe control is represented by a central supervisory system which can also be modeled as an interconnection of timers and logical blocks.

A special case of the switched continuous block is a timer (clock). The clock has two modes, running (at a certain rate) and stopped, which are switched by the mode input. An event input resets the clock to a (possibly condition-dependent) starting time x_0. A condition output indicates whether the clock is below or above a threshold, and an event output is generated at the time a threshold is reached. A clock is a condition/event system as far as the external behavior is concerned and can thus be connected to other c/e-systems.

Simulation of Hybrid Systems

As the mathematical analysis of hybrid dynamical systems is very complex, general models which include the continuous as well as the discrete behavior are most useful for simulation studies. It is only recently that simulation systems have become available which allow the implementation and simulation of the type of systems described above. Many simulation programs, such as *MATLAB/SIMULINK* which is frequently used for simulation in other fields but also for control-oriented simulations in chemical companies, currently provide limited features for event-driven changes of the model structure nor for logical subsystems. *MATLAB* code which can handle switched ODEs has been developed by Taylor (1995).

DYMOLA (Elmqvist et al., 1993) is a software package which allows modular modeling of dynamical systems including abrupt changes of the system structure and, e.g., hysteresis and stick-slip effects based on bond graph models. Its strength is the symbolic manipulation of modularly defined equations to produce a minimal description of the overall system. It has been successfully applied to large mechanical systems with state dependent changes of the system structure. There are, however, no mechanisms provided to define and to handle complex sequences and logical decisions. Another simulation tool that models abrupt switching dynamics is *dstool* (Back et al., 1993).

gPROMS (Barton 1992, Barton and Pantelides, 1994) constitutes a major progress in the simulation of general dynamical processes with structural changes and switching. It is a powerful simulation system which can handle large interconnected systems consisting of DAE-(sub)systems coupled with automata. The continuous system models are index 1 DAEs with mode changes of the Tavernini type (constant number of differential equations, continuous state variables). On a higher level, "tasks" and "schedules," which may change the system structure, including reinitialization of the state variables, can be defined.

Batches (Clark and Joglekar, 1992) and *BaSiP* (Engell and Wöllhaf, 1993, Engell et al., 1995) are simulation systems specifically designed to model recipe-driven batch processes. In *BaSiP*, the dynamics are described locally (in the technical functions of the plant) including the generation of discrete signals. Logical systems can also be defined locally (to model changes of the system structure on the subsystem level) as well as globally to perform the recipe-driven starting and stopping of operations, sequentially or in parallel. The dynamical state vector is generated dynamically by assembling the state vectors of the active components. In *BaSiP*, interactions of logical and continuous blocks as described above are simulated but defined only implicitly by the graphical configuration of plants and recipes. The continuous dynamics at present are described by ODEs.

Optimization

Based on modeling and simulation, the logical next step is optimization. Optimization of hybrid systems in general is a complex issue because the discrete states introduce non-differentiability of the cost function. A class of problems which can be handled relatively easily is when the switching sequence is known a priori, so only the durations of the switching intervals and continuous variables within the intervals are to be optimized as, e.g. in batch distillation with a fixed sequence of cuts (cf. Vassiliades et al., 1994). Infinite horizon trajectory optimization is discussed in (Branicki et al., 1994) where necessary conditions are given for a special class of problems.

Logical Systems Approach

An alternative view of hybrid systems has emerged from the purely discrete, or logical, view of dynamic systems. Logical models capture only the sequential behavior of the system evolution, often without reference to absolute time or continuous process variables. They have been applied to the discrete aspects of process control systems, and have formed the basis for programming languages that can be used for process control computers. Powerful computer tools have also been developed for verifying properties of logical systems, and control synthesis for logical systems has been an active area of control research for over a decade.

Methods have been proposed recently to extend purely discrete models to incorporate aspects of continuous dynamics. Although much of this work has been initiated by computer scientists, modeling and verification of process control systems has been one of the primary motivations for this work, and process control researchers have begun to explore the possibilities for exploiting the tools for discrete analysis to deal with hybrid problems.

Automata and Petri Nets

The most basic model of discrete dynamics is the finite state machine, or automaton, where transitions between discrete states are identified as instantaneous events. Formally, an automaton can be defined as $A = (Q, \Sigma, \delta, q_0)$, where Q is the finite set of discrete states, Σ is the set of events, $\delta: Q \times \Sigma \rightarrow Q$ is the state transition (partial) function, and $q_0 \in Q$ is the initial state. The state transition function is a partial function because it is defined only for the events that can occur when the automaton is in a given state. A state trajectory for the automaton is a sequence of states $q_0, q_1, \ldots \in Q$ satisfying $q_{k+1} = \delta(q_k, \sigma_k)$ for a corresponding event sequence $\sigma_0, \sigma_1, \ldots \in \Sigma$.

Note that the above model does not explicitly identify events as input or output signals. The assignment of these notions in a particular model depends upon the sources of the events: in some cases the system transitions are caused by external input events, and in other cases the state transitions occur due to the internal dynamics of the system being modeled.

Automata are state-based models of discrete dynamics. Associated with an automaton is the collection of all valid sequences of events consistent with the state transition function, denoted $L(A)$, known as the *language* of A. Different automata can share the same language, which is analogous to the existence of multiple state-space realizations for a given input-output behavior (transfer function) for a linear continuous dynamic system. When only the sequences of events are of interest, a process can be modeled by its language alone. Research on formal languages in computer science forms the basis for many methods for specification, verification, and synthesis of logical systems. In process control applications, it is usually more natural to develop a state-based model of the system dynamics.

If there are states from which more than one state can be reached for a single event, a *nondeterministic automaton* can be defined in which $\delta: Q \times \Sigma \rightarrow 2^Q$, and the state trajectories satisfy $q_{k+1} \in \delta(q_k, \sigma_k)$. A standard result in automata theory states that for any nondeterministic (finite state) automaton A, there always exists a deterministic automaton A_d such that $L(A_d) = L(A)$. This result is of limited value when the states in the automaton model have physical significance and nondeterminism represents an actual loss of observability of the state trajectories. In these cases, the nondeterminism in the model must be retained and can lead to significant increases in the computational complexity of analysis and synthesis of controls (Cieslak et al., 1988; Tsitsiklis, 1989).

Building modular models of systems using finite state automata requires a method of composition. One method for combining automata models of several system components is the *synchronous composition* in which the state space for the composite automaton is the product of the state spaces of each component automaton, and transitions (events) occur simultaneously in the component automata whenever they have the same event label. Otherwise, events can occur in the components independently without changing the states of other components.

Automata-based models even for small systems quickly become unmanageable. The number of states grows exponentially as more system components are added, and the model structure loses any intuitive meaning. Petri nets provide more compact and intuitive structured models for logical systems. A simple Petri net consists of *places* (circles) and *transitions* (bars) with arcs connecting places to transitions and transitions to places. The state of a Petri net model is given by the *marking* which identifies the number of so-called *tokens* in each place. State transitions occur when Petri net transitions *fire*. In a simple Petri net, a transition *can* fire (it does not have to) when it is *enabled*, which occurs when there is at least one token in each of its input places. When an enabled transition fires, a token is removed from each input place and a new token is placed in each of its output places. Two transitions are said to be in *conflict* if they are enabled for a given marking, and the firing of one of the transitions disables the other transition.

Like automata, Petri nets can be combined to form models from system components. Typically Petri nets are combined through common Petri net transitions and places. In contrast to the automaton created by the synchronous composition, the states of the subsystems retain their identity as separate places in the composite Petri net. Thus, the structure of the Petri net reflects the logical structure of the system being modeled. Moreover, the size of the Petri net grows linearly, rather than exponentially, with the number of subcomponents added to the system.

To analyze a system modeled by a Petri net, one can always define an automaton in which each state is associated with a (reachable) marking in the Petri net and the automaton state transitions are associated with the enabled Petri net transitions for each marking. This automaton is called the *reachability graph* for the Petri net. When Petri net analysis is based on construction of the reachability graph, the economy of the Petri net representation does not carry over to a savings in computation since the number of markings can grow exponentially with the size of the Petri net (or may even be infinite in the case of unbounded Petri nets; that is, when there is no limit on the number of tokens in the Petri net places). Computational savings are realized when methods for analysis or control synthesis are based directly on the Petri net structure, as in, e.g., (Holloway and Krogh, 1990; Li and Wonham, 1993, 1994; Giua and DiCesare, 1994).

Petri nets have been used extensively to model, simulate, and analyze discrete control systems. The reader is referred to (Murata, 1989) for a general survey of the Petri net literature. Extensions of the simple Petri net model described above to the so-called high-level Petri nets have made them even more attractive for modeling complex systems. Most of these extensions build on the concepts of colored Petri nets (Jensen, 1995) in which the tokens assume attributes with valuations that can be included in the transition firing rules, leading to even more compact representations of complex systems. Examples of the applications of Petri nets models to process control systems include (Brand and Kopainsky, 1988; Yamalidou and Kantor, 1990; Hanisch, 1992; Hanisch, 1993). An extension of the algebraic representation of Petri nets to a discrete-time, linear hybrid system model with real, integer and Boolean state components has been introduced by Antsaklis and Kantor (1995).

The internal discrete state dynamics of a c/e system can be represented as an automaton or Petri net (Sreenivas and Krogh, 1991b). In either case, the condition and event input signals define additional external conditions on the state transitions, with event signals being able to force state transitions. Output condition signals reflect the internal state of the system, whereas the output event signals are associated with state transitions.

Programming Languages

The earliest implementation of discrete control functions was by means of pneumatic or electrical relay systems. PLCs and DCSs have now replaced physical relays, allowing the implementation of much more sophisticated combinatorial and sequential switching logic. This has also increased the capability to integrate discrete and continuous control functions with the introduction of non-discrete functions in the PLC programming languages.

Because of the historical relationships between PLCs and physical relay systems, graphical relay ladder diagrams became a common programming language for PLCs (Wilhelm, 1985). The size and complexity of PLC applications in the early 1970's created the need for new PLC (and DCS) programming languages that offered a more transparent representation of sequential and concurrent control logic. It is also desirable to have features that facilitate the management of large programs, such as hierarchies, modularity, refinement, etc. Given the natural graphical representations of automata and Petri nets, it is not surprising that programming languages have been introduced for discrete control based on each of these formalisms.

Statecharts are a graphical method for representing hierarchical and interacting automata (Harel, 1987). With the ability to have concurrent states active simultaneously, statecharts avoid the state explosion encountered in the composition of simple automata. Continuous control functions can be incorporated into a statechart as modules associated with discrete states. For example, when a particular state is active in a statechart, an associated PID routine might operate to perform closed loop regulation with a specified set of gains. A transition to another state might change the gains in the feedback loop, or completely shut off the regulation function. Brave and Heymann (1993) have shown how the statechart structure can be exploited to reduce the computational complexity of control logic synthesis.

Grafcet, introduced in 1977 in France as a programming standard for PLCs, is based on a Petri net-like graphical structure (David and Alla, 1992; David, 1995). Grafcet *steps* correspond to Petri net places. When a step is active, the *action* associated with the step is performed. Actions can be sustained conditions, or impulse "one-shot" actions. These two types of actions correspond respectively to the condition and event signals in c/e systems. Grafcet transitions occur when all the steps preceding the transition are active and an additional condition associated with the transition (called the *receptivity*) is true. Unlike tokens in Petri net places, tokens cannot accumulate in a Grafcet step; a step is merely active or inactive. Like statecharts, Grafcet programs allow continuous control functions to be associated with the discrete steps in the program. Grafcet also allows hierarchy and refinement: steps can contain further Grafcet (sub)programs. Falcione and Krogh (1993) developed a method for translating simple relay ladder diagrams into Grafcet programs, thereby recovering the sequential and concurrent logic flow inherent in the relay ladder representation.

Grafcet has been accepted as an international standard PLC programming language under the label *sequential function charts* (SFC) (IEC, 1986). It is straightforward to translate SFCs into c/e-systems (Kowalewski et al., 1994). Grafcet has also been proposed as a graphical programming language for organizing high-level, rule-based programs for process control applications (Arzen, 1994).

Another group of programming languages for logical systems are synchronous languages, which have been developed for real-time embedded control applications, and have been based on computer science models emphasizing concurrency and synchronization (Hoare, 1985; Caspi, 1991). These programming languages, such as *Esterel* (Boussinot and de Simone, 1991), *Lustre* (Halbwachs et al., 1991), *Lucid* (Ashcroft and Wadge, 1985) and *Signal* (Le Guernic et al., 1991), are structured so that properties of the control program are enforced or can be verified by the formal model underlying the programming language. A primary application for these languages is to implement complex communication protocols, but they may become of interest for process control applications which are becoming increasingly distributed and strongly coupled.

Formal Verification of Process Control Systems

One of the principal motivations for developing mathematical models of logical systems is to demonstrate that the system satisfies the system specifications. There are two general classes of specifications for logical systems: *safety* specifications and liveness specifications. In this context, safety refers to specifications defined in terms of the reachable states or finite-length (but perhaps unbounded) strings of events. Liveness specifications, on the other hand, concern the behavior of system along trajectories of indefinite (infinite) length. In the case of process control systems, safety specifications are most often of concern. For example, it must be guaranteed a condition leading to an explosion never occurs. (Here the computer science notion of safety corresponds to safety in the usual sense of the word.)

The most straightforward way to check whether or not a safety specification is satisfied is to simulate all possible state trajectories for the system. In contrast to continuous dynamic models which usually must be simulated over an indefinite time horizon to evaluate a single state trajectory, all the trajectories for a finite state logical system can be evaluated with finite length simulations. Once the system returns to a previously visited state (which must occur since the number of states is finite), the simulation can be terminated and another trajectory can be initiated until all possible transitions from all states have been executed. This approach applies to automata or (bounded) Petri net models, but it is prohibitively time consuming for systems of any significant size.

The computational burden of brute force simulation is drastically reduced by *symbolic model checking* in which whole classes of system trajectories are evaluated through a single forward simulation (Burch et al., 1992; Kurshan, 1994). To perform symbolic model checking, a discrete-state model of the system dynamics is constructed, normally by the composition of simple automata-based models of the system components. The desired properties of the system (both safety and liveness specifications) are then defined using an appropriate language, and the computer then verifies whether the properties are satisfied. The verification is symbolic because the state transition relation is executed for entire classes of system states rather than exploring the reachable states by simulating single state trajectories.

The possibility of performing symbolic model checking for systems of realistic size has become possible by the availability of computers with increasing speed and storage, and by the introduction of efficient data structures for representing Boolean functions (Bryant, 1986; Clarke et al., 1994). Symbolic model checking has been applied widely in digital circuit and computer bus protocol verification. Applications of these tools to small process control examples can be found in (Moon et al., 1992) and (Probst et al., 1997).

Recently, there has been considerable interest in extending tools for discrete formal verification to systems with continuous dynamics. The simplest case to consider is a timed system where the discrete state space is augmented by a set of timers, or "clocks." State transitions are enabled by conditions on the timers, and the clocks can be reset when discrete state transitions occur. This is the model considered by Alur and Dill (1994) who demonstrated that seemingly simple verification problems can be undecidable (that is, there is no algorithm guaranteed to stop with a true or false verdict for some specifications) unless the ratios of the clock rates are rational numbers.

The clocked-automata model of Alur and Dill has been extended by various researchers and computer tools are now available to verify properties of discrete-state systems with simple scalar integrator dynamics and arbitrary linear inequalities determining the discrete state transitions conditions. HyTech (Henzinger et al., 1995), KRONOS (Nicollin et al., 1992) and UPPALL (Larsen, 1995) are representative tools for performing this type of system verification. A textual programming-type language is used to describe the system dynamics and reachability conditions can be checked. These programs will also compute ranges of threshold values that will guarantee the specified conditions remain valid. To date this and similar tools have been applied only to simple academic examples, but they are particularly interesting for process control problems where purely logical models are of limited applicability.

Automatic theorem proving is another approach to the verification of logical systems in the computer science literature (Caines and Wang, 1990; Henzinger et al., 1994; Manna and Pnueli, 1995). In this approach the system is described with logical *assertions*, and properties are verified as *theorems* in the proof system. The process and controller description can be refined through the addition of more assertions. Vitt and Hooman (1995) have applied this approach to a benchmark boiler control problem. Dyck and Caines (1995) have developed a system which controls an elevator using the theorem-proving approach on-line. This approach has not yet been applied to practical examples of significant size, due in part to the lack of effective computer tools. The Stanford TEmporal Prover (STEP) is being developed currently to address this problem (cf. Manna, 1992).

A quite different approach to the formal verification of process control systems is the one by Dimitriadis et al. (1995). In this work the basic reachability problem for a discrete-time hybrid model is reformulated as a mixed integer optimization problem. The idea is to search for an "optimal" control policy which leads the system to an undesired state. The controller is proved to be correct, if such a policy cannot be found.

Discrete Control Logic Synthesis

Process control is a problem of engineering synthesis: given the process dynamics and operating limits, a controller must be synthesized to guarantee the closed-loop system satisfies the control specifications. Ramadge and Wonham formulated the control synthesis problem for purely discrete systems (Ramadge and Wonham, 1989). The plant model to be controlled is an automaton for which some events can be inhibited by the controller (the *controllable events*). They considered two problem formulations. In the first problem, the specification for the closed-loop system is a language, that is, a set of acceptable sequences of events (Ramadge and Wonham, 1987a). In the second problem, the specification is a set of acceptable states (Ramadge and Wonham, 1987b). It can be shown that the first problem can be converted into the second problem through the introduction of a system memory (Li and Wonham, 1994). An application of the Ramadge and Wonham approach to batch process control synthesis is reported by Tittus et al. (1995).

For most process control applications, the simple Ramadge and Wonham (RW) model is inadequate: a real controller can normally force events to occur, not simply prohibit events from occurring. Moreover, the lack of an explicit input-output signal structure for the controlled automaton makes it difficult to model interacting controllers and processes intuitively. One way to resolve these problems is to formulate the control synthesis problem using c/e systems. In this framework, the RW type control problem can be formulated to include more realistic system behaviors, such as the forcing of controlled events (Krogh and Kowalewski, 1995). A further possibility to handle forced events has been proposed by Sanchez et al. (1995). In their approach the supervisor of the RW framework is replaced by a so-called *procedural controller* which is able to preempt uncontrollable plant events by controllable ones.

Controller synthesis for logical systems ought to take advantage of computational tools available for system verification. In (Krogh and Kowalewski, 1993) and (Balemi, 1992) Boolean approaches are proposed. Given the power of these tools, it is attractive to consider the possibility of applying them also to control synthesis problems for hybrid systems. Returning to the general model for a hybrid system presented earlier, we note that the signals connecting the discrete and continuous portions of the system are discrete (condition and event signals). If the continuous dynamics can be replaced by c/e systems which exhibit the same logical input-output behavior as the continuous dynamics, the control logic can be synthesized using logical tools alone.

Clearly it is impossible to exactly model arbitrary continuous dynamics with discrete-state c/e systems. An approach proposed in (Niinomi et al., 1995) is to construct discrete automata models which are an outer approximation of the input-output dynamics of the continuous system. It is shown that if a logical controller is synthesized using these approximate finite state automata models, the controller will perform correctly when applied to the actual system.

This approximation approach reflects what is typically done in practice — control logic is programmed to handle everything that could possibly happen without working with detailed models of the continuous system dynamics. The formal theory identifies the conditions that must be satisfied by the approximations to make this approach work. The application of this approach to process control problems of realistic size will require the development of computer tools to deal with the models of the hybrid systems of the form presented in the beginning of this paper.

One of the key problems in applying any of these formal methods is modeling, i.e. the derivation of correct untimed or timed discrete models of continuous plants. Experiences with c/e-system models for a realistic example were reported in (Kowalewski et al., 1995). The conclusion was that an automated "translation" of balance equations into discrete models should be performed because even with a modular approach, the complexity soon becomes quite large. A first step in this direction was discussed by Preisig (1993).

Conclusions

Interactions of discrete controls with continuous plants are ubiquitous in modern chemical plants, and are in fact more frequent than the well-studied interaction of continuous controllers with continuous plant dynamics. However, systematic methods for the verification of control programs, the analysis of mixed discrete-continuous dynamics and the synthesis of discrete controllers for realistic problems are not yet available. Even the use of programming tools which support a structured approach is not standard practice. There is a striking discrepancy between the efforts for fail-safe hardware and other measures to increase plant safety and reliability, and the efforts for improved discrete control software quality. With the continuing pressure of economic, environmental, and safety constraints, this situation cannot last forever.

In this paper we have reviewed important developments in hybrid systems modeling, discrete event dynamical systems theory and computer science. The most practically useful results of the research in discrete systems so far have been in the areas of simulation tools and programming languages. Most of the research results on verification and synthesis so far have been applied only to toy examples and it is not clear yet whether the scaling properties of the various approaches will allow their application to real industrial problems. Nonetheless, the theoretical progress and the available tools, in particular for symbolic model checking, promise that within the next five years, formal verification and maybe synthesis of control logic will become feasible for realistic problems.

Acknowledgments

The authors acknowledge helpful discussions with W.-P. de Roever, I. Hoffmann, Ch. Schulz and J. Taylor.

References

Alur, R.; and D. Dill (1994). The theory of timed automata. Theoretical Computer Science, 126, 183-235.

Antsaklis, P.J., J. A. Stiver, M. Lemmon (1993). Hybrid system modeling and autonomous control systems, *Hybrid Systems* (*LNCS* **736**), Springer, Berlin, 366-392.

Antsaklis, P.J., and J.C. Kantor (1995): Intelligent Control for High Autonomy Process Control Systems. ISPE '95, Snowmass, Colorado.

Arzen, K.-E. (1994). Grafcet for intelligent supervisory control applications. Automatica, **30**, 1513-1525.

Ashcroft, E.A., and W.W. Wadge (1985). *LUCID, the Data-Flow Language*. Academic Press, New York.

Back, A., J. Guckenheimer, and M. Myers (1993): A dynamical simulation facility for hybrid systems. *Hybrid Systems* (*LNCS* **736**), Springer, Berlin, 255-267.

Balemi, S. (1992). *Control of discrete event systems: theory and application*. Ph.D. Thesis, Swiss Federal Institute of Technology, Zurich.

Barton, P.I. (1992). *The Modeling and Simulation of Combined Discrete/Continuous Processes*, PhD Thesis, Univ. of London.

Barton, P.I., and C.C. Pantelides (1994). Modeling of combined discrete/continuous processes. AIChE Journal **40**, 966-979.

Boussinot, F., and R. de Simone (1991). The Esterel language. Proc. of the IEEE, **79**, 1293-1304.

Branicky, M., V. Borkar, and S. Mitter (1994): A unified framework for hybrid control. Proc. IEEE Conf. on Decision and Control, 4228-4234.

Brand, K.-P., and J. Kopainsky (1988). Principles and engineering of process control with Petri nets. IEEE Trans. on Automatic Control, **33**, 138-149.

Brave, Y., and M. Heymann (1993). Control of discrete event systems modeled as hierarchical state machines. IEEE Trans. on Automatic Control, **38**, 1803-1819.

Bryant, R.E. (1986). Graph-based algorithms for Boolean function manipulation. IEEE Trans. on Computers, **35**, 677-691.

Burch, J. R., E. M. Clarke, K.L. McMillan, D. L. Dill and L. J. Hwang (1992). Symbolic model checking: 10^{20} states and beyond. Information and Computation, **98**(2), 142-170.

Caines, P.E., and W. Wang (1990). COCOLOG: A conditional observer and controller logic for finite machines. Proc. IEEE Conf. on Decision and Control, 2845-2850.

Caspi, P. (1991). Models of discrete event systems in computer science. Proc. European Control Conference, 503-511.

Chase, C., J. Serrano, P. J. Ramadge (1993). Periodicity and chaos from switched flow systems. IEEE Trans. on Automatic Control, **38**, 70-83.

Cieslak, R., C. Desclaux, A. Fawaz, and P. Varaiya (1988). Supervisory control of discrete event processes with partial observations. IEEE Trans. on Automatic Control, **33**, 249-260.

Clark, S.M., and G.S. Joglekar (1992). Features of Discrete Event Simulation. Proc. NATO ASI on Batch Process Engineering.

Clarke, E.M., O. Grumberg, and D.E. Long (1994). Model checking and abstraction. ACM Trans. on Programming Languages and Systems, **16**, 1512-1542.

David, R. (1995). Grafcet: a powerful tool for specification of logic controllers. IEEE Trans. on Control Systems Technology, **3**, 253-268.

David, R., and H. Alla (1992). *Petri nets & Grafcet*, Prentice Hall, New York.

Dyck, D.N., and P. Caines (1995). The logical control of an elevator. *IEEE Trans. on Automatic Control* **40**, 480-486.

Elmqvist, H., F.E. Cellier, and M. Otter (1993). Object-oriented modeling of hybrid systems. *Proc. European Simulation Symposium*, Delft.

Engell, S., and K. Wöllhaf (1993). Dynamic simulation of batch plants. *Proc. ESCAPE-3*, Pergamon Press, 439-444.

Engell, S., M. Fritz, C. Schulz, and K. Wöllhaf (1995). Hybrid Simulation of Flexible Batch Plants. *Preprints IFAC Symposium DYCORD+*, 123-128.

Falcione, A., and B.H. Krogh (1993). Design recovery for relay ladder logic. *IEEE Control Systems Magazine*, vol. **13**, no. 2, 90-98.

Giua, A., and F. DiCesare (1994). Petri net structural analysis for supervisory control. *IEEE Trans. on Robotics and Automation*, **10**, 185-195.

Halbwachs, N. et al. (1991). The synchronous data flow programming language Lustre. *Proc. of the IEEE*, **79**, 1305-1320.

Hanisch, H.-M. (1992). Coordination control modeling in batch production systems by means of Petri nets. *Computers in Chemical Engineering*, **16**, 1-10.

Hanisch, H.-M. (1993). Analysis of place/transition nets with timed arcs and its application to batch process control. In: *LNCS* **691**, Springer, Berlin, 282-299.

Harel, D. (1987). Statecharts: a visual formalism for complex systems. *Sci. Computer Programming*, **2**, 231-274.

Henzinger, T.A., P-H Ho, and H. Wong-Toi, (1995). A user guide to HyTech. Dept. of Computer Science, Cornell University.

Henzinger, T.A., Z. Manna, and A. Pneuli (1994). Temporal proof methodologies for timed transition systems. *Inf. and Comp.* **112**, 273-337.

Herrmann, P., and H. Krumm (1994). Compositional specification and verification of high-speed transfer protocols. In: S. T. Voung, S. T. Chanson (eds.): *Protocol Specification, Testing, and Verification XIV*. Chapman and Hall.

Hoare, C.A.R. (1985). *Communicating Sequential Processes*, Prentice-Hall International.

Holloway, L.E., and B.H. Krogh (1990). Synthesis of feedback control logic for a class of controlled Petri nets. *IEEE Trans. on Automatic Control*, **35**, 514-523.

IEC 1131 (1986). Standard for programmable controllers. Part 3: Programming languages. International Electrotechnical Commission, Technical Committee 65: Industrial Process Measurement and Control, Geneva.

Jensen, K. (1995). *Coloured Petri Nets*. Springer, Berlin.

Kowalewski, S., R. Gesthuisen, and V. Roßmann (1994): Model-based verification of batch process control software. *Proc. IEEE Conf. on Systems, Man, and Cybernetics*, San Antonio, 331-336.

Kowalewski, S., S. Engell, M. Fritz, R. Gesthuisen, G. Regner, and M. Stobbe (1995). Modular discrete modeling of batch processes by means of condition/event systems. *Workshop Analysis and Design of Event-Driven Operations in Process Systems*, Imperial College, London, 10.-11. April.

Krogh, B.H. (1993). Condition/event signal interfaces for block diagram modeling and analysis of hybrid systems. *Proc. 8th Int. Symp. on Intelligent Control Systems*, 180-185.

Krogh, B.H., and S. Kowalewski (1993). Boolean condition/event systems: computational representations and algorithms. *Preprints IFAC 12th World Congress*, Sydney, vol. 3, 327-330.

Krogh, B.H., and S. Kowalewski (1995): State feedback control of condition/event systems. *Workshop Analysis and Design of Event-Driven Operations in Process Systems*, Imperial College, London. Also in *Mathematical and Computer Modeling* **23** (1996), no. 11/12, special issue on "Recent Advances in Discrete Event Systems," 161-174.

Kurshan, R.P. (1994). *Computer-aided Verification of Coordinating Processes: The Automata-Theoretic Approach*, Princeton University Press.

Larsen, K.G., P. Pettersson, W. Yi (1995). Model-checking for real-time systems. *LNCS*, **965**, Springer, Berlin.

Le Guernic et al. (1991). Programming real-time applications with Signal. *Proc. of the IEEE*, **79**, 1321-1336.

Li, Y., and M. Wonham (1993,1994). Control of vector discrete event systems: I - the base model, II - controller synthesis. *IEEE Trans. on Automatic Control*, **38**, 1214-1227, **39**, 512-531.

Manna, Z., and A. Pnueli (1995). *Temporal Verification of Reactive Systems: Safety*. Springer, New York.

Manna, Z. et al. (1992). Step: the Stanford temporal prover. Technical report, Dept. of Comp. Science, Stanford University.

Moon, I., G. J. Powers, J. R. Burch and E. M. Clarke (1992). Automatic verification of sequential control systems using temporal logic. *AIChE Journal* **38**, 67-75.

Murata, T. (1989). Petri nets: properties, analysis, applications. *Proceedings of the IEEE*, **77**, 541-580.

Nerode, A., and W. Kohn (1993). Models for hybrid systems. In: *Hybrid Systems* (LNCS, **736**), Springer, Berlin, 317-356.

Nicollin, X., J. Sifakis and S. Yovine (1992). Compiling real-time specifications into extended automata. *IEEE Tr. on SE* **18**, 794-804.

Niinomi, T., B.H. Krogh, and J. Cury (1995). Control of hybrid systems using finite state approximations. *Hybrid Systems III* (LNCS, **1066**), Springer, New York.

Pantelides, C. C. (1995). Modeling, simulation and optimisation of hybrid processes. *Proceedings Workshop Analysis and Design of Event-Driven Operations in Process Systems*, Imperial College, London, 10.-11. April.

Preisig, H. (1993). First principle-based event-discrete dynamic system models and EDD controller synthesis. *Preprints IFAC 12th World Congress*, Sydney, vol. 5, 275-278.

Probst, S.T., G.J. Powers, D.E. Long, I. Moon (1997). Verification of a logically controlled, solids transport system using symbolic model checking. *Comp. Chem. Engg.* **21**, no. 4, 417-429

Ramadge, P. J., and W. M. Wonham (1987a). Supervisory control of a class of discrete event processes. *SIAM J. Control Optim.*, **25**, 206-230.

Ramadge, P. J., and W. M. Wonham (1987b). Modular feedback logic for discrete event systems. *SIAM J. Control Optim.*, **25**, 1202-1218.

Ramadge, P.J., and W.M. Wonham (1989). The control of discrete event systems. *Proc. IEEE*, **77**, 81-98.

Sanchez, A., G. E. Rotstein, and S. Macchietto (1995). Synthesis of procedural controllers for batch chemical processes. Preprints *IFAC Symposium DYCORD+*, 415-420.

Schürmann, T., and I. Hoffmann (1995). The entropy of 'strange' billiards inside n-simplexes. *J. Phys. A.* **28**, 5033-5039.

Sreenivas, R.S., and B.H. Krogh (1991a). On condition/svent systems with discrete state realizations. *Discrete Event Dynamic Systems: Theory and Applications,* **1**, 209-236.

Sreenivas, R.S., and B.H. Krogh (1991b). Petri net based models for condition/event systems. *Proc. American Control Conf.*, Boston, 2899-2904.

Tavernini, L. (1987). Differential automata and their discrete simulators. *Nonlinear Analysis, Theory, Methods & Applications* **11**, 665-683.

Taylor, J.H. (1995). Rigorous handling of state events in MATLAB. *Proc. IEEE Conf. on Control Applic.*, Albany.

Tittus, M., M. Fabian, and B. Lennartson (1995). Controlling and Coordinating Recipes in Batch Applications. Proc. *34th IEEE Conf. on Decision and Contro*l.

Tsitsiklis, J.N. (1989). On the control of discrete-event dynamical systems. *Mathematics of Control, Singals, and Systems*, **2**, 95-107.

Vassiliadis, V.S., R.W.H. Sargent, and C.C. Pantelides (1994). Solution of a class of multistage dynamic optimization problems. I: Problems without path constraints, II: Problems with path constraints. *Ind. Eng. Chem. Res.* **33**, 2111-2133.

Vitt, J., and J. Hooman (1995). Specification and verification of the real-time part of a steam boiler control system. *Proc. 2nd European Workshop on Real-Time and Hybrid Systems*, Grenoble, 205-208.

Wilhelm, Jr., R.E. (1985). *Programmable Controller Handbook*, Hayden, Hasbrouck Heights, New Jersey.

Yamalidou, E.C., and J. Kantor (1990). Modeling and optimal control of discrete-event chemical processes using Petri nets. *Computers in Chemical Engineering*, **15**, 503-519.

SESSION SUMMARY: IMPACT OF COMPUTER SCIENCE

Lorenz T. Biegler
Chemical Engineering Department
Carnegie Mellon University
Pittsburgh, PA 15213

Summary

As control algorithms and characteristics become more sophisticated and more difficult control problems are addressed, computational needs in process control have started to receive growing consideration over the past decade. Computational needs occur in modeling the plant for process control, including dynamic behavior and/or discrete events, and the synthesis and implementation of control algorithms. For both tasks we need to consider issues in the solution of large sets on nonlinear equations, reliable handling of optimization problems and the efficient treatment of discrete and hybrid control systems.

With the development of more efficient processors and hardware, as well as advanced architectures, there have been spectacular advances in computational performance. This has had a strong impact on advances in modeling and optimization. This can be summed up by Moore's Law — that processing speed doubles every 18 months. Also, it was noted by John Perkins at the last FOCAPD meeting that the size of process problems that are routinely handled, grows by about three orders of magnitude every decade (or about double each year). As a result, we now see the simulation of dynamic systems with up to 100,000 differential-algebraic equations and real-time optimization with up to 500,000 variables.

However, computational effort does not scale linearly with problem size and faster hardware performance does not always translate into more efficient solution of larger and more difficult process problems. To realize these, there must also be advances in computational algorithms as well. These advances are often difficult to describe and rarely have the glamour of hardware or architectural advances. Nevertheless, it is clear that more efficient computational algorithms have played a leading role in today's advanced process control and monitoring environments.

This session is devoted to exploring these advances by considering the following questions:

- What is the state of the art of control-relevant algorithms for modeling, optimization and discrete events?

- What advantages and challenges are introduced in applying these powerful algorithms to process control?

- What new areas are opened up with these advances?

The first paper, by Dr. Stephen Wright, examines the role of numerical analysis for the development of new algorithms in process control. Special emphasis is given on linear and quadratic programming algorithms that are widely used in model predictive control. Moreover, the author contrasts the characteristics of conventional active-set strategies with more recent interior-point strategies. The latter show a lot of promise for large, structured nonlinear programming problems found in process control applications. The second paper, by Profs. Marlin and Hrymak, explores the development of real-time optimization in process engineering. Here, important topics include the development of efficient optimization algorithms, the interface of these methods to process control systems and the handling of model mismatch between the optimization model and the plant. Finally, the third paper, by Prof. Engell and colleagues, focuses on the challenging area of discrete event systems. Unlike continuous systems, there are still many open research questions for discrete and hybrid control systems. Here, the authors summarize the current status of control and verification for these systems, outline recent research results from computer science, and discuss their relevance to practical problems in the control of discrete event and hybrid processes.

CPC I - CPC V: 20 YEARS OF PROCESS CONTROL RESEARCH

Manfred Morari
Automatic Control Laboratory
ETH Zentrum
CH-8092 Zürich, Switzerland

Abstract

Data on process control research, publications, and funding are analyzed and related to statistics of overall research funding for chemical engineering, industrial investments in control and instrumentation, and the number of chemical engineering graduates. The two- to threefold increase in activity over the last two decades is somewhat larger than that in chemical engineering as a whole but can be justified by the expanded role of "process operations" for the chemical process industries. Industry is the main funding source. The control of nonlinear systems, process monitoring and diagnostics, model predictive control, and identification attracted many researchers during the last decade and continues to have future potential.

Keywords

Process control, Research funding, Publications, Investment, Research consortium, Cooperative research, Instrumentation.

Introduction

There is every indication that the level of research activity in the process control area has increased dramatically over the last 20 years. This perception has led to rumors whispered at conferences and loud criticism expressed in various discussion forums. Various preconceived notions are feeding these exchanges: there are too many papers published and presented with "epsilon modifications" and too few new ideas; too many Ph.D.'s are graduated in the control field who have done research on arcane academic topics; funding in process control is inadequate and misdirected. In the following paragraphs we will first look at the facts, then try to draw conclusions, and finally attempt some perspectives on the future of process control research.

The Facts

Publications

Two indicators of publication activity are the new journals which have started during the last two decades (Table 1) and the number of process control papers presented at the Annual AIChE Meeting and the American Control Conference (ACC), formerly the Joint Automatic Control Conference (Fig. 1).

These seemingly dramatic changes are put in perspective by noting that the increase in the number of publications in the process control area *parallels* that in chemical engineering (AIChE) and in control engineering (ACC) as a whole (Fig. 2).

Table 1. *Journals Where Process Control Papers are Published Started in the Time Period 1975-1995.*

Journal	Publisher	Year started
Computers & Chemical Engineering	Pergamon	1977
Int. J. of Adaptive Control and Signal Processing	Wiley	1987
Journal of Process Control	Elsevier	1991
IEEE Control Systems Magazine	IEEE	1981
IFAC Control Engineering Practice	Pergamon	1993
International Journal of Robust and Nonlinear Control	Wiley	1991
IEEE Transactions on Control System Technology	IEEE	1993
European Journal of Control	Springer	1995

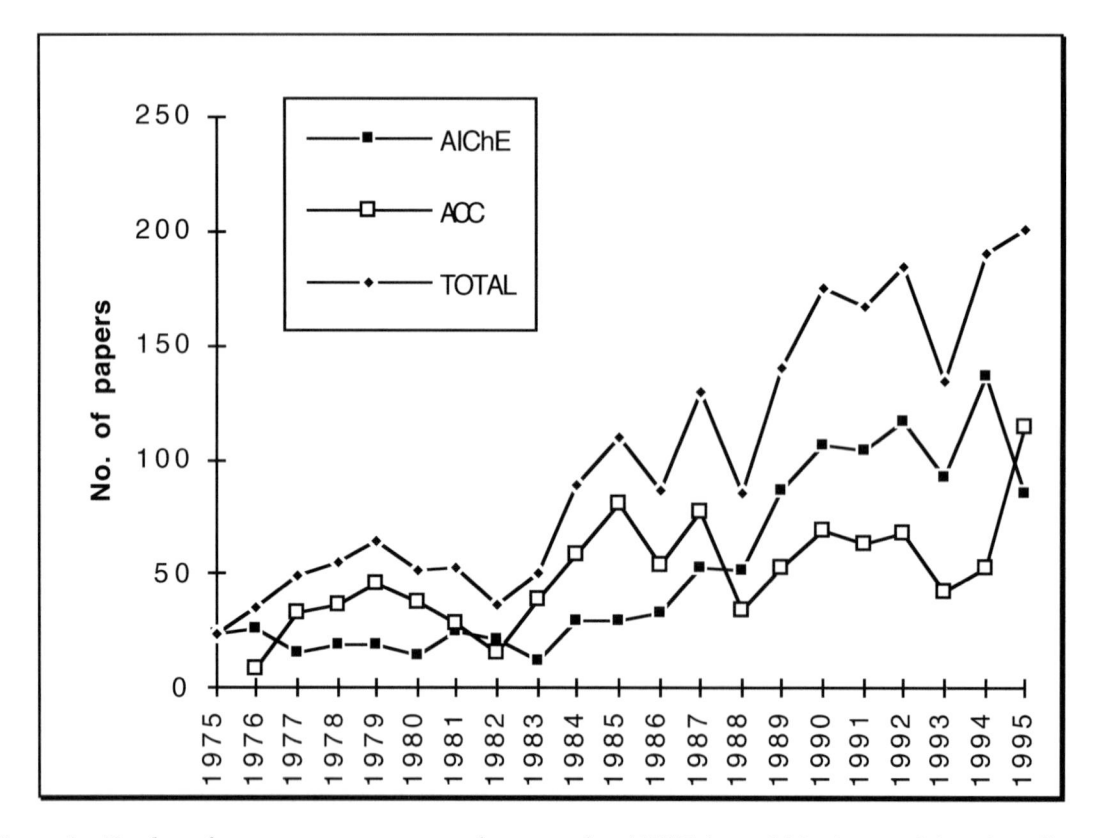

Figure 1. *Number of papers on process control presented at AIChE Annual Meetings and American Control Conferences (ACC).*

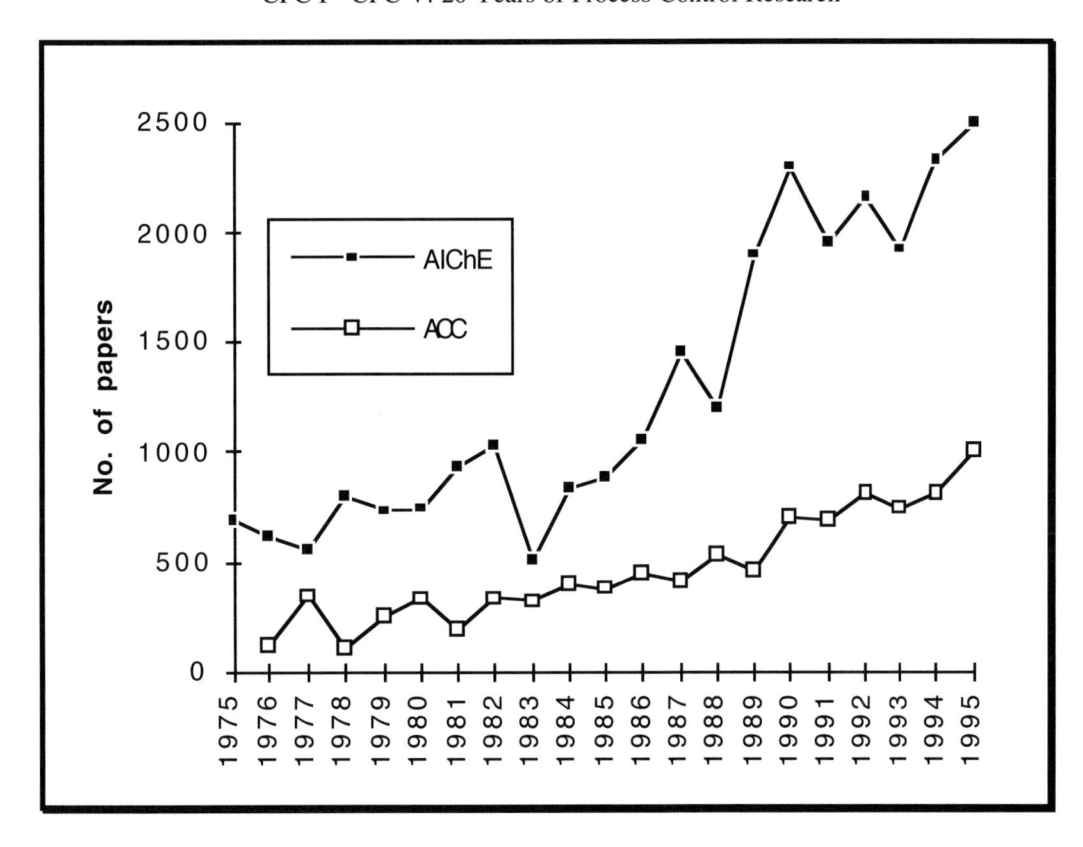

Figure 2. Total number of papers presented at AIChE Annual Meetings and American Control Conferences (ACC).

We find that this number is doubling every decade, resulting in an increase by a factor of four between 1975 and 1995. The fraction of process control papers at AIChE Annual Meetings is hovering around 4% with a slight uptrend in the last few years. At the ACCs the fraction is fluctuating wildly but the average has been around 12%.

Graduates

The number of chemical engineering Ph.D. graduates has increased by a factor of two from 1975 to 1993 (Fig. 3). Most of the research presented at conferences is done at universities. Thus it seems reasonable to assume a correlation between the number of presentations and the number of Ph.D. graduates. From this we may conclude that on the average researchers publish twice as much today as they did 20 years ago.

Because the fourfold increase in the number of *process control* presentations occurred largely during the last decade it is likely that the number of Ph.D. graduates in the control area has more than doubled during the last two decades, maybe increased by as much as a factor of three. While some of the new Ph.D.'s chose careers in academia, most of them took on positions in industry.

Investment

Between 1985 and 1996 the expenditures for instrumentation by the Hydrocarbon Processing Industry (HPI) almost doubled in the U.S. and increased by a factor of 2.6 worldwide (Fig. 4). This is remarkable when one considers that the total expenditures rose by only 35% for the U.S. and by 78% worldwide during the same time period (Fig. 5). Vendors and contractors like Honeywell show a somewhat smaller increase of 74% between 1983 and 1994.

One reviewer pointed out that the total industrial expenditures for instrumentation may not reflect the significance of the narrowly defined process control, but of the broader scope of process operations and control, including monitoring, analysis, scheduling, planning, logistics, etc.

Funding

It is probably fair to assume that most federal funding for process control is provided by NSF through the Division of Chemical and Transport Systems (formerly Chemical and Process Engineering). This funding has stayed essentially level through the 1980s and early 1990s and has increased substantially only during the last couple of years (Fig. 6)[1]. This increase is not unusual, however.

---

[1] During the last ten years an ever growing portion of the funding has gone to Presidential Young Investigators, Presidential Faculty Fellows and similar programs.

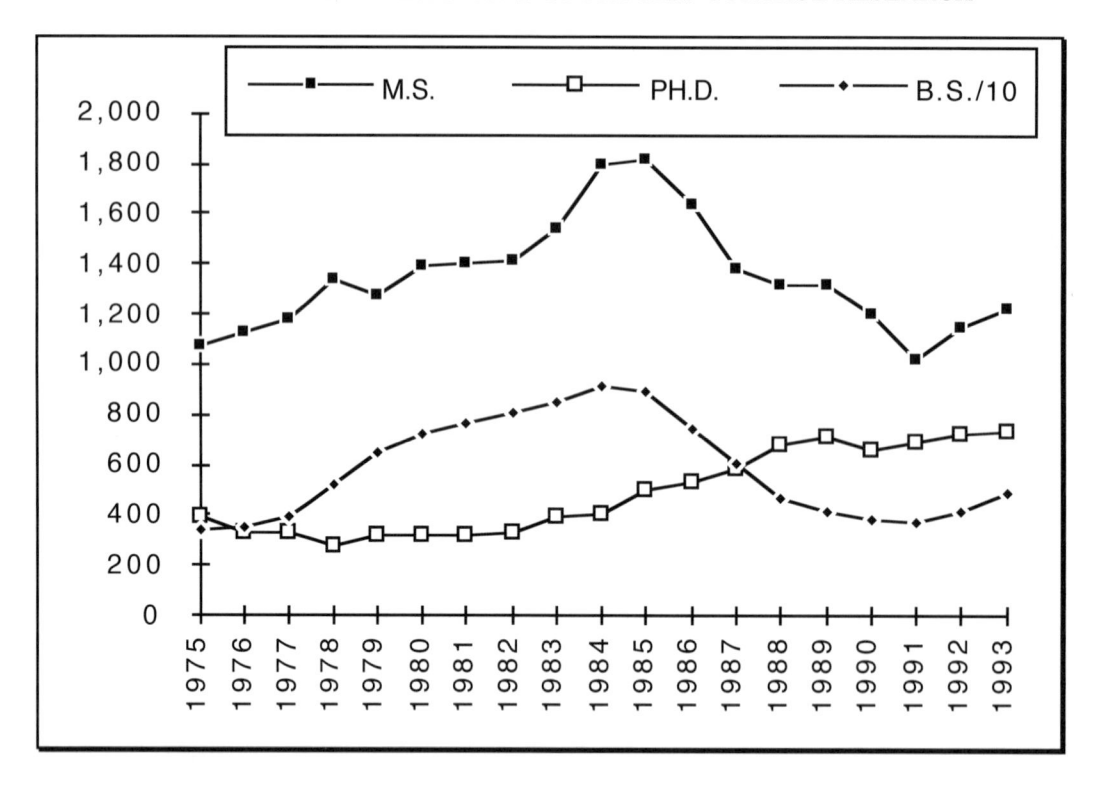

Figure 3. Chemical engineering degrees awarded in the U.S. Source: National Science Foundation, Division of Sciences Resources Study (http://www.nsf.gov/sbe/srs/s2893/start.htm).

Indeed, it is lagging behind the change in total federal funding of research and development in chemical engineering which more than doubled from 1983 to 1993 (110%) (Fig. 7). In addition it should be noted that chemical engineering fared slightly worse than engineering as a whole which increased by 114%. Total R&D support increased by 141% for chemical engineering and 136% for engineering reflecting the growing portion of funding from industry. An informal survey of nine research groups revealed that today the dominant portion of funds for process control research is provided by industry. In 1995 these nine consortia with focus on process design and control, i.e. broader than just "process control," raised close to $2.1 Mill. (about 30% more than the *total* NSF-CTS support for process control). In such consortia an average of 13 member companies pay annual fees ranging between $5,000 and $50,000.

Research Directions

Needless to say, the activity level in process control is not uniform across all subareas. The subareas tend to come in and go out of favor as time passes. These changes reflect fashion[2], new ideas and general interest which is usually expressed in the form of funding.

Topics with steadily increasing activity in the last decade were the control of nonlinear systems, identification and estimation, model predictive control, and monitoring and diagnostics (Fig. 8). While we barely find a couple of papers presented on each of these topics a decade ago, the numbers now range between ten and twenty.

In other areas interest has risen and faded within the last ten years (Fig. 9). The number of papers on adaptive control theory peaked in 1987 and has gone down steadily since. Neural networks showed a meteoric rise in 1990, but activity seems to have waned. Something similar can be said about Artificial Intelligence though the decrease is not as clear.

There are probably different explanations for these trends. Some of the tools developed originally under the label "artificial intelligence" have become parts of the routine tool kit in industry and cease to be the focus of research presentations, just as we do not see any papers on FORTRAN programming any more. Neural networks as well are now used routinely for certain classes of applications and little new theory is developed within the chemical engineering community.

The fading activity in adaptive control may actually express fading industrial interest as well as a lack of any conceptual breakthroughs in this complex and challenging area.

---

[2] According to a recent survey of the Japanese industry "popularity" and "industrial competition" were the main non-technical reason for choosing Model Predictive Control (Oshima, Ohno, Hashimoto, 1995).

Many other topics were surveyed but the numbers are too small to make any meaningful conclusions about trends. The area of design/control interactions together with control structure synthesis enjoys steady popularity but the number of papers presented each year fluctuates quite a bit. There is a positive development in the number of application papers (Fig. 10). Apart from the increasing number we find more applications ("Others") in non-traditional areas like biochemical engineering and semiconductor manufacturing.

Less encouraging is the fact that the fraction of papers where a person from industry appears as co-author shows a downward trend (Fig. 11).

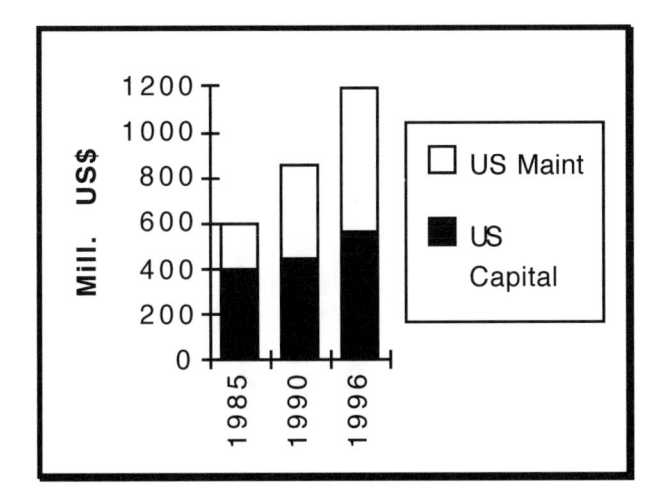

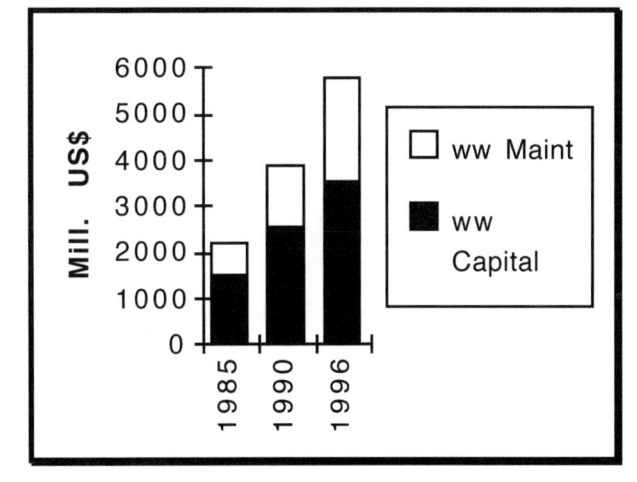

Figure 4. Capital and maintenance expenditures for instrumentation by the hydrocarbon processing industry in the U.S. and worldwide (HPI Market Data).

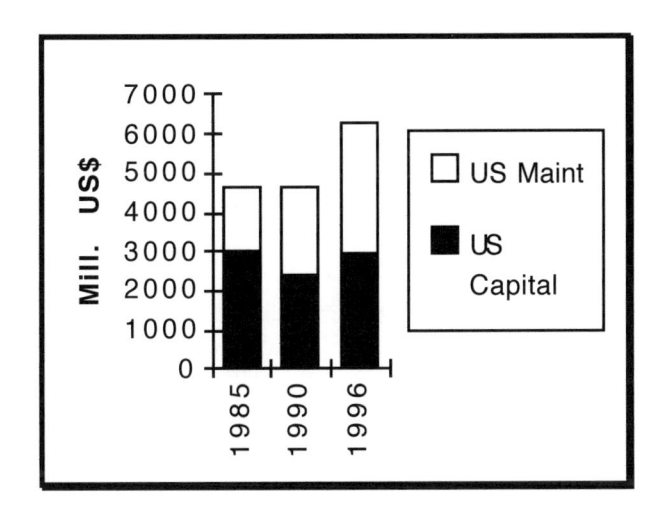

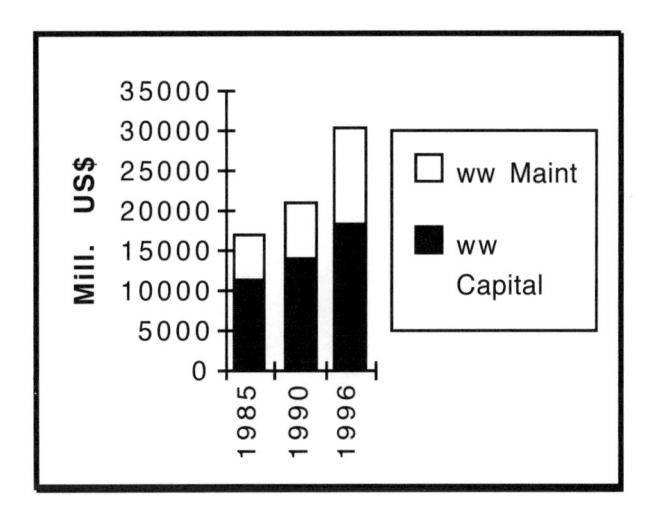

Figure 5. Total capital and maintenance expenditures by the hydrocarbon processing industry in the U.S. and worldwide (HPI Market Data).

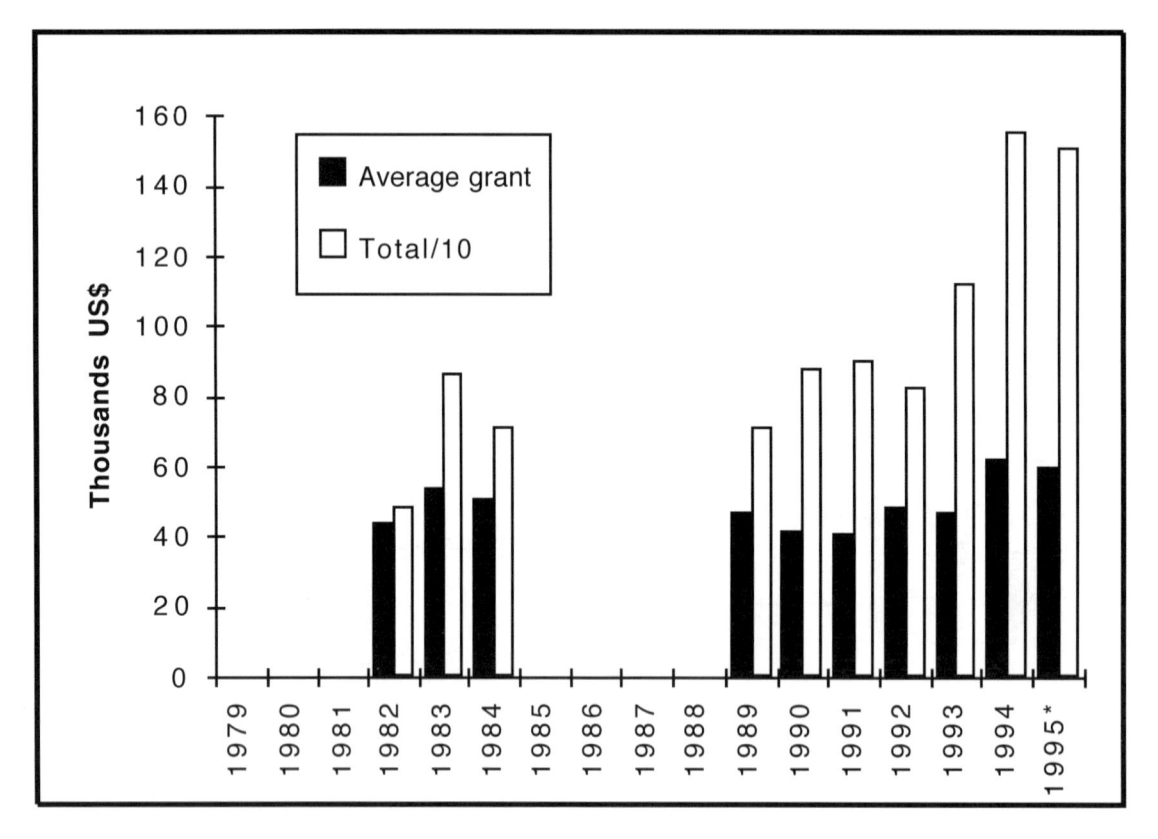

Figure 6. Funding in process control by the National Science Foundation through the Division of Chemical and Transport Systems (formerly Chemical and Process Engineering); Data through Nov. 1995; Source: NSF Annual Report and WWW.

Conclusions

The Virtual Increase in Productivity

Obviously, the fact that in chemical engineering the number of publications per researcher seems to have doubled over the last two decades does not imply that there is double the amount of publishable material. There are non-technical factors driving this process.

> "The twin devils ... are the perceived demands of funding agencies for quantifiable evidence of "productivity" and the present structure of the university reward system. The former is regrettable and damaging to scholarship, but is at least understandable; there is little excuse for the latter. I know of several chemical engineering departments where the measure of faculty effort and accomplishment rewards the *number* of refereed publications and I know of departments, where assistant professors are told the *number* of papers in refereed journals expected for tenure[3]. Is it any wonder that we are deluged by excessive publication of incomplete work, often in 'safe' areas, where results are assured?" (Denn, 1991).

The Real Increase

There are voices of concern that the real increased activity in process control research and the real increased number of Ph.D.'s is due to a lack of academic self control. This survey paper was approached exactly with that premise and statistics were *sought* to "prove" these conjectures. The facts, however, do not support these claims. First of all, the trends in process control are not greatly out of line with the overall trends in chemical engineering. Second, they are clearly not the result of some misdirected federal funding to carry out academically pleasing work which is irrelevant for industry. (The federal funding is too small for that.) The increase can only be explained by industrial interest.

This industrial interest is expressed in a variety of ways. Foremost is the increased investment in control and instrumentation by the HPI with parallel trend likely in other sectors of the Chemical Processing Industries. These investments are driven by the importance given to

---

[3] At one institution the Dean proposed the following criteria for promotion with tenure: Eight to ten archival and pre-reviewed publications; $50-100k per year of sustained external research support; supervision of two or more completed or nearly-completed Ph.D. dissertations. This proposal was rejected by the faculty. The Dean moved on to become President at another university.

the area of "operations," especially when the "chemistry" has matured. Naturally, new distributed control systems are rarely installed by Ph.D.'s. However, the accompanying research and development effort directed toward improved modeling and control strategies may be a significant factor to explain some of the increased industrial demand for Ph.D.'s in the controls area. Also the level of sophistication and complexity in the operations area has increased and with it has the skill level required for a certain task.

Obviously, industrial interest is also evident in the funding of academic research. There is, however, the - admittedly not clearly significant- downtrend in the industrial participation in scientific meetings (Fig. 11). Could this mean that for industry the product they are expecting from their financial support is the Ph.D. control engineer rather than the results of the Ph.D. research?

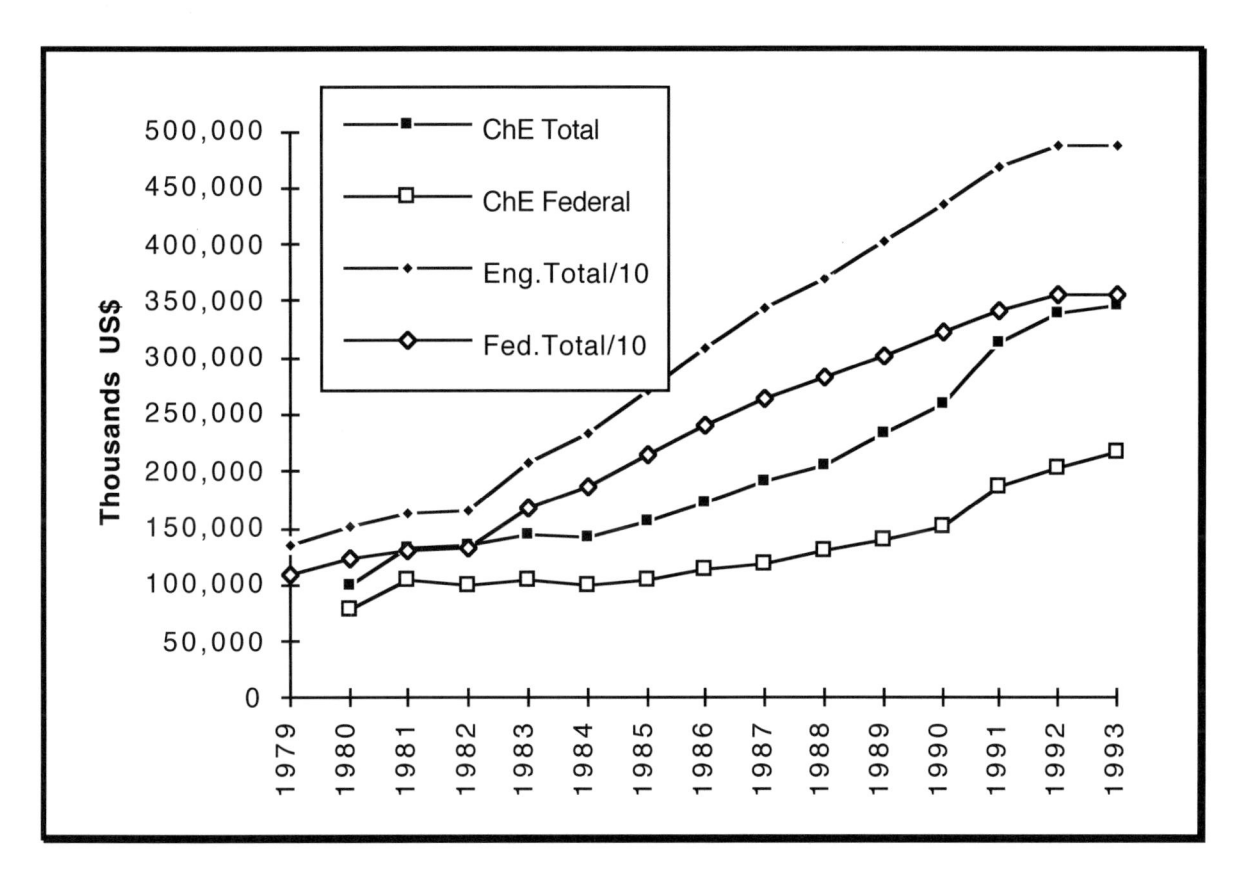

Figure 7. Research and development expenditures at U.S. universities and colleges. Source: NSF R&D Expenditures Survey (http://www.qrc.com/nsf/srs/rdexp/start.htm).

Some Future Perspectives

Production of Ph.D.'s in Process Control

It would be frivolous to make any predictions about the number of Ph.D.'s needed in the next decade. Industry tried to do that for B.S. chemical engineers in the early 1980s before the bottom dropped out of the job market with the cancellation of the synfuel activities. The result was a painful oversupply (Fig. 3) which took years to absorb. Process control Ph.D.'s trained in good graduate programs in the U.S. have a broad knowledge in applied mathematics, process dynamics, modeling and computer software. Therefore, they are flexible and are employed in a wide range of functions within and outside the process industries. They are not subjected to fluctuations in the popularity of a particular area. This is less true for graduates from European institutions where the Ph.D. is often very project oriented.

Eventually market forces will affect the supply and these forces are determined by the importance industry will place on "plant operations." Brosilow (1990) makes a very effective argument that as a mature industry which imports a large fraction of its raw materials the chemical process industries *must* concentrate on plant operations to stay competitive.

An example from the Far East serves to emphasize this point. Pohan Steel Company (POSCO) in Korea did not exist twenty years ago. Today it is the second largest steel company in the world. The technology was purchased mainly from the U.S. and Europe (mostly Germany), most of the raw materials are imported. Low labor costs and efficient manufacturing technologies give

POSCO the competitive edge. At the time POSCO was started, it was very difficult to raise capital. Nowadays investors are eager to put their money into Korea and skilled manpower is probably the only factor holding back Korean industry.

Funding

The amount of federal funding for process control research is small both in absolute and relative terms and unfortunately, it is unlikely to increase significantly in the foreseeable future. Thus its role must be more than ever to provide seed money to young promising researchers and to support fundamental and possibly high-risk research. It should not be used for joint University-Industry development projects. The money from NSF is so small that it could be easily provided by industry if the projects were of direct industrial benefit. The danger of the present large fraction of industrial funding is that academic process control research will become increasingly contract development which will lead to the demise of the current U.S. research dominance in this field. This is the situation in several European countries where overaged faculty are using neuro-fuzzy updates to theoretical developments of the 1960s in an attempt to solve industrial problems of the 1990s. The industrial funding of academic research can take a variety of forms depending if one or several universities and companies are involved.

The current standard is the single-university multiple-company scheme where the consortium members pay an annual fee either unrestricted or directed toward a particular project.

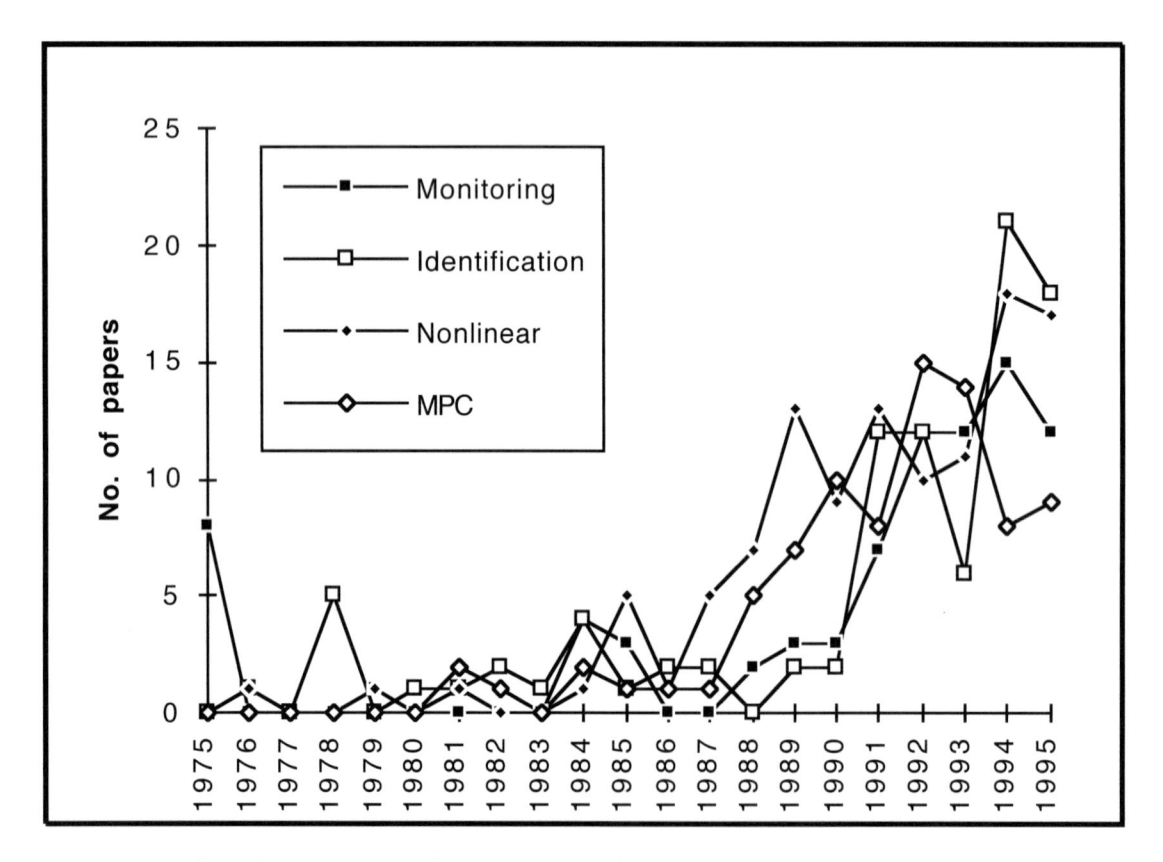

Figure 8. Number of papers presented at AIChE Annual Meetings in various subareas of process control. "Monitoring" includes fault detection and diagnostics. "Identification" includes state estimation. "Nonlinear" implies in general the use/development of a nonlinear controller for a nonlinear system. MPC = Model Predictive Control.

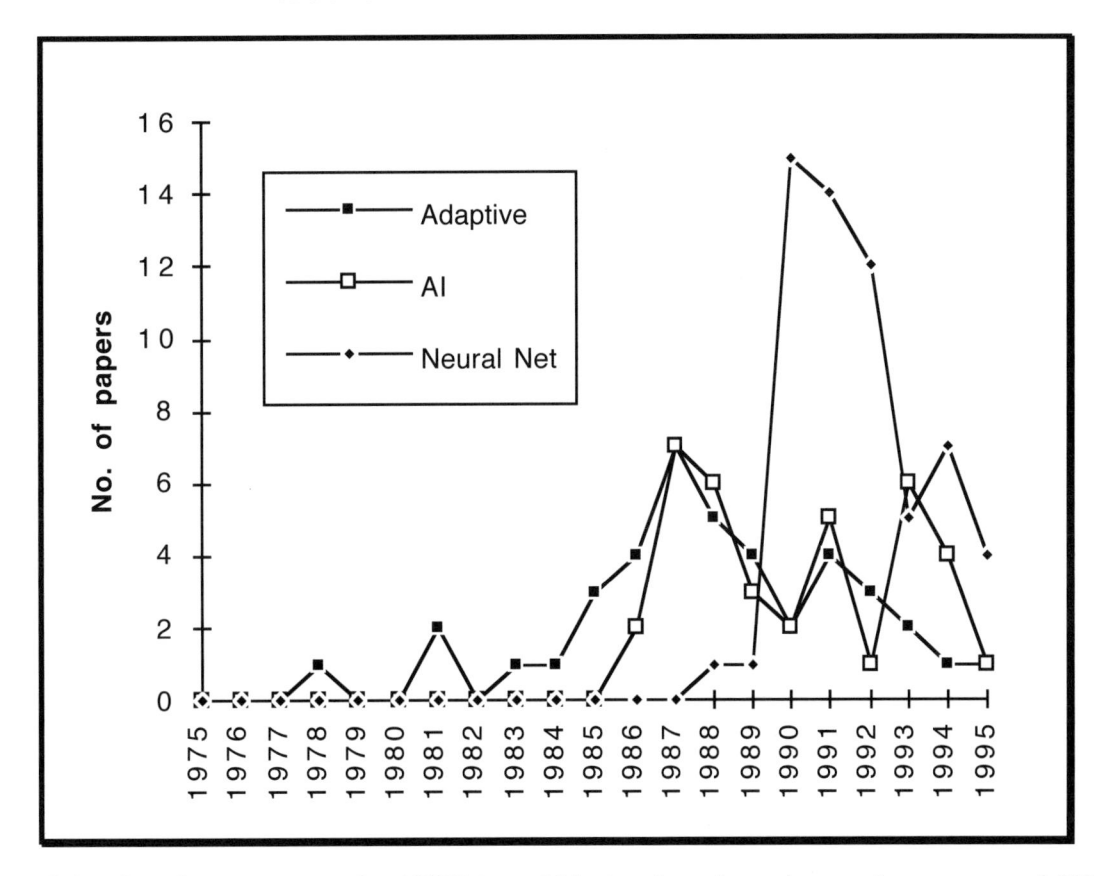

Figure 9. Number of papers presented at AIChE Annual Meetings in various subareas of process control. "Neural Net" includes fuzzy control and genetic algorithms. AI = Artificial Intelligence and knowledge-based systems.

The benefits are quick research access through meetings and reports and, in certain cases, royalty-free rights to inventions and software. In the multiple-university single-company scheme, a company would select a group of partners with the appropriate collection of skills to assist in a particular project.

For expensive development projects Brosilow (1990) proposes multi-company multi-university agreements similar to what SEMATECH is for the semiconductor industry. (He notes that of the 700 Electrical Engineering Ph.D.'s produced a year, 200 are supported by the Semiconductor Research Corporation-SRC). Projects which could be tackled in this manner are dynamic plant simulation, the development of an object-oriented framework to configure control systems including logic, and the utilization of new computer architectures for control. In all these cases the companies would want to limit their financial exposure and it would be difficult to find all the needed skills at a single university.

Another excellent industrial funding technique which has been used successfully by Shell Oil and DuPont is to offer postdoctoral fellowships and sabbatical leaves to faculty. This arrangement provides a very effective exchange of technology know-how and gives fresh Ph.D.'s a glance at real industrial problems which may affect their choice of research topics in academia.

Research Directions

Before we discuss research directions we have to define a value system. The value of process control research cannot only be judged by what fraction of the results is used directly by industry but by the insight it provides and the unifying view it offers.

The primary benefit of the research in optimal control, for example, are not the algorithms and the software but the insight into what can and cannot be achieved for a given system under precisely defined conditions. What can be accomplished by control is always quite limited and such an analysis reveals the process inherent limitations which can be affected only by changes in the design of the system itself.

Robust control, for example, cannot be dismissed as unimportant, because only very few chemical companies use MATLAB's Robust Control Toolbox and Mu Analysis and Synthesis Toolbox. Robust control has taught us how to quantify the effect of model uncertainty on stability and closed-loop performance and why multivariable systems can be much more difficult to control than single-variable systems.

When I spent a summer at Exxon in 1978 they had just generated extensive PID tuning tables based on numerical optimization of the Integral Time Absolute Error criterion for a wide range of plant models and inputs. While these tables were undoubtedly *useful* very few people would classify them as important research products. On the other hand, the derivations by Kestenbaum, Shinnar and Thau (1976) and Rivera, Skogestad and Morari (1986) provided analytic insight into the relationships between model and noise structure on one hand and controller structure and parameters on the other, a unifying view invaluable for education and further developments — despite the fact that the closed-loop performance is probably inferior than what can be obtained with the Exxon tables.

There is a sense of uneasiness in the research community, a searching for directions which may be attributable to the fact that at present there is no dominant research paradigm. There was the frequency domain / transfer function paradigm of the 1950s and the state space / linear quadratic regulator paradigm of the 1960s. In each case the new paradigm helped to obtain new insight into many old long standing problems which could not be tackled before. The H_∞ paradigm effectively bridged the gap between the classic frequency domain based design tools and the modern state space techniques.

The effect of model uncertainty could now be analyzed and understood in the multivariable context similar to what the classic work by Bode offered for single variable systems. There are many complex problems today but a paradigm is lacking. A paradigm implies that a problem is well defined, but that it is rich enough not to be "solved" quickly so that it attracts many researchers.

Model Predictive Control (MPC) may be the concept closest to a paradigm we have today. The formulation in the original papers by Richalet, et al. (1978) and Shell (Prett and Gillette, 1980) was not clean enough to form a paradigm, Internal Model Control (Garcia and Morari, 1982) was a step in this direction for unconstrained systems and the state space formulation of the constrained problem by Rawlings and Muske (1993) opened up many new research avenues.

The mathematicians are competing to reach a Kalman-like breakthrough in the nonlinear control field, but most people are pessimistic that any such general breakthrough is possible for the wide variety of nonlinear systems. Differential geometry has proven very useful for mechanical systems but has done relatively little for process control problems. We may have to get used to a time period without paradigm with a wide range of different problems and problem-specific solutions.

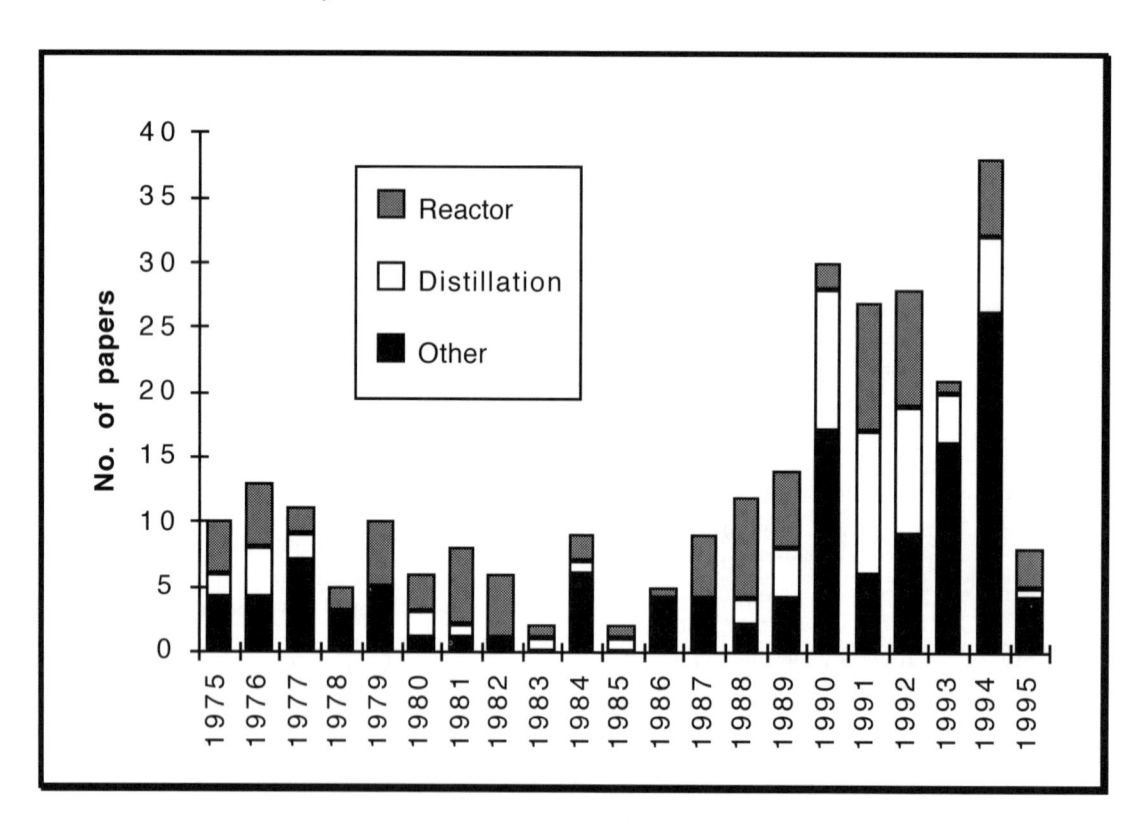

Figure 10. Number of papers on process control applications presented at AIChE Annual Meetings.

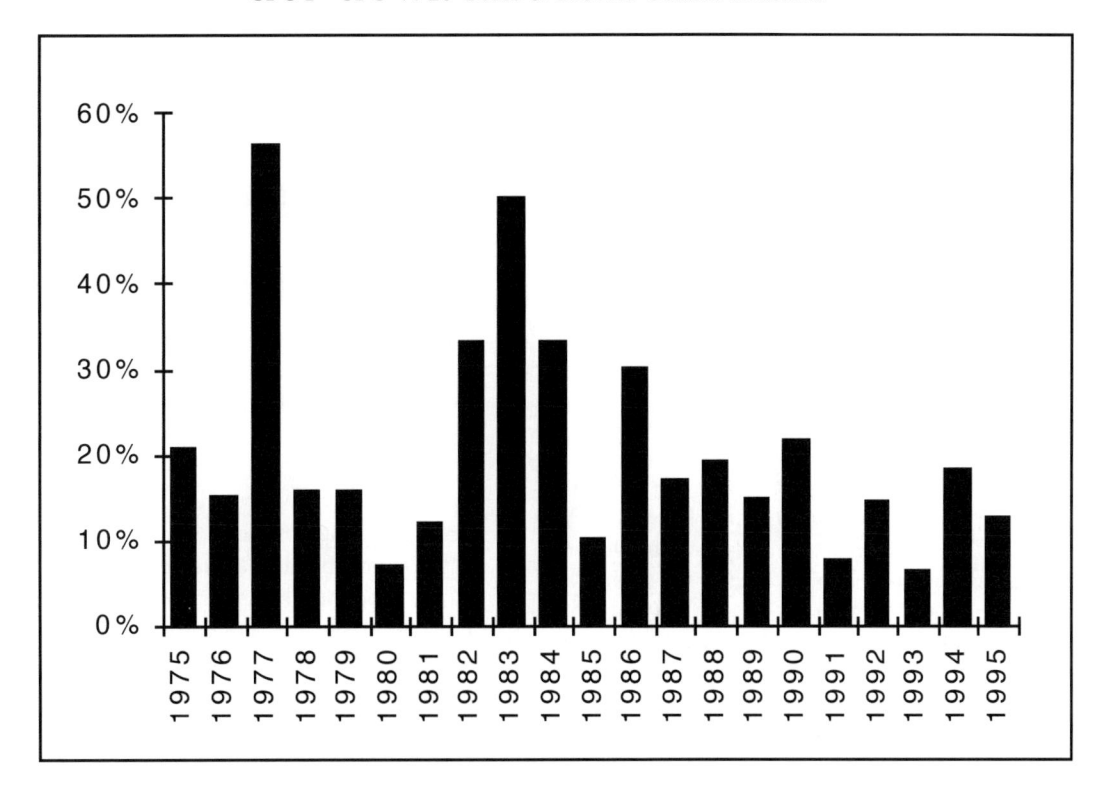

Figure 11. Fraction of process control papers with industrial co-author presented at AIChE Annual Meetings

Short of any obvious paradigm we offer the following thoughts regarding possible research directions:

- Take stock in the available theory.
- More process rather than system orientation.
- Modeling and identification.
- Monitoring and diagnostics.
- Nonlinear systems.
- Hybrid (continuous/discrete event) systems.

We will discuss these items in somewhat more detail in the following. Much theory was developed in the last decade, but little has been applied even to simulation examples. Before embarking on a new route it may be prudent to take stock and to look at what these new techniques offer. Unfortunately the funding situation and lack of an effective infrastructure makes it difficult to carry out experiments in the US.

The power of the system approach is sometimes overestimated. To apply some of the new tools from nonlinear system theory effectively it is essential to make the appropriate model simplifications which requires good knowledge of the underlying physical system. To make a control system work in practice even more knowledge of the technical details is necessary. The community is too much "tool driven" and not enough "process driven." It would benefit from focusing on a complete understanding of the dynamic behavior and control of traditional complex processes like azeotropic distillation and crystallization or new processes like semiconductor manufacturing with whatever tools are necessary to accomplish the task.

There is the obvious benefit from fundamental modeling for improved understanding and control, but the effort necessary for this task is often excessive. To fully benefit from the new nonlinear control techniques, better tools for building nonlinear models from input-output data are necessary. Robust control will remain at a conceptual level as long as we do not know how to determine model uncertainty descriptions from experimental data.

Apart from decades of work on fault tree analysis, monitoring and diagnostics have become the focus of attention only during the last few years (Fig. 8). The challenge is to find proper extensions of the traditional statistical quality control and digital signal processing techniques to multivariable systems.

With the improved computational capabilities the potential for the application of nonlinear control techniques has greatly increased. Most successful for process control applications has been Model Predictive Control. The synthesis of robust model predictive controllers is at its infancy.

All real controlled systems consist of a mixture of continuous and discrete elements. We have no effective tools for the representation, simulation, analysis, and design of such systems.

Conclusions

The number of Ph.D.'s graduating in the process control field may have tripled in the last two decades, but they appear to be absorbed by industry without significant problems. This and the millions of dollars of funding industry puts into universities for process control research is an indication of the value industry places on "process operations" to improve its competitive position.

At present there is no dominant research paradigm. To provide stimulation and discussion forums more than once every five years at CPCs it is suggested to organize invited sessions at AIChE meetings or ACCs as was the tradition 10-15 years ago. With the proper choice of topic and mix of people these sessions may have much more impact than the streamlined assembly line structure we have adopted today. An example would be the session on the Interface between Process Design and Control which Jim Douglas organized in 1982 with contributions from Foxboro, Taylor Instrument Co., Halcon SD Group, ICI, Carnegie Mellon and the U. of Wisconsin which provided a stimulus for many years of research.

I am grateful to a reviewer for pointing out, that "for the last few years, the number of Ph.D.'s hired for their thesis-based 'expertise', process control or other, is on the decline. Companies hire flexible and broadly educated individuals, who can accommodate more easily the rapidly shifting and project-oriented needs. The number of Ph.D.'s hired in traditional corporate research and development teams has been drastically reduced, with the very large majority being hired by "production-lines," "business-lines" and other groups all of which are very close to the production. This trend strengthens the significance of process control education and research but also the implications on the educational curriculum are significant and need to be discussed, since they suggest a possibly new paradigm.

An expanded scope of process control to include aspects of what is collectively referred to as Process Operations will benefit both traditional process control and the area of process operations. Researchers and practitioners should not be segmented by artificial delineations, while the methodological approaches should benefit from such integration."

Remarks Regarding the Survey of Presented Papers

Each AIChE paper was assigned to a single category. This assignment is clearly not unique and reflects personal judgment. Papers specifically on "operations" including Discrete Event Systems were not included in the count. Papers focusing solely on process dynamics and dynamic modeling (based on fundamentals) were also not counted. Poster sessions were included in the survey. The survey was not limited to presentations sponsored by CAST Group 10. In 1983 the total number of papers was smaller because of the AIChE Diamond Jubilee.

Remarks Regarding the Survey of Funding for Process Control Research

The reported funding from NSF is a lower bound. It does not reflect the money for process control from other NSF programs for jointly funded projects. Obviously, some control groups receive funds from other government agencies, but according to the author's superficial information the number of such groups is rather small.

As one reviewer pointed out, the industrial support measured by the contributions to nine academic research consortia may be misleading because the process control activity supported by these consortia is a small fraction of the total process systems engineering activities of these research labs or centers. While this is true, several other centers devoted exclusively to process control were not included in the statistics because financial data were not available.

Acknowledgments

Partial financial support from the National Science Foundation and the Department of Energy is gratefully acknowledged. I would also like to thank the following for their help in researching and processing the data on which this paper is based: Warren D. Seider (U. Penn.), Maria Burka (NSF), Chris Webb (Honeywell), Pierre Grosdidier (Setpoint); Simone De Oliviera, Thomas Güttinger, Myung Wan Han, Iftikhar Huq, Mayuresh Kothare, Carl Rhodes, Matthew Tyler (all at ETH).

References

Brosilow, C.B. (1990). Does American Industry Need Cooperative R&D in Automation and Control? *CAST Division Comm.*, **13** (1-Winter), 10-16.

Denn, M.D (1991). Distorting the Practice and Publication of Research. *AIChE J.*, **37**, 802.

Garcia, C.E. and M. Morari (1982). Internal Model Control - 1. A Unifying Review and Some New Results. *Ind. Eng. Chem. Process Des. & Dev.*, **21**, 308-323.

HPI Market Data (1985, 1990, 1996). *A Special Report on The Hydrocarbon Processing Industry Prepared Annually by Hydrocarbon Processing Magazine*. Gulf Publishing Company, Houston, Texas, USA.

Kestenbaum, A., R. Shinnar and F.E. Thau (1976), Design Concepts for Process Control. *Ind. Eng. Chem. Proc. Des. & Dev.*, **15**, 2.

Ohshima, M., H. Ohno and I. Hashimoto (1995). Model Predictive Control. Experiences in the University-Industry Joint Projects and Statistics on MPC Applications in Japan. *Int. Workshop on Predictive and Receding Horizon Control*.

Prett, D.M. and R.D. Gillette. (1980). Optimization and Constrained Multivariable Control of a Catalytic Cracking Unit. *Proc. Joint Autom. Control Conf.*

Rawlings, J.B. and K.R. Muske. (1993). The stability of constrained receding horizon control. *IEEE Trans. Auto. Cont.*, **38**(10), 1512-1516.

Richalet, J.A., A. Rault, J.L. Testud and J. Papon. (1978) Model Predictive Heuristic Control: Applications to an Industrial Process. *Automatica,* **14**, 413-428.

Rivera, D.E., S. Skogestad and M. Morari. (1986). Internal Model Control. 4. PID Controller Design. *Ind. Eng. Chem. Proc. Des. & Dev.,* **25**, 252-265.

THE FUTURE OF PROCESS CONTROL – A UK PERSPECTIVE

Roger Benson
ICI Manufacturing Technology
Wilton, UK

John Perkins
Centre for Process Systems Engineering
Imperial College
London SW7 2BY, UK

Abstract

Benefits of around US$ 220M are available from the full exploitation of process control. The authors define a "Winning Target" for the process industries and identify a significant research agenda in exploiting all the existing data, removing the need for trips and alarms and developing more aggressively the synergy between process design and control.

Keywords

Process control, Benefits, Foresight, UK innovative manufacture.

Introduction

The process industries are one of the world's most successful and important industries. The chemical industries alone had world-wide sales in 1993 of US$ 1250 billion (Ref. 1). The UK chemical industry accounts for 4% of world sales and is the country's most successful manufacturing exporter with a significant net positive balance of trade.

Growth of the industry and its present success have been built on a long standing partnership between the industry and the academic community. This partnership includes both the provision of properly trained graduates and the continuous exploitation of effective research.

Process control research has been an area of growing importance over the last thirty years. This paper has been stimulated by the suggestion that present factories, and "the factory of the future" do not require any further research in process control. There appears to be a feeling in some quarters that model based predictive control is capable of delivering all the potential benefits of process control. This paper addresses that question and suggests that the potential improvements in the future still far exceed those that have already been delivered.

These benefits will arise in two areas. First is the improvement in the operation of existing assets. Experience from companies in the UK and those reported in literature suggest that improving the process control in existing plants can bring about improvements in profit of the order of 5 to 10% of sales. Taking the lower figure would suggest that their remains a potential in the world of around US$ 60 billion. Second is the even greater potential for radical re-design and cost reduction of production assets based on exploiting the synergy between process design and process control.

Review

The process industries have some specific characteristics which do not make progress easy. The assets are expensive and difficult to modify, process modeling, especially of dynamic behavior, is regarded as difficult, the models tend to be significantly non-linear and the operations themselves are highly data rich. However, despite the large quantities of data available in modern process control systems, the amount of on-line information on critical quality variables is usually rather low. In spite of these difficulties significant progress has been made.

There is little doubt that simple three term control and the understanding provided by Bode, Ziegler, Nichols and others has made a significant contribution and is still the backbone of the operation of many existing processes. Sampled data control became important with the development of computer control and there are now very few plants and assets that operate in the world that do not have some form of computer control.

Model based predictive control has made rapid progress in terms of industrial acceptance in the past decade. There are currently at least five suppliers offering proven robust packages. There are now sufficient proven demonstrators around the world for even the most cynical industrialist to believe in the potential advantages and the evidence suggests that the technology will play a significant beneficial role. Model based predictive control is proving particularly suitable for continuous processes involving gases and liquids.

Analyzing the characteristics of the process control research and development that has achieved rapid adoption by industry suggests the importance of ease of understanding, robustness of the package, and marketing based on benefits that have been achieved. Research that has appeared to be more mathematical, less easy to understand, and where less effort has been put into the market, has tended not to be adopted by the process industries. Progress has however been less dramatic in areas of industry that involve batch processes, complex fluids and solids.

Are We Winning?

The answer to this question depends on how you define winning. For the purposes of this paper we would like to define winning in the process industries as follows:

> "Process plants that are achieving on-line equipment efficiency (OEE) in an excess of 95% with operators who are regularly in contact with the customers and where no alarms are visible and no safety, health or environmental issues arise."

Against this admittedly demanding target, there is still a great way to go. In the UK there has recently been published some manufacturing performance measures for the UK process industry (UK Tec Manufacturing Council, 1995). These are displayed in Figure 1.

The statistics illustrate the wide variation in performance achieved between the average and the top 10% in the UK process sector. They further suggest that the on-line equipment efficiency being achieved is roughly 70%. For our industry the on-line equipment efficiency is defined as: OEE = the actual rate of operation over the maximum rate achieved multiplied by the days on-line over 365 multiplied by the percent of 100% quality product first pass.

	Average Performance	Top 25%	Top 10%
Delivery Reliability	88%	97.5%	99%
Ex-Stock availability	86%	98.3%	100%
New product introduction over last 5 years (new product/product range)	2%	7%	55%
Scrap Rate	5.8%	1.3%	0.4%
Manufacturing added-value per manu-facturing employee (£000)	72	90	140
Total Stock turns	10	11	19
Capacity used for changeovers	8%	2%	0.95%
Average component set-up time (minutes)	188	15	10
Average assembly set-up time (minutes)	39	6	4
Training on job (days) Existing employees New employees Off job for existing employees	4 21 2	5 30 4	10 42 5
Absenteeism	3.6%	2.2%	1%

Figure 1. UK process sector.

Depending on the nature of the process the first pass percent quality is either the actual yield achieved over the theoretical yield or the on spec product percentage made.

The OEE is a very demanding measure which is sensitive to all aspects of process control. It measures the ability to control the plant to produce the desired product, and the capacity to start or shut down the plant quickly and reliably so that the plant availability remains high.

In our vision of winning we also feature the alarms on the plant. If process plants were under perfect control they would need very few trips and alarms, and would certainly not need the large number that are presently incorporated in most processes. In fact, one could regard the number of trips and alarms as a measure of the lack of confidence of the industry in the quality of its process control.

Finally we make the point in the winning statement that the operators would be in direct contact with the final customers and that they would be the people who were actually running the business.

Against this winning target industry still has some way to go, and there are considerable challenges remaining for the process control community.

Outstanding Research Issues with Existing Assets

The authors would suggest that an over-riding issue for the community is to continue to raise awareness of the significant contribution of process control. As an illustration of the current situation, it is interesting to compare the outcomes of two exercises recently conducted in the UK which sought to anticipate future research needs. First The Chemicals Industries Association developed a research agenda for the industry (CIA, 1995). Process control did not figure in that agenda, which was primarily driven by chemistry, materials and catalysis.

On a similar time frame the UK Government stimulated an exercise in Foresight (Technology Foresight Programme, 1995). The Foresight exercise had two main phases. First, fifteen sector-based panels developed a view of the research priorities for their areas, using scenario analysis as well as extensive consultations with their expert communities. The most relevant sector panels for the process control community are those in Chemicals, Food and Drink, Materials, and Manufacturing Production and Business Processes. In a second phase, the reports of the sector panels were considered by a Steering Group who sought to extract the generic technologies which appeared to be most significant. Figure 2 is a pictorial representation of the Steering Group's conclusions. It is worth noting that the Chemicals panel identified process engineering and control as a priority area for the future, and that the Steering Group report highlighted several topics of relevance to our community as key generic technologies.

In terms of the "Winning Target" defined earlier, a number of outstanding research challenges may be identified:

1. On all processes there is a challenge associated with the enormous amount of unused data that are available from the process and the trip and alarm scenarios. The exploitation of multivariate statistical methods indicates a potential contribution comparable to that of univariate statistical process control in discrete parts manufacture. However, given the complex nature of typical process operations, involving nonlinearities, dynamics, structured uncertainties and large numbers of interacting variables, significant research challenges remain. As well as developing the most appropriate techniques based on data alone, there will be a need for hybrid methods where data and *a priori* process knowledge may be combined to enhance interpretation.

2. Progress has been slow in the area of batch control and in the repeatable and reliable start up and shut down of processes. One can argue that a continuous process is no more than a batch process with relatively long periods between start ups and shutdowns. Examining the effort that is put in other industries to the whole area of start up, shutdown, grade change etc. suggests that the process industries have been backward in this area. Figures on the average performance of the UK Process Sector (Fig. 1) indicate that up to 8% of capacity is used for changeover. This points to a significant potential for improvement. To translate that figure into monetary terms using the worldwide sales volume quoted earlier, the output capacity of the world chemical industry might be increased by of order US$ 100 billion by improving the control of start ups, shutdowns and grade changes.

3. Conversations with many industrialists suggest a growing awareness of the importance of making significant improvements in the area of measurement. The rapid advances in sensors, computational power and interpretation means that this is now becoming a real possibility. To quote an extreme case, the pharmaceutical industry is having to expend enormous time and money to satisfy the requirements of validation of its products. If it were capable of measuring continuously all potential impurities in the product on line, then ultimately there would be no requirement to validate the process since the measurement would prove that the product being made was exactly the same as the product registered. As well as contributing to the development of "smart" sensors, there is an important role for process control in quantifying the potential of new measurement techniques in terms of contributing to improved operations. Sensors is a major research priority identified by the UK Foresight study.

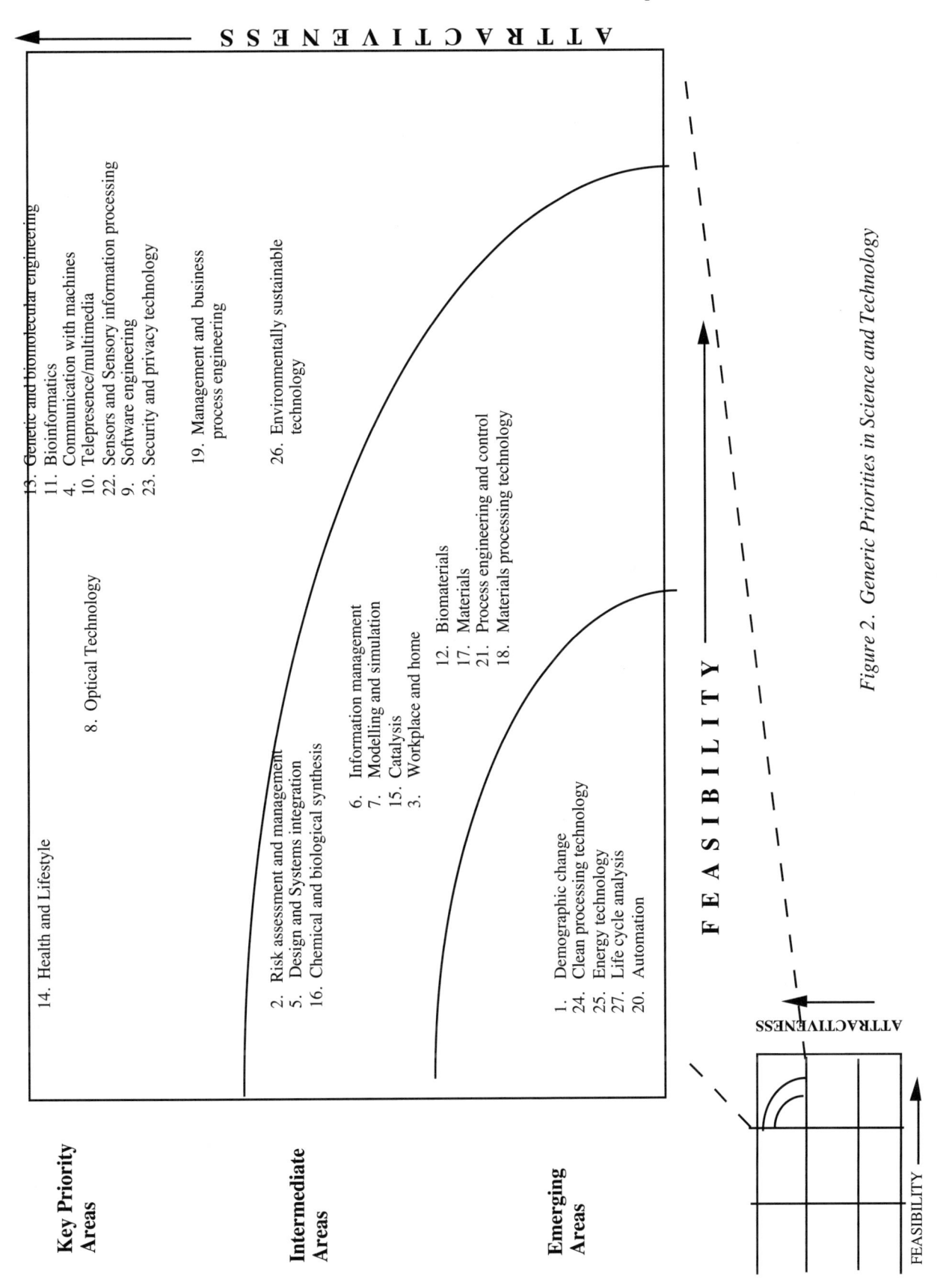

The Future of Process Control — A UK Perspective   195

Key Priority Areas

14. Health and Lifestyle

8. Optical Technology

13. Genetic and biomolecular engineering
11. Bioinformatics
4. Communication with machines
10. Telepresence/multimedia
22. Sensors and Sensory information processing
9. Software engineering
23. Security and privacy technology

19. Management and business process engineering

26. Environmentally sustainable technology

Intermediate Areas

2. Risk assessment and management
5. Design and Systems integration
16. Chemical and biological synthesis

6. Information management
7. Modelling and simulation
15. Catalysis
3. Workplace and home

12. Biomaterials
17. Materials
21. Process engineering and control
18. Materials processing technology

Emerging Areas

1. Demographic change
24. Clean processing technology
25. Energy technology
27. Life cycle analysis
20. Automation

ATTRACTIVENESS

FEASIBILITY

Figure 2. Generic Priorities in Science and Technology

Future Assets

The academic challenge in the future is to match the synergy in process design with that of process control to exploit each to the mutual benefits of the customer. The evidence suggests that this is rarely done at the moment yet it also suggests that where it is done the benefits are significant. Walsh (Walsh, 1993) shows that for an industrial chemical effluent treatment plant it is possible to reduce the capital cost of the equipment by 2/3rds by exploiting the synergy. In this application, the disturbance regime is rich, and uncertainties are large, so the benefits from an integrated approach might be expected to be significant. However, recent work by Mohideen (Mohideen, et al., 1995) suggests that worthwhile benefits are also available in milder disturbance regimes, with more modest uncertainties. Other non published evidence suggests that building pipeless plants for a certain multi product factory is capable of reducing the capital costs again by a factor of 60%. Take a current design of the process on a 3D package, and remove all the elements that are inherent in that design as a consequence of poor process control. For example all the trip and alarm equipment, the relief valve piping, the control room based on the assumption that men must be present, the access ways for people to reach the equipment when it fails etc. You will probably find as we have found that the actual investment often shrinks to the order of 30% of current values. While recognizing that this is an extreme case it does indicate the potential for capital reduction if perfect control were to be achieved.

UK industry typically spends 6% of sales on capital investment. On a world wide basis this would suggest world chemical investment at US$ 80 billion per year. If the synergy between process design and process control was to be achieved then this could possibly be cut by up to US$ 40 billion!

The UK view of the future is embodied in a research programme under the title of Innovative Manufacturing Initiative - Responsive Processing (SERC, 1994). This is based on the premise that the process industries have to respond to three major drivers:

- Responding to global markets.
- Responding to the demanding needs of the final customer.
- Responding to the rapid advances in science and technology, particularly biotechnology.

The consequences of this are becoming clear. Future processes will need to be very flexible, under tight control, low cost and need to be capable of being designed and built quickly and reliably to operate in any part of the world. This leads to the conclusion that they are likely to be smaller and distributed adjacent to the final customer's site. This implies that the plants would be small, have very little process capacity, and be totally predictable in their control in both normal and abnormal conditions.

The research challenges to meet this future vision are as much functional as technical. While chemical engineering and control engineering are seen as separate disciplines within the academic community it will remain difficult to bridge this gap.

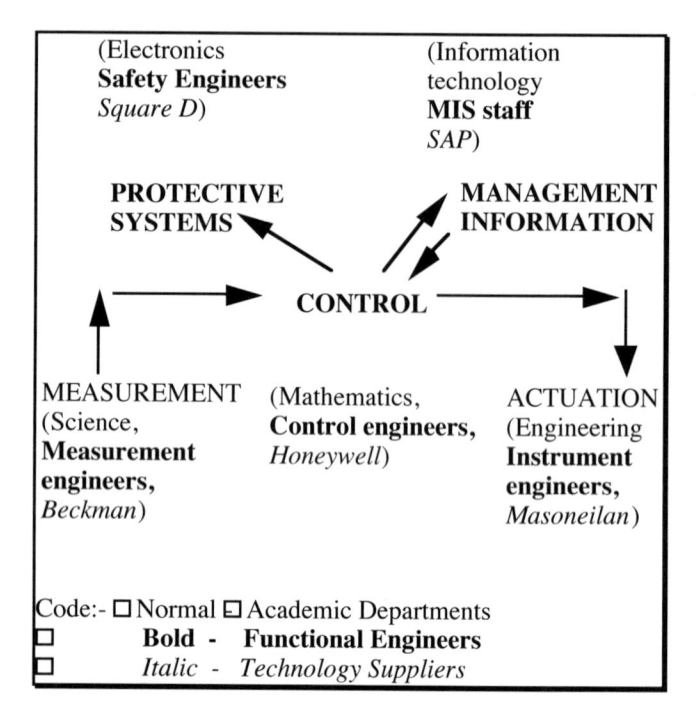

Figure 3. Functionalism in The Control Loop

Fig. 3 attempts to illustrate the high degree of functionalism built into the present control loop on academia, industry and the suppliers. The answer must lie in innovative design approaches and software. Imperial College have recently created the concept of a process design studio which will attempt to create an environment similar to that within the consumer product design area, where novel processes are rapidly developed with many alternatives to given process objectives. It is anticipated that this environment will stimulate the development of the appropriate tools to make the breakthrough required in this synergy between process design and process control. Achieving this synergy will require multi functional and multi centre ways of working.

Future Partnerships

As stated in the first paragraph, progress to date has been based on a long and successful partnership between the academic community and the industry. With the leading companies the overlap between those two partners has been significant (Fig. 4).

There are however a number of changes occurring. Industry is moving to a leaner structure and is removing many of the areas of expertise that have acted as the interface between academia and industry. Academia is under similar pressures in the UK in the form of research performance assessment and teaching performance assessment. The potential consequence of this is a gap arriving in the area of exploitation. A significant move in the last ten years has been that of the process equipment suppliers into the area providing process control solutions and in some cases taking financial and technical responsibility for the process control benefits to be delivered. This is illustrated in Figure 5 below where there are now three partners.

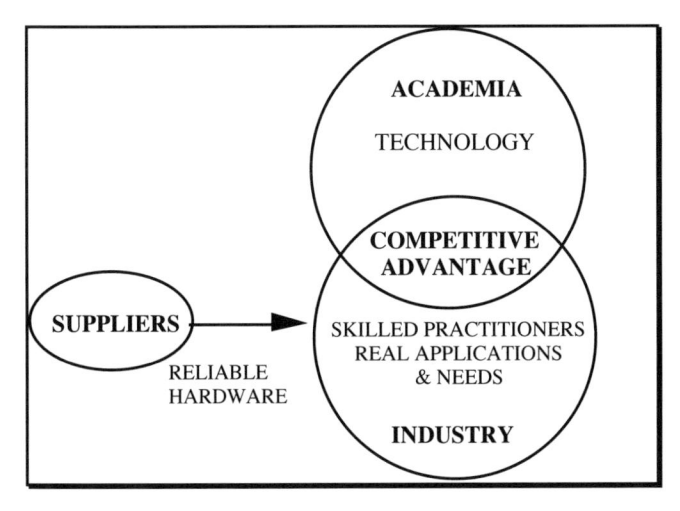

Figure 4. Academia/industry partnership.

As ever all the partners have different objectives. The role of the suppliers cannot be ignored given the nature of the changes in academia and industry. It is the authors' contention however that those companies that wish to remain at the leading edge of this technology because of the competitive advantage this brings, will wish to remain in very close harmony with the academic community. For those that are happy to be adopters rather than leaders the suppliers will become their route to new technology. Both are acceptable solutions and partnerships. They do however pose challenges. Developing the ongoing close relationship between academia and industry for the leaders will probably demand some form of clear exploitation route between the two groups. This could be in the form of university companies that rapidly take research through to demonstration and then to applications, it could be in the form of industry setting up intermediaries or of industry re-establishing centres of process control expertise. The industrial intermediary is considered unlikely because it would lead to one element of industry competing with others, hence the future favors the academic exploitation companies since these have the benefits of being independents. The establishment of these companies is a further challenge to the academic community.

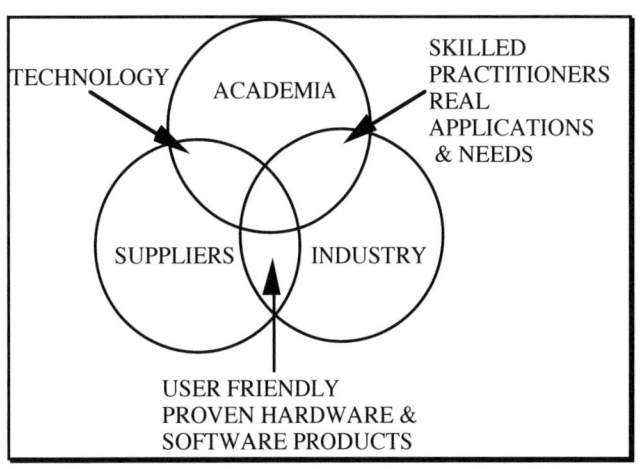

Figure 5. Academic/suppliers/industry partnership.

Summary of the Benefits of Process Control

This paper has suggested to the world chemical industry alone the benefits of improved control on existing processes are in the order of:-

- US$ 80 billion for straight control itself.
- US$ 100 billion for reducing start up, shutdown and changeover from 8% of capacity to a very low number.
- US$ 40 billion in capital by exploiting the full synergy of process design and control.

These are very large numbers indeed and should provide a spur to ongoing research. A way of making this happen is to agree a common vision of what a winning process industry will achieve.

Conclusions

Necessity is the mother of innovation. The process industries are facing that necessity both financially and in terms of license to operate. Agreeing a common vision of winning will provide a spur to substantial further research and progress in process control.

Acknowledgments

Roger Benson is a Chief Engineer within ICI and, a Visiting Professor at the Centre for Process Systems Engineering at Imperial College. He has recently spent two years within the Innovation Unit of the UK Department of Trade and Industry. John Perkins is the Director of the Centre for Process Systems Engineering at Imperial College and is a member of the Technology Foresight panel on Manufacturing, Production and Business Processes. They would wish to acknowledge that the views expressed in this paper are their own but have been much stimulated by the time and experiences they have had with the people in those respective organizations.

References

Chemical industry research priorities, CIA, 1995. ISBN 1-85897-028-8.

Innovative manufacturing — A new way of working. February 1994, SERC. ISBN 1-870669-79-7.

Manufacturing winners. UK Tec Manufacturing Council, 1995.

Mohideen, M.J., J.D. Perkins and E.N. Pistikopoulos (1995). *A framework for incorporating flexibility and controllability in process design.* Proc. ECC95, Rome, pp. 2677-2682.

Progress through partnership. Report from the Steering Group of the Technology Foresight Programme 1995. May 1995, HMSO, ISBN 0-11-430139-1.

Walsh, S.P.K. (1993). *Integrated design of chemical waste water treatment systems.* Ph.D. thesis, University of London.

SESSION SUMMARY:
PAST AND FUTURE DIRECTIONS OF PROCESS CONTROL RESEARCH

John Perkins
Centre for Process Systems Engineering
Imperial College
London SW7 2BY, UK

Summary

The tradition of assessing the current status of process control research and attempting to look forward to identify trends and opportunities goes back to the first Chemical Process Control (CPC) conference at Asilomar in 1976. Indeed, it is arguable that the conception of the whole CPC series arose from a concern about the direction that process control research was taking in the late 1960s and early 1970s, and about the perceived "large gap" between theory and practice in process control at that time. These themes have been revisited at each subsequent CPC, conference organizers taking a variety of different approaches to stimulate reflection and debate.

At this meeting, a formal conference session with papers looking back and forward across the whole spectrum of process control was the chosen option. The organizers were fortunate in being able to identify two authors who could do justice to the daunting brief of either reviewing twenty years of process control research (M. Morari), or foreseeing needs and opportunities for the next twenty years (R. Benson).

The paper by Morari presents an impressive range of facts and statistics related to the developments in process control research between 1975 and 1995. The perspective is largely North American, reflecting on the one hand the sponsorship of the meeting, but also the ready availability of data for the USA. The analysis is these data lead to some interesting conclusions, most notable of which, in the context of the history of the CPC series, is the growth in industrial support for process control research. It is fortunate that industry has demonstrated its interest in this way, given the surprisingly (at least to a European) low level of Government funding for this research area.

The predictions contained in the paper by Benson are informed by an analysis of the current performance of the (UK) process industries based on manufacturing benchmarks. It is argued that there is significant room for improvement, and that process control will have a key role in delivering these benefits. As well as identifying a number of research issues for the future, Benson argues that there are important organizational matters to be addressed both within industry and at the various interfaces between participants (academia, suppliers, industry) in the research/development/technology transfer system. The paper concludes with estimates of the possible economic benefits associated with different components of the overall strategy for the future.

Judging by the lively discussion in the session itself, and in the bar afterwards, the organizers of and contributors to this session were successful in stimulating the participants at CPC-V to think broadly about where they have come from and where they are heading.

RECENT ADVANCES IN MODEL PREDICTIVE CONTROL AND OTHER RELATED AREAS

Jay H. Lee* and Brian Cooley
Chemical Engineering Department
Auburn University
Auburn, AL 36849-5127

Abstract

Recent theoretical developments in Model Predictive Control (MPC) are reviewed. Various methods to ensure nominal stability of MPC algorithms are described and compared. MPC techniques for systems with parametric uncertainty are presented. Finally, some advances in related areas like state estimation and system identification are discussed.

Keywords

Model predictive control, Closed-loop stability, Optimal control, Robustness, Adaptive control, State estimation, Multi-variable identification.

Introduction

"In the last 10 years, ... the academic community devoted more attention to Model Based Predictive Control. Thus, it is to be noted with some satisfaction, that theory and practice supported each other quite efficiently in this methodology. Progress, which has consequently been steady, is now accelerating." (Richalet, 1993).

Model Predictive Control (MPC) is regarded by many as one of the most important developments in process control. The credit for its remarkable success is generally given to several industrialists, who outlined the basic algorithms and argued for its potential for industrial applications. Since the publication of their papers in the late 70s and early 80s, MPC has progressed rapidly, mainly in the petrochemical industries, building up an impressive track record. During the same period, MPC has also become a focal topic of academic research. Although the main idea of MPC existed in the literature since the early 60s, its potential for process control applications had been very much unexplored. The well-publicized success of MPC in the process industries fueled and in many ways revived the research in academia. The flexible formulation and ability to handle constraints in an optimal control setting attracted researchers from various disciplines. They not only analyzed and improved the industrial algorithms, but also tackled with MPC some important theoretical problems that were previously unsolvable with linear control techniques. It is only recently that we have begun to see some very powerful theoretical results derived from this line of MPC research and perhaps only the surface has been scratched.

Looking back at the papers presented at CPC-IV, one can sense that the major thrust by the process control community at the time was to interpret and generalize MPC techniques in the state-space optimal control setting. This represented a significant conceptual leap in several ways. An obvious benefit was that it generalized the existing MPC techniques considerably with respect to the model forms that can be used and types of disturbances that can be treated. More importantly, however, it established a framework to understand and evaluate various existing techniques in the optimal control context and to unify them. It also helped clear up various myths and misconceptions that existed so that the field can focus its attention on the right issues for further progress of the technology. The result is the past few years' prolific research, which established a firm, rigorous theoretical foundation for MPC.

The main purpose of this paper is to review the recent advances in MPC theory, put the results in perspective, and indicate some remaining issues. Our focus will be on the developments that occurred after CPC-IV, since many aspects of the field before then are well summarized in the papers that appeared at the conference. We will mainly concentrate

---
*To whom all correspondence should be addressed: phone (334)844-2060, fax (334)844-2063, e-mail JHL@ENG.AUBURN.EDU

on the results and issues that are pertinent to *linear* systems, as several other papers in this conference focus specifically on the issues for nonlinear systems. The only nonlinear issues that will be brought up here are those arising due to the constraints and coupling between uncertain parameters and states. As mentioned earlier, many exciting theoretical advances have come along during the past five years or so. The most notable development is that a clear framework for designing MPC achieving stability has been put forward. In addition, much effort has been devoted to the problem of designing MPC in the presence of well-defined model uncertainty.

On the practical application side, various early versions of MPC are continuing to find their applications in industry. To my knowledge, many important generalizations of MPC reported at CPC-IV are yet to find their ways into the commerical packages. This is somewhat disappointing, but can also be viewed as a healthy gap between academia and industry. In addition to the developments in the MPC theory, there has been some important headway made in related areas like state estimation and system identification, and this will undoubtedly play an important role in expanding the use of MPC in the future. For this reason, we will attempt to provide a brief summary of these developments as well.

The intended nature of this paper is that of a tutorial survey. This means that the goal is not just to collect and present various published results in a haphazard fashion, but to explain the results within a unified framework. This necessitated slight alteration or reinterpretation of some of the results in order to keep them in the context. In addition, as the paper attempts to cover advances in a very broad research area for a period of five years or so, theoretical rigor had to be compromised to a certain extent in the interest of maintaining the length of the paper at a reasonable level. References are provided for those who wish to see the original, detailed versions of the results.

With respect to the list of references, some comments are in order. MPC references provided in this paper are mainly those that are relevant to the specific topics discussed in the paper and primarily represent works that occurred (or were introduced to the process control community) during the past five years. Even so, no attempt was made to provide a complete list of the important publications on the topics that are discussed in this paper. Some important papers may have been left out simply because the author is unaware of them or he preferred the presentation of the same result in another paper. Several good review papers have been published over the past few years and most of them contain extensive lists of reference (If you want a really extensive list. It lists 391 references). Readers are strongly encouraged to compare the references in these papers to obtain a more complete and impartial list. In addition, even though every effort has been made to provide an objective review, some aspects and parts of the paper inevitably ended up otherwise. The authors' bias may have been reflected in several different ways, *e.g.*, in the list of topics covered, results cited for each topic and perspective given for each result. Once again,

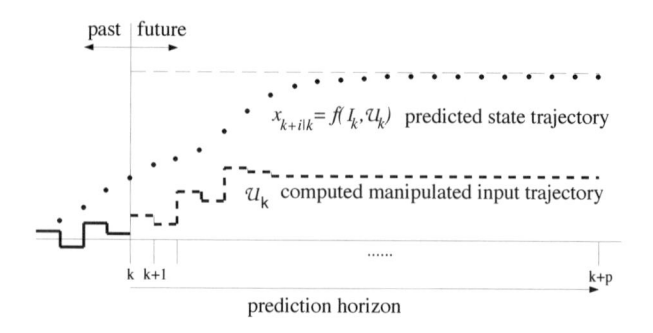

Figure 1. Basic idea of model predictive control.

we are fortunate to have several other recent review papers in this regard.

General Framework

What is MPC?

The basic concept of MPC is displayed in Fig. 1. At the k_{th} sample time, we receive information about the current state of the system. Based on the information and the system model, the dependence of the future states on the future manipulated inputs can be predicted. The future input trajectory is then determined according to some optimality criterion and implemented until the next sample time, when the entire process is repeated based on new information.

More specifically, let $\mathcal{I}_k$ represent the information vector summarizing the information available at the k_{th} sample time step. $\mathcal{I}_k$ may be as simple as the open-loop state estimate plus the current measurement (as in Dynamic Matrix Control) or substantially more complex like a vector parametrizing the probability distribution of the state vector (in the case of a nonlinear stochastic system). In addition, let $\mathcal{U}_k$ be a finite-dimensional vector parameterizing the future manipulated input trajectory. Then, the relationship between the future state trajectory $x(t)$ (either deterministic or stochastic) and $\mathcal{U}_k$ and $\mathcal{I}_k$ within a chosen predicition horizon ($k \cdot t_s < t < (k+p) \cdot t_s$ where t_s is the sample time) can be established. The following optimization is performed in order to determine $\mathcal{U}_k$:

$$\min_{\mathcal{U}_k} \phi(\mathcal{I}_k, \mathcal{U}_k) \qquad (1)$$

$$\psi(\mathcal{I}_k, \mathcal{U}_k) \leq 0 \qquad (2)$$

where functional ϕ is chosen so that the predicted future state trajectory (or its probability distribution) is shaped in a desired manner by minimizing it and vector field ψ is chosen so that the inequality represents the constraints of the system. Constraints may arise due to the limits of physical devices as well as safety and economic considerations. Even though the input trajectory for the entire prediction horizon is computed, it is implemented for only one sample interval and the procedure is repeated at the next sample time. This is referred to as receding horizon control. What results is a discrete *feedback* control law $u_k = f(\mathcal{I}_k)$, but the feedback law is not given explicitly, but implicitly through the optimization problem, which must be solved at each time step.

One noteworthy point is that the future input trajectory computed as above is a deterministic trajectory computed solely on the basis of the current information. Hence, a discrepancy exists in that *open-loop* control is assumed in the feedback control computation. This is a key feature distinguishing MPC from optimal feedback control. The optimal feedback control approach assumes an optimal relation between future input actions and future information but leads to an intractable dynamic programming problem (with a few notable exceptions like the LQG problem). The open-loop control assumption allows for a significant reduction in the complexity, but it can also lead to significantly worse performance than what is achievable with feedback control. The distinction is meaningless in the purely deterministic case or linear quadratic Gaussian case for which the separation holds, however. In the purely deterministic cases, MPC provides a computationally attractive way to solve an optimal control problem.

Problem Formulation

In this paper, we will assume that the underlying system is the following *discrete linear* system:

$$\begin{aligned} x_{k+1} &= Ax_k + Bu_k + w_k \\ y_k &= Cx_k + \nu_k \end{aligned} \quad (3)$$

x is the state vector including the disturbance states, y is the measured output vector, u is the manipulated variable vector (or the difference from the previous sample time step in the case that the model includes integrators for inputs or outputs). $\begin{bmatrix} w \\ \nu \end{bmatrix}_k$ is a zero-mean, independent vector sequence representing the unknown inputs to the system. The subscript $\{\cdot\}_k$ denotes the k_{th} sample time step. Without loss of generality, it will be assumed that the sample time is 1; hence, $t = k$ implies the k_{th} sample time step throughout this paper. (A, B) and (C, A) are assumed to be stabilizable and detectable pairs. The problem is to regulate the state vector sequence x_k (or certain linear combinations of x_k) to the origin, while satisfying the constraints

$$u_{\min} \leq u \leq u_{\max} \quad (4)$$
$$h_{\min} \leq Hx \leq h_{\max} \quad (5)$$

It is assumed that $u_{\min} < 0$, $u_{\max} > 0$, $h_{\min} < 0$, and $h_{\max} > 0$, and hence $u = 0$ and $Hx = 0$ are feasible solutions. Depending on the system, additional input constraints (*e.g.*, limit on its rate of change) may be imposed.

Comments:

1. The assumption of stabilizability and detectability of the unconstrained system is a reasonable one, but there are a few cases in which the assumption is violated and the problem is still meaningful. One such case is when the system includes integrators for constant disturbances or setpoint changes. These particular integrating states are not stabilizable and can also be undetectable if the number of measurements is insufficient. In this case, one can obtain an equivalent stabilizable / detectable system by differencing the model for two consecutive sample times. However, the system may still be unstabilizable if the number of outputs exceed the number of inputs. In addition, a system stabilizable with unconstrained input may not be globally stabilizable with constrained inputs. These issues will be discussed separately in the section "Application to the DMC Algorithm."

2. Although not stated explicitly, the problem of setpoint tracking can also be formulated as the above regulation problem by including the error terms as states.

3. In most of the paper, the system matrices A, B, C will be assumed to be constant, completely known matrices. However, in some parts of the paper, they may be assumed to be time varing and / or parametrized through a vector belonging to a compact set or a stochastic vector with known probability density evolution. Whenever the constant, fixed system matrix assumption is relaxed, the nature of these matrices will be defined.

MPC Design for Closed-Loop Stability

One of the most significant developments in the MPC theory during the past few years is that a clear framework for achieving closed-loop stability has been put forward. In this section, we review several published methods that ensure the stability of the closed-loop system. Throughout this section, it is assumed that the system matrices A, B, C are known, constant matrices.

Prototypical Algorithm

In this section, the following basic algorithm will be analyzed and modified so that closed-loop stability is guaranteed.

Let information vector $\mathcal{I}_k$ be composed of $x_{k|k}$, an estimate of x_k available from some state observer. Then, based on system model (3), one can develop the following prediction:

$$\begin{bmatrix} x_{k+1|k} \\ \vdots \\ x_{k+p|k} \end{bmatrix} = \mathcal{S}^x x_{k|k} + \mathcal{S}^u \begin{bmatrix} u_k \\ \vdots \\ u_{k+p-1} \end{bmatrix} \quad (6)$$

where $\mathcal{S}^x$ and $\mathcal{S}^u$ are constant matrices that can be computed from the system matrices. $x_{k+i|k}$ stands for prediction of x_{k+i} based on $\mathcal{I}_k$. Let the open-loop input trajectory to be computed be represented by $\mathcal{U}_k \triangleq [u_{k|k}^T, \cdots, u_{k+m-1|k}^T]^T$, thus requiring that the input trajectory returns to zero after m time steps, *i.e.*, $u_{k+i|k} = 0$ for $i \geq m$. The following optimization is solved at time k:

$$\min_{\mathcal{U}_k} \sum_{i=1}^{p} x_{k+i|k}^T Q x_{k+i|k} + \sum_{i=0}^{m-1} u_{k+i|k}^T R u_{k+i|k} \quad (7)$$

$$x_{k+i|k} \in \mathcal{X}, \; i = 1, \cdots, p \quad (8)$$

$$u_{k+i|k} \in \mathcal{V}, \; i = 0, \cdots, m-1 \quad (9)$$

Q and R are positive definite matrices.[1] $x_{k+i|k} \in \mathcal{X}$ and $u_{k+i|k} \in \mathcal{V}$ represent the hard constraints on the future states and manipulated inputs given by (4)–(5). We set

$$u_k = u_{k|k} \quad (10)$$

Then, the whole procedure (with k replaced by $k+1$) is repeated at the next sample time step. This gives an implicit feedback control law $u_k = f(x_{k|k})$ defined through the above QP.

Overview of Issues

It is well-known that the feedback control law defined through (7)–(9) and (10) is not necessarily stabilizing. In the unconstrained case for which the feedback law can be derived explicitly, the stability can be checked by examining the closed-loop eigenvalues. There are stability tests developed for the constrained case as well, but the procedure is quite complex, conservative and therefore ill-suited for practical use. What is desirable are condition(s) that directly tell the user how parameters like p and m should be selected in order to achieve stability regardless of how the rest of the tuning parameters are set. Furthermore, satisfying such conditions should not limit the achievable closed-loop performance in any significant way, and, preferably, one should be left with additional degrees-of-freedom that have consistent, predictable effects on the closed-loop performance. Hence, before one attempts to develop complex mathematical conditions for closed-loop stability, it would be of benefit to examine the fundamental reasons for undesirable closed-loop behavior like instability.

Since we are interested in asymptotic stability here, we assume that the MPC algorithm is used on an infinite time problem. In the prototypical MPC algorithm just described, there are three main discrepancies between control input computation and implementation.

1. *A finite moving horizon cost function is used to derive a control law for an infinite time problem.* This means not only that the objective used for the optimization is inappropriate for the underlying problem (because the finite horizon cost is used), but also that one is solving a different optimal control problem at each time step (because the horizon is moved).

---
[1] The assumption of positive definiteness of Q can be relaxed somewhat. Q can be assumed to be a positive semidefinite matrix chosen such that the unstable eigenspace of A does not overlap with the null-space of Q. This way, $Q^{1/2} x_k \to 0$ and $u_k \to 0$ together ensure that $x_k \to 0$. The stability results then remain valid.

2. *A finite dimensional parameterization of the input vector is assumed in the control computation.* In computing the input u_k at $t = k$, we restricted $u_{k+m} = u_{k+m+1} = \cdots = 0$. In computing the input u_{k+1} at $t = k+1$, however, we relax the restriction $u_{k+m} = 0$.

3. *Open-loop control is assumed to compute a feedback control law.*
   Notice that the input trajectory from k to $k+p$ computed at time k is a deterministic trajectory determined based on $x_{k|k}$. On the other hand, the inputs actually implemented depend on the future feedback measurements.

The third issue is not of relevance for unconstrained, linear time-invariant systems for which separation holds (see section "Extension of the Stability Proofs to the Output Feedback Case" for how the separation also extends to the constrained case in the stability context). The first and second problems can be overcome by stretching the prediction and control horizons out to infinity (i.e., by letting $p \to \infty$ and $m \to \infty$). However, this leads to an infinite dimensional programming problem (An exception is the unconstrained case where the problem can be solved explicitly using Dynamic Programming). Conversion into a finite dimensional problem is possible in principle, but one is still faced with a very large problem in general. In the next three sections, we will present several approximations that result in a fixed size problem.

Infinite Horizon MPC with Finite Number of Input Moves

Rawlings and Muske proposed an approximation to infinite horizon MPC where (7)–(9) is replaced by

$$\min_{\mathcal{U}_k} \sum_{i=1}^{\infty} x_{k+i|k}^T Q x_{k+i|k} + \sum_{i=0}^{m-1} u_{k+i|k}^T R u_{k+i|k} \quad (11)$$

$$x_{k+i|k} \in \mathcal{X}, \; i = 1, \cdots, \infty \quad (12)$$

$$u_{k+i|k} \in \mathcal{V}, \; i = 0, \cdots, m-1 \quad (13)$$

It is straightforward to show that the feedback law (10) defined by the above optimization leads to closed-loop stability regardless of the choice of tuning parameters as long as the optimization is solvable at $t = 0$. Asymptotic stability is proven using Lyanpuov's Theorem. We will discuss the proof for the *state feedback* case ($x_k = x_{k|k}$) only and defer the discussion of the output feedback case to a later section (see section "Extension of the Stability Proofs to the Output Feedback Case"). Define φ_k to be the optimal cost for (11). Since introducing an additional degree of freedom u_{k+m} in the optimization at $t = k+1$ can only lower the cost, it is clear that $\varphi_{k+1} \leq \varphi_k - x_{k+1}^T Q x_{k+1} - u_k^T R u_k$. Since $\varphi_0 < \infty$ by assumption and φ is bounded below by zero, φ_k reaches a constant as $k \to \infty$, which implies that $x_k \to 0$ and $u_k \to 0$.

There are two issues remaining to be discussed. The first is when the optimization is solvable (i.e., $\varphi_0 < \infty$) and the second is how to solve the optimization that includes infinite horizon cost and constraints. We will address the second concern first.

- **Solution of the infinite horizon problem** Let us ignore the state constraints for a moment. The restriction $u_{k+i|k} = 0$, $i \geq m$ means that the system is assumed to operate open-loop starting at time $k + m$. Hence, the cost from then on can be expressed as a quadratic weighting on the state vector at time $k + m$, i.e., $\sum_{i=m}^{\infty} x_{k+i|k}^T Q x_{k+i|k} = x_{k+m|k}^T \bar{Q} x_{k+m|k}$ where $\bar{Q}$ is the positive definite solution to the Lyapunov equation $A^T \bar{Q} A - \bar{Q} + Q = 0$. In order for the Lyapunov equation to have a positive definite solution, however, A must be stable. This, of course, means that one must constrain all the unstable modes to return to zero by time $k + m$ in order for the open-loop infinite horizon cost to be bounded.

When state constraints exist, one must find a constraint horizon $m + N$ large enough such that satisfaction of the constraints beyond the horizon is automatically implied by satisfaction of the constraints within the horizon. Rawlings and Muske used bounding arguments to show that such a choice of N always exists. Gilbert and Tan introduced the idea of *maximal output admissible set* and proposed an algorithm that can be used to compute such an N.

The above two ideas can be summarized as follows:

Let $\Gamma_s x$ and $\Gamma_u x$ be the projections of x onto the stable and unstable eigenspaces of A respectively. Then the infinite horizon problem (11)–(13) can be written as the following QP:

$$\min_{\mathcal{U}_k} \sum_{i=1}^{m-1} x_{k+i|k}^T Q x_{k+i|k} + \sum_{i=0}^{m-1} u_{k+i|k}^T R u_{k+i|k}$$
$$+ x_{k+m|k}^T (\Gamma_s)^T \bar{Q}_s \Gamma_s x_{k+m|k} \quad (14)$$
$$\Gamma_u x_{k+m|k} = 0 \quad (15)$$
$$x_{k+i|k} \in \mathcal{X}, \; i = 1, \cdots, m + N \quad (16)$$
$$u_{k+i|k} \in \mathcal{V}, \; i = 0, \cdots, m - 1 \quad (17)$$

where $\bar{Q}_s$ is the positive definite solution to Lyapunov equation $(\Gamma_s^\dagger)^T A^T \Gamma_s^T \bar{Q}_s \Gamma_s A \Gamma_s^\dagger - \bar{Q}_s + (\Gamma_s^\dagger)^T Q \Gamma_s^\dagger = 0$ with $\Gamma_s^\dagger \triangleq \Gamma_s^T (\Gamma_s \Gamma_s^T)^{-1}$.

- **Feasibility of the constraints** The other issue involves whether the equality and inequality constraints in (15)–(17) are feasible. The inequality constraints are further classified into manipulated variable constraints and state constraints. When the manipulated variable constraints are the only constraints, the optimization is always feasible. Addition of state constraints can make the optimization infeasible, but they are usually soft constraints that can be relaxed temporarily. Various ways to relax the state constraints without affecting the closed-loop stability are available in the literature. Rawlings and Muske, for instance, suggested removing the state constraints one by one starting from the initial point of the horizon until the optimization becomes feasible. Others proposed relaxing the infeasible constraints for the entire horizon by ϵ and penalizing its squared magnitude in the minimization.

The equality constraint (15), on the other hand, cannot be relaxed (if the stabilizing property is to be preserved) and can become infeasible. It is relevant, however, only for systems that include unstable modes. It is convenient to discuss the following two cases separately:

– *Unconstrained Input Case:* When the inequality constraints on the manipulated inputs are absent, the equality constraint is always feasible if (A,B) is a stabilizable pair and m is chosen to be larger than or equal to the state dimension. Depending on the system, however, significantly smaller m may be used. The stabilizability, on the other hand, is a necessary condition.

– *Constrained Input Case:* When the manipulated inputs are limited, the equality constraint may not always be satisfied. For the case where all the unstable modes are strictly on the unit circle (*e.g.*, integrating systems), there exists a result that the equality constraint is always feasible with a sufficiently large choice of m. However, the "sufficiently large" choice of m is not predefined and depends on the system and the initial state. Zheng and Morari presented a modification to the algorithm (14)–(17) so that a fixed choice of m can be used while preserving the globally stabilizing property. For systems whose unstable modes lie outside the unit circle, global stabilization is not possible with constrained inputs. Zheng and Morari presented a method to compute the upper and lower bounds of the region in the state-space within which the equality constraint is feasible (*i.e.*, if the initial state lies within, the unstable modes can be returned to zero within m time steps).

The idea of using the cost as a Lyapunov function for proving stability of MPC had existed in the literature before the work of Rawlings and Muske. For instance, Keerthi and Gilbert proved the stabilizing property of the optimal infinite horizon feedback laws for general nonlinear systems, using the same Lyapunov function approach. Since the optimal infinite horizon feedback laws are generally not computable, they proposed to approximate the infinite horizon feedback law with the finite horizon MPC with an end constraint, an approach that is described in a later section.

Infinite Horizon MPC with Fixed Linear Feedback Relation

An alternative to restricting $u_{k+m+i|k} = 0, i \geq 0$ is to assume a fixed linear feedback relation $u_{k+m+i|k} =$

$Kx_{k+m+i|k}, i \geq 0$, where K is a stabilizing gain. If K is chosen to be the optimal gain for the unconstrained problem, then the equality constraint and the terminal quadratic penalty term can be replaced with another quadratic penalty term $x_{k+m|k}^T Q^* x_{k+m|k}$ where Q^* is the positive-definite solution to the Riccati equation for the unconstrained infinite horizon LQ problem. The quadratic penalty term in this case represents the optimal *closed-loop* cost from time $k+m$ to ∞ assuming the solution starting from $t = k+m$ is *unconstrained*.

For general choice of m, in order to ensure that the inputs generated from the assumed linear feedback law do not violate the input constraints, one needs to add state constraints $Kx_{k+m+i|k} \in \mathcal{V}$, $i \geq 0$. These are hard constraints and cannot be softened. Hence, the problem of assuring the feasibility of the equality constraint (15) is replaced by that of assuring the feasibility of these state constraints. One has to use a large enough m such that these constraints are always feasible. Such a choice of m depends on many factors (e.g., size of disturbance) and may be difficult to compute *a priori*. Stability cannot be guaranteed if a smaller m is used.

Infinite Horizon MPC with End Constraint

Another way to approximate the infinite horizon MPC is to constrain the states to return to zero at the end of fixed prediction horizon. Since the states are constrained in such a manner, the infinite horizon objective is equivalent to the following finite horizon objective:

$$\min_{\mathcal{U}_k} \sum_{i=1}^{p-1} x_{k+i|k}^T Q x_{k+i|k} + \sum_{i=0}^{m-1} u_{k+i|k}^T R u_{k+i|k} \quad (18)$$

$$x_{k+p|k} = 0 \quad (19)$$
$$x_{k+i|k} \in \mathcal{X}, \; i = 1, \cdots, p-1 \quad (20)$$
$$u_{k+i|k} \in \mathcal{V}, \; i = 0, \cdots, m-1 \quad (21)$$

The closed-loop stability of the resulting feedback system can be proven similarly as before. This particular idea has existed in the literature since the 70's, but it was often presented as a special modification to finite horizon MPC instead of an approximation to infinite horizon MPC. Hence, the stability was discussed mostly in the context of an unconstrained system.

Note that it can be quite restrictive to demand that all the states return to zero after m time steps. For instance, even in the unconstrained case, in order for this to be possible with any finite choice of p, the system needs to be controllable (which is more restrictive than the stabilizability needed for the infinite horizon MPC). When the inputs are limited, the constraint can be even more restrictive and can require a very large choice of m. On the other hand, an advantage of the above formulation over the infinite horizon MPC is that the inequality constraints on the states need to be applied only within the prediction horizon. This avoids the need to compute the constraint horizon separately, a potentially cumbersome task. Furthermore, the method and the stability proof extend more readily to the nonlinear system case.

MPC with Contraction Constraint

Apart from using infinite horizon cost function, finite horizon MPC can be made stable by imposing a constraint that the state vector contracts by a prespecified factor before a new optimization begins. This idea originated from the work of Polak and Yang, who proposed the following algorithm:

$$\min_{\substack{u_{k|k}, \cdots, u_{k+p_k-1|k}, \\ p_k \in [1, p_{\max}]}} \sum_{i=1}^{p_k} x_{k+i|k}^T Q x_{k+i|k} + \sum_{i=0}^{p_k-1} u_{k+i|k}^T R u_{k+i|k} \quad (22)$$

$$\|x_{k+p_k|k}\|_P \leq \alpha \|x_{k|k}\|_P \quad (23)$$
$$x_{k+i|k} \in \mathcal{X}, \; i = 1, \cdots, p_k + N \quad (24)$$
$$u_{k+i|k} \in \mathcal{V}, \; i = 0, \cdots, p_k - 1 \quad (25)$$

where $\|\cdot\|_P$ denotes the Euclidean norm weighted by a positive-definite matrix P and $\alpha \in [0, 1)$. Note that the horizon length is also an optimization variable in the above algorithm. Unlike in the previous strategies, the entire input trajectory $u_k, \cdots, u_{k+p_k-1}$ must be implemented before the new optimization occurs (at time $k+p_k$). Assuming that the optimization is solvable every time, $\|x_{k_\ell}\|_P \leq \alpha^\ell \|x_o\|_P$ where $k_\ell = \sum_{i=0}^{\ell-1} p_i$ and the asymptotic stability ensues.

Obvious motivation for using the above algorithm is that the contraction constraint is less restrictive than the end constraint (19) in the previous algorithm. However, for systems that are not globally stabilizable, a theoretical complication is that the feasibility of the constraints at the initial time does not gurantee that the optimization will be solvable at subsequent time steps. The parameters to be chosen by the user are α and $p_{\max}$. Choosing α requires a trade-off. If α is chosen close to 1, the convergence may be slow. On the other hand, if α is chosen close to zero, convergence may be fast, but a large p may be required to satisfy the contraction constraint. This means larger computational demand and the system must be run without feedback for a longer period of time. $p_{\max}$ should be chosen large enough that the optimization is feasible, but it is difficult to say *a priori* what this should be. For *stable* systems, this problem can be somewhat alleviated. Let P be the solution to the Lyapunov equation $A^T P A - P = -S$ where $S > 0$. Then, it is easy to show that $\|x_{k+1}\|_P < \|x_k\|_P$ with $u_k = 0$. Hence, if contraction constraint (23) is not feasible, one can set $u_k = 0$ and try the optimization again at the next sample time. This way the algorithm is implementable with any choice of $p_{\max}$.

Zheng and Morari suggested the following variation to

the above algorithm:

$$\min_{\mathcal{U}_k} \sum_{i=1}^{p} x_{k+i|k}^T Q x_{k+i|k} + \sum_{i=0}^{m-1} u_{k+i|k}^T R u_{k+i|k} \quad (26)$$

$$\|x_{k+p_{\max}|k}\|_P \leq \alpha \|x_{k|k}\|_P \quad (27)$$

$$x_{k+i|k} \in \mathcal{X}, \; i = 1, \cdots, p+N \quad (28)$$

$$u_{k+i|k} \in \mathcal{V}, \; i = 0, \cdots, m-1 \quad (29)$$

In the above algorithm the horizon sizes p and m are fixed. After $u_{k|k}$ is computed, one checks whether $\|x_{k+1|k}\|_P \leq \alpha \|x_{k|k}\|_P$. If yes, the above optimization is repeated at the next sample time (with k replaced with $k+1$). If not, the following optimization is solved instead at time $k+1$:

$$\min_{\mathcal{U}_{k+1}} \sum_{i=1}^{p} x_{k+1+i|k}^T Q x_{k+1+i|k} + \sum_{i=0}^{m-1} u_{k+1+i|k}^T R u_{k+1+i|k} \quad (30)$$

$$\|x_{k+p_{\max}|k}\|_P \leq \alpha \|x_{k|k}\|_P \quad (31)$$

$$x_{k+1+i|k} \in \mathcal{X}, \; i = 1, \cdots, p+N \quad (32)$$

$$u_{k+1+i|k} \in \mathcal{V}, \; i = 0, \cdots, m-1 \quad (33)$$

The point to notice is that, unless the required contraction is achieved, one does not time-shift the contraction constraint. With the above procedure, the contraction constraint does not change until time $k+i$ where i is the smallest integer for which $\|x_{k+i|k}\|_P \leq \alpha \|x_{k|k}\|_P$.

Extension of the Stability Proofs to the Output Feedback Case

Thus far, the proof of closed-loop stability is established only for the case where $x_k = x_{k|k}$. In reality, perfect measurements of states are usually not possible. In addition, not all the states may be measurable and an observer may be required. For linear time-invariant systems, it is well-known that a stable linear state regulator can be combined with a stable linear state observer to produce a stable output feedback regulator. In addition, for the linear quadratic Gaussian problem, the optimality is achieved by combining the optimal observer (Kalman filter) with the optimal state feedback regulator (LQR). In the case of *constrained* linear time-invariant systems, however, the optimal regulator (or even a stabilizing regulator) is nonlinear and the standard result of separation no longer applies. Although the optimality result for the LQG problem no longer holds for constrained linear systems, it is possible to show that a combination of a stable linear observer and a stable nonlinear regulator satisfying a certain property indeed produce a stable closed-loop system.

The key point is that the stable MPC feedback laws we described earlier are Lipschitz continuous. Let the nonlinear feedback law generated by an MPC algorithm be represented by $u_k = f(x_{k|k})$. Assume that the feedback law is globally stabilizing. Lipschitz continuity implies that there exists a fixed constant K such that $\|f(x+\epsilon) - f(x)\| \leq K\|\epsilon\|$ for all ϵ and for all x. The closed-loop system can be written as follows:

$$\begin{aligned} x_{k+1} &= Ax_k + Bf(x_{k|k}) \\ &= Ax_k + Bf(x_k) + B[f(x_{k|k}) - f(x_k)] \end{aligned} \quad (34)$$

$\hat{x}_{k+1} = g(\hat{x}_k) \triangleq A\hat{x}_k + Bf(\hat{x}_k)$ is a dynamic system for which the origin is a globally asymptotically stable fixed point, since $u_k = f(\hat{x}_k)$ is a globally stabilizing feedback law for system $\hat{x}_{k+1} = A\hat{x}_k + Bu_k$. In addition, $g(x_k)$ is Lipschitz continuous since $f(x_k)$ is Lipschitz continuous. There exists a result in the literature that such systems are robust with respect to an exponentially decaying additive disturbance, *i.e.*, the origin remains an asymptotically stable fixed point for the perturbed system $x_{k+1} = g(x_k) + e_k$ where e_k is an exponentially converging sequence. What remains to show is that is $B[f(x_k) - f(x_{k|k})]$ is an exponentially converging sequence. This is indeed true since $(x_k - x_{k|k})$ is an exponentially converging sequence due to the assumption of a stable linear observer, and $f(x_k)$ is Lipschitz continuous. Hence, the stability results established for the state feedback case remain valid in the output feedback case as long as a stable linear observer is used.

Application to the DMC Algorithm

In this section, we apply the previously discussed stability results to the Dynamic Matrix Control (DMC) algorithm. There are two reasons for this exercise. First, the DMC or similar algorithms still represent the industry's standard, and it should be of practical value to translate the stability results into specific tuning rules for these algorithms. Second, it may expose some additional issues that need to be considered before the results are implemented into the general-purpose software. We base our discussion on the DMC algorithm for the truncated step response model. However, it should be noted that a similar algorithm can also be formulated using a state-space model. Since the finite impulse response (or, equivalently, truncated step response) assumption should be a reasonable one for all stable systems, one should be able to use the rules for similarly formulated state-space algorithms as well.

The step response model can be expressed in the following state-space form:

$$\begin{aligned} x_{k+1} &= Ax_k + B\Delta v_k + B^d \Delta d_k \\ y_k &= Cx_k + \nu_k \end{aligned} \quad (35)$$

where

$$A = \overbrace{\begin{bmatrix} 0 & I_{n_y} & 0 & \cdots & 0 \\ 0 & 0 & I_{n_y} & \cdots & 0 \\ \vdots & \vdots & \vdots & \ddots & \vdots \\ 0 & 0 & 0 & \cdots & I_{n_y} \\ 0 & 0 & 0 & \cdots & I_{n_y} \end{bmatrix}}^{N \cdot n_y};$$

$$B = \begin{bmatrix} S_1^v \\ S_2^v \\ \vdots \\ S_{N-1}^v \\ S_N^v \end{bmatrix}; \quad B^d = \begin{bmatrix} S_1^d \\ S_2^d \\ \vdots \\ S_{N-1}^d \\ S_N^d \end{bmatrix} \quad (36)$$

$$C = \overbrace{\begin{bmatrix} I_{n_y} & 0 & \cdots & 0 \end{bmatrix}}^{N \cdot n_y}$$

$$S_i^v = \begin{bmatrix} s_{1,1,i}^v & s_{1,2,i}^v & \cdots & s_{1,n_u,i}^v \\ s_{2,1,i}^v & s_{2,2,i}^v & \cdots & s_{2,n_u,i}^v \\ \vdots & \vdots & \ddots & \vdots \\ s_{n_y,1,i}^v & s_{n_y,2,i}^v & \cdots & s_{n_y,n_u,i}^v \end{bmatrix} \quad i = 1, \cdots, N$$

$$S_i^d = \begin{bmatrix} s_{1,1,i}^d & s_{1,2,i}^d & \cdots & s_{1,n_d,i}^d \\ s_{2,1,i}^d & s_{2,2,i}^d & \cdots & s_{2,n_u,i}^d \\ \vdots & \vdots & \ddots & \vdots \\ s_{n_y,1,i}^d & s_{n_y,2,i}^d & \cdots & s_{n_y,n_d,i}^d \end{bmatrix} \quad i = 1, \cdots, N \quad (37)$$

v is the manipulated variable and d is the disturbance. Δv_k and Δd_k are defined as $v_k - v_{k-1}$ and $d_k - d_{k-1}$ respectively. x_k is the system state vector and contains the unforced output response from $t = k$ to $k+p-1$, i.e., its elements are $y_k, \cdots, y_{k+N-1}$ achieved with $\Delta v_{k+i} = 0$, $\Delta d_{k+i} = 0$, and $\nu_{k+i} = 0$ for $i \geq 0$. N is the number of time steps required for all the step responses to settle. $s_{\ell,j,i}^v$ is the ith step response coefficient relating the jth manipulated input v to the ℓth output. $s_{\ell,j,i}^d$'s are the step response coefficients for disturbance d, defined in the same manner. n_v, n_d and n_y are the number of inputs, disturbances and outputs, respectively. Note that the above system model is in the same form as our basic state-space model (3) with $u_k = \Delta v_k$ and $w_k = B^d \Delta d_k$. Also note that (C, A) is a detectable pair and (A, B) is a stabilizable pair if $n_v \geq n_y$ and the steady-state gain matrix is nonsingular.

Again, the problem we consider is the regulation of y_k at the origin (Setpoint tracking can be studied in the same framework by defining the states with error terms $y_k - r_k$ instead). The following constraints are typically imposed:

$$v_{\min} \leq v_k \leq v_{\max} \quad (38)$$
$$0 > \Delta v_{\min} \leq \Delta v_k \leq \Delta v_{\max} > 0 \quad (39)$$
$$0 > y_{\min} \leq y_k \leq y_{\max} > 0 \quad (40)$$

For the step response model, it is easy to show that the infinite horizon MPC described in (11)–(13) (with $x_{k+i|k}^T Q x_{k+i|k}$ replaced with $y_{k+i|k}^T Q_y y_{k+i|k}$ where $Q_y > 0$) can be reformulated as the following finite horizon MPC:

$$\min_{\Delta v_{k|k}, \cdots, \Delta v_{k+m-1|k}} \sum_{i=1}^{m+N-1} x_{k+i|k}^T C^T Q_y C x_{k+i|k}$$
$$+ \sum_{i=0}^{m-1} \Delta v_{k+i|k}^T R \Delta v_{k+i|k}$$
$$(41)$$

$$y_{k+m+N|k} = 0 \quad (42)$$
$$y_{\min} \leq C x_{k+i|k} \leq y_{\max}, \quad i = 1, \cdots, m+N \quad (43)$$
$$v_{\min} \leq v_{k-1} + \sum_{j=0}^{i} \Delta v_{k+j|k} \leq v_{\max}, \quad i = 0, \cdots, m-1 \quad (44)$$
$$\Delta v_{\min} \leq \Delta v_{k+i|k} \leq \Delta v_{\max}, \quad i = 0, \cdots, m-1 \quad (45)$$

where $Q_y > 0$. It is also equivalent to the end constraint algorithm (18)–(21) with $p = m + N$. Even though $C^T Q_y C$ is not positive definite, the stability still ensues from the same argument as before (provided that the end constraint is feasible) since $Cx_k \to 0$ and $\Delta v_k \to 0$ together imply $x_k \to 0$. Hence, the infinite horizon MPC theory seems to provide a transparent way to tune the DMC algorithm such that the *nominal stability* is guaranteed: Set $p = m + N$ and constrain all the outputs to return to zero at the end of the prediction horizon.

However, there are at least a couple of reasons why the above algorithm may not be directly implementable in practice. First is the obvious case when $n_v < n_y$. In this case, the system is not stabilizable and the end constraint cannot be satisfied regardless of the choice of m. The second is the nature of the input constraint. For systems with all poles inside the *closed* unit disk, it was mentioned earlier that the system is globally stabilizable with magnitude-bounded input. However, notice that the input constraint $v_{\min} \leq v_k \leq v_{\max}$ is not a magnitude constraint on u_k, since $v_k = \sum_{i=1}^{k} u_i + v_0$ (Recall $u_k = \Delta v_k$). Physically, one can argue that, depending on the size of the perturbation, one or more inputs can saturate and then it is not possible to return all the outputs to zero. Again, the end constraint would be infeasible regardless of the choice of m in such a case. Since these are practically relevant cases, one should then ask what to do when such cases arise. In other words, how can we ensure that the system behaves nicely and the steady-state errors are minimized?

The simplest modification to (41)–(45) in order to address this problem appears to be the following two step algorithm:

Step 1: Calculation of steady-state target
Find $y^*_{k+m+N-1|k} = Cx^*_{k+m+N-1|k}$ by solving

$$\min_{\Delta v_{k|k}, \cdots, \Delta v_{k+m-1|k}} x^T_{k+m+N-1|k} C^T Q_y C x_{k+m+N-1|k} \quad (46)$$

$$y_{\min} \leq Cx_{k+i|k} \leq y_{\max}, \; i = 1, \cdots, m+N \quad (47)$$

$$v_{\min} \leq v_{k-1} + \sum_{j=0}^{i} \Delta v_{k+j|k} \leq v_{\max}, \; i = 0, \cdots, m-1 \quad (48)$$

$$\Delta v_{\min} \leq \Delta v_{k+i|k} \leq \Delta v_{\max}, \; i = 0, \cdots, m-1 \quad (49)$$

This provides the best achievable output at steady state.

Step 2: Dynamic optimization with steady-state target constraint
Solve the optimization below to further optimize $\Delta v_{k|k}, \cdots, \Delta v_{k+m-1|k}$.

$$\min_{\Delta v_{k|k}, \cdots, \Delta v_{k+m-1|k}} \sum_{i=1}^{m+N-2} x^T_{k+i|k} C^T Q_y C x_{k+i|k}$$
$$+ \sum_{i=0}^{m-1} \Delta v^T_{k+i|k} R \Delta v_{k+i|k} \quad (50)$$

$$y_{k+m+N-1|k} = y^*_{k+m+N-1|k} \quad (51)$$

$$y_{\min} \leq Cx_{k+i|k} \leq y_{\max}, \; i = 1, \cdots, m+N \quad (52)$$

$$v_{\min} \leq v_{k-1} + \sum_{j=0}^{i} \Delta v_{k+j|k} \leq v_{\max}, \; i = 0, \cdots, m-1 \quad (53)$$

$$\Delta v_{\min} \leq \Delta v_{k+i|k} \leq \Delta v_{\max}, \; i = 0, \cdots, m-1 \quad (54)$$

It is easy to show that $y^*_{k+m+N-1|k}$ will converge in finite time steps. Once $y^*_{k+m+N-1|k}$ converges, it can be shown that $y_k \to y^*_{k+m+N-1|k}$ and $\Delta v_k \to 0$.

It is interesting that some of the commercial algorithms already use a two step procedure similar to the above. There are some differences, however. For example, in the first step, LP is usually solved to find the best steady-state target. This optimization is done using a steady-state model only and often includes economic objectives. In the second step, the computed steady-state target is actually used as a reference value for the output throughout the prediction horizon. Input moves are also constrained such that the output reaches the computed steady-state target. This approach should work well as long as the steady-state optimization is performed at every time step and disturbances remain constant for long enough such that using the optimal achievable steady-state value as the transient reference value is justified.

Robust / Adaptive MPC

In this section, we examine the problem of designing a Model Predictive Controller for the following system:

$$\begin{aligned} x_{k+1} &= A(\theta_k)x_k + B(\theta_k)u_k + w_k \\ y_k &= Cx_k + \nu_k \end{aligned} \quad (55)$$

θ_k is a vector that parametrizes the system matrices and is not known exactly. It can be constant or time-varying. It can also be modelled as a deterministic vector that belongs to a compact set or a stochastic vector that follows a certain probability evolution. Similar constraints exist as before. Hence, the problem is to design a controller for a constrained linear system with parameter uncertainty.

To simplify the presentation, we assume that the relations between θ_k and system matrices are affine and x_k is perfectly measured (i.e., $y_k = x_k$).

Deterministic Formulation

Assume that $w_k = 0 \; \forall k$ for simplicity. Hence, the problem we treat here is bringing x to the origin starting from a nonzero initial condition in a purely deterministic setting.

Robust Control of Time-Varying Systems

Let us assume that θ_k is a time sequence and $\theta_k \in \Theta \; \forall k$ where Θ is a compact set. Furthermore, the time behavior of θ_k is assumed to be unknown, thus eliminating the possibility of adaptive control. The prediction equation for future states takes the same form as (6), but $\mathcal{S}^x$ and $\mathcal{S}^u$ are no longer constant matrices – they depend on the parameter vector sequence $\theta_k, \cdots, \theta_{k+p-1}$. Consequently, the prediction depends on the parameter vectors as well as the inputs.

The following min-max MPC formulation can be derived from game theory:

$$\min_{\mathcal{U}_k} \max_{\theta_{k|k}, \cdots, \theta_{k+p-1|k}} \sum_{i=1}^{p} x^T_{k+i|k} Q x_{k+i|k}$$
$$+ \sum_{i=0}^{m-1} u^T_{k+i|k} R u_{k+i|k} \quad (56)$$

$$x_{k+i|k} \in \mathcal{X}, \; i = 1, \cdots, p \quad (57)$$
$$u_{k+i|k} \in \mathcal{V}, \; i = 0, \cdots, m-1 \quad (58)$$
$$\theta_{k+i|k} \in \Theta, \; i = 0, \cdots, p-1 \quad (59)$$

The above computes an open-loop input trajectory that minimizes the "worst-case" cost for the chosen prediction horizon. It is to be noted that the state constraint (57) must be satisfied for all possible sequence of $\theta_{k+i|k}$. Although the optimization is nonconvex in general, in the case of a finite impulse response model or a step response model whose coefficients belong to a convex set, it can be shown to be convex. Thus the global optimum can be found with relative ease. The feedback control requires solving the above optimization at every sample time.

One can establish some stability results for the above min-max MPC by stretching the prediction horizon to ∞. The Lyapunov function argument can be used similarly as before to prove asymptotic stability, provided that the cost is bounded at the initial time step. However, this is true only for systems that are strictly stable. Even the presence of simple integrators (*e.g.*, those found in the step response model) makes the infinite horizon cost become unbounded. Zheng and Morari used an ∞-norm formulation instead and show that, in the case of a finite impulse response model, asymptotic rejection of constant disturbances can be proven for certain choices of prediction and control horizons.

An important point to realize is that the above formulation yields only a suboptimal *feedback* law, even as $p, m \to \infty$. The feedback control policy that minimizes the worst case cost of $\sum_{i=1}^{p} x_{k+i}^T Q x_{k+i+1} + u_{k+i-1}^T R u_{k+i-1}$ is determined from the following dynamic program:

$$V_{k+i}(x_{k+i}) = \min_{u_{k+i}} \max_{\theta_{k+i}} \left[x_{k+i+1}^T Q x_{k+i+1} \right.$$
$$\left. + u_{k+i}^T R u_{k+i} + V_{k+i+1}(x_{k+i+1}) \right] \quad (60)$$

where $V_{k+i}(x_{k+i})$ is the cost-to-go. Running the above recursion backward starting from $i = p-1$ with $V_{k+p} = 0$ to $i = 0$ gives the optimal feedback law $u_{k+i} = f_i(x_{k+i}), 0 \le i \le p-1$ for the finite horizon problem. However, analytical solution of the above dynamic program is not possible. Numerical computation of the optimal feedback law is also limited by the curse of dimensionality (Computational and storage requirement is proportional to q^n where n is the state dimension and q is the number of discretization points for each state). Even with today's powerful computers, there is little hope that such a calculation can be performed for a system of any reasonable size, even for a finite horizon problem. We mention this strategy only to caution against mistakenly believing that the mini-max MPC approaches the optimal feedback control in the worst-case sense as $p, m \to \infty$.

Robust / Adaptive Control of Time-Invariant Systems

Assume that θ is a *time-invariant* vector instead (*i.e.*, $\theta_k = \theta_{k+1} = \cdots$). Hence, it is an unknown constant vector that belongs to a compact set Θ. A key difference from the time-varying case is that the optimal feedback control strategy now involves adaptation of the feasible parameter set based on the measurement information. In other words, Θ needs to be updated each time as measurements can be used to reduce its size. Not updating it will clearly result in a more conservative control law. Another complication in the optimal feedback law computation is that the control input decision is intricately interconnected with the evolution of the feasible parameter set (The inputs provide the information for the future parameter set adaptation while the feasible parameter set affects the input decision). Veres and Norton studied this problem and proposed an algorithm for parameter set adaptation and worst-case predictive control calculation. However, their algorithm can be quite conservative as it does not rigorously account for the relationship between the control inputs and the feasible parameter set. In addition, some sort of bounding algorithm needs to be used for updating Θ as the shape of the feasible region can become very complex and difficult to parameterize. It is safe to say that the problem remains largely unsolved.

Several other researchers looked at this problem from a purely robust control standpoint. Campo and Morari proposed the following min-max MPC algorithm for a time-invariant system with a set membership uncertainty description:

$$\min_{\mathcal{U}_k} \max_{\theta \in \Theta} \sum_{i=1}^{p} x_{k+i|k}^T Q x_{k+i|k} + \sum_{i=0}^{m-1} u_{k+i|k}^T R u_{k+i|k} \quad (61)$$

$$x_{k+i|k} \in \mathcal{X}, \ i = 1, \cdots, p \quad (62)$$

$$u_{k+i|k} \in \mathcal{V}, \ i = 0, \cdots, m-1 \quad (63)$$

Although intuitively appealing, it turns out that the above MPC formulation does not even guarantee robust stability when implemented as a feedback law (no matter how the tuning parameters are chosen). The reason is that the above is an open-loop optimization, which does not account for the effect of future feedback measurements on the future input decisions. A rigorous account of this leads to the same dynamic programming as in (60). Zheng and Morari proposed to use Yang and Polak's contractive MPC in the context of this problem. They propose to solve (26)–(29) with the contraction constraint applied to the entire parameter set. Robust stability ensues assuming that the constraint can be satisfied every time. They show that, for an FIR system with a certain class of uncertainty descriptions, the problem can be cast as a QP of moderate size. Genceli and Nikolaou propose an ℓ_1-norm version of DMC for an FIR system with bounded coefficients. They propose to use the end condition (42) to achieve nominal stability and adjust the input weights in a special way to achieve robust stability. Since they constrain the structure of the control algorithm and adjust its parameters for robust stability, the performance can become limited. It must be remembered that all these robust MPC algorithms are inherently conservative for time-invariant systems, since adaptation of the feasible parameter set is missing.

Stochastic Formulation

Another way to set up the robust control problem is to model θ as a stochastic vector. Because θ_k is a stochastic vector sequence, x_k is also a stochastic vector sequence. An equivalent problem to the one we studied in the deterministic setting is to minimize the expected deviation of x_k from the origin. The stability here is defined as the probability density of the state vector converging to a point density at the origin starting from an arbitrary initial density (in the absence of any further disturbance).

Robust Control of Systems with Time-Indepent Parameter Uncertainty

Let θ_k be an independent, identically distributed (i.i.d.) sequence. The assumption implies that parameter variations show no temporal correlation, eliminating the need for adaptive control. w_k is also assumed to be a zero-mean i.i.d. sequence.

The following optimization computes the deterministic input trajectory that minimizes the expected deviation of the state vector from the origin for a chosen time horizon:

$$\min_{\mathcal{U}_k} E\left\{\sum_{i=1}^{p} x_{k+i}^T Q x_{k+i} \mid x_k\right\} + \sum_{i=0}^{m-1} u_{k+i|k}^T R u_{k+i|k} \quad (64)$$

$$E\{x_{k+i} \mid x_k\} \in \mathcal{X},\ i = 1, \cdots, p \quad (65)$$

$$u_{k+i|k} \in \mathcal{V},\ i = 0, \cdots, m-1 \quad (66)$$

$E\{\cdot \mid \cdot\}$ denotes the conditional expectation operator. Lee and Cooley evaluated the expectation explicitly and showed that the above is a QP of the same size as the constant, known parameter case. One can also establish some stability results by stretching the prediction horizon p to ∞. It can be shown that, in order for the infinite horizon cost to be bounded, a certain Lyanpunov-type equation should yield a positive-definite solution.

As pointed out previously, the above optimization does not describe the optimal feedback relation between x_k and u_k, even as $p, m \rightarrow \infty$. This is because the future inputs are optimized as deterministic variables in the above, while in fact they depend on the future x. The optimal feedback control policy that minimizes the cost $E\left\{\sum_{i=1}^{p} x_{k+i}^T Q x_{k+i} + u_{k+i-1}^T R u_{k+i-1}\right\}$ is again determined from the following dynamic program:

$$V_{k+i}(x_{k+i}) = \min_{u_{k+i}} E\left\{x_{k+i+1}^T Q x_{k+i+1}\right.$$
$$\left. + u_{k+i}^T R u_{k+i} + V_{k+i+1}(x_{k+i+1}) \mid x_{k+i}\right\} \quad (67)$$

Solving the above recursively from $i = p-1$ to $i = 0$ with the starting value of $V_{k+p} = 0$ gives the optimal feedback policy $u_{k+i} = f_i(x_{k+i})$ for $0 \leq i \leq p-1$. What is interesting is that, unlike in the deterministic problem, an analytical solution of the above dynamic program is possible in the unconstrained case. The optimal policy turns out to be a linear feedback law. The *infinite horizon* cost and the optimal feedback law can be computed by solving a Riccati-type equation. Although the unconstrained case may not be of practical interest, the result is important in that it enables comparison of MPC performance with that achievable through feedback. Lee and Cooley indeed found a significant difference in the closed-loop costs achieved by the two types of feedback laws. The above is one of the very few cases for which the stochastic optimal control law can be derived analytically (Another well-known case is the celebrated LQG problem).

Robust / Adaptive Control of Systems with Time-Correlated Parameter Uncertainty

Assume that θ evolves according to the following equation:

$$\theta_{k+1} = \theta_k + \varepsilon_k \quad (68)$$

ε_k is an i.i.d. Gaussian sequence. Assuming $\varepsilon_k = 0\ \forall k$ amounts to the constant parameter assumption. w_k is also assumed to be a Gaussian i.i.d. sequence with zero mean. In this case, because of the time correlation in the parameter vector, optimal feedback control requires parameter estimation. This can be done using the Kalman filter. The Kalman filter equations define the evolution of the conditional probability density of the parameter vector, *i.e.*, they express the conditional mean $\theta_{k|k}$ and covariance $P_{k|k}^\theta$ in terms of $\theta_{k-1|k-1}, P_{k-1|k-1}^\theta$ and x_k.

The computation of the optimal feedback control law requires the solution of the following dynamic program:

$$V_{k+i}(\mathcal{I}_{k+i}) = \min_{u_{k+i}} E\left\{x_{k+i+1}^T Q x_{k+i+1}\right.$$
$$\left. + u_{k+i}^T R u_{k+i} + V_{k+i+1}(\mathcal{I}_{k+i+1}) \mid \mathcal{I}_{k+i}\right\} \quad (69)$$

In the above, $\mathcal{I}_{k+i}$ represents the information vector at time $k+i$, which consists of $x_{k+i}, \theta_{k+i|k+i}$ and $P_{k+i|k+i}^\theta$. Unlike the previous case, an analytical solution is not possible. This is due to the interdependence between the parameter density evolution and the control computation. The coupling is very clear. The parameter density at time $k+i+1$ (which is a part of $\mathcal{I}_{k+i+1}$) depends on x_{k+i+1}, which in turn depends on u_{k+i} as well as $\mathcal{I}_{k+i}$. Hence, V_{k+i+1} (which is a function of $\mathcal{I}_{k+i+1}$) is related to u_{k+i} in a complex manner. On the other hand, the optimal choice of u_{k+i} clearly depends on V_{k+i+1} according to the above optimization. This coupling makes the optimization nonconvex. Numerical solution suffers from the curse of dimensionality and has been computed for only very simple systems. This is one of the few remaining theoretical challenges in adaptive control.

MPC provides a suboptimal, but computationally feasible way to compute the feedback law. The MPC formulation for the above is

$$\min_{\mathcal{U}_k} E\left\{\sum_{i=1}^{p} x_{k+i}^T Q x_{k+i}\right\} + \sum_{i=0}^{m-1} u_{k+i|k}^T R u_{k+i|k} \quad (70)$$

$$x_{k+i|k} \in \mathcal{X},\ i = 1, \cdots, p \quad (71)$$

$$u_{k+i|k} \in \mathcal{V},\ i = 0, \cdots, m-1 \quad (72)$$

Computing the feedback relation between u_k and x_k in this way is referred to as the "open-loop optimal feedback control" in the literature. In contrast to the optimal feedback calculation, the above reduces to a single stage deterministic optimization, once the expectation is evaluated. The computation can still be quite involved in general, but for FIR systems the problem simplifies considerably and can be cast as a QP of the same size as the constant, known parameter case. Chikkula *et al.* used the idea to develop a robust adaptive control strategy for a Volterra series model, for which parametric uncertainty can be very significant. A disadvantage is that the the above feedback law loses some attractive features (like active probing) that the optimal feedback law is known to possess. Hence, when the uncertainty is large, the above suboptimal feedback control law is likely to need some external probing scheme.

Moving Horizon Estimation

The success of MPC motivates us to seek a similar optimization-based technique for state estimation. Several researchers explored this possibility in various contexts. So-called "moving horizon state estimation" methods that can potentially offer better robustness and flexibility than traditional techniques like Kalman filtering, have appeared in the literature. Here, we briefly examine the main ideas behind moving horizon estimation.

Again, we assume that the underlying system is (3). The problem of state estimation is to extract the state sequence x_k from the measurement sequence $\hat{y}_k, \cdots, \hat{y}_1$ and initial estimate $x_{1|0}$. For this, the following batch optimization can be solved at each time step:

$$\min_{\substack{x^e_{1|k} \\ w_{1|k},\cdots,w_{k-1|k}}} \left\{ \sum_{i=1}^{k-1} w^T_{i|k}(P^w)^{-1} w_{i|k} \right.$$

$$+ \sum_{i=1}^{k} (\hat{y}_i - Cx_{i|k})^T (P^\nu)^{-1}(\hat{y}_i - Cx_{i|k})$$

$$\left. + (x^e_{1|k})^T (P^x_{1|0})^{-1} x^e_{1|k} \right\} \quad (73)$$

$$x_{1|k} = x_{1|0} + x^e_{1|k} \quad (74)$$
$$x_{i+1|k} = Ax_{i|k} + Bu_i + w_{i|k}, \quad i = 1, \cdots, k \quad (75)$$

Hence, one is looking for the error in the initial estimate x^e_1, state noise sequence $w_1, \cdots, w_{k-1}$, and measurement noise sequence $\nu_1, \cdots, \nu_k$, which are consistent with the observations $\hat{y}_1, \cdots, \hat{y}_k$. Since there exist infinite such sequences, these errors are weighed according to their expected sizes / correlations, and the squared sum of the weighted vectors is minimized. Once $x^e_{1|k}$ and $w_{1|k}, \cdots, w_{k-1|k}$ are determined, the state estimate $x_{k+1|k}$ can be constructed using the model equation (75).

It is well documented in the literature that this procedure yields the same state estimate as the nonsteady-state Kalman filter started with the initial estimate of $x_{1|0}$, initial covariance matrix of $P^x_{1|0}$, state noise covariance matrix of P^w and measurement noise covariance matrix of P^ν. The main advantage of the Kalman filter is that it is recursive and hence the required computational effort is minor. On the other hand, the advantage of the above optimization formulation is that it affords more flexibility. For example, the following constraints can be added to the estimation:

$$w_{\min} \leq w_{i|k} \leq w_{\max}, i = 1, \cdots, k-1 \quad (76)$$
$$\nu_{\min} \leq (\hat{y}_i - Cx_{i|k}) \leq \nu_{\max}, i = 1, \cdots, k \quad (77)$$
$$x_{\min} \leq x_{i|k} \leq x_{\max}, i = 1, \cdots, k+m \quad (78)$$

The above constraints come from prior knowledge and can speed up the convergence and improve the robustness if incorporated into the estimation. Another advantage of the optimization approach is that it extends straightforwardly to nonlinear systems, at least on a conceptual level. Robertson *et al.* give a probabilistic interpretation of the above constrained batch state estimation formulation. They interpret the Gaussian density with magnitude bounds as a truncated Gaussian density and show that the above calculation provides the joint maximum *a posteriori* estimate of the state trajectory.

A problem with (73)–(75) is that the size of optimization grows linearly with the number of measurements. The formulation can be modified to employ a fixed-size moving window in which the number of measurements that we base our estimate on (and hence the size of optimization) remains constant. In moving horizon state estimation, (73)–(75) is modified as follows:

$$\min_{\substack{x^e_{k-m+1|k} \\ w_{k-m+1|k},\cdots,w_{k-1|k}}} \left\{ (x^e_{k-m+1|k})^T (P^x_{k-m+1|k-m})^{-1} x^e_{k-m+1|k} \right.$$

$$+ \sum_{i=1}^{m} (\hat{y}_{k-m+i} - Cx_{k-m+i|k})^T (P^\nu)^{-1}(\hat{y}_{k-m+i} - Cx_{k-m+i|k})$$

$$\left. + \sum_{i=1}^{m-1} w^T_{k-m+i|k}(P^w)^{-1} w_{k-m+i|k} \right\} \quad (79)$$

$$x_{k-m+1|k} = x_{k-m+1|k-m} + x^e_{k-m+1|k} \quad (80)$$
$$x_{k-m+i+1|k} = Ax_{k-m+i|k} + Bu_{k-m+i} + w_{k-m+i|k}, \quad (81)$$
$$i = 1, \cdots, m$$

The above optimization with constraints (76)–(78) is a fixed-size QP. One additional issue that arises in the moving horizon formulation is the propagation of the initial estimate $x_{k-m+1|k-m}$ and weighting matrix $P^x_{k-m+1|k-m}$. A simple option is to use the Kalman filter equation to obtain $x_{k-m+1|k-m}$ and $P^x_{k-m+1|k-m}$ from $x_{k-m|k-m-1}$ and $P^x_{k-m|k-m-1}$. One can also propagate the constraints for x_{k-m}, w_{k-m} and ν_{k-m} as a feasible region for the initial state x_{k-m+1} in the next optimization. Nominal stability of such designed moving horizon estimator follows trivially from the stability of the Kalman filter. Compared to the Kalman filter, however, the constraints imposed within the window can improve the estimation and help avoid unrealistic estimates. On the other hand, from a probablistic standpoint, the distribution of the state vector is not truncated-Gaussian in general and the nice property of providing the-joint maximum *a posteriori* estimate is lost.

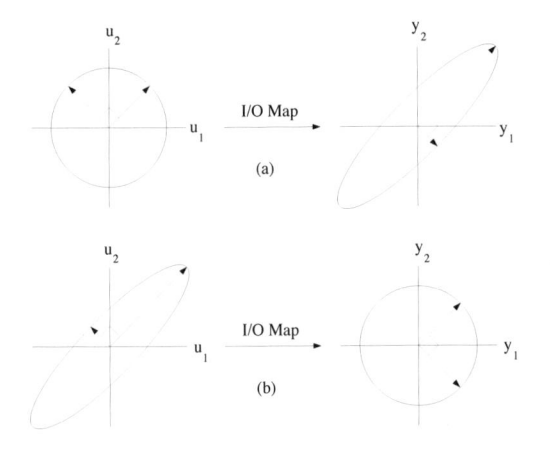

Figure 2. Mapping between the input and output spaces for 2 x 2 systems: (a) even excitation of the input space, (b) even excitation of the output space.

The moving horizon estimation is particularly appealing in the context of nonlinear estimation, for which there does not exist a well-established technique like the Kalman filter. Robertson et al. show that the moving horizon estimator is equivalent to the iterated extended Kalman filter with $m = 1$ and similar to the iterated linear filter-smoother with $m = 2$. The method affords the flexibility of choosing a compromise between performance and computational requirements.

Multivariable System Identification

There are two main issues in system identification. The first has to do with generating the input output data that contain relevant system information for closed-loop control. The second concerns the development of numerically stable and robust identification algorithms that do not require excessive prior knowledge. Both issues have been addressed to a satisfactory degree for single variable systems. Even though most identification methods developed in the single variable system context can be readily applied / extended to multivariable systems, the results have not always been satisfactory, especially when the multivariable system is highly interactive. For the last five years or so, much progress has been made in understanding the issues intrinsic to multivariable system identification, and many techniques that are specifically tailored for multivariable systems have appeared in the literature. Here we review some progresses that are relevant to MPC application.

Data Generation for Interactive Systems

For the simplicity of argument, let us consider for a moment a 2×2 static system (or a steady-state gain matrix). Fig. 2 shows the mapping between the input and output spaces. As displayed in Fig. 2a, exciting two orthogonal input directions evenly (or all input directions evenly for that matter) can produce output data that are highly collinear because of the gain difference. The degree of collinearity depends on the condition number of the matrix. The collinearity in the output data implies a poor signal-to-noise ratio for the low gain direction and makes the correct estimation of the gain for this direction very difficult. In addition to the measurement errors and disturbances, other factors like nonlinearity can adversely affect the estimation. While hardly visible in the open-loop response, the low gain direction can be just as important when the loops are closed. For instance, a model with a wrong sign of gain for this direction results in closed-loop instability under integral control.

The problem is more complex for dynamic systems for which the gain dependence on the input direction changes with frequency, but the fundamental idea remains the same. In the identification of dynamic models, another potential problem associated with the collinearity in the output data is its impact on the distribution of model bias. The practical evidence of the above-stated problem can be seen in the literature. For example, Andersen et al. present a case study involving a high-purity distillation column for which significant problems were found with conventional SISO or MISO type identification.

A perferable way to generate the data is to excite the inputs proportionally more in the low gain direction so that the gain difference can be offset (as shown in Fig. 2b). However, this requires prior knowledge of the weak direction and an estimate of the gain magnitude difference. Koung and MacGregor suggest iterating the open-loop identification in order to find this information. Others recommend generating data after closing a loop. Closing a loop for a 2 x 2 system enables one to restrict the direction of the output response in a specific manner. For example, putting a PI loop on the second output results in the output data that are mostly in $[1\ 0]^T$ direction. Switching the PI loop to the first output makes the outputs respond in $[0\ 1]^T$ direction instead. Hence, the loop closure facilitates the task of obtaining output data that show even level of excitation in orthogonal directions. A disadvantage is that the procedure can become quite cumbersome for systems with many inputs and outputs.

Other related methods and interpretations exist in the literature. Li and Lee, for instance, present an interesting analysis showing that closed-loop data can be used to identify the inverse of the system gain matrix directly. Combined with open-loop data, a model that accurately matches both the frequency matrix and the inverse in an element-wise sense can be constructed. The method proved to be effective in identifying a model for a nonideal high-purity distillation column.

Subspace Identification

Identification of multivariable polynomial models has been known to be difficult, giving rise to a numerically ill-conditioned, nonlinear estimation problem with possible local minima. In addition, significant prior knowledge (*e.g.*, the system order, the observability index) is needed to obtain a model parameterization. An alternative is to identify a state-space model directly, using so-called subspace identification methods. Significant developments have occurred in this area over the last few years and several subspace identification algorithms have appeared in the literature. Our goal here will be to introduce the basic concepts common among all the algorithms and identify some key issues for practical application.

Assume that the underlying system is given as in (3) where $\begin{bmatrix} w \\ \nu \end{bmatrix}_k$ is a zero-mean, i.i.d. Gaussian vector sequence. The system is assumed to be controllable (from $[u^T \ w^T]^T$) and observable. In addition, the stochastic part of the system is assumed to be stationary. The objective is to identify from input-output data a state-space model

$$\begin{aligned} \tilde{x}_{k+1} &= \tilde{A}\tilde{x}_k + \tilde{B}u_k + \zeta_k \\ y_k &= \tilde{C}\tilde{x}_k + \eta_k \end{aligned} \quad (82)$$

that matches (3) in an *input-output sense*. We will assume for the sake of simplicity that the input sequence u_k used in the identification is an independently designed, white noise sequence (having no temporal correlation).

To understand the main idea behind subspace identification, it is probably easiest to start from the following optimal multi-step prediction equation:

$$\begin{aligned} \begin{bmatrix} y_{k+1} \\ y_{k+2} \\ \vdots \\ y_{k+\bar{n}} \end{bmatrix} &= L_1 \begin{bmatrix} y_{k-\bar{n}+1} \\ y_{k-\bar{n}+2} \\ \vdots \\ y_k \end{bmatrix} + L_2 \begin{bmatrix} u_{k-\bar{n}+1} \\ u_{k-\bar{n}+2} \\ \vdots \\ u_k \end{bmatrix} \\ &+ L_3 \begin{bmatrix} u_{k+1} \\ u_{k+2} \\ \vdots \\ u_{k+\bar{n}-1} \end{bmatrix} + \begin{bmatrix} \varepsilon_{k+1|k} \\ \varepsilon_{k+2|k} \\ \vdots \\ \varepsilon_{k+\bar{n}|k} \end{bmatrix} \\ &= \begin{bmatrix} y_{k+1|k} \\ y_{k+2|k} \\ \vdots \\ y_{k+\bar{n}|k} \end{bmatrix} + \begin{bmatrix} \varepsilon_{k+1|k} \\ \varepsilon_{k+2|k} \\ \vdots \\ \varepsilon_{k+\bar{n}|k} \end{bmatrix} \end{aligned} \quad (83)$$

$\bar{n} > n$ where n is the system order. $y_{k+i|k}$ represents the optimal prediction of y_{k+i} based on data $y_{k-\bar{n}+1}, \cdots, y_k$ and $u_{k-\bar{n}+1}, \cdots, u_{k+\bar{n}-1}$. $\varepsilon_{k+i|k}$ denotes the respective prediction error. $L_1 \in \mathcal{R}^{n_y \cdot \bar{n} \times n_y \cdot \bar{n}}$, $L_2 \in \mathcal{R}^{n_y \cdot \bar{n} \times n_u \cdot \bar{n}}$ and $L_3 \in \mathcal{R}^{n_y \cdot \bar{n} \times n_u \cdot (\bar{n}-1)}$ are functions of system matrices and can be *identified* using the input and output data. The optimal prediction error $\varepsilon_{k+i|k}$, $i \leq 1 \leq \bar{n}$ is zero-mean and is independent of $y_{k-\bar{n}+1}, \cdots, y_k$ and $u_{k-\bar{n}+1}, \cdots, u_{k+\bar{n}-1}$. Hence, linear least squares identification provides unbiased, consistent (*i.e.*, asymptotically converging) estimates.

L_1, L_2 and L_3 are related to the system matrices and covariance matrices in a complex manner, and extracting the system matrices directly from L_1, L_2 and L_3 would involve a very difficult nonlinear optimization. It also requires a special parametrization of model matrices in order to avoid an identifiability problem. Hence, an alternative way to generate the system matrices is desired. This is where the idea gets interesting. We can rewrite the optimal predictions in (83) in terms of the Kalman filter estimate as follows:

$$\begin{bmatrix} y_{k+1|k} \\ y_{k+2|k} \\ \vdots \\ y_{k+\bar{n}|k} \end{bmatrix} = \begin{bmatrix} C \\ CA \\ \vdots \\ CA^{\bar{n}-1} \end{bmatrix} x_{k+1|k} + L_3 \begin{bmatrix} u_{k+1} \\ u_{k+2} \\ \vdots \\ u_{k+\bar{n}-1} \end{bmatrix} \quad (84)$$

$x_{k+1|k}$ represents an estimate of x_{k+1} that is obtained by running the nonsteady-state Kalman filter started with an initial estimate of $x_{k-\bar{n}+1|k-\bar{n}} = 0$ and initial covariance matrix representing the open-loop, steady-state covariance of x.[2] Comparing (84) with (83), one can conclude that

$$\begin{bmatrix} C \\ CA \\ \vdots \\ CA^{\bar{n}-1} \end{bmatrix} x_{k+1|k} = \begin{bmatrix} L_1 & L_2 \end{bmatrix} \begin{bmatrix} y_{k-\bar{n}+1} \\ \vdots \\ y_k \\ u_{k-\bar{n}+1} \\ \vdots \\ u_k \end{bmatrix} \quad (85)$$

Hence, the extended observability matrix $\begin{bmatrix} C \\ CA \\ \vdots \\ CA^{\bar{n}-1} \end{bmatrix}$ and $\begin{bmatrix} L_1 & L_2 \end{bmatrix}$ share the same image space and examinining the rank of the latter gives the system order.

In constructing a state-space model from input-output data, there clearly exists some extra degrees-of-freedom since the basis for the state vector can be chosen arbitrarily without affecting the input-output relation. This means that the extended observability matrix for the identified model (82) (denoted as Γ_o from this point on) can be *any* matrix (of dimension $(\bar{n} \cdot n_y) \times n$) that has the same image space as $\begin{bmatrix} L_1 & L_2 \end{bmatrix}$. Let the SVD of $\begin{bmatrix} L_1 & L_2 \end{bmatrix}$ be represented as follows:

$$\begin{bmatrix} L_1 & L_2 \end{bmatrix} = \begin{bmatrix} U_1 & U_2 \end{bmatrix} \begin{bmatrix} \Sigma_1 & 0 \\ 0 & 0 \end{bmatrix} \begin{bmatrix} V_1^T \\ V_2^T \end{bmatrix} \quad (86)$$

We choose $\Gamma_o = U_1 \Sigma_1^{1/2}$. This defines the basis for the state vector. Let $\tilde{x}$ denote x written in terms of the above-defined basis. We then rewrite the system equation in terms of $\tilde{x}$ as follows:

$$\begin{aligned} \tilde{x}_{k+1} &= \tilde{A}\tilde{x}_k + \tilde{B}u_k + \tilde{w}_k \\ y_k &= \tilde{C}\tilde{x}_k + \nu_k \end{aligned} \quad (87)$$

---

[2] This interpretation does not hold in the case of time-correlated input sequence since future inputs can then contribute to the estimation of past outputs. However, a similar interpretation can be developed and the theory extends straightforwardly with some modifications.

The form of the state-space model that will actually be identified is the following Kalman filter equation for the above system:

$$\tilde{x}_{k+2|k+1} = \tilde{A}\tilde{x}_{k+1|k} + \tilde{B}u_{k+1} + \underbrace{K_{k+1}\eta_{k+1}}_{\zeta_{k+1}}$$

$$y_{k+1} = \tilde{C}\tilde{x}_{k+1|k} + \eta_{k+1} \qquad (88)$$

$\tilde{x}_{k+1|k}$ and $\tilde{x}_{k+2|k+1}$ are two consecutive Kalman filter estimates generated from the same starting condition and K_{k+1} is the Kalman filter gain. Note that the above system equation is equivalent to (3) in an input-output sense.

Now that the identification problem is well-defined, we can discuss construction of the system matrices. In order to identify the system matrices using the relations in (88), we need data for the Kalman filter estimates $\tilde{x}_{k+1|k}$ and $\tilde{x}_{k+2|k+1}$. Let us define $\tilde{x}_{k+2|k+1}$ and $\tilde{x}_{k+1|k}$ as the estimates from the nonsteady-state Kalman filter for system (87), started with the initial estimate of $\tilde{x}_{k-\bar{n}+1|k-\bar{n}} = 0$ and initial covariance matrix representing the open-loop, steady-state covariance of $\tilde{x}$. Then, according to (84),

$$\begin{bmatrix} y_{k+1|k} \\ y_{k+2|k} \\ \vdots \\ y_{k+\bar{n}|k} \end{bmatrix} = \Gamma_o \tilde{x}_{k+1|k} + L_3 \begin{bmatrix} u_{k+1} \\ u_{k+2} \\ \vdots \\ u_{k+\bar{n}-1} \end{bmatrix} \qquad (89)$$

Hence, the data for $\tilde{x}_{k+1|k}$ can be found through the following formula:

$$\tilde{x}_{k+1|k} = \Gamma_o^\dagger \begin{bmatrix} L_1 & L_2 \end{bmatrix} \begin{bmatrix} y_{k-\bar{n}+1} \\ \vdots \\ y_k \\ u_{k-\bar{n}+1} \\ \vdots \\ u_{k-1} \end{bmatrix} \qquad (90)$$

The data for $\tilde{x}_{k+2|k+1}$ cannot be obtained by time-shifting the data for $\tilde{x}_{k+1|k}$, since this will result in the Kalman filter estimate for $\tilde{x}_{k+2}$ with a different starting estimate of $x_{k-\bar{n}+2|k-\bar{n}+1} = 0$. Instead, one must start from the prediction equation below and follow the same procedure as before:

$$\begin{bmatrix} y_{k+2|k+1} \\ y_{k+3|k+1} \\ \vdots \\ y_{k+\bar{n}|k+1} \end{bmatrix} = \hat{L}_1 \begin{bmatrix} y_{k-\bar{n}+1} \\ y_{k-\bar{n}+2} \\ \vdots \\ y_{k+1} \end{bmatrix} + \hat{L}_2 \begin{bmatrix} u_{k-\bar{n}+1} \\ u_{k-\bar{n}+2} \\ \vdots \\ u_{k+1} \end{bmatrix}$$

$$+ \hat{L}_3 \begin{bmatrix} u_{k+2} \\ u_{k+2} \\ \vdots \\ u_{k+\bar{n}-1} \end{bmatrix} \qquad (91)$$

Once the data for the optimal prediction vector $\begin{bmatrix} y_{k+2|k+1}^T & y_{k+3|k+1}^T & \cdots & y_{k+\bar{n}|k+1}^T \end{bmatrix}^T$ are obtained by least squares, the data for $\tilde{x}_{k+2|k+1}$ can be derived by multiplying them with the pseduo-inverse of $\hat{\Gamma}_o$ (which is Γ_o with the last n_y rows eliminated). Once the data for $\tilde{x}_{k+1|k}$ and $\tilde{x}_{k+2|k+1}$ are generated, one can find the system matrices using (88). Since ζ_{k+1} and η_{k+1} are zero-mean sequences that are independent of $\tilde{x}_{k+1|k}$ and u_{k+1}, the least squares identification gives unbiased, consistent estimates. The covariance matrix for $[\zeta^T \ \eta^T]^T$ can also be computed from the residual sequence. These estimates are biased, however, due to the fact that (88) is a nonsteady-state Kalman filter. The approximation error diminishes as $i \to \infty$.

The above is the main idea behind the N4SID algorithm proposed by Van Overschee and De Moor. A major strength of the algorithm is that it requires only numerically stable, noniterative linear algebra operations. Another advantage is that very little prior knowledge (an upper-bound of the system order) is needed to start up the algorithm. Other subspace algorithms available in the literature are based on similar principles, but differ on how the data for $\tilde{x}_{k+1|k}$ and $\tilde{x}_{k+2|k+1}$ are constructed. For instance, in Larimore's CVA method, the basis for the states are chosen by considering the statistical correlation betweeen $\begin{bmatrix} y_{k+1}^T & y_{k+2}^T & \cdots & y_{k+\bar{n}}^T \end{bmatrix}^T$ and $\begin{bmatrix} y_{k-\bar{n}+1}^T & \cdots & y_k^T & u_{k-\bar{n}+1}^T & \cdots & u_{k-1}^T \end{bmatrix}^T$. These algorithms (and their variants) have already appeared in commercial software packages (e.g., MATLAB System Identification Toolbox) and also been embedded into some of the identification packages used with MPC.

Despite the fact that these algorithms represent a major breakthrough, there are a few practical complications that should not be overlooked. First, although it is proven that the N4SID algorithm gives unbiased estimates of system matrices with infinite amount of data, very little can be said about the model quality obtained with finite data. In practice, one must always work with finite-length data sets. In addition, various nonideal factors like nonlinearity and nonstationarity make the residual sequence $\varepsilon_{k+i|k}$ in (83) become correlated with the regression data. Because of these reasons, L_1, L_2 obtained from the least squares identification (which are critical for determining the system order and data for the Kalman filter estimates) may be biased. Although expected errors in the estimates of these matrices can be quantified, it is difficult to say how these errors eventually affect the final model quality (e.g., prediction capability, frequency response, etc.). One implication is that, in general, one needs a large amount of data in order to have much success with these algorithms (which is only natural since these algorithms use very little prior knowledge). Clearly, more research is needed on the robustness of these algorithms. Another implication is that the above does not replace the traditional parametric identification, but complements it. For instance, it has been suggested that the subspace methods be used to provide a starting estimate for prediction error minimization.

Another issue arising from biased L_1 and L_2 estimates is that the SVD of $\begin{bmatrix} L_1 & L_2 \end{bmatrix}$ is likely to show many more nonzero singular values than the intrinsic system order. In order not to overfit the data, one has to limit the system order by eliminating the "negligible" singular values in forming the Γ_o matrix. In the context of model reduction, this is in the same line with the Hankel norm reduction. An alternative for deciding the system order and the basis for the states is to use the SVD of the matrix $\begin{bmatrix} L_1 & L_2 \end{bmatrix} Y$, where Y is the matrix whose columns contain the data for $\begin{bmatrix} y_{k-\bar{n}+1}^T & \cdots & y_k^T & u_{k-\bar{n}+1}^T & \cdots & u_{k-1}^T \end{bmatrix}^T$ (This is actually the approach advocated by Van Overschee and De Moor). In this case, the states are determined such that the output predictions best match the data (in the 2-norm sense). In the context of model reduction, this corresponds to a frequency-weighting with the input spectrum (for the deterministic part). This step of determining the model order and basis is somewhat subjective, but is often critical. For instance, one must consider the previously mentioned gain directionality issue carefully, since gain directions that are not important for open-loop dynamics can be important for closed-loop control.

Finally, the requirement that the stochastic part of the system should be stationary can be limiting. This means that the algorithms cannot be applied to systems with integrating-type disturbances. For such systems, differencing the input / output data before applying the algorithm may do the trick. This effectively takes out the system integrators, which can be added back into the model after the identification. In addition, it is noteworthy that the result that the algorithm being unbiased does not apply when the algorithm is used on closed-loop data. This is because the residual terms in (83) are now correlated with the regression data, and the least squares identification gives biased estimates.

Conclusion

The main focus of MPC research during the past five years has been on devising algorithms for which stability can be proven. There are now several methods to design Model Predictive Controllers that will deliver stability regardless of how they are tuned. When these methods are translated to typical industrial algorithms like DMC, they provide transparent ways to set the horizon size and constraints such that the stability can be guaranteed. On the other hand, the results should be applied in an intelligent manner, since there are many practical cases for which asymptotic stability is neither possible nor desired. The algorithms need to be expanded to deal with these cases.

The theory for designing model predictive controllers for systems with parametric uncertainty (both time-varying and time-invariant) does not appear to be as well established at this point. We examined both deterministic and stochastic formulations and reviewed some interesting results. A fundamental problem exists here, however, that, unlike the case of completely deterministic systems, MPC does not approximate the optimal feedback control very well, and the achievable performance with MPC can be far below the optimal feedback control peformance. Thus, from a theoretical standpoint, MPC (*i.e.*, open-loop optimal feedback control) appears not to be the best approach to the problem.

We also reviewed some progress in related areas like state estimation and multivariable system identification. In the state estimation area, a notable development is the so-called "moving horizon estimation" techniques, which can provide better robustness and more flexibility / extendability at the expense of added computational requirements. In multivariable system identification, much progress has been made in understanding how the multivariable-intrinsic issues like gain directionality affects the identification. Insight suggests that identification signals need to be generated in a special way for interactive systems, which may require iterative identification or closed-loop identification. Another significant achievement in this area is the development of so called subspace identification algorithms, which provide numerically stable, noniterative ways to identify state-space models with little prior knowledge. Again, these methods should not be understood as panacea to all identification problems and must be combined with intelligently designed identification experiments and other traditional methods like the prediction error method.

Acknowledgment

The author gratefully acknowledges the financial support from NSF NYI Program (CTS #9209808 and CTS #9357827), DuPont, Setpoint and Shell Development.

NONLINEAR MODEL PREDICTIVE CONTROL: AN ASSESSMENT

David Q. Mayne
Department of Electrical and Computer Engineering
University of California
Davis, CA 95616

Abstract

This paper examines the design of controllers with emphasis on their ability to handle state and control constraints and nonlinearity. A major motivation is the presence of hard constraints in most applications, not least in the petro-chemical industry where steady state optimization forces the operating point to lie on or near the boundary of the feasible set. Efficient handling of constraints requires nonlinear controllers, even if the system being controlled is linear, and design of nonlinear systems involves optimization. The interaction between nonlinear control and optimization is explored, particular attention being given to model predictive control of constrained nonlinear systems.

Keywords

Model based control, Predictive control, Nonlinear control, Constraints, Optimization.

Introduction

It is undoubtedly true that most control applications require satisfaction of hard constraints on controls and states; actuators saturate and safe operation requires limitations on states such as velocity, acceleration, temperature and pressure. Efficient handling of constraints requires nonlinear control whether the system being controlled is linear or nonlinear. The closed loop system is, therefore, nonlinear which is the reason for the slow development of an adequate theory. The value function $V^0(x,t)$ and optimal *feedback* control law $u^0(x,t)$ for the optimal control problem defined by the system
$$\dot{x} = f(x,u)$$
with control constraint $u(t) \in \Omega$ and cost $\int_0^T \ell(x(t), u(t), t)dt + F(x(T))$ may, in principle, be obtained by solving the Hamilton-Jacobi-Bellman partial differential equation

$$-\frac{\partial}{\partial t}V^0(x,t) = \min_{u \in \Omega} H(x, u, V_x^0(x,t))$$
$$H(x, u, \lambda) := \ell(x, u) + \lambda f(x, u)$$

with boundary condition $V^0(x,T) \equiv F(x)$. The feedback control law is given by

$$u^0(x,t) = \arg\min_{u \in \Omega} H(x, u, V_x^0(x,t))$$

Solution of the Hamilton-Jacobi-Bellman equation yields the optimal cost $V^0(x,t)$ and the optimal control $u^0(x,t)$ for each $(x,t) \in \mathbb{R}^n \times \mathbb{R}$ where n is the state dimension. This makes solution difficult, if not impossible, unless the system is linear and the cost quadratic or the state dimension n is low; in the linear, quadratic case, the value function is finitely parameterized by a symmetric matrix $P(t)$ ($V^0(x,t) = (1/2)x^T P(t)x$) and the Hamilton-Jacobi-Bellman partial differential equation reduces to an ordinary differential equation (the Riccati equation) for $P(t)$. This difficulty has, in general, precluded the development of procedures for designing nonlinear feedback controllers except in specific areas. One such area is adaptive control of linear systems; the adaptive controller is nonlinear but closed loop stability can be established by sophisticated arguments. A second area is model predictive control. Model predictive control avoids solving the Hamilton-Jacobi-Bellman partial differential equation by repeatedly solving online an *open-loop* optimal control problem for the current state, a considerably simpler task, and applying the minimizing control for a short time before repeating the process. In essence, instead of determining the value function $V^0(x)$ for each $x \in \mathbb{R}^n$, it is determined for the sequence of states $\{x[k] := x(k\Delta)\}$ actually encountered. An impossible problem is replaced by a soluble problem; the price is the complexity of the on-line controller (which has to have the capability of solving open-loop optimal control problems) which restricts its use to 'slow' plants.

This paper examines the interplay between optimization and control: how optimization may be employed to determine nonlinear control, and what restrictions this use imposes on achievable control.

Model Predictive Control

Introduction

Our purpose here is to explore model predictive control. We are interested in exposing limitations, both inherent and imposed by optimization. We will be concerned with the formulation of finite horizon, open loop, optimal control problem which is both implementable (soluble, at least approximately, by the online controller within appropriate time limits) and whose repeated solution online ensures satisfaction of the control objectives. We discuss the use of model predictive control for controlling constrained nonlinear systems. To do this, we examine existing MPC strategies, not because they are definitive, but because they reveal underlying problems. We do not pretend to survey the field; for recent surveys see (Garcia, Prett and Morari, 1989), (Bequette, 1991) and (Rawlings, Meadows and Muske, 1994).

The control objectives include stability, performance, implementability, constraint satisfaction, robustness, adaptivity, and operator transparency.

Control constraints arise from actuator limitations. State constraints are often imposed for safety. In process control, economic considerations often require the operating point to lie on, or near, the boundary of the feasible region (the subset of the state space satisfying the state constraints).

Efficient control of a constrained system requires nonlinear control even if the system being controlled is linear. Figure 1 shows the step response of a linear system satisfying an acceleration constraint; this forces the step response to lie below the indicated boundary. A linear zero velocity error controller forces area B to be equal to area A with the result that rapid settling necessitates large overshoot; conversely, low overshoot implies large settling time.

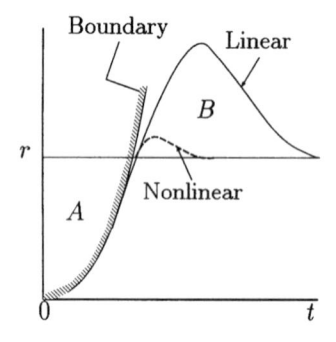

Figure 1. Linear vs. nonlinear control.

Far better performance (rapid response without excessive overshoot) can be achieved with nonlinear control. In contrast, nonlinear control is not advantageous if the system being controlled is linear and unconstrained and a H_∞ criterion is used.

Problem Formulation

The system to be controlled is assumed to be described by

$$\dot{x} = f(x, u) \quad (1)$$
$$y = h(x) \quad (2)$$

where $f : \mathbb{R}^n \times \mathbb{R}^m \to \mathbb{R}^n$ and $h : \mathbb{R}^n \to \mathbb{R}^r$ are twice continuously differentiable and

$$f(0, 0) = 0 \quad (3)$$

and to be subject to the hard control and state constraints

$$u(t) \in \Omega, \quad x(t) \in E \quad (4)$$

where $\Omega \subset \mathbb{R}^m$ is compact and convex and $E \subset \mathbb{R}^n$ is closed and convex; the origin lies in the interior of each of these sets. If the system is linear

$$f(x, u) = Ax + Bu \quad (5)$$

where $A \in \mathbb{R}^{n \times n}$ and $B \in \mathbb{R}^{n \times m}$. In optimal control, $u(\cdot)$ is often assumed to lie in the set $U := \{u(\cdot) \in L_\infty \mid u(t) \in \Omega\}$. To facilitate optimization, and other implementation issues, the control may be finitely parameterized. Among many options, the simplest, and most widely used, is the class of piecewise constant controls with period Δ and values in Ω; let $U_\Delta \subset U$ denote this class of controls. A control $u(\cdot)$ in U_Δ is characterized by the sequence $\{u[k]\}$ where $u[k] := u(k\Delta)$. A control $u(\cdot) \in U_\Delta$ satisfies $u(t) = u[k]$ for all $t \in [k\Delta, (k+1)\Delta)$.

We consider, for simplicity, the regulation problem. The state $x(t) \in \mathbb{R}^n$, the control $u(t) \in \mathbb{R}^m$, and the output $y(t) \in \mathbb{R}^r$ are chosen so that the origin is the set point. We also assume

For all $x \in \mathbb{R}^n$, all $u(\cdot) \in U$ there exists a unique solution $x^u(\cdot)$ to (1) satisfying $x^u(0) = x$.

When f is linear, this assumption is automatically satisfied.

Let $x^u(\cdot; x, s)$ denote the solution of (1), due to control $u(\cdot) \in U$, passing through (x, s) (i.e. satisfying $x^u(s; x, s) = x$). A control $u(\cdot) \in U$ (satisfying, therefore the control constraint $u(t) \in \Omega$) will be called *admissible*; similarly, a state trajectory $x^u(\cdot)$ which satisfies both the control and state constraints will be termed *admissible*.

Limitations

Constraints automatically impose limitations in what can be achieved, even if the system to be controlled is linear. Not every state of a controllable but constrained linear system can be steered to the origin. The set of states which *can* be steered to the origin in time T along admissible trajectories is X_T, defined, for all $T \geq 0$, by

$$X_T := \{x \in E \mid \exists u(\cdot) \in U \text{ such that}$$
$$x^u(t; x, 0) \in E, t \in [0, T]; x^u(T; x, 0) = 0\}$$

Under growth conditions, such as $\|f(x,u)\| \le M(1+\|x\|)$ for all $x \in \mathbb{R}^n$ all $u \in \Omega$, this set is bounded for all $T \in \mathbb{R}_+$. If the set $f(x, \Omega) := \{f(x,u) \mid u \in \Omega\}$, is convex for all x, then X_T is compact. If $f(\cdot)$ is linear, X_T is compact and convex (even if Ω is not convex) (Lee and Markus, 1967). If, in the definition of X_T, U is replaced by $U_\Delta \subset U$, smaller sets will be obtained.

The behaviour of X_T as $T \to \infty$ is of interest. If $f(\cdot)$ is linear, stable and controllable, and $E = \mathbb{R}^n$, then $X_T \to \mathbb{R}^n$ as $T \to \infty$. This result can be extended to include the case when A has simple eigenvalues on the imaginary axis. If A has unstable eigenvalues, then $X_T \to X_\infty$ where X_∞ is a convex, strict subset of $\mathbb{R}^n$. If all the eigenvalues of A are unstable, then X_∞ is compact and convex.

The concept of a *control invariant set*, was used in the analysis of unconstrained linear systems (Wonham, 1974). Gilbert and Tan (Gilbert and Tan, 1991) introduced the concept of an 'output admissible set' to analyze and synthesize linear controllers for constrained linear systems. For our purposes, the extension, by Aubin (Aubin, 1991), of the original definition is useful. A set $X \subset E$ is said to be control invariant (for the system $(f(\cdot), E, \Omega)$), if, for every $x \in X$, there exists a control $u \in U$ such that $x^u(t; x, 0) \in X$ for all $t \ge 0$. Roughly speaking, if X is control invariant, then, for every initial state x in X, there exists an *open-loop* control $u \in U$ which keeps any trajectory starting at x in X; this trajectory is admissible. Clearly the set X_T defined above is control invariant; every point $x(t)$ on a trajectory commencing at $x \in X_T$ at time 0 and terminating at the origin at time T can be steered to the origin along an admissible trajectory in a time not exceeding $T - t$; hence, $X_s \subset X_t$ for all s, t satisfying $s \le t$.

Thus, there is an inherent limitation on the set of states of a constrained system that can be steered to the origin in finite or infinite time. Any controller must keep the state x in X_∞ if not in X_T; this implied constraint should not be ignored.

Stability

In the process control literature, model predictive control (MPC) is conventionally formulated as solving online a finite horizon optimal control problem (FHP) of the form

$$\min\{V(x, t, u(\cdot)) \mid u(\cdot) \in U_\Delta;$$
$$x(s) \in E, s \in [t, t+T]\} \quad (6)$$

and applying the minimizing control $u^0(\cdot) \in U_\Delta$, to the plant over the interval $[k\Delta, (k+1)\Delta]$. The procedure is repeated indefinitely. In (6), $V(\cdot)$ denotes the cost function defined by

$$V(x, t, u(\cdot)) := \int_t^{t+T} \ell(x(s), u(s))ds + F(x(t+T)) \quad (7)$$

The minimizing control $u^0(t; x, t)$ is a time-invariant function $h(x)$ of the current state x, so MPC yields an implicit nonlinear control law even when the system being controlled is linear. When the system is linear, the cost quadratic, the sets Ω and E are specified by linear inequalities, FHP reduces to a quadratic program for which efficient software exists.

But *optimality does not imply stability*; the version of MPC described above does not guarantee closed loop stability. Stability can be achieved, at least when the system is linear and the cost quadratic, by increasing the terminal cost and/or adjusting the horizon T. Tuning (or fiddling) with parameters in this way in order to achieve stability is both undesirable (Bitmead, Gevers and Wertz, 1990) and unnecessary. Stability can, in fact, be achieved relatively simply (Thomas, 1975; Keerthi and Gilbert, 1988; Mayne and Michalska, 1990) by adding a stability constraint (SC) $x(t+T) = 0$ to the finite horizon control problem $P(x, t)$ as follows:

$$\min\{V(x, t, u(\cdot)) \mid u(\cdot) \in U;$$
$$x(s) \in E, s \in [0, T]; x(t+T) = 0\} \quad (8)$$

The terminal cost $F(\cdot)$ becomes irrelevant and can be discarded. For simplicity in presentation, the class of admissible controls is extended to U. With the additional stability constraint $x(t+T) = 0$, the value function $V^0(x)$ (the minimum value of V in (8)) can be shown to be a Lyapunov function for the closed loop system; for suitable $\ell(\cdot)$, $V(x) > 0$ if $x \ne 0$, and $\dot{V}^0(x) \le -\ell(x, h(x))$ where $h(\cdot)$ is the implicitly defined control law, which establishes stability using a standard Lyapunov argument; see (Mayne and Michalska, 1990) where differentiability of $V^0(\cdot)$ is established (a difficult task because of the terminal equality constraint). A proof, which does not require differentiability of the value function, is given in (Vinter and Michalska, 1991). See also (Michalska and Mayne, 1991).

The addition of the stability constraint $x(t + T) = 0$ does not unduly complicate implementation when the system is linear, the cost quadratic and the sets E and Ω are specified by linear inequalities; FHP is still a quadratic program, (with additional equality constraints) if U is replaced by U_Δ. It is interesting to note that in practice MPC is usually applied to stable plants and T is chosen sufficiently long to ensure the plant has settled to its equilibrium state at time $t + T$; this procedure implicitly ensures satisfaction of the stability constraint.

Subject to some modest conditions, it is possible to achieve closed loop stability by employing an infinite horizon but the resulting optimal control problem cannot be solved in general. But Keerthi and Gilbert (Keerthi and Gilbert, 1988) show how infinite horizon feedback laws may be approximated by finite horizon laws. An alternative to infinite horizon control laws is the ingenious procedure (Rawlings and Muske, 1993) in which FHP is replaced by a control problem with infinite horizon cost but finite horizon control; the resultant control problem can be solved in the linear quadratic case.

It is an interesting fact (Meadows, Henson, Eaton and Rawlings, 1995) that receding horizon control can stabilize

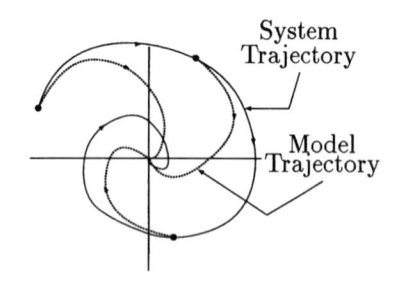

Figure 2. System and model trajectories.

systems which cannot be asymptotically stabilized with continuous feedback.

With this version of (classical) MPC, the trajectory of the controlled system may differ appreciably from the solution to each FHP; an operator cannot necessarily predict future plant behaviour (see Figure 2).

Model Predictive Control of Nonlinear Systems

Since MPC requires the repeated online solution of an optimal control problem, it is important to know whether the problem can be reliably solved. The big divide in nonlinear programming is not between linear and nonlinear but, rather, between convex and nonconvex. If an optimization problem is convex, as FHP is when the system being controlled is linear and $\ell(\cdot)$, $F(\cdot)$ and the state and control constraints are convex, and the feasible set is nonempty, optimization methods exist which ensure convergence to a global minimum (which is unique if the performance criterion is strictly convex). By exploiting duality, lower and upper bounds to the optimal value are easily computed, permitting cessation of the optimization procedure when a predetermined tolerance is satisfied.

Even if $\ell(\cdot)$, Ω and E are convex, but $f(\cdot)$ is not linear, the optimal control problem FHP will not, in general, be convex. A host of difficulties impacting on the implementability of the model predictive controller, ensue.

Computation

The purpose here is not to examine particular methods for solving the finite horizon optimal control problem $P(x,t)$ (see (Mayne, 1995; Santos, deOliveria and Biegler, 1995) and the references therein) but to draw some general conclusions arising from the nature of the problem. Standard descent methods (gradient and conjugate gradient, sequential quadratic programming, interior point methods ...), when applied to a nonconvex problem, yield local, rather than global, minima. Standard stability results, which rely on global optimality, are then inapplicable. Indeed, one cannot even guarantee a feasible solution (a control satisfying the control and state constraints and the stability constraint). In general, the state constraint set E will be characterized by a finite set of inequalities

$$E := \{x \in \mathbb{R}^n \mid q(x) \leq 0\} \quad (9)$$

where $q : \mathbb{R}^n \to \mathbb{R}^a$ is continuously differentiable. Because the state $x(t)$ is an implicit function of the control $u(\cdot)$, the state constraint (for a given initial state) is implicitly of the form

$$\psi(u(\cdot)) := \max\{q(x^u(s; x_0, t_0)) \mid s \in \mathcal{T}\} \leq 0 \quad (10)$$

where $\mathcal{T}$ is either the interval $[t, t+T]$ or a discrete subset thereof it we wish to avoid semi-infinite optimization. Since $\psi(\cdot)$ is, in general, nonconvex (*even* if q is linear in x), it will have local minima, and any *descent* algorithm which reduces $\psi(\cdot)$ to find an admissible control may jam at one of these minima. Hence, determination of a feasible control, let alone an optimal control, cannot be guaranteed. Infeasibility, or nonoptimality, of a control $u(\cdot)$ computed by the MPC controller destroys, of course, the theoretical properties of classical MPC; stability cannot be guaranteed.

Assuming an initial feasible control is available, can feasibility be maintained? This is possible in the absence of disturbances or model error. If the control $u_0(\cdot)$ steers the initial state x to the origin in time T along an admissible trajectory, then (in the absence of disturbances or model error) the control $u_0(\cdot)$, restricted to the interval $[\Delta, T]$, steers $(x(\Delta)$ to the origin in time $T - \Delta$ and so is feasible for the FHP $P(x(\Delta), \Delta)$. However, maintaining both feasibility *and* optimality may be difficult (see below).

The stability constraint is a terminal *equality* constraint which is nonlinear in the decision variable $u(\cdot)$. Satisfaction of nonlinear equality constraints (as contrasted with linear constraints) can only be achieved (if at all) asymptotically (in the sense that accumulation points of sequences generated by the optimization algorithm are feasible). In practice, computation must cease prior to exact satisfaction of the equality constraint. This requires that the MPC strategy be robust to such errors.

An additional, important, factor is the unpredictability of the computation time required to solve nonlinear optimal control problems. Even if the state dimension is relatively low, the dimension of the decision variable and the state constraint can be high (indeed, very high), resulting in complex optimization problems whose solution time can be very variable.

Although implementation of MPC requires the solution of many optimization problems, these problems are not randomly chosen, but strongly correlated. Indeed, the solution to $P(x(t), t)$ provides a feasible, albeit nonoptimal solution, to the successor problem $P(x(t + \Delta), t + \Delta)$. Can this feature be used to simplify online computation by 'tracking' the optimal solution? Indeed it can, but no guarantee against failure can be made. This is illustrated in Figure 3 which shows how a global minimum of $V(\cdot)$ can be transformed into a local minimum or, even, a global maximum, as the parameter x (the initial state) varies (Polak, 1993).

Consequently, it is possible, indeed probable, that, as time proceeds, an MPC controller, which commences with

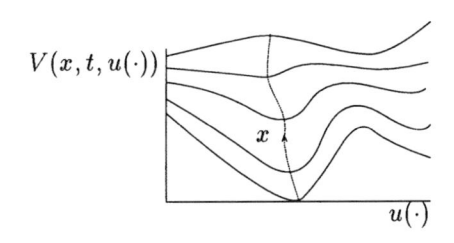

Figure 3. Transformation of global minimum.

a global minimum to $P(x(0), 0)$ and tracks the (local) minimum, will eventually generate a nonoptimal solution. The excellent properties of MPC may then be lost.

It seems, therefore, that nonlinear MPC requires the development of alternative approaches. We discuss, in turn, various possibilities before turning to an examination of robustness issues.

Global Optimization.

In the nonconvex case, ordinary descent algorithms will not, in general, solve the global optimization problem implicit in classical MPC. It is natural, therefore, to turn to global optimization algorithms. In the general case, establishing that a global minimum has been achieved involves excessive computation for most problems. However, there may exist a class of MPC problems for which global optimization may be successfully employed. Firstly, it is possible to discard the requirement of optimality, merely requiring feasibility of computed solutions (see discussion below on variable horizon algorithms); if the feasible set is not too small (relative to the space being explored), online determination of feasible controls (controls satisfying all constraints, including the stability constraint) may well be practical even if FHP is nonconvex. Secondly, there exists a class of problems for which globally *optimal* solutions may be computed online (Staus, Biegler and Ydstie, 1996; Sriniwas and Arkun, 1995); future research will, no doubt, widen this class.

Finite Response and Stable Systems

While, as indicated above and discussed more fully below, optimality may be discarded, achieving stability requires the online determination of feasible controls. It is interesting to note that there exist a class of problems, linear and nonlinear, for which the control and terminal stability constraints may be easily satisfied; these are problems involving *finite response* systems, i.e. those systems whose state returns to the origin in finite time under zero control. This includes FIR systems (linear systems with finite pulse response) and nonlinear systems whose input/output behaviour is described by a finite Volterra series (Genceli and Nikolaou, 1995). As Genceli and Nikolaou show, if the response time of the system is T_P, then the stability constraint is easily satisfied by choosing finite horizon control problem $P(x, t)$ in which the control horizon is restricted to T_C ($u(\tau) = 0$ for $\tau > t + T_C$) and the cost horizon T is chosen to satisfy $T > T_P + T_C$; the finite response ensures automatic satisfaction of the stability constraint $x(t+T) = 0$.

If the system being controlled is merely stable then approximate satisfaction of the stability constraint may be achieved by choosing the horizon T in $P(x, t)$ to be sufficiently large to ensure that $x(t + T)$ is 'small'; the MPC controller must be sufficiently robust (with respect to errors in satisfaction of the stability constraint). This appears to be done in practice.

Contractive MPC

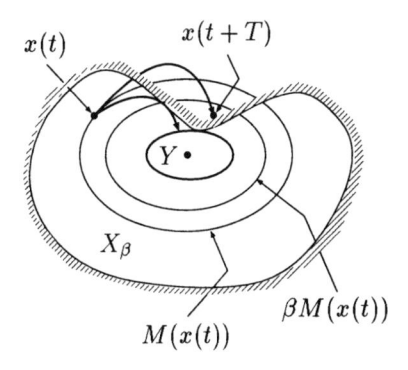

Figure 4. Contractive MPC.

An alternative form of MPC, derived from a procedure (Polak and Mayne, 1981) for the design of nonlinear controllers, is proposed in (Yang and Polak, 1993). This is a Lyapunov based approach in which a Lyapunov function $M(\cdot)$, chosen a-priori, is decreased, not continuously, but discretely; it is allowed to increase at other times. In this form of MPC, stability is achieved by replacing the terminal equality stability constraint by an inequality constraint. The resultant finite horizon optimal control problem $P(x, t)$ now takes the form

$$\min\{V(x, t, u(\cdot), T) \mid u(\cdot) \in S; T \geq 0;$$
$$x(s) \in E, s \in [0, T]; x(t + T) \in \beta M(x)\} \quad (11)$$

where $\beta \in (0, 1)$ and, typically,

$$M(x) := \{z \in \mathbb{R}^n \mid \langle z, Qz \rangle \leq \langle x, Qx \rangle\} \quad (12)$$

for some positive definite matrix Q; $\beta M(x)$ denotes the set $\{\beta z \mid z \in M(x)\}$; see Figure 4. Note that the decision variables are now the control $u(\cdot)$ and the horizon time T. Suppose that the state is x^k at time t^k and that $(u^k(\cdot), T^k)$ solves $P(x^k, t^k)$. The control $u^k(\cdot)$ is then applied to the system over the interval $[t^k, t^{k+1}]$ where $t^{k+1} := t^k + T^k$; at the end of this interval, the solution to FHP is recomputed. The behaviour of the system is easily understood; it is clear that $x^k \in \beta^k M(x(0))$ so that, under relatively mild conditions, stability ensues *provided* that $P(x^k, t^k)$ can be solved for each k. But solution of $P(x,t)$ cannot be guaranteed for every $x \in X$.

To see why this may arise, let $X_\beta \supset X_\infty$ denote the set of states which may be steered to the *set* $\beta M(x)$ along an admissible trajectory. Whereas X_T and X_∞, defined above, are clearly control invariant sets, it is by no means clear that X_β has this property. Given an arbitrary point x in X_β, the successor point x' generated by 'contractive' MPC will not necessarily lie in X_β as illustrated in Figure 4. Hence, FHP $P(x', t')$ is not necessarily solvable. As in classical MPC, the controller may generate states at which a solution to the finite horizon control problem cannot be found. However (Polak, private communication), there exists an $\alpha \geq 0$ such that $Y = \{x \mid \langle x, Qx \rangle \leq \alpha\} \subset X_\infty \subset X_\beta$. The FHP $P(x,t)$ can be solved for any point in this set. Hence, once the initial state of the system is steered to this set, contractive MPC may be safely employed. Since the initial state should be steered to Y in finite time, a strategy such as variable horizon MPC, discussed next, could be employed.

Variable Horizon MPC

The preceding discussion has shown that a satisfactory form of MPC for linear constrained systems is relatively easily obtained by the incorporation into FHP of a stability constraint. When the plant is nonlinear, a host of difficulties arise. Determining a globally optimal, or even a feasible, solution to FHP is, in general, problematic. Even if an initial global solution to FHP is available, it cannot necessarily be tracked as the state x varies. Suppose an MPC controller generates a sequence of states $\{x^i := x(t^i)\}$ where $t^i := i\Delta$. Unless global optimization is employed, it is generally impossible to guarantee that every problem in the sequence $\{P(x^i, t^i)\}$ can be solved. The *general* use of any MPC strategy relying on global optimality seems to be precluded. The *most* that can be hoped for in general is that every member of the problem sequence $\{P(x^i, t^i)\}$ can be solved if the initial problem $P(x^0, t^0)$ can be. MPC has this desirable property if both the problem and control horizons are infinite (and certain stabilizability and detectability assumptions are satisfied). Suppose $P(x^0, t^0)$ is solvable; under the above conditions, the optimal control u^0 for $P(x^0, t^0)$ steers x^0 to the origin. If $P(x^i, t^i)$ is solvable and the optimal control for this problem is $u^i(\cdot)$, then, by Bellman's principle of optimality, the control $u^i(\cdot)$, *restricted* to the interval $[t^{i+1}, \infty)$, is optimal for $P(x^{i+1}, t^{i+1})$. By induction, every member of the problem sequence $\{P(x^i, t^i)\}$ can be solved, and the minimizing control is, in each case, a restriction of $u^0(\cdot)$, the initial minimizing control. Other desirable properties ensue. The controlled trajectory coincides with the solutions to each $P(x^i, t^i)$, a property not possessed by the controllers previously described.

By permitting horizon time $T \in [0, T_{max}]$ to be a decision variable, as in (Michalska and Mayne, 1993), many difficulties are overcome. In its simplest form, variable horizon MPC has the following ingredients. The cost is T:

$$V(x, t, u(\cdot), T) := T \qquad (13)$$

so that FHP $P(x,t)$ becomes the minimum time problem

$$\min\{T \mid u(\cdot) \in U;\ T \in [0, T_{max}];$$
$$x(s) \in E, s \in [t, t+T]; x(t+T) = 0\} \qquad (14)$$

The minimizing control is applied to the plant for the period $\min\{\Delta, T\}$ and the process repeated. This version of MPC is similar to the infinite horizon version: if $P(x^i, t^i)$ is solvable and the optimal control for this problem is $u^i(\cdot)$, then, by Bellman's principle of optimality, the control $u^i(\cdot)$, *restricted* to the interval $[t^{i+1}, \infty)$, is optimal for $P(x^{i+1}, t^{i+1})$. By induction, every member of the problem sequence $\{P(x^i, t^i)\}$ can be solved if $P(x^0, t^0)$ can be solved, and the minimizing control is, in each case, a restriction of $u^0(\cdot)$, the initial minimizing control. *In the absence of disturbances or model error, the solution to every member of the sequence $\{P(x^i, t^i)\}$ of problems is simply obtained as a restriction of the solution to the original problem $P(x^0, t^0)$; only one optimal control problem $P(x^0, t^0)$ need be solved. Moreover, the controlled trajectory coincides with the solution to $P(x^0, t^0)$. Optimal variable horizon MPC is asymptotically stabilizing with a region of attraction $X_{max} := X_{T_{max}}$.* Stability can be established using standard Lyapunov arguments with the Lyapunov function either the value function $V^0(x)$ of (14), or $T^0(x)$, the minimizing value of the horizon T.

Even more can be gained; *optimality can be discarded*. Consider the following version of variable horizon MPC in which the controller seeks merely to improve the previous control rather than optimize. The finite horizon problem $P(x^i, t^i)$ to be solved by the controller at time t^i is now:

Determine a feasible control $\tilde{u}^i(\cdot)$ in U and horizon $\tilde{T}^i$ in $[0, T^{i-1} - \beta\Delta]$, $\beta \in (0,1)$, which satisfies

$$x^{\tilde{u}^i}(t^i + \tilde{T}^i; x^i, t^i) = 0 \qquad (15)$$

Determine a feasible control $u^i(\cdot)$ in U and horizon T^i in $[0, \tilde{T}^i]$ which satisfies

$$x^{u^i}(t^i + T^i; x^i, t^i) = 0 \qquad (16)$$

Apply the control $u^i(\cdot)$ to the system over the interval $[t^i, t^{i+1}]$, $t^{i+1} := t^i + \min\{\Delta, T^i\}$.

This strategy has the following properties:

- If (u^{i-1}, T^{i-1}) is feasible for $P(x^{i-1}, t^{i-1})$, then $(\tilde{u}^i, \tilde{T}^i)$, where $\tilde{u}^i$ is the restriction of u^{i-1} to $[t^i, t^{i-1} + T^{i-1}]$ and $\tilde{T}^i = T^{i-1} - \Delta$, is feasible for $P(x^i, t^i)$.

- The controller need do no more than accept the feasible control $(\tilde{u}^i, \tilde{T}^i)$.

- If it does, the system and model trajectories coincide.

- The horizon time decreases by $\beta\Delta$ at each iteration (except the last).

If a feasible solution (u^0, T^0) to the initial problem $P(x(0), 0)$ is known, then, by induction, every problem in the sequence $\{P(x^i, t^i)\}$ is (easily) soluble. See Figure 5. Online computation is substantially reduced. A global solution to an optimal control problem is *never* required. In the absence of model error or external disturbances, the solution (u^0, t^0) to the initial FHP provides a solution to all subsequent FHP's; if the original model trajectory provides a good path to the origin (because FHP incorporates all performance and stability objectives), why spoil it with further optimization?

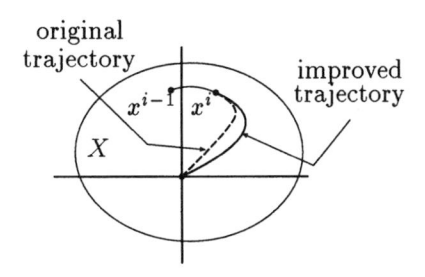

Figure 5. Variable horizon MPC.

Lyapunov analysis is complicated by the fact that the control is not unique. A direct proof of stability is, however, possible. Since the horizon time decreases by $\beta\Delta$ at each iteration, whatever control is used, the controller steers any point x in X_{max} to the origin in a time not exceeding T^0/β along a trajectory which remains in the bounded set X_{max}. *In the absence of model error or disturbance, variable horizon MPC is asymptotically stabilizing with a region of attraction X_{max}.*

Hybrid MPC

Hybrid MPC is a variant of variable horizon MPC in which the stability (terminal equality) constraint is replaced by an inequality constraint which is more easily satisfied. In general, an inequality constraint can be satisfied in a finite number of iterations of a suitable algorithm, whereas an equality constraint may be only satisfied asymptotically. To achieve this objective, a local stabilizing controller $h(\cdot)$ is employed near the equilibrium point (the origin). Suppose a stabilizing control law $h : W \to \Omega$, where $W \subset E$ is a positively invariant set for the system $\dot{x} = f(x, h(x))$, is known. The control law $h(\cdot)$ steers any initial state $x \in W$ to the origin along a trajectory which remains in W and satisfies the control and state constraints. The control law may be linear. The finite horizon problem $P(x^i, t^i)$ to be solved at time t^i becomes:

If $x^i \notin W$, determine a feasible control $\tilde{u}^i(\cdot)$ in U and horizon $\tilde{T}^i$ in $[0, \tilde{T}^i], \beta \in (0, 1)$, to satisfy

$$x^{\tilde{u}^i}(t^i + \tilde{T}^i; x^i, t^i) \in W \tag{17}$$

Determine a feasible control $u^i(\cdot)$ in U and horizon T^i in $[0, T^{i-1} - \beta\Delta], \beta \in (0, 1)$, *to satisfy*

$$V(x^i, t^i, u^i(\cdot), T^i) \leq V(x^i, t^i, \tilde{u}^i(\cdot), \tilde{T}^i)$$
$$x^{u^i}(t^i + T^i; x^i, t^i) \in W \tag{18}$$

Apply the control $u^i(\cdot)$ to the system over the interval $[t^i, t^{i+1}]$.
If $x \in W$, apply the control $h(x)$.

This controller shares many of the properties of its predecessor (16) and steers any initial state to the set W in a time not exceeding T^0/β; the local controller $h(\cdot)$ is employed in W. See Figure 6.

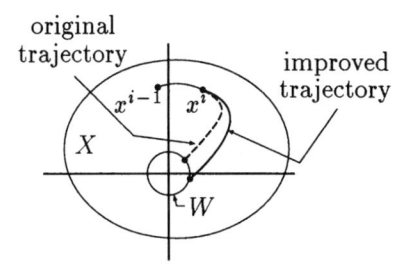

Figure 6. Hybrid MPC.

Hybrid MPC is asymptotically stabilizing with a region of attraction X_{max}.

Robust MPC

The discussion above assumes that the system being controlled is identical to the model employed by the controller. In reality, $f_R(\cdot) = f(\cdot) + d(\cdot)$, where the dynamics of the real system and the model are described by $\dot{x} = f_R(x, u)$ and $\dot{x} = f(x, u)$ respectively, and $d(\cdot)$ denotes the disturbance or model error. We distinguish three cases:

- $d = d(x, u)$ where $\|d(x, u)\| \leq c\|(x, u)\|$.
- $d = d(t)$ where $d(t) \to 0$ as $t \to 0$.
- $d = d(t)$ where $d(t) \in D$ for all t.

where $c \in (0, \infty)$, D is compact, and $0 \in \text{interior}(D)$. It is important to note that even an exponentially decaying disturbance can destabilize a nonlinear system, possibly resulting in finite escape time.

Parametric Model Error

Suppose parametric model error is present, so that the real system is described by

$$\dot{x} = f_R(x, u) \tag{19}$$

whereas the model is described, as before, by $\dot{x} = f(x, u)$. Assume also that the model error is bounded as follows:

$$\|f_R(x, u) - f(x, u)\| \leq c\|(x, u)\| \tag{20}$$

for all (x, u). The hybrid MPC controller employs, of course, the model in calculating solutions to FHP. The controller can be made robust by employing in $P(x,t)$ conservative constraint sets $E' \subset E$ and $W' \subset W$ in place of E and W respectively. This is illustrated in Figure 7. Consider robust stability. Suppose that, at time t^{i-1}, an admissible control u^{i-1} which steers the *model* from x^{i-1} to $z^i \in W'$ has been computed and that this control is applied to the plant for the period $[(i-1)\Delta, i\Delta)$, yielding the next *plant* state $x^i = x(t^i)$. Because of model error, $x^i \neq x_m^i$ which is the model state resulting from the same control being applied to the model. If the model error (i.e. c) is sufficiently small, time t^i can be chosen so that $x^i \in E'$. Because of model error, the control u^{i-1}, restricted to the period $[t^i, t^{i-1} + T^{i-1}]$, steers the model from x^i to the state $\tilde{z}^i \neq z^i$. If the model error is not too large, $\tilde{z}^i \in W$ and the local control law $h(\cdot)$ can steer the model from $\tilde{z}^i$ to W' in a time less than $\Delta/2$, yielding an admissible control-horizon pair $(\tilde{u}^i, \widetilde{T}^i)$ for the subsequent finite horizon control problem $P(x^i, t^i)$; $\tilde{u}^i$ is the restriction of u^{i-1} to the period $[t^i, t^{i-1} + T^{i-1}]$ (this control steers the model from x^i to $\tilde{z}^i$) concatenated with the control which steers the model from $\tilde{z}^i \in W$ to $z^{i+1} \in W'$ and $\widetilde{T}^i \leq T^{i-1} - \Delta/2$. The control $\tilde{u}^i$ thus steers the model from $x^i \in E'$ to W' in time $\widetilde{T}^i \leq T^i - \Delta/2$ so that $(\tilde{u}^i, \widetilde{T}^i)$ is an admissible control-horizon pair for $P(x^i, t^i)$. This control-horizon pair can be improved, if desired. (If model error or disturbance is too large for the above procedure to provide an admissible control-horizon pair, an admissible pair may be determined directly; however, this may require global optimization techniques). Hence again we have the property that, provided a feasible solution (u^0, T^0) to $P(x^0, t^0)$ is available, feasible solutions (u^i, T^i) to all subsequent problems $P(x^i, t^i)$ are easily obtained. Robust stability ensues (Michalska and Mayne, 1993). Let $X'_{max} \subset X_{max}$ denotes the set of states of the model that can be steered in time T_{max} to the set W' along trajectories satisfying the control and conservative state constraint.

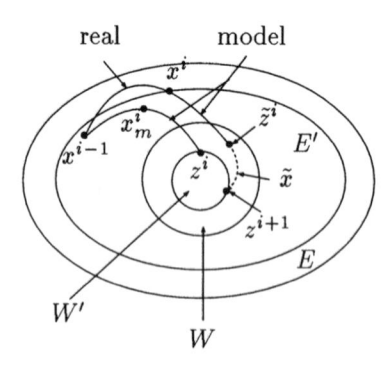

Figure 7. Robust MPC.

Again stability can be established by direct arguments, showing that the horizon time is reduced by at least $\Delta/2$ at each iteration, so that any initial state in X'_{max} is steered to the origin in a time not exceeding $2T^0$.

There exists a $c > 0$ such that robust hybrid MPC is asymptotically stabilizing with a region of attraction X'_{max} for every system satisfying (20).

Robustness of variable horizon MPC is, perhaps, of more interest. The situation is as depicted in Figure 7 if W' is regarded as a single point, the origin. Control $u^{i-1}(\cdot)$ steers the *model* from x^{i-1} to $z^i = 0$ in time T^{i-1}. Because of model error, the actual state at time t^i is $x^i \neq x_m^i$, and control $u^{i-1}(\cdot)$ (suitably restricted) steers the model from this state to $\tilde{z}^i$. Robustness depends on the ability to find a control which steers the model from $\tilde{z}^i$ to the origin in a time not exceeding $\Delta/2$.

For alternative approaches, see (Zheng and Morari, 1993; Genceli and Nikolaou, 1993).

Asymptotically Decaying Disturbance

This situation typically arises when an observer is used for a linear system (even if nonlinearly controlled) in which case $d(t) = A[x(t) - \hat{x}(t)]$ where $\hat{x}(t)$ is the estimated state. If the controlled system is exponentially stable, converse Lyapunov theorems can be employed to establish robustness (Vidyasagar, 1993; Scokaert and Rawlings, 1995); the latter have shown that the controlled system need merely be asymptotically stable if the perturbation decays exponentially.

Bounded Disturbance

Bounded deterministic, or stochastic, disturbances introduce a different problem. It is no longer possible to drive the state to the origin. If the disturbance $d(\cdot)$ is unknown but bounded, then the best that can be hoped for is that the state can be steered to a control invariant set. Consider the discrete time version of (1):

$$x[k+1] = g(x[k], u[k]) + d[k] \quad (21)$$

where $x[k] := x(kT)$ and the disturbance is subject to the constraint $d[k] \in D$ for all k; the constraint set $D \subset \mathbb{R}^n$ is compact and convex and contains the origin in its interior. The set $X \subset E$ is robust control invariant for $(g(\cdot), \Omega, E, D)$ if, for all $x \in X$ there exists a $u \in \Omega$ such that $g(x, u) + d \in X$ for all $d \in D$. If X is robust control invariant, and the initial state $x[0]$ of (21) is in X, then there exists a *feedback* control $u[k] = h(x[k])$ which maintains the state in X (despite the disturbance).

Because the system (21) cannot be controlled to the origin, MPC must be modified. An obvious possibility is to ignore the disturbance, i.e. to obtain a control by solving FHP for the undisturbed system

$$x[k+1] = g(x[k], u[k]) \quad (22)$$

Because possible future disturbances are neglected, performance can suffer (Rawlings et al., 1994). To overcome this deficiency, the use of min-max (differential game) MPC has been proposed; see (Zheng and Morari, 1993). For the sake of later comparison only, consider the following open-loop formulation of the min-max FHP for discrete time systems:

$$\min_{\{u[j]\}} \max_{\{d[j]\}} \{ \sum_{j=k}^{N-1} \ell(x[j], u[j]) + cd(x[j], E)$$
$$+ cd(x(N), E) \mid u[j] \in \Omega;\ w[j] \in W \}$$

where $d(x, E)$ denotes the distance of x from the set E and c is finite but sufficiently large to ensure satisfaction of the state constraint $x[j] \in E$ for all possible disturbance sequences. This formulation has two disadvantages. Firstly, it includes no stability constraint (the stability constraint $x(t+T) = 0$ cannot be included since the system cannot be steered to the origin by the control $\{u[j]\}$). Secondly, the worst case scenario corresponding to a nominal control can be considerably worse than that actually achieved with feedback control.

Output Feedback MPC

The controllers described above assume that the state is perfectly accessible ($y = x$). If this is not the case, state estimation has to be implicitly or explicitly introduced. There are several possibilities for dealing with this case.

(i) The simplest approach, widely used in adaptive control although not without danger, is 'certainty equivalence': estimate the state and use the estimated state $\hat{x}(t)$ in the model predictive controller. The observer can, under reasonable conditions, be designed to ensure that estimation error $\tilde{x}(t) := x(t) - \hat{x}(t) \to 0$ exponentially as $t \to \infty$. Since optimization is employed for the controller, it appears preferable to employ it also for the observer in place of weaker alternatives such as extended Kalman filtering.

(ii) The approach in (i) can lead to instability and even to finite escape time in certain nonlinear systems even if the estimation error is exponentially decaying. It appears necessary to couple the design of the controller and observer to avoid this possibility. One such approach is described in (Michalska and Mayne, 1995); the sampling frequency is increased as a potential finite escape time is approached so that the observer is no longer independent, as it is in certainty equivalence controllers, of control demands.

Observer Based MPC

Since optimization is employed for the controller, it appears preferable to employ it also for the observer in place of weaker alternatives such as extended Kalman filtering. This is what is proposed in (Michalska and Mayne, 1992; Rawlings and Muske, 1993; Robertson, Lee and Rawlings, 1994; Michalska and Mayne, 1995). Whereas the control looks at the future over the interval $[t, t+T]$, the observer's data is over the *moving horizon* $[t - T_E, t]$ in the immediate past where T_E is the estimation horizon which may differ from the control horizon. The estimation cost function $J(\cdot)$ is defined to be

$$J(x, t) := \int_{t-T_E}^{t} \|h(x^u(s; x, t)) - y(s)\|^2 ds$$

where $y(s)$ is the actual output $h(x(s))$ of the system, and $h(x^u(s; x, t))$ is the estimated output at time s assuming that the state of the system is x at time t (and the control is $u(\cdot)$). The function $p_t(\cdot) := J(\cdot, t)$ may be regarded as the state of the system at time t. If the system is linear, p_t is quadratic and may be written in the form

$$p_t(x) = (1/2)(x - \hat{x}(t))^T P^{-1}(t)(x - \hat{x}(t)) + a(t)$$

where $P(t)$ is positive definite if the system is observable, and $a(t) = 0$ if there is no process noise or model error. Under these ideal conditions, the estimate $\hat{x}(t) = \arg \min p_t(x)$ is the unique global minimizer of $p_t(x)$ and $\hat{x}(t) = x(t)$.

But, if the system is nonlinear, $p_t(x)$ may have the form shown in Figure 8 if there is no process noise or model error. Again, the estimate $\hat{x}(t) := \arg \min p_t(x)$ is the unique global minimizer of $p_t(x)$ and satisfies $\hat{x}(t) = x(t)$ if the nonlinear system is observable and there is no process noise or model error.

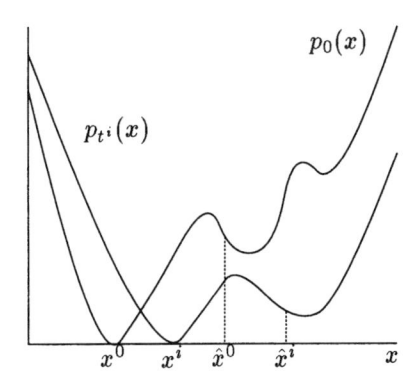

Figure 8. Observer-based MPC.

Finding $\hat{x}(t)$ is not a simple matter. Because of nonlinearity, global minimization may be required; see Figure 8.

Under these ideal conditions, if the system being controlled does not have finite escape time, a *conceptual* online observer may be defined as follows. The observer computes a sequence $\{\hat{x}^i\}$ of estimates of $\{x^i := x(t^i)\}$ to satisfy

$$p_{t^{i+1}}(\hat{x}^{i+1}) \leq \beta p_{t^i}(\hat{x}^i)$$

where $\beta \in (0, 1)$.

If, among other assumptions, the system is observable, $\hat{x}^i \to x^i$ exponentially as $i \to \infty$ (Michalska and Mayne, 1992; Michalska and Mayne, 1995). These estimators can be harnessed, using the 'certainty equivalence' strategy, to a robust model predictive controller; robustness permits the controller to cope with estimation error. Since even exponentially decaying inputs can destabilize a nonlinear system, fairly strong assumptions have to be made. Results of the following form are obtained for the control constrained control problem (Michalska and Mayne, 1995):

For any compact set Y in the interior of X_{max} and any bound on initial observer error, there exist observer parameters such that the observer-based moving horizon controller is asymptotically stabilizing in Y.

The assumption that $x(0) \in Y \subset \text{interior}(X_{max})$ and that the observer is sufficiently fast, is made to ensure that the state does not escape the set X_{max} of states which can be controlled to the origin; observer error implicitly reduces the set of states which can be successfully controlled.

The procedure can be extended to deal with systems with finite escape if the the sampling period Δ and the estimation horizon T_E are variable; the system is measured more frequently as a finite escape time is approached.

But results like this do not necessarily invite application. Figure 8 illustrates the type of difficulty which may arise; although the minima of $p_t(\cdot)$ vary with t there is no guarantee that the estimate $\hat{x}(t)$ will escape into the 'correct' valley (whose minimum is $x(t)$) before $x(t)$ moves out of the set X_{max}. The observer described above avoids this problem but only by implicitly assuming that global optimization is available and that there is no process or model error. If the initial estimate $\hat{x}(0)$ is in the 'correct valley', this problem may be avoided. Approximate observers, such as the extended Kalman filter, face this problem more acutely. If model error or disturbance is present, then $x(t)$ cannot be exactly determined, even with the help of global optimization. With noise, $p_t(x) > 0$ for all x; values of x corresponding to low values are more likely than those corresponding to high values. Using the minimizer of $p_t(x)$ as the state estimate together with certainty equivalence control does not necessarily provide satisfactory performance or, even, stability. As nonlinear H_∞ shows, the state of the system at time t is $p_t(\cdot)$, so, in principle, the control should depend on this function rather than its minimizer.

Adaptive MPC

If the system being controlled has unknown parameters, it may be modelled by

$$\dot{x} = f(x, u, \theta) \qquad (23)$$
$$y = h(x, \theta) \qquad (24)$$

where $\theta \in I\!\!R^p$ is a vector of unknown parameters. By adding the dynamics

$$\dot{\theta} = 0 \qquad (25)$$

we reduce the problem to one of observer based design in which the observer estimates the extended state (x, θ). But this reduction is not useful because the observability assumption is not appropriate for θ. (If the system is $\dot{x} = Ax + Bu, y = Cx$ and $\theta = (A, B)$, then $y(t) \equiv 0$ if $x(0) = 0$ and $\theta(0)$ is arbitrary.) Adaptive control theory does not, therefore, rely on convergence of $\hat{\theta}$ to θ whereas observer based control usually does requires convergence of $\hat{x}$ to x. While the early literature on generalized predictive control, like that on MPC, did not include a stability constraint, this is no longer the case (Clarke, Mosca and Scattaloni, 1991). However, these results rely on linear adaptive control theory precluding consideration of control and state constraints. Since the MPC controller is implicitly nonlinear, a new theory is required. In the ideal case (no model error or disturbance) it is possible, when the system is linear or linearly parameterized, to employ standard algorithms to provide an estimate of θ and to employ this to obtain an estimate of the state. A recurrent problem is that the current estimate $\hat{\theta}(t)$ of θ may not admit a solution to FHP even though $x(t) \in X_T$. This is overcome in (Mayne and Michalska, 1993) by using parameter perturbations which restore controllability without increasing estimation cost. Making use of a property of the innovation process, it is possible to establish, in the ideal case, that the MPC controller is stabilizing. The underlying idea is that if the innovation $e \equiv 0$, then $y \equiv \hat{y}$. Since the MPC controller steers $\hat{y}$ to zero, $y \to 0$. Using observability we deduce that $x \to 0$. These results require further development for application but give an indication of the computational effort required for adaptive control.

Alternatives to MPC

Nonlinear MPC will never achieve the simplicity of linear MPC. Unless a nonlinear MPC controller is exhaustively tested by simulation over the whole range of potential initial states (or observation sequences in the case of inaccessible states) doubt will always remain as to whether or not a state will be encountered for which an acceptable solution to the FHP can be found. It may be necessary to investigate alternative approaches, either to replace, or to be used in conjunction with, MPC to remove some of the disadvantages arising from limitations in what can be achieved from nonlinear optimization. Two possible alternatives, both of which involve a degree of feedback, are outlined below.

Feedback Min-Max MPC

It is possible to overcome some of the disadvantages of the min-max MPC controller, described above, by solving a more complex min-max problem online, one that takes into account future feedback. The penalty is increased complexity of the min-max problem; this complexity can be partly contained by course quantization of the disturbance.

If the disturbance sequence has a finite number of realizations $\{d^\ell[j]\}$ where $\ell \in \mathcal{L} := \{1, \ldots L\}$ indexes the realization, a 'feedback' version of MPC may be formulated as follows. The feedback FHP $P(x, k)$ is

$$\min_{(\{u^\ell[j]\}, N)} \max_{\ell \in \mathcal{L}} \{ \sum_{j=k}^{N-1} \ell(x^\ell[j], u^\ell[j]) + cd(x^\ell[j], E)$$
$$+ cd(x^\ell[N], Z) \mid u^\ell[j] \in \Omega;\ d^\ell[j] \in D;$$
$$u^{\ell_1}[j] = u^{\ell_2}[j] \text{ if } x^{\ell_1}[j] = x^{\ell_2}[j]\}$$

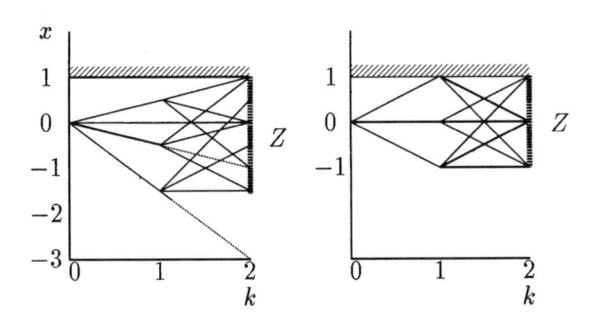

Figure 9. Min-max MPC.

where $x^\ell[j]$ denotes the solution of (20) due to control and disturbance sequences $\{u^\ell[i]\}_{i=k}^{N-1}$ and $\{d^\ell[i]\}_{i=k}^{N-1}$ and initial condition $x[k] = x$. An exact penalty with finite parameter c is employed to ensure satisfaction of the state constraint $x^\ell[j] \in E$ and satisfaction of the terminal stability constraint $x^\ell[N] \in Z$ for all ℓ, j. The set Z in the stability constraint is robust-control-invariant, preferably the smallest such set, and has to be pre-determined. It appears as if NL control actions $u^\ell[k]$ have to be determined, but the *causality* constraint $u^{\ell_1}[j] = u^{\ell_2}[j]$ for all j, ℓ_1, ℓ_2 such that $x^{\ell_1}[j] = x^{\ell_2}[j]$ enforces a single control action for each state, thereby reducing the number of independent variables. The FHP is a finite dimensional min-max problem for which local solutions can be found by standard algorithms. If the system is linear, the cost quadratic, and the sets Ω, D and Z specified by linear inequalities, FHP reduces to a quadratic program, but of much higher dimension than in the disturbance free case. It is interesting to note that variable horizon, min-max MPC inherits many of the advantages of its disturbance free predecessor; in the absence of errors and disturbances *not* already incorporated in the problem description, *the solution* $\{u^\ell[k] \mid \ell \in \mathcal{L}, k \in \{0, \ldots, N-1\}\}$ *to* $P(x[0], 0)$ *provides, suitably restricted, a solution to every problem* $P(x[k], k)$ *subsequently encountered*. Optimality can be discarded as before provided the stability constraint $x[N] \in Z$ is retained.

As an illustration, consider the system

$$x[k+1] = x[k] + u[k] + d[k], \quad x[0] = 0$$

with $\Omega = [-1, 1]$, $D = \{-1, 0, 1\}$, $E = (-\infty, 1]$ and $\ell(x, u) = x^2$. Figure 9 illustrates the two strategies of open-loop min-max MPC (described above) and feedback min-max MPC. For variable horizon, open loop min-max, the solution to $P(x[0], 0)$ is $\{u^0[k]\} = \{-0.5, -0.5\}, \{w^0[k]\} = \{1, 1\}$, yielding the worst case state trajectory $\{0, 0.5, 1\}$. At time $k = 0$, the predicted set of terminal states includes $\{1, -3\}$. At $k = 1$, using horizon $T = 1$, open loop min-max chooses controls $-0.5, 0.5, 1$ if the state is, respectively, $0.5, -0.5, -1.5$. Hence, the actual states at time $k = 2$ lie in the set $Z = [-1.5, 1]$. In fact, Z is a robust control invariant set for this controller and its size is a measure of performance; the set is larger than necessary, as shown below. If the open loop min-max controller employed a fixed horizon of 2, the resultant robust control invariant set would be $[-3, 1]$, even larger.

There are nine possible disturbance sequences $\{-1, -1\}, \{-1, 0\}, \{-1, 1\}, \ldots \{1, 0\}, \{1, 1\}$. Let state (x, k) denote state x at time k. Variable horizon, feedback min-max MPC generates control 0 for states $(0, 0)$ and $(0, 1)$, control -1 for $(1, 1)$ and control 1 for $(-1, 1)$. At time $k = 0$, the predicted set of terminal states includes $\{1, -1\}$ which lie in the minimal robust control invariant set Z= $[-1, 1]$ for this controller. At $k = 1$, the controller applies the precomputed controls $u = -1$ if $x = 1$, $u = 0$ if $x = 0$ and $u = 1$ if $x = -1$. The resultant worst case state trajectory is $\{0, 1, 1\}$. The robust control invariant set $Z = [-1, 1]$ for this controller is the minimal achievable.

Nonlinear Control of Constrained Linear Systems

While there are as yet no visible alternatives to MPC for constrained nonlinear systems, there is considerable activity in the field of nonlinear control of constrained linear systems (Gilbert and Tan, 1991; Gilbert, Kolmanovsky and Tan, 1994; Mayne and Schroeder, 1994). It is certainly possible that this activity may lead to the new proposals for controllers for nonlinear constrained systems which do not require online optimization.

One approach to nonlinear control of constrained linear systems is described below.

The system to be controlled is described by

$$\dot{x} = Ax + Bu \qquad (26)$$
$$y = x \qquad (27)$$

where $x(t) \in \mathbb{R}^n$ and $u(t) \in \mathbb{R}^m$. The control and state are subject to the following hard constraints

$$u(t) \in \Omega, \quad x(t) \in E \qquad (28)$$

where Ω is compact, E is closed, and each set contains the origin in its interior. The corresponding discrete-time system is described by

$$x[k+1] = Fx[k] + Gu[k] \qquad (29)$$

where $x[k] := x(k\Delta), u[k] := u(k\Delta), F := \exp(A\Delta)$ and $G := \int_0^\Delta \exp(As)B ds$ where Δ is the sampling period. It is subject to the control and state constraints $u[k] \in \Omega$ and $x[k] \in E$. Let V^0 denote the value function for the discrete time, minimum time problem, and X_k, $k \in \mathbb{N}$ the associated level sets defined now by:

$$X_k := \{x \mid V^0(x) \leq k\} \qquad (30)$$

Hence, X_k is the set of states that can be steered to the origin in k steps along trajectories satisfying the control and state constraints.

The sets $\{X_k\}$ are easily generated by the recursion $X_{k+1} = \{F^{-1}X_k - F^{-1}G\Omega\} \cap E$ together with the boundary condition $X_0 = \{0\}$. The value function has the constant value k in interior$(X_k \setminus X_{k-1})$ so that it is fully defined

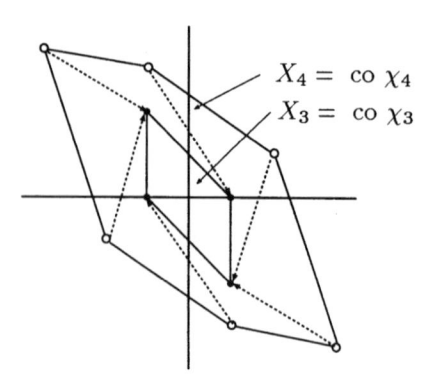

Figure 10. The sets X_k.

by the boundaries of the level sets X_k. If we impose the restriction that Ω is a polytope and E a polyhedron, then, as is easily shown, the sets X_k are all polytopes, and therefore finitely parameterizable (as the convex hull of a finite number of points, or, dually, as satisfying a finite number of linear inequalities) (Mayne and Schroeder, 1995a). However, minimum time controllers are notoriously sensitive, so we consider the perturbed system

$$x[k+1] = Fx[k] + Gu[k] + d[k] \quad (31)$$

where the disturbance $d[k]$ takes values in the polytope $D \subset \mathbb{R}^n$. The disturbance prevents control to the origin; the best that can be hoped for is control to a robust control invariant set X_0. Suppose the controller $u = Kx$ steers the unconstrained, unperturbed, discrete time system to the origin in s steps. Then

$$X_0 := D + F_K D + \ldots + F_K^{s-1} D \quad (32)$$

is a robust control invariant set if $X_0 \subset E$ and $KX_0 \subset \Omega$. The linear control $u = Kx$ keeps any state trajectory starting at $x \in X_0$ in X_0. Let V^0 denote the value function for the *robust* minimum time problem: determine the state feedback law which steers any initial state to X_0 in minimum time for all possible disturbance sequences, i.e.

$$V^0(x) = \inf_u \sup_w \{k \mid x^{u,w}[k; x, 0] \in X_0\} \quad (33)$$

where $u(\cdot)$ and $w(\cdot)$ are admissible state feedback controllers. The associated level sets $\{X_k\}$ are now generated by the recursion

$$X_{k+1} = \{F^{-1}[X_k \ominus W] - F^{-1}G\Omega\} \cap E \quad (34)$$

with boundary condition X_0, where

$$X \ominus W := \{x \in X \mid x + W \in X\} = \bigcap_{w \in W}(X - w) \quad (35)$$

Again, X_k is the set of states that can be *robustly* steered to X_0 in k steps. It can be shown (Bertsekas and Rhodes, 1971; Mayne and Schroeder, 1995b) that each set X_k is a polytope

and, hence, finitely parameterizable. It is thus possible to compute and store the sets X_k, at least for modest problems. Approximation, such as replacing the sets X_k by simpler inner approximants (polytopes or ellipsoids) permits consideration of more complex systems. The information encoded in the sets $\{X_k\}$, which are the level sets of the value function for the minimum time and robust minimum time problems, is sufficient to construct a controller. We present two alternatives below for the robust case.

One Step Model Predictive Controller

The controller determines, firstly, the minimum integer k such that the current state $x \in X_k \setminus X_{k-1}$, secondly, determines the control

$$u(x) = \arg\min\{\|u\| \mid u \in \Omega, \ Fx + Gu \in X'_{k-1}\}$$

and, thirdly, applies this (constant) control (to the discrete or continuous time system) for one step, after which the process is repeated. This controller requires the online solution of a low-dimensional linear program (the dimension is one plus the dimension of the control $u[k]$).

Simplex Controller

This makes use of the following result (see Figure 11). For any integer j, Σ_j is the convex set defined by:

$$\Sigma_j = \{\mu \in \mathbb{R}^j \mid \mu^i \geq 0, \sum_{i=1}^{j} \mu^i = 1\} \quad (36)$$

Suppose

(i) Control $u_i(\cdot)$ steers state x_i to state z_i in time T along a trajectory that satisfies both state and control constraints, $i = 1 \ldots n + 1$,

(ii) $x \in S_1$,

where the simplices S_1 and S_2 are defined by:

$$S_1 := \{\sum_{i=1}^{n+1} \mu^i x_i \mid \mu \in \Sigma_{n+1}\}, \quad (37)$$

$$S_2 := \{\sum_{i=1}^{n+1} \mu^i z_i \mid \mu \in \Sigma_{n+1}\}. \quad (38)$$

Then $x = \sum_{i=1}^{n+1} \mu^i x_i$ for some $\mu \in \Sigma_{n+1}$ and the control $u(\cdot) = \sum_{i=1}^{n+1} \mu^i u_i(\cdot)$ steers the state $x \in S_1$ to the the state $z = \sum_{i=1}^{n+1} \mu^i z_i \in S_2$ in time T along a trajectory that satisfies both state and control constraints.

To use this result for the robust minimum time problem, each polytope X_k is decomposed into a sequence $\{X_{kj} \mid j \in \{1 \ldots j_k\}\}$ of possibly overlapping simplices whose union covers $X_k \setminus X_{k-1}$. The simplex controller determines the simplex X_{kj} in which the current state x lies (where k is, as before, the least integer such that $x \in X_k \setminus X_{k-1}$), determines $\mu \in \Sigma_n$ such that $x = \sum_{i=0}^{n} \mu^i x_{kji}$, sets

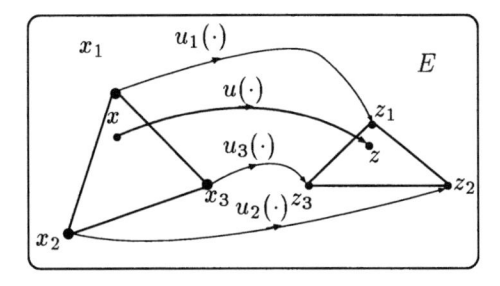

Figure 11. Simplex control.

$u(x) = \sum_{i=0}^{n} \mu^i u_{kji}$, and applies the constant control $u(x)$ to the discrete or continuous time system for time Δ. Here x_{kji}, $i = \{0 \ldots n\}$ are the vertices of X_{kj}, and, for each i, u_{kji} steers the discrete time system from x_{kji} to X_{k-1}; these controls are obtained as a byproduct of the algorithm for constructing the sets X_k. Hence, control $u(x)$ steers either system from state $x \in X_{km} \subset X_k$ to the set X_{k-1}.

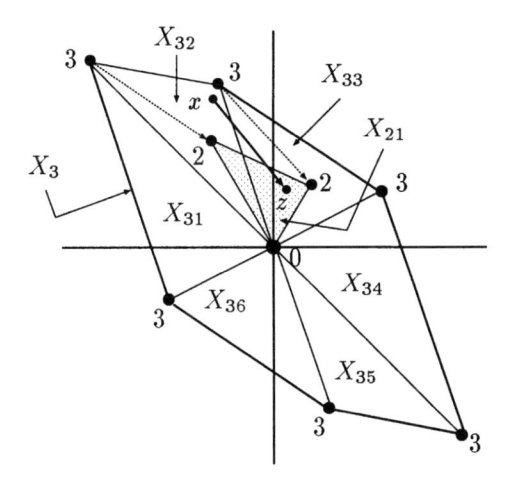

Figure 12. Nonlinear control.

Figure 12 shows how the controller steers a state x in the simplex $X_{32} \subset X_3$ to a state z in the simplex $X_{21} \subset X_2$.

Controller Properties

The robust one step model predictive and simplex controllers steer any state x in X_N to the invariant set X_0 along a trajectory satisfying the control constraints, and maintain the state in X_0 thereafter despite the disturbance.

Normalized step responses for a triple integrator system with control saturation and a six layer control are shown in Figure 13. The faster (slower) responses correspond to smaller (larger) step changes.

The procedure can be extended (Mayne and Schroeder, 1995b) to deal with the tracking problem.

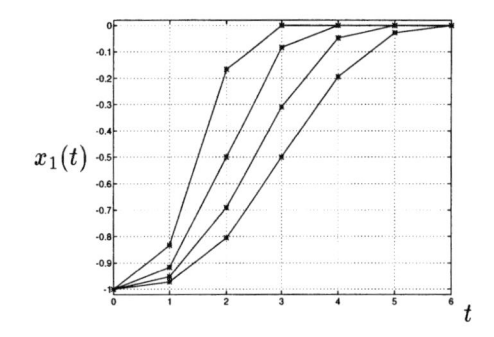

Figure 13. Step responses for triple integrator.

Conclusions

Efficient control of constrained systems, whether linear or nonlinear, requires the controller to be nonlinear.

Constraints impose limits on what can be achieved by any controller. The most important of these is the restriction of the set of states X_{max} which can be controlled. For model predictive controllers, this set is generally compact.

To ensure stability, FHP should include a stability constraint. If the disturbances decay to zero, this constraint takes the form of an equality constraint $x(t+T) = 0$ on the terminal state. If the disturbances are merely bounded, the stability constraint takes the form $x(t+T) \in Z$ where Z is a robust control invariant set (there exists a controller which can maintain the state in this set despite the disturbances).

Model predictive control of linear systems is relatively easily implemented because FHP (the finite horizon control problem solved online) is convex (usually a quadratic program).

Control of nonlinear systems raises formidable problems because FHP is then generally nonconvex.

Variants of MPC which require optimal solutions of FHP must then employ global optimization rather than standard descent methods. This is not generally practical, but there exist some problems, which may be extended by further research, for which global optimization is possible.

There exists a class of systems, namely finite response systems, for which satisfaction of the stability constraint is simple.

Optimal solutions of FHP are not necessary, merely feasible solutions. Obtaining global, feasible solutions to FHP is a considerably simpler task than computing global optimal solutions. Moreover, if variable horizon MPC is employed, feasible or optimal solutions to FHP are easily obtained from the initial solution, at least when disturbances are small or absent.

Variable horizon model predictive control achieves a region of attraction equal to X_{max} if state feedback is employed. Disturbances, state estimation error, parameter error, all reduce the achievable region of attraction. Other controllers generally have a smaller region of attraction.

MPC is not optimal when disturbances or model error

are present, because it does not optimize over the class of feedback controllers.

Alternative methods of controlling linear systems with constraints are becoming available. These require considerable memory, but less online computation.

Acknowledgements

I am grateful to Jim Rawlings for many helpful discussions, to the reviewers for their constructive comments, and to Pierre Scokaert for his assistance in revising the manuscript.

Research supported by National Science Foundation under grant ECS-9312922.

References

Aubin, J. P. (1991). *Viability theory*, Systems & control: foundations & applications, Birkhauser, Boston, Basel, Berlin.

Bequette, B. W. (1991). Nonlinear control of chemical processes: a review, *Ind. Eng. Chem. Res*. **30**: 1391–1413.

Bertsekas, D. P. and Rhodes, I. B. (1971). On the minimax reachability of target sets and target tubes, *Automatica* **7**: 233–247.

Bitmead, R. R., Gevers, M. and Wertz, V. (1990). *Adaptive Optimal Control—The Thinking Man's GPC*, Prentice Hall Int.

Clarke, D. W., Mosca, D. E. and Scattaloni, R. (1991). Robustness of an adaptive predictive controller, *Proceedings of the 30th IEEE Conference on Decision and Control*, Brighton, England, pp. 979–984.

Garcia, C. E., Prett, D. M. and Morari, M. (1989). Model predictive control: Theory and practice—a survey, *Automatica* **25**(3): 335–348.

Genceli, H. and Nikolaou, M. (1993). Robust stability analysis of constrained l_1 norm MPC, *AIChE J*. **39**(12): 1954–1965.

Genceli, H. and Nikolaou, M. (1995). Design of robust constrained model-predictive controllers with Volterra series, *AIChE J*. **41**: 2098–2106.

Gilbert, E. C., Kolmanovsky, I. and Tan, K. T. (1994). Nonlinear control of discrete-time linear systems with state and control constraints: a reference governor with global convergence properties, *Proceedings of the 33rd IEEE Conference on Decision and Control*, Lake Buena Vista, Florida, pp. 144–149.

Gilbert, E. G. and Tan, K. T. (1991). Linear systems with state and control constraints: the theory and application of maximal output admissible sets, *IEEE Transactions on Automatic Control* **AC-36**: 1008–1020.

Keerthi, S. and Gilbert, E. (1988). Optimal, infinite horizon feedback laws for a general class of constrained discrete time systems, *Journal of Optimization Theory and Applications* **57**: 265–293.

Lee, E. B. and Markus, L. (1967). *Foundations of Optimal Control Theory*, Wiley, New York.

Mayne, D. Q. (1995). Optimization in model predictive control, *in* R. Berber (ed.), *Methods of Model Based Process Control*, Vol. Proceedings of NATO Advanced Study Institute, Antalya, Turkey, August 7–17, 1994 of *Nato Advanced Study Institute series: E Applied Sciences 293*, Kluwer Academic Publishers, Dordrecht, The Netherlands.

Mayne, D. Q. and Michalska, H. (1990). Receding horizon control of non-linear systems, *IEEE Transactions on Automatic Control* **35**(5): 814–824.

Mayne, D. Q. and Michalska, H. (1993). Adaptive receding horizon control for constrained nonlinear systems, *Proceedings 32nd IEEE Conference on Decision and Control*, San Antonio, Texas, USA, pp. 1286–1290.

Mayne, D. Q. and Schroeder, W. R. (1994). Nonlinear control of constrained dynamic systems, *International Journal of Control* **60**: 1035–1043.

Mayne, D. Q. and Schroeder, W. S. (1995a). Nonlinear control of constrained dynamic systems, *Communications, Computing, Control and Signal Processing; Proceedings of a meeting in honor of Professor T. Kailath*, Kluwer, Stanford University.

Mayne, D. Q. and Schroeder, W. S. (1995b). Robust nonlinear control of constrained linear systems, *Technical Report UCD-ECE-SCR-95/3*, College of Engineering, University of California, Davis.

Meadows, E., Henson, M., Eaton, J. and Rawlings, J. (1995). Receding horizon control and discontinuous state feedback stabilization, *International Journal of Control* **62**: 1217–1229.

Michalska, H. and Mayne, D. Q. (1991). Receding horizon control of non-linear systems without differentiability of the optimal value function, *Systems & Control Letters* **16**: 123–130.

Michalska, H. and Mayne, D. Q. (1992). Moving horizon observers, *Proceedings IFAC Symposium on Nonlinear Control Systems Design*, Bordeaux, France, pp. 576–581.

Michalska, H. and Mayne, D. Q. (1993). Robust receding horizon control of constrained nonlinear systems, *IEEE Transactions on Automatic Control* **38**: 1623–1632.

Michalska, H. and Mayne, D. Q. (1995). Moving horizon observers and observer-based control, *IEEE Transactions on Automatic Control* **40**(6): 995–1006.

Polak, E. (1993). On the use of consistent approximations in the solution of semi-infinite optimization and optimal control problems, *Math. Programming* **62**: 385–414.

Polak, E. and Mayne, D. Q. (1981). Design of nonlinear feedback controllers, *IEEE Transactions on Automatic Control* **26**(3): 730–733.

Rawlings, J., Meadows, E. and Muske, K. (1994). Nonlinear model predictive control: a tutorial and survey, *Proceedings Adchem '94*, Kyoto.

Rawlings, J. and Muske, K. (1993). The stability of constrained receding horizon control, *IEEE Transactions on Automatic Control* **AC-38**: 1512–1516.

Robertson, D. G., Lee, J. H. and Rawlings, J. B. (1994). A moving horizon-based approach for least squares state estimation, Submitted for publication in *AIChE J*.

Santos, L., deOliveria, N. and Biegler, L. T. (1995). Reliable and efficient optimization strategies for nonlinear model predictive control, *in* J. B. Rawlings (ed.), *Proc. Fourth IFAC Symposium on Dynamics and Control of Chemical Reactors, Distillation Columns and Batch Processes (DYCORD '95)*, Helsingor, Denmark, Elsevier Science, Oxford, pp. 33–38.

Scokaert, P. O. M. and Rawlings, J. (1995). Stability of model predictive control under perturbations, *Proceedings of the IFAC symposium on nonlinear control systems design*, Lake Tahoe, California.

Sriniwas, G. R. and Arkun, Y. (1995). Optimization and convergence issues for MPC algorithms using polynomial ARX models, *Proceedings of the AIChE Meeting*, Miami, Florida. Paper 185c.

Staus, G. H., Biegler, L. T. and Ydstie, B. E. (1996). Adaptive control via non-convex optimization, *in* C. Floudas and

P. Pardolas (eds), *Nonconvex optimization and its applications*, Kluwer, pp. 1–19.

Thomas, Y. A. (1975). Linear quadratic optimal estimation and control with receding horizon, *Electronics Letters* **11**: 19–21.

Vidyasagar, M. (1993). *Nonlinear system analysis*, second edn, Prentice Hall, Englewood Cliffs, New Jersey.

Vinter, R. B. and Michalska, H. (1991). Receding horizon control for nonlinear time-varying systems, *Proceedings 30th IEEE Conference on Decision and Control*, Brighton, England, pp. 75–76.

Wonham, W. M. (1974). *Linear Multivariable Control: A Geometric Approach*, Springer-Verlag, Berlin, Heidelberg, New York.

Yang, T. H. and Polak, E. (1993). Moving horizon control of nonlinear systems with input saturation, disturbances and plant uncertainty, *International Journal of Control* **58**: 875–903.

Zheng, Z. Q. and Morari, M. (1993). Robust stability of constrained model predictive control, *Proceedings of the 1993 American Control Conference*, pp. 379–382.

AN OVERVIEW OF INDUSTRIAL MODEL PREDICTIVE CONTROL TECHNOLOGY

S. Joe Qin
Department of Chemical Engineering
University of Texas
Austin, TX 78712

Thomas A. Badgwell
Department of Chemical Engineering
Rice University
Houston, TX 77251

Abstract

This paper provides an overview of commercially available Model Predictive Control (MPC) technology, based primarily on data provided by MPC vendors. A brief history of industrial MPC technology is presented first, followed by results of our vendor survey of MPC control and identification technology. A general MPC control algorithm is presented, and approaches taken by each vendor for the different aspects of the calculation are described. Identification technology is then reviewed to determine similarities and differences between the various approaches. MPC applications performed by each vendor are summarized by application area. The final section presents a vision of the next generation of MPC technology, with an emphasis on potential business and research opportunities.

Keywords

Industrial survey, Model predictive control.

Introduction

Model Predictive Control (MPC) refers to a class of algorithms that compute a sequence of manipulated variable adjustments in order to optimize the future behavior of a plant. Originally developed to meet the specialized control needs of power plants and petroleum refineries, MPC technology can now be found in a wide variety of application areas including chemicals, food processing, automotive, aerospace, metallurgy, and pulp and paper.

Several authors have published excellent reviews of MPC theoretical issues, including the papers of García et al. (García, Prett and Morari, 1989), Ricker (Ricker, 1991), Morari and Lee (Morari and Lee, 1991), Muske and Rawlings (Muske and Rawlings, 1993) and Rawlings et al. (Rawlings, Meadows and Muske, 1994). The other papers in the present session by Mayne (Mayne, 1996) and Lee (Lee, 1996) summarize the very latest technical developments in MPC control theory. Froisy provides a vendor's perspective on industrial MPC technology and summarizes likely future developments (Froisy, 1994). A recent survey of MPC technology in Japan provides a wealth of information on application issues from the point of view of MPC users (Ohshima, Ohno and Hashimoto, 1995).

The primary purpose of this paper is to present an overview of commercially available MPC technology. A brief history of MPC technology development is presented first. A general MPC controller formulation is then described to provide a basis for discussion of the commercial products. Results of our industrial survey are then presented. The survey is not exhaustive in that several well-known companies either were not asked to participate, *chose* not to participate, or responded too late to be included in this paper. Nevertheless we believe that the products discussed here are sufficiently representative to allow us to draw conclusions regarding the current state of MPC technology. Significantly unique features of each offering are outlined and discussed. MPC applications to date by each vendor are summarized by application area. The final section presents a view of the next generation of MPC technology, emphasizing potential business and research opportunities.

A Brief History of Industrial MPC

This section presents an abbreviated history of industrial MPC technology. Control algorithms are emphasized here because relatively little published information is available on the identification technology.

The development of modern control concepts can be traced to the work of Kalman in the early 1960's, who sought to determine when a linear control system can be said to be optimal (Kalman, 1960a; Kalman, 1960b). Kalman studied a Linear Quadratic Regulator (LQR) designed to minimize an quadratic objective function. The process to be controlled can be described by a discrete-time, deterministic linear state-space model:

$$\begin{aligned}\mathbf{x}_{k+1} &= \mathbf{A}\,\mathbf{x}_k + \mathbf{B}\,\mathbf{u}_k \\ \mathbf{y}_k &= \mathbf{C}\,\mathbf{x}_k\end{aligned} \quad (1)$$

The vector $\mathbf{u}$ represents process inputs, or manipulated variables; vector $\mathbf{y}$ describes process output measurements. The vector $\mathbf{x}$ represents process states. Fig. 1 provides a schematic representation of a state space model. The state vector is defined such that knowing its value at time k and future inputs allows one to predict how the plant will evolve for all future time. Much of the power of Kalman's work relies on the fact that this general process model was used.

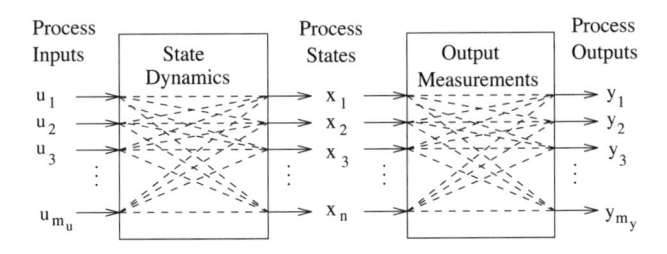

Figure 1. State-space process description.

The objective function to be minimized penalizes squared input and state deviations from the origin and includes separate state and input weight matrices $\mathbf{Q}$ and $\mathbf{R}$ to allow for tuning trade-offs:

$$J = \sum_{j=1}^{\infty}\left(\|\mathbf{x}_{k+j}\|_{\mathbf{Q}}^2 + \|\mathbf{u}_{k+j}\|_{\mathbf{R}}^2\right) \quad (2)$$

where the norm terms in the objective function are defined as follows:

$$\|\mathbf{x}\|_{\mathbf{Q}}^2 = \mathbf{x}^T \mathbf{Q} \mathbf{x} \quad (3)$$

Implicit in this formulation is the assumption that all variables are written in terms of deviations from a desired steady-state. The solution to the LQR problem was shown to be a proportional controller, with a gain matrix $\mathbf{K}$ computed from the solution of a matrix Ricatti equation:

$$\mathbf{u}_k = -\mathbf{K}\,\mathbf{x}_k \quad (4)$$

The infinite prediction horizon of the LQR algorithm endowed the algorithm with powerful stabilizing properties; it was shown to be stabilizing for any reasonable linear plant (stablizable and detectable) as long as the objective function weight matrices $\mathbf{Q}$ and $\mathbf{R}$ are positive definite. A dual theory was developed to estimate plant states from noisy input and output measurements, using what is now known as a *Kalman Filter*. The combined LQR controller and Kalman filter is called a Linear Quadratic Gaussian (LQG) controller. Constraints on the process inputs, states and outputs were not considered in the development of LQG theory.

Although LQG theory provides an elegant and powerful solution to the problem of controlling an unconstrained linear plant, it had little impact on control technology development in the process industries. The most significant of the reasons cited for this failure include (Richalet, Rault, Testud and Papon, 1978; García et al., 1989) :

- constraints
- process nonlinearities
- model uncertainty (robustness)
- unique performance criteria
- cultural reasons (people, education, etc.)

It is well known that the economic operating point of a typical process unit often lies at the intersection of constraints (Prett and Gillette, 1980). A successful industrial controller must therefore maintain the system as close as possible to constraints without violating them. In addition, process units are typically complex, nonlinear, constrained multivariable systems whose dynamic behavior changes with time due to such effects as changes in operating conditions and catalyst aging. Process units are also quite individual so that development of process models from fundamental physics and chemistry is difficult to justify economically. Indeed the application areas where LQG theory had a more immediate impact, such as the aerospace industry, are characterized by physical systems for which it is technically and economically feasible to develop accurate fundamental models. Process units may also have unique performance criteria that are difficult to express in the LQG framework, requiring time dependent output weights or additional logic to delineate different operating modes. However the most significant reasons that LQG theory failed to have a strong impact may have been related to the culture of the industrial process control community at the time, in which instrument technicians and control engineers either had no exposure to LQG concepts or regarded them as impractical.

This environment led to the development, *in industry*, of a more general model based control methodology in which the dynamic optimization problem is solved on-line at each control execution. Process inputs are computed so as to optimize future plant behavior over a time interval known as the *prediction horizon*. In the general case any desired objective function can be used. Plant dynamics are described by an explicit process *model* which can take, in principle, any required mathematical form. Process input and output constraints are included directly in the problem formulation so that future constraint violations are anticipated and prevented. The first input of the optimal input sequence is injected into the plant and the problem is solved again at the next time interval using updated process measurements. In addition to developing more flexible control technology, new process identification technology was developed to allow quick estimation of empirical dynamic models from test data, substantially reducing the cost of model development. This new methodology for industrial process modeling and control is what we now refer to as Model Predictive Control (MPC) technology.

In modern processing plants the MPC controller is part of a multi-level hierarchy of control functions. This is illustrated in Fig. 2, which shows a conventional control structure on the left for Unit 1 and a model predictive control structure on the right for Unit 2. Similar hierarchical structures have been described by Richalet et al. (Richalet et al., 1978) and Prett and García (Prett and García, 1988). At the top of the structure a plant-wide optimizer determines optimal steady-state settings for each unit in the plant. These may be sent to local optimizers at each unit which run more frequently or consider a more detailed unit model than is possible at the plant-wide level. The unit optimizer computes an optimal economic steady-state and passes this to the dynamic constraint control system for implementation. The dynamic constraint control must move the plant from one constrained steady state to another while minimizing constraint violations along the way. In the conventional structure this is accomplished by using a combination of PID algorithms, Lead-Lag (L/L) blocks and High/Low select logic. It is often difficult to translate the control requirements at this level into an appropriate conventional control structure. In the MPC methodology this combination of blocks is replaced by a single MPC controller.

Although the development and application of MPC technology was driven by industry, it should be noted that the idea of controlling a system by solving a sequence of open-loop dynamic optimization problems was not new. Propoi, for example, described a moving horizon controller in 1963 (Propoi, 1963). Lee and Markus (Lee and Markus, 1967) anticipated current MPC practice in their 1967 optimal control text:

> One technique for obtaining a feedback controller synthesis from knowledge of open-loop controllers is to measure the current control process state and then compute very rapidly for the open-loop control function. The first portion of this function is then used during a short time interval, after which a new measurement of the function is computed for this new measurement. The procedure is then repeated.

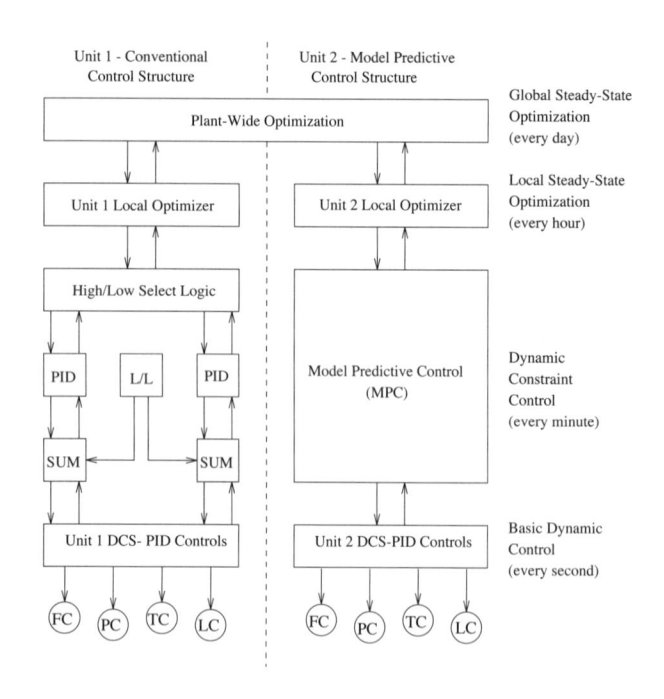

Figure 2. Hierarchy of control system functions in a typical processing plant. Conventional structure is shown at the left; MPC structure is shown at the right.

There is, however, a wide gap between theory and practice. The essential contribution of industry was to put these ideas into practice on operating units. Out of this experience came a fresh set of problems that has kept theoreticians busy ever since.

IDCOM

The first description of MPC control applications was presented by Richalet et al. at a 1976 conference (Richalet, Rault, Testud and Papon, 1976) and later summarized in a 1978 *Automatica* paper (Richalet et al., 1978). They described their approach as Model Predictive Heuristic Control (MPHC). The solution software was referred to as IDCOM, an acronym for Identification and Command. The distinguishing features of the IDCOM approach are:

- impulse response model for the plant, linear in inputs or internal variables

- quadratic performance objective over a finite prediction horizon

- future plant output behavior specified by a reference trajectory

- input and output constraints included in the formulation

- optimal inputs computed using a heuristic iterative algorithm, interpreted as the dual of identification.

Richalet et al. chose a black-box process representation of the process, shown in Fig. 3. From this point of view the process inputs influence the process outputs directly. Process inputs are divided into manipulated variables (MV's) which the controller adjusts, and disturbance variables (DV'S) which are not available for control. Process outputs are referred to as controlled variables (CV's). They chose to describe the relationship between process inputs and outputs using a discrete-time Finite Impulse Response (FIR) model. For the single input, single output (SISO) case the FIR model looks like:

$$y_{k+j} = \sum_{i=1}^{N} h_i \, u_{k+j-i} \quad (5)$$

This model predicts that the output at a given time depends on a linear combination of past input values; the summation weights h_i are the impulse response coefficients. The sum is truncated at the point where past inputs no longer influence the output; this representation is therefore only possible for stable plants.

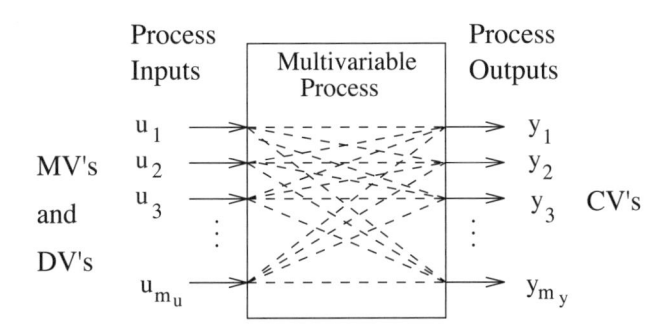

Figure 3. Input-output process description used in industrial MPC technology. Process inputs u_i consist of two types; manipulated variables (MV's) and disturbance variables (DV's). Process outputs are the controlled variables (CV's).

The finite impulse response was identified from plant test data using an algorithm designed to minimize the structural distance between the plant and model impulse responses in parameter space. The resulting iterative algorithm makes small adjustments to the coefficients at each step in such a way that the structural distance continuously decreases. The algorithm was shown to converge to unbiased parameter estimates in the face of noisy output measurements. The control problem was solved using the same algorithm by noting that control is the mathematical dual of identification. In the identification problem one knows the outputs and the inputs and wishes to estimate the coefficients; in the control problem one knows the desired output trajectory and the model coefficients and the goal is to estimate the required inputs. Because the output prediction appears as a dot product of input and coefficient vectors, the same algorithm can be used to find either one. The iterative nature of the control algorithm allows input and output constraints to be checked as the algorithm proceeds to a solution. Because the control law not linear and could not be expressed as a transfer function, Richalet et al. refer to it as *heuristic*. In today's context the algorithm would be referred to as a nonlinear MPC controller.

The MPHC algorithm drives the predicted future output trajectory as closely as possible to a reference trajectory, defined as a first order path from the current output value to the desired setpoint. The speed of the desired closed loop response is set by the time constant of the reference trajectory. This is important in practice because it provides a natural way to control the aggressiveness of the algorithm; increasing the time constant leads to a slower but more robust controller.

Richalet et al. make the important point that dynamic control must be embedded in a hierarchy of plant control functions in order to be effective. They describe four levels of control, very similar to the structure shown in Fig. 2:

- Level 3 - Time and space scheduling of production
- Level 2 - Optimization of setpoints to minimize costs and ensure quality and quantity of production
- Level 1 - Dynamic multivariable control of the plant
- Level 0 - Control of ancillary systems; PID control of valves.

They point out that significant benefits do not come from simply reducing the variations of a controlled variable through better dynamic control at level 1. The real economic benefits come at level 2 where better dynamic control allows the controlled variable setpoint to be moved closer to a constraint without violating it. This argument provides the basic economic motivation for using MPC technology. This concept of a hierarchy of control functions is fundamental to advanced control applications and seems to have been followed by many practitioners. Prett and García., for example, (Prett and García, 1988) describe a very similar hierarchy in their 1988 text *Fundamental Process Control*.

Richalet et al. describe applications of the MPHC algorithm to a Fluid Catalytic Cracking Unit (FCCU) main fractionator column, a power plant steam generator and a Poly-Vinyl Chloride (PVC) plant. All of these examples are constrained multivariable processes. The main fractionator example involved controlling key tray temperatures to stabilize the composition of heavy and light product streams. The controller adjusted product flowrates to compensate for inlet temperature disturbances and to maintain the level of a key internal tray. The power plant steam generator problem involved controlling the temperature and pressure of steam delivered to the turbine. This application is interesting because the process response time varied inversely with load on the system. This nonlinearity was overcome by executing the controller with a variable sample time. Benefits for the main fractionator application were reported as $150,000/yr, due to increasing the flowrate of the light product stream. Combined energy savings from two columns in the PVC plant were reported as $220,000/yr.

DMC

Engineers at Shell Oil developed their own independent MPC technology in the early 1970's, with an initial application in 1973. Cutler and Ramaker presented details of an unconstrained multivariable control algorithm which they named Dynamic Matrix Control (DMC) at the 1979 National AIChE meeting (Cutler and Ramaker, 1979) and at the 1980 Joint Automatic Control Conference (Cutler and Ramaker, 1980). In a companion paper at the 1980 meeting Prett and Gillette (Prett and Gillette, 1980) described an application of DMC technology to an FCCU reactor/regenerator in which the algorithm was modified to handle nonlinearities and constraints. Neither paper discussed their process identification technology. Key features of the DMC control algorithm include:

- linear step response model for the plant

- quadratic performance objective over a finite prediction horizon

- future plant output behavior specified by trying to follow the setpoint as closely as possible

- optimal inputs computed as the solution to a least-squares problem

The linear step response model used by the DMC algorithm relates changes in a process output to a weighted sum of past input changes, referred to as input moves. For the SISO case the step response model looks like:

$$y_{k+j} = \sum_{i=1}^{N-1} s_i \Delta u_{k+j-i} + s_N u_{k+j-N} \quad (6)$$

The move weights s_i are the step response coefficients. Mathematically the step response can be defined as the integral of the impulse response; given one model form the other can be easily obtained. Multiple outputs were handled by superposition. By using the step response model one can write predicted future output changes as a linear combination of future input moves. The matrix that ties the two together is the so-called *Dynamic Matrix*. Using this representation allows the optimal move vector to be computed analytically as the solution to a least-squares problem. Feedforward control is readily included in this formulation by modifying the predicted future outputs. In practice the required matrix inverse can be computed off-line to save computation. Only the first row of the final controller gain matrix needs to be stored because only the first move needs to be computed.

The objective of a DMC controller is to drive the output as close to the setpoint as possible in a least-squares sense with a penalty term on the MV moves. This results in smaller computed input moves and a less aggressive output response. As with the IDCOM reference trajectory, this technique provides a degree of robustness to model error. Move suppression factors also provide an important numerical benefit in that they can be used to directly improve the conditioning of the numerical solution.

Cutler and Ramaker showed results from a furnace temperature control application to demonstrate improved control quality using the DMC algorithm. Feedforward response of the DMC algorithm to inlet temperature changes was superior to that of a conventional PID lead/lag compensator.

In their paper Prett and Gillette (Prett and Gillette, 1980) described an application of DMC technology to FCCU reactor/regenerator control. Four such applications were already completed and two additional applications were underway at the time the paper was written. The overall FCCU control system was implemented in a multi-level hierarchy, with a nonlinear steady-state FCCU model at the top. At the start of each optimization cycle, parameters in the nonlinear model were estimated so as to match model predictions with measured steady-state operating data. The calibrated nonlinear model was then perturbed numerically to generate partial derivatives of each process output with respect to each process input (the matrix of partial derivatives is known as the Jacobian matrix in numerical analysis). The partial derivatives were then used in a Linear Program (LP) to compute a new economic optimal operating point for the FCCU, subject to steady-state process constraints. The optimal process input and output targets were then passed to a DMC algorithm for implementation. As soon as the DMC controller moved the unit to the new steady state the optimization cycle was repeated. This separation of the control system into constrained steady-state optimization and dynamic control is quite similar to the structure described by Richalet et al. and has since become standard in industrial control system design.

The DMC algorithm had the job of moving from the system from one optimal steady-state to another. Although the LP solution provided optimal targets for process inputs and outputs, dynamic disturbances could potentially cause the DMC algorithm to move inputs away from their optimal steady-state targets in order to keep outputs at their steady-state targets. Since moving one input away from its optimal target may be much more expensive than moving another, the control system should determine this trade-off in a rational way. The DMC algorithm was modified to account for such trade-offs by including an additional equation for each input in the process model. The new equation required that the sum of all moves for a particular input should equal the total adjustment required to bring that input to its optimal steady-state target. This allowed the inputs some freedom to move dynamically but required that the steady-state input solution be satisfied in a least-squares sense, with trade-offs determined by the appropriate objective function weights.

Prett and Gillette described additional modifications to the DMC algorithm to prevent violation of absolute input constraints. When a predicted future input came sufficiently close to an absolute constraint, an extra equation was added to the process model that would drive the input back into the feasible region. These were referred to as time variant constraints. Because the decision to add the equation had to be made on-line, the matrix inverse solution had to be recomputed at each control execution. Prett and Gillette developed a matrix tearing solution in which the original matrix inverse could be computed off-line, requiring only the matrix inverse corresponding to active time variant constraints to be computed on-line.

The initial IDCOM and DMC algorithms represent the *first generation* of MPC technology; they had an enormous impact on industrial process control and served to define the industrial MPC paradigm.

QDMC

The original IDCOM and DMC algorithms provided excellent control of unconstrained multivariable processes. Constraint handling, however, was still somewhat ad-hoc. Engineers at Shell Oil addressed this weakness by posing the DMC algorithm as a Quadratic Program (QP) in which input and output constraints appear explicitly. Cutler et al. first described the QDMC algorithm in a 1983 AIChE conference paper (Cutler, Morshedi and Haydel, 1983). García and Morshedi published a more comprehensive description several years later (García and Morshedi, 1986).

Key features of the QDMC algorithm include:

- linear step response model for the plant
- quadratic performance objective over a finite prediction horizon
- future plant output behavior specified by trying to follow the setpoint as closely as possible subject to a move suppression term
- optimal inputs computed as the solution to a quadratic program

García and Morshedi began with a clear and concise presentation of the unconstrained DMC algorithm, including an interesting discussion of tuning. Their experience showed that the DMC algorithm was closed loop stable when the prediction horizon was set long enough to include the steady-state effect of all computed input moves. This is supported by a rigorous proof presented by García and Morari (García and Morari, 1982) which shows that the DMC algorithm is nominally stabilizing for a sufficiently large prediction horizon.

García and Morshedi then show how the DMC objective function can be re-written in the form of a standard QP. Future projected outputs can be related directly back to the input move vector through the dynamic matrix; this allows all input and output constraints to be collected into a matrix inequality involving the input move vector. Although the QDMC algorithm is a somewhat advanced control algorithm, the QP itself is one of the simplest possible optimization problems that one could pose. The Hessian of the QP is positive definite for any reasonable problem and so the resulting optimization problem is convex. This means that a solution can be found readily using standard commercial optimization codes.

The default QDMC algorithm requires strict enforcement of input and output constraints at each point of the prediction horizon. Constraints that are strictly enforced are referred to as *hard constraints*. This is illustrated in Fig. 4. In practice García and Morshedi report that hard output constraints are typically required to be satisfied over only a portion of the horizon which they refer to as the *constraint window*. The constraint window generally starts at some point in the future and continues on to steady state. They report that if non-minimum phase dynamics are present, performance is improved by pushing the constraint window farther into the future. This amounts to ignoring hard output constraints during the initial portion of the closed loop response. It is interesting that Rawlings and Muske recently arrived at the same solution to ensure feasibility of their infinite-horizon algorithm (Rawlings and Muske, 1993) after a careful theoretical analysis. They show that output constraints can be made feasible by relaxing them for a finite time j_1, and they derive an upper bound for j_1.

García and Morshedi report another option for handling output constraints that may be useful when non-minimum phase dynamics are present. When output constraint violations are predicted to occur, one can attempt to minimize the violation in a least-squared sense. This is the *soft constraint* concept illustrated in the middle plot of Fig. 4. García and Morshedi described an approximate implementation of the soft constraint concept using a setpoint approximation, illustrated at the bottom of Fig. 4. The setpoint approximation idea is to guess a priori where the optimal solution will require a constraint violation, and penalize this deviation by adding a setpoint that forces the output to stick to the constraint boundary. One way to guess where output violations will occur is to examine output predictions based on the optimal input solution from the previous time step. Because it is difficult to guess where the true solution at the current time step will require a constraint violation, the setpoint approximation method is generally sub-optimal.

A true soft constraint can be implemented by adding a slack variable to an inequality constraint, and then adding the slack variable to the objective function to be minimized. This approach has been studied by several researchers, including Ricker et al. (Ricker, Subrahmanian and Sim, 1988), Zafiriou and Chiou (Zafiriou and Chiou, 1993), and Genceli and Nikolaou (Genceli and Nikolaou, 1993). Zheng and Morari (Zheng and Morari, 1995) recently analyzed an infinite horizon MPC algorithm with soft output constraints implemented in this way. They show that global asymptotic stability can be guaranteed provided that the plant is not un-

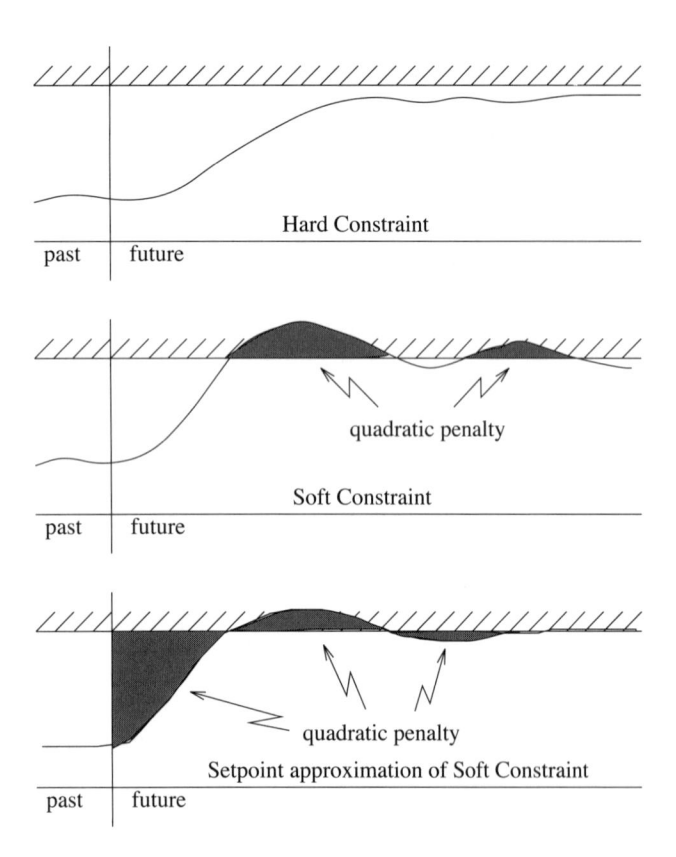

Figure 4. The three basic types of constraint; hard, soft and setpoint approximation. Hard constraints (top) should not be violated in the future. Soft constraints (middle) may be violated in the future, but the violation is penalized in the objective function. Setpoint approximation of constraint (bottom) penalizes deviations above and below the constraint. Shades areas show violations penalized in the dynamic optimization. Adapted from Froisy (Froisy, 1994).

stable.

García and Morshedi wrapped up their paper by presenting results from a pyrolysis furnace application. The QDMC controller adjusted fuel gas pressure in three burners in order to control stream temperature at three locations in the furnace. Their test results demonstrated dynamic enforcement of input constraints and decoupling of the temperature dynamics. They reported good results on many applications within Shell on problems as large as 12x12 (12 process outputs and 12 process inputs). They stated that above all, the QDMC algorithm had proven particularly profitable in an on-line optimization environment, providing a smooth transition from one constrained operating point to another.

The QDMC algorithm can be regarded as representing a *second generation* of MPC technology, comprised of algorithms which provide a systematic way to implement input and output constraints. This was accomplished by posing the MPC problem as a QP, with the solution provided by standard QP codes.

IDCOM-M, HIECON, and PCT

As MPC technology gained wider acceptance, and problems tackled by MPC technology grew larger and more complex, control engineers implementing second generation MPC technology ran into other practical problems. The QDMC algorithm provided a systematic approach to incorporate hard input and output constraints, but there was no clear way to handle an infeasible solution. For example it is possible for a feedforward disturbance to lead to an infeasible QP; what should the control do to recover from infeasibility? The soft constraint formulation is not completely satisfactory because it means that all constraints will be violated to some extent, as determined by the relative weights. Clearly some output constraints are more important than others, however, and should never be violated. Wouldn't it make sense then to shed low priority constraints in order to satisfy higher priority ones?

In practice, process inputs and outputs can be lost in real time due to signal hardware failure, valve saturation or direct operator intervention. They can just as easily come back into the control problem at any sample interval. This means that the structure of the problem and the degrees of freedom available to the control can change dynamically. This is illustrated in Fig. 5, which illustrates the shape of the process transfer function matrix for three general cases. The *square plant* case, which occurs when the plant has just as many manipulated variables (MV's) as controlled variables (CV'S), leads to a control problem with a unique solution. In the real world, square is rare. More common is the *fat plant* case, in which there are more MV's available than there are CV's to control. The extra degrees of freedom available in this case can be put to use for additional objectives, such as moving the plant closer to an optimal operating point. When valves become saturated or lower level control action is lost, the plant may reach a condition in which there are more CV's than MV's; this is the *thin plant* case. In this situation it will not be possible to meet all of the control objectives; the control specifications must be relaxed somehow, for example by minimizing CV violations in a least-squared sense.

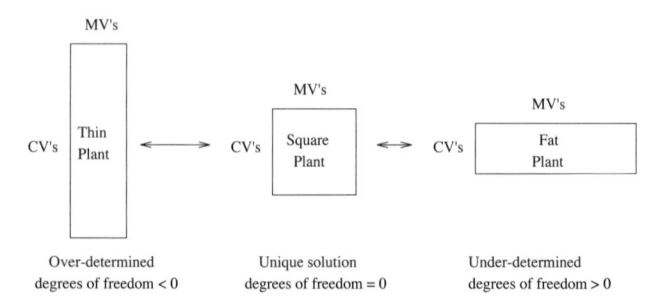

Figure 5. Process structure determines the degrees of freedom available to the controller. Adapted from Froisy (Froisy, 1994).

Fault tolerance is also an important practical issue. Rather than simply turning itself off as signals are lost, a practical MPC controller should remain online and try to make the best of the sub-plant under its control. A major barrier to achieving this goal is that a well conditioned multivariable plant may contain a number of poorly conditioned sub-plants. In practice an MPC controller must recognize and screen out poorly conditioned sub-plants before they result in erratic control action.

It also became increasingly difficult to translate control requirements into relative weights for a single objective function. Including all the required trade-offs in a single objective function means that relative weights have to be assigned to the value of output setpoint violations, output soft constraint violations, inputs moves, and optimal input target violations. For large problems it is not easy to translate control specifications into a consistent set of relative weights. In some cases it does not make sense to include these variables in the same objective function; driving the inputs to their optimal targets may lead to larger violation of output soft constraints, for example. Even when a consistent set of relative weights can be found, care must be taken to avoid scaling problems that lead to an ill-conditioned solution. Prett and García commented on this problem in their text on *Fundamental Process Control* (Prett and García, 1988):

> *The combination of multiple objectives into one objective (function) does not allow the designer to reflect the true performance requirements.*

These issues motivated engineers at Adersa and Setpoint, Inc. to develop a new version of the IDCOM algorithm. The version marketed by Setpoint was called IDCOM-M (the M was to distinguish this from a single input/single output version called IDCOM-S), while the Adersa version was referred to as HIECON (Hierarchical Constraint Control). The IDCOM-M controller was first described in a paper by Grosdidier et al. (Grosdidier, Froisy and Hammann, 1988). A second paper presented at the 1990 AIChE conference describes an application of IDCOM-M to the Shell Fundamental Control Problem (Froisy and Matsko, 1990) and provides additional details concerning the constraint methodology. Distinguishing features of the IDCOM-M algorithm include:

- linear impulse response model of plant

- controllability supervisor to screen out ill-conditioned plant subsets

- multi-objective function formulation; quadratic output objective followed by a quadratic input objective

- controls a single future point in time for each output, called the coincidence point, chosen from a reference trajectory

- a single move is computed for each input

- constraints can be hard or soft, with hard constraints ranked in order of priority

The IDCOM-M controller retains the linear impulse response plant model used by the original IDCOM algorithm. The IDCOM-M controller allows the capability to include purely integrating responses, however. These are assumed to describe the response of the first order derivative of the output with respect to time.

The IDCOM-M algorithm includes a controllability supervisor which decides, based on the current set of available inputs and outputs, which outputs can be independently controlled. The selection is based on the effective condition number of the plant gain matrix; a list of controllability priorities is used to determine which outputs to drop from the problem if an ill-conditioned set is encountered.

An important distinction of the IDCOM-M algorithm is that it uses *two separate objective functions*, one for the outputs and then, if there are extra degrees of freedom, one for the inputs. A quadratic output objective function is minimized first subject to hard input constraints. Each output is driven as closely as possible to a desired value at a single point in time known as the *coincidence point*. The name comes from the fact that this is where the desired and predicted values should coincide. The desired output value comes from a first order reference trajectory that starts at the current measured value and leads smoothly to the setpoint. Each output has two basic tuning parameters; a coincidence point and a closed loop response time, used to define the reference trajectory.

In many cases the solution to the output optimization is not unique. When additional degrees of freedom are present a second input optimization is performed. A quadratic input objective function is minimized subject to equality constraints that preserve the outputs found in the output optimization. The inputs are driven as closely as possible to their *Ideal Resting Values* (IRV's) which may come, for example, from an overlying steady-state optimizer. By default the IRV for a given input is set to its current measured value. The input optimization makes the most effective use of available degrees of freedom without altering the optimal output solution.

The IDCOM-M calculation is greatly simplified by computing a single move for each input. This *input blocking* assumption results in a loss of performance but provides additional robustness to modeling errors. In practice this has been an acceptable trade-off. Badgwell has analyzed the robustness properties of input-blocking for the SISO case (Badgwell, 1995).

In the IDCOM-M context, constraints are divided into hard and soft categories, with the understanding that hard constraints must be ranked in order of priority. When the calculation becomes infeasible, the lowest priority hard constraint is dropped and the calculation is repeated. One can specify several constraints to have the same priority, and it is possible to require that the control turn itself off and notify the operator if constraints above a given priority level cannot be enforced.

Grosdidier et al. (Grosdidier et al., 1988) describe the flow of a typical calculation:

- Determine available process inputs and outputs
- Determine the list of controllable outputs (controllability supervisor)
- Output Optimization
- Input Optimization

Grosdidier et al. provide simulation results for a representative FCCU regenerator control problem. The problem involves controlling flue gas composition, flue gas temperature, and regenerator bed temperature by manipulating feed oil flow, recycle oil flow and air to the regenerator. The first simulation example demonstrates how using multiple inputs can improve dynamic performance while reaching a pre-determined optimal steady-state condition. A second example demonstrates how the controller switches from controlling one output to controlling another when a measured disturbance causes a constraint violation. A third example demonstrates the need for the controllability supervisor. When an oxygen analyzer fails, the controllability supervisor is left with only flue gas temperature and regenerator bed temperature to consider. It correctly detects that controlling both would lead to an ill-conditioned problem; this is because these outputs respond in a very similar way to the inputs. Based on a pre-set priority it elects to control only the flue gas temperature. When the controllability supervisor is turned off the same simulation scenario leads to erratic and unacceptable input adjustments.

Engineers at Profimatics addressed similar issues in the development of their PCT algorithm (Predictive Control Technology). This alogrithm also uses constraint prioritization to recover from infeasibility. Another interesting feature is known as *predict-back*, in which unmeasured disturbances that enter lower level PID loops are estimated and used for feedforward control. This feature is very useful when a lower level PID loop has slow dynamics, where the predict-back estimate helps the MPC controller respond to an unmeasured disturbance much faster.

The IDCOM-M algorithm is one of several that represent a *third generation* of MPC technology; others include the PCT algorithm sold by Profimatics, the RMPC controller developed by Honeywell, and the PFC algorithm developed by Adersa. This generation distinguishes between several levels of constraints (hard, soft, ranked), provides some mechanism to recover from an infeasible solution, addresses the issues resulting from a control structure that changes in real time, and allows for a wider range of process dynamics and controller specifications.

Survey of MPC Technology Products

Commercial MPC technology has developed considerably since the introduction of third generation technology nearly a decade ago. We recently surveyed five MPC vendors in order to assess the current status of commercial MPC technology. We believe that this list is representative in that the technology sold by these companies represents the industrial state of the art; we fully recognize that we have omitted some major MPC vendors from our survey. Some companies were not asked to participate, some *chose* not to participate, and some responded too late to be included in the paper. Only companies which have documented successful MPC applications were asked to participate.

It should be noted that several companies make use of MPC technology developed in-house but were not included in the survey because they do not offer their technology externally (Shell, Exxon, etc.). These MPC packages are either well-known to academic researchers (e.g., QDMC from Shell Oil) or not known at all for proprietary reasons.

The companies surveyed and their product names and acronyms are listed in Table 1. Initial data were collected from industrial vendors of MPC technology using a written survey. Blank copies of the survey are available upon request from the authors. Survey information was supplemented by published papers, product literature (DMC Corp., 1994; Setpoint, Inc., 1993; Honeywell, Inc., 1995), and personal communication between the authors and vendor representatives. Results of the survey are summarized in Tables 2, 3 and 4. In presenting the survey results our intention is to highlight the important features of each algorithm; it is not our intent to determine the superiority of one product versus another. The choice of an appropriate MPC package for a particular application is a complex question that must be answered on a case by case basis; such issues are beyond the scope of this paper.

We focus here on the main aspects of the control and identification technology. We fully understand that a sound industrial offering must address many needs not necessarily related to the mathematics of the algorithms; these include software and hardware compatibility, user interface requirements, personnel training, and configuration and maintenance issues. It should also be clear that the descriptions given here are necessarily incomplete, since every MPC product has proprietary features. With this understanding in mind, we first discuss the process models at the core of MPC technology and then describe the details of a typical MPC calculation. Subsequent sections describe how different MPC vendors approach the different aspects of implementing MPC technology.

Table 1. Companies Surveyed, their Product Names and Acronyms

Company	Acronym	Product Name (Function)
Adersa	HIECON	Hierarchical Constraint Control
	PFC	Predictive Functional Control
	GLIDE	(Identification package)
DMC Corp.	DMC	Dynamic Matrix Control
	DMI	Dynamic Matrix Identification
Honeywell	RMPCT	Robust Model Predictive Control Technology
Profimatics	PCT	Predictive Control Technology
Setpoint Inc.	SMCA	Setpoint Multivariable Control Architecture
	SMC-Idcom	(Multivariable control package)
	SMC-Test	(Plant test package)
	SMC-Model	(Identification package)
Treiber Controls	OPC	Optimum Predictive Control

Table 2. Summary of Reported MPC Vendor Applications by Areas (estimates based on vendor survey; estimates do not include applications by companies who have licensed vendor technology)

Area	DMC Corp.	Setpoint Inc.	Honeywell Profimatics	Adersa	Treiber Controls	Total
Refining	360	320	290	280	250	1500
Petrochemicals	210	40	40	-	-	290
Chemicals	10	20	10	3	150	193
Pulp and Paper	10	-	30	-	5	45
Gas	-	-	5	-	-	5
Utility	-	-	2	-	-	2
Air Separation	-	-	-	-	5	5
Mining/Metallurgy	-	2	-	7	6	15
Food Processing	-	-	-	41	-	41
Furnaces	-	-	-	42	-	42
Aerospace/Defense	-	-	-	13	-	13
Automotive	-	-	-	7	-	7
Other	10	20	-	45	-	75
Total	600	402	377	438	416	2233
First App	DMC:1985	IDCOM-M:1987 SMCA:1993	PCT:1984 RMPCT:1991	IDCOM:1973 HIECON:1986	OPC:1987	
Largest App	603x283	35x28	28x20	-	24x19	

Process Models

The mathematical form of the process model defines the scope of an MPC algorithm. Tables 3 and 4 show that a wide variety of model forms are used in industrial MPC algorithms. All of the control and identification algorithms described here use time-invariant models. A general nonlinear discrete-time state space model may be described as

$$\mathbf{x}_{k+1} = \mathbf{f}(\mathbf{x}_k, \mathbf{u}_k, \mathbf{v}_k, \mathbf{w}_k) \quad (7)$$
$$\mathbf{y}_k = \mathbf{g}(\mathbf{x}_k) + \xi_k \quad (8)$$

where $\mathbf{u}_k \in \Re^{m_u}$ is a vector of m_u MV's, $\mathbf{y}_k \in \Re^{m_y}$ is a vector of m_y CV's, $\mathbf{x}_k \in \Re^n$ is a vector of n state variables, $\mathbf{v}_k \in \Re^{m_v}$ is a vector of m_v measured DV's, $\mathbf{w}_k \in \Re^{m_w}$ is a vector of m_w unmeasured DV's or noise.

The PFC algorithm is the only one considered in this survey that allows for nonlinear and unstable linear internal models. Nonlinear dynamics can be entered in the form of the nonlinear state space model shown above. The PFC algorithm uses transfer functions or ARX models to describe linear unstable dynamics. The remaining MPC products are designed based on Linear Time-Invariant (LTI) process models with stable or integrating dynamics. Nonlinearities may be accounted for in an approximate way by using a local linear model or nonlinear transformation of a specific CV. The SMC-Idcom algorithm, for example, allows the model gains to be adjusted on-line.

A very general discrete-time LTI model form is the linear state space model:

$$\mathbf{x}_{k+1} = \mathbf{A}\mathbf{x}_k + \mathbf{B}_u\mathbf{u}_k + \mathbf{B}_v\mathbf{v}_k + \mathbf{B}_w\mathbf{w}_k \quad (9)$$
$$\mathbf{y}_k = \mathbf{C}\mathbf{x}_k + \xi_k \quad (10)$$

An equivalent transfer function model in the form of matrix fraction description(Kailath, 1980) can be written as:

$$\mathbf{y}_k = [\mathbf{I} - \mathbf{\Phi}_\mathbf{y}(\mathbf{q}^{-1})]^{-1}[\mathbf{\Phi}_\mathbf{u}(\mathbf{q}^{-1})\mathbf{u}_k + \mathbf{\Phi}_\mathbf{v}(\mathbf{q}^{-1})\mathbf{v}_k + \mathbf{\Phi}_w(q^{-1})\mathbf{w}_k] + \xi_k \quad (11)$$

where q^{-1} is a backward shift operator. The output error identification approach (Ljung, 1987) minimizes the measurement error ξ_k, which results in nonlinear parameter estimation. Multiplying $[\mathbf{I} - \mathbf{\Phi}_\mathbf{y}(\mathbf{q}^{-1})]$ on both sides of the above equation results in an *autoregressive model with exogenous inputs* (ARX),

$$\mathbf{y}_k = \mathbf{\Phi}_y(q^{-1})\mathbf{y}_k + \mathbf{\Phi}_u(q^{-1})\mathbf{u}_k + \mathbf{\Phi}_v(q^{-1})\mathbf{v}_k + \mathbf{\Phi}_w(q^{-1})\mathbf{w}_k + \zeta_k \quad (12)$$

where

$$\zeta_k = [\mathbf{I} - \mathbf{\Phi}_\mathbf{y}(\mathbf{q}^{-1})]\xi_\mathbf{k} \quad (13)$$

The equation error identification approach minimizes ζ_k, which is colored noise even though the measurement noise ξ_k is white. For a stable system, a Finite Impulse Response (FIR) model can be derived as an approximation to the transfer function model:

$$\mathbf{y}_k = \sum_{i=1}^{N_u} \mathbf{H}_i^u \mathbf{u}_{k-i} + \sum_{i=1}^{N_v} \mathbf{H}_i^v \mathbf{v}_{k-i} + \sum_{i=1}^{N_w} \mathbf{H}_i^w \mathbf{w}_{k-i} + \xi_k \quad (14)$$

This model form is used by the SMC-Idcom, HIECON, and OPC algorithms. Typically the sample time is chosen so that from 30 to 120 coefficients are required to describe the full open loop response. An equivalent velocity form is useful in identification:

$$\Delta \mathbf{y}_k = \sum_{i=1}^{N_u} \mathbf{H}_i^u \Delta\mathbf{u}_{k-i} + \sum_{i=1}^{N_v} \mathbf{H}_i^v \Delta\mathbf{v}_{k-i} + \sum_{i=1}^{N_w} \mathbf{H}_i^w \Delta\mathbf{w}_{k-i} + \Delta\xi_k \quad (15)$$

An alternative model form is the finite step response model (FSR) (Cutler et al., 1983), given by:

$$\mathbf{y}_k = \sum_{i=1}^{k} \mathbf{S}_i^u \Delta\mathbf{u}_{k-i} + \sum_{i=1}^{k} \mathbf{S}_i^v \Delta\mathbf{v}_{k-i} + \sum_{i=1}^{k} \mathbf{S}_i^w \Delta\mathbf{w}_{k-i} + \xi_k \quad (16)$$

where $\mathbf{S}_0 = \mathbf{0}$ and $\mathbf{S}_j = \mathbf{S}_N$ for $j > N$. Note that the summation goes from the initial time to the current time k. The FSR model is used by the DMC and RMPCT algorithms. The FIR model is related to the FSR model through:

$$\mathbf{H}_i = \mathbf{S}_i - \mathbf{S}_{i-1} \quad (17)$$

The SMC-Idcom and RMPCT algorithms also provide the option to enter Laplace transfer function models. All of the algorithms allow control of processes with integrating dynamics, either by modeling the time derivative of the output response or by using a modified feedback procedure.

A General MPC Control Calculation

MPC controllers are designed to drive the process from one constrained steady state to another. They may receive an optimal steady-state operating point from an overlying optimizer, as shown in Fig. 2, or they may compute an optimal operating point using an internal steady-state optimizer. The general objectives of an MPC controller, in order of importance, are:

1. prevent violation of input and output constraints

2. drive the CV's to their steady-state optimal values (dynamic output optimization)

3. drive the MV's to their steady-state optimal values using remaining degrees of freedom (dynamic input optimization)

4. prevent excessive movement of MV's

5. when signals and actuators fail, control as much of the plant as possible.

The translation of these objectives into a mathematical problem statement involves a number of approximations and trade-offs that define the basic character of the controller. Like any design problem there are many possible solutions; it is no surprise that there are a number of different

MPC control formulations. Table 3 summarizes how each of the MPC vendors has accomplished this translation.

Fig. 6 illustrates the flow of a representative MPC calculation at each control execution. The first step is to read the current values of process inputs (DV's and MV's) and process outputs (CV's). In addition to their numerical values, each measurement carries with it a sensor status to indicate whether the sensor is functioning properly or not. Each MV will also carry information on the status of the associated lower level control function or valve; if saturated then the MV will be permitted to move in one direction only. If the MV controller is disabled then the MV cannot be used for control but can be considered a measured disturbance (DV). The following sections describe other aspects of the calculation in greater detail.

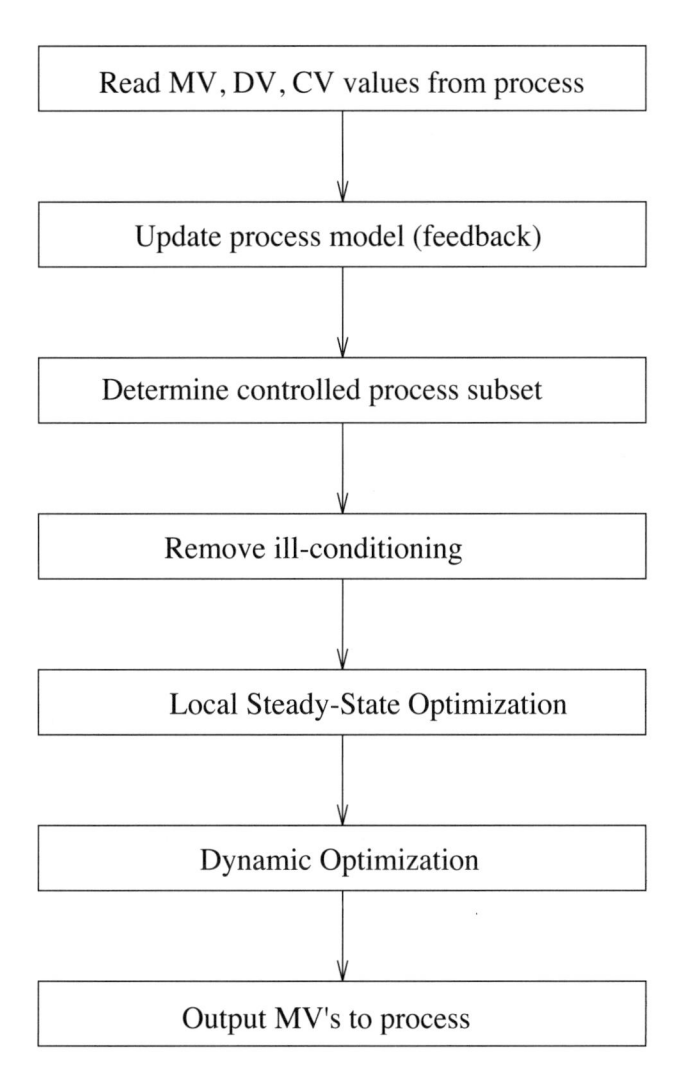

Figure 6. Flow of MPC calculation at each control execution.

Flow chart:
- Read MV, DV, CV values from process
- Update process model (feedback)
- Determine controlled process subset
- Remove ill-conditioning
- Local Steady-State Optimization
- Dynamic Optimization
- Output MV's to process

Output Feedback

The model update step is where feedback enters the loop. All of the industrial MPC algorithms surveyed here use the same form of feedback for stable processes, based on comparing the current measured process output $\mathbf{y}_k^m$ to the current predicted output $\mathbf{y}_k$:

$$\mathbf{b}_k = \mathbf{y}_k^m - \mathbf{y}_k \qquad (18)$$

The bias $\mathbf{b}_k$ term is added to the model for use in subsequent predictions:

$$\mathbf{y}_{k+j} = \mathbf{g}\left(\mathbf{x}_{k+j}\right) + \mathbf{b}_k \qquad (19)$$

This form of feedback is equivalent to assuming an output disturbance that remains constant for all future time (Morari and Lee, 1991; Lee, Morari and García, 1994). Rawlings et al. (Rawlings et al., 1994) show that this method of feedback removes steady-state offset, which provides theoretical support for its use.

Variations of this basic feedback approach are used for the case of integrating dynamics. In RMPCT, for example, both a model bias and the rate of change of the bias are used for CV's that have integrating elements. DMC uses a rotation factor to combine a contribution from the bias term with a contribution from the rate of change of the bias term.

Additional practical details of the bias term calculation should be noted. The bias calculation may be filtered to remove high frequency noise; the RMPCT and SMC-Idcom algorithms provide this option. In some cases the CV measurement may not be available at each control execution; this may happen, for example, when the CV measurement is provided by an analyzer. In this case one can skip the bias update for the affected CV for a number of control intervals. A counter is provided to disable control of the CV if too many executions go by without feedback. The DMC and SMC-Idcom algorithms provide this feature.

Determining the Controlled Sub-process

Once the model has been updated the controller must determine which MV's can be manipulated and which CV's should be controlled. In general, if the measurement status for a CV is good, and the operator has enabled control of the CV, then it should be controlled. An MV must meet the same criteria to be used for control; in addition, however, the lower level control functions must also be available for manipulation. If the lower level controller is saturated high or low, one can add a temporary hard MV constraint to the problem to prevent moving the MV in the wrong direction. If the lower level control function is disabled, the MV cannot be used for control. In this case it should be treated as a DV. From these decisions a controlled subprocess is defined at each control execution. In general the shape of the subprocess changes in real-time as illustrated in Fig. 5.

The RMPCT and DMC algorithms provide an additional mechanism to prevent low level control saturation by including the low level control outputs (e.g., valve position) in the control formulation as additional CVs. These CV's are then forced to stay within high and low saturation limits by treating them as *range* or *zone* control variables. In this configuration, the number of MV's is typically less than the number of CV's, which include both range CV's and setpoint CV's. However, the number of setpoint CV's is typically less than or equal to the number of MV's. If a range CV is well within the saturation constraints, it has no effect on the objective function. In this case, the control configuration reduces to a typical fat or square plant. It should be noted that other CV's which need not to be controlled tightly at a setpoint value (e.g., surge tank level) are also treated as range CV's. When these CV's are within constraint limits, no MV action is required for these CV's. This releases additional degrees of freedom to drive remaining CV's or MV's to their targets.

The DMC, RMPCT, and SMC-Idcom algorithms distinguish between a *critical* CV failure and a *non-critical* CV failure. If a non-critical CV fails, the DMC controller completely removes it from the control calculation. The RMPCT and SMC-Idcom algorithms continue control action by setting the failed CV measurement to the model predicted value, which means there is no feedback for the failed CV. If the non-critical CV fails for a specified period of time, RMPCT drops this CV from the control objective function.

If a critical CV fails, the DMC and RMPCT controllers turn off immediately. The SMC-Idcom algorithm, however, tries to maintain control for the part of the process that is not affected by the critical CV failure. The SMC-Idcom algorithm also allows a sensor to be temporarily turned off for calibration without interrupting control.

The OPC algorithm does not distinguish between critical and non-critical CV's; if a CV fails, its measurement is replaced with a model estimate.

In most MPC products, sensor faults are limited to complete failure that goes beyond pre-specified control limits. Sensor faults such as significant bias and drifting that are within normal limits are generally not detected or identified in these products.

Removal of Ill-conditioning

At any particular control execution, the process encountered by the controller may require excessive input movement in order to control the outputs independently. This problem may arise, for example, if two outputs respond in an almost identical way to the available inputs. Consider how difficult it would be to independently control adjacent tray temperatures in a distillation column, or to control both regenerator and cyclone temperature in an FCCU. It is important to note that this is a feature of the *process* to be controlled; any algorithm which attempts to control an ill-conditioned process must address this problem. For a process with gain matrix G, the condition number of $G^T G$ provides a measure of process ill-conditioning; a high condition number means that small changes in the future error vector will lead to large MV moves.

Although the conditioning of the full control problem will almost certainly be checked at the design phase, it is nearly impossible to check all possible sub-processes which may be encountered during future operation. It is therefore important to examine the condition number of the controlled sub-process at each control execution and to remove ill-conditioning in the internal model if necessary. Three strategies are currently used by MPC controllers to accomplish this; singular value thresholding, controlled variable ranking and input move suppression.

The Singular Value Thresholding (SVT) method used by the RMPCT controller involves decomposing the process model using a singular value decomposition. Singular values below a threshold magnitude are discarded, and a process model with a much lower condition number is then reassembled and used for control. The neglected singular values represent the direction along which the process hardly moves even if a large MV change is applied; the SVT method gives up this direction to avoid erratic MV changes. This method solves the ill-conditioning problem at the expense of neglecting the smallest singular values. If the magnitude of these singular values is small comparing to model uncertainty, it may be better to neglect them anyway. After thresholding, the collinear CV's are approximated with the principal singular direction. In the case of two collinear CV's, for example, this principal direction is a weighted average of the two CV's. Note that the SVT approach is sensitive to output weighting. If one CV is weighted much more heavily than another, this CV will represent the principal singular direction and the results will be approximately equivalent to the priority approach used in SMC-Idcom.

The SMC-Idcom algorithm addresses this issue using a user-defined set of CV controllability ranks. When a high condition number is detected, the controller drops low priority CV's until a well-conditioned sub-process remains. The sub-process will be controlled without erratic input movement but the low priority CV's will be uncontrolled. Note, however, that if a low priority CV is dropped because it's open loop response is close to that of a high priority output, it will follow the high priority CV and will therefore still be controlled in a loose sense. In the case of two collinear CV's having no differentiable priority, it may be desirable to use an weighted average of the two.

Controllers that use input move suppression, such as the DMC and OPC algorithms, provide an alternative strategy for dealing with ill-conditioning. Input move suppression factors increase the magnitude of the diagonal elements of the matrix to be inverted in the least squares solution, directly lowering the condition number. The move suppression values can be adjusted to the point that erratic input movement is avoided for the commonly encountered sub-processes. In the limit of infinite move suppression the condition number becomes one for all sub-processes. There probably exists a set of finite move suppression factors which guarantee that all sub-processes have a condition number greater than a desired threshold value. In the case of two collinear CV's, the move suppression approach gives up a little bit on moving each CV towards its target. The move suppression solution is similar to that of the SVT solution in the sense that it tends to minimize the norm of the MV moves.

Local Steady-State Optimization

The DMC, SMC-Idcom, and RMPCT controllers split the control calculation into a local steady-state optimization followed by a dynamic optimization. Optimal steady-state targets are computed for each input and output; these are then passed to a dynamic optimization to compute the optimal input vector. This should not be confused with the more comprehensive nonlinear optimization that takes place above the MPC algorithm in the plant control hierarchy (see Fig. 2). The local steady-state optimization uses a linear steady-state model which may come from linearizing a comprehensive nonlinear model at each control execution or may simply be the steady-state version of the linear dynamic model used in the dynamic optimization.

The DMC controller uses an LP to do the local steady-state optimization. The optimization is carried out subject to hard input and soft output constraints. The LP is used primarily to enforce steady-state input and output constraints and to determine optimal steady-state input and output targets for the thin and fat plant cases of Fig. 5. Input optimization in the fat plant case is accomplished using economic factors which describe the cost of using each input. In the thin plant case the output error trade-offs are evaluated using slack variable weights.

The RMPCT and PFC algorithms provide an additional level of flexibility by allowing for both linear and quadratic terms in the steady-state objective function. They also includes hard input and output constraints.

The SMC-Idcom algorithm solves the local steady-state optimization problem using a sequence of quadratic programs. CV's are ranked by priority such that control performance of a given CV will never be sacrificed in order to improve performance of a lower priority CV. The prediction error can be spread across a set of CV's by grouping them together at the same priority level. The calculation proceeds by optimizing the highest priority CV's first, subject to hard and soft output constraints on the same CV's and all input hard constraints. Subsequent optimizations preserve the future trajectory of high priority CV's through the use of equality constraints. Likewise inputs can be ranked in priority order so that inputs are moved sequentially towards their optimal values when extra degrees of freedom permit.

Dynamic Optimization Objectives

At the dynamic optimization level, an MPC controller must compute a set of MV adjustments that will drive the process to an optimal steady-state operating point without violating constraints. All of the controllers discussed here can be described (approximately) as minimizing the following dynamic objective function:

$$J = \sum_{j=1}^{P} \|\mathbf{e}_{k+j}^y\|_{\mathbf{Q}_j}^2 + \sum_{j=0}^{M-1} \|\Delta \mathbf{u}_{k+j}\|_{\mathbf{S}_j}^2 + \sum_{j=0}^{M-1} \|\mathbf{e}_{k+j}^u\|_{\mathbf{R}_j}^2 \quad (20)$$

subject to a model constraint:

$$\mathbf{x}_{k+j} = \mathbf{f}(\mathbf{x}_{k+j-1}, \mathbf{u}_{k+j-1}) \quad \forall j = 1, P$$
$$\mathbf{y}_{k+j} = \mathbf{g}(\mathbf{x}_{k+j}) + \mathbf{b}_k \quad \forall j = 1, P$$

and subject to inequality constraints:

$$\underline{\mathbf{y}}_j \leq \mathbf{y}_{k+j} \leq \bar{\mathbf{y}}_j \quad \forall j = 1, P$$
$$\underline{\mathbf{u}} \leq \mathbf{u}_{k+j} \leq \bar{\mathbf{u}} \quad \forall j = 0, M-1$$
$$\Delta \underline{\mathbf{u}} \leq \Delta \mathbf{u}_{k+j} \leq \Delta \bar{\mathbf{u}} \quad \forall j = 0, M-1$$

The objective function 20 involves three conflicting contributions. Future output behavior is controlled by penalizing deviations from a desired response, defined by $\mathbf{e}_{k+j}^y$, over a prediction horizon of length P. Future input behavior is controlled using input penalties defined by $\mathbf{e}_{k+j}^u$ over a control horizon of length M. Rapid input changes are penalized with a separate term involving the moves $\Delta \mathbf{u}_{k+j}$. The relative importance of the objective function contributions are controlled by setting the time dependent weight matrices $\mathbf{Q}_j$, $\mathbf{S}_j$, and $\mathbf{R}_j$; these are assumed to be positive semi-definite.

The solution is a set of M input adjustments:

$$\mathbf{u}^M = (\mathbf{u}_k, \mathbf{u}_{k+1}, \ldots, \mathbf{u}_{k+M-1}) \quad (21)$$

The first input $\mathbf{u}_k$ is injected into the plant and the calculation is repeated.

The DMC, SMC-Idcom, RMPCT, PFC and OPC algorithms use a single quadratic objective function similar to 20. The DMC and OPC algorithms penalize the last input in order to drive the system towards the optimal steady state:

$$\mathbf{R}_j = \mathbf{0}; \forall j < M-1 \quad (22)$$
$$\mathbf{R}_{M-1} \neq \mathbf{0}$$

If the final input weight is large enough and the process is stable, it is approximately equivalent to having a terminal state constraint. If the dynamic solution is significantly different from the LP targets, which means the terminal states are not effectively constrained, the DMC controller will be turned off. This setting may provide nominal stability for DMC controller; further analysis is necessary to firmly establish this result. The final input weights are also applicable to integrating processes where the derivative of the integrator is driven to zero.

The SMC-Idcom, RMPCT, HIECON and PFC controllers do not penalize input moves directly. The SMC-Idcom, HIECON and PFC algorithms use a predefined reference trajectory to avoid aggressive MV moves. The RMPCT controller defines a funnel, which will be described later in the paper, and finds the optimal trajectory and optimal MV moves by minimizing:

$$(\mathbf{u}^M, \mathbf{y}^r) = \arg\min \sum_{j=1}^{P} \|\mathbf{y}_{k+j} - \mathbf{y}_{k+j}^r\|_{\mathbf{Q}}^2 + \|\mathbf{u}_{k+M-1} - \mathbf{u}_{ss}\|_{\mathbf{R}}^2 \quad (23)$$

subject to the funnel constraints. The relative priority of the two terms is set by the two weighting matrices. In the case that the first term is completely satisfied, which is typical due the funnel formulation, the CV error will vanish and the minimization is in fact performed on the second term only. In this case the results will be similar to having two separate objectives on CV's and MV's. In the case of an infinite number of solutions, which is also typical due to "relaxing" the trajectory, a minimum norm solution to the MV's is obtained due to the use of singular value thresholding. This provides a similar effect to the move suppression used in DMC.

Using a single objective function means that trade-offs between the three contributions must be resolved using the the relative weight matrices $\mathbf{Q}_j$, $\mathbf{S}_j$, and $\mathbf{R}_j$. The HIECON algorithm resolves conflicting dynamic control objectives by solving a sequence of multiple optimization problems. The decision is made, a priori, that CV errors are more important than MV errors. A quadratic output optimization problem is solved first, similar to 20 but including only the $\mathbf{e}^y_{k+j}$ terms. For the thin and square plant cases this will provide a unique solution and the calculation terminates. For the fat plant case there are remaining degrees of freedom that can be used to optimize the input settings. For this case the controller solves a separate quadratic input optimization problem, similar to 20 but including only the $\mathbf{e}^u_{k+j}$ terms. The input optimization includes a set of equality constraints that preserve the future output trajectories found in the output optimization. This eliminates the need to set weights to determine the trade-off between output and input errors, at the cost of additional computation.

Constraint Formulations

There are basically two types of constraints used in industrial MPC technology; hard and soft. These are illustrated in Fig. 4. Hard constraints, shown in the top of Fig. 4 are those which should never be violated. Soft constraints, shown in the middle of Fig. 4 are those for which some violation may be allowed; the violation is generally subjected to a quadratic penalty in the objective function.

All of the MPC algorithms allow hard MV maximum, minimum, and rate of change constraints to be defined. These are generally defined so as to keep the lower level MV controllers in a controllable range, and to prevent violent movement of the MV's at any single control execution. The PFC algorithm also accommodates maximum and minimum input acceleration constraints which are useful in mechanical servo control applications. The SMC-Idcom, RMPCT, HIECON and OPC algorithms perform rigorous optimizations subject to the hard input constraints. The DMC and PFC algorithms, however, enforce input hard constraints in a sub-optimal manner in the sense that the solution does not generally satisfy the Karush-Kuhn-Tucker (KKT) conditions. In the DMC algorithm, when an input is predicted to violate a maximum or minimum limit it is set equal to the limit and the calculation is repeated with the MV removed. The PFC algorithm performs the calculation without constraints and then clips the input values if they exceed hard constraints. Both of these techniques will prevent violation of hard input constraints but will, in general, involve a loss of performance that is difficult to predict.

The SMC-Idcom, RMPCT, HIECON and PFC algorithms enforce output soft constraints in an optimal manner as part of the dynamic optimization. The DMC and OPC algorithms, however, use a setpoint approximation to enforce soft constraints. This is illustrated at the bottom of Fig. 4. Setpoints are defined for each soft constraint, resulting in objective function penalties on both sides of the constraint. The output weight is adjusted dynamically, however, so that the weight becomes significant only when the CV comes close to the constraint. When a violation is predicted the weight is increased to a large value so that the control can bring the CV back to its constraint limit. As soon as the CV is within the constraint limit, the LP target is used as the setpoint instead.

The SMC-Idcom and RMPCT algorithms consider hard output constraints only in the local steady-state optimization. This prevents problems which may occur due to enforcing output constraints early in the prediction horizon of the dynamic optimization (see the discussion in Zafiriou's paper, for example (Zafiriou, 1990)). Only the HIECON and PFC algorithms consider hard output constraints in the dynamic optimization. Hard output constraints are ranked in order of priority so that low priority constraints can be dropped when the problem becomes infeasible. The PFC algorithm also considers hard output constraints in the steady-state optimization.

Output and Input Trajectories

Industrial MPC controllers use four basic options to specify future CV behavior; a setpoint, zone, reference trajectory or funnel. These are illustrated in Fig. 7. The shaded areas correspond to the $\mathbf{e}^y_{k+j}$ and $\mathbf{e}^u_{k+j}$ terms in 20. All of the controllers provide the option to drive the CV's to a fixed setpoint, with deviations on both sides penalized in the objective function. In practice this type of specification is very aggressive and may lead to very large input adjustments, unless the controller is detuned in some fashion. This is particularly important when the internal model differs significantly from the process. The DMC and OPC algorithms use move suppression factors for this purpose.

All of the controllers also provide a CV zone control option, designed to keep the CV within a zone defined by upper and lower boundaries. One way to implement zone control is to define upper and lower soft constraints. Other implementations are possible, however. The DMC algorithm, for example, uses a *dynamic weighting* strategy to implement the zone using objective function penalties. When the CV is predicted to lie within its zone, its weight is set to zero so the controller will ignore it. If the CV is near one edge of the zone, its weight increases gradually depending on how close it is to the constraint.

Table 3. Comparison of Industrial MPC Control Technology

Company	DMC Corp	Setpoint Inc	Honeywell IAC	Adersa	Adersa	Treiber Controls
Algorithm	DMC	SMC-Idcom	RMPCT	HIECON	PFC	OPC
Model Forms[1]	SR L,S,I	IR,TF L,S,I	SR,TF,ARX L,S,I	IR L,S,I	SS,TF,ARX L,N,S,I,U	IR L,S,I
Feedback[2]	CD, ID	CD, ID	CD, ID	CD, ID	CD, ID	CD, ID
Rem Ill-cond[3]	IMS	RCV	SVT	-	-	IMS
SS Opt Obj[4]	L[I,O]	Q[O]...Q[I],R	Q[I,O]	-	Q[I,O]	-
SS Opt Const[5]	IH,OS	IH,OH,OS,R	IH,OH	-	IH,OH	-
Dyn Opt Obj[6]	Q[I,O,M],S	Q[I,O]	Q[I,O]	Q[O],Q[I]	Q[I,O],S	Q[I,O,M],S
Dyn Opt Const[7]	IH,OS	IH,OS	IH,OS	IH,OH,OS,R	IC,OH,OS,R	IH,OS
Output Traj[8]	S,Z	S,Z,RT,F	S,Z,F	S,Z,RT	S,Z,RT	S,Z
Output Horiz[9]	FH	FH,MP	FH	FH	FH,MP	FH
Input Param[10]	MM	SM	MM	SM	BF,SM	MM

[1] Model Form: (IR) finite Impulse Response, (SR) finite Step Response, (TF) Laplace Transfer Function, (SS) State-Space, (ARX) Auto-Regressive Exogenous input, (L) Linear, (N) Nonlinear, (S) Stable, (I) Integrating, (U) Unstable
[2] Feedback: (CD) Constant output Disturbance, (ID) Integrating output Disturbance
[3] Removal of Ill-conditioning: (RCV) Ranked Controlled Variables, (SVT) Singular Value Thresholding, (IMS) Input Move Suppression
[4] Steady-State Optimization Objective: (L) Linear, (Q) Quadratic, (I) Inputs, (O) Outputs, (...) multiple sequential objectives, (R) outputs Ranked in order of priority
[5] Steady-State Optimization Constraints: (IH) Input Hard maximum, minimum and rate of change constraints, (OH) Output Hard maximum and minimum constraints
[6] Dynamic Optimization Objective: (Q) Quadratic, (I) Inputs, (O) Outputs, (M) Input Moves, (S) Sub-optimal solution
[7] Dynamic Optimization Constraints: (IH) Input Hard maximum, minimum and rate of change constraints, (OH) Output Hard maximum and minimum constraints, (OS) Output Soft maximum and minimum constraints, (IC) Input Clipped maximum, minimum and rate of change constraints (sub-optimal), (R) constraints Ranked in order of priority
[8] Output Trajectory: (S) Setpoint, (Z) Zone, (RT) Reference Trajectory, Funnel (F),
[9] Output Horizon: (FH) Finite Horizon, (MP) Multiple Point
[10] Input Parameterization: (SM) Single Move, (MM) Multiple Move, (BF) Basis Functions

Table 4. Comparison of Industrial MPC Identification Technology

Product	Test Protocol	Model Form	Estimation Method	Uncertainty Bound
DMC	multi-steps	(velocity) FIR	modified LS	No
SMC-Model	PRBS, step	FIR, ARX	output error LS[1]	Yes
RMPCT	PRBS, step	FIR, ARX	LS, G-N[2]	Yes
Glide	non-PRBS	G(s)	GD[3],G-N, GM[4]	Yes
OPC	PRBS, step, pulse, operating data	FIR, ARMAX	LS	Yes

[1] LS: Least Squares
[2] G-N: Gauss-Newton
[3] GD: Gradient Descent
[4] GM: Global Methods

If the CV violates the constraint, the weight is set to a large value. This causes CV's which violate or almost violate constraints to be driven towards the zone boundaries.

The SMC-Idcom, HIECON, and PFC algorithms provide a CV reference trajectory option. The reference trajectory idea is to bring the CV up to the setpoint more slowly, in order to avoid overshoot. A first order curve is drawn from the current CV value to the setpoint, with the speed of the response determined by a trajectory time constant. Future CV deviations from the reference trajectory are penalized. In the limit of a zero time constant the reference trajectory reverts back to a pure setpoint; for this case, however, the controller would be sensitive to model mismatch unless some other strategy such as move suppression is also being used. A drawback of the reference trajectory formulation is that it penalizes the output when it happens to drift too quickly towards the setpoint, as might happen in response to an unmeasured disturbance. If the CV moves too quickly due to model mismatch, however, the reference trajectory is beneficial in that it will slow down the CV and minimize overshoot. The reference trajectory can be interpreted mathematically as a filter in the feedback path, similar to the robustness filter recommended by IMC theory (Morari and Zafiriou, 1989). In general, as the reference trajectory time constant increases, the controller is able to tolerate larger model mismatch.

The RMPCT algorithm attempts to keep each CV within a user defined zone, with setpoints defined by setting the maximum and minimum zone limits equal to each other. When the CV goes outside the zone, the RMPCT algorithm defines a CV funnel, shown at the bottom of Fig. 7, to bring the CV back within its range. The slope of the funnel is determined by a user defined performance ratio, defined as the desired time to return to the zone divided by the open loop response time. A weighted average open loop response time is used for multivariable systems.

The SMC-Idcom algorithm uses a variation of the funnel when a zone CV falls out of it's range. In this case a funnel is defined using a one-sided reference trajectory that terminates just inside the zone boundary.

A potential advantage of the funnel is illustrated in Fig. 8. Consider the case of a reference trajectory, shown on the left side of Fig. 8. If a disturbance causes the predicted future CV to reach the setpoint more quickly than the reference trajectory allows, the controller will take action to bring the CV back down to the defined trajectory. In the same situation the funnel shown on the right side of Fig. 8 would take no action at all.

All of the MPC algorithms surveyed here provide MV setpoints to drive the inputs towards their optimal values when there are sufficient degrees of freedom. The SMC-Idcom algorithm also provides an option to ramp the MV along a straight line to it's optimal value.

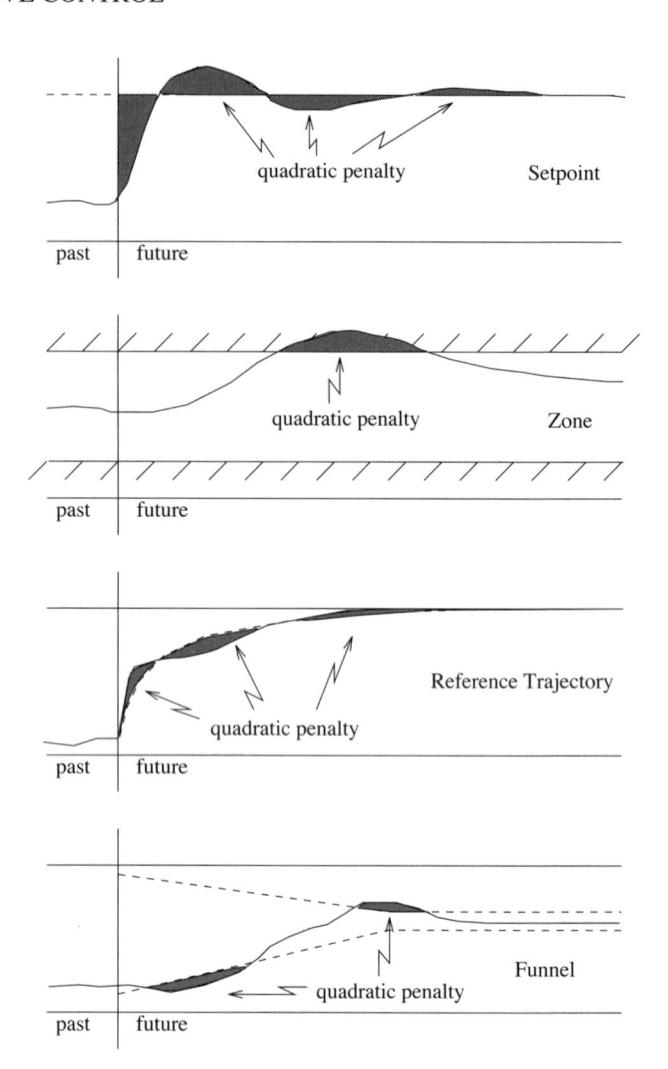

Figure 7. Four options for specifying future CV behavior; setpoint, zone, reference trajectory and funnel. Shaded areas show violations penalized in the dynamic optimization.

Output Horizon and Input Parameterization

Industrial MPC controllers generally evaluate future CV behavior over a finite set of future time intervals called the *prediction horizon*. This is illustrated at the top of Fig. 9. The finite output horizon formulation is used by all of the algorithms discussed in this paper. The length of the horizon P is a basic tuning parameter for these controllers, and is generally set long enough to capture the steady-state effects of all computed future MV moves.

The SMC-Idcom and PFC controllers allow the option to simplify the calculation by considering only a subset of points in the prediction horizon. This concept is illustrated at the bottom of Fig. 9. These points are called *coincidence points* because the desired and predicted future outputs are required to coincide at these points. A separate set of coincidence points can be defined for each output, which is useful when one output responds quickly relative to another. The full finite output horizon can be selected as a special case.

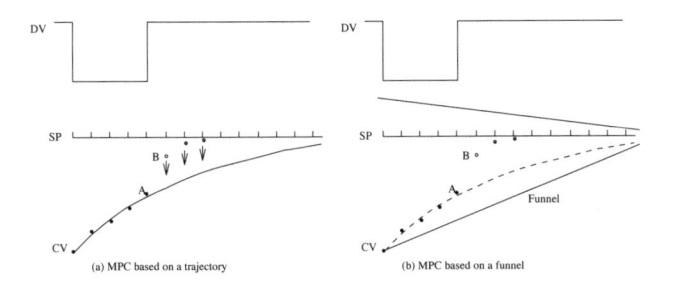

Figure 8. MPC based on a funnel allows a CV to move back to the setpoint faster than a trajectory would require if a pulse disturbance releases. A trajectory based MPC would try to move away from the setpoint to follow the trajectory.

Industrial MPC controllers use three different methods to parameterize the MV profile; these are illustrated in Fig. 10. The DMC, RMPCT, and OPC algorithms compute a set of future moves to be spread over a finite *control horizon*, as shown at the top of Fig. 10. The length of the control horizon M is another basic tuning parameter for these controllers. Increased performance is obtained as M increases, at the expense of additional computation.

The SMC-Idcom and HIECON algorithms compute a single future input move, as shown in the middle of Fig. 10. This greatly simplifies the calculation for these algorithms, which is helpful because they solve a series of optimization problems at each control execution. The use of a single move involves a sacrifice of closed loop performance that is difficult to quantify, however.

The PFC controller parameterizes the input function using a set of polynomial basis functions. A possible solution is illustrated at the bottom of Fig. 10. This allows a relatively complex input profile to be specified over a large (potentially infinite) control horizon, using a small number of unknown parameters. This may provide an advantage when controlling nonlinear systems. Choosing the family of basis functions establishes many of the features of the computed input profile; this is one way to ensure a smooth input signal, for example. If a polynomial basis is chosen then the order can be selected so as to follow a polynomial setpoint signal with no lag. This feature is important for mechanical servo control applications.

Control Design and Tuning

The MPC control design and tuning procedure is generally described as follows (DMC Corp., 1994; Honeywell, Inc., 1995; Setpoint, Inc., 1993):

1. From the stated control objectives, define the size of the problem, and determine the relevant CV's, MV's, and DV's

2. Test the plant systematically by varying MV's and DV's; capture and store the real-time data showing how the CV's respond

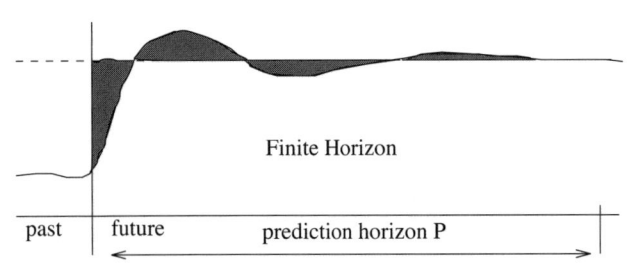

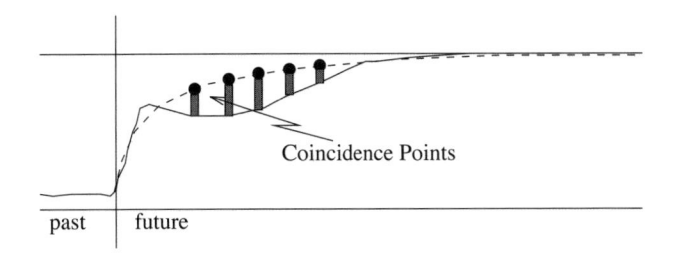

Figure 9. Output horizon options. Finite horizon (top) includes P future points. A subset of the prediction horizon, called the coincidence points (bottom) may also be used. Shaded areas show violations penalized in the dynamic optimization.

3. Derive a dynamic model from the plant test data using an identification package

4. Configure the MPC controller and enter initial tuning parameters

5. Test the controller off-line using closed loop simulation to verify the controller performance.

6. Download the configured controller to the destination machine and test the model predictions in *open-loop* mode

7. Commission the controller and refine the tuning as needed.

All of the MPC packages surveyed here provide software tools to help with the control design, process identification and closed loop simulation steps. A significant amount of time is currently spent at the closed loop simulation step to verify acceptable performance of the control system. Typically tests are performed to check the regulatory and servo response of each CV, and system response to violations of major constraints is verified. The final tuning is then tested for sensitivity to model mismatch by varying the gain and dynamics of key process models. However, even the most thorough simulation testing usually cannot exhaust all possible scenarios.

Controller tuning is always required not only for stability, but also for trade-offs between performance and robustness. It is true that a high performance controller will not be robust with respect to model mismatch, but a low performance controller is not necessarily robust. Most MPC

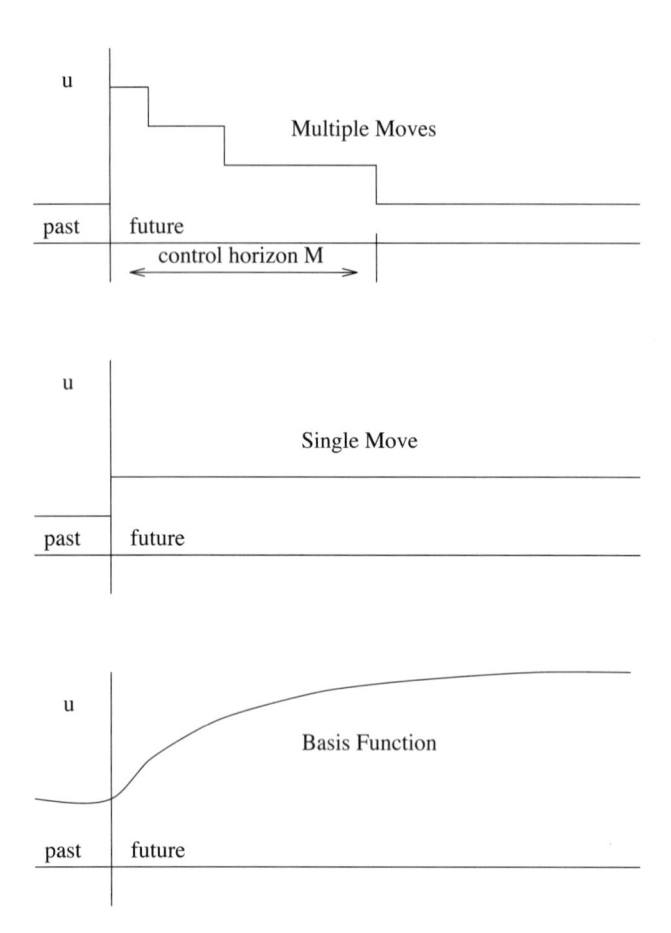

Figure 10. Input parameterization options. Multiple move option (top), single move option (middle), basis function parameterization (bottom).

products provide tuning "knobs" to de-tune the controller. In DMC and OPC, two types of tuning parameters are used: (1) move suppression factors, which are weights on $\Delta \mathbf{u}$ and (2) equal concern error factors, which are the inverse of output weights. The move suppression factors change the aggressiveness of the controller, while the equal concern error factors normalize the importance of each CV.

The SMC-Idcom, HIECON, and PFC controllers use the time constant of the reference trajectory as the main tuning parameter. Smaller time constants demand more aggressive control, while larger time constants result in less aggressive action. One may start with the trajectory time constant equal to the open loop time constant of the CV, then refine the tuning based on performance/robustness trade-offs.

A performance ratio is provided in RMPCT which is defined as the ratio of the required closed loop settling time to the weighted-average open loop settling time. The performance ratio is used to determine the length of the funnel, which is somewhat similar to the settling time of a setpoint trajectory. A performance ratio equal to one means that the closed loop settling time is equal to the open loop settling time. A performance ratio less than one results in a more aggressive controller. Only one tuning parameter per CV needs to be specified. Independent tuning is available in RMPCT for feedforward control, which allows the user to achieve faster response in feedforward control than in setpoint tracking.

The RMPCT package provides a min-max design procedure in which the user enters estimates of model uncertainty directly. Tuning parameters are computed to optimize performance for the worst case model mismatch. Robustness checks for the remaining MPC controllers are performed by closed loop simulation.

All of the MPC packages provide a way to test final controller performance by closed loop simulation. It is particularly important to check the response of the final controller with respect to key constraint violations. It is almost impossible, however, to test all possible situations of active constraints for a realistic problem. The problem becomes even more intractable if one wishes to test performance in the presence of model mismatch. This is one place where academic research can help industrial MPC practice significantly. Rawlings and Muske (Rawlings and Muske, 1993), for example, have shown that their infinite horizon MPC algorithm is nominally stabilizing in the presence of constraints if the initial optimization problem is feasible. Additional work is needed to extend this result to the case of an imperfect plant model. Vuthandam et al. (Vuthandam, Genceli and Nikolaou, 1995) have done this already for a modified QDMC algorithm; their results should be useful to many industrial practitioners.

Identification Test Protocol

Test signals are required to excite both steady-state (low frequency) and dynamic (medium to high frequency) dynamics of a process. A process model is then identified from the process input-output data. DMCC believes that the plant test is the single most important phase in the implementation of DMC controllers. To prepare for a formal plant test, a pretest is usually necessary for three reasons: (i) to step each MV and adjust existing instruments and PID controllers; (ii) to obtain the time to steady state for each CV; and (iii) to obtain data for initial identification.

Most identification packages test one (or at most several) manipulated variables at a time and fix other variables at their steady state. This approach is valid as long as the process is assumed linear and superposition works. A few packages such as DMI and SMC-Test allow several MV's to change simultaneously with uncorrelated signals for different MV's. For DMI, the plant test is run 24 hours a day with engineers monitoring the plant. Each MV is stepped 8 to 15 times, with the output (CV) signal to noise ratio at least six. The plant test may take up to 10 to 15 days, depending on the time to steady state and number of variables of the unit. Two requirements are imposed during the test: (i) no PID configuration or tuning changes are allowed; and (ii) operators may intervene during the test to avoid critical situations, but no synchronizing or correlated moves are allowed. One may merge data from multiple test periods, which allows the user to cut out a period of data which may be corrupted with disturbances.

If the lower level PID control tuning changes significantly then it may be necessary to construct a new process model. A model is identified between the input and output, and this is combined by discrete convolution with the new input setpoint to input model. The SMC-Model package provides a convenient interface for such calculations.

It appears that PRBS or PRBS-like stepping signals are the primary test signals used by the identification packages. The GLIDE package uses a binary signal in which the step lengths are optimized in a dedicated way. Others use a step test or pulse test in combination with PRBS (e.g., OPC, SMC-Test, and RMPCT). As a special case, OPC allows one to use operating data as the ultimate source to build steady-state models for quality variables (Lines, Hartlen, Paquin, Treiber, de Tremblay and Bell, 1993). These variables may be measured by an analytical sensor or laboratory analysis. The following steady-state model is used for analyzer variables:

$$\mathbf{y}_k = \mathbf{y}_{k-1} + \mathbf{K}(\mathbf{u}_k - \mathbf{u}_{k-1}) + \mathbf{K}_f(\mathbf{y}_k^m - \mathbf{y}_{k-d}) \quad (24)$$

where $\mathbf{K}$ is the steady-state gain matrix identified from the operating data. The vector $\mathbf{y}_k^m$ represents current analyzer measurements which have a time delay represented by d. $\mathbf{K}_f$ is an adjustable feedback gain matrix to correct the model estimation. The entire relation may be thought of as a steady-state Kalman filter. The same mechanism can be applied to dynamic models as well.

Model Forms and Parameter Estimation

The model forms used in identification are generally more diversified than those used for control. Table 4 lists various model forms used in different MPC products. The identification approaches in the MPC products are mainly based on minimizing the following least squares criterion,

$$J = \sum_{k=1}^{L} \|\mathbf{y}_k - \mathbf{y}_k^m\|^2 \quad (25)$$

using either an equation error approach or an output error approach (Ljung, 1987). The major difference between the equation error approach and the output error approach appears in identifying ARX or transfer function models. In the equation error approach, past output *measurements* are fed back to the model in Eqn. 12,

$$\mathbf{y}_k = \mathbf{\Phi}_y(q^{-1})\mathbf{y}_k^m + \mathbf{\Phi}_u(q^{-1})\mathbf{u}_k + \mathbf{\Phi}_v(q^{-1})\mathbf{v}_k \quad (26)$$

while in the output error approach, the past model output *estimates* are fed back to the model,

$$\mathbf{y}_k = \mathbf{\Phi}_y(q^{-1})\mathbf{y}_k + \mathbf{\Phi}_u(q^{-1})\mathbf{u}_k + \mathbf{\Phi}_v(q^{-1})\mathbf{v}_k \quad (27)$$

The equation error approach formulates a linear least squares problem, but the estimates are biased even though the measurement noise ξ in Eqn. 11 is white. The output error approach is unbiased given white measurement noise. However, the ARX model parameters appear nonlinearly in the model, which requires nonlinear parameter estimation. One may also see that the equation error approach is a one-step ahead prediction approach with reference to $\mathbf{y}_k^m$, while the output error approach is a long range prediction approach since it does not use $\mathbf{y}_k^m$.

Using FIR models results in a linear-in-parameter model and an output error approach, but the estimation variance may be inflated due to possible overparametrization. In DMI, a least squares method is used to estimate FIR model parameters in velocity form (Eqn. 15). The advantage of using the velocity form is to reduce the effect of a step-like unmeasured disturbance (Cutler and Yocum, 1991). However, the velocity form is sensitive to high frequency noise. Therefore, pre-smoothing for the process data is done in DMI before fitting models to the data. The FIR coefficients are then converted into FSR coefficients for control.

RMPCT adopts a three-step approach: (i) identify an FIR model using least squares; (ii) fit the FIR coefficients to a low-order ARX model to smooth out induced variance due to possible overparametrization in the FIR model. The output error approach is used to fit the ARX model and a Gauss-Newton method is used to estimate the parameters; and (iii) convert the ARX models into Laplace transfer functions. When the model is used in control, the transfer function models are discretized into FSR models based on a given sampling interval. The advantage of this approach is that one has the flexibility to choose different sampling intervals than that used in data collection. OPC provides options to identify FIR and ARMAX models directly from data. The ARMAX model is used on a single-input-single-output basis. Note that OPC estimates the noise dynamics by using moving average terms in the model. These models are finally converted into FIR models for control.

It appears that different philosophies exist in choosing the model forms for identification. DMC believes that complex dynamics can only be identified with high-order FIR models and worries little about overparametrization. A few other products are more concerned with overparametrization which may induce unrealistic variance, and use low order transfer function models as an alternative.

Model uncertainty bounds are provided in several products such as OPC, RMPCT, and SMC-Model. In GLIDE, continuous transfer function models are identified directly by using gradient descent or Gauss-Newton approaches. Then model uncertainty is identified by a global method, which finds a region in the parameter space where the fitting criterion is less than a given value. This given value must be larger than the minimum of the criterion in order to find a feasible region.

Most products apply nonlinear transformations to variables that exhibit significant nonlinearity. For example, a logarithm transformation is often performed on composition variables.

Applications Summary

Table 2 summarizes the reported applications experience of each MPC vendor surveyed. Note that this is a count of completed MPC *applications* reported by each vendor; for a particular problem one vendor may use a single large controller while another may use several smaller controllers. In some cases a single controller is designed and then subsequently used in thousands of copies; this may happen with an automobile application, for example. Note also that this is a count of MPC applications performed by the vendors themselves; this does not include in-house applications performed by licensees of vendor technology. Vendors were given wide latitude to report the numbers in any distribution that they wished to use.

The total number of reported MPC applications is currently over 2200. All of the vendors report a considerable number of applications in progress so it is likely that this number will continue to increase rapidly. Table 2 shows that MPC technology can now be found in a wide variety of application areas. The majority of applications (67 %) are in refining, one of the original application areas, where MPC technology has a solid track record of success. A significant number of applications can also be found in petrochemicals and chemicals, although it has taken longer for MPC technology to break into these areas. Significant growth areas include the chemicals, pulp and paper, food processing, aerospace and automotive industries.

The DMC Corporation reports the largest total number of applications, at 600, with the other four vendors reporting between 350 and 450 applications each. Table 2 shows that DMC, Setpoint, Honeywell and Treiber Controls are highly focused in refining and petrochemicals, with a handful of applications in other areas. Adersa appears to have the broadest range of experience with applications in the food processing, aerospace and automotive areas, among others.

The bottom of Table 2 shows that largest applications to date by each vendor, in the form of (outputs)x(inputs). The numbers show a difference in philosophy that is a matter of some controversy. The DMC Corporation recommends solving a large control problem with a single controller whenever possible; they report an olefins application with 603 outputs and 283 inputs. Other vendors prefer to break the problem up into meaningful sub-processes. This is an issue that needs further attention from the academic community.

Limitations of Existing Technology

Morari and Lee pointed out several problems with existing technology in their CPC-IV paper (Morari and Lee, 1991). Muske and Rawlings have also pointed out limitations of existing industrial MPC technology (Muske and Rawlings, 1993). These include:

- impulse and step response models are over-parameterized and limit application of the algorithm to strictly stable processes

- sub-optimal solution of the dynamic optimization

- sub-optimal feedback (constant output disturbance assumption)

- tuning is required to achieve nominal stability

- model uncertainty is not addressed adequately

Impulse and step response models are known to be over-parameterized. The dynamics of a first order process, for example, can be described by three numbers using a parametric model (gain, time constant and deadtime). An impulse response will typically require from 30 to 120 coefficients to describe the same dynamics. These difficulties can be overcome at the identification step by first identifying a low order model and then calculating the impulse response coefficients. However the impulse response model still requires more storage space than is necessary. A potentially more significant problem with the impulse and step response models is that they are limited to strictly stable processes. While it is certainly possible to modify the algorithms to accommodate a pure integrator, these modifications may lead to other problems, such as adding the derivative of a noisy output signal into the feedback path. It is not possible, in general, to represent an unstable process using an impulse response model. All of these problems can be overcome by using an auto-regressive parametric model form such as an ARX or state-space model.

Sub-optimal solutions to the dynamic optimization 20 are used in several of the packages, presumably in order to speed up the solution time. This seems difficult to justify for the case of refining and petrochemical applications, where the controllers run on the order of once each minute, unless it can be shown that the sub-optimal solution is always very nearly optimal. For high speed applications where the controller must execute in a few milliseconds, such as tracking the position of a missile, it may not be feasible to solve a QP at every control execution. For this case a good sub-optimal solution may be the only option.

The bias update feedback technique used by industrial MPC controllers can be interpreted as assuming an output disturbance that remains constant for all future time (Morari and Lee, 1991; Muske and Rawlings, 1993). This is probably the best assumption that can be used in the absence of detailed disturbance information, but better feedback is possible if the distribution of disturbances can be characterized more carefully. Morari and Lee have shown how to extend MPC technology to achieve better feedback while still retaining the step response model (Morari and Lee, 1991). Muske and Rawlings have demonstrated how better performance can be achieved by using a state space model and an optimal state observer (Muske and Rawlings, 1993).

Tuning MPC controllers for stable operation in the presence of constraints may be difficult, even when the process model is perfect. This is why so much effort is spent on closed loop simulation prior to commissioning a controller. Rawlings and Muske addressed this issue directly in the development of their infinite horizon MPC algorithm (Rawlings and Muske, 1993). They used the LQR controller of Kalman (Eqn. 2) as a starting point. The Kalman LQR is stable for any reasonable choice of tuning parameters, due to the use of an infinite prediction horizon. They developed an infinite horizon MPC controller that is guaranteed to be closed loop stable, for the nominal case, *in the presence of constraints*. Feasibility of the initial QP is enough to guarantee constrained stabilizability. When the QP is infeasible, Muske and Rawlings propose giving up on all of the constraints for a short period of time, much like the constraint window idea used by the QDMC algorithm (García and Morshedi, 1986). Other options for recovering feasibility are also possible, such as dropping low priority constraints. Zheng and Morari (Zheng and Morari, 1995) recently analyzed an infinite horizon MPC algorithm which uses soft output constraints to avoid infeasibility. They show that the closed loop system using output feedback is stable for a strictly stable plant.

Model uncertainty is not addressed adequately by current MPC technology. While most of the identification packages provide estimates of model uncertainty, only one vendor (Honeywell) provides a convenient way to use this information in the control design. All of the MPC algorithms provide a way to detune the control to improve robustness, but the trade-off between performance and robustness is generally not very clear. Until this connection is made, it will not be possible to determine when a model is accurate enough for a particular control application. This is one area where academic research can help. Vuthandam et al. for example, have recently presented robust stability conditions for a modified QDMC algorithm (Vuthandam et al., 1995). More research is needed in this area.

Next-Generation MPC Technology

MPC vendors were asked to describe their vision of next-generation MPC technology. Their responses were combined with our own views and the earlier analysis of Froisy (Froisy, 1994) to come up with a composite view of future MPC technology.

Basic Controller Formulation

Because it is so difficult to express all of the relevant control objectives in a single objective function, next-generation MPC technology will probably utilize multiple objective functions. The infinite prediction horizon has beneficial theoretical properties and will probably become a standard feature. Output and input trajectory options will include setpoints, zones, trajectories and funnels. Input horizons will include options for multiple moves or parameterization using basis functions.

Nonlinear MPC

MPC using nonlinear models is likely to become more common as users demand higher performance and new software tools make nonlinear models more readily available. Developing adequate nonlinear empirical models may be very challenging, however. Test signals such as PRBS that are adequate for linear models are not likely to provide adequate excitation of nonlinear systems. Also there is no model form that is clearly suitable to represent general nonlinear processes. Froisy (Froisy, 1994) points out that second order Volterra models may bridge the gap between linear empirical models and nonlinear fundamental models in the near future. Genceli and Nikolaou, for example, have studied the use of second order Volterra series with a modified QDMC controller (Genceli and Nikolaou, 1995). However, nonlinear empirical models such as Volterra series or neural networks do not seem to extrapolate well.

An alternative approach would be to use first-principles models developed from well known mass, momentum, and energy conservation laws. However, the cost of developing a reasonably accurate first-principles model is likely to be prohibitive until new software tools and validation procedures become available. Hybrid models that integrate steady state nonlinear first-principles models with dynamic empirical models (linear or nonlinear) may prove most promising for the near future. Gain-scheduling with linear dynamic models is an example of this approach.

From a theoretical point of view using a nonlinear model changes the control problem from a convex QP to a non-convex Non-Linear Program (NLP), the solution of which is much more difficult. There is no guarantee, for example, that the global optimum can be found. Bequette describes several approaches to solving the general nonlinear MPC problem in his review of nonlinear control for chemical processes (Bequette, 1991). Although solving the nonlinear MPC problem at each time step is much more difficult, Rawlings et al. (Rawlings et al., 1994) have shown that the nominal Lyapunov stability argument presented for linear models carries over to the general nonlinear case with minor modifications. The most important change is that a constraint must be included to zero the states at the end of the input horizon. They point out, however, that nonlinear MPC may require unexpected input adjustments. They present an interesting example in which a simple nonlinearity in the process model leads to a discontinuous feedback control law. This implies that tuning nonlinear MPC controllers may be very difficult, particularly for the case of model mismatch.

Adaptive MPC

A few adaptive MPC algorithms such as the GPC algorithm introduced by Clarke et al. have been proposed (Clarke, Mohtadi and Tuffs, 1987) but only a single adaptive MPC algorithm has reached the marketplace (STAR from Dot Products (Dollar, Melton, Morshedi, Glasgow and Repsher, 1993)). This is despite the strong market incentive for a self-tuning MPC controller. This reflects the difficulty of doing adaptive control in the real world. Barring a theoretical breakthrough, this situation is not likely to change in the near future.

On the other hand, adaptive and on-demand tuning PID controllers have been very successful in the marketplace. This suggests that adaptive MPC controllers may emerge for SISO loops as adaptive PID technology is generalized to handle more difficult dynamics.

Robust Stability of MPC

With one exception (Honeywell), industrial MPC controllers rely solely on brute-force simulation to evaluate the effects of model mismatch. Robust stability guarantees would significantly reduce the time required to tune and test industrial MPC algorithms. It is likely that the powerful theoretical results recently presented for MPC with a perfect model (Muske and Rawlings, 1993; Zheng and Morari, 1995) will be extended to include model mismatch in the near future. This has already been accomplished for a modified QDMC algorithm using an impulse response model (Vuthandam et al., 1995). New robust stability guarantees will then be combined with uncertainty estimates from identification software to greatly simplify design and tuning of MPC controllers.

Robust Identification Schemes

The use of FIR models results in overparametrization. In practice this leads to some kind of engineering modification, such as a smoothness factor to "regularize" the model parameters. Although these modifications are proprietary, regularization (smoothness) and biased regression algorithms (ridge regression or partial least squares) are common approaches to dealing with overparametrization. An alternative approach would be to identify the process model based on parametric models, such as transfer function models or state space representation, then convert the model into desired form for the controller to use.

Conclusions

MPC technology has progressed steadily in the twenty two years since the first IDCOM and DMC applications. Survey data reveal approximately 2200 applications to date, with a solid foundation in refining and petrochemicals, and significant penetration into a wide range of application areas from chemicals to food processing.

Current generation MPC technology offers significant new capabilities but the controllers still retain, for the most part, an IDCOM-like or a DMC-like personality. The SMC-Idcom and HIECON algorithms are IDCOM-like controllers which have evolved to use multiple objective functions and ranked constraints. The DMC, RMPCT and OPC algorithms are DMC-like controllers that use a single dynamic objective function to evaluate control and economic trade-offs using weighting factors. The PFC controller inherits some of the IDCOM personality but is significantly different in that it can accommodate nonlinear and unstable processes and uses basis functions to parameterize the input function.

An important observation is that industrial MPC controllers almost always use empirical dynamic models identified from test data. The impact of identification theory on process modeling is perhaps comparable to the impact of optimal control theory on model predictive control. It is probably safe to say that MPC practice is one of the largest application areas of system identification. The current success of MPC technology may be due to carefully designed plant tests.

Another observation is that process identification and control design are clearly separated in current MPC technology. Efforts towards integrating identification and control design may bring significant benefits to industrial practice. For example, uncertainty estimates from process identification could be used more directly in robust control design. Ill-conditioned process structures could be reflected in the identified models and also used in control design.

Choosing an MPC technology for a given application is a complex question involving issues not addressed in this paper. It is the opinion of the authors that for most applications, a knowledgeable control engineer could probably achieve acceptable control performance using any of the packages discussed here, although the time and effort required may differ. If the process is nonlinear or unstable, or needs to track a complex setpoint trajectory with no offset, the PFC algorithm may offer significant advantages. If a vendor is to be selected to design and implement the control system, it would be wise to weigh heavily their experience with the particular process in question.

Research needs as perceived by industry are mostly control engineering issues, not algorithm issues. Industrial practitioners do not perceive closed loop stability, for example, to be a serious problem. Their problems are more like: Which variables should be used for control? When is a model good enough to stop the identification plant test? How do you determine the source of a problem when a controller is performing poorly? When can the added expense of an MPC controller be justified? How do you design a control system for an entire plant? How do you estimate the benefits of a control system? Answering these questions could provide control practitioners and theoreticians with plenty of work in the foreseeable future.

Several technical advances have not yet been incorporated into industrial MPC technology. These include using an infinite prediction horizon to guarantee nominal closed loop stability, and using linear estimation theory to improve output feedback. In addition, robust stability conditions have been developed for a modified QDMC algorithm. It would seem that the company which first implements these advances will have a significant marketing and technical advantage.

The future of MPC technology is bright, with all of the vendors surveyed here reporting significant applications in progress. Next-generation MPC technology is likely to include multiple objective functions, an infinite prediction horizon, nonlinear process models, better use of model uncertainty estimates, and better handling of ill-conditioning.

Acknowledgments

The authors wish to thank the vendor representatives who responded to our MPC survey for their time and effort. We also wish to thank one anonymous reviewer and the following people for their constructive criticism of the initial draft: Jim Downs, Bjarne Foss, Brian Froisy, Pierre Grosdidier, Rob Hawkins, Evelio Hernandez, Joseph Lu, Jim Rawlings, Jaques Richalet and Steve Treiber.

References

Badgwell, T. A. (1995). Robust stability conditions for a SISO model predictive control algorithm. Manuscript in preparation.

Bequette, B. W. (1991). Nonlinear control of chemical processes: A review, *Ind. Eng. Chem. Res.* **30**: 1391–1413.

Clarke, D. W., Mohtadi, C. and Tuffs, P. S. (1987). Generalized predictive control—Part I. The basic algorithm, *Automatica* **23**(2): 137–148.

Cutler, C., Morshedi, A. and Haydel, J. (1983). An industrial perspective on advanced control, *AIChE Annual Meeting*, Washington, D.C.

Cutler, C. R. and Ramaker, B. L. (1979). Dynamic matrix control—a computer control algorithm, AIChE National Meeting, Houston, TX.

Cutler, C. R. and Ramaker, B. L. (1980). Dynamic matrix control—a computer control algorithm, *Proceedings of the Joint Automatic Control Conference*.

Cutler, C. R. and Yocum, F. H. (1991). Experience with the DMC inverse for identification, in Y. Arkun and W. H. Ray (eds), *Chemical Process Control—CPC IV*, Fourth International Conference on Chemical Process Control, Elsevier, Amsterdam, pp. 297–317.

DMC Corp. (1994). [DMC]TM: Technology Overview. Product literature from DMC Corp.

Dollar, R., Melton, L. L., Morshedi, A. M., Glasgow, D. T. and Repsher, K. W. (1993). Consider adaptive multivariable predictive controllers, *Hydrocarbon Process.* pp. 109–112.

Froisy, J. B. (1994). Model predictive control: Past, present and future, *ISA Trans.* **33**: 235–243.

Froisy, J. B. and Matsko, T. (1990). Idcom-m application to the shell fundamental control problem, AIChE Annual Meeting.

García, C. E. and Morari, M. (1982). Internal model control. 1. A unifying review and some new results, *Ind. Eng. Chem. Proc. Des. Dev.* **21**: 308–323.

García, C. E. and Morshedi, A. M. (1986). Quadratic programming solution of dynamic matrix control (QDMC), *Chem. Eng. Commun.* **46**: 73–87.

García, C. E., Prett, D. M. and Morari, M. (1989). Model predictive control: Theory and practice—a survey, *Automatica* **25**(3): 335–348.

Genceli, H. and Nikolaou, M. (1993). Robust stability analysis of constrained l_1–norm model predictive control, *AIChE J.* **39**(12): 1954–1965.

Genceli, H. and Nikolaou, M. (1995). Design of robust constrained model-predictive controllers with volterra series, *AIChE J.* **41**(9): 2098–2107.

Grosdidier, P., Froisy, B. and Hammann, M. (1988). The IDCOM-M controller, in T. J. McAvoy, Y. Arkun and E. Zafiriou (eds), *Proceedings of the 1988 IFAC Workshop on Model Based Process Control*, Pergamon Press, Oxford, pp. 31–36.

Honeywell, Inc. (1995). RMPCT Concepts Reference. Product literature from Honeywell, Inc.

Kailath, T. (1980). *Linear Systems*, Prentice-Hall, Englewood Cliffs, New Jersey.

Kalman, R. E. (1960a). Contributions to the theory of optimal control, *Bull. Soc. Math. Mex.* **5**: 102–119.

Kalman, R. E. (1960b). A new approach to linear filtering and prediction problems, *Trans. ASME, J. Basic Engineeering* pp. 35–45.

Lee, E. B. and Markus, L. (1967). *Foundations of Optimal Control Theory*, John Wiley and Sons, New York.

Lee, J. H. (1996). Recent advances in model predictive control and other related areas, *Chemical Process Control—CPC V*, CACHE, Tahoe City, California.

Lee, J. H., Morari, M. and García, C. E. (1994). State-space interpretation of model predictive control, *Automatica* **30**(4): 707–717.

Lines, B., Hartlen, D., Paquin, F., Treiber, S., de Tremblay, M. and Bell, M. (1993). Polyethylene reactor modeling and control design, *Hydrocarbon Process.* pp. 119–124.

Ljung, L. (1987). *System Identification: Theory for the User*, Prentice-Hall, Inc., Englewood Cliffs, New Jersey.

Mayne, D. Q. (1996). Nonlinear model predictive control: An assessment, *Chemical Process Control—CPC V*, CACHE, Tahoe City, California.

Morari, M. and Lee, J. H. (1991). Model predictive control: The good, the bad, and the ugly, in Y. Arkun and W. H. Ray (eds), *Chemical Process Control—CPC IV*, Fourth International Conference on Chemical Process Control, Elsevier, Amsterdam, pp. 419–444.

Morari, M. and Zafiriou, E. (1989). *Robust Process Control*, Prentice-Hall, Englewood Cliffs, New Jersey.

Muske, K. R. and Rawlings, J. B. (1993). Model predictive control with linear models, *AIChE J.* **39**(2): 262–287.

Ohshima, M., Ohno, H. and Hashimoto, I. (1995). Model predictive control: Experiences in the university-industry joint projects and statistics on mpc applications in japan, International Workshop on Predictive and Receding Horizon Control, Korea.

Prett, D. M. and García, C. E. (1988). *Fundamental Process Control*, Butterworths, Boston.

Prett, D. M. and Gillette, R. D. (1980). Optimization and constrained multivariable control of a catalytic cracking unit, *Proceedings of the Joint Automatic Control Conference*.

Propoi, A. I. (1963). Use of linear programming methods for synthesizing sampled-data automatic systems, *Automn. Remote Control* **24**(7): 837–844.

Rawlings, J. B., Meadows, E. S. and Muske, K. R. (1994). Nonlinear model predictive control: A tutorial and survey, *ADChEM '94 Proceedings, Kyoto, Japan*.

Rawlings, J. B. and Muske, K. R. (1993). Stability of constrained receding horizon control, *IEEE Trans. Auto. Cont.* **38**(10): 1512–1516.

Richalet, J., Rault, A., Testud, J. L. and Papon, J. (1976). Algorithmic control of industrial processes, *Proceedings of the 4th IFAC Symposium on Identification and System Parameter Estimation*, pp. 1119–1167.

Richalet, J., Rault, A., Testud, J. L. and Papon, J. (1978). Model predictive heuristic control: Applications to industrial processes, *Automatica* **14**: 413–428.

Ricker, N. L. (1991). Model predictive control: State of the art, *in* Y. Arkun and W. H. Ray (eds), *Chemical Process Control—CPC IV*, Fourth International Conference on Chemical Process Control, Elsevier, Amsterdam, pp. 271–296.

Ricker, N. L., Subrahmanian, T. and Sim, T. (1988). Case studies of model-predictive control in pulp and paper production, *in* T. J. McAvoy, Y. Arkun and E. Zafiriou (eds), *Proceedings of the 1988 IFAC Workshop on Model Based Process Control*, Pergamon Press, Oxford, pp. 13–22.

Setpoint, Inc. (1993). SMC-Idcom: A State-of-the-Art Multivariable Predictive Controller. Product literature from Setpoint, Inc.

Vuthandam, P., Genceli, H. and Nikolaou, M. (1995). Performance bounds for robust quadratic dynamic matrix control with end condition, *AIChE J.* **41**(9): 2083–2097.

Zafiriou, E. (1990). Robust model predictive control of processes with hard constraints, *Comput. Chem. Eng.* **14**(4/5): 359–371.

Zafiriou, E. and Chiou, H.-W. (1993). Output constraint softening for SISO model predictive control, *Proceedings of the 1993 American Control Conference*, pp. 372–376.

Zheng, A. and Morari, M. (1995). Stability of model predictive control with mixed constraints, *IEEE Trans. Auto. Cont.* **40**: 1818–1823.

SESSION SUMMARY: PREDICTIVE CONTROL

James B. Rawlings
Chemical Engineering Department
University of Wisconsin
Madison, WI 53706-1691

Summary

The model predictive control (MPC) session at CPC V consisted of three papers and a panel discussion.

In the first paper, Jay Lee provided a broad summary of academic research supporting the implementation of predictive control with linear models. He addressed numerous issues, such as stability and optimality of predictive control under uncertainty, predictive control for stochastic disturbances, moving horizon state estimation, and some recent approaches to state space model identification.

David Mayne then provided a stimulating discussion of methods for applying predictive control with nonlinear models. A major theme in this presentation was the computational complexity caused by the nonconvex programs generated by applying model predictive control with nonlinear models. Several novel ideas were presented for making this computational problem tractable in real time while maintaining the excellent closed-loop stability properties of the optimal controller.

The final presentation, a joint effort by Joe Qin and Tom Badgwell, summarized the industrial technology currently available for implementation of model predictive control. In preparing their paper, the authors made a detailed survey of vendors and users of model predictive control technology. The several tables summarizing the features of different approaches provided a wealth of information for practitioners wishing to evaluate the different commercial products.

The panel discussion followed the three presentations, but also followed several days of other spirited and informal discussions of model predictive control. In particular, Jacques Richalet led a thought provoking afternoon discussion of MPC that was well attended. Evelio Hernandez joined the four authors for the panel discussion that covered a wide variety of topics and addressed questions from the audience.

In summary, the session organizer feels the participants were indeed fortunate to have experienced for themselves some of the best interactions in the process control field. The industrial need for advanced control has been expressed clearly through the vehicle of model predictive control. Academic researchers have been in tune to this need and have attempted to provide a sound and practical theory to support a flexible and high-level implementation of MPC. Diverse groups have interacted in very positive ways, and the interactions have increased everyone's understanding of the issues.

As to what the future holds, model predictive control will continue to spark the imagination as long as it maintains a position as one of the key methods for making the best operational decisions in real time using the best available process information. In this larger context, model predictive control is the descendant of a long tradition of variational problems and optimal control theory. It is connected to a long history of scientific achievement, and its future may turn out to be just as bright as its past.

VARIATIONAL SYSTEMS: A BASIS FOR NEW PRODUCTION-SUPPORT PROGRAMMING

Bruce C. Moore
Data Refining Technologies, Inc.
Plaquemine, LA   70765-0893

Abstract

This paper gives a brief overview of a production-support applications paradigm together with a rigorous design basis for automated information processes which serve that paradigm. The paradigm encompasses both process monitoring and process control. Relational (data base) theory is extended with mathematics by which the half-open value intervals [a,b) on the reals can be utilized as a value domain for relation attributes. These extensions are packaged as *variational relations*, together with a "device," the *illuminator*, by which loop structures can be programmed to partition and combine relation records. An abstract class of *reduction processes* is shown to enable diverse functionality relating to signal filtering and dispersion measurement. A core concept, *assigned neighborhood characteristic*, is applied with the illuminator to create a seven-component design framework.

Keywords

Variational mathematics, Relational mathematics, Interval mathematics, Signal processing, Dispersion measures, Illuminator, Reduction process, Assigned neighborhood characteristic.

Introduction

In the applications paradigm, depicted in Fig. 1, there is an organization of job functions responsible for operating a production facility for profit, subject to environmental and safety constraints. Associated with this operation are people, materials, equipment, and markets, with many measured or estimated streams of time-dependent data, and many degrees of freedom to be exercised through actions, either by people or their programmed agents (e.g. control loops).

Capital resources, including production equipment, instrumentation, and computer systems, function within an *informed action feedback loop* where people and their programmed agents act to satisfy constraints and optimize economic return in the face of changing conditions.

Digital process control systems fall naturally within this paradigm. There is not a digital loop in all of industry for which no job-function is responsible. In production plants, the job function of a board-operator is to take informed action as appropriate for safe, efficient operation of the production facility.

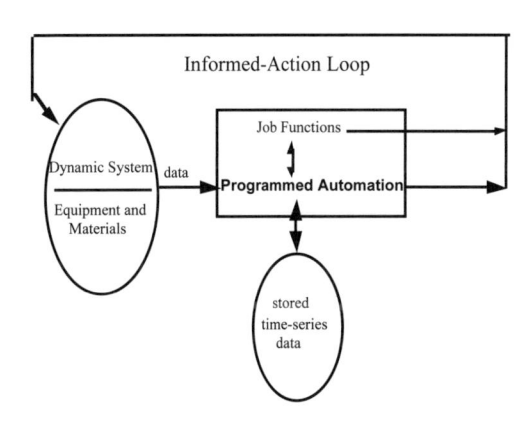

Figure 1. Applications paradigm.

An operator may put a controller in manual, but this does not break the informed-action feedback loop the controller was serving. Rather, in doing so, the operator moves a level deeper into the details of the informed-action loop.

There are three significant focal points of this applications paradigm:

1. *programming*: this is seen as a primary activity in the creation of applications. Within an applications area, a vital role of theoretical work is to provide a rigorous design basis for programming activities.
2. *job functions*: the success of specific programmed automation depends on how well it serves the target job-functions. Job-functions need not be fixed input to the design process. Instead, they can be treated as a variable component in the design of programmed automation.
3. *stored time-series data*: Time-series data (recent and historical lumped together here) is potentially very valuable within informed-action systems.

This paper explicitly addresses the first of these focal points by describing the basis for *variational processes*. A set of such processes in an application is called a "variational processing system," or *variational system*. Two diverse examples of applications functionality will be illustrated, one in the area of signal processing, the other in dispersion measurement. Detailed derivations and expanded discussion are given in Moore (1995).

This works stands on the theoretical fundamentals of sets and relations, using the following basic terminology. A *relation* can be thought of as a table containing data pertaining to attributes of a set. Each row of the table (*record* in the relation) corresponds to one of the set elements, and each column corresponds to one *attribute* of the relation. Whereas a relation is an actual table of data, a relation *schema* gives the structural definition of the data table. The set of attribute names for a relation is its *sort*.

Variational Relations

A *variational schema* is a relational schema whose attributes are restricted to one of three types: continuous, symbolic, and cumulative. The type of an attribute is determined by the value domains associated with it. For continuous and symbolic attributes there are two value domains, a *point domain* and a *variational domain*.

- *continuous*: The point domain is the set of reals on a specified interval [min,sup). The variational domain is a set of half open intervals of the form [a,b), described below in a discussion of analog-to-digital conversion.
- *symbolic*: The point domain is a finite set of distinct symbols **S**. The variational domain is the power set **P(S)**.
- *cumulative*: There is only one value domain, the positive reals, but it simplifies discussion to think of it as having a point domain and a variational domain which are identical.

For a given variational schema (*v-schema*) there are two kinds of relations: an *interface relation*, where attribute values are taken from the point domains, and a *variational relation* (*v_relation*), where values are taken from the variational domains.

An analog-to-digital conversion process maps an input interface relation into a corresponding v-relation. Superimposed on the point domain is an interval scale: a non-intersecting sequence of *basis intervals* $\{[a_i,b_i), i=1,...,N\}$ which are contained in, and collectively cover [min,sup). Every point value in [min,sup) is mapped into exactly one basis interval. The variational value domain induced by an interval scale is its closure in the following sense. For every pair of basis intervals [w,x), [y,z), the induced value domain contains [min(w,y), max(x,z)).

There are two special attributes of variational schemas: *Time*, which is a continuous attribute and *Norm*, which is cumulative. An *time-series v-schema* contains attribute *Time*, and a *measurable v-schema* contains attribute *Norm*. For a measurable time-series v-schema, the value of attribute *Norm* is the duration of the *Time* attribute: i.e. for time $[t_1, t_2)$, the corresponding *Norm* value is $t_2 - t_1$. An *ordered* time-series v-relation has the property that no two records have intersecting time attributes. In this paper all time-series v-relations are assumed to be ordered.

The *relation-norm* (or just *norm*) of a measurable v-relation is the sum of its values for attribute *Norm*. The norm of a measurable relation is zero if and only if it has no records.

Pure Reduction Processes

A fundamental operation on a v-relation is *natural reduction* which reduces it to a simple v-relation of the same sort. Natural reduction of an empty relation give the empty relation, but otherwise the result contains a single record whose attribute values are determined as follows: For cumulative attributes, the result is the sum of the attribute values. For symbolic attributes, it is the smallest element of **P(S)** containing all symbols in the attribute values. For continuous attributes, it is the interval of least range which includes all of the attribute values.

Natural reduction is the charter member a class called *pure reduction processes,* whose members operate on a v-relation and produce a v-relation of the same sort, usually containing fewer records. Pure reduction processes are structured by a recursive skeleton given below. At the bottom of the recursion is natural reduction.

- *partition*: the input v-relation records are partitioned into subsets.
- *reduce*: a reduction process is applied to each subset in the partition.

- *insert*: the reduced records for selected subsets are inserted in the output v-relation

To this skeleton is added muscle: mathematics packaged as an *illuminator*, a "device" used in controlled loop operations during the partition step of the reduction process. An illuminator is applied to a *subject v-relation* and is controlled by an *illumination control v-relation*. The control relation v-schema is not arbitrary; rather, each attribute is either a continuous or symbolic attribute of the subject. Furthermore, the scale (continuous) or symbol set (symbolic) for each attribute of the control relation is the same as that of the selected control attribute.

The essence of a *k-set illuminator* is its *illumination map* which assigns to every (control, subject) record pair (Rec_C, Rec_S) one symbol from a distinct set of *illumination symbols* $\{S_0, S_1,..., S_k\}$. Members of $\{S_1,..., S_k\}$ are called *bright symbols* and S_0 is the *dark symbol*. For a given control record, the effect of this map is to assign every subject record to one of k illumination subsets, each of which qualifies as a (possibly empty) v-relation of the same sort as the subject.

For a measurable subject, there is also a well defined *illumination distribution relation* whose attributes are *Norm*, $\%\_S_0$, $\%\_S_1$,..., $\%\_S_k$, and an *illumination distribution map* which maps (Rec_C, Rec_S) into a single record. This distribution record gives the norm of the subject together with the distribution of that norm over the illumination symbols; that is, $\%\_S_0$ gives the percentage (by norm) of the subject relation mapped into the dark symbol. By convention the distribution map assigns an empty subject a distribution record with norm zero with $\%\_S_0$ set equal to 100%.

In practice one works with single-attribute component illumination maps and combines these into a composite illumination map. A k-set component map is created by selecting k disjoint binary relations (called *bright relations*) on the value-domain of the attribute. The following set of thirteen relations, first reported in Birdwell and Moore (1994) using slightly different labels, are such that exactly one is true for any two intervals [w,x], [y,z] in the domain of a continuous attribute. In the table below, these relations are presented as one symmetric relation and six dual pairs:

- Equals
- LessThanSeparate   • GreaterThanSeparate
- LessThanAdjacent   • GreaterThanAdjacent
- LessThanOverlaps   • GreaterThanOverlaps
- LessThanCovered   • GreaterThanIsCovered
- LessThanIsCovered   • GreaterThanCovers
- Covers   • isCovered

These fundamental relations, a generalization of $\{=,<,>\}$ on the reals, can be combined with the "OR" operator to create various sets of bright relations.

A common illuminator setup uses the INCLUDES relation, the logical OR of { Equals, Covers }, to form a single bright set.. If the control interval includes the corresponding attribute value for a subject record, then that record belongs to the bright set. Otherwise, the subject record belongs to the dark set.

An illuminator has two modes of operation: forward and inverted. In *forward illumination*, the control relation has a single record which is manipulated at the beginning of each iteration of a loop. The illuminator assigns each subject record to one of the k+1 illumination sets, which are subsequently processed by other loop operations. With a continuous control attribute the "slider sweep" is a common operation. The control attribute value is assigned fixed range so that it acts like a fixed-width slider which is swept from a starting position to a final position during the iterations of a loop.

Inverted illumination is applicable only to measurable v-relations, and is used in *measurement reduction processes*. This goes the other direction -- illumination *produces* a control relation which is either empty or contains one record. Inverted illumination is controlled by a set of *distribution specifications* (such as $\%\_S_1 > 95\%$), and a *selection specification*, collectively called the *measurement specifications*. Because finite variational domains are used for continuous attributes, the inverse of the illumination distribution map is well behaved. The (possibly infinite) set of records satisfying the distribution constraints are mapped into a finite set of control records. If this set is not empty, the selection specification is used to select a final control record. Otherwise, the final control relation is empty record, and the inverted illumination operation is said to have failed.

In a measurement reduction process, inverted illumination is first applied to the subject using measurement specifications, as described in the preceding paragraph. If this operation is successful, then the resulting control record is used in forward illumination applied to the subject and the bright sets are reduced and inserted into the output.

Figure 2 illustrates the result of measurement process DispMeas_drti9401 which measures dispersion in one continuous attribute of a v-relation. Shown in this figure are three things:

- relative distribution of the subject attribute values by norm on value domain [0,100).
- dispersion measure equal to the mean value +- 2.35 standard deviations
- attribute interval value from the result of a measurement reduction process using a single INCLUDES bright relation, with distribution specification $\%S_1>98$.

For a large population with normal Gaussian distribution, the interval which is 2.35 standard deviations either side of the mean would also correspond (roughly) to the minimal interval containing at least 98% of those values.

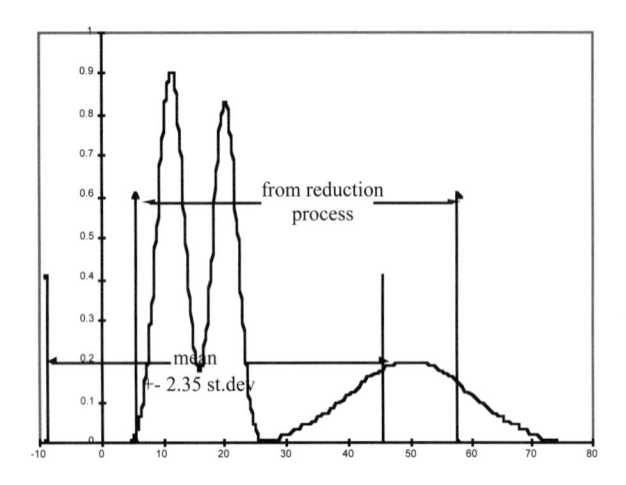

Figure 2. Dispersion measurement

Assigned Neighborhood Characteristic

Figure 3 shows a time-series interface relation as a chart image, with a time-index used instead of actual time. This relation contains 4000 records and will be used to illustrate the process SigFilt_drti9301, a filtering reduction process consisting of two pure reduction processes in series, each designed using the seven-component ANC design method.

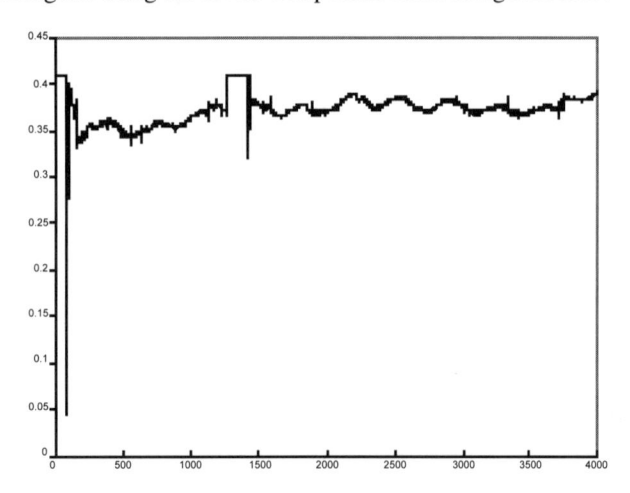

Figure 3. Interface time-series relation

The ANC method applies forward illumination using *the assigned neighborhood characteristic* concept. The chart image in Figure 4 shows the reduced time-series v-relation produced by SigFilt_drti9301, for which there is only a "time-scale" parameter (no threshold parameter). Note that this v-relation, which has only 146 records, retains much of the character of the input signal, including timing and magnitude of large transient swings. SigFilt_drti9301 has been in successful daily operation for over two years, reducing tens of thousands of signals and storing them for downstream processing.

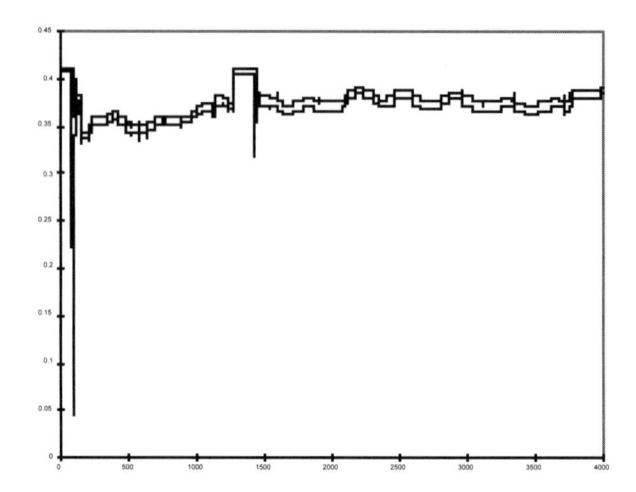

Figure 4. Reduced time-series v-relation

Space limitations permit only a brief sketch of the ANC design method. It provides a structured seven-component method for specifying a process which executes the partition step by building a *characteristic map* from the subject (input) v-relation. This map assigns to each record of the subject a *characteristic record*. Records assigned the same characteristic form one subset of the partition. The characteristic map is necessarily a function of the subject.

A neighborhood in this context is defined directly in terms of illumination. It is assumed that a *1-set illuminator* (design component 1) is applied to the subject and executed for a sequence of control records in a loop (dc 2), producing a bright set at each iteration. Each of the bright sets illuminated by the loop is a *neighborhood* (induced by the control record) which is assigned a *trial characteristic* (dc 3) by a *trial map* (dc 4). The trial map is not a function of the subject. Every record in the subject has a list of assignable characteristics, which is initialized by a *default characteristic map* (dc 5). At each iteration of the loop, the trial characteristic is added to the assignable list some neighborhood records, based on a *update criterion* (dc 6). After the illumination loop is done, a *final assignment process* (dc 7) builds the characteristic map by assigning each subject record one characteristic from its list of assignable characteristics.

References

Birdwell, J.D., and B. Moore (1994). Condensation of information from signals for process modeling and control. Workshop on Hybrid Systems and Autonomous Control, Cornell University, October, 1994.

Moore, B.C. (1995). Variational relation processes: extended relational theory with diverse applications. DRTI report TF50223c, to be submitted for publication.

THE GAME OF FATE — A MODELING PARADIGM FOR MANUFACTURING OPERATIONS

Robert E. Young
Exxon Chemical Co.
Baytown, TX 77522

Abstract

The "Game of Fate" is a framework originally proposed by Moore (1986) for the development and application of process monitoring technology. In this paper, the "Game of Fate" abstraction is applied to the manufacturing organizations at Exxon's Baytown Chemical Plant. The typical structure of these groups and their morning monitoring and analysis routine is presented. These specific instances provide insight into enhancements to this process monitoring paradigm. The game boundaries are extended to include the manufacturing and technical-support personnel involved and typical control and data historization systems. The goals are modified to more closely match the needs and capabilities of these manufacturing organizations. These enhancements also reflect the industry-wide trend in downsizing that has left numerous opportunities in the process monitoring arena. An industrial example illustrates the paradigm. Finally, challenges faced by education, academic research, and industrial application are posed in the framework of the "Game of Fate."

Keywords

Process monitoring, Data reconciliation, Fault detection, Blending processes, Dynamic modeling.

Introduction

Every morning in control centers everywhere in the world, the day begins with a new "Game of Fate." Simply stated, the goal of the original game developed by Moore (1986) is to find all unplanned changes in the manufacturing process.

At the time, the challenges of global economics motivated the process industry to stop "keeping the plant running" and start "running the plant wisely." This operating philosophy change coupled with work-force reductions has inspired many process monitoring developments throughout the last decade. The process monitoring problem is now widely recognized as multivariable, often nonlinear, and dynamic. As a result, academic activities have focused in areas such as multivariate statistical analysis, data reconciliation, estimation, and fault detection. As powerful as these techniques are, relatively few academic activities have produced industrial applications.

The motivations for the development of the "Game of Fate," global economics and resource reductions, have only intensified in the last five years. These effects and challenges are examined by applying the "Game of Fate" structure to manufacturing organizations typical to Exxon's Baytown Chemical Plant.

The Original Game

A summary of the original seminar is presented here as a basis for further discussion. Moore's description of the original game is a very visual decomposition of a manufacturing plant into simple building blocks or cartons. The six faces and the inside of the carton categorize a piece of process equipment or a process unit operation as presented in Fig. 1. Note that this framework recognizes the difference between material and energy flows (tubes) and the measurements of these flows (wires). The ambient and the constraint cartons are two special cartons that contain information about ambient conditions, raw material supply, product demand, cost, etc., all determined by fate knob settings.

Equipped with this carton decomposition of the plant, the game master generates a historical database containing the sampled field signals changing fate knobs on the ambient and constraint cartons only. Then, the game master generates another database containing sampled field signals generated by a fixed number of fate knob changes in the plant cartons. The carton decomposition, equipment data, tube and wire descriptions, and signal databases are supplied to the players according to a game schedule. At the conclusion of the game, the players must submit a list of changes that include the time of the change, the carton that contained the change, and the source of the change — field signal variation or variation in tube flows. Scores are based on accuracy and computer resource costs.

The Game Applied to Manufacturing

Some variant of this game is played in manufacturing plants everywhere. Every morning, process and control engineers sift through reams of data, review operator logs, and view trend after trend of data from the day before. Console operators are constantly trolling through process schematics, reviewing process alarm summaries, making setpoint changes. All of these activities are focused on pursuing the hand of fate. In the original game, the organization on the end of the wires included the teams of players, the computer with a real-time historical database, graphic terminals, and high speed printers. Except to motivate the need for the paradigm, the impact of the plant's organizational structure on the "rules, goals, and scoring" was not explored.

Consider two companies playing the game instead of two groups of players. If both companies have the same manufacturing process with the same carton decomposition of the plant, the single source of competitive advantage is the organizational structure on the other side of the wires. At Exxon's Baytown Chemical Plant (BTCP), there are four major control centers. Each control center contains multiple consoles. Each of these consoles is the "other end of the wires" for the BTCP. A console is composed of control/computing systems hardware, on-shift technicians, and support engineer staff. The key individual for each console is the console operator (CO). The CO is responsible for the ongoing plant operation and relies on the rest of the organization for field support, technical expertise, and data such as sensor readings, histories, trends, and alarms. The typical CO has many years of experience in the plant and has seen many of the possible effects of the hand of fate. However, most of the console operators lack in-depth training or skills in mathematics, statistics, modeling, and computing. An average console has two or three process/operations engineers and one or two control/systems engineers dedicated to the console. Other engineering disciplines are not typically dedicated to a single console.

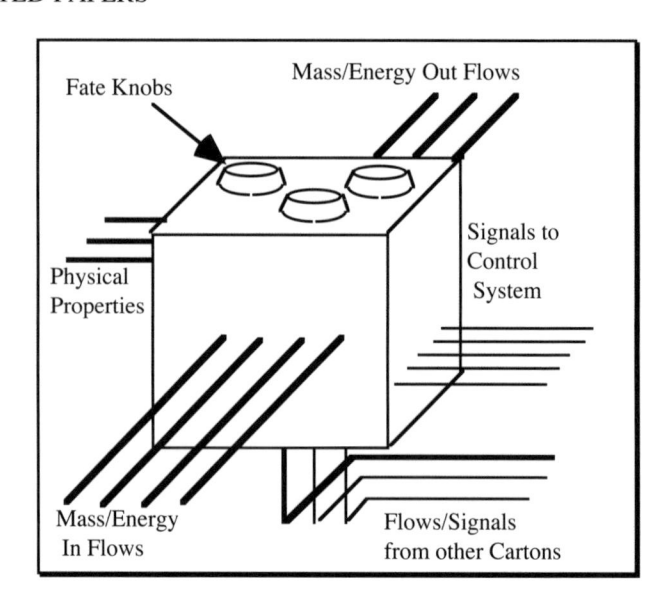

Figure 1. Carton decomposition visualization.

Most consoles at BTCP operate from Honeywell TDC 3000 distributed control systems coupled with VAX computers. Each console typically controls 200 to 400 manipulated variables, depending on the degree of automation and on the degree of digital integration. Each console is connected to between 1000 and 3000 field signals. The CO can typically access four to six TDC 3000 screens and a desktop computer. These systems also supply process data to databases accessible to each engineers desktop computer.

Relevant Observations

Console resources have undergone three significant changes in the last five to ten years. First, the number of engineering support people has been reduced and more emphasis placed on the on-shift personnel's ability to identify and correct deviations from normal operation. This general trend observed throughout the process industries has been also seen at the BTCP. Besides simply reducing the number people looking at the data, the amount of technology applied manufacturing problems has declined. This trend motivated the original "Game of Fate" in 1986, and has only intensified since that time, and is unlikely to swing back soon.

Second, the plants have implemented significant upgrades to the control systems. These upgrades tend to offset some of the personnel reductions. The control system can detect some instrument failures, provides more access to history with better trending, and significant improvement in the alarm subsystems. The integration of the control and computer systems provides a ripe platform for the automation of process monitoring applications.

Finally, manufacturing organizations have developed a strong, well-defined goal orientation. Manufacturing defines goals in the areas of safety, production, fixed costs, and environmental impacts to be consistent with corporate plans. Progress towards these goals is measured by both traditional and contemporary metrics. Traditional metrics include profit, fixed costs, production, and injury ratios while more contemporary metrics include environmental losses, process capability (Cpk), and overall equipment efficiency ratio (OEER). This goal orientation serves to prioritize the organization.

Process Monitoring Insights

These observations lead to several insights into the goals of the "Game of Fate." Most importantly, all process changes are equal. Identifying those changes with the greatest impact on the manufacturing goals is more important that finding every process change that occurred. For example, finding every level instrument zero error in the plant is not as important as identifying the one flow instrument zero error that will make many thousands of product off-prime quality — affecting the plant's Cpk and OEER. Note that safety is rarely a motivation for process monitoring since safety systems are required to be separate from the regulatory control system. However, a process monitoring application can reduce the probability of unplanned safety shutdown and avoid the associated costs.

Process monitoring automation is necessary to keep up with the organizational changes. Their development must focus on the on-shift personnel without avoiding the multivariate, dynamic, and nonlinear nature of the analysis. There is no need to duplicate control system functions such as alarms for failed instruments. With the rich graphics capability and the alarm subsystem available on current control systems, these complex applications can be integrated into the standard console interface.

Blending Application Example

A single major upset in the blending operation shown in Fig. 2 can make it impossible for one BTCP console to achieve its product quality goal as measured by Cpk. There are adequate measurements to identify abnormal operation before a major upset. Specifically, the three inlet flows are measured where w1 is pure component 1, w2 is pure component 2, and w3 is a mixture of component 1 and a diluent. The level, the outlet flow and the drum compositions are also measured. Historically, upsets occur because of flow meter or analyzer failures that go undetected. Once an upset has detected, the source of the failure is unknown and every instrument is often serviced.

This problem can be modeled with fundamental models or correlation models such as principle components (PC) or neural networks (NN). Challenges include the consistent non-steady state operation, the asynchronous analyzer sampling and the highly correlated, closed-loop nature of the process.

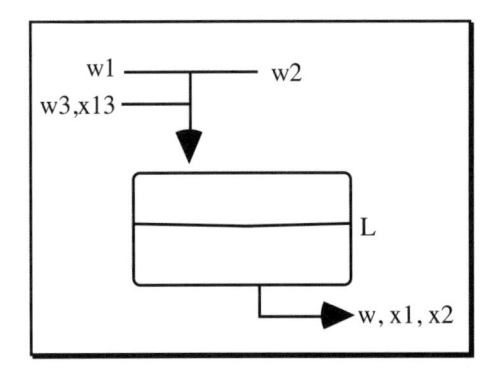

Figure 2. Blending unit schematic.

The tubes in and out of the carton are easy to identify and all of these quantities have an associated field signal wires. For this process, detecting inconsistent blending operation is the key factor related to the plant goals. Changes resulting from outlet demand flow changes and composition setpoint changes are expected and do not impact the plant goals. Composition setpoint deviations and high/low levels are handled by control system alarms. Detecting poor blending operation translates to detecting inconsistent correlations among the process data and the model. A fundamental model is best suited to this requirement. The model is composed of an overall and two component dynamic mass balances:

$$\frac{dM}{dt} = w1 + w2 + w3 - w$$
$$\frac{dMx1}{dt} = w1 + x13 * w3 - x1 * w \quad (1\text{-}3)$$
$$\frac{dMx2}{dt} = w2 - x2 * w$$

Note that the mass, M, is an algebraic function of the level, L. The energy balance is not required since the drum temperature is tightly controlled to a constant setpoint.

Dynamic data reconciliation (DDR) of this model and the process data readily identifies inconsistencies. The DDR application computes values for the model variables by minimizing the weighted squared deviation between the measurements and the model variables subject Eqns. (1-3). Large values of the objective function correspond to inconsistent blending.

Typical normalized process data is presented in Fig. 3. Note that for the first 425 minutes, the blend rate, inventory, and blend compositions are reasonably constant with only a slight drift in the x13 composition. At 425 minutes, a disturbance affects both the inlet and outlet analyzers. At 450 mins, the demand rate changes abruptly, causing a change in inventory that affects the blend rate. The individual inlet flows track together as a result of tight ratio control to maintain the drum compositions.

Observation of the data indicates that the disturbance at 425 minutes is an obvious abnormality. However, it is difficult to determine if the slow drift in the x13 composition and the data after 450 minutes are consistent with the model.

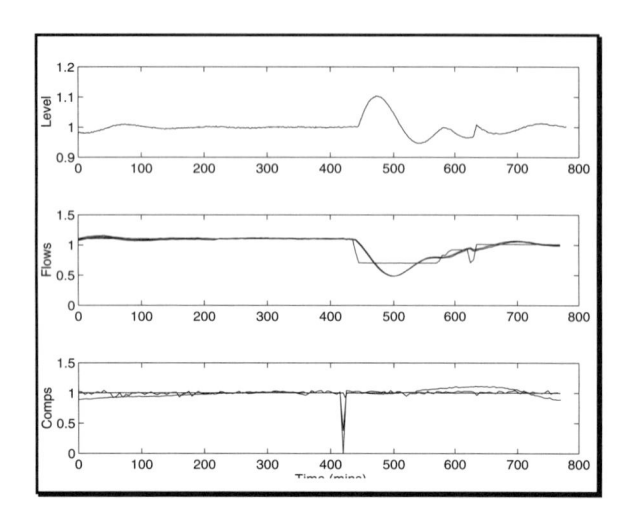

Figure 3. Typical normalized blending data.

The sum of the weighted squared deviations determined by sequentially applying the DDR application to the data in Fig. 3 is less than 0.1 for all times except those near the disturbance at 425 minutes where the value is approximately 1. Hence, the inlet analyzer drift and the non-steady state data are consistent, indicating good blending operation except at approximately 425 minutes.

The deviations computed with the reconciliation application around the disturbance are smeared across several variables. This smearing effect makes determination of the inconsistency from the computed deviations difficult. However, examination of the material balance residual presented in Fig. 4 indicates that there is no inconsistency

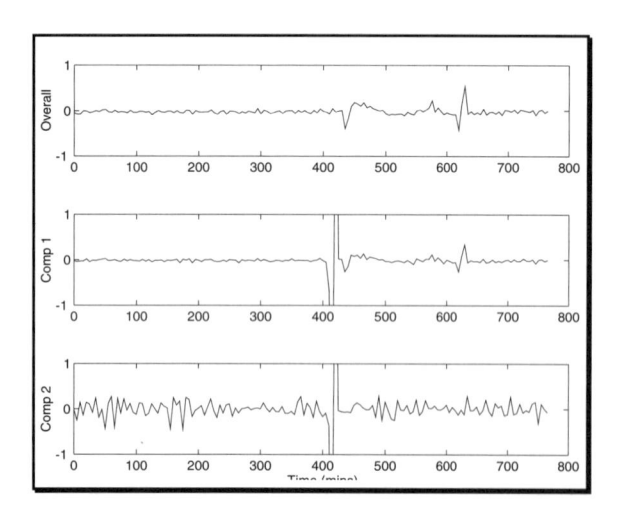

Figure 4. Normalized mass balance residuals.

in the overall balance, just in both component balances. Consequently, the level and the flow data are consistent and the analyzer data is inconsistent. Since both components are involved, the outlet analyzer is definitely inconsistent and the inlet analyzer may be. Since the inlet flows were not affected, the composition of the blend was not really inconsistent. However, the analyzer data was inconsistent and should be investigated. Prioritization of the most probable device error via this type of fault analysis is feasible for a problem with only seven instruments and can be nicely integrated into the console operator's schematics.

To an experienced technician, the abnormality at 425 minutes was obviously a carrier gas bottle swap that effected both analyzers. In many cases, simply identifying an abnormal condition through the alarm subsystem is sufficient to avoid process upsets that impact the plant's goals.

Conclusions

This paper has examined the "Game of Fate" as a modeling paradigm for manufacturing operations. A new focus on the goals has been motivated based on organizational structure. Examination of a typical organization at Exxon's Baytown Chemical plant leads to the following recommendations:

- focus on changes that impact the manufacturing goals.
- focus on tools for the on-shift personnel.
- focus on tools that solve problems that can not be handled by the existing control system.
- focus on scarce resources (rarely computing hardware).

An industrial example illustrates how these insights relate to the development of on-line monitoring applications.

Appropriately focused, the "Game of Fate" provides educators, researchers, and practitioners with a framework to embrace the future of manufacturing. Educators are challenged to combine the existing educational focus on design with a focus on manufacturing to produce engineers better prepared to contribute to a manufacturing organization. Researchers are clearly challenged to use the game to focus technology advances on significant problems, avoiding problems that can be solved with existing on-line systems. Finally, the practitioners are challenged to recognize the value of complex academic solutions and to translate these solutions into applications that focus on their particular manufacturing organization. Perhaps most importantly, the "Game of Fate" is a framework that facilitates communication among these diverse groups.

References

Moore, B. (1986) Industrial Challenges and "The Game of Fate." Seminar presented to Department of Electrical and Computer Engineer, University of California, Santa Barbara, CA.

GEOMETRIC INTERPRETATION OF SVD, RANK, MEAN CENTERING AND SCALING IN APPLYING MULTIVARIATE STATISTICAL ANALYSIS METHODS

Tsewei Wang, Srinivas Vedula, and Atul Khettry[1]
Department of Chemical Engineering
The University of Tennessee
Knoxville, TN 37996-2200

Abstract

A geometric interpretation of the singular value decomposition (SVD) of a matrix, its rank determination, along with the preprocessing steps of mean centering and scaling in applying multivariate statistical analysis methods is presented. The SVD of a matrix is viewed from the perspective of sphere to ellipsoid transformation. The effect of mean centering and scaling on the SVD of the resultant matrix, and thus the rank determination is discussed. Examples of 2-dimensional data matrices are used to illustrate the visualization.

Keywords

SVD, Rank, Multivariate statistical analysis, Geometric, Mean centering, Scaling.

Introduction

Within the last decade, multivariate statistical analysis (MVS) methods, such as principal component analysis (PCA) or regression (PCR), have been increasingly employed by engineers, especially chemical engineers in performing multivariable process modeling for the purpose of fault detection, preventive maintenance, and multivariable statistical process control (MacGregor et al, 1991; Sarma et al, 1993; Wise et al, 1990).

At the heart of the multivariate statistical analysis methods is the determination of the number of factors to keep (i.e., the effective rank of the data matrix). In determining the effective rank, a variety of analytical methods (though not always consistent) have been used, such as cross validation. The bottom line is that the number of factors kept determines the degree of fit of the data by the MVS model, as well as the sensitivity of the MVS model to small changes in the data. If enough factors are not retained, the resultant MVS model would not fit the data well and will leave a relatively large residual. Although including too many factors would result in a relatively small residual, it would render the resultant model too sensitive to slight changes in the measured data. Therefore, the ultimate determination of the effective factors (i.e., the effective rank of the data matrix) is one of the most important decisions to be made in the MVS modeling approach.

Before the number of factors is to be determined from the data matrix, the raw data matrix is usually column mean centered and scaled, such that the mean of each column is zero, and the standard deviation or the sum of squares of each column is one. In textbooks and tutorials, this preprocessing is given as a matter of fact and is rarely given with any sort of justification. Pell and colleagues (Pell et al, 1992) and Seasholtz and Kowalski (Seasholtz and Kowalski, 1992) gave an analytical test to see if mean centering of data matrix is required. Those authors concluded that if the rank of a data matrix is reduced by one or more after mean centering, then mean centering is required. They further showed that if there is closure of the matrix (the row or column sum adds up to a constant number, such as 100%), then mean centering is always required. We present here a geometric visualization of what

---

[1] Present address: Perten Instruments, Chatham, Illinois.

mean centering and scaling do to a data matrix, using 2-d examples. Extension into higher dimension is straightforward. Further, we present a geometric test method to see if mean centering and scaling are required. Our geometric criteria for mean centering is of course the same as that of Pell et al, in that rank reduction if mean centered, is the determining factor. In this communication, we also present a test for the need for scaling. The criteria is that if scaling (after mean centering) leads to a rank increase, then scaling is strongly advised.

Geometric Interpretation of the Singular Value Decomposition of a Matrix

The singular value decomposition (SVD) is the preferred numerical method for rank determination of a matrix (Strang, 1988). The condition number, or the ratio of the largest to the smallest singular value, is a gauge of how near singular the matrix is. The bigger the condition number, the more near singular the matrix is. The number of nonzero singular values determines the nominal rank, and the ratio of the largest to the smallest nonzero singular value will give a measure of the effective rank of the matrix. The effective rank, r, of a matrix is to be loosely defined by the authors as the rank of the matrix such that the ratio between the largest (the first) and the rth (and the last nonzero) singular value is not too large.

Consider a rectangular matrix, **A**, of size mxn, with m>n and rank of n. The economical singular value decomposition of **A** can be expressed as the product of three matrices, **U**, **Σ**, and **V**, each with a distinct property. Thus **A=UΣV'**, where **U** and **V** have orthonormal columns and are of dimensions mxn, and nxn respectively. **V'** is the transpose of **V** and **Σ**, a diagonal matrix of dimension nxn, contains the positive singular values arranged in descending order. **U** and **V** are referred to as the left and right singular vectors and span the column and row space of **A** respectively.

The SVD of A can be visualized as shown below.

$$\begin{bmatrix} & n & \\ & A & \\ m & & \end{bmatrix} = \begin{bmatrix} & n & \\ & U & \\ m & & \end{bmatrix} \begin{bmatrix} n \\ \Sigma \\ n \end{bmatrix} \begin{bmatrix} n \\ V' \\ n \end{bmatrix}$$

Geometrically speaking the matrices, **U**, **Σ**, and **V** can be viewed as n-dimensional spanning ellipsoids. Let's consider an mxn **A** with n<m, as representing a linear transformation matrix, that transforms a vector **x**, from the n-dimensional domain space to a vector **b**, in the m-dimensional recipient space. This transformation can be represented as **Ax = b**, or visually

$$\begin{bmatrix} & n & \\ & A & \\ m & & \end{bmatrix} \begin{bmatrix} 1 \\ x \\ n \end{bmatrix} = \begin{bmatrix} 1 \\ b \\ m \end{bmatrix}$$

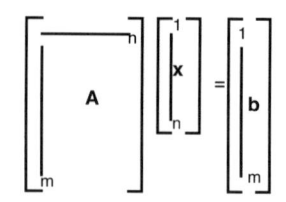

Consider an array of unit-length vectors, say a hundred of them spread around an n-dimensional sphere in the domain space, being transformed by **A** into one hundred corresponding **b** vectors. One asks the question of what the one hundred **b** vectors look like in the recipient space; how are they oriented and what are their lengths. Let's take a 2-d example. If one takes one hundred vectors of all unit lengths from the domain space and transform them by a 2x2 matrix A, as shown below, into a hundred **b** vectors,

$$A = \begin{bmatrix} 2 & 1 \\ -1 & 6 \end{bmatrix}$$

then the one hundred **b** vectors will span an ellipse in the recipient space, as shown in Fig. 1-a. The singular value decomposition of **A** is given below:

$$A = U\Sigma V' = \begin{bmatrix} -0.126 & 9.994 \\ -0.992 & -0.209 \end{bmatrix} \begin{bmatrix} 6.123 & 0 \\ 0 & 2.123 \end{bmatrix} \begin{bmatrix} 0.111 & 0.992 \\ -0.994 & 0.127 \end{bmatrix}'$$

Geometrically speaking then, the major axis of the ellipse spanned by the one hundred **b** vectors points in the direction of u_1, the first column of the left singular matrix **U**, above; and its minor axis points in the direction of u_2, the second column of **U** above. The two vectors from the domain space that are transformed by **A** into these two **b** vectors come from the first and second columns of the right singular matrix **V** above. Further, the lengths of the major and minor axes correspond to the first and second singular values, respectively, as given by the diagonal elements of the singular value matrix **Σ** above. In short, **A** transforms v_1 into u_1, v_2 into u_2, v_i into u_i, each **u** vector being amplified by the corresponding σ_i, the ith singular value.

Extensions into higher dimension **A** is straight forward in that an n-dimensional spheroid of unit vectors is mapped by **A** into an n-dimensional ellipsoid, with its major and minor axes having the lengths given by the n singular values respectively. Viewed from this point, the mapping is decoupled, in that each **v** vector is mapped into a corresponding **u** vector, amplified by a corresponding singular value. **Ax=b**, or **Ax=UΣV'x** can also be interpreted geometrically as **x** changing its representation to that of **V**, (**V'x**), followed by stretching of each component by the corresponding singular value (**Σ(V'x)**, which now has representation with respect to **U**), and finally changing the representation back from that of U to that of normal (**U(ΣV'x)**).

Geometric Interpretation of the Rank of A Matrix

The key step at the heart of MVS approach is the determination of the effective rank of the data matrix. In other words, how many dimensions of independent information is really contained in the data? Let's take a look at this notion of rank and view it from the geometric point of view.

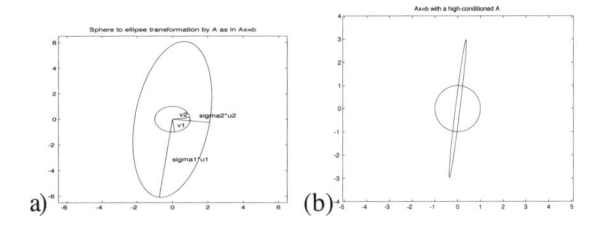

Figure 1(a) and (b). (a) Sphere to ellipsoid transformation by A, as in Ax=b; (b) A has effective rank of only 1

Let's consider a matrix A of say 3 by 2 dimension. The column space of a matrix A is viewed as the space spanned by all possible linear combinations of the columns of A. The dimension of the column space of A then is the dimension spanned by all possible linear combinations of the columns of A. Imagine now the case in which although the two columns of A are nominally linearly independent, but just barely, so that the two vectors represented by the two columns in the 3-dimensional space appear very close to each other, then all linear combination of these two vectors would span nominally a 2-d column space, although in reality, more like a solid 1-d space. The second dimension would be very weak. From the perspective of the circle to ellipse mapping as presented above, a circle of vectors would be mapped into a long and thin "baguette" like ellipse, with a very short minor axis compared to its major axis (i.e. the condition number of **A** is high). Fig. 1(b) shows such a case. Therefore, the presence of a weak dimension of the ellipsoid always signifies the lack of true linear independence among the columns of **A**, thus the effective rank of **A** is less than the nominal dimension of the ellipsoid. The effective rank can be viewed as equal to the number of the strong major and minor axes of the ellipsoid. A flat disc in a 3-d space implies an effective rank of 2; whereas a long baguette in 3-d implies an effective rank of only 1.

Geometric Interpretation of SVD in PCA or PCR with respect to Mean Centering

In multivariate statistical analysis (MVS) methods, such as PCA / PCR, the data matrix is usually rectangular, with the longer dimension much longer than the other. Usually, the m rows represent the samples, and the n columns represent the process variables, that are correlated and contain noise. In reality, this mxn data matrix contains a much smaller dimensional but mutually independent information. The SVD of the data matrix should then give an indication of the effective rank represented by the data, thus the number of factors, or principal components to keep. In MVS, column mean centering is usually the first step performed on a data matrix followed by SVD of the centered matrix that leads to the determination of the effective rank, or the number of factors to keep. Let's take a look of the SVD of a data matrix with and without mean centering.

Let's first consider a nominally two dimensional data matrix, with **A** given by a 21x2 matrix. Fig. 2 shows these 21 points (the 21 rows) given by **A**. It can be seen that these points lie more or less on a straight line, i.e. a linear regression line can be drawn through them with a statistically significant nonzero slope. However, the SVD of the unmean-centered **A** above gives the singular values as 326 and 37 (a condition number of 8.8) not exactly considered to be highly conditioned enough to warrant an effective rank of 1. Based on the sphere-ellipsoid interpretation above, an ellipse with major and minor axes of 326 and 37 can be drawn so as to "circumscribe" these 21 points. (Fig. 2, the bigger ellipse.) Thus with **A** not mean centered, the effective rank is considered a solid 2, even though it is really 1 (from construction), because there is an nonzero intercept present.

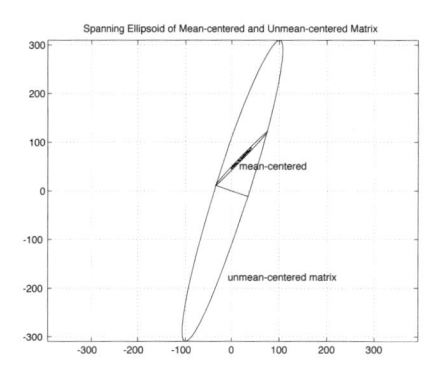

Figure 2. The spanning ellipsoid of mean-centered, as well as unmean-centered matrix.

When the SVD of the mean-centered **A** above is taken, the singular values become 78 and 2, with a condition number of 39, a much larger number from that of before. Fig. 2 also shows the ellipse "circumscribing" the mean-centered points (the smaller ellipse). The new ellipse is long and skinny, representing effectively, a one dimensional span, much closer to the true span of the original points.

With mean centering, such that the column means are all zero, the SVD of the above gives a different story. Mean centering moves the center of gravity of the cluster of points to that of origin, thus removing the intercept, so as to render the relationship between the columns truly linear, i.e. y=mx, instead of just affine, i.e. y=mx+b. As a consequence, a linear vector space results, such that the rules from linear algebra, such as the four fundamental spaces, linear transformation, singular value decomposition, would apply. Thus the SVD of the mean centered matrix would truly reflect its rank.

Geometric Interpretation of Scaling with respect to PCA or PCR

A second data preprocessing step usually taken following mean centering is scaling, or normalization, such that the standard deviation or the sum of the squares of each column is 1. The rationale given for this step is that if the process variables that correspond to the n columns of the data matrix **A**, have different physical units, then the nominal SVD of the unscaled but mean centered data matrix will reflect this fact, and sometimes the condition number will change drastically when the units are changed. One desires an expression of the data matrix such that its SVD is unit invariant. As a result, scaling each column by its sum of squares has the advantage of treating the span of each process variable equally, regardless of the units used. One may also scale the columns differently so as to reflect any desired weighting of the variables. In addition to rendering the SVD invariant, scaling also lends to the extraction of how much contribution each of the principal component makes, with respect to the total "dispersion" present in the original data matrix. If each column of the mean centered matrix is scaled to 1, then the sum of the squares of the singular values of the scaled **A** becomes equal to the number of nonzero singular values of the scaled **A**.) Thus the ratio of the ith singular value squared to the sum of the squares of all the singular values, would give the relative contribution made by the ith principal component. Thus, the mean centered columns need to be scaled. The SVD will reflect the true or the effective rank only when the matrix is scaled.

Let's consider a matrix with two columns which are about 60° apart, thus not singular at all. This data is mean centered and scaled with respect to the standard deviations of the columns. One notices that the numbers in the second column is about 1 to 2 orders of magnitude higher than that of column 1. Such inequity in the scaling would affect the nominal condition number of the matrix. The condition number is nominally high (around 93), signifying a near singular matrix with effective rank of only 1. After scaling however, the singular value matrix has an effective rank of 2 which is reflected by the small value of the condition number (around 2.8), Fig. 3(a) gives the geometric visualization. Before scaling, the points appear to be dispersed in only 1 dimension, in which the ellipse appears as a long and skinny "baguette." However after scaling such that the standard deviation of each column is 1, the points appear to cover a solid 2 dimensional space, as shown in Fig. 3(b). In summary, if scaling increases the effective rank, then scaling is strongly recommended, so as to strongly reflect the true dimensionality of the data.

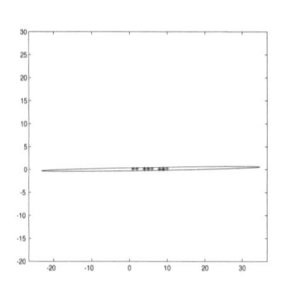

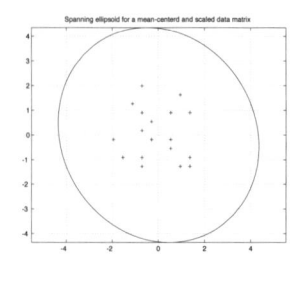

Figure 3(a) and (b). The spanning ellipsoid of (a) centered but unscaled; and (b) the same centered but scaled data matrix.

Conclusion

We presented a geometric interpretation of the singular value decomposition of a matrix, its rank and condition number from the perspective of mapping of a spheroid to an ellipsoid. The relative aspect ratio of the resulting ellipsoid gives a visual determination of the effective rank of the matrix. The true rank of a matrix is brought only after mean centering and scaling.

References

MacGregor, J.F., T.E. Marlin, J. Kresta, and B. Skagerberg (1991). Multivariate statistical methods in process analysis and control, *Proc. of the CPC-IV Conf.*, South Padre Island, Texas, Feb. 18-22.

Pell, R.J., M. B. Seasholtz, and B. Kowalski (1992). The relationship of closure, mean centering and matrix rank interpretation, *J. of Chemometrics*, **6**, 57-62.

Sarma, P., R. Horn, D. Rangel, and T.W. Wang (1993). Intelligent multivariate monitoring and diagnosis of a continuous exothermic reactor, AIChE Annual Meeting, Session 150, St. Louis, MO.

Seasholtz, M.B., and B. R. Kowalski (1992). The effect of mean centering on prediction in multivariate calibration, *J. of Chemometrics.*, **6**, 103-111.

Strang, G. (1988). Linear Algebra and its applications, 3rd ed., Saunders Publishing, New York.

Wise, B. M., N .L. Ricker, D.F. Veltkamp, and B.R. Kowalski (1990). A theoretical basis for the use of principal component models for monitoring multivariate processes, *Process Control and Quality*.

MULTI-WAY ANALYSIS IN PROCESS MONITORING AND MODELING

Barry M. Wise and Neal B. Gallagher
Eigenvector Research, Inc.
Manson, WA 98831

Abstract

Continuous chemical processes with multiple measurements produce data that is logically arranged as a 2-way matrix (second order tensor). Data from batch processes are logically arranged as a 3-way matrix (third order tensor) where data from each batch is a "slice" of the data "cube." Increasingly, sensing systems are capable of producing a second order response for each sample analyzed. When these instruments are used to monitor chemical processes the result is data that is naturally arranged as a n-way matrix, with n ≥ 3. It has been shown that it is possible to take advantage of this higher order data provided that the proper data models and numerical methods are employed. For instance, analytical instruments that produce second order (2-way) data (such as GC-MS) have the ability to quantify species of interest in the presence of previously unseen interferents. This is not possible with first order instruments. In this manuscript we review some of the methods for analyzing multi-way data and show how the methods can be used in chemical process monitoring and modeling.

Keywords

Multi-way data analysis, Process monitoring, Process modeling, Multivariate curve resolution, Trilinear decomposition, PARAFAC, Tucker models.

Introduction

Analytical chemistry has recently undergone a paradigm shift. Analytical instruments used to be classified in terms of the physics governing the method, e.g. spectroscopy vs. chromatography. Now, however, instruments are often classified in terms of the type of data they produce (Booksh, 1994). An instrument that generates a single datum per sample, e.g. a pH electrode, is a zero order instrument since a single number is a zero order tensor. An instrument that generates a vector (first order tensor) for each sample is a first order instrument, e.g. a spectrometer.

The reason behind this shift in emphasis is that the order of the data produced determines what information may be extracted from it. It has been shown, for example, that it is not possible to quantify any particular chemical species in the presence of interfering species with a zero order instrument, but it is possible with a first order instrument (e.g. a spectrometer) provided that a model was developed with data which includes the interferent. With a second order instrument (e.g. GC-MS) it is possible to quantitate in spite of a previously unseen interferent (Sanchez 1990). The ability to extract relevant information from higher order systems in spite of interferences is known as the "advantage of order."

Today's chemical processes are typically multivariate, many measurements are made at regular intervals, and the data is logically ordered by time. Thus a continuous process produces a 2-way (second order) data matrix. Batch processes, where the same "recipe" is followed many times, produce a 2-way data matrix for each run. These matrices are logically arranged as a 3-way data cube. The combination of a higher order instrument with a process can lead to logical data arrangements that are n-way, where n is ≥ 3.

To realize of the advantage of order, the proper data modeling method must be used. Fortunately, the potential advantages have lead to the development of model forms and numerical methods for extracting the information from multi-way data sets. It is quite likely that these methods could be very useful in the context of chemical process monitoring and modeling. The purpose of this document is to introduce the reader to some of these methods and to provide some examples of how they might be used in practice.

Multi-way Methods

In this section a brief description of the theory behind the analysis of n-way analysis is given. Several of the more common methods are reviewed. Note that the term "chemical system" can refer to an analytical instrument or a chemical process. Likewise, the term "input" can refer to an analyte or a process input.

Weakly Multi-way Methods

Applications of Multi-way Principal Components Analysis (MPCA) and Multi-way Partial Least Squares (MPLS) have been reported in the literature (Nomikos, 1994 and 1995). These methods are not explicitly multi-way as they rely on a rearrangement of the n-way data (where n is usually 3) into a conventional 2-way matrix prior to analysis. Because of this rearrangement we refer to these methods as weakly multi-way.

In MPCA the 3-way matrix $\underline{\mathbf{X}}$ is decomposed as the summation of the product of score vectors ($\mathbf{t}$) and loading matrices ($\mathbf{P}$) plus a residual array $\underline{\mathbf{E}}$ that is minimized in a least squares sense.

$$\underline{\mathbf{X}} = \sum_{r=1}^{R} \mathbf{t}_r \otimes \mathbf{P}_r + \underline{\mathbf{E}} \qquad (1)$$

This decomposition is accomplished by arranging the elements of each 2-way slice of $\underline{\mathbf{X}}$ in a single row of a matrix, then performing conventional PCA. Several applications of MPCA and MPLS have been very successful, particularly for batch process monitoring. Note, however, that MPCA models confound two of the orders (usually variables and time) into the $\mathbf{P}_r$ matrices, which, in general, are of arbitrary rank.

Bilinear Data

The simplest case of multi-way data is the bilinear case, which is also known as the rank 1 case (Sanchez, 1988a). In this instance, the matrix of data produced by the chemical system $\mathbf{N}_k$ in response to a single input k can be modeled as the outer product of the system's response in each of the two orders:

$$\mathbf{N}_k = \mathbf{x}_k c_k \mathbf{y}_k^T + \mathbf{E} \qquad (2)$$

where $\mathbf{x}_k$ (a column vector) is instrument response in the first order, $\mathbf{y}_k^T$ (a row vector) is the instrument response instrument in the second order, c_k is the magnitude of the input (e.g. the analyte concentration or process input) and $\mathbf{E}$ is the measurement error. In such a case, the response matrix $\mathbf{N}_k$ has an effective rank of 1, however, the true mathematical rank of the data will be greater than 1 due to noise. Examples of second order bilinear analytical systems include GC/MS and LC/UV. Linear processes with multiple outputs will produce a rank 1 response provided that only one process state is affected by the input.

The response to multiple inputs $\mathbf{N}_m$ can be modeled as the sum of responses to the single inputs

$$\mathbf{N}_m = \sum_{k=1}^{m} \mathbf{x}_k c_k \mathbf{y}_k^T + \mathbf{E} = \mathbf{X}\mathbf{C}_m \mathbf{Y}^T + \mathbf{E} \qquad (3)$$

where m is the number of inputs, $\mathbf{X}$ and $\mathbf{Y}$ are the matrices of responses in the first and second order respectively, and $\mathbf{C}_m$ is a diagonal matrix of input magnitudes. In this instance the response matrix has an effective rank of m.

Generalized Rank Annihilation Method

It has been shown that the Generalized Rank Annihilation Method (GRAM) can be used to determine the magnitude of a single input (e.g. concentration of an analyte of interest) in a multiple input response $\mathbf{N}_m$ given the response $\mathbf{N}_c$ of a calibration sample of known input magnitude (Sanchez, 1988b) where

$$\mathbf{N}_c = \mathbf{X}\mathbf{C}_c \mathbf{Y}^T + \mathbf{E} \qquad (4)$$

This leads to the generalized eigenvalue/eigenvector problem

$$\mathbf{N}_c \mathbf{Z} \mathbf{C}_m = \mathbf{N}_m \mathbf{Z} \mathbf{C}_c \qquad (5)$$

where $\mathbf{Z}$ is a matrix of eigenvectors. Several algorithms have been proposed for performing the GRAM procedure, however, the state of the art appears to be the method proposed by Wilson (1989). GRAM is a relatively straightforward method and has utility in a large number of applications; many second order analytical methods are bilinear.

Multivariate Curve Resolution

One difficulty with GRAM is that it is affected by shifts in either order due to process drift. These difficulties can often be resolved with Multivariate Curve Resolution (MCR). This technique is also known as self modeling curve resolution (SMCR).

Simply put, the goal of MCR is to extract the number of inputs or species, their time profiles and single input responses with as few assumptions about the data as possible (Tauler, 1993a, 1993b). The most common application of MCR is the extraction of concentration time profiles and pure component spectra in chemical processes or second order instruments. The following discussion refers to a process spectrometer which monitors a batch chemical process. Restated mathematically in this context, given a response matrix $\mathbf{N}_m$ (not mean centered) that is the product of concentration profiles **C** and pure component spectra **A**,

$$Nm = C\,A \quad (6)$$

we wish to obtain physically meaningful **C** and **A**.

MCR makes use of the fact that both concentrations and spectra are necessarily non-negative. Using this constraint, it is possible to extract pure component spectra through a procedure of alternating and constrained least squares optimization. The result of the procedure is an estimate of pure component responses in both instrument orders. These estimates can provide powerful diagnostic tools for experimental investigation and process analysis. As and additional benefit, if standards are used, MCR can be used as a calibration technique for second order methods.

An advantage of MCR is the ability to incorporate additional information about the problem at hand. It is possible to incorporate this information into the iteration as a constraint. For instance, it is often assumed that the concentration profiles are unimodal. In other cases, closure or some known stoichiometry may be used. When additional known (correct) constraints are used, the quality of the solution obtained generally improves.

It is also possible to combine data sets from different analytical samples or multiple runs of a process under consideration simply by appending matrices from subsequent runs. Combining data from different runs can also contribute to the stability of the solutions obtained.

Parallel Factor Analysis

GRAM has the unique and powerful ability to perform quantitation using a single reference sample. However, it is also restricted to use with a single calibration sample. Therefore, the signal to noise advantage gained when multiple calibration samples are used is lost. Parallel Factor Analysis (PARAFAC) is a curve resolution technique that estimates responses in each of the orders and can be used with multiple samples (Smilde, 1991, 1994). The PARAFAC model of a third order tensor **D**, consisting of a "stack" of second order system responses is

$$d_{ijk} = \sum_{q=1}^{Q} a_{iq} b_{jq} c_{kq} + e_{ijk} \quad (8)$$

where Q is chosen such that **E** with elements e_{ijk} has small norm. This is the trilinear model. If **A** is defined such that its qth column is a_q, and likewise for **B** and **C** then the outer product of a_q, b_q and c_q is the qth triad of the PARAFAC model.

One very important property of the PARAFAC decomposition is that, under very mild assumptions, it is unique. The result is that **A**, **B** and **C** give pure component responses and relative concentration estimates directly. This is a very powerful advantage of the method.

Like MCR, PARAFAC uses an alternating least squares approach in the decomposition, which, unfortunately, often does not converge to chemically meaningful solutions unless proper constraints are applied. Examples of constraints are non-negativity of spectra and unimodal GC response curves. A challenge in the implementation of PARAFAC and Multivariate Curve Resolution is the automation of the incorporation of reasonable constraints.

The Trilinear Decomposition

The Trilinear Decomposition (TLD) can also be used with multiple calibration samples TLD estimates pure component instrument responses in both orders through solution of an eigenvalue/eigenvector problem (Sanchez, 1990). Concentrations are then estimated by least squares. This allows for a direct, exact solution and provides signal averaging. Early implementations of TLD led to non-meaningful imaginary solutions. Recent modifications to the original TLD algorithm, however, have rectified this problem and have lead to significantly improved concentration estimates. TLD provides essentially the same information and advantages of PARAFAC.

Tucker Models

The Tucker model (Tucker, 1963) is another important tensor decomposition. Given $\underline{D}$ with elements d_{ijk}, the Tucker model is

$$d_{ijk} = \sum_{p=1} \sum_{q=1} \sum_{r=1} a_{ip} b_{jq} c_{kr} g_{pqr} + e_{ijk} \quad (9)$$

and is a quadrilinear model. The g_{pqr} form a three-way array $\underline{G}$ (P by Q by R) which is much smaller than $\underline{D}$, and is known as the core array. P, Q and R are parameters to be decided upon and chosen in such a way that $\underline{E}$ has a sufficiently small norm. An alternating least squares procedure has been developed to estimate the Tucker model parameters. One difficulty with the application of the Tucker model is rotational ambiguity of the solutions. Often **A**, **B**, and **C** are restricted to be orthonormal matrices. In this instance, physically meaningful solutions are not obtained. Methods for constraining the solution to a physically reasonable one are yet to be explored. Note that PARAFAC is a special case of Equation 9, with $g_{pqr} = 1$ when $p = q = r$, and $g_{pqr} = 0$ otherwise. In this respect PARAFAC is a Restricted-Tucker Model (RTM).

In general, an RTM is a Tucker model with some of the elements of $\underline{G}$ set to zero (Kroonenberg, 1980). The elements of $\underline{G}$ can be set to zero using either a theoretical understanding of the chemical system. Another way to restrict the Tucker model is by explicitly using prior information on the loading matrices. For instance, it may be known that **A**, **B**, and **C** are non-negative and parts of the matrices may be known explicitly. In such instances the estimation procedure becomes a combination of alternating least squares and non-negative least squares. Theory regarding the conditions which assure the uniqueness of the RTM solution have yet to be developed. RTMs have a great potential for application to systems where the response of the system to each input is of medium rank (~2-5). To realize the potential power of the method more applications must be attempted.

Some (Brief) Examples

As an example of MPCA consider a proprietary batch process where a reacting system is being monitored with a spectrometer and many conventional engineering variables are also measured. In the reactor a reactive gas strips layers with differing compositions off the product part.

In this example, the data set consists of 37 samples (batches). The total number of variables is 6206 and 28 time steps are taken per batch. Thus, the 3-way array is 37 by 6206 by 28, for a total of 6,429,416 elements. When MPCA was performed on this data, it was found that 51% of the variation in the data was captured by the first PC; over 72% of the variation was captured by the first 3 PCs. The MPCA scores of the data are shown in Figure 1. The data clearly split into two groups due to a change between lots which had slightly different upstream processing conditions.

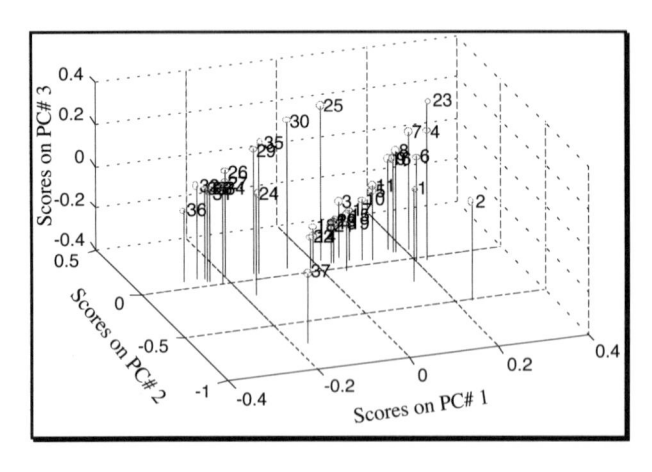

Figure 1. MPCA Scores of Proprietary Batch Chemical Process.

MCR was used on the spectral data from the batch process. The estimated concentration profiles reactant gas and analytes from the layers being stripped are shown in Figure 2. It was found that an additional, unexpected, analyte appeared in the batch. This analyte may be a previously unseen reaction intermediate. Pure component spectra were also recovered and are being used to help improve other system monitoring devices. Given the pure component spectra of the analytes it is possible to pick optimal regions of the spectra for monitoring purposes.

Conclusions

This document serves as a very brief introduction to multi-way methods and is far from exhaustive. These methods are an as yet untapped, but potentially very powerful approach to modeling chemical processes. Many connections between the multi-way models and dynamic process models have yet to be explored. For instance, the relationship between restricted Tucker models and state-space process models may prove quite enlightening. It is the authors' belief that the application of the methods in the chemical process domain will lead to new process insights and models that are very useful for process monitoring and modeling for control purposes.

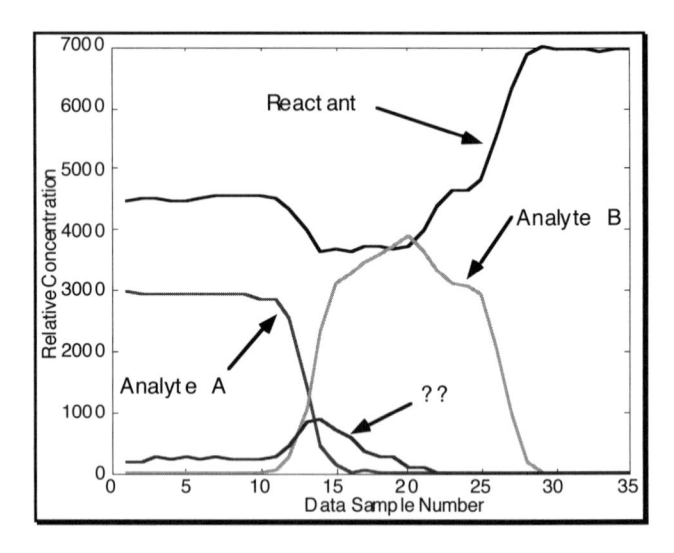

Figure 2. Concentration Profile Estimates from MCR of Proprietary Batch Chemical Process.

References

Booksh, K. and B.R. Kowalski (1994). *J. Chemometrics*, 8, 45-63.
Kroonenberg, P.M. and J. de Leeuw (1980). *Psychometrika*, 45, 69.
Nomikos, P. and J.F. MacGregor (1994). *AIChE Journal*, **40**(8), 1361-1375.
Nomikos, P. and J.F. MacGregor (1995b). *Technometrics*, **37**(1).
Sanchez, E. and B.R. Kowalski (1988a). *J. Chemometrics*, 2, 247-263.
Sanchez, E. and B.R. Kowalski (1988b). *J. Chemometrics*, 2, 265-283.
Sanchez, E. and B.R. Kowalski (1990). *J. Chemometrics*, 4, 29-46.
Smilde, A.K. and D.A. Doornbos (1991). *J. Chemometrics*, 5, 345-360.
Smilde, A.K., Y. Wang and B.R. Kowalski (1994). *J. Chemometrics*, 8, 21-36.
Tauler, R. and B. Kowalski (1993a). *Anal. Chem.*, 65, 2040-2047.
Tauler, R., A. Izquierdo-Ridorsa and E. Casassas (1993b). Chemo. and Intell. Lab. Sys., 18, 293-300.
Tucker, L. (1963) Problems in Measuring Change, University of Wisconsin Press, Madison, WI.
Wilson, B.E., E. Sanchez and B.R. Kowalski (1989). *J. Chemometrics*, 3, 493-498.

IDENTIFICATION OF PAPER MACHINE FULL PROFILE DISTURBANCE MODES USING ADAPTIVE PRINCIPAL COMPONENT ANALYSIS

Apostolos Rigopoulos and Yaman Arkun
School of Chemical Engineering
Georgia Institute of Technology
Atlanta, GA 30332-0100

Ferhan Kayihan and Eric Hanczyc
Weyerhaeuser Technology Center
Tacoma, WA 98477

Abstract

The significant disturbance characteristics of a paper machine profile are identified using adaptive Karhunen-Loève expansion. The full profile under analysis is assumed to be either measured by stationary full-width sensors or reconstructed from measurements coming from a moving scanner. Different criteria to select the number of disturbance modes are discussed and used in simulations.

Keywords

Paper machine, Identification, Estimation, Modes, Principal component analysis, Karhunen-Loève expansion.

Introduction

In paper machines the profiles of different paper properties (e.g. basis weight, thickness, moisture) vary both in cross (CD) and machine directions (MD) which result in a two-dimensional distributed system that poses challenging estimation and control problems. Most of the early works have focused on the separability of CD and MD variations from scanned measurements and their independent controllability (e.g. Wilhelm, 1984; Taylor, 1991). Adopting the traditional assumption that CD changes occur more slowly than MD, Wang et al. (1993b) are able to model and estimate these variations separately using Kalman filtering in parallel with an exponential forgetting and resetting algorithm. The same authors also implement their estimator using fixed array sensors and note their potential advantages (Wang et al., 1993a). Recently Tyler and Morari (1995) have implemented a periodic Kalman filter to reconstruct the cross-directional profile by using the "lifting" approach. They compare the estimator performance for scanning and stationary sensors using different disturbance and noise characteristics. Rawlings and Chen (1994) and Bergh and MacGregor (1987) also design periodic Kalman filters but they do not use the "lifting" approach.

Whether working with scanners or stationary sensors, one is confronted with large sets of correlated data which have to be efficiently handled during estimation and closed-loop control. In this work we are interested in identifying the most significant modes of the underlying full profile. We develop an on-line implementation of a particular principal component analysis technique, known as the Karhunen-Loève (KL) expansion which does data compression and filtering. Once such a model is made available, it can be used for closed-loop control where the controller responds only to the most relevant disturbance profile at any given time.

Karhunen-Loève Analysis

The Karhunen-Loève analysis is a Principal Component Analysis technique applied to random processes. Consider a paper machine with N CD positions, and M MD positions where $N < M$. It is assumed that the paper properties are measured by either a fixed array sensor or a scanner. In the case of the stationary sensor we have access to N CD measurements taken in M time intervals. These measurements are stored in a data matrix Z of size $M \times N$ where each row represents data taken at the same time (same MD position) across the sheet. In the case of the moving scanner the data matrix Z is constructed similarly by replacing the missing CD measurements with their estimates which can be obtained in different ways including the Kalman filter. Given a batch of data the first step in the KL analysis is to mean center them by subtracting the temporal average from each column of the data matrix Z to form a new matrix $\hat{Z}$. Next the elements of the two-point spatial correlation matrix $\mathbf{K}$ are computed from:

$$K(i,j) = \frac{1}{M}\sum_{k=1}^{M}\hat{Z}(k,i)\hat{Z}(k,j) \quad i,j = 1:N$$

The finite KL expansion of the data set over our finite sample space can be written as a linear combination of the eigenvectors of $\mathbf{K}$, $\{\phi^{(n)}\}$, where the coefficients $\{a_n\}$ of the expansion are uncorrelated random variables on that space i.e. $\hat{z}(k) = \sum_{n=1}^{N} a_n(k)\phi^{(n)}$. The column vector $\hat{z}(k)$ consists of N CD properties at time k. The stochastic evolution of this CD profile is attributed to the temporal modes a_n which can be computed from $a_n(k) = \hat{z}(k)^T \phi^{(n)}$. Among the class of all linear decompositions, the KL method is optimal in the sense that the KL expansion on a subspace of lower dimension $P < N$ retains the most average energy possible. In other words the (% variance), $100 \times \sum_{n=1}^{P}\lambda_n / \sum_{n=1}^{N}\lambda_n$, is maximized where each eigenvalue λ_i of $\mathbf{K}$ represents the mean energy of its corresponding mode. Rigopoulos and Arkun (1996) have used the KL expansion to model disturbance profiles given a batch of data from stationary sensors.

In on-line applications the KL procedure has to account for the nonstationary nature of disturbances. To accomplish this the KL expansion is repeatedly applied to a fixed size moving window of data in which exponential weighting may be used to eliminate the influence of old data. As new data are collected, the local statistical information contained in the particular window is used to identify and select the significant modes in an adaptive fashion. The important decision is how to choose the number of dominant modes. In order of increasing complexity and accuracy we propose three selection criteria below and illustrate their use in the examples.

(% variance) Criterion. Here the number of significant modes is selected to satisfy a desired value of (% variance) as shown in Fig. 1. It is simple to implement on-line and works very well when the data do not contain significant additive white noise, or when the measurement noise is already filtered out. However, as the S/N decreases, it becomes increasingly more difficult to specify the cut-off value for the desired percent variance. A too small a value filters the noise but misses some significant modes, while a too large a value would start modeling the noise.

Scaled Eigenvalues Criterion. For additive white noise, the (% variance) varies almost linearly with the number of modes, since each noise mode contributes almost equally to the total variance. To separate these modes from the rest, the (% variance) is differenced and plotted against the number of modes. Differencing amplifies the contribution of

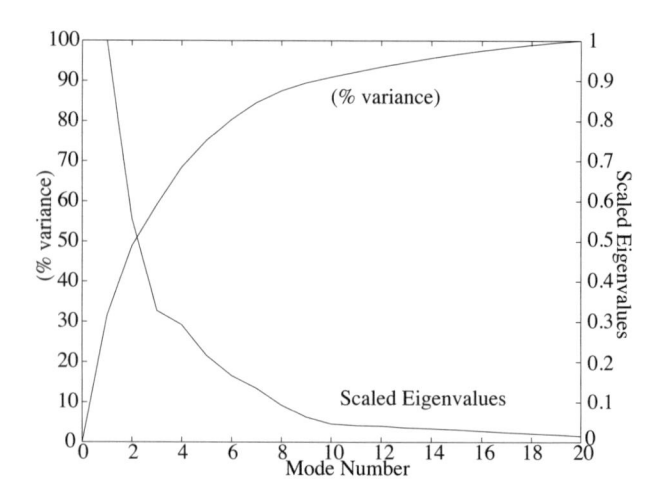

Figure 1. A typical plot of (% variance) and scaled eigenvalues versus the number of modes.

the modes which correspond to the noise-free signal, thus making the cut-off point less ambiguous. When normalized by the (% variance) contribution of the most significant mode, this plot is equivalent to the distribution of the scaled eigenvalues λ_i/λ_1. The distribution curve starts with a large slope (due to the most significant modes) and reaches an almost straight line with a small slope. So the first few dominant modes are retained in such a way that the remaining ones (due to noise) are aligned on a straight line in the least square sense (by presetting the coefficient of determination, r^2, of the line.).

Whiteness of Temporal Modes Criterion. The first two criteria are simple to use and certainly justified in many cases. However, if one is interested in identifying the number of modes which do not correspond to white noise more precisely, they may not be suitable when the choice of the cut-off values ((% variance) or r^2) is not obvious. This last criterion checks the autocorrelation of the temporal eigenvectors to detect if they correspond to white noise. Because of their orthogonality, each temporal mode can be analyzed independently, thus significantly reducing the computational load. Thus by using a hypothesis test based on the χ^2 distribution (Ljung, 1987), the modes that do not describe a particular pattern are rejected. For the remaining modes, either of the first two criteria are used to select the significant ones.

Results and Discussion

The property profile under study is generated from a proprietary paper machine simulator which is developed from first-principles (Kayihan and Hanczyc, 1995). The profile has 20 CD positions and data are taken at 3000 time intervals. It includes four different sections with smooth transitions:

Identification of Paper Machine Full Profile Disturbance Modes Using Adaptive Principal Component Analysis 277

- Section 1: MD Positions 1 to 300. The profile is assumed to be around its pre-specified mean with only measurement noise added to it.

- Section 2: MD Positions 300 to 1800. Here, MD and CD variations of increasing magnitude constitute the profile along with measurement noise.

- Section 3: MD Positions 1800 to 2500. The CD variations are minimized leaving only the MD frequencies and the noise.

- Section 4: MD Positions 2500 to 3000. The CD variations return, only their magnitude is smaller than what it was for section 2.

Simulations Based on Full-scan Measurements

Here, we assume that we have access to full-scan measurements from a stationary array scanner. After a warm-up period equal to the window size selected, KL expansion is repetitively performed on the most recent data comprised within a window of size W.

At first, all data are weighted equally. Fig. 2 illustrates the number of modes selected for a window size $W = 100$. The plots of criteria 2 and 3 are very similar in shape and magnitude whereas criterion 1 is completely different. The reason for this is that criterion 1 (with a pre-set (% variance) cut-off value of 95%) does not take into consideration the magnitude of the measurement noise. What is important to note is that the number of modes selected based on criteria 2 and 3 matches the description of the different sections of the property profile. So, initially, where only measurement noise is present, both criteria indicate that no modes are needed. For section 2 the number of modes increases to a maximum of eight, then it drops to two or three for sec-

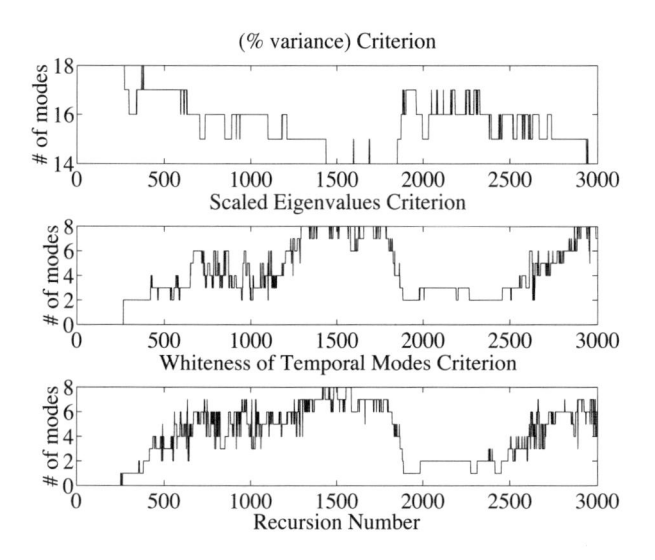

Figure 2. Number of modes selected based on the three criteria as a function of the window number.

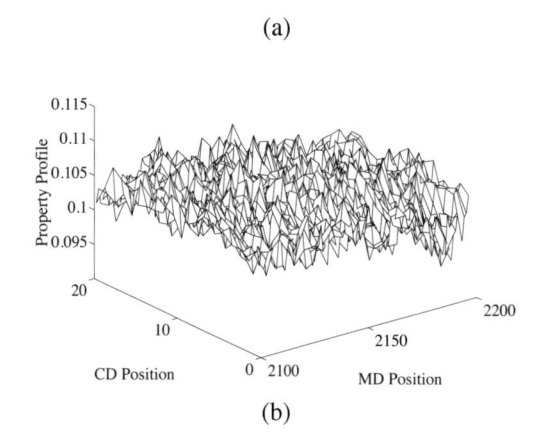

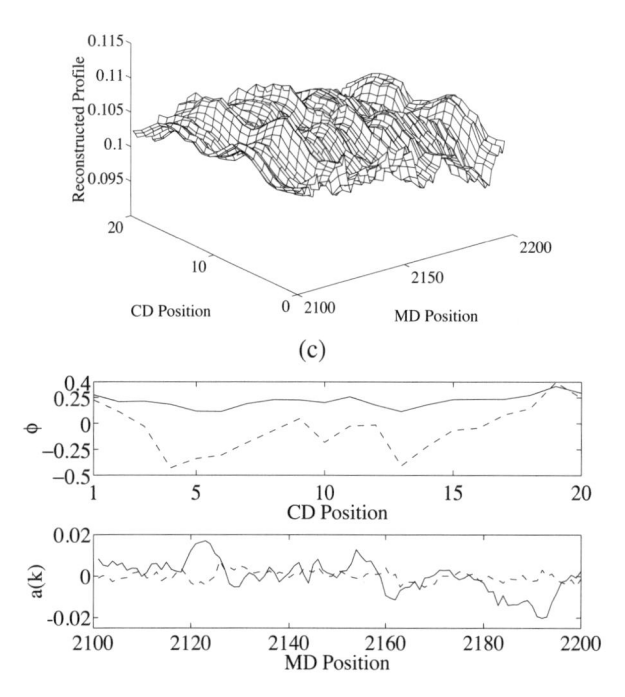

Figure 3. Original profile (a), and reconstructed profile (b) using two modes as selected by criterion 3. The reconstructed profile is almost identical to the noise-free original simulated profile with a normalized sum of square reconstruction error equal to 4.6×10^{-7}. The first (solid) and second (dashed) spatial and temporal eigenvectors for the window at time $k = 2200$ is shown at the bottom plot (c).

tion 3, and finally it increases again for section 4. Fig. 3(a) illustrates the original property profile of a part of section 3, whereas the reconstructed profile adaptively evaluated retaining 95% of (% variance) of the significant modes (after the noise modes have been excluded using criterion 3) is shown in Fig. 3(b). Two modes are enough to explain the profile to the given accuracy. They are shown in Fig. 3(c).

If the window size increased from 100 to, say, 200 one would notice two main differences from the plots of Fig. 2. First, the number of modes per recursion would be larger (on the average one more mode would be needed). This should

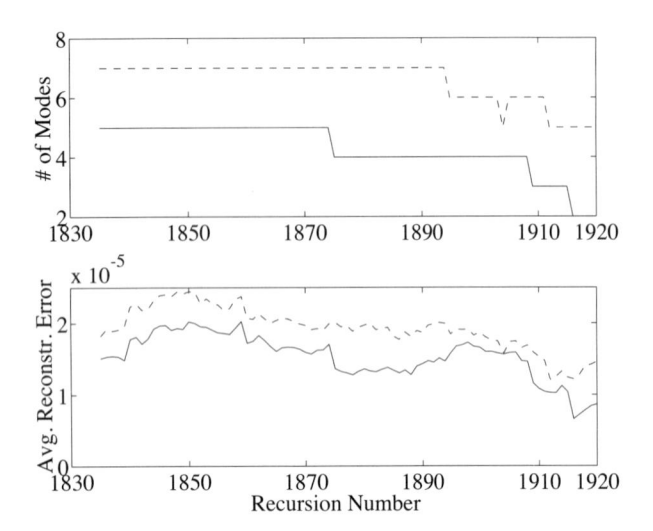

Figure 4. Comparison of modes and average reconstruction error in a transition region for exponentially (solid line) and equally (dashed line) weighted KL with $W = 200$, $w = 20$.

be expected since the number of patterns to be captured is larger for the larger window size. Second, the plots would look less jagged. Since the new information is weighted less for large window size, there must be an important change in the property profile to justify a sharp change in the λ_i value, and as a consequence, criteria 2 and 3 will be less eager to change the number of modes. In the limit, when $W = M$, batch KL is implemented. In this case, application of criteria 1, 2, and 3 would indicate that fifteen, eight, and eight modes, respectively are necessary. The advantage of the adaptive KL method with a relatively small window size is now obvious, since depending on the window to be analyzed at a particular time, fewer modes may be needed as shown in Fig. 2.

Until now, we assumed that within the window selected all data are equally weighted. However, since we are interested in representing the most recent data more accurately to quickly adapt to changing conditions, we could apply a weighting function to the data so that older data are valued less. For a given window size W, we keep the most recent w CD profiles untouched, whereas the remaining $W - w$ profiles are multiplied by an exponentially decreasing weighting function and KL is applied to this modified profile. An improvement over the equal weights case can be observed during a transition from section 2 to 3 where the number of modes changes due to slowly disappearing CD variations. Fig. 4 shows that the use of the weighted data allows KL to adapt to the mode changes faster, and as a result the normalized sum of square errors for the last 20 CD profiles for each window between noise-free original profile and its reconstructed one is smaller for the exponentially weighted than the equally weighted case.

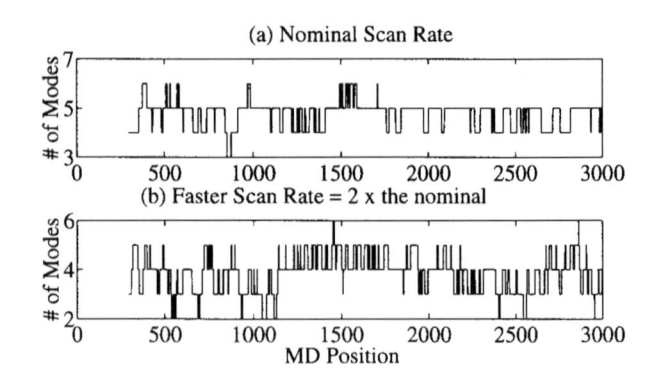

Figure 5. Number of modes evaluated using criteria 1 and 3 for the nominal and faster scan rates.

Simulations Based on Moving Scanner Measurements

If instead of a stationary array of sensors, a moving scanner is available, the missing CD data at each time k can be estimated through a periodic Kalman filter. The profile generated in this way is then used by KL adaptively to select the number of modes as before. In this case, however, it is important to note that since the measurement noise is already filtered, sole application of criterion 1 with a 95% cut-off value is enough to get good results.

Fig. 5(a) shows the number of modes selected using criterion 1 with a sampling rate of 1 time unit and $W = 100$. The first profile section which corresponds to noise has been ommitted from the calculations since it does not include any significant modes to be identified. Comparison of this plot against the middle and bottom plots of Fig. 2 illustrates that KL is not able to effectively differentiate between the different regions because of the significant estimation error due to the missing CD data. If the sampling rate is decreased to one half time units however (Fig. 5(b)), then the number of modes selected by KL comes closer to what was calculated using full-scan data.

Conclusions

We have presented an adaptive implementation of the Karhunen-Loève (KL) expansion to identify the significant disturbance modes of a paper machine profile. Different criteria to select the number of modes are discussed and used in a simulation example. The weighted version of KL has proven to be effective during transition regions in which the modes change. The technique can be used with either stationary or moving scanners. In the latter case, a Kalman filter is used to reconstruct the full profile which is then subjected to the KL analysis. The effect of scanning rates is also briefly discussed.

Acknowledgment

The first two authors gratefully acknowledge the financial support of Weyerhaeuser Co.

References

Bergh, L.G. and MacGregor, J.F. (1987). Spatial control of sheet and film forming processes, *Can. J. Chem. Eng.*, **65**, 148-155.

Kayıhan, F. and Hanczyc, E. (1995). Paper machine model and full sheet property estimation, *Weyerhaeuser Control Workshop on Kamyr Digesters and Paper Machines*, June, Weyerhaeuser Technology Center, Federal Way, WA.

Ljung, L. (1987). *System Identification: Theory for the User*, PTR Prentice Hall.

Rawlings, J.B. and Chien, I.-L. (1994). Gage control of sheet and film forming processes, Accepted in *AIChE J*.

Rigopoulos, A. and Arkun, Y. (1996). Principal components analysis in estimation and control of paper machines, *ESCAPE-96*, Rhodes, Greece, submitted.

Taylor, B.F. (1991). Optimum separation of machine-direction and cross-direction product variations. The scanning measurement challenge, *Tappi J.*, 87-92.

Tyler, M.L. and Morari, M. (1995). Estimation of cross directional properties: Scanning versus stationary sensors, *AIChE J.*, **41**, 846-854.

Wang, X.G. and Dumont, G.A. and Davies, M.S. (1993a). Basis weight estimation from array measurements, *Proc. IFAC World Congress*, Sydney, Australia, **6**, 55-59.

Wang, X.G. and Dumont, G.A. and Davies, M.S. (1993b). Estimation in paper machine control, *IEEE Control Systems Magazine*, **13**(4), 34-42.

Wilhelm, R.G. (1984). On the controllability of cross-direction variations of sheet properties. *Engineering Conference, Proceedings of the Technical Association of the Pulp and Paper Industry*, 621-629.

EXPERIMENTAL EVALUATION OF NEURAL NONLINEAR MODELING TECHNIQUES

John D. Bomberger and Dale E. Seborg
University of California
Santa Barbara, CA 93106

Gordon Lightbody and George W. Irwin
The Queen's University of Belfast
Belfast BT9 5AG, Northern Ireland

Abstract

In this paper three nonlinear neural modeling techniques, multi-layer perceptron networks (MLP), B-spline networks, and radial basis function networks (RBFN), are compared in an experimental application for a laboratory scale pH neutralization process at UCSB. The neural modeling techniques are evaluated on the basis of model accuracy, practicality, and ease of implementation. It will be shown that accurate models can be obtained using any of the networks, and that RBFN and B-Spline networks have the benefit of local basis function support. Furthermore, the effects of the "Curse of Dimensionality" on the three networks types will be discussed in terms of computational difficulty and practicality.

Keywords

Nonlinear identification, Neural networks, Radial basis function network, Multi-layer perceptron, B-Spline network.

Introduction

In the past several years, there has been considerable interest in using neural network models for process identification and control. However, many of the published papers are based on simulation studies for low dimensional, simplified mathematical models instead of realistic chemical processes or experimental data. Furthermore, comparisons between different types of neural network models for actual processes are few.

In this paper three different types of neural network models are examined for use in the identification of the laboratory scale pH neutralization process at UCSB. The three types of networks are: multi-layer perceptron, B-spline, and radial basis function networks. All three SISO models can be represented as nonlinear, auto-regressive models with exogenous inputs (NARX), where the model output at time k, $\hat{y}_k$, is expressed as

$$\hat{y}_k = F(\mathbf{x}_k) \quad (1)$$

and $\mathbf{x}_k$ is the network input vector,

$$\mathbf{x}_k = \begin{bmatrix} y_{k-1} & \cdots & y_{k-m} & u_{k-1-\theta} & \cdots & u_{k-n-\theta} \end{bmatrix}^T \quad (2)$$

y_k and u_k are the process output and input, respectively, m is the number of past outputs, n the number of past inputs, and θ the number of samples time delay.

Multi-layer Perceptron Networks

Multi-layer perceptron networks (MLP) are perhaps the most widely used networks for process identification in chemical engineering. The modeling capabilities of these networks have been demonstrated in numerous publications, and several well known training algorithms have been developed, most notably backpropagation.

In this paper, the Broyden-Fletcher-Goldfharb-Shanno (BFGS) method, a conjugate gradient technique, is used for network training (Battiti and Masulli, 1990). Conjugate gradient techniques can show a substantial acceleration of the training process compared to gradient descent based training algorithms like backpropagation (Charalambous, 1992).

B-spline Networks

The B-spline network can be represented as a linear combination of multivariate basis functions. Each multivariate basis function is generated based on univariate polynomial spline functions: the outputs of all the univariate splines generated for each scalar element of the network input vector, each of which is nonlinearly mapped onto a higher dimensional space according to a set of "knots" (Lightbody and Irwin, 1995). Due to the large number of parameters that need to be approximated and the sparseness of the vector of basis function outputs, normalized least mean squares is used for identification (Harris et al., 1993; Brown and Harris, 1994).

Radial Basis Function Networks

Radial basis function networks (RBFN), like B-spline networks, are linear combinations of weighted basis functions. In this case however, the basis functions are radially symmetric maps from the space containing the network input vector to a scalar. Often, Gaussian radial basis functions are used,

$$\phi(r_{ki},\beta) = \exp\left(-\frac{\|\mathbf{x}_k - \mathbf{c}_i\|_2^2}{\beta}\right) \quad (3)$$

where β is a shape parameter, and $\mathbf{c}_i$ is the center of the RBF.

The most noteworthy methods for identifying RBFN are k-means clustering (Moody and Darken, 1989; Leonard and Kramer, 1991) and some form of linear regression, *e.g.* stepwise regression used by Pottmann and Seborg (1992). Both of these methods attempt to place RBF centers and then approximate the linearly appearing connection weights using some variant of least squares estimation.

Localized Basis Functions

Localization of the basis functions can be advantageous for network models, especially if the model must be updated or adapted to accommodate changing process conditions. Networks with localized basis functions can improve model accuracy in one region without affecting other regions. In contrast, if networks using globally activated basis functions are adapted to improve accuracy within a particular operating region, the network model obtained for previous operating regions may be lost.

It has been suggested that because of the global nature of the activation function within the MLP, this type of network is not readily suitable for on-line adaptive modeling (Lightbody and Irwin, 1995). For B-spline networks, basis function localization is determined by the number of knots and the order of the polynomial and thus these networks are typically very localized. Radial basis functions can be made to appear localized because they decrease exponentially to zero as the radius increases. The rate at which they decrease is determined by the shape parameter β; small values of β cause a rapid decrease in the value of the RBFs, and hence localized RBFs, while large β values result in broader RBFs that act more globally, because the RBF values approach zero only well outside the expected range of the network inputs.

Experimental Results

The apparatus used in this study is essentially the same pH neutralization process as that described by Hall and Seborg (1989) and used in many other studies (Seborg, 1994). In this study, a single continuous stirred tank reactor configuration is used. The reactor inputs are an acid stream, a buffer stream, and a base stream: their respective flowrates are Q_1, Q_2, and Q_3. There is a single effluent stream. The measured process output is effluent pH, which has an time delay of approximately 30 s due to placement of the pH sensor downstream of the reactor. The process input is considered to be the base flowrate Q_3, with Q_1 and Q_2 acting as disturbances.

Three sets of experimental open loop data were recorded at a sampling time of 15 s, and used for evaluation of the neural network models. There are numerous unmeasured disturbances present in the data sets in the form of acid flowrate changes and feed tank concentration changes. Set 1 is a portion of the data set used for training the network models: the entire training set consisted of 2352 data points. Sets 2 and 3 are validation data sets, *i.e.* they were not used for training the models and as such represent a measure of the models' prediction abilities. Two types of predictions were made: one-step-ahead prediction, and multiple-step-ahead prediction from initial conditions and process inputs, where $\hat{y}$ was substituted for y in Eqn. (2).

Neural Network Models

The networks were all trained on the same data set, using two past outputs, two past inputs, and a delay of two samples in the network input vector in Eqn. (2). The low dimensionality of the input vector is a concession to the limitations of B-spline networks. The training data set was also chosen based on the requirements of the B-spline network; specifically, the very large number of parameters and the sparseness of the vector of basis function outputs.

The MLP network was identified by a time consuming process of cross-validation to select the number of iterations (to prevent over-fitting), the model size, and the initial weights. This process involved 80 different networks and about 24 hours of computer time. The selected model size was a single hidden layer of nine hyperbolic tangent nodes which was identified after 544 presentations of the data set. The network was trained using the BFGS algorithm.

The B-spline network used third-order polynomial splines and six interior and four exterior knots placed uniformly over the range 4 - 22 ml/s for the input and 0.7 - 13.3 pH for the output. Normalized least mean squares was used for identification, for almost 1000 iterations of the training set.

The RBFN was trained using k-means clustering with 20 clusters. No optimization was attempted. The clusters were determined after 14 iterations, taking 41.4 s, a dramatic time savings over the other two network types. Stepwise regression could not be attempted because of the large size of the training data set. The P-nearest-neighbors heuristic (Leonard and Kramer, 1991) with P=2 was used to determine the parameter β_i, which was different for each RBF in the model.

Fig. 1 shows the one-step-ahead prediction for all three networks compared to Set 1, which is a portion of the training data set, and Table 1 summarizes some information about each model. All the networks provided approximately the same accuracy over the training set, as calculated by the error index I, given as a percentage,

$$I = \frac{(\mathbf{y} - \hat{\mathbf{y}})^T (\mathbf{y} - \hat{\mathbf{y}})}{(\mathbf{y} - \bar{\mathbf{y}})^T (\mathbf{y} - \bar{\mathbf{y}})} \times 100 \quad (4)$$

where $\bar{y}$ is the sample mean for the data set.

The networks were each tested on the validation sets as well. Table 2 summarizes the results. Figure 2 shows the one-step-ahead prediction (1 SAP) for each of the three models compared to Set 2. Figure 3 shows the multiple-step-ahead prediction (m SAP) over the same data set. Finally, Fig. 4 compares the multiple-step-ahead prediction for all three models to Set 3.

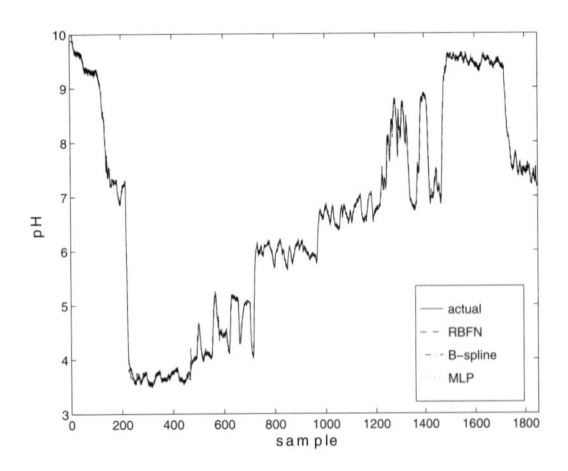

Figure 1. Comparison of neural network models to Set 1 using one-step-ahead prediction.

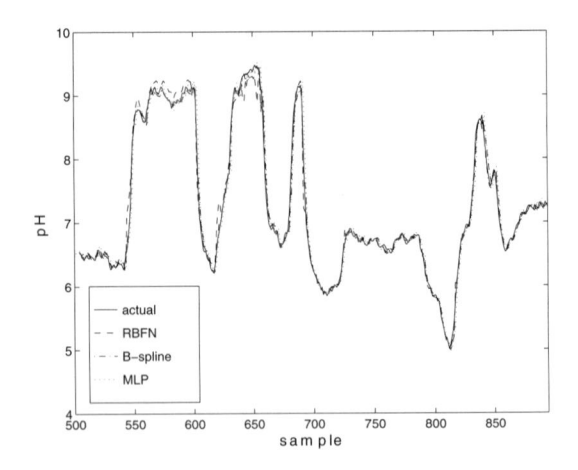

Figure 2. One-step-ahead prediction comparison of neural network models to Set 2.

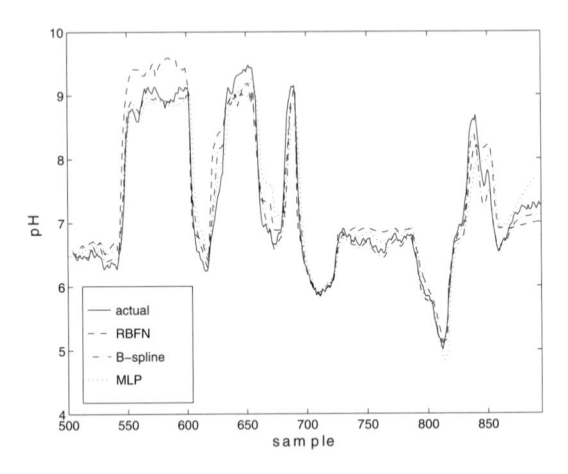

Figure 3. Multiple-step-ahead prediction comparison of neural network models for Set 2.

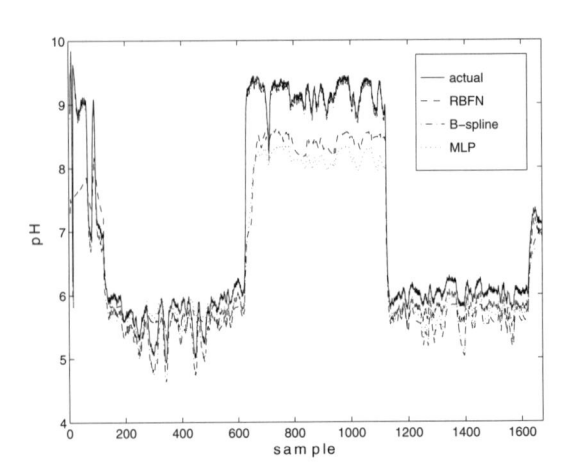

Figure 4. Multiple-step-ahead prediction comparison of neural network models for Set 3.

Table 1. Network Identification Information.

Network	No. Nodes	No. Param.	$I(\%)$
MLP	4–9–1	55	0.0881
B-spline	4–6561–1	6561	0.0734
RBFN	4–20–1	61	0.1389

Table 2. Network Validation.

Network	Set 2: 1 SAP	Set 2: m SAP	Set 3: m SAP
MLP	0.427	6.55	19.20
B-spline	0.764	3.43	1.64
RBFN	1.540	10.95	17.76

Discussion

Each network was able to adequately approximate the experimental data for one-step-ahead predictions. For multiple step ahead predictions, only the B-spline network performed very well.

The key drawback of B-spline networks is that the number of basis functions and hence the number of weights depend exponentially on the size of the input vector $\mathbf{x}_k$. In order to obtain a sufficient approximation accuracy, a large number of knots may be required for each input, and even utilizing low order univariate splines may result in a prohibitively large number of basis functions. This problem is commonly known as the "Curse of Dimensionality," and results in practical limits on the order of the system being identified (Lightbody and Irwin, 1995), *i.e.*, the number of past outputs and inputs included in the network input vector in Eqn. (2).

Unlike B-spline networks, the number of basis functions required for RBFNs identified using k-means clustering or linear regression does not increase exponentially with the order of the system, since RBF centers are typically placed only where they are most needed for approximation accuracy. Hence the "Curse of Dimensionality" is not as severe, and experience shows that RBFN may be used successfully for higher order systems. Some authors (Sanner and Slotine, 1992) utilize mesh-based center placement for the radial basis functions. As was the case for B-spline networks, the number of parameters then increases exponentially with the order of the system, limiting the usefulness of this method.

Conclusions

Three different types of neural network models have been identified from experimental data from the pH neutralization process at UCSB. Results demonstrate that all the networks have comparable accuracy for one-step-ahead prediction, but vary widely in the amount of time required for successful implementation, with RBFN requiring the least effort by far. Furthermore, the relative merits of each network type with regard to basis function localization and the "Curse of Dimensionality" have been discussed.

Acknowledgments

D. E. Seborg would like to acknowledge financial support from the National Science Foundation and Dupont. G. Lightbody would like to acknowledge Dupont (UK) at Maydown, N. Ireland.

References

Battiti, R. and F. Masulli (1990). BFGS optimisation for faster and automated supervised learning. *Proc. Int. Neural Conf.*, **2**, 757-760.

Brown, M. and C. J. Harris (1994). *Neurofuzzy Adaptive Modelling and Control*. Prentice-Hall, London.

Charalambous, C. (1992). Conjugate gradient algorithm for efficient training of artificial neural networks. *IEE Proc. G*, **139**, 3.

De Boor, C. (1978). *A Practical Guide to Splines*. Springer-Verlag, New York.

Hall, R. C. and D. E. Seborg (1989). Modeling and self-tuning control of a multivariable pH neutralization process. Part 1: modeling and multiloop control. *Proc. of the American Control Conf.*, Pittsburgh. 1822-1827.

Harris, C. J., C. G. Moore, and M. Brown (1993). *Intelligent Control: Aspects of Fuzzy Logic and Neural Nets*. World Scientific Press, Singapore.

Leonard, J. A. and M. A. Kramer (1991). Radial basis function networks for classifying process faults. *IEEE Control Systems Mag.*, **11**, 3, 31-38.

Lightbody, G. and G. W. Irwin (1995). Adaptive neural internal model control. *EURACO Workshop*, Florence, Italy. 440-463.

Moody, J. and C. J. Darken (1989). Fast learning in networks of locally-tuned processing units. *Neural Computation*, **1**, 281-294.

Pottmann, M. and D. E. Seborg (1992). Identification of non-linear processes using reciprocal multiquadric functions. *J. Process Control*, **2**, 4, 189-203.

Sanner, R. M. and J.-J. E. Slotine (1992). Gaussian networks for direct adaptive control. *IEEE Trans. on Neural Networks*, **3**, 6, 837-863.

Schumaker, L. (1981). *Spline Functions: Basic Theory*. John Wiley and Sons, New York.

Seborg, D. E. (1994). Experience with nonlinear control and identification. In *Control '94*, IEE Conf. Pub. No. 389, IEE, London. 879-886.

IDENTIFICATION AND CONTROL OF INDUSTRIAL-SCALE PROCESSES VIA NEURAL NETWORKS

Masoud Nikravesh[‡], Anthony R. Kovscek and Tad W. Patzek[‡]
Earth Sciences Division
Lawrence Berkeley National Laboratory
Berkeley, CA 94720

[‡]Department of Materials Science and Mineral Engineering
University of California
Berkeley, CA 94720

Masoud Soroush
Department of Chemical Engineering
Drexel University
Philadelphia, PA 19104

Abstract

This article surveys a series of successful applications of neural networks for identification and model-based control of industrial-scale processes. These include electrolyzers, petroleum reservoirs, contaminant removal processes, and chemical reactors, whose complex dynamics are little known in general. Neural network models have been able to predict the complex behavior of these processes accurately. Furthermore, application of these models for model-based control has led to significant improvement in the control of these processes.

Keywords

Neural network, Model identification, Model-based control, Industrial implementation, Waste management, Electrolyzer, Petroleum reservoir, Chemical reactor.

Introduction

During the last decade, application of neural networks for identification and control has increased exponentially (Hunt et al., 1992; Widrow and Lehr, 1990). Artificial neural networks have been used in nonlinear process identification (Chen et al., 1990), in internal model control (Nahas et al., 1992), in adaptive control (Boskovic and Narendra, 1995), and in hybrid first-principles neural-network models (Psichogios and Ungar, 1992; Nikravesh et al., 1996a). This widespread application has been due to several attractive features of neural networks. For example, neural networks have the ability to approximate arbitrary nonlinear functions, they can be trained easily by using past data records from the system under study, they are readily applicable to multivariable systems, and they do not require specification of structural relationship between input and output data. From a control theory viewpoint, the ability of neural networks to deal with nonlinear systems is perhaps the most significant feature. Given sufficient input-output data, neural network models can easily be obtained to predict outputs as a function of inputs (process model) and inputs as a function of outputs (process model inverse). Because of these features, neural networks have made possible model-based control of the processes for which there are no other accurate models.

Currently, a great number of neural network canned software packages are available, which have made the process of developing a neural network model very easy. However, developing a proper model that is an *"accurate"* representation of the process may be an arduous task, which requires

- sufficient experience with the qualitative effect of the structural parameters of neural network models, such as number of nodes in each layer and the dynamics associated with each node, on the overall behavior of the model,
- sufficient insight into the behavior of the process under consideration.

The organization of this article is as follows. First, application of neural networks in predicting the future distribution of contaminants in contaminated sites and in achieving predictive control of contaminants is surveyed. Next, neural-network model-based control of an industrial-scale Chlorine/Caustic electrolyzer is described. Application of neural networks in identification and model-based control of petroleum reservoirs is touched on, followed by that of chemical reactors.

Contaminant Removal Processes

In a contaminated site, it is of great importance to characterize how the contaminants move and spread. However, data from waste sites are often difficult to analyze because a physical relationship cannot be established to show how the data are correlated. In a study, we developed a series of neural network models to analyze actual field data from a contaminated site in Alameda County, California. A neural network model was first developed to reconstruct the topology of the earth as a function of location, (x, y), at that site. The network had 2 nodes in the input layer (dimensionless location, x and y), 25 nodes in the hidden layer with a nonlinear transfer function, and one node in the output layer (elevation above sea level, dimensionless) with a nonlinear transfer function. Based on the topology of the earth and a series of neural network models, the geology of the earth in this area was reconstructed. The model provided the composition of the soil infill, the soil color and the pH of the soil in this site as a function of the earth topology over a period of time. Other network models were then developed to reconstruct the distribution of the waste in this site as a function of the topology and geology of the site, soil infill composition, soil color, and pH of the soil. The goal was to find a trend in the contaminant movement during a certain time period. The neural network models were also used to predict future distribution of the contaminant, thus allowing predictive control and preventing contamination of natural resources.

To improve the accuracy of the neural network models, the site was divided into distinct regions based on existing knowledge of the site. The networks developed earlier as general models for the entire site were further adapted for smaller regions. Results showed improvement in the accuracy of the adapted model over the general model. In this case study, knowledge-based neural networks provided a "more accurate" model of the process than black-box neural networks.

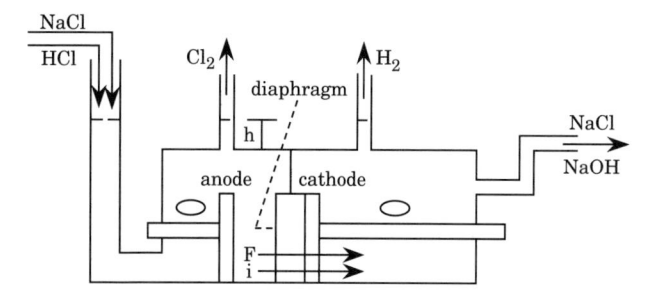

Figure 1. Schematic of the diaphragm-type chlorine/caustic cell (Nikravesh et al, 1995).

Electrochemical Processes

Here we limit our survey to the application of neural networks for multivariable model-based control of a industrial-scale Diaphragm-Type Chlorine/Caustic electrolyzer. Fig. 1 depicts a schematic of this process. The electrolyzer produces caustic and chlorine by electrolysis of brine. These two chemicals ranked seventh and eighth respectively in terms of annual mass production among all US industrial chemicals in 1990. Among the major features of this process are their complex nonlinear behavior (Nikravesh et al., 1995), their relatively large time constants and dead times, and their tight operation (there are hard constraints on several process variables). Moreover, in these cells, the current density, i, is a major process disturbance.

For this specific process, the caustic effluent concentration and anolyte pH were chosen as controlled outputs, and the brine and HCl feed flow rates as manipulated inputs. The values of both controller outputs and both manipulated inputs must always remain within desirable ranges.

For this process, a dynamic neural network model was developed first to predict the behavior of the process. The developed model was used to synthesis a dynamic matrix controller. For each manipulated variable, a control horizon of two was used. The prediction horizons were chosen to be ten. The neural-network dynamic matrix controller provided superior servo and regulatory performance over conventional dynamic matrix control (with step response model) and PID control.

The implementation of the neural-network dynamic matrix controller led to (a) a decrease in the effective energy consumption, (b) a decline in energy waste by returning the process to the desired operating point or to within control limits in an optimal manner, (c) a decrease in the production of off-specification products during the transition periods, and (d) an increase in the life of the diaphragm.

Oil Recovery from Petroleum Reservoirs

Petroleum reservoirs are known to exhibit inherently complex, nonlinear, and nonstationary behavior (Nikravesh et al., 1996a). Fluids such as water or steam are injected into reservoirs to maintain pressure and displace oil. Currently, engineers develop injection policy on the basis of past expe-

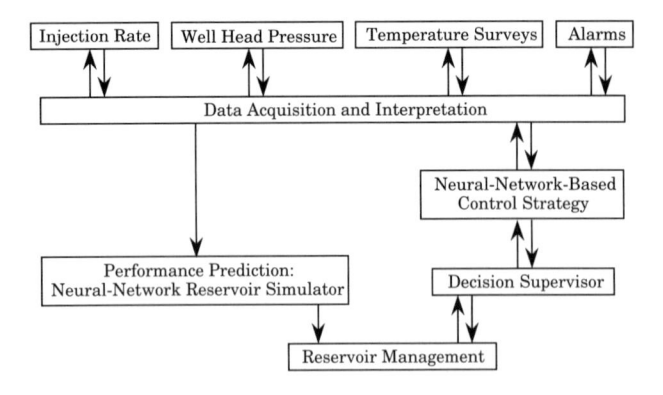

Figure 2. Simplified schematic representation of optimal injection policy.

rience, partial knowledge of the state of the reservoir stress, production history, and limited predictions of future reservoir performance from numerical simulation. Although an oil field is a complex and highly coupled system, injectors are usually controlled individually, with constant setpoints, and without feedback among neighboring patterns. In this project, we are creating an innovative Computer Assisted Operations (CAO) tool for developing and maintaining optimal injection policy for water or steam. Fig. 2 shows a simplified schematic of optimal fluid injection policy. First, the CAO system will continuously acquire data from the injectors (Wellhead Pressures (WHPs), rates, and temperatures) and producers (WHPs, rates, etc.). Second, the pattern-by-pattern responses are inverted to "back-out" the evolution of reservoir properties such as permeability and the extent of fracturing. A simplified model of reservoir behavior that maintains first order physics is used for rapid, ideally real-time inversion (Kovscek et al., 1995). Third, a neural network is used to "learn" the reservoir response and recognize symptoms of inefficient injection, so that control of the field improves with time. Fourth, feedback from neural network is used simultaneously to decide how to adjust the coupled network of well-head controllers. They are also used as input to a sophisticated reservoir simulator, used to make long term reservoir management decisions. The final goal of this project is to bring the neural network-model based technology to the point of commercialization.

In this study, the diatomaceous fields of California which represent a $42 billion resource, have been chosen to develop and demonstrate the neural network technology for optimal fluid injection policy. However, this methodology will be applicable directly to fluid injection into tight fractured reservoirs such as the Austin Chalk and the West Texas Carbonates. Using historical data from steamdrive pilots, a neural network model has been developed to predict wellhead pressure as function of injection rate, and vice versa. In addition, the network was taught the reservoir injection/production/temperature response and reservoir behavior (even though steam drive dynamics in this project are complex and all mechanisms affecting injection and production are not elucidated (Nikravesh et al., 1996a). The neural network model has been trained to model the temperature responses of the observation wells and predicts these responses into the future. This neural network has then been used in conjunction with a mechanistic reservoir model (Kovscek et al., 1995), to predict the extensions of hydro- and natural fractures as a function of the steam injection pressure. This approach allows predictive control.

Likewise, historical data from a waterflood in the South Belridge and Lost Hills Diatomite (Kern County, CA) have been used to develop a neural network model. There are two approaches to analyze and predict the performance of waterfloods. In the first approach, the behavior of each injector or producer is considered independently and modeled with a neural network. For an injector or producer, historical data consisting of flow rate and well pressure are used along with the assumption that the same strategy of well operation will continue into the future. This approach uses a very simple neural network model and is easy to train and implement based on a minimum amount of information from the field (Nikravesh et al., 1996a). In the second approach, the behavior of the waterflood is considered as a coupled, highly nonlinear system of injectors and producers. The field-wise objective is to meet a given production goal with the minimum amount of injectant. The oil field is divided into sections with similar characteristic behavior and each well within a section and its interactions with the surrounding wells are studied. The model helps to improve waterflood management and the design of recovery strategies (Nikravesh et al., 1996b).

In summary, neural network models can match and then predict complicated reservoir behavior when historical databases are available. To capture even small extensions of injection hydrofractures in tight rocks, the historical data must be acquired at 30-60 second intervals. This necessitates the use of Computer-Assisted Operations (CAO) systems, mass data storage devices and fast data transfer protocols. With current computer hardware and networks, these requirements can be satisfied at a relatively low cost and the potential savings in terms of otherwise forfeited oil production may be huge. The neural networks are capable of making accurate predictions even when all mechanisms affecting injection or production behavior are not known. Further, neural networks provide a way to incorporate disparate information because a structural relationship between input and output data is not required.

In addition, using steam and water injection data from field operations in the South Belridge Diatomite, we have demonstrated that neural networks are capable of predicting injection rate as a function of wellhead pressure and vice versa. These networks used are simple and are based on accepted neural network designs and training algorithms. The networks match field data very well and have exceptional generalization properties. They are able to accurately predict tip extensions of injection hydrofractures and provide us with a means of preventing unwanted fracturing. Finally, with regard to steam injection, neural networks used in conjunction with reservoir simulation provide a novel tool for predicting, visualizing, and screening various steam injection strategies. Thus, predictive simulation can be achieved even when mechanisms of injection and oil displacement are not fully understood.

Chemical Reactors

Chemical reactors are a class of complex nonlinear processes, which often exhibit nonstationary behavior because of factors such as catalyst deactivation and heat-transfer-coefficient fouling. Examples include polymerization reactors whose dynamics are dominated by the complex kinetics of polymerization reactions.

Compared to pilot-scale reactors, industrial-scale reactors have a more stochastic nature and their residence-time distributions are usually much more complex. Several factors, such as those two, have made the development of first-principle models for industrial-scale reactors very challenging, if not impossible.

In the case of polymerization reactors, although the kinetics of the polymerization reactions of many monomers are well-understood, for an industrial-scale polymerization reactor—because of the above-mentioned factors—a sufficiently accurate, pure first-principle model may be difficult to find.

We have studied the application of neural networks for identification and model-based control of several chemical reactors including (a) a reactor with a classical reaction and a decreasing reactor-jacket overall heat-transfer coefficient, (b) a catalytic reactor with catalyst deactivition, and (c) a methyl methacrylate batch polymerization reactor with a decreasing reactor-jacket overall heat-transfer coefficient.

In the case of the reactor (a), a neural-network model-based controller was able to provide a satisfactory servo and regulatory performance in the presence of the fouling in the heat transfer coefficient. In the case of the reactor (b), the performance of the controller was significantly affected by the catalyst deactivation. However, the controller was robust enough to perform well over a wide range of operating conditions. The performance was improved further by using a model developed via neural networks in conjunction with recursive least squares (Azimi and Liou, 1992).

For the polymerization reactor, a neural network model was developed to describe the relationship between the heat input to the jacket and the reactor temperature, by using on-line data on the reactor and jacket temperatures and the heat input. By using the developed model, a geometric model-based controller (Soroush and Kravaris, 1992) was synthesized and then implemented to achieve a tight temperature control of the batch reactor. The neural-network-based controller was able to provide the desired tight control in the presence of the heat-transfer coefficient fouling.

Acknowledgments

We thank CalResouces LLC, formerly Shell Western E&P and Mobil U.S., for providing the water and steam injector data and Forsenic Management Associates for the contamination data. Support for the work discussed in Oil Recovery from Petroleum Reservoirs was provided by the Office of Laboratory Technology Applications, Office of Energy Rersearch of the U.S. Department of Energy under contract No. DE-ACO3-76FS00098 to the Ernest Orlando Lawrance Berkeley National Laboratory of the University of California.

References

Azimi, M. S., and R. J. Liou (1992). Fast learning process of multilayer neural networks using recursive least squares method. *IEEE Trans. on Sig. Process.*, **40**, 446-449.

Boskovic, J. D., and K. S. Narendra (1995). Comparison of linear, nonlinear and neural-network-based adaptive controllers for a class of fed-batch fermentation processes. *Automatica*, **31**, 817-840.

Chen, S., S. A. Billings, and P. M. Grant (1990). Non-linear system identification using neural networks. *Int. J. of Control*, **51**, 1191-1214.

Hunt, K. J., D. Sbarbaro, R. Zbikowski, and P. J. Gawthrop (1992). Neural networks for control systems – A survey. *Automatica*, **28**, 1083-1112.

Kovscek, A. R., R. M. Johnston, and T. W. Patzek (1995). Evaluation of rock/fracture interaction during steam injection through vertical hydrofractures. SPE Paper No. 29622.

Nahas, E. P., M. A. Henson, and D. E. Seborg (1992). Nonlinear internal model control strategy for neural network models. *Comput. & Chem. Engng.*, **16**, 1039-1057.

Nikravesh, M., A. R. Kovscek, T. W. Patzek, and R. M. Johnston (1996a). Prediction of formation damage during fluid injection into fractured, low permeability reservoirs via neural networks. SPE Paper No. 31103, SPE Formation Damage Symposium, Lafayette, LA.

Nikravesh, M., A. R., Kovscek, A. R., Murer, and T. W. Patzek (1996b). Field-Wise waterflood management in low permeability, fractured oil reservoirs with neural networks, SPE Paper No. 35721, Western Regional Meeting of SPE, Anchorage, Alaska.

Nikravesh, M., A. E. Farell, C. T. Lee, and J. W. Van Zee (1995). Dynamic matrix control of diaphragm-type chlorine/caustic electrolyzers. *J. of Process Control*, **5**, 131-136.

Psichogios, D. C., and L. H. Ungar (1992). A hybrid neural network-first principles approach to process modeling. *AIChE J.*, **38**, 1499-1511.

Soroush, M., and C. Kravaris (1992). Discrete-time nonlinear controller synthesis by input/output linearization. *AIChE J.*, **38**, 1923-1945.

Widrow, B. and M. A. Lehr (1990). 30 Years of adaptive neural networks: perceptron, medaline, and back-propagation,*Proc. of IEEE*, **78**, 1441-1445.

DESIGN OF A COMPOSITION ESTIMATOR FOR INFERENTIAL CONTROL OF HIGH-PURITY DISTILLATION COLUMNS

Joonho Shin and Sunwon Park
Department of Chemical Engineering
KAIST
Taejon 305-701, Korea

Moonyong Lee
Department of Chemical Engineering
Yeungnam University
Kyongsan 712-749, Korea

Abstract

In distillation column control, temperature as a secondary measurement is widely used in order to infer product composition. This paper addresses the design of a static PLS estimator using multiple temperatures for estimating the product compositions of the distillation columns. Design guidelines on the development of the composition estimator using PLS regression are presented. It is shown that the loading vector space of the estimator exactly reflects the direction of temperatures due to corresponding input variables by considering an inverse estimation problem. We also discuss the effects of data pre-processing and variable transformation. The estimator based on the guidelines is robust to sensor noise and has a good predictive power.

Keywords

Composition estimator, PLS (Partial-Least-Squares) regression.

Introduction

Product quality measurement is one of the major difficulties associated with the composition control of distillation columns. In distillation columns, the product composition control is performed by using process analyzers such as Gas Chromatography(GC), but the analyzers suffer from large measurement delays, high investment/maintenance costs and low reliability. For these reasons, many workers (Joseph and Brosilow, 1978; Mejdell and Skogestad, 1991; Kresta, Marlin and MacGregor, 1994; Piovoso and Kosanovich, 1994) have studied the inferential model, which estimates the product composition of the distillation column. It was reported that the Partial-Least-Squares(PLS) provides models with good predictive power and robustness to the process noise and sensor failure (Kresta et al., 1994).

In developing the composition estimator using PLS regression, we have to consider the following issues: (1) the determination of the number of factors in PLS regression; (2) the selection of the secondary measurements to be used; (3) the selection of the most effective variable transformation or scaling. These estimator design issues are discussed by analyzing the results of simulation studies on a high-purity binary column example.

Problem Definition

The linear estimator inferring the product compositions of the distillation column may be written by

$$\hat{\mathbf{y}} = \mathbf{K}\theta \qquad (1)$$

where $\hat{\mathbf{y}}$ is the estimated composition and θ is the secondary measurement vector. The problems concerned in this work are the design problems stated in the introduction.

Evaluation Criteria

The Explained Prediction Variance(EPV) (Mejdell and Skogestad, 1991) in percent is used to evaluate the performance of the estimator:

$$EPV(k) = 100 \times \left(1 - \frac{MSEP(k)}{MSEP(0)}\right) \quad (2)$$

where the mean squared error of prediction $MSEP(k) = 1/N_{set} \sum_{i=1}^{N_{set}} (\hat{y}_i(k) - y_i)^2$ and N_{set} is the number of data sets(here, N_{set}=64).
The Prediction Error Sum of Squares(PRESS) (Kresta et al., 1994) is also used to evaluate the absolute performance:

$$PRESS = \sum_{i=1}^{N_{set}} (\hat{y}_{D,i} - y_{D,i})^2 \quad (3)$$

Example Column

The steady state simulation of the binary column of normal-hexane and cyclo-hexane with 40 theoretical stages is performed using the rigorous steady state simulator, Aspen-Plus. The feed stream enters the column at stage 20 as saturated liquid. The nominal operation conditions and the variation of the inputs of the binary column are given in Table 1.

Table 1. *Simulation Conditions for the Binary Distillation Column.*

	Base case conditions	Variation in steady state reference set
F	1000.0 kmol/hr	Constant
T_F	346.57 K	Constant
z_f	0.5	0.25~0.75
D	500.0 kmol/hr	Vary
Q_B	0.130736×10^8 cal/sec	Vary
P	1 atm	Constant
y_D	0.99	0.97~0.997
x_B	0.01	0.003~0.03

Variable Transformation

Logarithmic transformation of the product compositions was used by many investigators (Joseph and Brosilow, 1978; Mejdell and Skogestad, 1991; Kresta et al., 1994). Several transformation methods were tested by Mejdell and Skogestad (1991). They presented that the logarithmic transformation of both the composition and the temperatures improved the estimation performance. In this work, the same transformation methods are adopted and evaluated to select the most effective variable transformation method.

Multivariate Statistical Methods

Various multivariate statistical methods such as Multiple Linear Regression(MLR), Principal Component Regression(PCR) and PLS are used to build the linear static estimator inferring the product composition in this work. For the convenience of the discussion, only the results of PLS are presented in this paper.

PLS Regression

This method is a variation of the PCR which recently has become popular among analytical chemists (Geladi and Kowalski, 1986). The latent variables are determined in order to have the largest covariance with the dependent variables. The PLS estimator (Mejdell and Skogestad, 1991) based on k factors is given by

$$\mathbf{K_{PLS}} = \mathbf{Q}\left(\mathbf{P^T W}\right)^{-1}\mathbf{W^T} \quad (4)$$

where $\mathbf{Q}^{p \times k}$ is the loading matrix for the dependent variables (e.g. the product compositions), $\mathbf{P}^{q \times k}$ is the loading matrix for the independent variables (e.g. temperatures and flow rates), and $\mathbf{W}^{q \times k}$ is the matrix formed in the PLS algorithm.

Results of a Case Study

As many researchers (Mejdell and Skogestad, 1991; Kresta et al., 1994) reported, the performance of the inferential model using a single temperature measurement is very sensitive to noise (e.g. EPV decreases by 32 % when the noise 0.1 °C is added to the temperature measurement). In order to reduce the sensitivity of the inferential models using a single temperature, we should use the inferential model based on multiple temperature measurements. The design guidelines on the development of the composition estimator will be presented by analyzing the binary column case study in this section. The results presented in this work are based on all tray temperature measurements.

Determination of the Number of Factors in PLS Estimator

There is an optimal model dimension which minimizes the prediction error (Kresta et al., 1994). Therefore, when we build an inferential model, the optimal model dimension (here the number of factors in PLS estimator) should be considered.

Unique temperature profiles are obtained by specifying different values of feed composition z_f, distillate composition y_D and bottom product composition x_B (or z_f, reflux ratio R, and reboiler duty Q_B) at constant pressure. That means the number of degrees of freedom is three for the general binary columns assuming that the unmeasured disturbance is only the feed composition: the binary column system is fully defined by specifying three variables such as z_f, y_D, and x_B (or z_f, R, and Q_B). Therefore the number of directions of the temperature profile affected by the independent variables are clearly three for the distillation column of this case study. To verify this, let's consider the following inverse estimation problem:

$$\hat{\theta} = \mathbf{A}\left[z_f \ y_D \ x_B\right]^T \quad (5)$$

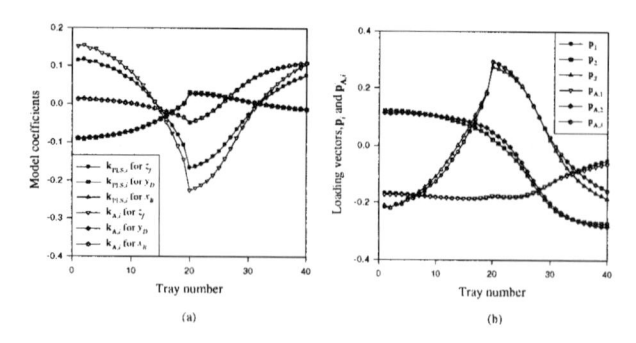

Figure 1. Comparison of (a) $\mathbf{K_{PLS}}$ and $\mathbf{K_A}$ and (b) loading vectors of $\mathbf{A}$ and $\mathbf{K_{PLS}}$.

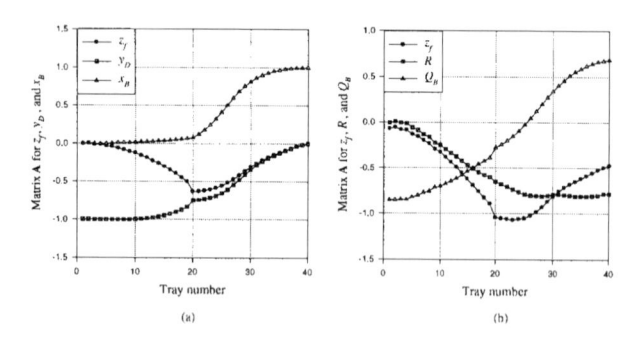

Figure 2. Plot of $\mathbf{A}$ for input variables (a) z_f, y_D, and x_B and (b) z_f, R, and Q_B.

One can estimate the temperature profile by the above equation and the inverse problem of it is the estimation of the compositions :

$$[\hat{z}_f \; \hat{y}_D \; \hat{x}_B]^T = \mathbf{K_A} \, \theta \qquad (6)$$

From Eqs. 5 and 6, one can easily find

$$\mathbf{K_A} = (\mathbf{A}^T \mathbf{A})^{-1} \mathbf{A}^T \qquad (7)$$

Matrix $\mathbf{A}$ is calculated using z_f, y_D and x_B with unit variance scaling, and the EPV for temperature is 93.87. The EPV value shows that the estimation of the temperature is good using z_f, y_D, and x_B. Fig. 1(a) compares matrix $\mathbf{K_A}$ with matrix $\mathbf{K_{PLS}}$ of which the number of factors k is three.

We can expect that the SVD(Singular Value Decomposition) of $\mathbf{A}$ may give us the same eigenvectors as the loading vectors of Θ which is the matrix of the temperatures for calibration. The SVD of matrix $\mathbf{A}$ is performed and the largest three eigenvectors ($\mathbf{p_{A,1}}$, $\mathbf{p_{A,2}}$, and $\mathbf{p_{A,3}}$) are plotted with loading vectors($\mathbf{p_1}$, $\mathbf{p_2}$, and $\mathbf{p_3}$) from the PLS in Fig. 1(b). It is clearly shown that the directions of the temperature profile have a one-to-one correspondency with the input variables z_f, y_D, and x_B.

The plot of matrix $\mathbf{A}$ in Fig. 2(a) shows the direct relation between the temperatures and z_f, y_D, and x_B. One may try to use R and Q_B instead of y_D and x_B. But EPV using z_f, R, and Q_B is only 50.4 and the performance of temperature estimation is not good. Therefore the directions of the temperature profile cannot be explained in terms of z_f, R,

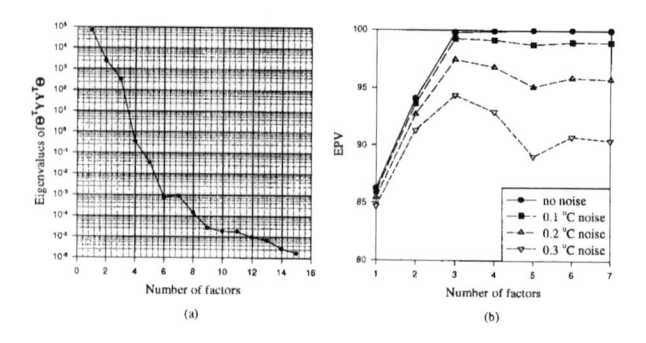

Figure 3. (a) Plot of eigenvalues of $\Theta^T \mathbf{YY}^T \Theta$ and (b) Effect of measurement noise on EPV.

and Q_B in the linear form. The plot of the matrix $\mathbf{A}$ for input variables z_f, R, and Q_B is shown in Fig. 2(b).

Additionally the eigenvalues of $\Theta^T \mathbf{YY}^T \Theta$ for PLS are plotted in Fig. 3(a) where $\mathbf{Y}$ is the matrix of product composition y_D for calibration runs. The magnitudes of eigenvalues drastically decrease as the number of factors increases. One can easily find that the condition number of $\Theta^T \mathbf{YY}^T \Theta$ is directly related to the sensitivity of the PLS estimator with corresponding factors. One should select the first three factors as the condition number increases by the order of 10^3 if the number of factors increases from 3 to 4.

We checked the sensitivity of the composition estimator by the cross-validation procedure (Mejdell and Skogestad, 1991). As shown in Fig. 3(b), the EPV increases monotonically as the number of factors increases. However, when the inferential model with more than 4 factors is applied to noise corrupted data, the EPV decreases. As we expect, the EPV without noise is saturated when the number of factors are three and the estimator with $k=3$ is robust to the noise. We can conclude that the optimal number of factors in the sense of the robustness and the prediction ability is three for our binary column and the factors describe the effects of z_f, y_D and x_B. This results can be extended for multicomponent distillation columns. The recommended number of factors is equal to the number of independent variables (e.g. z_f, y_D, x_B, and column pressure P) which have influence on the temperature profile. This guideline for determination of the number of factors is verified by other examples of the multi-components columns. The number of factors for BTX column (Kresta et al., 1994) and for a 4 component column is 4 and 5 respectively assuming that the unmeasured disturbance is only the feed composition.

Effects of data pre-processing and variable transformation

The results with unit-variance scaling are better than those with no scaling. Unit variance scaling makes the temperature measurements, which are collinear with T_1, have the same information as T_1(see Fig. 2(a).). The unit variance scaling of the temperature measurements provides us the useful information about T_1 or T_N highly related with y_D or x_B (note that $y_D = f(T_1)$ for the binary column). But the estimator with unit variance scaling is sensitive to noise because the weight on the temperature measurement, which has small variance (e.g. T_1), is large. The performances of the inferential models based on the scaled variables are almost the same except for the case of the logarithmic transformation on the composition only. For k=3, EPV=98.048 and PRESS=0.575×10^{-4} without variable transformation and scaling. When the transformation $\ln(y_D/(1-y_D))$ on y_D is applied, the relationship between $\ln(y_D/(1-y_D))$ and T_i is linear around the middle of the column. But the performance of the transformation is not good because the linear relationship between $\ln(y_D/(1-y_D))$ and T_i around the feed tray becomes distorted when the feed disturbances such as z_f is introduced. The logarithmic transformation on the composition is not desirable when the change of z_f is frequent. The performance of the logarithmic transformation is not good (EPV=92.698, PRESS=0.531×10^{-3} with k=3) because the change in z_f is very large for our column. The performance of the estimator is improved by using the logarithmic transformation both on the composition and temperature by $\ln(y_D/(1-y_D))$ and $\ln((T_i-T_L^b)/(T_H^b-T_i))$ (Mejdell and Skogestad, 1991). The EPV with no noise is close to 100 % after only 3 factors due to the linearizing effect of the transformation (EPV=99.943, PRESS=0.958×10^{-6} with k=3). But the transformation is somewhat sensitive to noise because $(T_i-T_L^b)/(T_H^b-T_i)$ term becomes zero or infinite at the end of the column. Estimation without variable transformation is best from the view points of prediction accuracy and robustness for our binary column.

Use of Measured Inputs

The relationship between the auxiliary measured inputs(e.g. R and Q_B) and the tray temperatures cannot be described in linear forms. Therefore the estimation using the auxiliary measured inputs is not desirable. Actually the estimation performance decreases when we use R and Q_B in addition to all tray temperatures. The results using all tray temperatures, R, and Q_B with three factors were EPV=99.119 and PRESS=0.263×10^{-4} with unit variance scaling while EPV=99.805 and PRESS=0.588×10^{-5} were obtained when only the temperatures were used. Similar results were also reported by previous researchers (Mejdell and Skogestad, 1993; Piovoso and Kosanovich, 1994).

The projection estimator (Joseph and Brosilow, 1978) has been reported to be sensitive to noise and modeling error (Mejdell and Skogestad, 1993). The R and Q_B are used to infer the composition in the projection estimator. It is clear that the performance of the estimator using the measured inputs is not good due to the modeling errors. Additionally, the projection estimator cannot handle the collinear measurements properly and provide the advantage of the multiple temperature measurements (e.g. robust to the process noise).

Conclusions

The guidelines on the design of composition estimator via PLS have been presented: The recommended number of factors is equal to the number of independent variables(e.g. z_f, y_D, x_B, and column pressure P) which have influence on the temperature profile. It has been shown that the loading vector space of the estimator exactly reflects the direction of temperatures due to corresponding input variables. The relationship between the auxiliary measured inputs (R and Q_B) and the product composition cannot be described in the linear form. Thus, the estimation using the measured inputs is not desirable. The performance of the logarithmic transformation on the composition is not good if the change of z_f is frequent. The transformation $\ln(y_D/(1-y_D))$ and $\ln((T_i-T_L^b)/(T_H^b-T_i))$ is effective but somewhat sensitive to measurement noise. When all tray temperature measurements are used, unit variance scaling enhances the estimation performance by making the collinear measurements with T_1 have the same information as T_1, but makes the estimators sensitive to noise.

Acknowledgment

Financial support from the Korea Science and Engineering Foundation through the Automation Research Center at Postech is gratefully acknowledged. This paper was also supported in part by Korea Research Foundation.

References

Geladi, P. and B.R. Kowalski. (1986). Partial least-squares regression: A tutorial, *Analytica Chimica Acta*, **185**(3), 1-17.

Joseph, B. and C.B. Brosilow. (1978). Inferential control of processes : Part I. Steady state analysis and design, *AIChE*, **24**(3), 485-492.

Kresta, J.V., T.E. Marlin and J.F. MacGregor. (1994). Development of inferential process models using PLS, *Computers Chem. Eng.*, **18**(7), 597-611.

Mejdell, T. and S. Skogestad. (1991). Estimation of distillation compositions from multiple temperature measurements using partial-least-squares regression, *Ind. Eng. Chem. Res.*, **30**, 2543-2555.

Mejdell, T. and S. Skogestad. (1993). Output estimation using multiple secondary measurements: high-purity distillation, *AIChE J.*, October, **39**(10), 1641-1653.

Piovoso, M.J. and K.A. Kosanovich. (1994). Application of multivariate statistical methods to process monitoring and controller design, *Int. J. Control*, **59**(3), 743-765.

ON-LINE INFERENCE SYSTEM OF TUBE-WALL TEMPERATURE FOR AN INDUSTRIAL OLEFIN PYROLYSIS PLANT

Manabu Kano, Toshihiro Shiren, and Iori Hashimoto
Kyoto University
Kyoto 606-01, Japan

Masahiro Ohshima
Miyazaki University
Miyazaki 889-21, Japan

Umetaro Okamo
Toyo Engineering Corporation
Chiba 275, Japan

Shyoji Aoki
Osaka Petro-Chemical Industries, LTD.
Osaka 592, Japan

Abstract

The period of decoking operation is crucial for effective functioning of the industrial pyrolysis plants. In this paper, an on-line inferential system is developed to predict the tube-wall temperature, which is called skin temperature, and help the operator to determine the timing of decoking operation. The inferential models used in the system are constructed by incorporating the first principles models with a PLS (Partial Least Squares) regression model. The application of the proposed system is illustrated using operating data obtained from an industrial olefin pyrolysis plant. The results show its effectiveness in practical use.

Keywords

Pyrolysis plant, Inferential control, Inferential model, Partial least squares, Cross-validation.

Introduction

In olefin plants, good performance of the pyrolysis plants is one of the key elements to an efficient operation of the entire facility. A major hindrance to efficient functioning of the industrial pyrolysis plants, where naphtha is cracked, is coking. This phenomenon, coking, decreases the olefin yield and simultaneously increases the wall temperature of the tubes. In order to protect the tube itself and maintain efficient pyrolysis functioning, a decoking operation is needed.

Ideally, the decoking operation should be commenced, when the tube-wall temperature exceeds a certain level, determined by the material of the tube. However, due to the lack of a suitable on-line sensor for the tube-wall temperature as well as coke thickness, the commencement of decoking operation is determined based purely on experience, with time-vs.-temperature data, collected from several off-line measurements during the pyrolysis operation, and extrapolated over time. Noting that the starting moment for decoking operation is crucial for effective functioning of the plant, such estimation of timing for commencement of decoking operation is far from being optimal. Moreover, in order to lessen the burden of the operators, it is required to reduce the frequency of manual measurements of the tube-wall temperature.

This is one of the typical control problems where the inferential control is the only solution to estimate the primary output from the secondary measured process variables and past off-line measured primary output values. Most of the papers related to the inferential control (Brosilow, 1972; Lee et al., 1992) were mainly focusing on the control structure but not on the structure of the models. Any inferential scheme of estimating the primary output from secondary measured process variables requires a model, which is the so-called inferential model, that can describe the dynamic relationship between the primary output and secondary measured process variables. The control performance of the inferential control system strongly depends on the prediction accuracy of the model. Therefore, for any plant to which inferential schemes are applied, the inferential model should be intensively developed before designing the controller.

Coking in a commercial pyrolysis tube is a highly complex process that has not been modeled in precise mathematical terms. Therefore, few applications of the inferential schemes to the pyrolysis have been reported in the process systems engineering field. In our previous work (Ohshima et al., 1994), an on-line inference system of tube-wall temperature was developed. Using an inferential model that simply describes how the temperature is affected by the pyrolysis operating conditions, the system predicts the tube-wall temperature as well as coke thickness from the available on-line measured process variables, such as gas temperature, fuel gas flowrate and naphtha feed flowrate. The inferential model was built upon the first principles of chemical engineering. The system could predict the temperature accurately for the most operating conditions but not for all.

This paper further investigates the inferential system and presents some advancements, where the inferential model is rebuilt by incorporating a PLS (Partial Least Squares) regression model with the physical model so as to improve the prediction accuracy for any operating conditions.

Process Description

The pyrolysis studied here is one of ten pyrolysis units at the Osaka Petro-Chemical Industries and its schematic is illustrated in Fig. 1. The naphtha is mixed with a certain amount of a diluent steam and then fed to the upper part of the pyrolysis unit. The feed is separated into four tubes. The pyrolysis itself consists of three chambers, which are a convection, a crossover and a radiant chamber. By burning fuel gas, the tubes are heated up as they pass through those chambers. The temperature is measured on-line at each chamber as illustrated in Fig. 1. The flowrate of naphtha and diluent steam are monitored for each tube. The fuel gas flowrate also is monitored and used to regulate the temperature of the effluent cracked naphtha.

As the cracking proceeds, a layer of coke adheres to the internal surface of the tubes. This layer changes the heat transfer coefficients at the tube surface. As a result, the skin temperature increases gradually during the operation because the effluent temperature is regulated at a specific value as the overall heat transfer coefficient of the tube surface decreases. In order to make sure that the skin temperature is below a certain temperature limit, which is determined by the tube material, an operator occasionally measures the temperature manually using a pyrometer. The extra work in frequent measurement of the skin temperature is impractical. As a result of not closely monitoring the skin temperature, the decoking operation is often performed while the temperature is far below the limit. This reduces the pyrolysis efficiency.

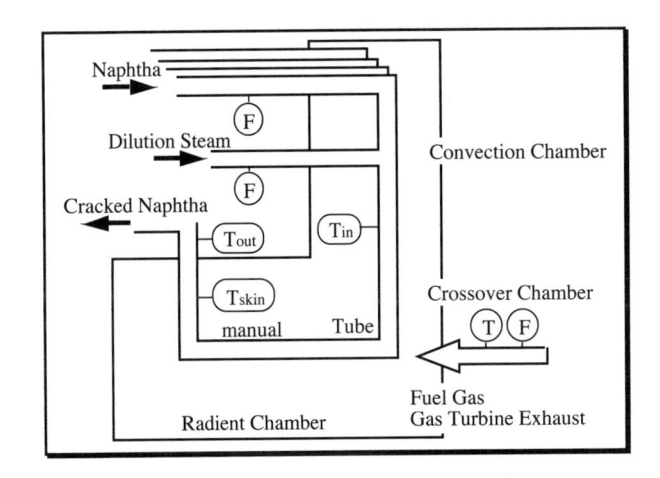

Figure 1. Schematic of the pyrolysis.

Inferential Model

In order to perform the decoking operation timely, the skin temperature of the pyrolysis tube should be monitored on-line. That is, an inferential system of the skin temperature is very much needed to monitor the temperature on-line. The system should be constructed on the basis of a certain inferential model that can describe how the temperature is affected by the pyrolysis operating conditions.

One can distinguish two principally different approaches to developing the inferential model: physical and empirical modeling approach. In the physical modeling approach, the model is built by applying the first principles to, at least conceptually, describe the process dynamics. One of the key characteristics in this approach is that a first principles model can be developed without recourse to measured data from the process.

In the empirical modeling approach, a generic input-output or state space models are assumed. Then certain estimation technique is used to determine the parameter values from measured process data. If the pyrolysis is operated at a particular set of conditions, then an empirical linear plant model will often suffice. However, the pyrolysis is operated in a long-term batch wise operation. Furthermore, the growth of coke layer keeps pyrolysis heat balance changing. The model developed for this application should be valid over a large range of skin temperature. Thus, linear empirical models alone are not sufficient for this application. The approach taken in developing the proposed inferential model is a combination of both methods.

Physical Model

A physical model is developed in this paper by extending the simple coking growth rate model and the basic heat transfer model proposed by Ohshima et al. (1994). (The derivations are left out because of the limited space.)

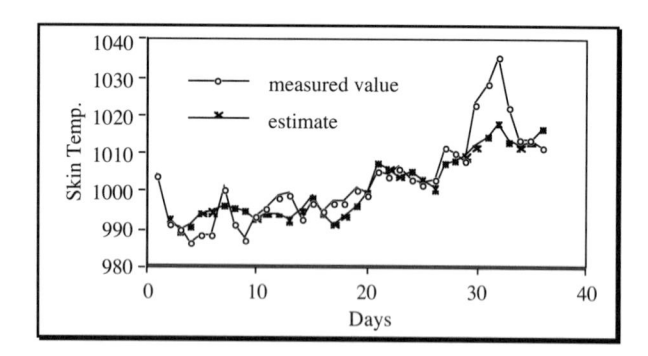

Figure 2. On-line inference of skin temp. using physical model.

The Revised Coking Growth Rate Model

$$d(t+1) = d(t) + C_1 \frac{W_f(t)^{0.8}}{(D_i - 2d(t))^{1.8}} \frac{W_n(t)}{W_f(t)} \quad (1)$$

where d(t) denotes the thickness of the coke layer at time t and unit time is one day. D_i is the inside diameter of the tube. W_n and W_f are the mass flowrate of naphtha and total flow rate. C_1 is a constant parameters to be estimated from measured process variables. The ratio of $W_n(t)$ to $W_f(t)$ is referred to as the diluent steam ratio, which was not taken into account in the previous model.

Heat Transfer Model of the Skin Temperature

$$T_{skin}(t) = T_p(t) + q(t)C_2 \left\{ \alpha \left(\frac{D_i - 2d(t)}{W_f(t)}\right)^{0.8} + \ln\left(\frac{D_i}{D_i - 2d(t)}\right) + \beta \right\}$$

$$\alpha = \frac{2\pi k_c}{0.0877} \frac{\mu^{0.4}}{k^{0.6} C_p^{0.4}}, \quad \beta = \frac{k_c}{k_w} \ln\left(\frac{D_o}{D_i}\right) \quad (2)$$

where T_p denotes the temperature of process flow and q denotes the heat flow rate. α and β are constant parameters to be determined from physical properties of naphtha and design parameters of the tube. μ, k, and C_p represent viscosity, thermal conductivity, and specific heat of the fluid running in the tube, respectively. k_c and k_w are thermal conductivity of the coke and tube. D_o is the outside diameter of the tube. C_1 and C_2 are constant parameters to be estimated from measured process variables.

In order to estimate the skin temperature by the proposed model, the thickness of the coke layer at the first day, d(1), must be given as an initial value. When the previous decoking operation was perfectly performed, one could assume d(1) to be zero. However, it is not often the case. Therefore, d(1) is considered as a parameter which is estimated from available measurement of process variables and off-line measured skin temperature.

Application of Physical Model for Temperature Inference

The developed physical model is tested on the data obtained from an industrial pyrolysis. C_1 and C_2 are determined beforehand, based on another data set. The initial value, d(1), is estimated using the nonlinear least squares method from first two off-line measurements of skin temperature. An estimation result is illustrated in Fig. 2. Figure 2 shows the good agreement between estimated and actual skin temperature. However, the fairly large error is observed around 30 days.

Combination of Physical and Empirical Models

The physical model alone could predict the temperature accurately for the most operating conditions but not for all. Two alternative approaches are available to further improve the prediction accuracy, depending on the assumption (Mellichamp et al., 1992): (i) The parameters of the physical model are not appropriate; (ii) some dynamic effects, which the physical model cannot explain, are present in the process. In the case (i), the approach to take is to formulate model identification as an adaptive problem, in which some model parameters are adapted using available measurements. In the second case, the approach to take is either developing a more precise physical model or employing an empirical model.

In the previous paper, based on the assumption (i), the Extended Kalman Filter (EKF) is employed to update the parameter values as well as the state variables. However, some extra measurements of skin temperature are needed.

The approach suggested in this paper is a variation of approach (ii). Namely, the dynamics is represented by integrating the developed physical model with an empirical model. A PLS regression model (Glen et al., 1989) is employed to represent the dynamic effects that the physical model alone could not represent.

PLS Regression Model

Since the PLS regression model to be developed can describe dynamic effects that the physical model cannot describe, the output of PLS regression model should be the estimation error of the physical model and the inputs are on-line measured process variables. In this study, twelve process variables are considered as input variables. The PLS analysis is applied to their normalized data.

The number of latent variables used in the PLS model is determined in the cross-validation test. In the cross validation test, the following steps are performed assuming that N data points are available; 1) one of the data point is left out from the considered data set. 2) a PLS regression model is developed from the remaining N-1 data points. For the selected data point, the squared prediction error is calculated using the developed model. This step is repeated by changing the selected data point. Then, the total prediction error of the model is calculated by summing up the errors over all N data points. 3) The step 1) and 2) is repeated by changing the number of latent variables employed in the PLS regression model.

Figure 3 illustrates the results of the total prediction error along the number of latent variables. The value of the error at zero indicates the total prediction error for the case that no PLS model is used, i.e., the developed physical model alone is used.

In Fig. 3, the number of latent variables showing the minimum prediction error is eleven. However, it is also shown that the prediction accuracy is not improved drastically after the two latent variables. For this reason, two latent variables are employed in this study.

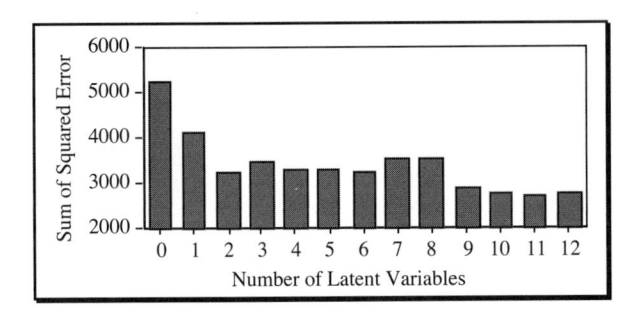

Figure 3. Total prediction error vs. number of latent var.

Procedure of On-line Inference

The estimate of the skin temperature, T_{est}, is given by $T_{est}(t) = T_{skin}(t) - T_{err}(t)$. T_{skin} is the estimate by the physical model and T_{err} is the estimate given by the developed PLS model. Consequently, the procedure of inferring the skin temperature on-line is summarized as follows:

1. Parameter values of α and β are determined from the physical property data as well as the tube design parameters.
2. Parameter values of C_1 and C_2 for a pyrolysis are estimated from past measured data at the pyrolysis unit.
3. The initial value of the coke layer thickness, $d(1)$, is determined from first two measurements of skin temperature and available process measurements by using the nonlinear least squares method so that the prediction error is minimized at two measuring points.
4. The estimation of the skin temperature is successively carried out by using the measured process variables.

Application Results

The inferential model combining physical and PLS model is applied to the same data that is shown in Fig. 2. Note that the parameter C_1 and C_2 as well as the PLS model is not built on the data. A result is shown in Fig. 4. It is shown that the error observed around 30 days in previous simulation can be decreased. Furthermore, by using the PLS model in addition to the physical model, the maximum estimation error is also decreased to 12 °C from 18 °C.

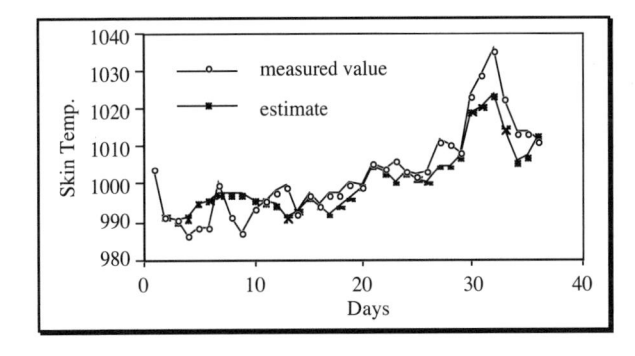

Figure 4. On-line inference of skin temp. using physical and PLS regression model

Conclusions

An on-line inferential system for the industrial pyrolysis units is developed using nonlinear physical model combined with a PLS regression model. The results show that, by taking two manual measurements of the skin temperature at the beginning of the cracking operation, the proposed system can predict the tube-wall temperature successfully for almost all operating conditions. As can be seen through the application, the proposed modeling approach, where nonlinear physical model is incorporated with an empirical model, provides more accurate prediction and is effective in practical use.

References

Weber, R. and C. Brosilow (1972). The use of secondary measurements to improved control. *AIChE J.*, **18**, 614-623.

Lee, J.H., M.S. Gelormino and M. Morari (1992). Model predictive control of multi-rate sampled-data systems: a state-space approach. *Int. J. of Control*, **55**(1), 153-191.

Kemna, A. H. and D. A. Mellichamp (1992). Identification of combined physical/empirical models with application to an industrial process. AIChE mtg. Miami

Ohshima. M, H. Nakagawa, I. Hashimoto, U. Ohkamo, G. Suzuki, and Y. Kawabata (1994). On-line inference of tube-wall temperature in an industrial olefins pyrolysis plant. *Proc. of PSE'94*, 627-633.

Glen, W.G., W.J.Dun II, and D.R. Scott (1986). Principal component analysis and partial least squares regression. Tetrahedron Computer Methodology, **2**(6), 329-376.

SOFTWARE SENSORS AND ADAPTIVE CONTROL OF BIOCHEMICAL PROCESSES

Denis Dochain
Maître de Recherches FNRS, Cesame
Université Catholique de Louvain
1348 Louvain-La-Neuve, Belgium

Abstract

This paper presents a brief survey about model-based monitoring and control approaches of bioprocesses, based on material balances and which accounts for the uncertain knowledge about the process kinetics.

Keywords

Software sensors, Linearizing control, Adaptive control, Bioprocesses.

Introduction

Automatic control of industrial biotechnological processes is still developing slowly. There are two main reasons for this : the internal working and dynamics of these processes are as yet badly grasped; another essential difficulty lies in the absence, in most cases, of cheap and reliable instrumentation suited to real-time monitoring. The classical monitoring and control methods do not prove very efficient to tide over these basic difficulties. In this paper, it is shown how to incorporate the well-known knowledge about the dynamics of biochemical processes (basically, the reaction network and the material balances) in monitoring and control algorithms which are moreover capable of dealing with the process uncertainty (in particular on the reaction kinetics) by introducing, for instance in the control algorithms, an adaptation scheme). A key reference to this paper is (Bastin and Dochain, 1990).

General Dynamical Model

A biotechnological process can be defined as a set of M biochemical reactions involving N components. The dynamical model of a bioprocess in a stirred tank reactor can be deduced from mass balance considerations and written in the following compact form :

$$\frac{d\xi}{dt} = -D\xi + Kr + F - Q \qquad (1)$$

where ξ is the vector of the bioprocess component (g/L) $(\dim(\xi) = N)$, D is the dilution rate (h^{-1})(i.e. the ratio of the influent flow rate over the medium volume), K is the yield coefficient matrix, r is the reaction rate vector $(g/(L.h))$, F $(g/(L.h))$ is the feed rate vector and Q $(g/(L.h))$ the gaseous outflow rate vector $(\dim(F) = \dim(Q) = N)$. The model (1) has been called the *General Dynamical Model* for stirred tank bioreactors (see (Bastin and Dochain, 1990)).

Example #1: PHB Producing Process

Let us first consider the production of poly-β-hydroxybutyric acid (PHB), which is a biodegradable polymer. The PHB can be produced in an aerobic culture of *Alcaligenes eutrophus*, and the production may follow two metabolic pathways : the first one is growth associated (with three limiting substrates : oxygen, a source of carbon (e.g. fructose or glucose) and nitrogen (ammonia)) and characterized by a very low yield; the second one is non-growth associated, where the biomass plays simply the role of a catalyst, and is completely inhibited by nitrogen :

$$S + N + C \longrightarrow X + P_1 \qquad (2)$$
$$S + C + X \longrightarrow X + P_1 \qquad (3)$$

where S, C, N, X and P_1 represent the carbon source, the oxygen, the nitrogen, the biomass and the PHB, respectively. In the first reaction scheme, the biomass X is an autocatalyst and is only present on the right hand side; the presence of X on both sides of the arrow in the second reaction scheme means that X simply plays the role of a catalyst. The dynamical equations of the PHB process can be written in the above matrix form (in which $\Delta Q_{O2} = Q_{in} - Q_{out}$) :

$$\frac{d}{dt}\begin{bmatrix} S \\ C \\ N \\ X \\ P_1 \end{bmatrix} = -D \begin{bmatrix} S \\ C \\ N \\ X \\ P_1 \end{bmatrix} + \begin{bmatrix} DS_{in} \\ \Delta Q_{02} \\ DN_{in} \\ 0 \\ 0 \end{bmatrix} + \begin{bmatrix} -k_1 & -k_2 \\ -k_3 & -k_4 \\ -k_5 & 0 \\ 1 & 0 \\ k_6 & 1 \end{bmatrix} \begin{bmatrix} r_1 \\ r_2 \end{bmatrix} \quad (4)$$

where S_{in} and N_{in} the influent carbon source and nitrogen concentrations, Q_{in} and Q_{out} the inlet and outlet gaseous oxygen flowrates, r_1 and r_2 the reaction rates of the reactions (2) and (3)) respectively and k_i (i = 1 to 6) the yield coefficients.

Example #2: Anaerobic Digestion

Anaerobic digestion is a biological wastewater treatment process which produces methane. A simplified model of this complex process considers two important steps : an acidogenesis step in which the organic matters are decomposed into volatile fatty acids (VFA) and CO_2 by acidogenic bacteria, and a methanization step in which VFA are transformed into methane and CO_2:

$$S_1 \longrightarrow X_1 + S_2 + P_2 \quad (5)$$
$$S_2 \longrightarrow X_2 + P_1 + P_2 \quad (6)$$

where S_1, S_2, X_1, X_2, P_1 and P_2 are respectively organic matters, VFA, acidogenic bacteria, methanogenic bacteria, methane and CO_2. The dynamical model of the anaerobic digestion process in a stirred tank reactor can be described within the above formalism (1) by using the following definitions :

$$\xi^T = \begin{bmatrix} X_1 & S_1 & X_2 & S_2 & P_1 & P_2 \end{bmatrix}$$

$$K = \begin{bmatrix} 1 & 0 \\ k_1 & 0 \\ 0 & 1 \\ k_3 & -k_2 \\ 0 & k_4 \\ k_5 & k_6 \end{bmatrix}, F = \begin{bmatrix} 0 \\ DS_{in} \\ 0 \\ 0 \\ 0 \\ 0 \end{bmatrix}, Q = \begin{bmatrix} 0 \\ 0 \\ 0 \\ 0 \\ Q_1 \\ Q_2 \end{bmatrix}$$

$$r^T = \begin{bmatrix} r_1 & r_2 \end{bmatrix} = \begin{bmatrix} \mu_1 X_1 & \mu_2 X_2 \end{bmatrix}$$

where μ_1, μ_2 are the specific growth rates (h^{-1}) of reactions (5), (6), respectively, and S_{in}, Q_1 and Q_2 represent respectively the influent glucose concentration and the gaseous outflow rates of CH_4 and CO_2.

Model Order Reduction

Consider that, for some i, the dynamics of the component ξ_i are to be neglected. The dynamics of ξ_i are described by equation (1) :

$$\frac{d\xi_i}{dt} = -D\xi_i + K_i r + F_i - Q_i \quad (7)$$

where K_i is the line of K corresponding to the component ξ_i. The simplification is then achieved by setting ξ_i and $d\xi_i/dt$ to zero i.e. by replacing the differential equation (7) by the following algebraic equation :

$$K_i r = -F_i + Q_i \quad (8)$$

It has been shown by singular perturbation arguments that the above model order reduction rule is valid for low solubility products (Bastin and Dochain, 1990), (Dochain, 1994) and for bioprocesses with fast and slow reactions. In the latter, the above order reduction (which is indeed a quasi steady-state approximation) rule (8) applies to substrates of fast reactions (as long as they intervene only in fast reactions).

Example: Anaerobic Digestion

First of all, it is well-known that methane is a low solubility product. Furthermore, assume that for instance the first acidogenic path (reaction (5)) is limiting, i.e. that the methanization (6) is fast and (5) is slow. We can then apply the model order reduction rule to the VFA concentration S_2 and the dissolved methane concentration P_1. By setting their values and their time derivatives to zero, we reduce their differential equations to a set of algebraic equations. If one considers the expressions of the reaction rate r_1 which can be drawn from these algebraic equations, and introduce them in the dynamical equation of the glucose concentration S_1, this one can be rewritten as follows :

$$\frac{dS_1}{dt} = -DS_1 - k_7 Q_1 + DS_{in} \quad (9)$$

where k_7 is defined as follows :

$$k_7 = \frac{k_1 k_2}{k_3 k_4} \quad (10)$$

The equation (9) will be used for the design of an adaptive linearizing controller of the organic matter concentration.

Monitoring of Bioprocesses : Software Sensors

Design of Asymptotic Observers

We shall now concentrate on the design of software sensors, i.e. algorithms based on the general dynamical model to estimate on-line the unknown parameters (like the reaction rates) and the unmeasured components from the few available on-line measurements. The objective of this section is to develop a software sensor (called *asymptotic observer*) for the process component concentrations without the knowledge of the process kinetics being necessary. The derivation of the asymptotic observer equations is based on the following assumptions:

1. M (the number of reactions) components are measured on-line.
2. The feedrates F, the gaseous outflow rates Q and the dilution rate D are known either by measurement or by choice of the user.
3. The yield coefficient matrix K is known.
4. The reaction rate vector r is unknown.
5. The M reactions are irreversible and independent, i.e. rank(K) = R = M

From assumption 1, we can define the following state partition :

$$\xi = \left[\begin{array}{c} \xi_1 \\ \xi_2 \end{array} \right] \quad (11)$$

where ξ_1 and ξ_2 hold for the measured component concentrations and the unmeasured ones, respectively (with the yield coefficient matrix K_1 corresponding to ξ_1 being full rank). Let us consider the following state transformation :

$$\zeta = \xi_2 - K_2 K_1^{-1} \xi_1 \quad (12)$$

The dynamics of ζ are independent of the reaction rate $r(\xi)$:

$$\frac{d\zeta}{dt} = -D\zeta + (F_2 - Q_2) - K_2 K_1^{-1}(F_1 - Q_1) \quad (13)$$

The equations (12)(13) are the basis for the derivation of the asymptotic observer. The dynamical equations of ζ are used to calculate an estimate of ζ on-line, which is used, via equation (12) and the on-line data of ξ_1, to derive an estimate the unmeasured component ξ_2 :

$$\frac{d\hat{\zeta}}{dt} = -D\hat{\zeta} + (F_2 - Q_2) - K_2 K_1^{-1}(F_1 - Q_1)$$
$$\hat{\xi}_2 = [\hat{\zeta} + K_2 K_1^{-1} \xi_1]$$

Remarks

1) The above software sensor has been called an *asymptotic observer* because it has been theoretically shown (Bastin and Dochain, 1990) that the error on the unmeasured process components' on-line reconstruction, $\xi_2 - \hat{\xi}_2$, *asymptotically* tends to zero as long as the dilution rate D is positive (or more precisely, as long as D does not remain equal to zero for too long).

2) The state transformation (12) is closely related to the notion of reaction invariants ((Gavalas, 1968)).

Example #1: PHB Producing Process

In practice the PHB process is operated in two sequential steps : a first step for biomass growth *in presence of nitrogen*, and a second step for non-growth associated production of PHB *in absence of nitrogen*; i.e. the process can be viewed as a sequence of two reactions. Only one measured component is then necessary to reconstruct the state of the bioreactor. Here we have designed an asymptotic observer for each step on the basis of oxygen data. Fig. 1 shows the estimation results for X, P_1, S and N : note the good correspondence of the software observation with the off-line analyses, and that the switching from step 1 to step 2 is efficiently driven by the estimate of nitrogen $\hat{N}$.

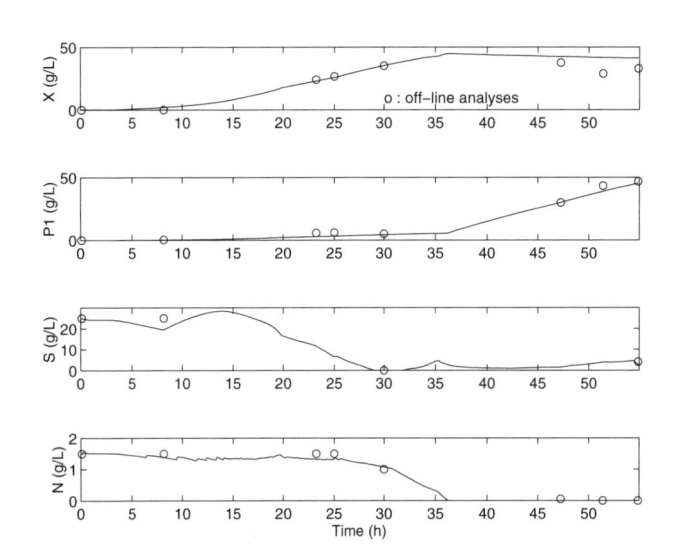

Figure 1. Estimation results of the asymptotic observer of the PHB process.

Adaptive Linearizing Control of Bioprocesses

Design of the Adaptive Linearizing Controller

We shall now concentrate on the design of model-based controllers for bioreactors. Since the model is generally nonlinear, the model-based control design will result in a *linearizing* control structure, in which the on-line estimation of the unknown variables (component concentrations) and parameters (reaction rates and yield coefficients) are incorporated. The resulting controller will be an *adaptive linearizing controller*. One interesting aspect of the control design in the example below will be to show how to eliminate the (unknown) kinetics terms by incorporating the gas measurements into the controller. Let us consider here the objective of controlling the concentration of one reactant component under the following practical constraints in addition to assumptions 2 and 4 :

6. the controlled components are assumed to be measured on-line; the concentrations of the other components (particularly of the biomass) are not available for on-line measurement;
7. the yield coefficients are unknown;

By defining y the controlled component, the dynamics of y are simply the equation of y in model (1) and can be rewritten as follows :

$$\frac{dy}{dt} = -Dy + K_y r + F_y - Q_y \quad (14)$$
$$= f(F_y, Q_y, K_y r) + g(F_y, y)u \quad (15)$$

where the index y holds for the rows of K, F and Q corresponding to the controlled output y. Assume now that we wish to have a linear stable closed-loop dynamical behaviour, i.e.:

$$\frac{dy}{dt} = C_1(y* - y), \ C_1 > 0 \quad (16)$$

By combining equations (15)(16), we readily obtain the control law:

$$u = g(F_y, y)^{-1}[C_1(y* -y) - f(F_y, Q_y, K_y r)] \quad (17)$$

Since the kinetics and most of the yield coefficients are assumed to be unknown, they are replaced by on-line estimates of selected parameters.

Example #2: Anaerobic Digestion

In anaerobic digestion processes, control strategies have particularly to concentrate on the control of the substrate concentration (which characterizes the pollution level and the presence of acids). COD (Chemical Oxygen Demand) is a priori a very good candidate: an adaptive linearizing controller has been designed, theoretically analyzed and experimentally validated on a pilot anaerobic digester ((Renard, Dochain, Bastin, Naveau and Nyns, 1988)). The control design is based on the reduced order model (9) and results in the following control law:

$$D = \frac{C_1(S_1^* - S_1) + \hat{k}_7 Q_1}{S_{in} - S_1} \quad (18)$$

The unknown parameter is now k_7. It can be estimated on-line by using an appropriate updating equation (see (Bastin and Dochain, 1990)), for instance here a "Lyapunov design" estimation equation:

$$\frac{d\hat{k}_7}{dt} = C_2(S_1^* - S_1), \qquad C_2 > 0 \quad (19)$$

One of the experimental result of the above controller implemented on a 60 liter CSTR (continuous stirred tank reactor) pilot reactor is shown in Fig. 2. In this experiment, the influent organic matter concentration S_{in} has been doubled over a period of 8 hours. Note the ability of the controller to maintain the effluent substrate concentration S_1 close to its desired value ($S_1^* = 8.9\ gCOD/L$) despite of the variation of S_{in} (the regulation error $S_1 - S_1^*$ was kept below 2.7%).

Discussion

One of the key feature of the adaptive linearizing can be illustrated on the basis of the anaerobic digestion example: it allows to incorporate the well-known characteristics of the process dynamics, while keeping the usual features of classical controllers:

- *proportional action*: via the term $C_1(S_1^* - S_1)$;
- *integral action*: via the adaptation mechanism (19);
- *feedforward action*: via the presence of S_{in}. Indeed the controller is capable of anticipating the effect of a variation of the influent substrate concentration: for instance, an increase of S_{in} will result in a decrease of the control input D, inversely proportional to this increase (see also Figures 2a and 2c).

Besides the controller contains a state "estimate" via the term $\hat{k}_7 Q_1$. If the process is in good working conditions, in particular if the biomass is in a good state, then the process

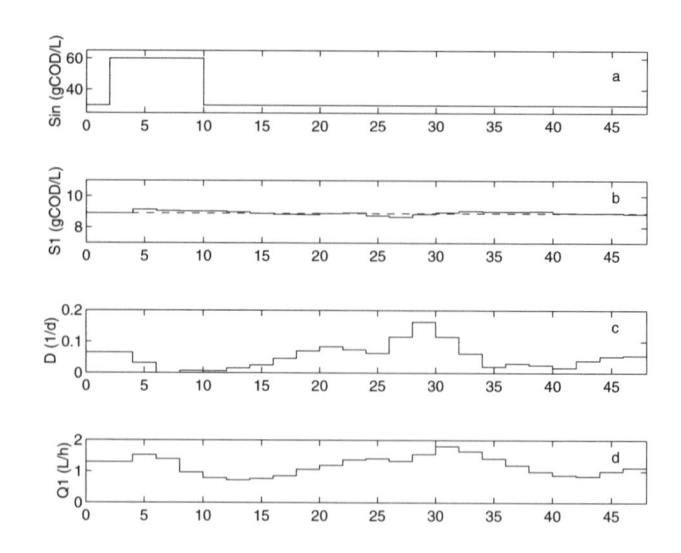

Figure 2. Experimental results of the control of COD in an anaerobic digestion process (Fig. 2b : - - - : S_1^*).

produces a lot of methane. Then it is possible to treat high amounts of organic matter, since the control action D is proportional to Q_1 (compare also Figures 2c and 2d).

Finally note that beside the integral action, the estimation of "physical" parameters (here k_7, which is indeed a conversion coefficient of COD organic matter into methane) has the further advantage of giving useful information, which can be used for monitoring the process, and possibly also for analyzing the internal working of the process.

Conclusions

The objective of this paper was to present a brief survey about model-based monitoring and control approaches for bioreactors. The proposed methodologies can be extended to chemical reactors (Dochain, 1995) or to non completely mixed reactors like fixed bed reactors (Dochain, 1994).

Acknowledgment

This paper presents results of the Belgian Programme on Interuniversity Poles of Attraction initiated by the Belgian State, Prime Minister's Office, Science Policy Programming. The scientific responsibility rests with its authors.

References

Bastin, G. and Dochain, D. (1990). *"On-line Estimation and Adaptive Control of Bioreactors"*, Elsevier, Amsterdam.

Dochain, D. (1994). *Contribution to the Analysis and Control of Distributed Parameter Systems with Application to (Bio)chemical Processes and Robotics*, Thèse d'Aggrégation de l'Enseignement Supérieur, UCL, Belgium.

Dochain, D. (1995). Design of adaptive linearizing controllers for non-isothermal reactors, *Int. J. Control* **59 (3)**: 689–710.

Gavalas, G. (1968). *"Nonlinear Differential Equations of Chemically Reacting Systems"*, Springer Verlag, Berlin.

Renard, P., Dochain, D., Bastin, G., Naveau, H. and Nyns, E. (1988). Adaptive control of anaerobic digestion processes. a pilot-scale application, *Biotechnol. Bioeng.* **31**: 287–294.

CONTROL OF NONLINEAR DISTRIBUTED PARAMETER SYSTEMS: RECENT RESULTS AND FUTURE RESEARCH DIRECTIONS

Panagiotis D. Christofides and Prodromos Daoutidis
Department of Chemical Engineering and Materials Science
University of Minnesota
Minneapolis, MN 55455

Abstract

This paper reviews recent developments on control of nonlinear partial differential equations (PDEs), outlines our results on control of quasi-linear hyperbolic and parabolic PDEs, and discusses directions for future research on control of nonlinear distributed parameter systems (DPS).

Keywords

PDE systems, Nonlinear control, Transport-reaction processes.

Introduction

Chemical processes are inherently distributed in nature, owing to the presence of convection, diffusion and phase-dispersion phenomena, time-delays etc., and are naturally modeled by DPS in the form of partial differential equations, integro-differential equations and delay-differential equations. The conventional approach to the control of DPS involves the discretization of the original model in space followed by the application of control methods for ordinary differential equations (ODEs). Although this approach may lead to satisfactory control quality in processes with mildly distributed nature, it usually leads to poor performance in processes with strongly distributed nature. This realization motivates research on the development of distributed control algorithms for DPS. The literature on control of *linear DPS* is really extensive. Excellent surveys of both theoretical and application papers on this topic can be found in (Ray, 1978; Balas, 1982; Keulen, 1993). However, the range of applicability and the efficiency of linear control methods are significantly restricted by the presence of severe nonlinearities in chemical processes. Motivated by this, there is an emerging research activity on the development of control methods for *nonlinear DPS*, that account for both nonlinearities and spatial variations. Most of this research has focused on systems modeled by nonlinear PDEs, which typically arise in practice as models of convection-reaction and diffusion-convection-reaction processes.

Convection-reaction processes (e.g., plug-flow and pressure swing reactors) can be adequately described by systems of quasi-linear first-order *hyperbolic* PDEs. The distinct feature of these systems is that all the eigenmodes of the spatial differential operator contain the same or nearly the same amount of energy, and thus an infinite number of modes is required to accurately describe their dynamic behavior. This property prohibits the application of modal decomposition techniques to derive reduced-order ODE models that approximately describe the dynamics of the PDE system and suggests addressing the control problem on the basis of the infinite dimensional model. Following this approach, distributed control algorithms have been developed employing optimal control (Ray, 1981), sliding-mode control (Sira-Ramirez, 1989; Hanczyc and Palazoglu, 1995), geometric control (Christofides and Daoutidis, 1996a) and Lyapunov-based control (Alonso and Ydstie, 1995; Christofides and Daoutidis, 1996c).

Diffusion-convection-reaction processes (e.g., packed-bed and fluidized-bed reactors) are naturally modeled by quasi-linear *parabolic* PDEs. In contrast to hyperbolic PDEs, parabolic PDEs are characterized by a finite-number of dominant modes, which implies that their dynamic behavior can be approximately described by finite-dimensional ODE systems. Thus, the standard approach for the control of parabolic PDEs utilizes eigenfunction expansion techniques to obtain an ODE system that approximately reproduces the dynamics of the original PDE system, which is then used for controller design purposes (Ray, 1981; Chen and Chang, 1992). However, there are two key problems associated with this approach. First, the number of modes that should be retained to derive an ODE model that yields the desired degree of approximation may be very large, leading to high dimensionality of the resulting controllers (Gay and Ray, 1995). Second, there is a lack of a systematic way to characterize the discrepancy between the solutions of the PDE model and the reduced-order ODE model in finite time, which is essential for characterizing the transient performance of the closed-loop system. To address these problems, a general methodology was developed in (Christofides and Daoutidis, 1995) for the design of *minimal-order* output feedback controllers for systems of quasi-linear parabolic PDEs. The methodology involves the derivation of accurate minimal-order ODE models, through combination of nonlinear Galerkin's method and a novel procedure for the construction of approximate inertial manifolds. The transient performance of the closed-loop system is precisely characterized using singular perturbation theory for infinite-dimensional systems.

Finally, other work on control and analysis of parabolic PDEs has focused on: *i*) the synthesis of state feedback controllers through combination of the theory of symmetry groups and sliding mode techniques (Hanczyc and Palazoglu, 1996), *ii*) the use of concepts from thermodynamics to construct a Lyapunov function candidate and derive conditions that guarantee asymptotic stability of the closed-loop system under boundary proportional-integral-derivative control (Ydstie and Krishnan, 1994), *iii*) the use of singular perturbations to reduce the PDE model that describes a fixed-bed reactor with strong diffusive phenomena to an ODE one (Dochain and Bouaziz, 1993), and *iv*) the use of concepts from infinite-dimensional systems theory to determine the extent to which linear boundary proportional control influences the dynamic and steady-state response of the closed-loop system (Byrnes et al., 1994; Byrnes et al., 1995).

Control of Nonlinear PDEs

Hyperbolic PDEs

Geometric control: We consider first-order hyperbolic PDEs of the form:

$$\begin{aligned}\frac{\partial x}{\partial t} &= A\frac{\partial x}{\partial z} + f(x) + g(x)b(z)u(t) \\ y(t) &= \int_a^b c(z)h(x)dz \\ R(t) &= C_1 x(a,t) + C_2 x(b,t)\end{aligned} \quad (1)$$

where $x(z,t) = [x_1(z,t) \cdots x_n(z,t)]^T \in \mathcal{H}^n[(a,b), \mathbb{R}^n]$ denotes the vector of state variables, $z \in [a,b] \subset \mathbb{R}$, denotes position, $u(t)$ denotes the vector of manipulated inputs, $y(t)$ denotes the vector of controlled outputs, $b(z)$ is a function which determines how the control action is distributed in space, $c(z)$ is a function which depends on the desired performance specifications, $f(x)$ and $g(x)$ are vector functions, $h(x)$ is a scalar function, $R(t)$ is a column vector, and A, C_1, C_2 are constant matrices with A being real symmetric and nonsingular.

The developed framework for output feedback controller synthesis for systems of the form of Eq.1 entails the following steps (Christofides and Daoutidis, 1996a):

1. Distributed nonlinear state feedback controllers are synthesized on the basis of the PDE model to enforce output tracking and guarantee closed-loop stability, by following an approach conceptually similar to the one used in the geometric control of ODE systems.

2. Distributed nonlinear output feedback controllers are synthesized through combination of the developed state feedback controllers with appropriate distributed state observers.

Application to a plug-flow reactor: Consider a nonisothermal plug-flow reactor where an elementary series reaction $A \xrightarrow{k_1} B \xrightarrow{k_2} C$ takes place. The process is described by three quasi-linear hyperbolic PDEs:

$$\begin{aligned}\frac{\partial C_A}{\partial t} &= -v_l \frac{\partial C_A}{\partial z} - k_{10}e^{\frac{-E_1}{RT_r}}C_A \\ \frac{\partial C_B}{\partial t} &= -v_l \frac{\partial C_B}{\partial z} + k_{10}e^{\frac{-E_1}{RT_r}}C_A - k_{20}e^{\frac{-E_2}{RT_r}}C_B \\ \frac{\partial T_r}{\partial t} &= -v_l \frac{\partial T_r}{\partial z} + \frac{(-\Delta H_{r_1})}{\rho_m c_{pm}}k_{10}e^{\frac{-E_1}{RT_r}}C_A \\ &+ \frac{(-\Delta H_{r_2})}{\rho_m c_{pm}}k_{20}e^{\frac{-E_2}{RT_r}}C_B + \frac{U_w}{\rho_m c_{pm}V_r}(T_w - T_r) \\ & \quad (2)\\ C_A(0,t) &= C_{A0}(t), \; C_B(0,t) = 0, \; T_r(0,t) = T_{A0}(t)\end{aligned}$$

where C_A, C_B are the concentrations of the species A and B in the reactor, T_r is the temperature of the reactor, v_l is the fluid velocity, ρ_m, c_{pm} are the density and heat capacity of the fluid, U_w is the heat transfer coefficient, V_r is the reactor volume, and $k_{10}, E_1, \Delta H_{r_1}, k_{10}, E_1, \Delta H_{r_2}$ denote the pre-exponential constants, activation energies and enthalpies of the two reactions. The control objective is the regulation of the concentration of the desired product C_B throughout the reactor by manipulating the wall temperature $T_w(t)$. The controlled output is of the form $y(t) = \int_0^1 C_B(z,t)dz$.

Fig. 1 (Christofides and Daoutidis, 1996a) shows the evolution of C_B throughout the reactor for a 30% increase in the set-point. The controller achieves excellent control at all positions and times, regulating C_B at each point of the reactor to a new steady-state value, which is 30% higher than the one of the original steady-state.

Lyapunov-based control: A framework for the design of distributed robust nonlinear controllers was developed in (Christofides and Daoutidis, 1996c), for systems of hyperbolic PDEs of the form of Eq.1 with time-varying uncertain variables and unmodeled dynamics. The derived con-

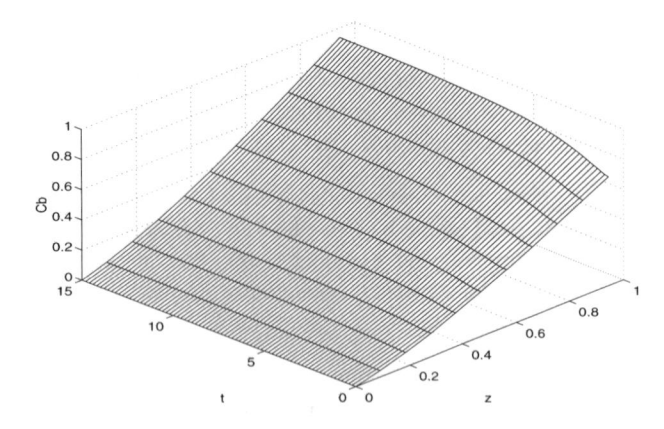

Figure 1. Profile of evolution of concentration of species B throughout the reactor for a 30% increase in the set-point (C_B (mol/lt), z (m), t (min)).

trollers guarantee boundedness of the state and achieve robust output tracking with arbitrary degree of asymptotic attenuation of the effect of uncertainty on the output of the closed-loop system, provided that the unmodeled dynamics are asymptotically stable and sufficiently fast. The controller design is performed constructively using Lyapunov's direct method, and requires that a structural condition in the manner in which the uncertain variables enter the system is satisfied and that there exist known bounding functions that capture the magnitude of the uncertain variables. The methodology was successfully applied to a fixed-bed reactor where the reactant wave propagates through the bed with significantly larger speed than the heat wave, and the heat of reaction is unknown.

Parabolic PDEs

We consider systems of parabolic PDEs of the form:

$$\frac{\partial x}{\partial t} = Lx + cb(z)u + f(x)$$

$$y(t) = \int_\Omega c(z)kx dz \quad (3)$$

$$Ax(z,t) + B\frac{\partial x}{\partial z}(z,t) = R(t), \text{ on } \Gamma$$

where $x(z,t) \in \mathcal{H}[\Omega, \mathbb{R}^n]$, $z = [z_1\ z_2\ z_3]^T$ denotes the vector of spatial variables, Γ denotes the boundary of the spatial domain Ω, L is a linear spatial differential operator which involves first- and second-order spatial derivatives, A, B are constant matrices, and c, k are constant vectors.

The developed framework for the synthesis of minimal-order output feedback controllers for systems of the form of Eq.3 entails the following steps (Christofides and Daoutidis, 1995):

1. Initially, nonlinear Galerkin's method is employed to transform the model of Eq.3 into an interconnection of a finite-dimensional (possibly unstable) system with an infinite dimensional stable system. To this end, two appropriate sets of orthogonal basis functions, which completely span the space $\mathcal{H}$, are selected and used to derive a two-time-scale system, modeled within the mathematical framework of singular perturbations, whose slow subsystem is of dimension equal to the number of slow modes of the system operator, and the fast subsystem is infinite dimensional and stable.

2. A novel procedure is then employed for the construction of an approximate inertial manifold, which is used for the derivation of a minimal-order slow subsystem which yields approximate solutions of arbitrary degree of accuracy. Given the fact that the fast subsystem is exponentially stable, a finite dimensional nonlinear output feedback controller is synthesized on the basis of the minimal-order slow subsystem, to enforce output tracking and guarantee exponential stability of the closed-loop slow subsystem.

3. Finally, the resulting infinite-dimensional closed-loop system (PDE model and controller) is analyzed utilizing singular perturbation theory to derive a lower bound on the degree of separation of the slow versus the fast modes that guarantees the exponential stability of the closed-loop system and that the output satisfies a relation of the form:

$$y(t) = y^s(t) + O(\epsilon), \ t \geq 0 \quad (4)$$

with $y^s(t)$ being the output of the closed-loop slow subsystem. $O(\cdot)$ denotes the usual order of magnitude notation and ϵ is a small positive parameter which depends on the degree of approximation of the original parabolic PDE model from the slow system and tends to zero as the dimension of the ODE system increases.

Application to a packed-bed reactor: Consider a non-isothermal packed-bed reactor (Chen and Chang, 1992), where an elementary reaction of the form $A \overset{k_1}{\rightarrow} B$ takes place. The process is described by two quasi-linear parabolic PDEs:

$$\rho_m c_{pm}\frac{\partial T_r}{\partial t} = -\rho_m c_{pm} v_l \frac{\partial T_r}{\partial z} + K\frac{\partial^2 T_r}{\partial z^2}$$
$$+(-\Delta H_{r_1})k_{10} exp^{-\frac{E_1}{RT_r}} C_A + \frac{U_w}{V_r}(T_w - T_r)$$

$$\frac{\partial C_A}{\partial t} = -v_l \frac{\partial C_A}{\partial z} + D\frac{\partial^2 C_A}{\partial z^2} - k_{10} exp^{-\frac{E_1}{RT_r}} C_A$$

$$z = 0, \quad \rho_m c_{pm} v_l (T_r - T_f) = K\frac{\partial T_r}{\partial z},$$

$$v_l(C_A - C_{A_f}) = D\frac{\partial C_A}{\partial z}; \quad z = 1, \quad \frac{\partial T_r}{\partial z} = \frac{\partial C_A}{\partial z} = 0 \quad (5)$$

where K, D denote the conductivity and diffusivity, and C_{A_f}, T_f denote some reference values for concentration and

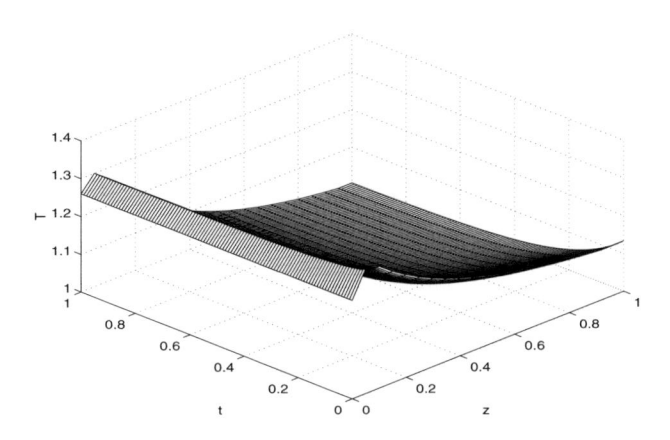

Figure 2. Profile of evolution of dimensionless reactor temperature for an 8% decrease in the set-point ($T = \dfrac{T_r - T_f}{T_f}, z\ (m), t\ (hr)$)

temperature. The control objective is the regulation of the temperature profile along the length of the reactor by manipulating the wall temperature T_w. The controlled output is of the form: $y(t) = \int_0^1 \dfrac{T_r(z,t) - T_f}{T_f} dz$. A four-dimensional ODE model was constructed and used for the synthesis of the controller (Christofides and Daoutidis, 1996b). Fig. 2 displays the profile of the temperature of the reactor for an 8% decrease in the set-point. The controller achieves excellent control at all positions and times, regulating the temperature at each point of the reactor to a new steady-state value, which is almost 8% lower than the one of the
original steady-state.

Future Research Directions

The results on quasi-linear PDEs provide the first step towards developing a general and practical framework for the analysis and control of nonlinear distributed parameter chemical processes. Main unresolved theoretical and practical problems that need to be addressed include:

1. The study of control problems in more general classes of nonlinear PDEs, such as models arising in fluid dynamics (e.g., Navier-Stokes equations) and pattern formation (they contain differential operators of fourth order in space, e.g., Kuramoto-Sivashinsky equation). It is well-known (Foias et al., 1989; Bangia et al., 1996) that such systems exhibit low-dimensional dynamic behavior, which implies that the proposed framework for parabolic PDEs can be used, in principle, to address their control.

2. The development of control algorithms for nonlinear delay-differential equations, as well as DPS modeling disperse-phase processes (e.g., crystallizers, emulsion polymerization reactors), such as nonlinear integro-differential equation systems.

3. The development of robust control algorithms in order to deal with disturbances and parametric uncertainty, and the derivation of design guidelines for the developed control algorithms, that will allow handling of manipulated input constraints.

4. The control-relevant modeling and analysis (e.g., selection of manipulated inputs and measurements, as well as optimal locations for sensors and actuators), and the experimental implementation of the control algorithms to industrially important DPS.

References

Alonso, A. A. and E. B. Ydstie (1995). Nonlinear control, Passivity and the second law of thermodynamics, AIChE Ann. Mtg., paper 181i, Miami Beach, FL.

Balas, M. J. (1982). Trends in large scale structure control theory: Fondest hopes, wildest dreams, *IEEE Trans. Automat. Contr.*, **AC-27**, 522.

Bangia, A. K. and P. F. Batcho and I. G. Kevrekidis and G. E. Karniadakis (1996). Unsteady 2-D flows in complex geometries: Comparative bifurcation studies with global eigenfunction expansion, submitted.

Byrnes, C. A. and D. S. Gilliam and V. I. Shubov (1994). Global Lyapunov stabilization of a nonlinear distributed parameter system, *Proc. of 33rd IEEE Conference on Decision and Control*, Orlando, FL, 1769.

Byrnes, C. A. and D. S. Gilliam and V. I. Shubov (1995). On the dynamics of boundary controlled nonlinear distributed parameter systems, *Proc. of Symposium on Nonlinear Control Systems Design '95*, Tahoe City, CA, 913.

Chen, C. C. and H. C. Chang (1992). Accelerated disturbance damping of an unknown distributed system by nonlinear feedback, *AIChE J.*, **38**, 1461.

Christofides, P. D. and P. Daoutidis (1995). Feedback control of parabolic PDE systems, AIChE Ann. Mtg., paper 180b, Miami Beach, FL.

Christofides, P. D. and P. Daoutidis (1996a). Feedback control of hyperbolic PDE systems, *AIChE J.*

Christofides, P. D. and P. Daoutidis (1996b). Nonlinear control of diffusion-convection-reaction processes, *Comp. Chem. Engng.*, Special issue for ESCAPE-6, to appear.

Christofides, P. D. and P. Daoutidis (1996c). Robust control of hyperbolic PDE systems, *Chem. Eng. Sci.*, submitted.

Dochain, D. and B. Bouaziz (1993). Approximation of the dynamical model of fixed-bed reactors via a singular perturbation approach, *Proc. of IMACS International Symposium MIM-S2 '93*, 34.

Foias, C. and M.S. Jolly and I.G. Kevrekidis and G.R. Sell and E.S. Titi (1989). On the computation of inertial manifolds, *Physics Letters A*, **131**, 433.

Gay, D. H. and W. H. Ray (1995). Identification and control of distributed parameter systems by means of the singular value decomposition, *Chem. Eng. Sci.*, **50**, 1519.

Hanczyc, E. M. and A. Palazoglu (1995). Sliding mode control of nonlinear distributed parameter chemical processes, *I & EC Res.*, **34**, 455.

Hanczyc, E. M. and A. Palazoglu (1996). Use of symmetry groups in sliding mode control of nonlinear distributed parameter systems, *Int. J. Contr.*, in press.

Keulen, B. (1993). H_∞-control for distributed parameter systems: A state-space approach, Birkhauser, Boston.

Ray, W. H. (1978). Some recent applications of distributed parameter systems theory - A survey, *Automatica*, **14**, 281.

Ray, W. H. (1981). Advanced process control, McGraw-Hill, New York.

Sira-Ramirez, H. (1989). Distributed sliding mode control in systems described by quasilinear partial differential equations, *Syst. & Contr. Lett.*, **13**, 177.

Ydstie, B. E. and P. V. Krishnan (1994). From "Thermodynamics to a macroscopic theory for process control," AIChE Ann. Mtg., paper 228a, San Francisco, CA.

OPERATING REGIME-BASED CONTROLLER STRATEGY

Karlene A. Kosanovich and James G. Charboneau
Department of Chemical Engineering
University of South Carolina
Columbia, SC 29208

Michael J. Piovoso
E.I. duPont de Nemours and Company
Wilmington, DE 19880-0101

Abstract

A hybrid, dynamical supervisory control strategy is introduced to handle transition control of multi-product processes. The scheme involves the selection of the best setpoint controller, from an existing family of feedback controllers so as to cause the output to track the desired setpoint. The performance of the supervisory control strategy is demonstrated on two examples. For both, linear controllers are designed for the operating regions, and the role of the supervisory control strategy will be to determine the switching logic as different scheduling policies are demanded.

Keywords

Switching signal, Grade changeovers, Transition control, Multiple product processes.

Introduction

To meet the growing demands of a global economy, many industries strive to remain competitive by producing multiple grades of a large volume, high value product using the same process equipment. One such example, is the production of different molecular weight grades of polyethylene, others include styrene and olefin production. Chemical reactors, in particular polymer reactors, are typically multivariable, exhibit nonlinear characteristics, and often have significant time delays. Lost production capacity and off-spec product, associated with reactor grade transitions, are major costs associated with polymer production.

Because of the multi-product nature, the inherent complexity of chemical processes, and multiple objectives such as safety and environmental constraints, minimum creation of off-spec product, downtime, feedstock waste, and costly energy consumption, the control of multi-product processes is difficult. This necessitates the design of a robust controller strategy that can regulate the process successfully, not only at the particular operating point but also in the regions of transition, since poor transition control leads to periods of transient operation that can be costly.

Recently, (Banerjee, Arkun, Pearson and Ogunnaike, 1994) proposed controlling a 2-state continuous-stirred tank reactor (CSTR) using multiple linear models and a scheduling strategy. (Schott and Bequette, 1995) suggest the use of a model-based control strategy that incorporates a set of models plus controllers to address the control of nonlinear processes with widely varying dynamics. Their technique is similar to that of Banerjee et al., in that a probability function is used to discriminate among models to determine one that best fits the current operation. Finally, (Morse, 1993) discusses the design of a supervisory control scheme that selects the best feedback controller, from a set of linear controllers to cause a single-input single-output linear system (SISO), to track a setpoint exactly even if bounded, constant process disturbances are present. The feedback controller strategy to be investigated here will build upon the works cited above and on the development of a dynamic, supervisory scheme for switching to appropriate controllers.

Problem Formulation

The objective of this work is to develop a methodology which will be able to switch properly into feedback control, one from a set of potentially applicable controllers, to control a process over a wide range of operating conditions. It is proposed to achieve this by: (1) defining a set of appropriate linear dynamic time-invariant systems, each of which models the process within a certain operating region, that together are sufficient to represent the process performance over the entire operating space; (2) designing a setpoint controller, one for each nominal model, that will control the process to a desired set of specifications *within* the corresponding valid region; (3) defining a state space model, that is capable of simultaneously producing the controller output and an estimate of the plant/model mismatch for each of the models, so that control performance can be assessed without actually putting each controller in feedback with the process; and (4) defining an appropriate measure of the plant/model mismatch error: the difference between the actual nonlinear process output and a *filtered*, predicted output.

Consider, that the multi-product process can be represented by a suitable set of local, linear models $C(p)$ or linear in the parameter models centered at each operating point and that there is overlap between the receptive field of each local model. Let the complete representation be in the union of a number of subclasses defined over a finite, real-valued parameter set, P, $C(P) = \bigcup C(p)$, $p \in P$, the parameter space of $C(p)$'s. Each subclass contains a nominal process model $\nu_p = \alpha_p(s)/\beta_p(s)$, about which the subclass is centered and there exists for each subclass a controller, $\kappa_q = \gamma_q(s)/\rho_q(s)$, which would solve the positioning control problem were the process sufficiently well-modeled by a member of that subclass. The pair, $\{\nu_p, \kappa_q\}$, defines a system of two outputs: the pth model output, y_m, and the qth controller output, v_q.

Define the norm of the pth plant/model mismatch error to be $\|e_p\| = \|\hat{y}_p - y\|$ where $\hat{y}_p$ is the filtered predicted response one for each value of $p \in P$ and y is the true nonlinear performance response. This error will be used to assess the appropriateness of the pth model to represent the current process operation. The outputs of the controllers together with the actual process output are used by a suitably constructed supervisor controller to assess the potential closed-loop performance should the qth controller be switched into feedback with the process. Such a supervisor controller is designed to be a specially structured causal, dynamic system whose output switching signal, σ, and whose inputs are the states, x_c, of an expanded linear system that contains the information about the plant/model mismatch errors and the controller output. This intelligent design is intended to avoid an exhaustive or round robin search or intentional excitation to a current operating process to discover which setpoint controller, κ_q, is most appropriate during transitions at or near the new operating point.

Simple Example

A simple example is outlined to illustrate the performance of the proposed control strategy. Consider the following feedback SISO system shown in Fig. 1. The transfer function models, ν_p, that represent the time-invariant, linear process at two different operating points are

$$\frac{\alpha_1}{\beta_1} = \frac{s+5}{(s+4)(s+2)} \quad \frac{\alpha_2}{\beta_2} = \frac{s+5}{(s+1)(s+7)} \quad (1)$$

where it is assumed that β_p is monic and that α_p and β_p are coprime. A linear feedback controller, κ_q, can be designed, one at each operating point, so that if model ν_p is the best representation of the process, then its associated feedback controller κ_q is the one that should be placed in feedback so that the output of the process tracks the reference signal. Assume two such controllers are:

$$\frac{\gamma_1}{\rho_1} = \frac{10(s+2)}{s+6.5} \quad \frac{\gamma_2}{\rho_2} = \frac{s+7}{s+10} \quad (2)$$

and κ_q is a reduced and proper rational function. Integral action, a common part of both controllers is implemented separately as shown in Fig. 1.

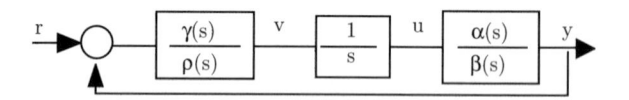

Figure 1. Feedback system.

The controllers are designed using a pole-placement method (Aström and Wittenmark, 1984) where the control law (error feedback) is given as

$$u = (r - y)\frac{\gamma_q(s)}{\rho_q(s)} \quad (3)$$

and r is the reference input, y is the controlled output, and $\gamma(s)$ and $\rho(s)$ are polynomials to be determined so that a pre-specified closed-loop transfer function

$$\frac{B_m(s)}{A_m(s)} = \frac{\alpha_p(s)\gamma_q(s)}{s\beta_p(s)\rho_q(s) + \alpha_p(s)\gamma_q(s)} \quad (4)$$

is obtained. The control law given by Equation (3) will be *proper* if and only if the degree of the denominator polynomial is greater than or equal to the degree of the numerator polynomial, that is, $\delta(\rho_q(s)) \geq \delta(\gamma_q(s))$ where $\delta(\cdot)$ symbolizes the degree of the polynomial.

A realizable, pole-placement controller design with good disturbance rejection properties, places restrictions on the choices of the polynomials, $\gamma(s)$ and $\rho(s)$. Factor $\alpha_p(s)$ into unstable and stable zeros respectively, $\alpha_p(s) = \alpha_p^-(s)\alpha_p^+(s)$. Observe that $\alpha_p^-(s)$ must remain a factor of $B_m(s) = \alpha_p^-(s)B_m'(s)$, to avoid generating unstable poles in the controller. Since $\alpha_p^+(s)$ will be canceled by the poles of the controller (see Equation (4)), it must be a factor of $\rho_q(s) = \alpha_p^+(s)\rho_q'(s)$. Making these substitutions, Equation (4) becomes

$$\frac{\alpha_p^-(s)B_m'(s)}{A_m(s)} = \frac{\alpha_p^-(s)\alpha_p^+(s)\gamma_q(s)}{\alpha_p^+(s)\left(s\beta_p(s)\rho_q'(s) + \alpha_p^-(s)\gamma_q(s)\right)} \quad (5)$$

In addition to $\alpha_p^+(s)$, there may be other factors in the numerator and denominator of the right hand side

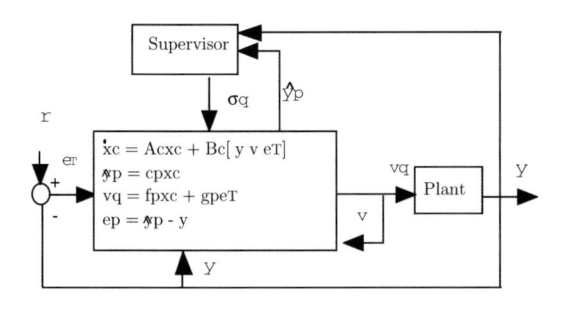

Figure 2. Supervisory controller scheme.

of Equation (5) that may have common factors that will cancel; let this term be $A_0(s)$. Following Åström and Wittenmark (1984) this is called the *observer polynomial*. The conditions on the choices of $A_m(s)$, $B_m(s)$, and $A_0(s)$ are given by the following theorem.

Theorem 1: There exists a causal solution to the pole-placement design if
(1) $\delta(A_m(s)) - \delta(B_m(s)) \geq \delta(s\beta_p(s)) - \delta(\alpha_p(s))$
(2) $\delta(A_0(s)) \geq 2\delta(s\beta_p(s)) - \delta(A_m(s)) - \delta(\alpha_p^+(s)) - 1$
where $A_m(s)$ defines the desired closed loop poles and $B_m(s)$ the zeros. The zeros of $\beta_p(s)$ are the process poles, those of $\alpha_p(s)$ are the process zeros, $A_0(s)$ is the observer polynomial, and $\alpha_p^+(s)$ defines the canceled stable zeros.

Intuitively, the theorem implies that any delays in the closed-loop system must be at least as large as the delay in the process, and the degree of the observer polynomial must be sufficiently high if a causal control law is to be found. The proof can be found in (Kosanovich, Charboneau and Piovoso, 1996).

Next construct a transfer function with the following properties: (1) under ideal conditions, if the transfer function of the process was the same as the nominal process model transfer function, ν_p, then the plant/model mismatch error, e_p, tends to zero as fast as $e^{-\mu_S t}$, where μ_S is a positive number that typifies the largest zero of a designed polynomial, and (2) the transfer function between the tracking error, e_T, and the qth controller output, v_q, under feedback connection will be κ_q after cancellation of common factors.

Let $T_{pq}(s)$ be that transfer function matrix that has as inputs $[y\ e_T\ v]'$, and produces as outputs, e_p, the plant/model mismatch error and v_q, the qth candidate control signal,

$$T_{pq}(s) = \begin{pmatrix} \frac{-s\beta_p\theta_p}{\omega_\nu} & \frac{\alpha_p\theta_p}{\omega_\nu} & 0 \\ 0 & \frac{\omega_\kappa - \rho_q\psi_q}{\omega_\kappa} & \frac{\gamma_q\psi_q}{\omega_\kappa} \end{pmatrix} \quad (6)$$

For a unique realization, $\omega_\nu(s)$ and $\omega_\kappa(s)$ are chosen to be stable, monic polynomials such that the $\delta(\omega_\nu(s))$ is at least the $\delta(s\beta_p\theta_p)$, and the $\delta(\omega_\kappa(s))$ is at least the $\delta(\rho_q\psi_q)$. $\theta_p(s)$ and $\psi_q(s)$ are the observer polynomials required to obtain a proper transfer function.

A minimal realization of $T_{pq}(s)$ is obtained by first selecting SISO controllable pairs $\{A_\nu, b_\nu\}$, $\{A_{\nu\kappa}, b_{\nu\kappa}\}$, and $\{A_\kappa, b_\kappa\}$ in such a way that $\omega_\nu(s)$ and $\omega_\kappa(s)$ are the characteristic polynomials of A_ν and A_κ respectively; while the *least common multiple* (LCM) of $\{\omega_\nu(s), \omega_\kappa(s)\}$ is the characteristic polynomial of $A_{\nu\kappa}$.

Next, define A_c to be a block diagonal matrix, $\{A_\nu, A_{\nu\kappa}, A_\kappa\}$, and B_c to be a three column matrix defined by the block diagonal matrix, $B_c = [\{b_\nu, 0, 0\}\ \{0, b_{\nu\kappa}, 0\}\ \{0, 0, b_\kappa\}]$. This defines a realization with $[y\ v\ e_T]'$ as inputs and $[\hat{y}_p\ v_q]'$ as outputs,

$$\begin{aligned} \dot{x}_c &= A_c x_c + B_c [y\ v\ e_T]' \\ \hat{y}_p &= c_p x_c \\ v_q &= f_p x_c + g_p e_T \\ e_p &= \hat{y}_p - y \end{aligned} \quad (7)$$

where x_c is a shared control state that circumvents the exhaustive evaluation of every controller in feedback. Lastly, vectors $[c_p\ f_q]$ and scalar g_p are found as unique solutions to

$$\begin{pmatrix} c_p \\ f_q \end{pmatrix}(sI - A_c)^{-1} B_c + \begin{pmatrix} -1 & 0 & 0 \\ 0 & 0 & g_p \end{pmatrix} = T_{pq}(s) \quad (8)$$

for each process and associated controller. The solution to Equation (8) is a well-known problem in elementary algebra where the idea is to find two polynomials X and Y that satisfy the *Diophantine equation*[†].

The overall goal is to develop a finite dimensional, dynamical controller, whose output is a piecewise-constant switching signal, σ, that takes values in p, with inputs x_c, the shared controller state, and the measured output of the process, y. The value of σ at each instant of time indicates which controller, κ_q, should be used as the feedback controller.

The supervisor controller generates a signal that rates the expected closed-loop performance of the qth controller, one for each $p \in P$. This is accomplished by determining which of the process models is most closely associated with the current plant performance. The switching logic that is employed is a function of two additional parameters, a dwell time, T_D, and a computation time, T_C, that is at most T_D. Define the event time, ΔT, the elapsed time between T_C and T_D, as the time when the closed-loop performance is evaluated and used to find a candidate σ_q that would provide the best closed-loop performance during the time interval $\Delta T < t < T_D$.

Two possible outcomes may result. First, if a different controller produces an improvement over the current controller then this new controller is switched into feedback to replace the existing controller. Second, if there is no improvement, initiate another evaluation of the closed-loop performance. Observe that this process involves both continuous and discrete time scales; hence the *hybrid* nature of the finite dimensional, dynamical controller. Fig. 2 illustrates the proposed closed-loop supervisory control scheme.

For the simple problem defined in Equation (1), $\delta(\omega_\kappa(s)) = \delta(\rho_q(s)) = 1$ and $\delta(\omega_\nu(s)) = \delta(\beta_p(s)) + 1 =$

[†] Aryabhatta's identity

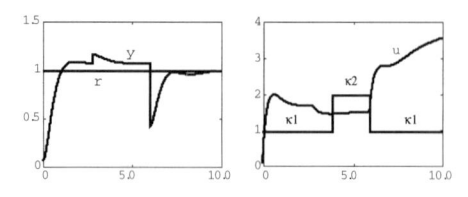

Figure 3. $\nu_1 \to \nu_2 \to \nu_3$.

3. This implies that $\theta_p(s)$ and $\psi_q(s)$ both have degree 0 and Theorem 1 is satisfied. The closed-loop performance using this switching control strategy is demonstrated on the simple example. A third operating point in addition to the two previously defined in Equation (1) is included. No controller is designed for this third operating point where the transfer function is $\nu_3 = \frac{s+0.5}{s^2+2s+2}$.

Fig. 3 illustrates the process performance for a transition policy of $\{\nu_1 \to \nu_2 \to \nu_3\}$ at $t = 2.75$ and $t = 6$ time units, respectively. Observe that κ_2 is selected as the best controller that minimizes the performance function at operating point 2 but when operating point 3 is specified, the supervisor selects κ_1 over κ_2 as the best controller, which is the correct choice.

3-State Exothermic Reactor

The open-loop behavior of a 2-state exothermic continuous stirred tank reactor is known to exhibit output multiplicities (Uppal, Ray and Poore, 1974). In more recent work, (Russo and Bequette, 1992) have shown the importance of including jacket dynamics for this system. With this additional state, the reactor has been observed to exhibit chaotic behavior. This work uses the 3-state system (ECSTR) studied by (Russo and Bequette, 1992) to demonstrate the proof-of-concept of this transition control approach.

The dimensionless component and energy balances that describe this system are given by

$$\begin{aligned}
\dot{x}_1 &= -\phi x_1 \mathcal{K}(x_2) + q(x_{1f} - x_1) \\
\dot{x}_2 &= \beta\phi x_1 \mathcal{K}(x_2) - (q+\xi)x_2 + \xi x_3 + q x_{2f} \\
\dot{x}_3 &= \xi_1 u(x_{3f} - x_3) + \xi\xi_1\xi_2(x_2 - x_3) \\
\mathcal{K}(x_2) &= exp(\frac{x_2}{1+\frac{x_2}{\gamma}})
\end{aligned} \quad (9)$$

The product of this 3-state system can be placed in the framework of transition control if one considers the three operating regimes as producing three different product grades, as reflected by the conversion. In this work, the parameters ($\phi = 0.072, \gamma = 20, \beta = 8, \xi = 0.3, \xi_1 = 1, \xi_2 = 1, x_{1f} = 1, x_{2f} = 0$, and $x_{3f} = -1$ are identical at the three operating points) produce approximate conversion levels of 10%, 47%, and 80%. From a conversion viewpoint, the third operating point is attractive, however, this operation is at or very near the safety design constraints on the cooling system. Hence, the middle operating regime is desirable, even though it is open-loop unstable.

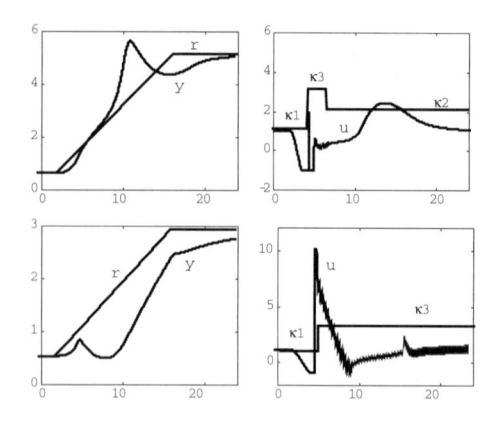

Figure 4. top: 10% $\longrightarrow$ 80%; bottom: 10% $\longrightarrow$ 47%.

The transformed linearized system about the three operating points yields third order transfer functions. Using pole placement techniques, appropriate linear controllers are designed. The hybrid dynamical supervisor controller is implemented for a scheduling policy of 10% $\longrightarrow$ 80% and 10% $\longrightarrow$ 47%. Fig. 4 illustrates the process and controller performance for both transitions. In the first changeover (top panel), the supervisor switches from the controller designed for the 10% conversion to the controller designed for the 47% conversion to the controller designed for the 80% conversion. In the second case (bottom panel), while there is a 6% error in the controlled variable, y, the supervisor correctly switches to the controller designed for the 47% conversion case. There is however, considerable chattering in the manipulated variable which may be removed by improving the linear controller design.

References

Aström, K. J. and Wittenmark, B. (1984). *Computer-Controlled Systems: Theory and Design*, 2nd ed., Prentice-Hall, Engelwood Cliffs, NJ.

Banerjee, A., Arkun, Y., Pearson, R. K. and Ogunnaike, B. A. (1994). Robust nonlinear control by scheduling multiple model based controllers, *AIChE Nat. Mtg.*, paper no. 230a, San Francisco, CA.

Kosanovich, K. A., Charboneau, J. G. and Piovoso, M. J. (1996). Operating regime-based controller strategy for multi-product processes. Submitted to *J. Proc. Contr.*

Morse, A. S. (1993). Supervisory control of linear set-point controllers – part I: Exact matching. Submitted to *IEEE Trans. on Auto Contr.*

Russo, L. P. and Bequette, B. W. (1992). CSTR performance limitations due to cooling jacket temperature dynamics, *DYCORD$^{+'}$92 IFAC Symp.*, College Park, MD, June.

Schott, K.D. and Bequette, B. W. (1995). Control of chemical reactors using multiple-model adaptive control (mmac), *DYCORD$^{+'}$ IFAC Symp.*, Helsingør, Denmark, June, 345-350.

Uppal, A., Ray, W. H. and Poore, A. B. (1974). On the dynamic behavior of continuous stirred tank reactors, *Chem. Eng. Sci.*, **29**, 967.

APPLICATION OF MULTI-UNIT CONTROL ANALYSIS AND DESIGN TO A REACTOR/SEPARATION PROCESS

Yudi Samyudia, Peter L. Lee, Ian T. Cameron
Chemical Engineering Department
The University of Queensland
Brisbane, Queensland 4072, Australia

Michael Green
Department of Engineering and Department of Systems Engineering
Research School of Information Sciences and Engineering
The Australian National University, Australia

Abstract

This paper addresses the control analysis and design of a reactor/separation process. It is shown that the application of single-unit controllers to this process results in poor overall closed-loop performance due to the presence of inter-unit interactions through recycle streams. In this paper, a multi-unit control analysis is proposed to systematically determine the best control strategy for this process.

Keywords

Control system design, Gap metric, Decentralized control.

Introduction

The applications of numerous innovations and advancements in model-based control have been primarily aimed at single processing units. However, any modern chemical process plant is made up of multiple processing units (Downs and Vogel, 1993; Ramachandra et al 1992). Multi-unit processing plants are commonly coupled and interactive because of material and energy connections between units through recycle streams. Current strategies for controlling a coupled multi-unit processing plant rely upon treating the whole plant as a *large* single unit. As a result, a centralized controller is produced to control the whole plant. This might not be the only way of controlling a multi-unit processing plant. As indicated in (Luyben, 1993), a more systematic multi-unit control analysis and design is required to determine the best control strategy for such processes.

Recently, a systematic procedure to multi-unit control analysis and design has been developed (Samyudia et. al.

1994; Samyudia et. al. 1995a). This procedure uses some quantitative indicators (Samyudia et. al. 1995b) to predict the stability and achievable closed-loop performance of alternative control configurations. Based on these indicators, the best control strategy can be determined. This paper presents the application of the developed procedure to a reactor/separation process.

A Reactor/Separation Process

A reactor-separation process shown in Fig. 1 is studied in this paper. The process is comprised of the reactor system and the separation system. The reactor system is a continuous stirred tank reactor (CSTR), and is responsible for facilitating the desired reaction. The separation system is made up of an extractor unit, a flash-drum unit, and a distillation column. The separation system is responsible for extracting the by product and separating the product from the reactant.

A first-order, endothermic, irreversible reaction A+B → P+G occurs in the liquid phase in the reactor at 80°C. It is assumed that the component B is all consumed, while some of the component A is unreacted. The reactor feeds are the pure B feed, F_2 (15 kmol/hr) and the recycle stream, which is a mixing of the recycle component A from the bottoms of the distillation column and the pure A feed, F_1 (15 kmol/hr). The recycle flowrate F_4 at steady state is 22.5 kmol/hr, and its composition is x_{A4} (mole fraction A). The specific reaction rate k_r is 1 hr^{-1}, and the reactor hold-up is controlled by a perfect controller at a steady state value of 75 kmol. Since the reaction is endothermic, the reaction can only occur if the required heat is supplied to the reactor. In this process, steam is circulated through the reactor jacket to give the required heat. The steam flowrate S at steady state is 4654.2 m^3/hr, and its inlet temperature T_{Jo} is 280°C. The jacket volume V_J is 2.5m^3, and its temperature T_J at steady state is 154°C.

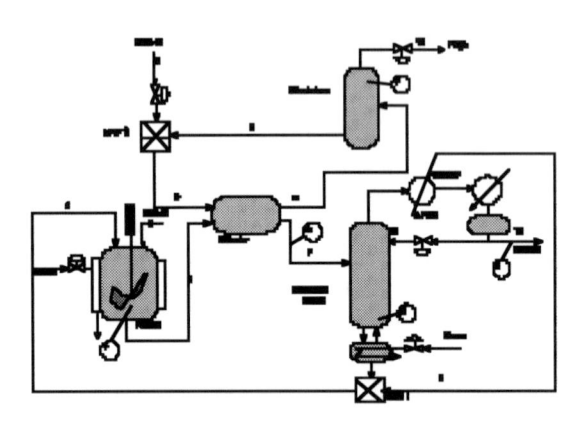

Figure 1. A Reactor-separation process.

The reaction produces a by-product G in addition to the product P. The by-product G needs to be extracted and a single stage extractor is used. The solvent C is used to extract the by-product G in the extractor. After extraction, the by product G is then separated out as a purge in the flash-drum unit. The solvent C is recycled back and mixed with the fresh make-up, replacing the small amount of C lost with the purge.

The extractor raffinate F_7 can be treated as a binary mixture of A and P. The binary mixture is passed through the distillation column to separate the product P from the reactant A. The vapor coming out at the top of the column is used to preheat the fresh A feed before the vapor is condensed by a condenser. The product P is drawn as the distillate with flowrate F_{10} (30 kmol/hr), its composition at steady state is 0.9999 (mole fraction P). The bottoms product is rich in the reactant A. Its flowrate and composition at steady state is 7.5 kmol/hr and 0.9991 (mole fraction A).

In this study, five controlled variables are considered : the reactor temperature; the raffinate composition; the distillate composition; the bottoms composition; and the flash-drum's pressure. The manipulated used are : the steam flowrate; the make-up solvent flowrate; the reflux flowrate; the boil-up rate; and the purge flowrate. The control design objective is to have a settling time within 1.5 hr for all controlled variables.

A Procedure for Multi-Unit Control Analysis and Design

A three step procedure is proposed for multi-unit control analysis and design. The first step is to decompose a multi unit processing plant into subsystems of smaller dimension. This approach was motivated by the high dimensionality and complexity of multi-unit processing plants. Additionally, this approach produced a decentralised control system for the whole plant, which is popular in the chemical process industries. The second step is to analyze the achievable closed-loop performance when the controller is designed for this plant decomposition. The third step is to design the controller for each subsystem such that the overall performance objectives are satisfied.

To implement the proposed procedure, two approaches to plant decomposition are proposed. The first technique is based on the physical unit operations. By using this technique, the effect of recycle streams on the overall performance can be analyzed such that the performance of single-unit controllers applied to a multi-unit processing plants can be investigated. The second technique is based on the dynamics of the variables to be controlled, with no regard to the physical units in which these variables occur. Hence, alternative plant decompositions might be produced for a multi-unit processing plant from which the best plant decomposition could be determined.

Furthermore, a method for analyzing interactions is required to analyze the achievable performance of a decentralized controller design for a plant decomposition. In this work, the method of (Samyudia et. al. 1995b) is used for this purpose. Two indicators, namely the maximum stability margin b_{max} and the gap β, are calculated for screening alternative plant decompositions. By observing these indicators, the stability and achievable performance of decentralized control could be predicted. Hence, the best plant decomposition could be determined simply by observing whether or not the gap β is less than b_{max}, and a lower value of β is better.

The application of this procedure to a multi-unit processing plant would lead to the best control strategy using a decentralized structure in which the overall closed-loop performance is comparable to a centralized controller.

Analysis of Control Configurations for the Reactor/Separation Process

In this study, three plant decompositions are considered. The first is based on the physical unit operation (SA Mode) : the reactor; the extractor; the distillation column;

and the flash drum. The others are decomposition across units (DAU1 and DAU8 Modes). In the DAU1 mode, there are two subsystems: the reactor as the first subsystem; the extractor, the flash drum, and the distillation column as the second subsystem. In the DAU8 mode, three subsystems are produced : the reactor as the first subsystem; the extractor and the flash-drum as the second subsystem; and the distillation column as the third subsystem.

In terms of controller complexity, the SA mode produces three SISO controllers and a 2x2 multivariable controller, while the DAU1 mode has a SISO controller and a 4x4 multivariable controller. In the DAU 8 mode, a SISO controller and two 2x2 multivariable controllers are applied to the process. Hence, the DAU1 mode produces the more complex controller than the other modes.

The analysis results using the indicators, β and b_{max} for these plant decompositions are presented in Table 1. These results indicate that the DAU1 and DAU8 modes are predicted to produce a better performance than the SA mode. This implies that single-unit approach to design the controller for this process does not produce the best overall performance. Also, this analysis demonstrates that the performance of the DAU1 mode is similar to that of the DAU8 mode. This means that increasing the controller complexity in a subsystem does not always guarantee a better overall performance. Rather, the way of decomposing the system would lead to the best control strategy.

Table 1. Results of Interactions Analysis.

Indicator	SA	DAU1	DAU8
β	0.883	0.232	0.232
b_{max}	0.382	0.382	0.382

Further, these results also indicate that interactions between the extractor and the flash-drum units are strong. Separating these units in the controller design significantly deteriorates the overall performance (compare the DAU8 mode with the SA mode). In contrast, the interactions between the reactor and other units are not strong. Thus, even though it is separated from other units, it does not deteriorate overall performance (compare the DAU8 mode and the DAU1 mode). The multi-unit control analysis can therefore identify whether or not a unit operation can be separated into a different subsystem.

Hence, the DAU8 mode is selected as the best control strategy for this process.

The closed-loop responses to 20% change in the A Feed flowrate are given in Fig. 2 to confirm the performance prediction, where comparison is made to a control structure (centralized control) which treats the entire plant as one unit to be controlled.

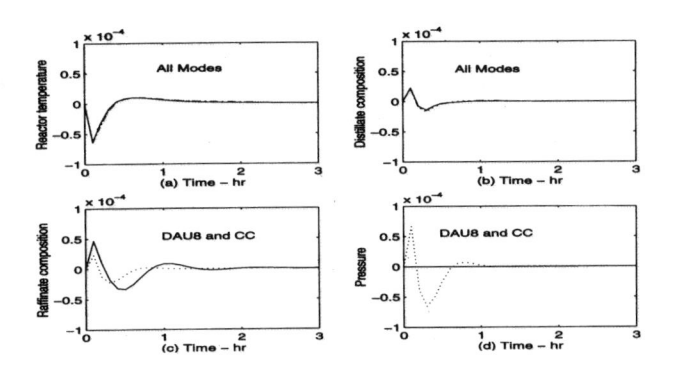

Figure 2. Response to 20% step change in A Feed flowrate. (a and b) Responses of all modes perfectly fit that of a centralized controller; (c and d) Responses of DAU1, DAU8 and a centralized controller (__); responses of SA Mode (....).

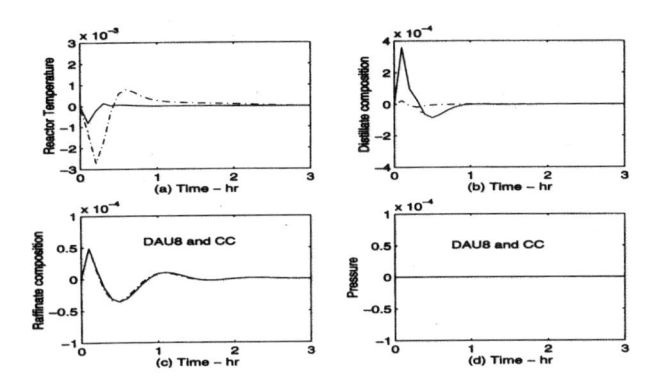

Figure 3. Response to 20% step change in A Feed flowrate when a less tight controller is applied to the reactor temperature. (a and b) Responses of DAU8 (-.-.) and centralized controller (__); (c and d) Responses of DAU8 and a centralized controller are identical.

Furthermore, the control study aims to investigate the effect of the reactor temperature's controller on the overall performance. For the DAU8 mode, a less tight controller is applied to the reactor temperature. The multi-unit analysis produces the gap β value of 0.996 for this change. This means that a poor overall performance is produced when a less tight controller is applied to the reactor temperature. Hence, a tight controller must be used for the reactor temperature.

The closed-loop simulations are shown in Fig. 3. These responses show that the overall performance of the DAU8 mode deviates significantly from a centralized control system as predicted by the proposed analysis method.

Conclusions

This paper has investigated alternative feedback control configurations for the reactor/separation process. It has been shown that the overall performance of a decentralized controller can be comparable to a centralized controller if the right plant decomposition is applied to the process and the performance specification is posed realistically. The developed procedure for multi-unit control analysis and design has been demonstrated to be a useful method that could systematically determine the best control strategy for the reactor/separation process.

References

Downs, J.J., and E.F. Vogel (1993) A plant wide industrial control problem. *Comp. Chem. Eng, 17, 245-255.*

Green, M. and D.J.N. Limebeer (1995) *Linear Robust Control*, Prentice Hall, NJ.

Luyben, W. (1993) Dynamics and control of recycle systems. *Ind. Eng. Chem. Res.,32, 466 -475.*

Ramachandra, B, J.B. Riggs, H.R. Heichelheim, A.F. Seibert, and J.R. Fair (1992) Dynamic simulation of supercritical fluid extraction processes. *Ind. Eng. Chem. Res., 31, 281-290.*

Samyudia, Y., P.L. Lee, and I.T. Cameron, (1994) A methodology for multi-unit processing plants. *Chem. Eng. Sci, 49, 23, 3871-3882.*

Samyudia, Y., P.L. Lee, I.T. Cameron, and M. Green, (1995a) Control of multi-unit processing plants : A new approach *AIChE Miami Beach Annual Meeting, Nov 13-17, Florida, USA.*

Samyudia, Y., M. Green, P.L. Lee, and I.T.Cameron, (1995b) A new approach to decentralised control design. *Chem. Eng. Sci., 50, 11, 1695-1706.*

OPERABILITY OF BATCH REACTORS: TEMPERATURE PROFILE FEASIBILITY

B. Wayne Bequette
Chemical Engineering Department
Rensselaer Polytechnic Institute
Troy, NY 12180

Abstract

General operability characteristics for a batch reactor with a single nth order exothermic reaction are presented. The focus is on how the desired reactor temperature profile determines the required jacket temperature profile. Analytical expressions are developed to find conversion and jacket temperature profiles, including the time and value of a jacket temperature "peak," when a reactor temperature is ramped to its final desired value. These results can be used to assess the operability of reactors that have jacket temperature constraints. A pharmaceutical reaction is used to illustrate the approach.

Keywords

Batch reactors, Scale-up, Operability, Feasibility, Reactor design.

Introduction

The motivation for this paper is the scale-up of jacketed exothermic batch chemical reactors from the laboratory (including heat-flow calorimeters, such as the Mettler RC1), to pilot-plant and full-scale production. Temperature control for laboratory reactors is typically easy because of high heat transfer area to reactor volume ratios, which do not require large driving forces for heat transfer from the reactor to the jacket. Pilot and full-scale reactors often have a limited heat transfer capability. A process development engineer will often have a choice of reactors when moving from the laboratory to the pilot plant. Kinetic and heat of reaction parameters obtained from the laboratory reactor, in conjunction with information on the heat transfer characteristics of each pilot plant vessel, can be used to select the proper pilot plant reactor.

Similarly, when moving from the pilot plant to manufacturing, a process engineer will either choose an existing vessel or specify the design criteria for a new reactor. A necessary condition for operation with a specified reactor temperature profile is that the required jacket temperature is feasible. As in the case of continuous process systems, the integration of design and control (or dynamic operability) is extremely important.

The objective of this paper is to provide a methodology to determine if a particular reactor can handle a desired temperature profile. The techniques developed can be used to determine, for example, the maximum rate that a reactor temperature can be ramped and still meet jacket temperature constraints. In this work we develop analytical expressions for jacket temperature and reactor conversion profiles, given the desired reactor temperature profile. The technique presented can be used as a shortcut approach to determine reactor temperature feasibility at the scale-up stage.

We make the following assumptions to obtain analytical expressions:

- Perfectly mixed liquid-phase reactor and jacket
- Single, nth order exothermic reaction
- Constant physical properties
- The jacket temperature is directly manipulated
- The "thermal mass" of the reactor wall is "lumped" and at the reactor fluid temperature

Dimensionless Model

It is convenient to work with a dimensionless model where the state variables are conversion and temperature

$$\frac{dx}{d\tau} = \kappa(\theta)(1-x)^n \quad (1)$$

$$Le\frac{d\theta}{d\tau} = \beta\kappa(\theta)(1-x)^n - \alpha(\theta - \theta_j) \quad (2)$$

and the dimensionless parameters and variables are defined in the notation section.

Reactor Temperature Trajectory

A common operating procedure is to ramp the batch reactor from an initial charge temperature to a final temperature where it is maintained constant (isothermal operation) for most of the batch time, as shown in Fig. 1. After the desired conversion is reached the temperature is then ramped down.

A successful batch reactor design assures that a desired temperature trajectory leads to a feasible jacket temperature trajectory. In this section we derive the required jacket temperature using a "perfect control" assumption.

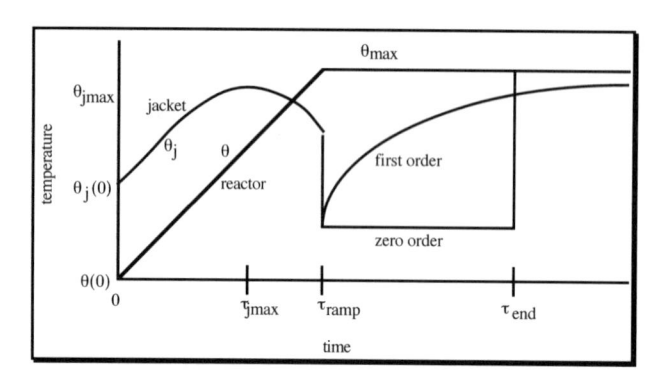

Figure 1. Typical temperature profiles for a linear ramp in reactor temperature.

A superscript (*) is used to represent a variable obtained assuming an ideal temperature trajectory. The desired temperature profile, $\theta^*(\tau)$, is specified, and the conversion, $x^*(\tau)$, is obtained by integrating Eqn. (1). The required jacket temperature, $\theta_j^*(\tau)$, is obtained by solving (from Eqn. (2))[1]

---

[1] Note the similarity between Eqn. (3) and reference trajectory approaches for control (Bequette, 1991; Kravaris and Kantor, 1990). It can be shown that this approach is identical to using input/output linearization with a perfect derivative controller (Bequette, 1995).

$$\theta_j^*(\tau) = \theta^*(\tau) - \frac{\beta}{\alpha}\frac{dx^*}{d\tau} + \frac{Le}{\alpha}\frac{d\theta^*}{d\tau} \quad (3)$$

The rate of change of the jacket temperature can be found by differentiating Eqn. (3).

For a linear ramp (from $\theta^*(0)$ at $\tau = 0$ to θ_{max}^* at $\tau = \tau_{ramp}$) followed by a constant temperature ($\theta^* = \theta_{max}^*$), the temperature profile is specified as $\theta^*(\tau) = \theta^*(0) + b\tau$ for $\tau \le \tau_{ramp}$ and $\theta^*(\tau) = \theta_{max}^*$ for $\tau \ge \tau_{ramp}$ where b is the desired slope. We assume that $\gamma \gg \theta^*$ (the positive exponential approximation) so $\kappa(\theta^*) \approx \exp(\theta^*)$ yielding the approximate relationship

$$\frac{dx^*}{d\tau} = \exp(\theta^*(0))\exp(b\tau)(1-x^*)^n \quad (4)$$

which can be integrated analytically. The reactor conversion and jacket temperature profiles are derived in Bequette (1995) and are not presented here due to space constraints. Typical temperature profiles are shown in Fig. 1. There are discontinuities in the jacket temperature at $\tau = 0$ (the jacket temperature is not equal to the initial reactor temperature) and $\tau = \tau_{ramp}$ (where the jacket temperature must instantaneously decrease). We use $\theta_j^*(\tau_{ramp-})$ and $\theta_j^*(\tau_{ramp+})$ to denote the values of jacket temperature before and after the discontinuities at τ_{ramp}. A peak in the jacket temperature of θ_{jmax}^* occurs at $\tau = \tau_{jmax}$.

For a maximum or minimum to occur in the jacket temperature, the derivative condition, $d\theta_j^*/d\tau = 0$, must be satisfied. Assuming minimal conversion (or a zero-order reaction), an extrema occurs at $\tau = \tau_{jmax}$, where

$$\tau_{jmax} = \frac{(\ln(\alpha/\beta) - \theta^*(0))\tau_{ramp}}{\theta_{max}^* - \theta^*(0)} \quad (5)$$

For τ_{jmax} to occur between $\tau = 0$ and $\tau = \tau_{ramp}$, then

$$\beta\exp(\theta_{max}^*) > \alpha > \beta\exp(\theta^*(0)) \quad (6)$$

If $\alpha < \beta\exp(\theta^*(0))$ then the heat released by the reaction is so great at the initial batch temperature that cooling must be initiated immediately. This is a particularly dangerous situation and requires redesign of the heat exchanger area (or decreasing the reactor size). Alternatively, semibatch operation may be used, where the reactant is added slowly to control the temperature release. For industrial-scale reactors $\beta\exp(\theta_{max}^*) > \alpha$ is generally satisfied, unless the heat of reaction is small (or negative, i.e. an endothermic reaction) or the heat transfer coefficient is large.

The maximum jacket temperature is

$$\theta^*_{jmax} = \ln(\alpha/\beta) + Le\, b/\alpha - 1 \qquad (7)$$

It may be desirable to specify a constraint on θ^*_{jmax}. For example, to satisfy $\theta^*_{jmax} \leq \theta^*_{max}$ the ramp time requirement is

$$\tau_{ramp} \geq \frac{Le(\theta^*_{max} - \theta^*(0))}{\alpha\,[\theta^*_{max} + 1 - \ln(\alpha/\beta)]} \qquad (8)$$

The minimum jacket temperature during the isothermal phase occurs as a discontinuous drop at $\tau = \tau_{ramp+}$

$$\theta^*_{jmin} = \theta^*_{max} - \frac{\beta}{\alpha}\exp(\theta^*_{max})\left.\frac{dx^*}{d\tau}\right|_{\tau=\tau_{ramp+}} \qquad (9a)$$

$$\left.\frac{dx^*}{d\tau}\right|_{\tau=\tau_{ramp+}} = \exp(\theta^*)(1 - x^*(\tau_{ramp}))^n \qquad (9b)$$

$$x^*(\tau_{ramp}) = 1 - \left(1 + \frac{(n-1)\exp(\theta^*(0))}{b}\{\exp(b\tau_{ramp}) - 1\}\right)^{1/1-n}$$

If a constraint on θ^*_{jmin} is specified, we can solve for the required τ_{ramp} from the following sequence

$$\theta^*_{jmin} \longrightarrow \left.\frac{dx^*}{d\tau}\right|_{\tau=\tau_{ramp+}} \longrightarrow x(\tau_{ramp}) \longrightarrow b \longrightarrow$$

τ_{ramp}

The only way to increase θ^*_{jmin} for a zero-order reaction is to increase α (by increasing the heat transfer area, for example) or decrease the desired isothermal temperature, θ^*_{max}. The ramp time can be used to change θ^*_{jmin} for first or higher order reactions.

Landau et al. (1994) suggested a temperature profile that is a cubic function of time and does not require discontinuities in the jacket temperature. Approximate analytical expressions for the jacket temperature extrema for a cubic reactor temperature profile are developed in Bequette (1995). A disadvantage is that a cubic profile with the same trajectory time as a linear profile must have a larger maximum jacket temperature.

Example

The batch reaction studied involves the attack of a cyclic sulfite by sodium cyanide in the presence of catalyst and solvent, yielding a hydroxynitrile compound, an intermediate which ultimately leads to a drug currently under investigation by Merck. Landau et al. (1994) used a Mettler RC1 laboratory heat flow calorimeter to study a 0.12 kg system. They used heat transfer information obtained from pilot plant data to simulate the required jacket temperature profiles for a 154.7 kg pilot plant reactor. The parameters for this first-order reaction are given in Landau et al. and Bequette (1995).

The initial and final temperatures are 20°C and 90°C. The minimum, maximum and end-of-ramp jacket temperatures as a function of ramp time for a linear ramp are shown in Fig. 2 for the reaction calorimeter (the maximum jacket temperature occurs at the end of the ramp) and Fig. 3 for the pilot plant reactor.

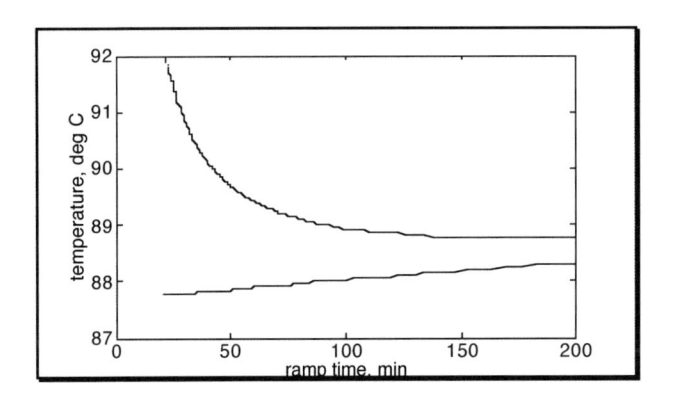

Figure 2. Temperature extrema for calorimeter as a function of linear ramp time.

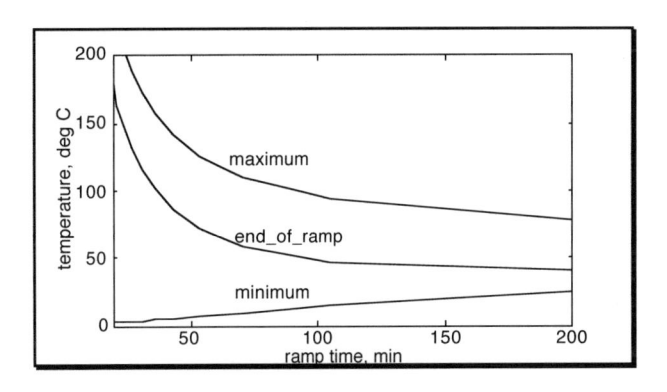

Figure 3. Temperature extrema for pilot plant reactor as a function of linear ramp time.

The minimum and maximum jacket temperature estimates as a function of trajectory time for a cubic trajectory are compared in Fig. 4. If the maximum jacket temperature constraint is 100 °C then a trajectory time of at least 140 minutes is required.

The linear and cubic trajectory responses for a 150 minute trajectory time are shown in Fig. 5. Notice that the cubic profile has a larger jacket temperature maximum, while the linear profile requires a lower minimum jacket temperature.

A manufacturing scale reactor typically has less favorable heat transfer capability, requiring a larger temperature gradient between the reactor and jacket. In this case a semi-batch reactor should be used. The operability of semi-batch reactors is presented in Bequette (1996).

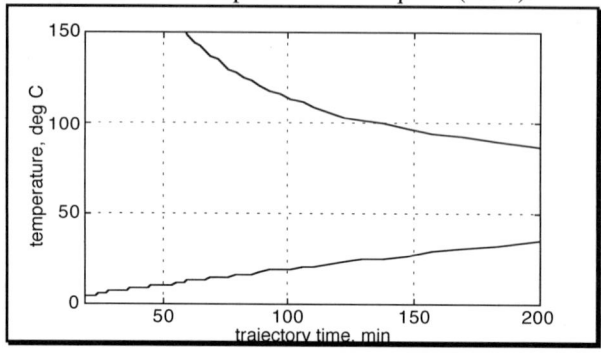

Figure 4. Temperature extrema for a cubic trajectory, pilot plant reactor.

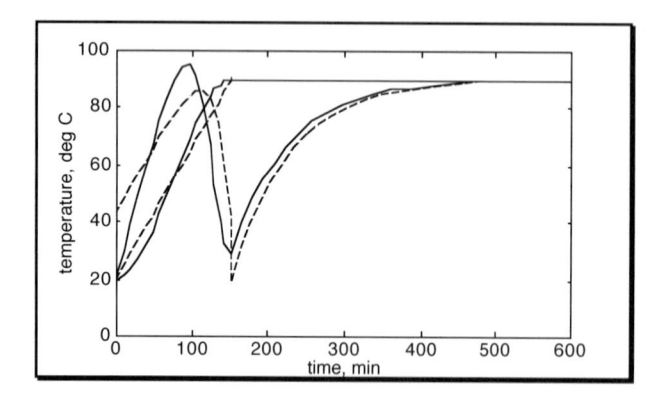

Figure 5. Comparison of linear (dashed) and cubic (solid) temperature profiles for a 150 minute ramp (or trajectory) time, pilot plant reactor.

Conclusions

In this paper we have shown how a desired batch reactor temperature trajectory determines the jacket temperature trajectory, assuming a perfect model. The approach provides first approximations to the minimum and maximum jacket temperatures, and hence the feasibility of a desired reactor temperature profile. The relationships can be used to scale-up to a reactor size that yields a desired jacket temperature profile characteristic. The methodology developed is meant to be a screening tool for batch reactor operability. Since analytical expressions are developed, it is easy to analyze the sensitivity of the jacket temperature profiles to uncertainties in the reactor system parameters.

A major disadvantage to a linear reactor temperature profile is that the required jacket temperature profiles are discontinuous. The use of a cubic temperature profile assures continuous jacket profiles, at the expense of a higher maximum jacket temperature than a linear profile with the same average slope.

Acknowledgment

This paper was written while on sabbatical at Merck Research Laboratories in Rahway, NJ. I gratefully acknowledge support from Merck and the NSF through the GOALI program, and the Petroleum Research Fund.

Notation

A_o = pre exponential factor
b = temperature ramp slope = $(\theta^*_{max} - \theta^*(0))/\tau_{ramp}$
C_A = reactor concentration
E_a = activation energy
$-\Delta H$ = heat of reaction
$k(T)$ = reaction rate constant = $A_o \exp(E_a/RT)$
Le = Lewis number = $1 + (mC_p)_w/(mC_p)_r$
$(mC_p)_r$ = heat capacity, reactor fluid
$(mC_p)_w$ = heat capacity, reactor wall
R = ideal gas constant
t = time
T = reactor temperature
T_o = reference temperature for dimensionless analysis
UA = heat transfer coefficient*area
V = reactor volume
V_j = jacket volume
x = conversion = $(C_{Ao} - C_A)/C_{Ao}$
α = dim. heat tran. coeff= $UA/((mC_p)_r k(T_o) C_{Ao}^{n-1})$
β = dim. heat of reaction = $\dfrac{-\Delta H \, V \, C_{Ao}}{(mC_p)_r} \dfrac{\gamma}{T_o}$
γ = dim. activation energy = E_a/RT_o
$\kappa(\theta)$ = dim rate constant = $\exp(\theta/(1 + \theta/\gamma))$
τ = dim. time = $k(T_o) \, t \, C_{Ao}^{n-1}$
τ_{ramp} = ramp time (for linear ramp)
τ_{jmax} = time for maximum jacket temperature
θ = dim. reactor temperature = $(T - T_o)\gamma/T_o$
θ_j = dim. jacket temperature = $(T_j - T_o)\gamma/T_o$

References

Bequette, B.W. (1991) Nonlinear Control of Chemical Processes: A Review. *Ind. Eng. Chem. Res.* **30**(7), 1391-1413 (1991).

Bequette, B.W. (1995). Understanding the Behavior of Batch Reactors. 1. Temperature Profile Feasibility, Dept. of Chem. Eng., Rensselaer Polytechnic Institute.

Bequette, B.W. (1996). Operability Analysis of an Exothermic Semi-Batch Reactor, *Comp. Chem. Eng.*, ESCAPE-6 special issue, Rhodes (in press, 1996).

Kravaris, C. and J.C. Kantor (1990). Geometric Methods for Nonlinear Process Control. I. Background, II. Controller Synthesis. *Ind. Eng. Chem. Res.*, **29**, 2295-2324.

Landau, R.N., D.G. Blackmond and H.-H. Tung (1994). Calorimetric Investigation of an Exothermic Reaction: Kinetic and Heat Flow Modeling. *Ind. Eng. Chem. Res.*, **33**, 814-820.

PROCEDURAL CONTROL OF CHEMICAL PROCESSES

A. Sanchez, G.E. Rotstein, N. Alsop, and S. Macchietto
Centre for Process Systems Engineering
Imperial College
London SW7 2BY, UK

Abstract

Procedural control is a formalism recently introduced to consider feedback control concepts in event–driven processes and to facilitate the automatic generation of sequential control code. This work summarises the control paradigm and a method for the synthesis of procedural controllers. It is shown how a provably correct and "optimal" control law (i.e. a procedural controller) is synthesised from discrete-event models of the process and desired behaviour. The translation of the control structure into sequential control code is also discussed. The synthesis method is applied to the automation of a CIP operation in a batch pilot plant located at Imperial College.

Keywords

Procedural control, Automation, Discrete event systems.

Introduction

The operation of chemical plants involves a large number of event–driven activities including plant start–up and shut–down, execution of emergency procedures and equipment interlocking. In automated plants, these activities are normally executed by control devices with logic and numerical processing capabilities (such as PLCs or DCSs). The generation of the control logic and the associated programming code required by these devices is a key stage in the automation of a process. The software must be able to guarantee both the correct normal operation of the plant and the safe handling of abnormal situations. Despite the extent to which procedural control software is used in practice, limited theoretical frameworks exist to support its formal analysis and design. The need for such a framework as well as formal methods and tools for synthesis, verification and validation has been stressed by academics and industrialists (Schuler, Allgower and Gilles, 1991). The methodologies should help to guarantee that the control logic and software performs according to specification and is free from errors that compromise the safe operation of the system. This is particularly relevant in flexible production environments, characterised by frequent changes in product recipes, production modes and equipment configuration. This paper outlines a framework for the synthesis of logic controllers for event–driven processes. The framework relies on the concept of *procedural controller*, a formal device for the representation of logic controllers (Rotstein, Sanchez and Macchietto, 1995). The process interacts with a procedural controller in feedback mode issuing responses corresponding to state and time events while the controller sends control commands. The procedural controller is synthesised using

- A process input–output model in which only event-driven actions are considered

- A description of the desired process behaviour captured as a set of logic statements

Both the normal and abnormal operation are considered in the synthesis procedure. The procedural controller thus synthesised can be viewed as an abstract representation of the control code required for the implementation device. Thus this approach complements current software engineering methods for control code generation. A realistic sized example is used to illustrate the concepts and method. Directions for future work are are also discussed.

Modelling Framework

The process is modelled as a labelled finite state machine (*FSM*)(Sanchez, 1996):

$$P = \{Q, V^{n_v}, \Sigma, \delta, q_0, Q_m\} \quad (1)$$

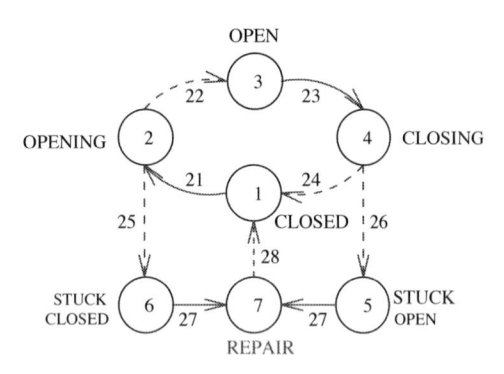

Figure 1. FSM logical model corresponding to an onoff valve.

where Q is the set of states ($q \in Q$) and V^{n_v} is the set of state-variables ($\{(v_j)_q ; j = 1 \ldots n_v$ (number of state-variables defining state q)$\}$). A state-variable describes an elementary component of the process (e.g. the position of an on/off valve or the status of a pump) and is characterised by a domain of possible values (e.g. {open, closed}). A system state is fully described at any instant by an ordered set of n_v state-variables. Σ is the set of transitions ($\sigma \in \Sigma$). Transitions are instantaneous events leading from a source state to a destination state which change only one state-variable value. Each transition σ is labelled as either controllable ($\sigma \in \Sigma_c$) or uncontrollable ($\sigma \in \Sigma_u$). Controllable transitions are associated with inputs to the process (e.g. sending a command to open a valve) while uncontrollable transitions are associated with the system responses (e.g. a signal indicating the position of the valve). δ is the state transition function defined as the partial function $\delta : \Sigma \times Q \to Q$. is the unique initial *state* and the set of marked states Q_m distinguishes states of special significance for the system. Events in process systems can, in principle, take place in parallel. However, in this work, events are considered to occur instantaneously and asynchronously. Parallelism is modelled by interleaving. That is, all possible sequences of events are considered. As an example consider an onoff valve with a FSM model as shown in fig 1. From the point of view of the controller, the valve can take the following logical values: valve *open* or *closed*, in the process of *opening* or *closing*, *stuck open* or *stuck closed* and *under repair*. From the valve open state, a control command (a continuous arrow) is issued to instruct the valve to close. Afterwards, the valve may either fail to open or fulfill the control instruction issuing the corresponding response (dashed arrows). If it fails and becomes stuck open, the next control command will drive the valve to the repair state but with the valve still open. Following repair, the valve returns to the closed state. Similar behaviour can arise when the valve is closed. If the valve fails to execute the opening action, it will reach a state in which it is stuck closed and a control command must be issued to repair it, eventually reaching the closed state again.

A process model of this nature is amenable to analytical manipulations supported by FSM Theory which are exploited in the synthesis of the controller. Although standard FSM theory does not support the notion of quantitative time, the proposed modelling framework has proven to be sufficiently rich to capture a considerable part of the phenomena encountered in event-driven processes. User requirements or process specifications within which the controlled process must be contained are expressed using logic formalisms. A consistent state-variable notation is introduced for use in both the FSM and logic domains. Methods for model construction either of the process or the user requirements have been developed (Sanchez, 1996) and are illustrated in the poster.

The Control Paradigm

In the feedback control setting employed here, a control device (the *procedural controller*) responds to a sequence of process events (outputs) by asynchronously issuing control commands (inputs) to the process. Thus, a procedural controller is modelled as a FSM $C = \{X, \Sigma, \gamma, x_0, X_m\}$, in which for each $x \in X$ such that $\gamma(\sigma, x)!$, one of the following is true:

1. $\sigma \in \Sigma_u \wedge (\forall \sigma_c \in \Sigma_c, \ \gamma(\sigma_c, x)$ is undefined).
2. $\sigma \in \Sigma_c \wedge (\forall \sigma'|(\sigma' \neq \sigma) \in \Sigma, \ \gamma(\sigma', x)$ is undefined). In other words, a procedural controller can either be in: 1) a state in which it is waiting for one of a set of uncontrollable transitions to take place (i.e. a "response" from the controlled process) or 2) a state in which it immediately forces the execution of the only defined controllable transition, anticipating the occurrence of any uncontrollable transition that can occur from that process state. The possible exclusion of behaviour generated by uncontrollable transitions from the closed–loop behaviour relies on the assumption that controllable transitions can be executed *before* uncontrollable transitions take place. Even though this assumption may appear to be strong, it is largely a modelling issue (as illustrated in the valve example). That is, even if a controllable transition has slow dynamics, it can always be divided into two transitions including a fast controllable transition representing the control command and an uncontrollable transition representing the slow system response to that command. The theoretical foundations of procedural control are presented elsewhere (Rotstein, Sanchez and Macchietto, 1995). A comparison with Supervisory Control Theory (Wonham, 1988) can be found in (Sanchez, Rotstein and Macchietto, 1995).

The Synthesis Method

The Procedural Control paradigm has been used successfully for the specification and synthesis of logic controllers (Sanchez and Macchietto, 1995). Optimal controller performance and implementation issues have also been considered (Rotstein, Sanchez, Alsop, Camillocci and Macchietto, 1995). The method comprises the following stages:

Process Modelling: Abstraction of system features relevant for logic control as elementary components modelled as FSMs (e.g. measurement devices and actuators). Elementary components are then combined to create the process model (Sanchez, 1996).

Specification: Description of the properties that the closed–loop system must meet including undesired situations that are to be avoided. Predicate and temporal logics are employed (Sanchez, 1996).

Controller Synthesis: The generation of a feedback control law that forces the system to follow the desired behaviour in spite of disturbances. For this task we adopt a "model–based" approach, in which the control law is derived from the process model by using the proposed control framework (Rotstein, Sanchez and Macchietto, 1995).

Optimization: Often, several procedural controllers satisfying the specified properties exist. It is then possible to define a performance measure, and select the controller that maximises the proposed measure (Rotstein and Macchietto, 1996).

Implementation and Testing: The formal abstraction obtained must be translated into implementable forms for execution on industrial control systems and for testing the closed loop performance of the system. Two platforms have been addressed so far including the gPROMS dynamic simulator (Barton and Pantelides, 1994) in which a FSM is used to drive a plant simulation (Sanchez, 1996) and an industrial control system used to drive a batch pilot plant (Alsop, Camillocci and Macchietto, 1995).

Case Study

A case study demonstrating the specification, modelling and synthesis techniques has been presented elsewhere (Alsop, Camillocci, Sanchez and Macchietto, 1995). In this study, procedural controllers are synthesised for a Cleaning In Place (CIP) operation (Liu and Macchietto, 1993) on the batch pilot plant at Imperial College (Macchietto, 1992). The multipurpose batch pilot plant consists of 5 interconnected 100L vessels, 12 pumps, 3 heat exchangers and CIP circuits and is highly suitable for foods processing. The plant is controlled by an industrial distributed control system (ACCOS 30) from APV Baker. The CIP operation is a complex sequence of discrete activities for cleaning a vessel and its associated pipework. The operation is broadly divided into four distinct phases including a pre–rinse (for removal of residual solids from the interior of a tank and associated pipe work by bursts of high pressure water), detergent service (the preparation of an inventory of hot concentrated caustic solution at a CIP station), detergent rinse (cleansing of the interior surfaces by high pressure bursts of hot concentrated caustic solution) and a post–rinse (final rinsing with water to remove residual detergent and to yield the equipment suitable for food contact). The synthesis method is applied to the pre–rinse phase only. A simplified P&I diagram of the equipment employed during the pre–rinse phase is shown in Fig. 2. Water from the mains is fed via a complex series of valves (collectively labelled "feed route"), pump P6, and valve SSV1-3 to tank T1. T1 is fitted with a proximity switch PS1-1 which detects the position of the lid. For safety reasons, the rinsing procedure must be halted should the lid be opened during operation. T1 is also fitted with a continuous level sensor IT1-1. Spent water from tank T1 is drained via valve AV1-41, pump P1 and a complex series of single and double seat valves (collectively labelled "return route") to drain.

Pre–rinse is effected by successive drain and fill cycles. High pressure water from P6 is sprayed into T1 until a volume of 20 litres is achieved. Pump P1 drains the residue to a volume of 6 litres after which the cycle repeats. As P6 delivers at a much greater rate than P1, P1 need not be switched off during the fill cycle. However to avoid cavitation, P1 is deenergised if the level in tank T1 falls below 3 litres. Space restrictions do not permit the presentation in detail of each method's stage. The interested reader is referred to the indicated bibliography. A FSM model of the process was constructed from the 8 identified elementary process models. Process specifications were elicited from the user requirements and expressed in predicate and temporal logic statements. A controller was then synthesised for the pre–rinse phase according to the design principles outlined above. Finally, the formal control structure was trans-

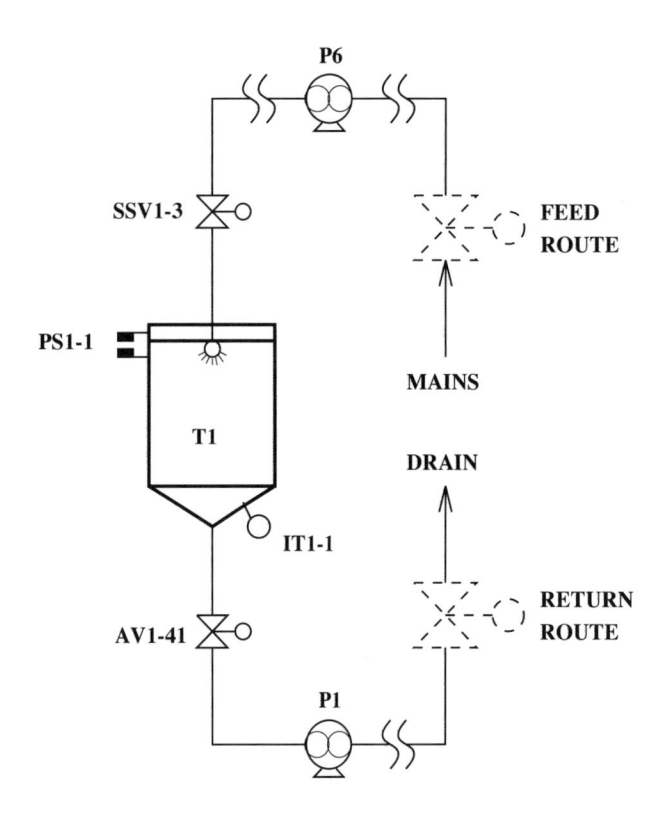

Figure 2. Simplified Process Diagram

lated into a form suitable for implementation on the batch pilot plant computer. Similarly, controllers were synthesised for each phase of the CIP process for implementation on the batch pilot plant.

Significance

This work provides a process systems approach to the conceptual design of control logic for chemical processes modelled as discrete event systems. A generic mathematical framework is presented which supports modelling and synthesis of discrete processes and controllers. In particular, we demonstrate a specific instance of the generic formulation in an application on the batch pilot plant. This approach is intended to complement software engineering methods in the creation of control code for chemical processes which is consistent with accepted standards. The advantage of this method is that it is mathematically rigorous and generates controllers which by construction are correct with respect to the process model and specifications. Other benefits follow from the systematic development including improved traceability and documentation of the design and greater potential for code reuse. Currently, studies are being carried out on hierarchical and decentralisation issues. At the same time, the techniques presented above are being tested on industrial case studies in collaboration with a number of major chemical and automation companies.

References

Alsop, N., Camillocci, L. and Macchietto, S. (1995). Automated synthesis of procedural control code, *Preprints 4th. conference on Advances in Process Control. 27–28 Sept. York, U.K.*

Alsop, N., Camillocci, L., Sanchez, A. and Macchietto, S. (1995). Synthesis of procedural controllers - Application to a batch plant. To be presented at ESCAPE6, May 27-29, Rhodes, Greece.

Barton, P. I. and Pantelides, C. C. (1994). The modelling of combined discrete/continuous processes, *AIChE Journal* **40**: 966–979.

Liu, Z. and Macchietto, S. (1993). Cleaning In Place policies for a food processing batch pilot plant, *Food and Bioproducts Processing* **71**(C3).

Macchietto, S. (1992). Automation research on a food processing pilot plant, *IChemE Symp. Series, no. 126*, pp. 179–189.

Rotstein, G. E., Sanchez, A. and Macchietto, S. (1995). Procedural control of discrete event systems. In preparation for IEEE trans. Aut. Cont.

Rotstein, G. and Macchietto, S. (1996). Synthesis of stable and optimal procedural controllers. Submmitted to Automatica.

Rotstein, G., Sanchez, A., Alsop, N., Camillocci, L. and Macchietto, S. (1995). Synthesis and implementation of procedural controllers and the automatic generation of control code, *Workshop on the Analysis and Design of Event–Driven Operations, London, U.K.*

Sanchez, A. (1996). *Formal Specification and Synthesis of Procedural Controllers for Process Systems*, Vol. 212, Lecture Notes on Control and Information Sciences, Springer–Verlag.

Sanchez, A. and Macchietto, S. (1995). Design of procedural controllers for chemical processes, *Computers and Chemical Engineering* **19**(Suppl.): S381–S386.

Sanchez, A., Rotstein, G. E. and Macchietto, S. (1995). Synthesis of procedural controllers for Batch Chemical Processes, *Proc. DYCORD+'95. 4th IFAC Symposium on Dynamics and Control of Chemical Reactors, Distillation Columns and Batch Processes. Helsingor, Denmark. June*, pp. 415–420.

Schuler, H., Allgower, F. and Gilles, E. D. (1991). Chemical Process Control: Present status and future needs. The view from the European industry, *in* Y. Arkun and W. H. Ray (eds), *CPC IV*, Elsevier, pp. 29–52.

Wonham, W. M. (1988). A control theory for discrete–event systems, *in* M. Denham and A. Laub (eds), *Advanced Computing Concepts and Techniques in Control Engineering*, Springer Verlag, pp. 129–169.

DYNAMIC OPERATIONS IN THE PLANNING AND SCHEDULING OF MULTI-PRODUCT BATCH PLANTS

Tarun Bhatia and Lorenz T. Biegler
Department of Chemical Engineering
Carnegie Mellon University
Pittsburgh, PA 15213

Abstract

Dynamic processing freedom of batch units is considered in design and scheduling to improve profitability under strict constraints. The long term planning and scheduling formulation of Birewar and Grossmann (1989b) is extended to include dynamic processing. Orthogonal collocation is used to solve the optimal control profile for individual tasks. An example is solved with the extended formulation and significant advantages are presented.

Keywords

Batch processing, Collocation, Design, Scheduling.

Introduction

Despite inherent dynamic processing freedom in batch units, an integration of flexible aspects while planning, e.g. (Voudouris and Grossmann, 1993) often considers processing along fixed recipes. Operating considerations have been addressed in process development (Allgor, Barrera, Bartona and Evans, 1994) and plant design (Salomone, Montagna and Irribarren, 1994; Mujtaba and Macchietto, 1994) through sequential, nested or bi-level solution approaches.

This work includes dynamic process models of tasks within the design and scheduling formulation for a special class of batch operations. Collocation over finite elements permits a simultaneous solution to the DAE process models. Optimal control profiles are then compared with best constant control levels (best recipes) in improving several planning objectives. The combinatorial scheduling issue is resolved by considering flowshop plants with Unlimited Intermediate Storage (UIS) and Zero Wait (ZW) transfer policies and one unit per stage, through the formulation of (Birewar and Grossmann, 1989b).

Problem Formulation

Consider the representation in Fig. 1. A total of J processing stages, in the flowshop sequence, produce N_p products with batches of size B_i for each product i. Variables $z_{i,j}^0$ and $z_{i,j}^f$ denote initial and final batch states for product i at stage j. State trajectories $z_{i,j}(t)$, between these boundary states, depend on control profiles $u_{i,j}(t)$ for product i at stage j.

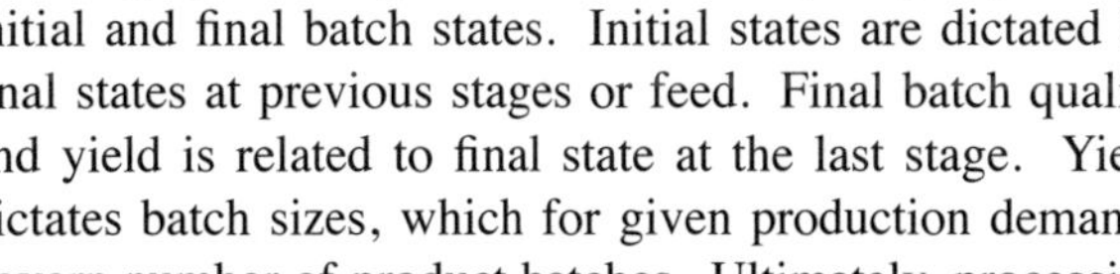

Figure 1. Problem representation.

Stage processing time is dictated by stage dynamics, initial and final batch states. Initial states are dictated by final states at previous stages or feed. Final batch quality and yield is related to final state at the last stage. Yield dictates batch sizes, which for given production demands govern number of product batches. Ultimately, processing times and batch sizes directly affect planning.

The method of collocation over finite elements permits efficient handling of state path constraints, essential for quality issues. State and control profiles in each element are approximated by Lagrange polynomials of appropriate order. Coefficients in these polynomials become decisions and the overall problem becomes an NLP (Logsdon, Diwekar and Biegler, 1990). Collocation thus permits a simultaneous solution for processing profiles within planning. The overall formulation used is (M1).

$$\max_{z_{ijlm}, u_{ijlm}, p_j, \delta\tau_{ijl}} \psi_{profit}(CT_j, V_j, Q_i)$$
s.t.
$$\dot{z}_{ij}(\tau_{lm}) - \delta\tau_{ijl} f(z_{ijlm}, u_{ijlm}, p) = 0 \quad \forall i, j, l, m$$
$$g(z_{ijlm}, u_{ijlm}, p) \leq 0 \quad \forall i, j, l, m$$
$$g_e(z_{ijl}, u_{ijl}, t_{ij}, p) \leq 0 \quad \forall i, j; \quad l = ne$$
$$h(z_{ijlm}, u_{ijlm}, p) = 0 \quad \forall i, j, l, m$$
$$h_e(z_{ijl}, u_{ijl}, t_{ij}, p) = 0 \quad \forall i, j; \quad l = ne$$
$$z^L_{ijlm} \leq z_{ijlm} \leq z^U_{ijlm}$$
$$u^L_{ijlm} \leq u_{ijlm} \leq u^U_{ijlm}$$
$$t_{ij} = \sum_l \delta\tau_{ijl} \quad \forall i, j$$
$$z^0_{ij} = z_{ij10} \quad \forall i, j$$
$$z_{ij,l+1,0} = \sum_{m'=0}^{ncol} z_{ijlm}\phi_m(1) \quad \forall i, j, l; \quad l \neq ne$$
$$z^f_{ij} = \sum_{m'=0}^{ncol} z_{ijlm}\phi_m(1) \quad \forall i, j, l; \quad l = ne$$
$$z^0_{ij} = f(z^f_{i,j-1}) \quad \forall i, j; j \neq 1$$
$$B_i = f(z^f_{ij}) \quad \forall i, j = J$$
$$V_j \geq S_{ij} B_i \quad \forall i, j$$
$$n_i = \frac{Q_i}{B_i} \quad \forall i$$
$$\sum_{k=1}^{N_p} NPRS_{ik} = n_i \quad \forall i$$
$$\sum_{i=1}^{N_p} NPRS_{ik} = n_k \quad \forall k$$
$$NPRS_{ii} \leq n_i - 1 \quad \forall i$$
$$CT_j = \sum_{i=1}^{N_p}\left(n_i t_{ij} + \sum_{k=1}^{N_p} SL_{ikj} NPRS_{ik}\right) \quad \forall j$$
$$CT_j \leq H \quad \forall j$$
$$SL_{ik,j} + t_{k,j} = t_{i,j+1} + SL_{ik,j+1} \quad \forall ik, j; \quad j < J$$
$$V_j \geq 0, n_i, B_i \geq 0, NPRS_{ik} \geq 0,$$
$$t_{i,j} \geq 0, SL_{ikj} \geq 0 \qquad (M1)$$

where:

ψ	is the objective,
$(i, k), j$	are product and stage indices,
l, m	are element and collocation point indices,
z, u	represent state and control profiles,
$\tau, \delta\tau_l$	are normalized time and element size
f, g, h	represent stage dynamics and constraints
g_e, h_e	represent end point constraints
p	are continuous parameters,
t_{ij}	is processing time for i in j,
L, U	represent lower and upper bounds,
$\phi_m(t)$	is Lagrange basis function,
V_j	are continuous equipment size parameters,
S_{ij}	are size factors,
n_i	are number of batches,
Q_i	is production capacity,
$NPRS_{ik}$	are no. of $i-k$ pairs in the schedule,
SL_{ikj}	are ZW slacks for product pairs ik,
H	is planning horizon.

Ideally n_i and $NPRS_{ik}$ must be integers. Although this is guaranteed under special conditions (Birewar, 1989) for recipe based processes, the formulation treats them as continuous.

Example

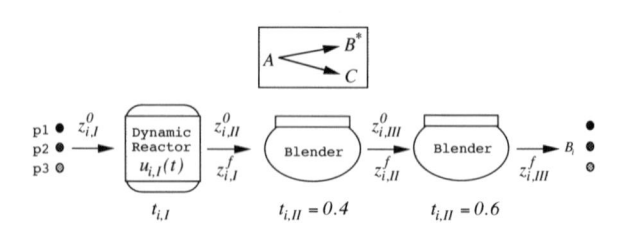

Figure 2. Example problem.

Three similar flowshop products requiring dynamic processing in a reactor followed by two recipe based blenders (Fig. 2) are considered. Process models for products differ in reaction kinetic parameters. A transformed temperature control profile $u_{i,I}(t)$ exists for each product i in the reactor. Initial feed charge in the reactor is fixed for each run. Dynamic control is expected to improve desired conversion for the competing mechanism, both in time and extent. Four planning scenarios are considered with constant and transient reactor control profiles, for both UIS and ZW policies.

- **P1** - Minimize Operating Cost - to achieve a production target in an existing facility.

- **P2** - Minimize Fixed Equipment Costs - to achieve a production target in a specified horizon.

- **P3** - Maximize Production Revenues - from operating in an existing facility for a specified horizon.

- **P4** - Maximize Overall Profit - a function of fixed costs, operating costs and revenues.

Solution Strategy

The reactor control profile (transformed temperature) is resolved using four equal sized finite elements spanning the processing time for each operation. Two collocation points are used within each element.

Problems P1 - P4 are solved successively, for both policies, with the solution for one problem providing an initialization to the next. Constraints are added and dropped as required by the problems. P1 is initialized by solving individual reactor subproblems that maximize desired conversion for each product in 2 hour.

Results

For all problems, UIS performance is better than the more constrained ZW policy. Dynamic control profiles improve all cases beyond operating at the best constant level.

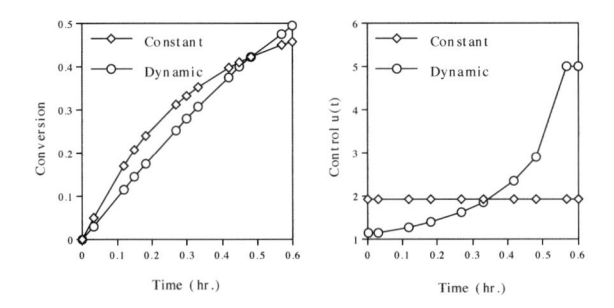

Figure 3. Optimal reactor control profiles (product 1, problem P1, ZW policy).

Reactor control profiles for a typical case are shown in Fig. 3. Optimal dynamic control gives higher desired conversion (x_B^f) towards the end, than best constant operation (best recipe), producing larger batches in equal time.

Problem P1 - Minimize Operating Cost

Problem P1	ZW	
	const. ψ=**26,728**	dyn. ψ=**24,922**
CT_j (hr.)	190.9, 190.9, 190.9	178.0, 178.0, 178.0
x_B^f	0.459, 0.492, 0.414	0.495, 0.527, 0.442
$t_{i,I}$ (hr.)	0.6, 0.6, 0.6	0.6, 0.6, 0.6
Problem P1	UIS	
	const. ψ=**23,539**	dyn. ψ=**22,026**
CT_j (hr.)	153.9, 143.0, 214.5	145.7, 132.8, 199.2
x_B^f	0.387, 0.411, 0.412	0.422, 0.445, 0.435
$t_{i,I}$ (hr.)	0.360, 0.343, 0.594	0.377, 0.359, 0.581

Table 1. Results for Problem P1.

With ZW policy, operating costs reduce by 6.8% for the dynamic case (Table 1). The reactor processes each product for 0.6 hours with both constant and transient control. This eliminates slacks from two stages out of three, for all product pair combinations. Optimal control produces larger batches for all products and thus fewer batches suffice to meet production targets. Stage cycle times thus reduce, even though reactor processing times in both cases are equal.

With UIS policy, 6.4% is saved in the dynamic case. Unequal stage cycle times are a result of decoupling between successive stages for UIS policy. Processing times fall below 0.6 hours as any improvement in time benefits the UIS cycle time. Batch sizes are thus smaller than in ZW case due to shorter processing. Optimal control still produces larger batches although with slightly longer reactor operations. The reduction in number of batches that satisfy demands however compensates for the increase in processing times. Cycle time is thus reduced in all stages for the dynamic case, although not as significantly as in the ZW case.

Problem P2 - Minimize Fixed Costs

Problem P2	ZW	
	const. ψ=**33,669**	dyn. ψ=**33,239**
V_j (l.)	83.3, 86.8, 87.9	82.0, 85.6, 86.6
x_B^f	0.455, 0.470, 0.382	0.464, 0.468, 0.377
$t_{i,I}$ (hr.)	0.6, 0.6, 0.6	0.6, 0.6, 0.6
Problem P2	UIS	
	const. ψ=**33,589**	dyn. ψ=**33,185**
V_j (l.)	83.1, 86.5, 87.7	81.9, 85.4, 86.4
x_B^f	0.451, 0.449, 0.4	0.459, 0.450, 0.393
$t_{i,I}$ (hr.)	0.585, 0.515, 0.689	0.582, 0.519, 0.686

Table 2. Results for Problem P2.

Fixed costs reduce by a mere 1.28% for ZW policy with a dynamic profile (Table 2). This is a result of higher desired conversion per unit volume with transient control of the reactor profile. Smaller equipment is thus able to address planning constraints.

UIS shows an improvement of 1.2% for problem P2 with dynamics. In this case too, desired conversion to product per unit volume is higher for two products out of three, for operating with a transient operating profile.

Problem P3 - Maximize Revenues

Problem P3	ZW	
	const. ψ=**45,330**	dyn. ψ=**48,704**
$n_i B_i$ (lb.)	4800, 4800, 5426	4800, 4800, 6419
x_B^f	0.459, 0.492, 0.414	0.495, 0.527, 0.442
$t_{i,I}$ (hr.)	0.6, 0.6, 0.6	0.6, 0.6, 0.6
Problem P3	UIS	
	const. ψ=**45,767**	dyn. ψ=**49,086**
$n_i B_i$ (lb.)	4800, 4800, 5555	4800, 4800, 6531
x_B^f	0.434, 0.459, 0.471	0.467, 0.492, 0.491
$t_{i,I}$ (hr.)	0.504, 0.479, 0.798	0.506, 0.483, 0.758

Table 3. Results for Problem P3.

An improvement of 7.4% is observed with dynamics for ZW policy (Table 3). Conversion and processing time decisions are same as for corresponding P1 problems. The specified horizon is larger than stage cycle time values in P1. Demand specifications are first met for all products at the earliest. In remaining time, product 3 (highest value) is produced until the horizon constraint becomes active.

For UIS, increased production with dynamics improves the objective by 7.3%. Higher conversions are achieved with slightly longer reactor processing, although product 3 behaves otherwise. This provides more efficient processing of product 3.

Problem P4 - Maximize Overall Profit

Problem P4	ZW	
	const. ψ=**23,967**	dyn. ψ=**27,875**
CT_j (hr.)	397.8, 397.9, 397.9	371.2, 371.2, 371.2
V_j (l.)	93.1, 96.5, 98.3	100.8, 104.5, 106.4
x_B^f	0.459, 0.492, 0.414	0.495, 0.527, 0.441
$t_{i,I}$ (hr.)	0.6, 0.6, 0.6	0.6, 0.6, 0.6
Problem P4	UIS	
	const. ψ=**30,010**	dyn. ψ=**33,391**
CT_j (hr.)	352.8, 280.8, 421.2	334.4, 260.6, 390.9
V_j (l.)	93.7, 97.1, 98.9	100.9, 104.6, 106.5
x_B^f	0.411, 0.435, 0.437	0.448, 0.472, 0.463
$t_{i,I}$ (hr.)	0.429, 0.408, 0.676	0.447, 0.426, 0.668

Table 4. Results for Problem P4.

Overall profit increases by 16.3% when reactor dynamics are considered for ZW policy (Table 4). Production reaches the upper bound for all products in both cases. A higher conversion is achieved for all products in the dynamic reactor. As a result, fewer batches with smaller cycle times exist for the dynamic case, at the cost of larger equipments to accommodate the bigger batches.

With UIS policy, overall profit increased by 11.3% due to smaller variations in cycle time or equipment sizes. With optimal reactor control, higher conversions are achieved with longer processing for products, except product 3. The result is an improved plan with few larger sized batches.

Derivation of exact product sequences for all problems studied is simple. UIS problems are sequence independent. ZW problems process all products in the reactor (bottleneck stage) for the same time. All products thus become identical in processing time requirements. This renders the ZW case to be sequence independent too. Determination of the exact makespan from the family of schedules is similarly simplified.

Problem	ZW		UIS	
	const.	dyn.	const.	dyn.
P1	0.390	0.510	0.430	0.400
P2	0.450	0.420	0.380	0.380
P3	0.440	0.450	0.400	0.450
P4	0.400	0.450	0.390	0.380

Table 5. CPU Seconds for Solving Example 1 Using the NLP Solver CONOPT Through the GAMS Modeling System on an HP 9000/700 Workstation.

The accuracy of the profile was verified using LSODE (Hindmarsh, 1983). This motivating example reflects some benefits in including dynamic process models in planning.

Conclusions

Dynamics can contribute to planning in less expected ways. In the chosen example, product sequencing was simplified. With more dynamic stages, interesting load shifts that improve schedule perfomance can be expected. At a bottleneck stage, extra units must be considered after assessing the role of dyanmics. The overall formulation with process modeling can be extended to other aspects such as plant superstructure, unit task assignments, etc. The formulation has been extended to include inventory planning and shows significant benefits for a larger example (Bhatia and Biegler, 1995). The collocation formulation embedded within the overall framework, serves these needs with modest computational expense.

The inherent assumption of perfect dynamic models is more limiting than the class of batch operations considered. The challenge in exploiting processing freedom lies in process model reliability. Developing perfect models or techniques that address processing with imperfect models would be critical in adding confidence to process integration in planning. This would be the focus of future research.

Acknowledgments

Funding from the Department of Energy is gratefully acknowledged.

References

Allgor, R., Barrera, M., Barton, P. and Evans, L. (1994). Optimal batch process development, *PSE*, 153-157.

Bhatia, T. and Biegler, L. (1995). Dynamic operations in the planning and scheduling of multi-product batch plants, submitted to *Ind. Eng. Chem. Res.*

Birewar, D. (1989). *Design, Planning and Scheduling of Multiproduct Batch Plants*, Ph.D. thesis, Department of Chemical Engineering, Carnegie Mellon University.

Birewar, D. and Grossmann, I. (1989b). Incorporating scheduling in the optimal design of multiproduct batch plants, *Comp. and Chem. Engng.*, **13**, 141-161.

Hindmarsh, A. (1983). Odepack, a systematic collection of ode solvers, *Scientific Computing*, North-Holland, Amsterdam.

Logsdon, J., Diwekar, U. and Biegler, L. (1990). On the simultaneous optimal design and operation of batch distillation columns, *Trans. I. Cheme.*, **68**(A), 434-444.

Mujtaba, I. and Macchietto, S. (1994). Optimal design of multiproduct batch distillation column – single and multiple separation duties, *PSE*, 179-184.

Salomone, H., Montagna, J. and Irribarren, O. (1994). Dynamic simulation in the design of batch processes, *Comp. Chem. Engng.*, **18**(3), 191-204.

Voudouris, V. and Grossmann, I. (1993). Optimal synthesis of multiproduct batch plants with cyclic scheduling and inventory considerations, *Ind. Eng. Chem. Res.*, **32**(2), 1962-1980.

THE CURRENT STATUS OF SHEET AND FILM PROCESS CONTROL

Richard D. Braatz
Chemical Engineering Department
University of Illinois
Urbana, IL 61801

Abstract

Sheet and film processes, which include coating, papermaking, and polymer film extrusion processes, pose a challenging multivariable control problem whose solution is of substantial industrial interest. Effective control of sheet and film processes translates to significant financial benefits arising from: reduced material consumption, increased production rates, improved product quality, elimination of product rejects, and reduced energy consumption. Existing techniques for sheet and film process control are reviewed, with particular attention to the effectiveness of these techniques for addressing the critical aspects of sheet and film processes. Besides reviewing the classical work in this area, the review includes a summary of the substantial developments made over the past five years which exploit the inherent structure of sheet and film processes. A conclusion is that, although classical and generic modern control approaches do not effectively address all of the characteristics of these processes, recent efforts show promise. Directions for future research opportunities are identified.

Keywords

Cross directional control, Large scale systems, Sheet and film, Polymer film extrusion, Papermaking, Coating, Robust control, Model Predictive Control, Identification, Control relevant identification.

Introduction

The control of sheet and film processes, which include coating, paper manufacturing, and polymer film extrusion processes, is of substantial industrial interest. Coating processes are of great importance to manufacturing, especially in the photographic, magnetic and optical memory, electronic, and adhesive industries. The total capitalization of industries which rely on coating technology has been estimated to be over $500 billion worldwide (Cohen, 1990). Paper manufacturing is the mainstay of the pulp and paper industries; and polymer film extrusion is used to make a variety of products from the manufacturing of plastic films for windshield safety glass to blow molding for making large plastic bags (Martino, 1991; Callari, 1990). Improved control of sheet and film properties can mean significant reductions in material consumption; greater production rates for existing equipment; improved product quality despite inexperienced operators resulting from a high turnover rate in the work force; elimination of product rejects; and reduced energy consumption (Atkins, Rodencal and Vickery, 1982; Wallace, 1981; Wallace, 1986).

Sheet and film processes pose a challenging multivariable control problem. The processes are truly large scale, in that the numbers of inputs and outputs can range from 100 to 1000. The processes operate at high speeds, with measurements taken approximately every 0.05 – 0.1 seconds. In order to handle the process dimensionality within the computational constraints, existing industrial control algorithms implemented on these processes are very simple. In particular, the sampling time for the control algorithms is chosen to be 100 to 1000 times longer than the time between individual sensor readings, and sensors are grouped into sensor "blocks" to reduce the dimensionality (Braatz, 1995). Existing industrial sheet and film control algorithms address model inaccuracies and actuator constraints in only an indirect manner, e.g., through excessive detuning, or by clipping the control actions resulting from an unconstrained (typically, LQG) controller. This simplicity is associated with reduced product quality and a loss of flexibility.

This article reviews existing techniques for sheet and film process control, with particular attention to the effectiveness of these techniques for addressing the critical as-

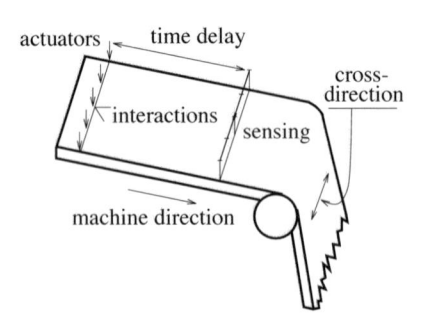

Figure 1. A generic sheet and film process.

pects of sheet and film processes, namely, the large dimensionality, the constraints on actuator movements, and the inaccuracies of their associated models. Directions for future research are identified.

Characteristics of Sheet and Film Processes

Fig. 1 is a diagram for a generic sheet and film process. Models for sheet and film processes usually consist of scalar dynamics multiplied by a constant interaction matrix which is symmetric or near-symmetric (Wilhelm, Jr. and Fjeld, 1983). The characteristics which make the effective control of these processes especially interesting and challenging are as follows.

- The *large process dimensionality* makes control challenging because: (1) most off-the-shelf software for the synthesis of modern (e.g., H_∞, μ) controllers have numerical difficulties for processes with a large number of inputs and outputs (Hovd, Braatz and Skogestad, 1994); (2) algorithms which perform optimizations on-line (e.g., Dynamic Matrix Control, Model Predictive Control, IDCOM) are too computationally demanding for these high-speed large-scale applications (Chen and Wilhelm, Jr., 1986); and, (3) plant matrices tend to be more poorly conditioned for processes of large dimension (Braatz, 1995; Braatz and Morari, 1994).

- Unknown disturbances and inaccurate values for the parameters of the physical system make it impossible to determine an exact model, either phenomenologically or via input-output identification. This *model inaccuracy* must be taken into account for high performance reliable control.

- *Actuator positions are constrained*. "Clipping" control actions, which is common in practice, can cause disastrous performance or even instability (Braatz, 1995). Constraint-handling will be needed when the disturbances are sufficiently large and have sharp spatial variations across the sheet or film.

Another characteristic of sheet and film processes is that measurements are usually taken from a limited number of sensors which continuously travel back and forth transverse to the movement of the sheet or film. The problem of estimating the sheet or film profile from these measurements for use by the control algorithm has been heavily studied (Bergh and MacGregor, 1987; Campbell and Rawlings, 1994; Rigopoulos, Arkun, Kayihan and Hanczyc, 1996; Tyler and Morari, 1995) and will not be discussed further here, as the focus of this manuscript is on controller design.

Previous Research

The large-scale, uncertain, constrained nature of sheet and film processes makes their control especially interesting and challenging. Here we summarize previous research on sheet and film process control, and the ability of developed techniques to address the critical aspects of sheet and film processes.

Linear Control

The plastic film industry began implementing multivariable control in the early 1970's (Wallace, 1986). Such control began to be implemented on paper machines in the mid-1970's; however, widespread use of such systems did not occur until the 1980's (Wilhelm, Jr. and Fjeld, 1983). The two multivariable control techniques reported in the literature before 1988 were linear-quadratic-optimal (LQ) (Boyle, 1978; Wilhelm, Jr. and Fjeld, 1983) and model inverse-based control (Wilkinson and Hering, 1983). Mostly steady state models were used. New control actions were often taken only after steady state was reached. The weaknesses of these linear control approaches are that constraints and model inaccuracies are not explicitly addressed.

Model Predictive Control

The application of *Model Predictive Control* has been considered for the control of paper machines (Boyle, 1977), coating processes (Braatz, Tyler, Morari, Pranckh and Sartor, 1992), and polymer film extruders (Campbell and Rawlings, 1994). In Model Predictive Control, the control objective is optimized on-line subject to the constraints. A linear or quadratic optimization is solved at each sampling instance, and off-the-shelf software is available for performing these calculations. These optimization problems can be very large—over 200 variables and 200 constraints for steady-state control of a medium-size sheet and film process (Braatz, 1995). Although recent efforts have been directed towards reducing the solution time for these optimization problems (Doyle III, Pekny and Dave, 1995), for high speed processes it is currently not feasible for the fastest workstations to perform the control computations within the measurement sampling interval of 0.05-0.1 seconds.

Robust Control

The function, μ, is a nonconservative measure for system robustness. Due to its computational complexity, no researcher has ever proposed the direct application of μ-theory to design controllers for sheet and film processes. In particular, it has been proven that *any* algorithm must take an exorbitant amount of computation time to compute μ with high precision (Braatz, Yound, Doyle and Morari, 1994). Although tight approximations allow computation which has polynomial growth as a function of process dimension, the order of this polynomial is large ($\mathcal{O}(n^4)$) (Braatz, 1995). Published examples (Hovd, Braatz and Skogestad, 1993; Laughlin, Morari and Braatz, 1993) illustrate that existing robust control software is inadequate for application to sheet and film processes.

Exploiting Structure in Robust Control

The multivariable interaction matrices for sheet and film processes typically have one of the following three structures (Laughlin et al., 1993): Toeplitz symmetric, circulant symmetric, and centrosymmetric. Most models found in the literature are Toeplitz symmetric; this structure follows from the assumption that changes observed downstream from one actuator caused by adjustments at the nearest neighboring actuators are independent of position across the machine. Circulant symmetric matrices represent interactions for circular sheet or film processes, as are used in some polymer extrusion applications (Callari, 1990; Martino, 1991). Centrosymmetric models include the other structures as special cases and can also take into account different effects near the edges. In recent years, attention has been directed to exploiting these model structures to reduce the computational expense associated with robust control.

Circulant matrix theory has been used to develop methods for designing conservative robust multivariable controllers based on the design of only *one* single loop controller (Laughlin et al., 1993). Circulant symmetric, Toeplitz symmetric, and centrosymmetric models are all covered by the theory. Other efforts use the properties of unitary-invariant norms to design nonconservative robust multivariable controllers via DK-iteration performed on an $m \times m$ diagonal plant, where m is the number of interaction parameters which is typically much smaller than the process dimension n (Hovd et al., 1994). Though the computational load can be substantially reduced using these dimensional reduction methods, the disadvantages of these methods include: (1) edge effects are not always allowed; (2) non-square systems are not always allowed; (3) constraints are not addressed; and, (4) methods of representing plant/model mismatch are limited.

Linear Control and Static Nonlinear Elements

The traditional method for dealing with constraints is to use simple static nonlinear elements (for example, selectors and overrides) which modify the linear control system only when necessary (Buckley, 1971). Two advantages of this method are that: (1) well-developed linear control methods can be applied to design the linear controller; and, (2) such constraint-handling methods are almost as easy to implement as a purely-linear controller. The static nonlinear elements are simple operations requiring very little computational effort and are already standard in industrial control.

The disadvantages of using static nonlinear elements are that they can cause severe performance degradation such as limit cycles and increased variance (Bialkowski, 1988). *Ad hoc* design methods have been developed for avoiding some of these problems, but all of these techniques perform poorly (or may even lead to instability) in some situations.

One such method which has had some success at handling constraints is referred to as *directionality compensation*. When the output of the linear controller cannot satisfy the constraints on the actuator movements, the directionality compensator scales back the linear control output *while keeping the same direction* until the control action becomes feasible. The directionality compensator has been shown to perform nearly as well as Model Predictive Control for some industrial scale adhesive coaters (Braatz et al., 1992); however, it is expected that MPC would provide better performance for most sheet and film processes.

Future Directions

Although existing techniques for explicitly addressing model uncertainty and actuator constraints are too computationally burdensome for direct application to sheet and film processes, recent methods to reduce computational expense by exploiting the inherent structure of sheet and film processes look promising. Future efforts should extend these approaches to handle actuator constraints while removing potentially restrictive assumptions regarding edge effects or the model uncertainty description. As the quantity of on-line data available for sheet and film processes is low, efforts should be directed to developing control relevant identification algorithms which exploit model structure to extract as much process information as possible for use by the control algorithms (Braatz and Featherstone, 1995; Featherstone and Braatz, 1995).

Over the past five years, several companies have initiated research programs on sheet and film process control. Substantial progress can be made relatively quickly through joint efforts by individuals from academia, government, and industry. Several academic process control researchers have begun to study these industrially-relevant control problems, with data for testing the developed algorithms provided by collaborating industrial control engineers (Braatz, Ogunnaike and Featherstone, 1996; Rigopoulos et al., 1996). The role of the government agencies should be to encourage such joint efforts. The prevalence and importance of sheet and film process operations provide opportunities for substantially improved process efficiencies with little capital investment.

Conclusions

A significant amount of research has been conducted on sheet and film process control over the past few decades. Although the developed techniques have addressed many of the important aspects of these processes, no technique directly addresses both actuator constraints and model uncertainty in a computationally feasible manner. It is argued that the key to the solution of the sheet and film identification and control problems will be to exploit the structure inherent in these processes. This research area provides an excellent opportunity for individuals from academia, industry, and the government agencies to work together to develop innovative practical solutions for an industrially important problem.

References

Atkins, J. W., Rodencal, T. E. and Vickery, D. E. (1982). Correcting a poor moisture profile improves productivity, *Tappi*, **65**, 49-52.

Bergh, L. G. and MacGregor, J. F. (1987). Spatial control of sheet and film forming processes, *Can. J. of Chem. Eng.*, **65**, 148-155.

Bialkowski, W. L. (1988). Integration of paper machine operations, *Tappi*, **71**, 65-71.

Boyle, T. J. (1977). Control of cross-direction variations in web forming machines, *Can. J. of Chem. Eng.*, **55**, 457-461.

Boyle, T. J. (1978). Practical algorithms for cross-direction control, *Tappi*, **61**, 77-80.

Braatz, R. D. (1995). Control of sheet and film processes, *Third SIAM Conf. on Control and its Applications*, St. Louis, Missouri.

Braatz, R. D. and Featherstone, A. P. (1995). Identification and control of large scale paper machines, *Weyerhaeuser Workshop on Modeling and Control of Kamyr Digesters and Paper Machines*, Tacoma, Washington.

Braatz, R. D. and Morari, M. (1994). Minimizing the Euclidean condition number, *SIAM J. on Control and Optim.*, **32**, 1763-1768.

Braatz, R. D., Ogunnaike, B. A. and Featherstone, A. P. (1996). Identification, estimation and control of shet and film processes, *Proc. of the IFAC World Congress*, San Franscisco, paper 7c-01-1.

Braatz, R. D., Tyler, M. L., Morari, M., Pranckh, F. R. and Sartor, L. (1992). Identification and cross-directional control of coating processes, *AIChE J.*, **38**, 1329-1339.

Braatz, R. D., Young, P. M., Doyle, J. C. and Morari, M. (1994). Computational complexity of μ calculation, *IEEE Trans. on Auto. Control*, **39**, 1000-1002.

Buckley, P. S. (1971). Designing overide and feedforward controls, *Control Eng.*, **18**, 48-51.

Callari, J. (1990). New extrusion lines run tighter, leaner, *Plastics World*, 28-31.

Campbell, J. C. and Rawlings, J. B. (1994). Gage control of film and sheet forming processes, *AIChE Annual Meeting*, San Francisco, paper 224c.

Che, S.-C. and Wilhelm, Jr., R. G. (1986). Optimal control of cross-machine direction web profile with constraints on the control effort, *Proc. of the American Control Conf.*, Seattle, Washington, 1409-1415.

Cohen, E. D. (1990). Coatings: Going below the surface, *Chem. Eng. Prog.*, **86**, 19-23.

Doyle III, F. J., Pekny, J. F. and Dave, P. (1995). Application of large scale MPC to paper machines, *Weyerhaeuser Workshop on Modeling and Control of Kamyr Digesters and Paper Machines*, Tacoma, Washington.

Featherstone, A. P. and Braatz, R. D. (1995). Control relevant identification of sheet and film processes, *Proc. of the American Control Conf.*, Seattle, Washington, 2692-2696.

Hovd, M., Braatz, R. D. and Skogestad, S. (1993). On the structure of the robust optimal controller for a class of problems, *Proc. of the IFAC World Congress*, vol. IV, Sydney, Australia, 27-30.

Hovd, M., Braatz, R. D. and Skogestad, S. (1994). SVD controllers for H_{2-}, $H_{\inf-}$, and μ-optimal control, *Proc. of the American Control Conf.*, Baltimore, Maryland, 1233-1237.

Laughlin, D., Morari, M. and Braatz, R. D¿. (1993). Robust performance of cross-directional basis-weight control in paper machines, *Automatica*, **29**, 1395-1410.

Martino, R. (1991). Motor-driven cams actuator flexible-lip automatic die, *Modern Plastics*, **68**, 23.

Rigopoulos, A., Arkun, Y., Kayihan, F. and HAnczyc, E. (1996). Esitimation of paper machine full profile properties using Kalman filtering and recursive principal component analysis, *Proc. of the Fifth Int. Conf. on Chemical Process Control (CPC-V)*, Tahoe City, California.

Tyler, M. L and Morari, M. (1995). Estimation of cross directional properties: Scanning versus stationary sensors, *AIChE J.*, **41**, 846-854.

Wallace, B. W. (1981). Economic benefits offered by computerized profile control of weight, moisture and caliper, *Tappi*, **64**, 79-83.

Wallace, M.D. (1986). Cross-direction control: Profitablity and pitfalls, *Tappi*, **69**, 72-75.

Wilhelm, Jr., R. G. and Fjeld, M. (1983). Control algorithms for cross directional control, *Proc. of the 5th IFAC PRP Conf.*, Antwerp, Belgium, 163-174.

Wilkinson, A. J. and Hering, A. (1983). A new control technique for cross machine control of basis weight, *Preprints of the 5th IFAC PRP Conf.*, Antwerp, Belgium, 151-155.

ON INFEASIBILITIES IN MODEL PREDICTIVE CONTROL

P.O.M. Scokaert and J.B. Rawlings
Department of Chemical Engineering
University of Wisconsin
Madison, WI 53706-1691

Abstract

This paper discusses the problem of infeasibility in model predictive control, caused by state inequality constraints. We review the current approaches and show that only partial solutions are available. In particular, we analyze the methods that minimize the duration of constraint violations and the methods that minimize the peak size of constraint violations. We highlight advantages and disadvantages of these and other approaches and illustrate our points with a simulated example.

Keywords

Model predictive control, State constraints, Soft constraints, Feasibility.

Introduction

Over the past two decades, model predictive control (MPC) has become established as a control tool of choice for advanced applications, and has become one of the most prominent control strategies used in the chemical industries. It is interesting that the popularity of the method with practitioners has come in spite of serious theoretical questions to which no answers were available, even at the most recent CPC IV meeting. Since that meeting, researchers have made much progress on MPC theory: for example, the features of MPC formulations that lead to stabilizing control have been identified. The problem of stability, however, is closely linked to the problem of feasibility, which has received much less attention. In this paper, we review two recent contributions in this area, and outline some reasons why further research is warranted.

A Perspective on Stability and Feasibility Questions

In the context of finite horizon MPC, Zafiriou showed that the addition of state constraints can make an otherwise stabilizing control law unstable (Zafiriou, 1991), regardless of tuning. The use of an infinite horizon solves the stability problem, because MPC laws defined on an infinite horizon are guaranteed to be stabilizing, under a feasibility assumption (Rawlings and Muske, 1993). However, the examples which cause instability for finite horizon laws, cause infeasibility for infinite horizon laws. Consequently, the problems that arise with the use of hard state constraints cannot be resolved without a discussion of feasibility.

Two basic approaches have been proposed to deal with infeasibility. The first suggests removal of the state constraints over the early portion of the horizon, in order to enforce them at later times (Rawlings and Muske, 1993). Feasibility is assessed at every sample, and the full constraints are therefore re-introduced into the control law as soon as that is possible. The second method relies on the use of soft rather than hard constraints (Ricker, Subrahmanian and Sim, 1988; Zheng and Morari, 1995). In this approach, violations of state constraints at all times are allowed, but an additional term is introduced into the control objective, penalizing the constraint violations.

Before considering the details of particular solutions to infeasibility, we first discuss the feasibility problem in a broader context. Performance limitations are identified that are inherent to the process and independent of the chosen control method. The trade-offs that must be made by any controller are then highlighted. This discussion sets the stage for the ensuing study of potential solution methods. A simulated example is presented that illustrates the main points of the discussion.

The Problem of Infeasibility

We consider linear, time invariant, discrete-time systems of the form

$$x_{t+1} = Ax_t + Bu_t, \qquad u_{\min} \leq u_t \leq u_{\max} \qquad (1)$$

where $x_t \in \real^n$ and $u_t \in \real^m$ are the state and input vectors at time t, $u_{\min} \in \real^m$ and $u_{\max} \in \real^m$ denote physical input limits and $A \in \real^{n \times n}$ and $B \in \real^{n \times m}$ are the state transition and input distribution matrices. The state constraints we consider take the form

$$Hx_{j|t} \leq h, \qquad \forall j > t, \qquad (2)$$

where, denoting by n_h the number of constraints, $H \in \real^{n_h \times n}$ denotes constraint transition matrix, $h \in \real^{n_h}$ contains the constraint levels, $u_{j|t} \in \real^m$ is the control postulated at time t for time j, and

$$x_{j+1|t} = Ax_{j|t} + Bu_{j|t}, \qquad (3)$$

with $u_{\min} \leq u_{j|t} \leq u_{\max}$, for all $j \geq t$, and $x_{t|t} = x_t$.

In order to ensure that steady state at zero is feasible, we restrict the elements of $u_{\min}$ to be strictly negative and those of $u_{\max}$ and h to be strictly positive. This does not however guarantee that there always exists a sequence of controls within the input limits that satisfies the state constraints. The condition that arises when no such sequence exists is called infeasibility, the focus of this paper.

When the state constraints cannot all be satisfied, the predicted constraint violations should be reduced in some way. Different measures of constraint violation may be used. In this discussion, we consider the "size" of the constraint violation as one measure and the "duration" of the constraint violation as another. Although alternative measures could be equally appropriate, these two allow us to demonstrate the multi-objective nature of reducing state constraint violations, which is our main purpose in this section. Size of violation may be measured in any L_p norm, but the L_∞ is the traditional choice in the MPC literature. The duration of a violation may be measured in different ways as well: duration may mean the time after which the state constraints are satisfied or the time after which no constraints are even active. These distinctions do not matter in this section.

In many plants, the simultaneous minimization of size and duration of state constraint violations is not a conflicting objective. The optimal way to handle infeasibility is then simply to minimize both size and duration; regulator performance may then be optimized, subject to the "optimally relaxed" state constraints.

However, in some cases, such as non minimum-phase plants, a reduction in size of violation can only be obtained at the cost of a large increase in duration of violation, and vice-versa. The optimization of constraint violations then becomes a multi-objective problem. For a given system, the optimal size/duration curves can be plotted for different initial conditions, as in Fig.1. The user must then decide where

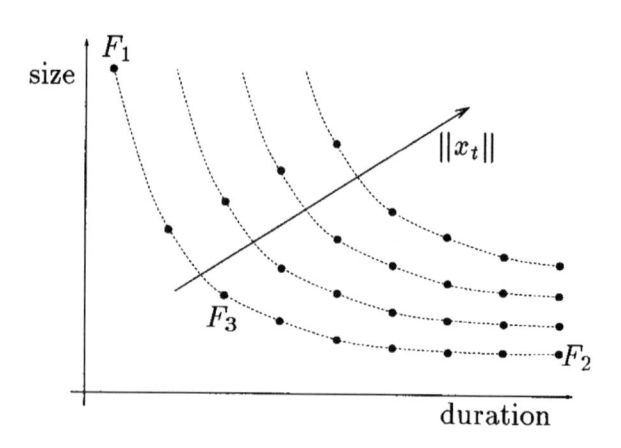

Figure 1. Pareto optimal constraint violation size/duration curves for varying state x_t.

in the size/duration plane, the plant should operate at times of infeasibility. Desired operation may lie on the Pareto optimal curve, since points below this curve cannot be attained and points above it are inferior, in the sense that they correspond to larger size and/or duration than are required.

In some applications, the best operation is to have constraint violations of the smallest possible duration. For instance, if the quality of the product is unacceptable and the product must be discarded during times of constraint violation, then optimal operation is to minimize duration. Assume the current state of the process corresponds to the bottom curve in Fig.1. In this situation, the controller should steer operation towards point F_1. In other applications, it may be critical to minimize the size of constraint violations, so that process shutdown or other exception conditions do not occur. Desirable control behavior is then close to F_2. In many applications, both the size and duration of violations are important. Then F_3 may be the best compromise for the controller to aim for.

The question now arises how to implement a control law that steers operation to a user-defined size/duration compromise. This issue must be addressed by any stabilizing controller.

Implementation of Infeasibility Solutions

The previous section highlights some possible ways in which the problem of infeasibility can be resolved on-line. One approach is to minimize a weighted sum of size and duration. However, this approach requires solution of a mixed-integer linear program and can be computationally taxing. Instead, the user can choose a state dependent value for duration; this choice identifies, for every state, the time of constraint violation to allow in the control optimization. Although this approach determines open-loop predictions that have the desired compromise, the closed-loop behavior can be quite different from the open-loop predictions, and closed-loop performance can be poor. This feature is a well-known and undesirable consequence of receding-horizon implementations of some optimal open-loop strategies. The problem can be resolved by choosing at each time step the minimal value of duration; the mismatch between open and closed-loop then disappears. This approach is discussed in (Rawlings and Muske, 1993), and we expand on that discussion below. An alternative to the minimal duration approach is to minimize size of violation, as measured by some chosen norm. This is the approach of soft-constrained MPC, which we also discuss.

Minimal Duration Approach

In (Rawlings and Muske, 1993), the control algorithm identifies the smallest horizon beyond which the state constraints can be enforced on the infinite horizon. Rawlings and Muske suggest enforcing the state constraint only after that horizon. They do not penalize early constraint violations, and this can lead to unnecessarily large transient violations. A simple modification of the method is to minimize the constraint violations until the hard constraint is enforced. The regulation optimization is then be performed subject to the relaxed constraint over an initial horizon and the hard constraint beyond that.

Calculation of the minimum constraint horizon may require the solution of several linear programs. It is, however, easy to show that the minimum constraint horizon is finite for all finite x_t, and there exists a finite upper bound on the computation involved in calculating it. This technique is clearly the most appropriate in the case mentioned above, when constraint violation implies manufacture of product with no value.

In some examples, even the minimal early violations can be large. This undesirable situation can be viewed as a limitation of the process, rather than the controller. Even in these examples, if the duration of the constraint violations is the critical factor, the control method leads to the best performance that can be achieved. However, if the duration of constraint violation is not the only concern at times of infeasibility, this technique can lead to unsatisfactory performance.

Our assessment is that the minimum duration approach is valid and is optimal in some cases. Its limitations should be understood and the method should be implemented only when it is appropriate. A further disadvantage is that the method may give rise to discontinuous $u(x)$. The perturbed stability of the method should therefore be established in order to address the case of state estimation, coupled with MPC regulation.

Soft Constraints

As mentioned above, in some cases the duration of the constraint violations is not the only concern. The large transients that can be induced by enforcing the hard state constraints at the earliest possible time may then be undesirable and some systematic means is required to reduce the size of the constraint violations. A natural solution appears to be soft constrained MPC. In this approach, a penalty term is added to the control objective, that penalizes a weighted norm of the control violations. The weight applied to the violation norm is scaled so as to give a desirable size/duration compromise, but operation may take place at an inferior point above the Pareto optimal curve. Increasing the constraint weight "hardens" the constraints and reduces the size of the violations. The performance that can be obtained by tuning soft constrained MPC therefore ranges from unconstrained to hard constrained control, if that is feasible.

An attractive feature of the method is that the soft constrained MPC control requires only the solution of a positive definite quadratic program. Besides the computational simplicity of the approach, the control law is Lipschitz continuous in the state and perturbed stability follows.

∞-*norm solutions:* The popular norm in the MPC literature is the ∞-norm, used by Ricker (Ricker et al., 1988) and Zheng and Morari (Zheng and Morari, 1995); this norm measures the largest violation. Although intuitively appealing, this measure of constraint violation inherently introduces mismatch between the open and closed-loop, which can lead to poor closed-loop performance. With non minimum-phase systems, as the weight on the constraint violation term in the cost is increased, the result can be an increasingly slow return of the state to its allowed region. In other words, as the user attempts to make the constraint harder, the duration of violations increases and the size does not necessarily decrease. This non-intuitive property of the method makes it difficult to tune. The advantage of the approach is that only one new decision variable, the peak violation, must be computed on line.

p-norm solutions: In view of the difficulties encountered with the ∞-norm approach, it is interesting to consider the use of other norms. It appears that the ∞-norm is the only one for which the open/closed-loop mismatch is inherent. The performance obtained with other norms is therefore intuitive and it can be shown that open and closed-loop behaviors converge as the control horizon in the MPC law is increased. The 1 and 2-norm solutions are most attractive, since they are the only ones for which the MPC optimization remains a quadratic program. A combination of the 1 and 2-norms may be the best choice, because the 1-norm allows an exact penalty effect to be obtained (Scokaert and Rawlings, 1996), and the 2-norm emphasizes the penalty on large violations.

To summarize, adding the ∞-norm of the constraint violations to the regulator cost adds only a single degree of freedom to the control optimization. The use of other norms adds N more degrees of freedom, but use can be made of the problem structure to reduce the numerical complexity of the optimization to essentially the same level as with the ∞-norm solution.[1]

---

[1] The authors wish to thank Dr. Steve J. Wright for pointing this out.

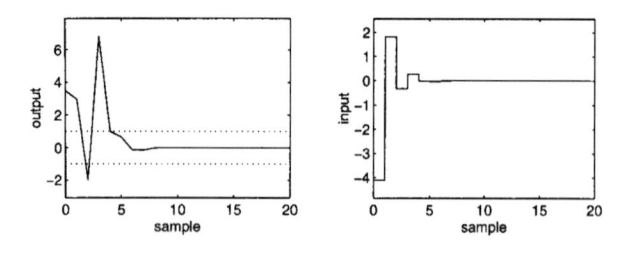

Figure 2. Minimal duration approach

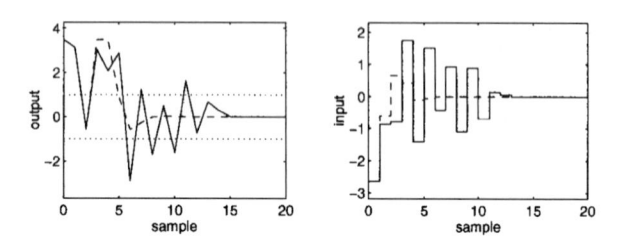

Figure 3. Soft constrained MPC: ∞-norm solution

Illustrative Example

We consider the process described by

$$A = \begin{bmatrix} 1.4 & -0.61 & 0.084 & -0.0036 \\ 1 & 0 & 0 & 0 \\ 0 & 1 & 0 & 0 \\ 0 & 0 & 1 & 0 \end{bmatrix}, B = \begin{bmatrix} 1 \\ 0 \\ 0 \\ 0 \end{bmatrix},$$

with the output given by $y_t = [0.1 \quad 1 \quad -2 \quad 1.48]x_t$. There is no input constraint and the state constraint is defined as $|y_t| \leq 1$. The initial condition we consider is $x_0 = [6 \quad 6 \quad 6 \quad 6]^T$ and we implement the MPC law described in (Rawlings and Muske, 1993), with $Q = I$, $R = 1$ and $N = 12$. The process in non minimum-phase and the initial condition is a steady state that violates the constraint. The hard constrained MPC optimization is therefore initially infeasible. We investigate different ways of relaxing the state constraint.

First, we implement the minimal duration approach and obtain the results displayed in Fig.2. Figs.3 and 4 show the soft constraint simulations, with ∞ and 2-norms, respectively, and and a weight on these of 100. In Figs.2-4, the solid line represents actual closed-loop results and the dashed line represents the open-loop predictions made at time 0. The dotted lines indicate the state constraint levels.

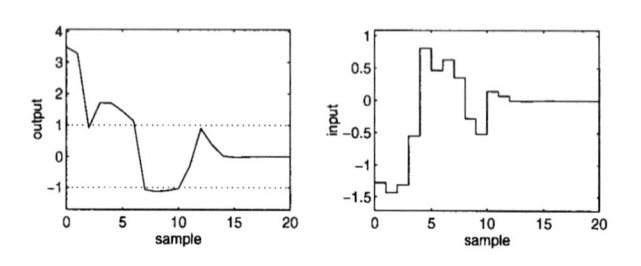

Figure 4. Soft constrained MPC: 2-norm solution

As expected, the results obtained with the minimal duration solution feature fairly large transient violations. However, constraint violations occur for only 4 samples and closed-loop behavior is very close to the open-loop predictions. With the ∞-norm soft constrained solution, the worst constraint violation is smaller and the last violation is observed at sample 11. However, closed-loop behavior is quite different from the open-loop predictions and the mismatch does not decrease as N increases, which makes the approach difficult to tune without extensive trial and error simulations. With the 2-norm solution, the open/closed-loop mismatch is again small enough that open and closed-loop results are indistinguishable. The behavior of the loop is therefore intuitive and easily tuned. The worst constraint violation is no worse than with the ∞-norm solution and constraint violations are not observed after sample 10.

Conclusion

The problem of infeasibility in MPC was investigated and the limitations of current solution methods were discussed. The minimal duration solution was found to give good results and soft constrained approaches were found to be interesting alternatives. The minimal duration and p-norm soft constrained solutions both have their strengths, with violations of low duration for the former and low amplitude for the latter. The use of the ∞-norm approach appears to cause some tuning difficulties. The 1-norm has the attractive feature that it allows implementation of exact penalties (Scokaert and Rawlings, 1996). Finally, it is noted that 1 and 2-norm soft constrained MPC need not be more computationally taxing than the ∞-norm formulation.

References

Rawlings, J. B. and Muske, K. R. (1993). Stability of constrained receding horizon control, *IEEE Trans. Auto. Cont.* **38**(10): 1512–1516.

Ricker, N. L., Subrahmanian, T. and Sim, T. (1988). Case studies of model-predictive control in pulp and paper production, in T. J. McAvoy, Y. Arkun and E. Zafiriou (eds), *Proceedings of the 1988 IFAC Workshop on Model Based Process Control*, Pergamon Press, Oxford, pp. 13–22.

Scokaert, P. O. and Rawlings, J. B. (1996). Infinite horizon linear quadratic control with constraints, *Proceedings of the 13th IFAC World Congress*, San Francisco, California.

Zafiriou, E. (1991). On the effects of tuning parameters and constraints on the robustness of model predictive controllers, in Y. Arkun and W. H. Ray (eds), *Chemical Process Control–CPC IV*, Fourth International Conference on Chemical Process Control, Elsevier, Amsterdam, pp. 363–393.

Zheng, A. and Morari, M. (1995). Stability of model predictive control with mixed constraints, *IEEE Trans. Auto. Cont.* **40**(10): 1818–1823.

LINEAR MODEL PREDICTIVE CONTROL OF INPUT-OUTPUT LINEARIZED PROCESSES WITH CONSTRAINTS

Michael J. Kurtz and Michael A. Henson
Department of Chemical Engineering
Louisiana State University
Baton Rouge, LA 70803-7303

Abstract

An input-output linearization strategy for nonlinear processes with input and/or output constraints is proposed. The nonlinear control system is comprised of: (*i*) an input-output linearizing controller that compensates for process nonlinearities; (*ii*) a constraint mapping procedure that transforms the original constraints into constraints on the feedback linearized system; and (*iii*) a linear model predictive controller that regulates the resulting constrained linear system. The approach combines the computational simplicity of input-output linearization and the constraint handling capability of model predictive control. Simulation results for a continuous stirred tank reactor demonstrate the superior performance of the proposed strategy as compared to conventional model predictive control techniques.

Keywords

Nonlinear processes, Constrained control, Feedback linearization, Model predictive control.

Introduction

Nonlinear model predictive control (NMPC) and input-output linearizing control (IOLC) are the most widely studied nonlinear control techniques for chemical process applications. NMPC offers many of the appealing features of linear model predictive control, including explicit compensation for input and output constraints (Meadows, Henson, Eaton and Rawlings, 1995). As compared to NMPC, IOLC offers several important advantages including transparent controller tuning and low computational requirements (Kravaris and Kantor, 1990). However, conventional feedback linearization techniques do not have constraint handling capabilities (Rawlings, Meadows and Muske, 1994). This has motivated the development of several modifications of the basic input-output linearization approach (Balchen and Sandrib, 1995; Kendi and Doyle, 1995). The most significant disadvantages of these techniques is that they only provide indirect compensation for input constraints and cannot address output constraints whatsoever.

In this paper, an input-output linearization strategy for constrained nonlinear processes is presented. The linearizing controller is designed by neglecting constraints on input and output variables. At each sampling instant, the original input constraints are mapped into constraints on the manipulated input of the feedback linearized system. This transformation yields a *linear* dynamic system with constant output constraints and *time-varying* input constraints. The constraints are handled explicitly by applying linear model predictive control (LMPC) to the constrained linear system. It is important to note that a similar technique has been developed independently by Morari and co-workers (Oliveira, Nevistic and Morari, 1995).

Constrained Input-Output Linearization

The key idea of the proposed strategy is to map the actual constraints into corresponding constraints in the feedback linearized space. Note that the presence of constraints precludes feedback linearization in the traditional sense. The constraint mapping procedure yields a constrained linear system, which necessarily leads to a nonlinear control problem. Our motivation is that the *linear* model predictive controller is much easier to design and implement than the *nonlinear* model predictive controller that is required for the original constrained nonlinear system.

Nonlinear Controller Design

The nonlinear process model has the form,

$$\dot{x} = f(x) + g(x)u \quad (1)$$
$$y = h(x)$$

where u and y are the manipulated input and controlled output, respectively. We assume that the state vector x is available for feedback. The input-output linearizing control law can be written as (Isidori, 1989),

$$u = \frac{v - L_f^r h(x)}{L_g L_f^{r-1} h(x)} \quad (2)$$

We assume that the relative degree r is well defined throughout the region of operation. As discussed below, v is the manipulated input for the feedback linearized system.

Under this control law, there exists a nonlinear coordinate transformation $[\xi^T, \eta^T] = \Phi^T(x)$ such that (1) can be represented as a partially linear system (Isidori, 1989),

$$\dot{\xi} = A\xi + Bv$$
$$\dot{\eta} = q(\xi, \eta) \quad (3)$$
$$y = C\xi$$

We assume that the $(n-r)$-dimensional nonlinear subsystem in (3) is bounded-input, bounded-state stable with respect to the ξ variables as inputs. This assumption ensures that the nonlinear system (1) is stabilized if the r-dimensional linear subsystem in (3) is stabilized.

Constraint Mapping

The nonlinear process is subject to the following input and output constraints:

$$u_{min} \leq u \leq u_{max} \quad y_{min} \leq y \leq y_{max} \quad (4)$$

First, the linear subsystem is discretized to facilitate the subsequent LMPC design. Exact discretization yields,

$$\xi(k+1) = A_d \xi(k) + B_d v(k) \quad (5)$$
$$y(k) = C\xi(k)$$

where the pair (A_d, B_d) is controllable, but the eigenvalues of A_d are located on the unit circle.

The next step is to map the constraints on the original nonlinear system (1) into constraints on the discretized linear system (5). The output constraints for the two systems are identical since the output y is not transformed as part of the IOLC design. By contrast, the constraints on the input u must be mapped into constraints on the new input v. This transformation must be performed at each sampling instant because the mapping is state dependent. Moreover, the transformation must be performed for the entire control horizon of the predictive controller. Thus, at each time step k the objective is to find constraints of the form,

$$v_{min}(k+j|k) \leq v(k+j|k) \leq v_{max}(k+j|k) \quad (6)$$

where $0 \leq j \leq N-1$, and N is the control horizon of the LMPC controller.

The state-dependent relation between $u(k)$ and $v(k)$ follows from (2):

$$v(k) = L_f^r h[x(k)] + L_g L_f^{r-1} h[x(k)] u(k) \quad (7)$$
$$\equiv b[x(k)] + a[x(k)] u(k)$$

Constraint mapping is problematic because estimates of future values of the input and state are not available until the LMPC problem is solved, and the LMPC problem cannot be solved until the input constraints are specified. As a result, exact solution of the LMPC problem requires nonlinear programming or iterative techniques (Oliveira et al., 1995), both of which are computationally expensive. Therefore, it is necessary to approximate the constraints $v_{min}(k+j|k)$ and $v_{max}(k+j|k)$ for $j \geq 1$. In a previous paper (Henson and Kurtz, 1994), we considered holding the constraints constant over the control horizon. This approach is simple, but it can lead to calculated constraints that are significantly different than the actual constraints. The incorrect constraints can result in implemented control moves that are unnecessarily conservative or aggressive.

An alternative approach is to use the inputs calculated at the *previous* sampling time to determine future constraints at the current sampling time. Our experience is that this method yields constraints that are very close to the actual constraints. The constraints are computed using the solution of the LMPC problem at the previous time step:

$$V(k-1|k-1) = [v(k-1|k-1) \cdots v(k+N-2|k-1)]^T \quad (8)$$

The first input is used to calculate the implemented input $u(k-1)$, while the remaining inputs are employed as an estimate of the control sequence at the current sampling time,

$$V(k|k-1) = [v(k|k-1) \cdots v(k+N-2|k-1) \; v_a]^T \quad (9)$$

where the value v_a is arbitrary since it is not actually utilized. The current measurement $x(k)$ is used to calculate the transformed state variables $\xi(k)$ and $\eta(k)$ via the nonlinear change of coordinates $\Phi(x)$. Using these values as initial conditions, the normal form (3) is integrated with the piecewise constant input sequence $V(k|k-1)$ to yield predicted values of the transformed state variables.

These predicted values and the inverse transformation $\Phi^{-1}(\xi, \eta)$ are used to compute estimated values of the state variables,

$$X(k|k) = [x^T(k|k) \cdots x^T(k+N-1|k)]^T \quad (10)$$

where $x(k|k) = x(k)$. The transformed constraints are calculated using the predicted state variables $x(k+j|k)$. For example, the lower constraints are computed as,

$$v_{min}(k+j|k) = \min_u \; b[x(k+j|k)] + a[x(k+j|k)]u \quad (11)$$

where: $u_{min} \leq u \leq u_{max}$. This procedure yields the input constraints $v_{min}(k+j|k)$ and $v_{max}(k+j|k)$ that are used in the LMPC optimization problem. The algorithm is initialized by holding the constraints constant over the control horizon. It is important to note that the first set of constraints, $v_{min}(k|k)$ and $v_{max}(k|k)$, map exactly to the actual input constraints since they are calculated using the current state measurement. Therefore, the implemented input,

$$u(k) = \frac{v(k|k) - L_f^r h[x(k)]}{L_g L_f^{r-1} h[x(k)]} \quad (12)$$

necessarily satisfies the actual constraints.

Linear Model Predictive Controller Design

The LMPC design is based on the linear model (5), the input constraints (6), and the output constraints (4). At each time step the linear model is initialized with the current state measurement. Because the matrix A_d is unstable, we utilize a LMPC technique specifically developed for unstable systems (Muske and Rawlings, 1993),

$$\min_{V(k|k)} \sum_{j=0}^{\infty} [\xi(k+j|k) - \xi_s]^T Q [\xi(k+j|k) - \xi_s] + s [v(k+j|k) - v(k+j-1|k)]^2 + r [v(k+j|k) - v_s]^2 \quad (13)$$

where: ξ_s and v_s are target values for ξ and v, respectively; $r > 0$ and $s \geq 0$ are scalar tuning parameters; and Q is a positive semidefinite tuning matrix. The decision vector is: $V(k|k) = [v(k|k) \cdots v(k+N-1|k)]^T$. To obtain a finite set of decision variables, inputs beyond the control horizon are set equal to the target value: $v(k+j|k) = v_s \; \forall \; j \geq N$.

A necessary condition for the LMPC problem to have a solution is that the unstable modes are driven to their steady-state values by the end of the control horizon: $\xi(k+N|k) = \xi_s$. The targets ξ_s and v_s are calculated from the steady-state form of (5) under the condition that $y = y_s$, where y_s is the setpoint. Although not discussed here, a disturbance model is used to shift the targets such that offset caused by plant/model mismatch is eliminated.

The infinite horizon LMPC problem (13) can be formulated as a quadratic program that is efficiently solved with standard software (Muske and Rawlings, 1993). There exists a control horizon N such that the quadratic program is feasible if the linear system is constrained stabilizable. If the constraints are constant, feasibility at $k = 0$ implies feasibility at all future times (Rawlings and Muske, 1993). This property no longer holds when input constraints vary with time. We address this problem by successively dropping constraints at the end of the control horizon until feasibility is achieved. Also, the constraint mapping strategy is modified so unconstrained inputs are not used for constraint prediction. In this case, we extend the last constrained input over the control horizon to generate the input sequence $V(k|k-1)$.

Table 1. Nominal Operating Conditions.

q	100 L/min	$\frac{E}{R}$	8750 K
C_{Af}	1 mol/L	k_0	7.2 x 10^{10} min^{-1}
T_f	350 K	UA	5 x 10^4 J/min·K
V	100 L	T_c	300 K
ρ	1000 g/L	C_A	0.5 mol/L
C_p	0.239 J/g·K	T	350 K
$-\Delta H$	5 x 10^4 J/mol		

Simulation Study

Consider an irreversible, first-order chemical reaction $A \rightarrow B$ which occurs in a constant volume, continuous stirred tank reactor. The process model can be written as:

$$\dot{C}_A = \frac{q}{V}(C_{Af} - C_A) - k_0 \exp\left(-\frac{E}{RT}\right) C_A \quad (14)$$

$$\dot{T} = \frac{q}{V}(T_f - T) + \frac{(-\Delta H)}{\rho C_p} k_0 \exp\left(-\frac{E}{RT}\right) C_A + \frac{UA}{V \rho C_p}(T_c - T)$$

The nominal conditions in Table 1 correspond to an unstable operating point. The manipulated input and controlled output are the coolant temperature (T_c) and reactor temperature (T), respectively. The manipulated input is constrained: 280 K $\leq T_c \leq$ 380 K. Output constraints are not considered in this example.

We compare the proposed control strategy to linear and nonlinear MPC techniques. The linear model obtained from feedback linearization is discretized with a sampling period $T = 0.05$ min, and used for LMPC design. The resulting controller is called "IOL-MPC." The linear model predictive controller is designed by linearizing (via Taylor series expansion) the CSTR model at the unstable operating point in Table 1; this controller is called "LMPC." The nonlinear model predictive controller is obtained by designing an LMPC controller for the CSTR model successively linearized at the current operating point (Garcia, 1984) This controller is called "SLMPC." The MPC controllers are tuned as follows: IOL-MPC (N=10, q=2, r=1, s=1); LMPC (N=16, q=1, r=4, s=1); SLMPC (N=16, q=4, r=0.3, s=0.3).

The performance of the three controllers for a +35 K disturbance in the feed temperature (T_f) is shown in Figures 1–2. LMPC yields a very sluggish response and unnecessarily large control actions. SLMPC provides significantly improved disturbance rejection, but generates oscillatory control moves. IOL-MPC yields an output response that is very similar to that produced by SLMPC, but with much smoother control moves. The controllers are compared for a –25 K setpoint change in Figure 3. LMPC produces a very sluggish response since the linear model used is not valid at the new setpoint. SLMPC had to be detuned with r=s=0.5 to obtain a stable closed-loop response. Even with detuning, it is unable to attain the new setpoint and the con-

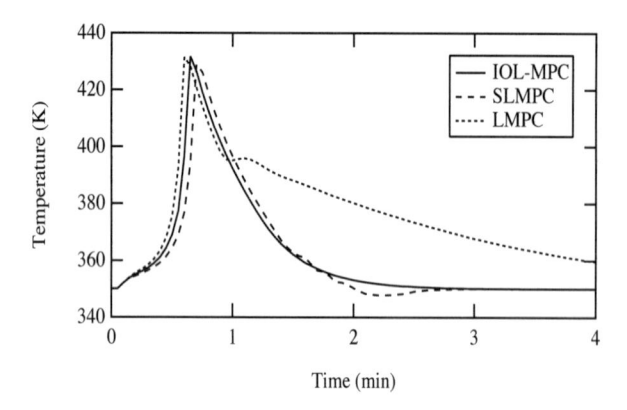

Figure 1. Response for a T_f disturbance.

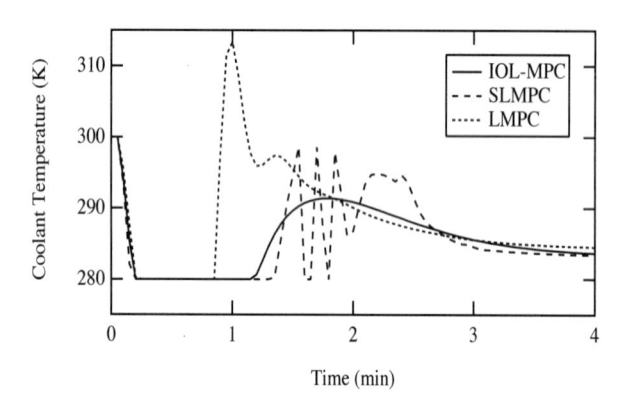

Figure 2. Inputs moves for Figure 1.

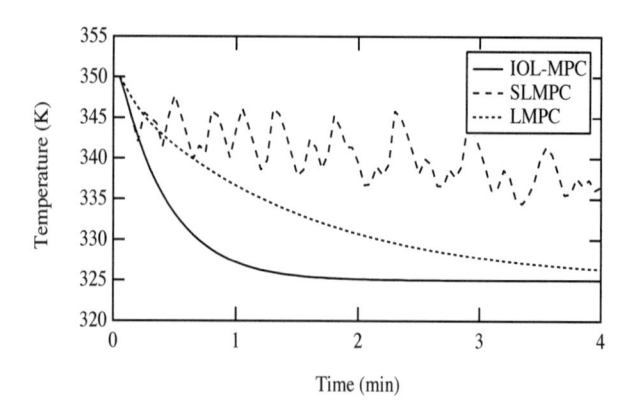

Figure 3. Response for a setpoint change.

trol moves (not shown) are extremely erratic. This behavior is attributable to the successive linearization repeatedly switching between stable and unstable models. By contrast, IOL-MPC yields a fast, smooth setpoint response with little control effort.

Conclusions

An input-output linearizing control strategy for constrained nonlinear processes has been developed and evaluated. The proposed method retains the computational simplicity of input-output linearization while providing the constraint handling capability of model predictive control. Simulation results for a continuous stirred tank reactor show that the proposed method provides significantly improved performance as compared to model predictive control techniques based on local and successive model linearization.

Acknowledgments

The authors would like to acknowledge the National Science Foundation (CTS-9501368) for support of this research.

References

Balchen, J. G. and Sandrib, B. (1995). Input saturation in nonlinear multivariable processes resolved by nonlinear decoupling, *Model. Ident. Control* **16**: 95–106.

Garcia, C. E. (1984). Quadratic dynamic matrix control of nonlinear processes: An application to a batch reactor process, *AIChE Annual Mtg.*, San Francisco.

Henson, M. A. and Kurtz, M. J. (1994). Input-output linearization of constrained nonlinear processes, *AIChE Annual Mtg.*, San Francisco, CA.

Isidori, A. (1989). *Nonlinear Control Systems*, Springer-Verlag, New York, NY.

Kendi, T. A. and Doyle, F. J. (1995). An anti-windup scheme for input-output linearization, *Proc. European Control Conf.*, Rome, Italy, pp. 2653–2658.

Kravaris, C. and Kantor, J. C. (1990). Geometric methods for nonlinear process control: 2. Controller synthesis, *Ind. Eng. Chem. Res.* **29**: 2310–2323.

Meadows, E. S., Henson, M. A., Eaton, J. W. and Rawlings, J. B. (1995). Receding horizon control and discontinuous state feedback stabilization, *Int. J. Control* **62**: 1217–1229.

Muske, K. R. and Rawlings, J. B. (1993). Linear model predictive control of unstable processes, *J. Process Control* **3**: 85–96.

Oliveira, S. L., Nevistic, V. and Morari, M. (1995). Control of nonlinear systems subject to input constraints, *Proc. IFAC Symposium on Nonlinear Control Systems Design*, Tahoe City, CA, pp. 15–20.

Rawlings, J. B., Meadows, E. S. and Muske, K. R. (1994). Nonlinear model predictive control: A tutorial and survey, *Proc. IFAC Symposium on Advanced Control of Chemical Processes*, Kyoto, Japan, pp. 203–214.

Rawlings, J. B. and Muske, K. R. (1993). The stability of constrained receding horizon control, *IEEE Trans. Autom. Control* **AC-38**: 1512–1516.

STABILITY OF NN-BASED MPC IN THE PRESENCE OF UNBOUNDED MODEL UNCERTAINTY

Alexandros Koulouris and George Stephanopoulos
Laboratory for Intelligent Systems in Process Engineering
Massachusetts Institute of Technology
Cambridge, MA 02139

Abstract

Uncertainty associated with neural network models used in control is unstructured, can be unbounded and is also difficult to be characterized in an a priori sense using robustness measures. Feedback limitation is an "emergency" technique that alleviates in an on-line way the stability problems inflicted by unbounded model uncertainty. Like relay autotuning, feedback limitation forces the closed-loop through an oscillation, thus producing additional process information suitable for hierarchical, localized model adaptation.

Keywords

Neural networks, Model predictive control, Stability, Feedback limitation, Uncertainty, Wave-Net.

Introduction

Model Predictive Control (MPC) algorithms based on nonlinear models have recently appeared in the control literature. The shift to nonlinear models for control is fueled by the need for more realistic and, therefore, more accurate process representations. Neural networks (NN) with their ability to learn from historical data and represent any nonlinear mapping offer an attractive alternative to laborious first-principles models.

Although the universal approximation property of NNs guarantees the existence of at least one NN that can represent a given plant with the desired accuracy, it does not suggest ways on how this NN can be obtained. It is, therefore, difficult to derive a priori estimates of the generalization accuracy of a NN model constructed out of finite empirical data. The weights of a NN are empirical coefficients which bear no (functional or conceptual) resemblance to the parameters of the underlying physical model. Although attempts to characterize the uncertainty of NN models with respect to their weights have been reported, the approach does not make much sense in view also of the fact that the number of nodes in a NN and their structural arrangement (i.e., the NN architecture) are defined in ad hoc empirical terms. Finally, due to the fact that most NNs use local or squashing (i.e., saturating) activation functions, they are unable to capture infinite-diverging dynamics even for simple linear plants.

In control, all these facts amount to the conclusion that when NNs are used as process models, modeling uncertainty is unstructured, can be unbounded, and most importantly, difficult to be characterized in an *a priori* way. That, in turn, means that even if we neglect the mathematical difficulties, stability theory cannot handle the robust stability issue for a NN-based MPC controller in an a-priori-guaranteed framework. Absolute stability (Xiao-xin, 1993), although general enough to apply to unstructured and unbounded uncertainty, does not encompass control problems where uncertainty cannot be well-characterized before controller implementation.

This situation is not unique, however, to NN models. In many practical situations, even for the simplest linear models, underestimated uncertainty bounds can lead to unstable behavior even when robustness issues have been taken into account. Such cases force a paradigm shift from the a-priori-guaranteed-stability concept to stability achieved in real-time. The latter requires the use of "emergency" tools that are activated when instability of the closed-loop is observed and whose only purpose is to force the closed-loop to return to the regions of expected stable behavior. In this paper, we propose the use of the *feedback limitation* (Carrier, 1994) technique as a tool for stabilizing NN-based controllers.

Model adaptation is the natural thing to do in order to avoid re-occurrence of the destabilizing implications of uncertainty. In this framework, the overstated but underused ability of NNs to learn in an on-line way becomes really important. Adaptive control techniques with NN models have indeed been reported, and theoretical results on stability have been constructed. The disadvantage of those techniques, however, is that they confine model uncertainty to the NN weights, an assumption difficult to justify in practice. In this paper, the hierarchical multiresolution framework of *Wave-Nets* (Koulouris, 1995), employing wavelets as the basis functions of the NN, is used to perform both parametric and structural model adaptation.

NN-based MPC

Fig. 1 presents the generic MPC closed-loop with a NN process model. The functionality and importance of the **F** block will be explained in the next section. The behavior of a NN-based controller is characterized by two features: a) localization, and, b) possible unboundedness of modeling error. As it is shown in Koulouris (1995), nonlinear closed-loops exhibit very different stability patterns of behavior at various operating regimes, with the set-point value behaving like a "bifurcation" parameter. This is partially due to the uneven distribution of error in the nonlinear model space. This localization of uncertainty is particularly evident for NN models whose saturation outside their training range can result in unbounded errors. The following example will illustrate the effects of unbounded uncertainty.

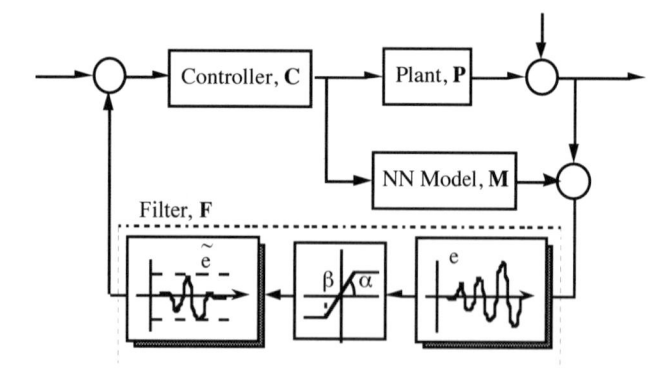

Figure 1. Schematic of NN-based MPC.

Example 1

Consider the following nonlinear discrete-time plant:

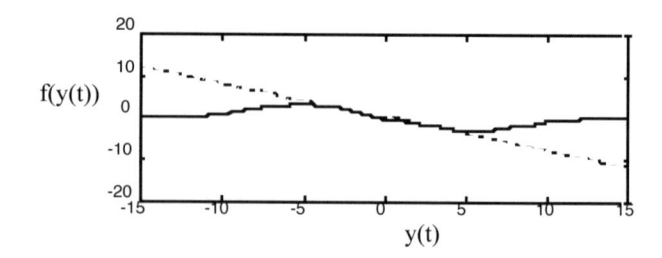

Figure 2. Plant vs. model nonlinearity.

$$y(t+1) = f(y(t)) + u(t) = -0.8\frac{y^3(t)}{1+y^2(t)} + u(t) \quad (1)$$

Fig. 2 compares the plant nonlinearity, $f(y(t))$, (dashed line) with a *Wave-Net* model (solid line) constructed out of 100 data in the range [-5,5]. Outside the training range the model settles to zero, thus failing to capture the infinite-diverging process dynamics. Fig. 3 presents the closed-loop behavior of a deadbeat MPC (i.e., unit horizon) for a set-point change to $r=6$. The unstable behavior is attributed to the unboundedness of the error.

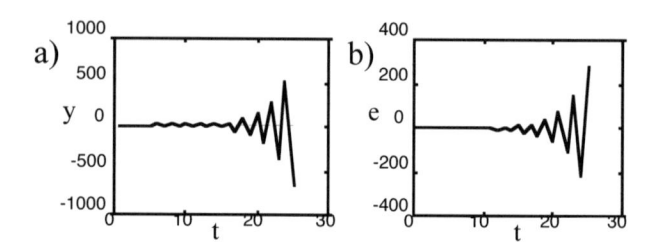

Figure 3. Closed-loop response to r=6.

Feedback Limitation

As clearly shown in Fig. 3b, in MPC the magnitude of the modeling error is reflected on the values of the feedback signal. Assuming that an upper bound on the magnitude of the disturbances exists, the feedback signal can effectively be used as a sensor of large modeling uncertainties and approaching instabilities. It should be noted that uncertainty-induced instabilities are the only type of instabilities that we are concerned with in this paper.

Feedback limitation or *clipping* is the natural thing to do when the magnitude of the feedback signal exceeds the bounds established during the course of "normal" operation. In its simplest form, the feedback limitation filter is given mathematically by the following function:

$$\widetilde{e}(t) = \begin{cases} \alpha\, e(t) & |e(t)| < \beta \\ \alpha\beta\, sign(e(t)) & |e(t)| > \beta \end{cases} \quad (2)$$

where, $e(t)$, is the feedback signal and α, β are adjustable parameters. In the extreme case of complete feedback *elimination* (i.e., β=0), the policy corresponds to operating the system in open-loop, a strategy usually implemented by operators in the event of instabilities (Carrier, 1994).

Following Economou et al. (1986), a sufficient condition for BIBO stability of the above closed-loop is:

$$g(F)g(C)g(P-M) < 1 \qquad (3)$$

where, $g()$, represents the gain of the corresponding operator. It is easy to show that, $g(F) = \alpha$. Therefore, for given stable C and P-M, a filter F with appropriately small gain can be constructed so that condition (3) is always satisfied. For a clipped IMC-type closed-loop with a stable plant and controller, it is trivial to show that BIBO stability is always guaranteed since all signals in the loop are bounded.

Filter Tuning

Despite its ability to contain unbounded instabilities, feedback limitation does not necessarily provide *asymptotic* stabilization of the closed-loop. Asymptotic robust stability and performance considerations will, therefore, determine the values of the filter parameters that will ensure acceptable closed-loop performance in real-time.

In designing the clipping filter, α can initially be set to 1 (no signal distortion within the passing range) and β equal to the worst-case error bound estimated during the process identification phase. With these settings, the filter is expected to operate only as an emergency tool when the encountered model uncertainty exceeds the estimated bound. Decreasing the values of both filter parameters can provide additional robustness to the closed-loop. By varying the magnitude of the clipping parameters over the different operating ranges, we can accommodate the localization of the modeling error and avoid conservatism.

The next example illustrates the different types of closed-loop behavior that can be obtained while varying the values of the filter parameters.

Example 1 (revisited)

Fig. 4 presents the closed-loop responses of the plant given by Eqn. (1) to a set-point change to r=6 in the presence of two different clipping filters with β=0 and β=10 respectively. In both cases, α=1. As expected, feedback limitation is able to stabilize the output in a BIBO sense, it creates, however, offset whose magnitude increases with the clipping parameter, β, decreasing. On the other hand, tighter clipping results in better control of the output oscillations.

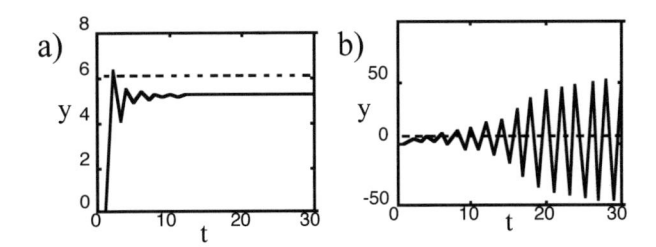

Figure 4. Output responses with feedback clipping. a) β=0, b) β=10.

Multiscale Clipping

The feedback limitation filter is a special case of the stabilizing filter proposed by Economou et al. (1986) for nonlinear IMC. Its effect on the closed-loop is also not very different from the actuator constraints whose presence, as observed by Li and Biegler (1988), induces BIBO stability to the closed-loop. The frequency-domain character of the nonlinear IMC filter and the intuitiveness of the time-domain input constraints can be effectively combined in the multiresolution extension of the feedback limitation idea presented in this section.

Fig. 5 (Carrier, 1994) graphically illustrates *multiscale* clipping. The idea is to apply different clipping thresholds to the different frequency bands resulting from the application of the wavelet transform on the feedback signal. Multiscale clipping allows the selective isolation and treatment of phenomena living at different ranges of the frequency spectrum. For example, low frequency components of the feedback that are responsible for integral action should be allowed to pass intact, thus eliminating offset. Similarly, by properly selecting the clipping thresholds and allowing process disturbances of variable amplitude at various frequencies to pass through the filter, effective disturbance rejection and noise elimination are also possible.

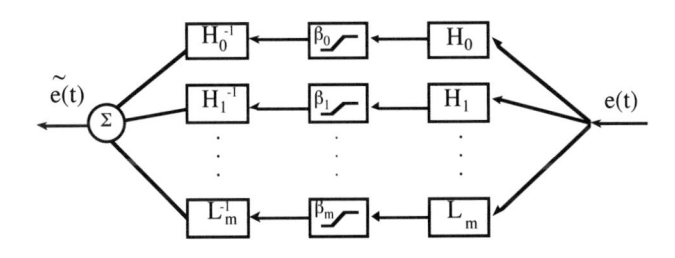

Figure 5. Multiscale clipping.

Example 2

Consider the plant given by the following equation:

$$y(t+1) = \frac{y(t)}{1+y^2(t)} + 0.2\sin(5y(t)) + u(t) \qquad (4)$$

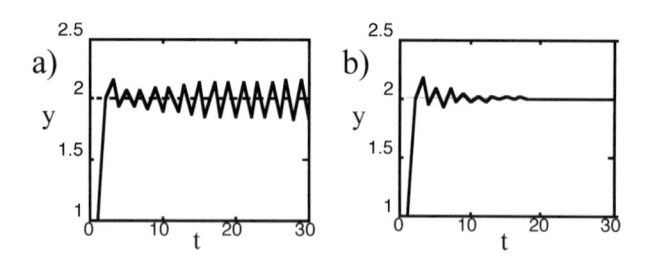

Figure 6. Output responses for a) unclipped and, b) clipped feedback.

A *Wave-Net* model of the process is constructed with the data indicating an upper bound of 0.25 on the modeling error. With a deadbeat MPC controller, the unclipped closed-loop response to a set-point change to $r=2$ is unstable, as shown in Fig. 6a. The clipped response (Fig. 6b) is, however, stable and, most importantly, offset-free. To achieve this performance, multiscale clipping was implemented with β set at the error bound (i.e., 0.25) at all levels except the top one where the need to eliminate high frequency oscillations indicated a value of $\beta=0$.

Clipping and Adaptation

As it was pointed out earlier, feedback clipping does not necessarily lead to asymptotic stabilization of the closed-loop. It is well-known (e.g., Campo and Morari, 1990) that the presence of saturation nonlinearities results in closed-loop oscillations, and clipping is no different in that regard. The only difference is that oscillations are desirable in this case, since the additional process information created can be used for model adaptation. In this respect, feedback limitation is related to *relay autotuning* proposed for PID controllers.

To accommodate the spatial localization of the modeling error, a localized adaptation procedure should be constructed. These properties are encapsulated in the *Wave-Net* learning mechanism (Koulouris, 1995) which implements both parametric (global) and structural (local) adaptation. The benefits of localized adaptation are demonstrated in the following example.

Example 1 (final visit)

If we implement the unclipped closed-loop of Example 1 for a set-point change to $r=10$, unbounded unstable behavior is observed similar to the one shown in Fig. 3a. With clipping only ($\beta=10$), boundedness is achieved (Fig. 7a), however, large amplitude oscillations are still present at the output. If we allow on-line model adaptation with the use of the data produced by clipping, the result is asymptotic stabilization of the closed-loop (Fig. 7b). The result is not surprising if we look at the model (Fig. 8) after adaptation. Learning with the use of the *Wave-Net* algorithm allowed the local extension of the model's validity to a new range of values, without, however, distorting the model in its original training area.

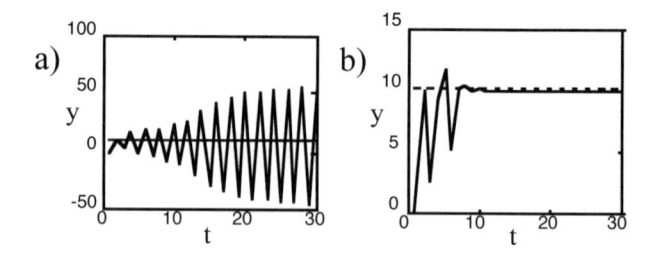

Figure 7. Output responses with a) clipping only and b) clipping and adaptation.

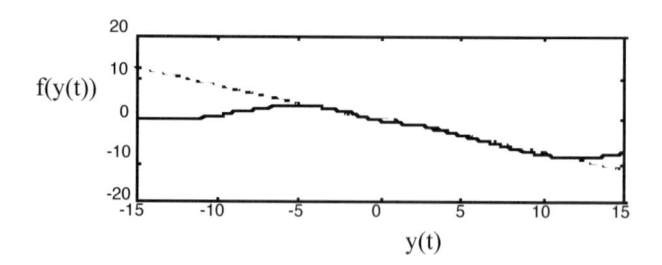

Figure 8. Plant vs. model after adaptation.

Conclusions

The uncertainty of NN models used in control is difficult to be accurately characterized in an a priori way. When robust stability cannot be guaranteed, feedback limitation can effectively serve to sense and reject uncertainty-induced instabilities. Through localized tuning and selective clipping of different frequency ranges, it is possible to achieve offset-free disturbance rejection without imposing conservatism in the performance. By reducing model uncertainty, localized model adaptation can deactivate clipping, effectively disallowing the re-occurrence of instabilities caused by unbounded uncertainty.

References

Campo, P. J., and M. Morari (1990). Robust stability of processes subject to saturation nonlinearities. *Computers chem. Engng*, **14**, 343-358.

Carrier, J. (1994). *The Application of Time-Frequency Techniques to Identification and Control*. PhD thesis. Massachusetts Institute of Technology.

Economou, C. G., M. Morari, and B. O. Palsson (1986). Internal Model Control. 5. Extension to nonlinear systems. *Ind. Eng. Chem. Process Des. Dev.*, **25**, 403-411.

Koulouris, A. (1995). *Multiresolution Learning in Nonlinear Dynamic Process Modeling and Control*. PhD Thesis. Massachusetts Institute of Technology.

Li, W. C., and L. T. Biegler (1988). Process control strategies for constrained nonlinear systems. *Ind. Eng. Chem. Res.*, **27**, 1421-1433.

Xiao-xin, L. (1993). *Absolute Stability of Nonlinear Control Systems*. Kluwer Academic Publishers.

MIXED OBJECTIVE OPTIMIZATION FOR ROBUST PREDICTIVE CONTROLLER SYNTHESIS

Kostas Hrissagis and Oscar D. Crisalle
Chemical Engineering Department
University of Florida
Gainesville, FL 32611

Abstract

A new design framework for synthesis of robust predictive regulators with various time-domain performance specifications is proposed. The controller robustly stabilizes a nominal plant model corrupted by unstructured uncertainty, and satisfactorily rejects exogenous disturbances affecting the plant output. Disturbances may be certain fixed sequences or alternatively be modeled as persistent or energy bounded signals. Adequate nominal performance in the regulation mode is obtained by optimizing a convex objective that minimizes the worst-case amplification of the disturbance inputs. Robust stability is guaranteed by satisfying classical frequency-domain conditions through an H_∞ problem. The achievable limits of performance in the time domain using a linear controller are quantified in this framework. The proposed mixed objective methodology avoids *ad hoc* procedures and rules of thumb, in a systematic two-stage design process.

Keywords

Robust predictive control, Disturbances, Constrained control, l_1/H_∞ design.

Introduction

Although predictive control has been an active research area for almost two decades, the design of robust predictive controllers subject to input and output constraints has proven to be a most challenging problem. An early study is that of Zafiriou (1990), and more recent results are given by Genceli and Nikolaou (1993), Zheng and Morari (1993), and also Vuthandam et al. (1995). All these methods apply to FIR plant representations. A number of theoretical results have appeared for predictive control systems using state-space (Rawlings and Muske, 1993) or transfer-function plant representations (*cf.* Rossiter et al., 1995; Yoon and Clarke, 1995; Scokaert and Clarke, 1994), but these techniques do not address the problem of *constrained robust control*.

Taking advantage of recent developments in mixed objective optimization theory, new venues for synthesizing robust predictive regulators for processes under process constraints and various performance specifications are proposed. In this framework transfer function models are used to represent the process so that unstable systems can be included in the design. The plant model is corrupted by unstructured uncertainty and subject to bounded but possibly persistent exogenous disturbances. The resulting optimization is not computationally expensive, and the method is applicable to conventional quadratic functionals.

The objectives of this paper are (i) to promote a discussion on the state-of-the-art on the design of robust constrained predictive controllers for linear models, and (ii) to identify new opportunities for continued research. It is hoped that the perspective offered by the authors will foster the emergence of novel venues to robustify and improve the performance of predictive controllers.

Controller Parameterization

The essence of predictive control design is the minimization of the quadratic performance functional

$$J(t) = \sum_{i=1}^{N_y}\left[r(t+i) - y(t+i|t)\right]^2 + \lambda \sum_{i=1}^{N_u}\left[\Delta u(t+i)\right]^2 \quad (1)$$

where $\{r(t+i)\}$ is the sequence of future set point values, $\{y(t+i|t)\}$ is the sequence of predicted future values of the output, $\{\Delta u(t+i)\}$ is the sequence of future control increments, λ is the move-suppression parameter, and parameters N_y and N_u are the prediction and control horizons, respectively. The terms in the first summation penalize future predicted errors, and the terms in the second summation penalize excessive control energy. If modeling uncertainties, process constraints, and the presence of exogenous disturbances are ignored, the following *nominal* predictive control law results as a consequence of the minimization of (1):

$$\frac{R(z)}{z^n} u(z) = T(z) r(z) - \frac{S(z)}{z^n} y(z) \quad (2)$$

where $R(z)$, $S(z)$, and $T(z)$ are polynomial operators. The nominal closed-loop system with the RST configuration (2) is shown in Fig. 1. where $g(z) = B(z)/A(z)$ is the nominal plant. For the purpose of this paper it is useful to remark that polynomials $R(z)$ and $S(z)$ are of degree n, i.e., of degree equal to that of the nominal plant polynomial $A(z)$. Explicit formulas for the three predictive control polynomials can be found in Crisalle et al. (1989) and in Hrissagis et al. (1995, 1996a).

It is possible to parameterize a stabilizing nominal RST predictive controller (2) using the Wiener-Hopf theory to yield a family of nominally stabilizing predictive controllers (Hrissagis et al. 1995, 1996a). We deploy the resulting parameterized structure as shown in Fig. 1b, where the Youla parameter $Q(z)$ is an arbitrary proper and stable transfer function. Furthermore, the stable and proper transfer functions $N(z)$, $M(z)$, $X(z)$, and $Y(z)$ shown in Fig. 1b satisfy the Diophantine equation $N(z)X(z) + M(z)Y(z) = 1$, and are functions of the nominal predictive controller and the nominal plant through the fractional representations

$$M(z) := \frac{A(z)}{A_1(z)}, \quad N(z) := \frac{B(z)}{A_1(z)} \quad (3)$$

$$Y(z) := \frac{R(z)}{A_2(z)}, \quad X(z) := \frac{S(z)}{A_2(z)} \quad (4)$$

where polynomials $A_1(z)$ and $A_2(z)$ have real coefficients, are both of degree n, and are obtained by factoring the characteristic closed-loop polynomial in the form $A(z)R(z) + B(z)S(z) = A_1(z)A_2(z)$. Finally, the operator $Z(z) = z^n A_2(z)/T(z)$ shown in Fig. 1b is the set-point advancement transfer function.

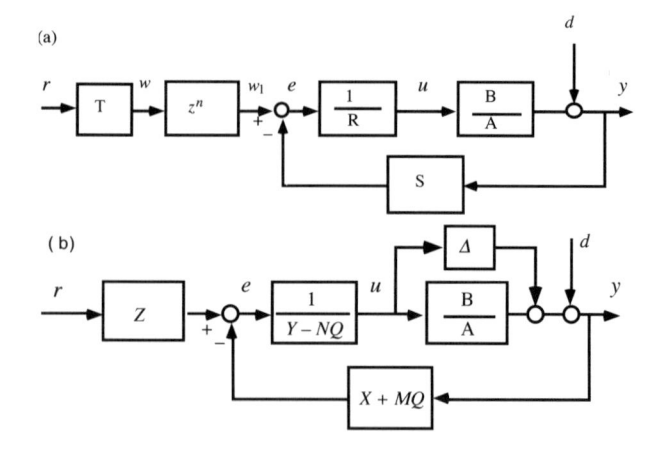

Figure 1. (a) Structure of a nominal predictive controller. (b) Structure of the parameterized predictive controller featuring the Youla parameter $Q(z)$ and the uncertainty block $D(z)$.

Note that Fig. 1b also features an additive uncertainty $\Delta(z)$ which satisfies $|\Delta(e^{i\omega})| \leq |W(e^{i\omega})|$ $\forall \omega$, where $W(z)$ is a known weight. Other uncertainty descriptions can be treated analogously. The following theorem characterizes the robust stability of the parameterized control loop.

Theorem 1. *A necessary and sufficient condition for the robust stability of the closed-loop system of Figure 1b is the inequality condition*

$$\|W(z)C(z)S_l(z)\|_\infty < 1 \quad (5)$$

where the operator $C(z)$ and the loop sensitivity $S_l(z)$ are given by

$$C(z) := \frac{X(z) + M(z)Q(z)}{Y(z) - N(z)Q(z)} \quad (7)$$

$$S_l(z) := M(z)\left[Y(z) - N(z)Q(z)\right] \quad (8)$$

The proof of Theorem 1 makes use of standard robustness results (Francis, 1987). The dependence of the robust stability condition on the parameter $Q(z)$ is made more evident by writing (5) in the model-matching form

$$\|T_1(z) - T_2(z)Q(z)\|_\infty < 1 \quad (9)$$

where

$$\begin{aligned} T_1(z) &= W(z)X(z)M(z) \\ T_2(z) &= -W(z)M(z)^2 \end{aligned} \quad (10)$$

It is more convenient for optimization purposes to further manipulate (5) into the form

$$\|\tilde{R}(z) - Q(z)\|_\infty \leq \gamma \quad (11)$$

which is obtained from (9) through a series of norm preserving transformations (Hrissagis et al. 1996a). The notation $R^{\sim}(z) = R(z^{-1})$ represents the reciprocal operation.

Hence, a robustly stabilizing predictive controller may be obtained by solving the model-matching problem (9) for $Q(z) \in H_\infty$, and then deploying the controller as in Fig. 1b (*cf.* Hrissagis et al., 1995). The resulting parameterized controller is shown to preserve the nominal servo (set-point tracking) performance of the original RST predictive controller. On the other hand, the nominal regulation response is a function of the Youla parameter. Hence one can select the appropriate $Q(z)$ parameter that simultaneously achieves two objectives: (i) robust stability with respect to the family of perturbations $\Delta(z)$, and (ii) optimal regulation performance with respect to disturbances $d(t)$.

This dual specification problem is addressed in the following section by combining the H_∞ problem (9) with a convex optimization problem that accounts for nominal regulation performance using various measures such as the l_∞, l_1, and H_2 norms. Thus the synthesis procedure can be cast as a mixed-objective optimization problem.

Mixed Objective Optimization

We propose a two-stage design procedure for synthesizing predictive controllers that satisfy given nominal performance specifications, and are robust in the face of modeling uncertainty. The first stage consists of designing a predictive controller for the nominal unconstrained system using any of the currently available and well established methods. Then we parameterize the predictive controller with respect to a Youla parameter $Q(z)$ as discussed in the previous section. In the second stage the design is optimized in order to satisfy the performance specifications (such as operational constraints, disturbance rejection, etc.), and to achieve a desirable level of robustness against unstructured perturbations.

A general mixed-objectives optimization problem for the second stage in the design of robust predictive controllers can be cast as the following optimization problem:

$$\min_{Q(z) \in H_\infty} \varphi(Q(z)) \tag{12}$$

such that

$$\left\| R^{\sim}(z) - Q(z) \right\|_\infty \leq \gamma \tag{13}$$

where $\varphi(\cdot)$ is a performance functional. Clearly equation (13) yields the robustness condition (11) with g<1, and equation (12) addresses the nominal performance requirements. The mixed-objective problem (12)-(13) affords systematic solutions when the performance functional is concave. In general, the performance functional may be nondifferentiable.

The explicit form of the functional $\varphi(\cdot)$ depends on the design specifications. In particular $\varphi(\cdot)$ can take the form of a system or signal norm as discussed in three cases of interest shown below.

l_∞/H_∞ *Design:*

$$\min_{Q \in H_\infty} \left\| v(k) \right\|_{l_\infty} \tag{14}$$

such that $\left\| R^{\sim}(z) - Q(z) \right\|_\infty \leq \gamma$

where $v(k) = \mathcal{Z}^{-1}\{v(z)\}$, $v(z) = T_{vd}(z;Q)d(z)$, and where $T_{vd}(z;Q)$ represents the transfer function between signals $d(z)$ and $v(z)$ and is an affine function of $Q(z)$. Note that the l_∞ signal norm is the maximal time-domain absolute peak of the signal. Hence, in this formulation, the l_∞ optimization (14) seeks to minimize the maximum amplitude of a signal $v(k)$ which in turn depends on given (known) disturbance signal $d(k)$. Cases of particular interest are those where signal $v(k)$ is the output of the controller $u(k)$ or of the process $y(k)$. In this fashion, input and/or output saturation constraints are directly addressed in the l_∞/H_∞ problem.

l_1/H_∞ *Design:*

$$\min_{Q \in H_\infty} \left\| T_{vd}(z;Q) \right\|_1 = \min_{q \in l_1} \sum_{k=0}^{\infty} |t(k;q)| \tag{15}$$

such that $\left\| R^{\sim}(z) - Q(z) \right\|_\infty \leq \gamma$

where $t(k;q)$ are the impulse response coefficients of the transfer function $T_{vd}(z;Q)$, and q represents the impulse-response sequence of $Q(z)$. Note that the l_1 operator norm represents the "peak-to-peak" gain between the input and output signals. Hence the l_1 optimization problem (15) seeks to minimize the amplification of peaks in signal $v(k)$ caused by signal $d(k)$. Due to the definition of the operator 1-norm, the class of disturbances considered is the entire family of signals $\|d\|_\infty \leq 1$; that includes all persistent but bounded signals. The l_1/H_∞ method is thus a worst case design because it yields performance guarantees for all possible bounded signals. Again, if $v(k)$ is chosen to represent either $u(k)$ or $y(k)$, the design method permits the synthesis of controllers in the presence of saturation constraints. A comprehensive discussion can be found in (Hrissagis et al., 1996b).

H_2/H_∞ *Design:*

$$\min_{Q \in H_\infty} \left\| T_{vd}(z;Q) \right\|_2 = \min_{q \in l_1} \left(\sum_{k=0}^{\infty} t(k;q)^2 \right)^{1/2} \tag{16}$$

such that $\left\| R^{\sim}(z) - Q(z) \right\|_\infty \leq \gamma$

The optimization problem (16) can have two interpretations. First, from a stochastic point of view, (16) seeks to minimize the RMS value of signal $v(k)$ subject to a white noise disturbance signal $d(k)$. Second, using the definition (Dahleh and D.-Bobillo, 1995)

$$\|T_{vd}(Q)\|_2 := \sup_{\|d\|_2 \leq 1} \|v\|_\infty$$

another interpretation of (16) is that it minimizes the maximum amplitude of signal $v(k)$ subject to the worst-case energy bounded disturbance $d(k)$. As before, interpreting $v(k)$ as either $u(k)$ or $y(k)$, the H_2/H_∞ design method permits the explicit design for input or output constrained systems.

Decoupled Problem— Control Synthesis

The three mixed objective optimization problems formulated in the previous section involve a semi-infinite convex optimization problem coupled with a standard H_∞ problem. This remains a most challenging problem. On the other hand, a suboptimal solution for these kinds of problems can be obtained using the method described in Sznaier (1994a). The common approach is to truncate the norm of the performance objective after a finite horizon N, and introducing constraints that ensure adequate performance at times beyond N. This simplification has great computational advantages because the truncated performance problem and the H_∞ problem can be decoupled and solved sequentially. The decoupled problem then can be posed as follows:

$$\min_{\substack{q \in \mathbb{R}^N \\ \|L(R^\sim, q)\|_2 \leq \gamma}} \|t(k;q)\|_i \qquad (17)$$

where $i = 1, 2, \infty$, as needed, and the constraint $\|L(R^\sim, q)\|_2 \leq \gamma$ (Sznaier, 1994a) is introduced to guarantee the existence of a solution to the robustness problem (18). Once an optimal solution q to (17) is found, the operator $Q_F = \sum_{i=0} q_i z^{-i}$ is formed and used to pose the unconstrained H_∞ problem

$$\|R(z)^\sim - (Q_F(z) + z^{-N} Q_S(z))\|_\infty \leq \gamma \qquad (18)$$

where the only unknown is $Q_S(z)$. Finally, an appropriate $Q_S(z)$ is found using any standard method for unconstrained H_∞ optimization (Rotstein and Sideris, 1992), and the optimal mixed-objective predictive controller is synthesized using $Q(z) = Q_F(z) + z^{-N} Q_S(z)$ and constructing the operators shown in Fig. 1b.

Conclusions

A design methodology for robustifying predictive regulators is proposed for control system design in the presence of performance constraints based on mixed objective optimization. Experience with the l_1/H_∞ approach shows that good control is achieved. In general, high order controllers are obtained; however, large controller orders are often deemed acceptable in predictive control schemes where high-order FIR models are used to represent the nominal plant. The robust design approaches based on the l_∞/H_∞ and H_2/H_∞ methods have not yet been extensively studied in the context of predictive control. It appears that the mixed-objective design methodology is inherently capable of delivering controllers of practical utility and merits further investigation.

References

Crisalle, O.D., D. E. Seborg, and D. A. Mellichamp (1989). Theoretical analysis of long-range predictive controllers. American Control Conference, Pittsburgh PA, 570-576.

Dahleh M. A., and I. Diaz-Bobillo (1995).*Control of Uncertain Systems*, Prentice Hall.

Francis, B. A. (1987).*A Course in H_∞ Control Theory*, Springer Verlag.

Genceli, H., M. Nikolaou (1993). Robust stability analysis of constrained l_1-norm model predictive control. *AIChE Journal*, **39**, 12, 1954-1965.

Hrissagis, K., O. Crisalle, and M. Sznaier (1995). Robust design of unconstrained predictive controllers. American Control Conference, Seattle WA, pp. 3681-3687.

Hrissagis, K., O. Crisalle, and M. Sznaier (1996a). Robust predictive control design with guaranteed nominal performance. To appear *AIChE Journal*.

Hrissagis, K., O. Crisalle, and M. Sznaier (1996b). Robust predictive control with constraints via l_1/H_∞ design. To appear *IFAC World Congress*, San Francisco, CA.

Rawlings, J. B., and K. Muske (1993). The stability of constrained receding horizon control. *IEEE Transactions on Automatic Control*, **38**, 2.

Rossiter, J.A., B. Kouvaritakis, and J.R. Gossner (1995). Feasibility and stability results for constrained stable generalized predictive control. *Automatica*, **31**, 6,. p. 863-877.

Rotstein, H., and A. Sideris (1992). Discrete-time H_∞ control: The one-block case. *Proceedings of the IEEE Conference on Decision and Control*, Tucson AR.

Sznaier, M. (1994a). Mixed l_1/H_∞ controllers for SISO discrete time systems. *Systems and Control Letters*, **23**, 9, 179-186.

Sznaier, M. (1994b). An exact solution to general SISO mixed H_2/H_∞ problems via convex optimization. *IEEE Trans. Autom. Control*, **39**, 12, 2511-2517.

Vuthandam, P., H. Genceli, and M. Nikolaou (1995). Performance bounds for robust quadratic DMC with end condition. *AIChE Journal*, **41**, 9, 2083-2098.

Yoon, T.-W., and D. W. Clarke (1995). Observer design in receding-horizon predictive control. International Journal of Control, **61**, 1, 171-191.

Zafiriou, E., (1990). Robust Model Predictive Control of Processes with Hard Constraints. *Computers Chem. Engin.*, **14**, 4, 359-371.

Zheng, Z. Q., and M. Morari (1993). Robust stability of constrained model predictive control. *American Control Conference*, San Francisco CA, 379-383.

M_p TUNING OF MULTIVARIABLE UNCERTAIN PROCESSES

Karel Stryczek and Coleman B. Brosilow
Chemical Engineering Department
Case Western Reserve University
Cleveland, OH 44106

Abstract

Mathematical models of processes are commonly used to design controllers that must perform well with real processes and not just with the models. For design and tuning purposes a real process can be represented by a linear model whose real parameters vary within specified intervals. The controller is then tuned to work well with the whole set of linear processes. A maximum peak (Mp) procedure for tuning multiple input multiple output (MIMO) internal model control (IMC) systems under parametric uncertainty is presented. This technique finds parameters of an IMC controller such that a closed-loop performance criterion is satisfied for each process from the set of possible process models, and for each process variable. Two examples of two input, two output processes show how to apply Mp tuning.

Keywords

Robustness, Tuning, Uncertain systems, Internal model control, Multivariable control systems.

Introduction

Mp tuning uses frequency domain methods to achieve time domain performance objectives. It differs from other H_∞ methods in that multiple objectives are used to tune the multivariable system. These objectives are to obtain the fastest possible closed loop responses given the maximum allowable overshoot of each process variable to a step change in its setpoint, and the maximum allowable response of any process variable to a step change in the set point of another process variable (i.e. the maximum allowable interactions).

Multivariable Mp Tuning Criteria

The time domain objectives given above convert easily into the following frequency domain objectives for the maximum allowable overshoot of a process variable to a step change in its set point.

$$\max_{\omega, \mathbf{x}} \left| \mathbf{T}_{ii}(\omega, \varepsilon, \mathbf{x}) \right| = \mathbf{Mp}_{ii} \quad (1) \text{ s.t. } \varepsilon_i > 0, \omega \geq 0, \mathbf{x}_L \leq \mathbf{x} \leq \mathbf{x}_U,$$

where:

$\mathbf{T}$ = Closed Loop Transfer Function Matrix between Outputs and Set Points. $\mathbf{Mp}_{ii}$ = Frequency domain equivalent of the time domain specification
$\mathbf{x}$ = vector of uncertain parameters with upper and lower limits of $\mathbf{x}_U$ and $\mathbf{x}_L$ respectively.
ε = vector of IMC controller filter time constants.

For example, specification of a maximum overshoot of ten percent, corresponds to a $\mathbf{Mp}_{ii}$ of 1.05. The relationship between $\mathbf{Mp}_{ii}$ and other overshoot specifications can be found in Seborg et al. (1989).

The conversion of time domain maximum interaction specifications to the frequency domain is slightly more complicated. We assume that the off diagonal terms in $\mathbf{T}(s)$ can be approximated as $\kappa \tau s / ((\tau s)^2 + 2\zeta \tau s + 1)$ cascaded with a possible dead time. The values of κ and ζ determine the peak of the interaction in the time domain. The maximum interaction can be estimated by fitting the foregoing second order system to $\max_{\mathbf{x}} \left| \mathbf{T}_{ij}(\omega, \varepsilon, \mathbf{x}) \right|$.

These estimates are usually somewhat conservative, since the curve of the maximum of $\mathbf{T}_{ij}$ over the parameters, $\mathbf{x}$, vs frequency, ω, includes parts of the frequency responses of several process.

Mp Tuning Algorithm

For MIMO control systems, it is not always possible to achieve an arbitrary set of specifications on the closed loop input-output responses. Therefore, the first task of the tuning algorithm is to determine if the current specifications are feasible. If they are not, then the engineer may either adjust his specifications, or try to find a new model which allows the specifications to be achieved. To determine whether a set of specifications are achievable, we claim that it is sufficient to investigate the situation where all filter time constants are very large relative to the unforced dynamics of the process. In this case, only the process and model gains influence closed loop behavior, and the closed loop frequency response can be closely approximated by:

$$\mathbf{T}_\infty(j\omega,\varepsilon) = \mathbf{P}(0)\mathbf{M}^{-1}(0)\mathbf{F}(j\omega,\varepsilon)(\mathbf{I}+(\mathbf{P}(0)-\mathbf{M}(0))\mathbf{M}^{-1}(0)\mathbf{F}(j\omega,\varepsilon))^{-1} \quad (2)$$

where
$\mathbf{T}_\infty(j\omega,\varepsilon)$ = Matrix of closed loop frequency responses
$\mathbf{P}(0) = \mathbf{P}(\mathbf{x},0)$ Matrix of uncertain process gains
$\mathbf{x}$ = A vector of uncertain parameters
$\mathbf{M}(0)$ = Matrix of model gains
$\mathbf{F}(s,\varepsilon)$ = Diagonal matrix of controller filters
$\mathbf{Q}(s)$ = IMC Controller Matrix $\cong \mathbf{M}^{-1}(0)\mathbf{F}(s,\varepsilon)$ for large ε

If there exists a vector of large filter time constants such that $\max_{\omega,\mathbf{x}} |\mathbf{T}_{\infty ii}(\omega,\varepsilon,\mathbf{x})| \leq \mathbf{Mp}_{ii}$ for all i, then we can find at least one vector of filter time constants which satisfies (1). (Note the parameters $\mathbf{x}$ in the foregoing are only the variable gains). In addition to satisfying (1), however, we also would like to have the property that if we increase any filter time constant beyond that which satisfies (1), then all maximum magnitudes either decrease, or stay the same. We call such behavior consistent tunability because it is consistent with intuition that the maximum of the closed loop frequency response should not increase when the filter time constant increases. Consistent tunability is necessary for our algorithm to find filter time constants which satisfy interaction specifications, as well as set point response specifications. The multivariable **Mp** tuning algorithm follows:

Step 1. **Check for existence of the solution..** Compute the upper bounds of $|\mathbf{T}_{\infty ii}|$ as defined in (2) for all n set points to output transfer functions and **equal** values of filter time constants. If a maximum peak of an upper bound is larger than specified, then see if specifications can be met by increasing the associated filter time constant. However, our limited experience to date indicates that if the overshoot specifications are not met with large equal filter time constants, they will **not** be met by increasing individual filter time constants. If no filter time constants can be found to meet specifications, then either change the model gains such that specifications are met, or change the specifications. Limited experience to date indicates that changing the model gains so as to decrease the model condition number reduces **Mp**ij when the model is ill conditioned.

Step 2. **Tune the control system for setpoint step changes.** The tuning algorithm follows later in this section and yields a vector of filter time constants $[\varepsilon_1^*, \varepsilon_2^*, ..., \varepsilon_n^*]$ for which the performance criteria on all setpoint - output responses are satisfied.

Step 3. **Compute the upper bounds of all interactions** and fit the upper bounds with second order systems.

Step 4. **If the interactions are acceptable, terminate the procedure.** If an interaction to a process variable i from setpoint step change in variable j is not acceptable (large peak in the time domain, oscillations, long settling time or the variable i is more important to maintain at its setpoint than the other variables) we need to slow down the response of the variable j to its setpoint, i.e. we need to increase the corresponding filter time constant ε_j.

Step 5. **Determine if filter time constant ε_j can be increased to suppress interactions.** Compute the upper bounds of all diagonal transfer functions of the complementary sensitivity transfer function matrix for the vector of filter time constant obtained in Step 2 except for $\varepsilon_j \to \infty$. This step tells whether we can increase the filter time constant ε_j without violating the tuning criteria on the diagonal transfer functions.

Step 6. **Increase ε_j to suppress interactions,** if possible. If not, tuning criteria yield an infeasible set of constraints. Go to Step 4 assuming modified performance criteria on interactions. Go to Step 1 if performance criteria on setpoint step changes or parametric uncertainty or the model were modified. Note, that the procedure always terminates at Step 4 when the system is tuned for setpoint step changes and all possible interactions are acceptable.

There is a trade-off when tuning for interactions. Only interactions from less important to more important variables can be partially suppressed using the above procedure. Reducing interactions from more important to less important variables requires sacrificing the speed of response of the more important process variable to set point changes and to disturbances.

The tuning algorithm for step set point changes is:

Step 1. **Find lower bound on ε.** Find a vector of filter time constants, $\underline{\varepsilon}$, such that no transfer function of the IMC controller Q amplifies noise by more than a selected factor (e.g. $|\mathbf{Q}_{ij}(\infty)/\mathbf{Q}_{ij}(0)| \leq 20$). This step corresponds to a MIMO IMC controller design for a perfect model. The vector, $\underline{\varepsilon}$, provides a lower bound on all sought filter time constants. Each element of the vector must be strictly greater than zero.

Step 2. **Supply an upper bound on ε.** Supply a vector of very large and equal filter time constants, $\bar{\varepsilon}$. We suggest choosing $\bar{\varepsilon}$ to be 10^3 times the value of the largest process time constant.

Step 3. **Initialization 1.** For the *i-th* diagonal transfer function of the complementary sensitivity transfer function matrix choose the lower bound value of $\underline{\varepsilon}_i$ and the upper bound values of the other filter time constants as initial guess of vector of filter time constants.

Step 4. **Tuning 1.** Perform a SISO IMC tuning procedure for the *i-th* diagonal transfer function to satisfy its specified $\mathbf{Mp}_{ii}$ by adjusting ε_i. This step yields **new lower bound for** ε_i and other filter time constants as steps 3 and 4 are performed *n* times separately for each diagonal transfer function.

Step 5. **Initialization 2.** Choose the lower bound values of all filter time constants as initial guess of vector of filter time constants.

Step 6. **Tuning 2.** Perform a SISO IMC tuning procedure for the *i-th* diagonal transfer function to satisfy its specified $\mathbf{Mp}_{ii}$ by adjusting ε_i. This step yields **new upper bound for** ε_i and other filter time constants as steps 5 and 6 are performed *n* times separately for *each diagonal transfer function*.

Step 7. **End.** If each $(\bar{\varepsilon}_i - \underline{\varepsilon}_i)$ is smaller than desired accuracy the algorithm terminates. If not go to Step 3. The maximum peaks found are considered to be the global optima and corresponding filter time constants and the MIMO IMC controller satisfy performance requirements.

Example 1

Control problem definition: Find a vector of filter time constants $[\varepsilon_1, \varepsilon_2]$ such that the maximum peak of magnitudes of diagonal terms of complementary sensitivity functions is equal to 1.05 for some frequency (frequencies) and less than 1.05 for other frequencies.

$$P = \begin{bmatrix} \dfrac{x(1)}{x(2)s+1}\exp(-x(3)s) & \dfrac{x(4)}{x(5)s+1}\exp(-x(6)s) \\ \dfrac{x(7)}{x(8)s+1}\exp(-x(9)s) & \dfrac{x(10)}{x(11)s+1}\exp(-x(12)s) \end{bmatrix} \quad (4)$$

Lower bound of uncertain parameters $\mathbf{x}_-$:
[3, 1, 0.5, 0.5, 0.5, 1, 1, 3, 0.3, 2, 3, 0.5]'
Upper bound of uncertain parameters $\mathbf{x}_+$:
[5, 3, 1.5, 1.5, 1.5, 3, 3, 5, 0.7, 4, 6, 1.5]'

$$M = \begin{bmatrix} \dfrac{4}{2s+1}\exp(-s) & \dfrac{1}{s+1}\exp(-2s) \\ \dfrac{2}{4s+1}\exp(-2.5s) & \dfrac{3}{5s+1}\exp(-2s) \end{bmatrix} \quad (5)$$

The IMC controller inverts all of the invertible part of the model.

We tuned the control system such that the maximum peak of magnitudes of frequency responses of both $\mathbf{T}_{11}$ and $\mathbf{T}_{22}$ transfer functions of the complementary sensitivity function $\mathbf{T}$ is $\mathbf{Mp}_{11} = \mathbf{Mp}_{22} = 1.05$. The resulting filter time constants are $\varepsilon_1^* = 8.3$ and $\varepsilon_2^* = 4.8$.

In this case as we increase any filter time constant from the values of $\varepsilon_1^*, \varepsilon_2^*$ the performance criteria on $|\mathbf{T}_{11}|$ and $|\mathbf{T}_{22}|$ will not be violated and the system will be slowed down. Fig. 1 shows the contours for $\mathbf{Mp}_{ii} = 1.05$.

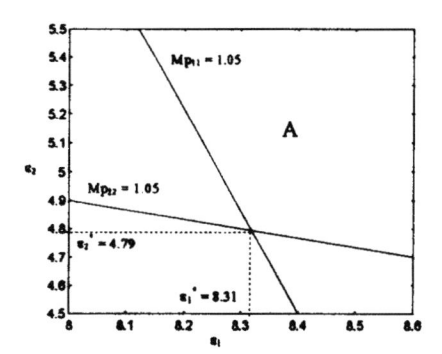

Figure 1. Contour plot of functions $Mp_{11} = Mp_{11}(\varepsilon_1, \varepsilon_2)$ and $Mp_{22} = Mp_{22}(\varepsilon_1, \varepsilon_2)$

If any vector of filter time constants from the area marked "A" is chosen both criteria on maximum peaks in the frequency domain will be satisfied as the functions $\mathbf{Mp}_{ii} = \mathbf{Mp}_{ii}(\varepsilon_1, \varepsilon_2)$ decrease to the value of one for $\varepsilon_1 \to \infty$ and/or $\varepsilon_2 \to \infty$. In this case we are allowed to increase any filter time constant by any amount and not violate the Mp criterion on either closed loop diagonal transfer function.

Consider the $\mathbf{T}_{12}$ transfer function of the closed loop transfer function matrix. This transfer function represents response of variable y_1 to change in setpoint in variable y_2. The upper bound of frequency responses of $|\mathbf{T}_{12}|$ for $\varepsilon_1^* = 8.3$ and $\varepsilon_2^* = 4.8$ is given in Fig. 2.

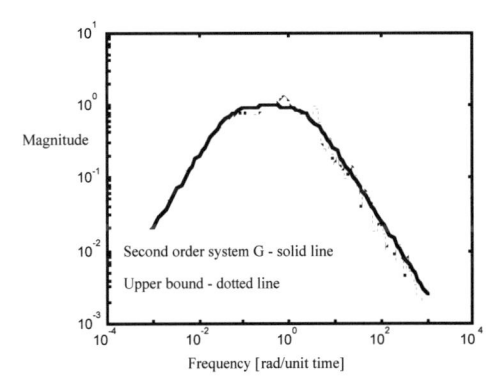

Figure 2. Upper bound of frequency responses fit by an overdamped second order system.

The second order system that fits the upper bound is

$$G = \frac{20s}{(20s+1)(0.4s+1)} \quad (6)$$

The estimated maximum peak of the interaction in the time domain is:

$$y_{max} = 20 \frac{1}{20-0.4}\left[\left(\frac{0.4}{20}\right)^{\frac{0.4}{20-0.4}} - \left(\frac{0.4}{20}\right)^{\frac{20}{20-0.4}}\right] = 0.923 \quad (7)$$

We can obtain the actual maximum peak of the interactions only by simulating the closed loop system for setpoint step change in output y_2. Simulating the behavior of the system using parameters :
x = [5 1 1.5 0.5 0.5 1 1 3 0.3 4 3 1.5]
gives the graphs shown in Fig. 3. The above parameters cause $|T_{11}|$ to reach a maximum peak of 1.05. This process also results in the largest interaction between process variables in terms of maximum peak in the time domain (~ 0.5).

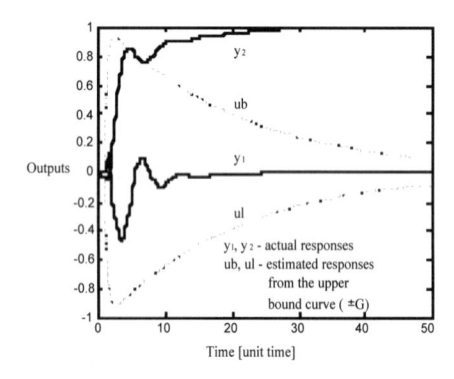

Figure 3. Closed loop responses to set point step change in variable y_2.

Our estimate of the maximum peak is somewhat conservative. In all cases that we solved, overdamped second order systems fit best the upper bounds. Of course, it is possible to get the maximum peak in the time domain directly by simulating the system using the process parameters associated with the maximum of the sensitivity function.

If it were necessary to decrease the interaction of y_2 on y_1, the designer could increase the filter time constant, ε_2^*, from 4.8 up to any value necessary to achieve the desired level of interaction, at the price of slowing the response of y_2.

Example 2

The process (with uncertainty $-0.5 \leq x_1, x_2 \leq 0.5$)

$$P = \begin{bmatrix} \frac{7.22+1.33x_1}{19s+1} & \frac{4.38+3.11x_2}{33s+1}e^{-20s} \\ \frac{5.88+0.59x_1}{50s+1}e^{-28s} & \frac{4.05+2.11x_2}{50s+1}e^{-27s} \end{bmatrix} \quad (8)$$

The mid-range model of the process ($y_i = 0$ for all i)

$$M = \begin{bmatrix} 1 & 0 \\ 0 & e^{-27s} \end{bmatrix} \begin{bmatrix} \frac{7.22+1.33y_1}{19s+1} & \frac{4.38+3.11y_2}{33s+1}e^{-20s} \\ \frac{5.88+0.59y_3}{50s+1}e^{-s} & \frac{4.05+2.11y_4}{50s+1} \end{bmatrix} \quad (9)$$

Using a decoupling IMC controller with the above model yields minimum Mp's of 2.2 and 3.2 for the closed loop transfer functions when the filter time constants are very large. However, if we select a model with $y_1 = y_4 = 0.5$ and $y_2 = y_3 = -0.5$, then the maximum Mp's are 1.0 for very large filter time constants. The condition numbers for the two models are 35 and 5 respectively.

The set point responses are shown in Fig. 4 and Fig. 5, for the case that $Mp_{ii} = 1.05$ for the low condition number model ($\varepsilon_1^* = 2.0$, $\varepsilon_2^* = 1.8$) and $Mp_{11} = 1.09$ and $Mp_{22} = 1.33$ for the mid-range model ($\varepsilon_1 = \varepsilon_2 = 10$). Notice, that the controller for the mid-range model is not consistently tuned.

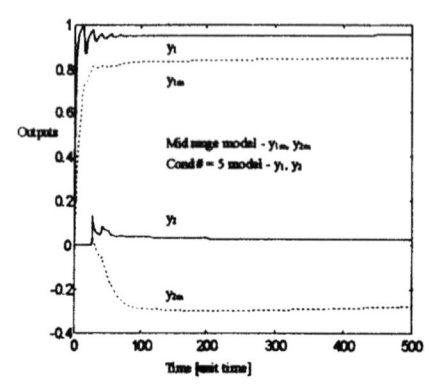

Figure 4. Closed loop responses for a unit step change in set point for y_1.

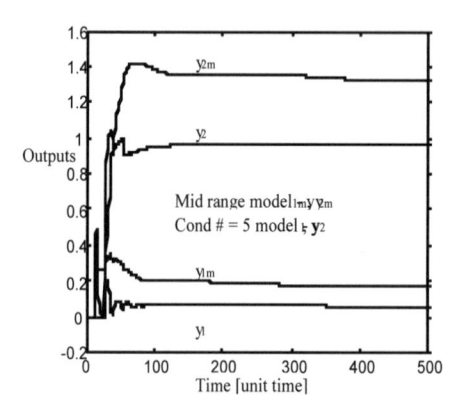

Figure 5. Closed loop responses for a unit step change in set point for y_2.

Conclusions

Mp tuning of low dimensional multivariable uncertain processes is feasible and provides good insights to control system performance. Stryczek (1996) has also applied Mp tuning to a three input, three output process.

References

Seborg, Edgar and Mellichamp (1989). *Process Dynamics and Control*. John Wiley, New York.

Stryczek, K. (1996). *Mp Tuning of Internal Model Control Systems Under Parametric Uncertainty*. Ph.D. Thesis, Case Western Reserve University.

SESSION SUMMARY: POSTER SESSION

Richard D. Braatz
Chemical Engineering Department
University of Illinois
Urbana, IL 61801

Introduction

The poster session papers cover many of the major control issues beginning with the process data and leading to the development of advanced supervisory control and optimization strategies. Dispersed throughout are the general themes of model development and refinement to represent accurately the nonlinear chemical process and to account for constraints and parametric uncertainties. A variety of processes are represented, including chemical reactors, large scale petrochemical operations, papermaking, sheet and film production, and batch processing.

Moore describes an abstract paradigm for developing automated data processing systems in support of the operation, maintenance, and improvement of industrial production facilities. Young discusses the application to a chemical plant of the abstract organizational structure for the manufacturing process and data acquisition systems.

Data analysis naturally follows to provide a sound basis for model validation, monitoring, and control. Wang et al. discuss the geometric interpretation of the principal components analysis and the preprocessing steps of mean centering and scaling. Wise and Gallagher review some of the methods for analyzing multi-way (e.g., batch) data and show how the methods can be used in chemical process monitoring and modeling. Rigopoulos et al. apply adaptive Principal Component Analysis to identify the critical disturbance modes in paper machine profiles.

Nonlinear, linear, and hybrid models are constructed from process data. Bomberger et al. compare the performance of different classes of artificial neural network models through experimental application to a pH neutralization process. Nikravesh et al. survey a series of successful applications of artificial neural networks models to industrial-scale processes, including electrolyzers, oil recovery from petroleum reservoirs, contaminant recovery, and chemical reactors. Shin et al. use Partial Least Squares to construct an inferential model for product composition based on multiple temperature measurements. Kano et al. incorporate first principles modeling with Partial Least Squares to construct a hybrid inferential model for tube wall temperature in an industrial olefins pyrolysis plant.

Models are used in linear and nonlinear control strategies. Dochain provides an overview of monitoring and control approaches for biochemical reactors, using biodegradable polymer production and biological waste treatment as illustrative examples. Christofides and Daoutidis describe the development of nonlinear control strategies for distributed parameter systems. Kosanovich et al. describe the synthesis of a hybrid dynamic supervisory control strategy for transition control. Samyudia et al. describe the development of control strategies for coupled multi-unit processes. Bequette studies the control and operability characteristics of batch exothermic reactors. Sanchez et al describe the procedural control formalism for representing feedback control concepts in event-driven processes. Bhatia and Biegler integrate dynamic optimization into the design and scheduling of multi-product batch plants.

Several papers describe techniques for the control of processes under constraints and/or uncertainties. Braatz reviews the applicability existing control techniques for addressing these issues for sheet and film processes. Scokaert and Rawlings review existing solutions to the problem of infeasibilities arising in model predictive control due to the use of state constraints. Kurtz and Henson propose a control strategy for nonlinear processes with input and/or output constraints that couples input-output linearization with linear model predictive control. Koulouris and Stephanopoulos use feedback limitation to guarantee closed loop stability with neural network-based model predictive controllers in the presence of model uncertainty. Hrissagis and Crisalle propose a framework for the design of robust predictive controllers which satisfy a range of time domain performance specifications. Stryczek and Brosilow describe the use of Mp tuning for the design of internal model controllers for multivariable systems with parametric uncertainty.

AUTHOR INDEX

A
Allgower, Frank 24
Alsop, N. 319
Aoki, Shyoji 292
Arkun, Yaman 60, 275

B
Badgwell, Thomas A. 232
Barton, Paul I. 102
Benson, Roger 192
Bequette, B. Wayne 315
Bhatia, Tarun 323
Biegler, Lorenz T. 177, 323
Bomberger, John D. 280
Braatz, Richard D. 327, 352
Brosilow, Coleman B. 347
Butler, Stephanie W. 133

C
Cameron, Ian T. 311
Charboneau, James G. 307
Christofides, Panagiotis D. 302
Cooley, Brian 201
Crisalle, Oscar D. 343

D
Daoutidis, Prodromos 302
Dochain, Denis 297
Doyle III, Francis K. 24, 115

E
Edgar, Thomas F. 133
Engell, Sebastian 165

G
Gallagher, Neal B. 271
Green, Michael 311

H
Hanczyc, Eric 275
Hashimoto, Iori 292
Henson, Michael A. 335
Hernandez, Evelio 1
Hrissagis, Kostas 343
Hrymak, Andrew N. 156

I
Irwin, George W. 280

K
Kayihan, Ferhan 117, 275
Kano, Manabu 292
Khettry, Atul 267
Kosanovich, Karlene A. 307
Koulouris, Alexandros 339
Kovscek, Anthony R. 284
Kowalewski, Stefan 165
Kowalski, Bruce R. 97
Kozub, Derrick J. 83
Krogh, Bruce H. 165
Kurtz, Michael J. 335

L
Lau, Henry K. 1
Lee, Jay H. 201
Lee, Moonyong 288
Lee, Peter L. 311
Lightbody, Gordon 280

M
Macchietto, Sandro 71, 319
Marlin, Thomas E. 156
Mayne, David Q. 217
Moore, Bruce C. 259
Morari, Manfred 81, 179

N
Nikolaou, Michael 61
Nikravesh, Masoud 284

O
Ogunnaike, Babatunde A. 46
Okamo, Umetaro 292
Oshima, Masahiro 292

P
Park, Sunwon 288
Park, Taeshin 102
Patzek, Tad W. 284
Perkins, John 192, 199
Piovoso, Michael J. 307

Q
Qin, S. Joe 232

R
Ramaker, Brian L. 1
Rawlings, James B. 257, 331
Rigopoulos, Apostolos 275
Rotstein, G. E. 319

S
Samyudia, Yudi 311
Sanchez, A. 319
Scokaert, P.O.M. 331
Seborg, Dale E. 145, 280
Shin, Joonho 288
Shiren, Toshihiro 292
Soroush, Masoud 284
Stephanopoulos, George 339
Stryczek, Karel 347

V
Vedula, Srinivas 267

W
Wang, Tsewei 267
Wise, Barry M. 271
Wright, Raymond A. 46
Wright, Steven J. 147

Y
Ydstie, B. Erik 9
Young, Robert E. 263

SUBJECT INDEX

A
Adaptive control 9, 201, 297
Assigned neighborhood characteristic 259
Automation 319

B
B-Spline network 280
Batch
    control 133
    processing 323
    reactors 315
Beer production 71
Benefits 192
Bioprocesses 297
Blending processes 263
Business drivers 71

C
CD control 117
Chemical reactor 284
Chips 117
Closed-loop stability 201
Coating 327
Collocation 323
Combined discrete/continuous systems 102
Composition estimator 288
Computer-aided process engineering 61
Constrained control 335, 343
Constraints 217
Control 102
    future vision 1
    research challenges 1
    business drivers 1
    relevant identification 327
    system design 311
Controller performance
    diagnosis 83
    monitoring 83
Controller synthesis 165
Cooperative research 179
Cross
    directional control 327
    validation 292

D
Dairy production 71
Data reconciliation 156, 263
Decentralized control 311
Design 323
Differential
    algebra 24
    geometry 24
Digesters 117
Discrete event systems 165, 319
Dispersion measures 259
Disturbances 343
Dynamic modeling 117, 263

E
Economic optimization 156
Electrolyzer 284
Equipment models 133
Estimation 275

F
Fault detection 263
Feasibility 315, 331
Feedback
    limitation 339
    linearization 335
Food manufacturing control 71
Finite state machines 165
Flowsheeting 156
Formal verification 165
Foresight 192

G
Gap metric 311
Geometric 267
Grade changeovers 307

H
Headbox 117
Hybrid systems 102, 165

I

Identification 275, 327
Illuminator 259
Industrial
    implementation 284
    process control 46
    survey 232
Industry-academia partnering 1
Inferential
    control 292
    model 292
Internal model control 347
Information research challenges 1
Instrumentation 179
Interior-point methods 147
Interval mathematics 259
Investment 179

K

Kappa number 117
Karhunen-Loève expansion 275

L

l_1/H_∞ design 343
Large-scale systems 327
Lime kilns 117
Linearizing control 297

M

Manufacturing execution system 71
Mean centering 267
Microelectronics manufacturing 133
Model
    based control 217, 284
    identification 284
    predictive control 46, 147, 201, 232, 327, 331, 335, 339
Modeling 102, 156, 165
Modes 275
Multi-layer perceptron 280
Multi-way data analysis 271
Multiple product processes 307
Multivariable
    control systems 347
    identification 201
Multivariate
    curve resolution 271
    statistical analysis 267

N

Neural network 46, 280, 284, 339
Nonlinear
    analysis 24
    control 24, 46, 217, 302
    H_∞ 24
    identification 280
    processes 335
    system 9
Nonlinearity measure 24

O

On-line scheduling 71
Operability 315
Optimal control 201
Optimization 147, 217
    economic 156
    real-time 156
Optimizing control 156

P

Paper
    control 117
    machine 117, 275
Papermaking 327
PARAFAC 271
Parameter estimation 156
Partial-least-squares regression 288, 292
PDE systems 302
Petroleum reservoir 284
pH control 46
Phase equilibrium 102
Polymer film extrusion 327
Predictive control 217
Principal component analysis 275
Procedural control 319
Process
    analytic chemistry 97
    analyzer 97
    modeling 271
    monitoring 263, 271
Process control 9, 24, 179, 192
    education 1
Profile control 117
Publications 179
Pulp control 117
Pyrolysis plant 292

R

Radial basis function network 280
Rapid thermal processing 133
Rank 257
Reactor design 315
Real-time optimization 156
Reduction process 259
Refiners 117
Relational mathematics 259
Research
    consortium 179
    funding 179
Results analysis 156
Robust
    control 327
    predictive control 343
Robustness 9, 201, 347

S

Scale-up 315
Scaling 267
Scheduling 323
Session summary 60, 81, 115, 145, 177, 199, 257, 352
Sheet and film 327
Simulation 102, 165
Signal processing 259
Snack food 61
Soft constraints 331
Software sensors 297
Stability 9, 339
State
    constraints 331
    estimation 201
Stochastic control 83
SVD 267
Switching signal 307

T

Time series analysis 83
Transition control 307
Transport-reaction processes 302
Trilinear decomposition 271
Tucker models 271
Tuning 347

U

UK innovative manufacture 192
Uncertain systems 347
Uncertainty 339

V

Variational mathematics 259
Verification 102
    formal 165

W

Waste management 284
Wave-Net 339